W9-CNX-860

ANNUAL REVIEW OF ENTOMOLOGY

ANNUAL REVIEW OF ENTOMOLOGY

VOLUME 48, 2003

MAY R. BERENBAUM, *Editor*
University of Illinois, Urbana-Champaign

RING T. CARDÉ, *Associate Editor*
University of California, Riverside

GENE E. ROBINSON, *Associate Editor*
University of Illinois, Urbana-Champaign

www.annualreviews.org   science@annualreviews.org   650-493-4400

ANNUAL REVIEWS
4139 El Camino Way • P.O. Box 10139 • Palo Alto, California 94303-0139

ANNUAL REVIEWS
Palo Alto, California, USA

International Standard Serial Number: 0066-4170
International Standard Book Number: 0-8243-0148-X
Library of Congress Catalog Card Number: A56-5750

♾ The paper used in this publication meets the minimum requirements of American National Standards for Information Sciences—Permanence of Paper for Printed Library Materials, ANSI Z39.48-1992.

TYPESET BY TECHBOOKS, FAIRFAX, VA
PRINTED AND BOUND BY MALLOY INCORPORATED, ANN ARBOR, MI

PREFACE

It's my turn to write the Preface for the latest volume of the *Annual Review of Entomology*. It is the tradition of the *Annual Review of Entomology* Editorial Committee to invite the outgoing member to write the Preface for that year's volume. So here goes.

The first thing on my mind is, "Does anybody read these prefaces?" Most people probably don't even know there is a Preface. If they're like me, people don't read the *Annual Review of Entomology* like a book, lovingly opening the front cover and leafing past the forward and the title page until they get to Chapter One. I delve into each volume somewhere in the middle after quickly scanning the Table of Contents to find a particular summary of an entomology research field that I need to bone up on. Each article is almost always an interesting and comprehensive analysis, put together in a way you can't find anywhere else. That's the beauty of this publication and probably one of the reasons why it's so widely read and cited. It summarizes the work going on in a huge field in an engaging and useful way, each and every year it is published.

From my five years on the Editorial Committee, I've learned why this is so. The work that the Editorial Committee does is really an interesting process that involves a combination of first working within pre-selected broad fields (e.g., insect physiology) that always need updating each year, and then selecting the most critical-needs topic from within that broad field. Authors submit proposals on their own, volunteering to review something in their area that hasn't been reviewed in a while, or perhaps even never before. Others in that field also suggest the names of potential topics and authors. Finally there is the process of sheer brainstorming on the part of the Editorial Committee during its annual meeting, where topics are often identified that have previously been completely overlooked and need addressing. These might be hot new research areas, or else a research theme from several old areas that nobody has conceived of before to shape into a compelling review topic.

From the articles each year we thus get a solid idea of exactly how dynamic and productive the science of Entomology truly is. I enjoyed very much being a part of the *Annual Review of Entomology* Editorial Committee and taking part in the process of displaying to the world the incredible variety of research that is being conducted in the field I have loved since childhood. Entomology continues to be a vibrant and ever-changing science. We need to be vigilant in keeping it strong, dutiful in getting our findings out to where they can be read, appreciated, and effective in elevating our collective intellect. The *Annual Review of Entomology* is a major vehicle for this, the advancement of entomological knowledge, and its applicability to human society. I am proud to have been a part of this effort, and grateful for the opportunity to have served.

Thomas C. Baker
For the Editorial Committee

Annual Review of Entomology
Volume 48, 2003

CONTENTS

ERRATA

An online log of corrections to *Annual Review of Entomology* chapters may be found at http://ento.annualreviews.org/errata.shtml

Related Articles

From the ***Annual Review of Ecology and Systematics***, Volume 33 (2002)

Saproxylic Insect Ecology and the Sustainable Management of Forests, Simon J. Grove

Troubleshooting Molecular Phylogenetic Analyses, Michael J. Sanderson and H. Bradley Shaffer

The Causes and Consequences of Ant Invasions, David A. Holway, Lori Lach, Andrew V. Suarez, Neil D. Tsutsui, and Ted J. Case

The (Super)Tree of Life: Procedures, Problems, and Prospects, Olaf R.P. Bininda-Emonds, John L. Gittleman, and Mike A. Steel

Homogenization of Freshwater Faunas, Frank J. Rahel

The Role of Animals in Nutrient Cycling in Freshwater Ecosystems, Michael J. Vanni

The Evolution and Maintenance of Androdioecy, John R. Pannell

Mast Seeding in Perennial Plants: Why, How, Where? Dave Kelly and Victoria L. Sork

Phylogenies and Community Ecology, Campbell O. Webb, David D. Ackerly, Mark A. McPeek, and Michael J. Donoghue

The Quality of the Fossil Record: Implications for Evolutionary Analyses, Susan M. Kidwell and Steven M. Holland

Herbivore Offense, Richard Karban and Anurag A. Agrawal

Estimating Divergence Times from Molecular Data on Phylogenetic and Population Genetic Timescales, Brian S. Arbogast, Scott V. Edwards, John Wakeley, Peter Beerli, and Joseph B. Slowinski

From the ***Annual Review of Genetics***, Volume 36 (2002)

Directions in Evolutionary Biology, R. C. Lewontin

Genetics of Motility and Chemotaxis of a Fascinating Group of Bacteria: The Spirochetes, Nyles W. Charon and Stuart F. Goldstein

Meiotic Recombination and Chromosome Segregation in Drosophila Females, Kim S. McKim, Janet K. Jang, and Elizabeth A. Manheim

Chromosome Rearrangements and Transposable Elements, Wolf-Ekkehard Lönnig and Heinz Saedler

Transvection Effects in Drosphila, Ian W. Duncan

Estimating F-Statistics, B. S. Weir and W. G. Hill

From the ***Annual Review of Microbiology***, Volume 57 (2003)

Climate and Infectious Disease, Joan B. Rose

Natural Born Killers: Theory and Applications of Biological Control, Raghavan Charudattan

Termite Gut Microbiology, John A. Breznak

From the ***Annual Review of Phytopathology***, Volume 40 (2002)

Evolutionary Ecology of Plant Diseases in Natural Ecosystems, Gregory S. Gilbert

Making an Ally from an Enemy: Plant Virology and the New Agriculture, Gregory P. Pogue, John A. Lindbo, Stephen J. Garger, and Wayne P. Fitzmaurice

Gene Expression in Nematode Feeding Sites, Godelieve Gheysen and Carmen Fenoll

Phytochemical Based Strategies for Nematode Control, David J. Chitwood

ANNUAL REVIEWS is a nonprofit scientific publisher established to promote the advancement of the sciences. Beginning in 1932 with the *Annual Review of Biochemistry*, the Company has pursued as its principal function the publication of high-quality, reasonably priced *Annual Review* volumes. The volumes are organized by Editors and Editorial Committees who invite qualified authors to contribute critical articles reviewing significant developments within each major discipline. The Editor-in-Chief invites those interested in serving as future Editorial Committee members to communicate directly with him. Annual Reviews is administered by a Board of Directors, whose members serve without compensation.

Annual Reviews of

Anthropology
Astronomy and Astrophysics
Biochemistry
Biomedical Engineering
Biophysics and Biomolecular Structure
Cell and Developmental Biology
Earth and Planetary Sciences
Ecology and Systematics
Entomology
Environment and Resources
Fluid Mechanics
Genetics
Genomics and Human Genetics
Immunology
Materials Research
Medicine
Microbiology
Neuroscience
Nuclear and Particle Science
Nutrition
Pharmacology and Toxicology
Physical Chemistry
Physiology
Phytopathology
Plant Biology
Political Science
Psychology
Public Health
Sociology

SPECIAL PUBLICATIONS
Excitement and Fascination of Science, Vols. 1, 2, 3, and 4

M. Locke

Annu. Rev. Entomol. 2003. 48:1–27
doi: 10.1146/annurev.ento.48.091801.112543

First published online as a Review in Advance on August 19, 2002

SURFACE MEMBRANES, GOLGI COMPLEXES, AND VACUOLAR SYSTEMS

Michael Locke
Department of Zoology, University of Western Ontario, London, Ontario, Canada, N6A 5B7; e-mail: mlocke@uwo.ca

Key Words epidermis, fat body, cell remodeling, ferritin, peptide routing

■ **Abstract** In the absence of fossils, the cells of vertebrates are often described in lieu of a general animal eukaryote model, neglecting work on insects. However, a common ancestor is nearly a billion years in the past, making some vertebrate generalizations inappropriate for insects. For example, insect cells are adept at the cell remodeling needed for molting and metamorphosis, they have plasma membrane reticular systems and vacuolar ferritin, and their Golgi complexes continue to work during mitosis. This review stresses the ways that insect cells differ from those of vertebrates, summarizing the structure of surfacce membranes and vacuolar systems, especially of the epidermis and fat body, as a prerequisite for the molecular studies needed to understand cell function. The objective is to provide a structural base from which molecular biology can emerge from biochemical description into a useful analysis of function.

CONTENTS

0066-4170/03/0107-0001$14.00

INTRODUCTION

This paper reviews insect cell membrane systems, that is, the inward and outward moving membranes spanning the space between the nuclear envelope and the plasma membrane. It describes the generality of membrane systems based especially on studies of the epidermis, fat body, and other cells of *Calpodes ethlius* and *Rhodnius prolixus* (summarized in Table 1). The principle that I have followed is that very often questions cannot be asked without knowing the structure involved (41). For example, microvesiculation of the nuclear envelope had to be observed to allow for the possibility of *trans*-envelope transport (52). Molecular biology can only emerge from biochemical description into useful analysis of function by working from a base of discovered structure. This review summarizes part of the structural foundation prerequisite for the molecular studies needed to understand cell function.

A problem for developmental biology is to understand the mechanism by which genetic codes are translated into three-dimensional cell architecture (31). *Drosophila* genetics combined with insect cell structural studies give us a clear base to work from, but one that for the most part has yet to be realized. A second objective emphasizes membrane systems as a component of three-dimensional cell architecture.

Workers on vertebrates have claimed vertebrate cells as the general animal eukaryote model. Because they have been reluctant to include insect work in their thinking, there has been little attempt to distinguish vertebrate specializations from common ancestral features (74). The two lines separated in the Proterozoic Eon nearly a billion years ago (15), making many generalizations drawn from vertebrate cells inappropriate for insects (46, 50, 53, 54, 59, 61, 83). This review stresses the ways that insect cells differ from those of vertebrates. The original literature must be consulted for supporting experiments and the visual picture. Terms used are defined in (28, 39).

SURFACE MEMBRANES

Fat body, oenocytes, prothoracic glands, pericardial cells, and other tissues interface with the hemolymph but live within extracellular lymph spaces (36). A clue to the properties of these spaces comes from the observation that they are absent in hemocytes, the only tissue not surrounded by a basal lamina. The spaces contain lymph, that is, hemolymph that has had its components selected for size and charge (Figure 1). Lymph spaces are intercellular and reticular system compartments separated from the hemolymph by negatively charged sieves preventing the penetration of molecules larger than 15 nm in diameter (1, 53, 72). They form either from the fusion of plasma membrane processes to create reticular systems, extracellular compartments "within" cells, or from the meshing of processes from nearby cells, elaborating the compartment between them (Figure 2).

Plasma Membrane Reticular Systems and Intercellular Lymph Spaces

Vertebrates are for the most part large three-dimensional solid-tissue builders requiring high-pressure fluid movements for bulk exchange. Insect complexity, on the other hand, is on a small scale, relying on simple epithelia and individual or loosely interacting cells (*pace* nerves and muscles). Insect cells have exploited the efficiency that comes from rapid diffusion over short distances by evolving compartments that minimize the distances for exchange between cells and the medium around them. These compartments are osmotically separate from the hemolymph and the cells. They occur both between cells (intercellular lymph spaces) and within intricately folded cell surfaces (reticular systems), creating reticular galleries, extra "internal" surfaces similar to those that we imagine gave rise to endoplasmic reticulum (ER) in evolution (30). Reticular systems enormously increase the surface area of plasma membranes and allow the cell to live within its own environment of lymph (36). They form by the extension of surface processes that enlarge and fuse, except for entry pores (17, 30).

TABLE 1

Compartment	Epidermis	Fat body	Other cells
Plasma membranes			
Reticular systems	Basal surface of wax-secreting cells, bristles, and lenticles	Outside and between cells in much fat body	Outer face of oencytes, prothoracic glands and many other cells
Intercellular lymph spaces	Lateral and basal surfaces	Between cells	Between many cells
Perimicrovillar spaces	Formed around apical microvilli		
Tokus membrane vesicles			
Pretransition region			
Nuclear envelope	Trans envelope transport vesicles	Not seen but probably general when nuclear pore transport pre-empted	—
Nuclear envelope	Distended in protein reserve vacuoles in *Tenebrio*		
Rough endoplasmic reticulum, RER	Lamellate or distended	Contains apoferritin or holoferritin	Lamellate or distended, abundant holoferritin in athrocytes and midgut
Concretions usually dilated RER		Enlarged NE or RER cisternae used for iron or other metal storage	Calcium and other ion stores in midgut and Malpighian tubules
Smooth endoplasmic reticulum, always tubular, never reticulate	Only in wax-secreting cells	Usually absent, present in termites	Extensive in oenocytes, prothoracic glands and midgut
Confronting cisternae	On lateral and basal membranes	Occasional	Prothoracic glands and other cells
Perioxosomes	In cells derived from epidermis	Formed initially from RER vesicles. Grow while confronting RER cisternae. Cyclically present in relation to urate metabolism	In oenocytes in relation to lipid metabolism
Posttransition region			
Golgi complex	Transfer vesicles pass through GC bead rings to make the outer lamella in all cells. Staining always shows one blank compartment as though processing is a batch process		
Golgi complex primary derivatives	Smooth-surfaced secretory vesicles, 1^0 lysosomes, isolation envelopes		
Autophagic vacuoles	Isolation envelopes and 1^0 lysosomes used cyclically in all cells		

Heterophagic vacuoles	Coated endocytic vesicle transport from apical and basal faces to apical MVB sorting vacuole	Massive heterophagy for hemolymph turnover	Massive heterophagy of hemolymph in athrocyte MVBs
Multivesicular bodies and sorting vacuole	Usually apical, ? transcellular transport vacuole, apical membrane turnover	Perinuclear, precursors to urate and protein storage vacuoles	General
Cell and membrane phagocytic vacuoles	Phagosomes for gap junctions and whole cells in apoptosis. Smooth-surface microvesicles for metamorphic basal membrane turnover into lameller bodies	Basal lamina and membrane turnover in phagosomes	
Lamellar bodies	Basal endocytic vesicles for turnover of membranes after resorption of feet	Endocytic vacuoles for turnover of excess membrane after release of tyrosine vacuoles	Probably occur generally
Glycogen-associated vacuoles	Phenolic glucoside storage vacuoles continuously present	Phenolic glucoside storage vacuoles continuously present near glycogen	Probably occur generally
Tyrosine storage vacuoles	Soluble carbohydrate and phenolic store? related to glycogen-associated vacuoles	Formed cyclically at molting from plasma membrane	
Urate storage		Vacuoles in cockroaches, granules derived from MVBs in caterpillars, intermediate in locusts	
Symbiont-containing vacuoles		Endocytic within plasma membrane vacuoles in many insects	Commonly in midgut
Inceretae sedis			
Protein storage vacuoles	Locust epidermis		
Apical vacuoles reacting to a UDP-N-acetyl glucosamine phosphatase	? chitinogenic vacuoles		
Vacuoles appearing after epicuticular abrasion	Supply of precursors for wound repair		
Pigment-containing vacuoles	Contain red ommochromes. Blue insecticyanins are in transport vacuoles		
Polarized subapical vesicular blebs	? basis for antero-postero epithelial polarity and polarized control of growth as in tracheae		

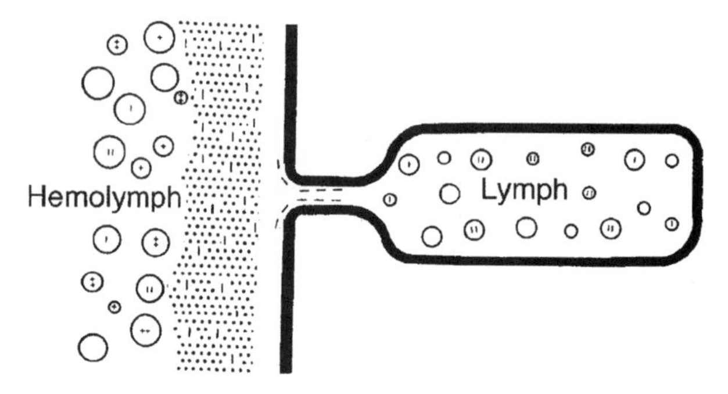

Figure 1 Plasma membrane reticular system and intercellular lymph spaces are separated from the hemolymph by negatively charged barriers tending to exclude large, positively charged particles. Both the basal lamina and reticular system entrances are negatively charged sieves (1).

Perimicrovillar Spaces and Cuticle Secretion

The epidermis has evolved an apical chamber from perimicrovillar spaces below the cuticular compartment (Figure 2). Fibrous cuticle arises and is oriented in the assembly zone on the plaques at the tips of the microvilli (25, 34, 37, 41, 51). The space between the microvilli is characterized by its emptiness, in spite of being a receptacle for apical secretion. Cuticle peptides are missing but abundant in the cuticle above (56). The absence of cuticle peptides may be explained if the perimicrovillar space is a compartment for mixing precursors that pass quickly

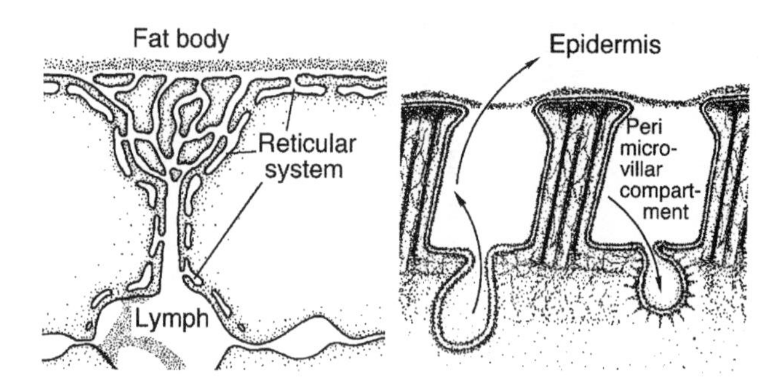

Figure 2 Many insect cells interface with their environment through compartments of their own creation. *Left*: Fat body cells elaborate the lymph spaces between them by the extension of surface processes. They also fold their surface membranes inward to create reticular systems (see Figure 9). *Right*: The apical surface of epidermal cells forms a compartment around the microvilli separate from that of the cuticle. Cuticle precursors traverse this compartment on their way to the assembly zone. Peptides destined for basal secretion pass into the compartment before being endocytosed (see Figure 11).

from the emptying vesicles to strong affinity binding sites (71) in the assembly zone, working like the spacer gel in electrophoresis. Apically secreted noncuticle peptides, on the other hand, are present in the perimicrovillar space (56). Such peptides are also basally secreted, suggesting that the perimicrovillar space may be a sorting compartment that separates cuticle precursors from peptides destined to be turned around for basal secretion, probably through the apical multivesicular body (MVB) acting as a sorting vacuole.

Thus, many insect cells have separate plasma membrane–lined compartments at their surfaces. The compartments differ in their structure and functions. The effective external surfaces of many cells appose lymph rather than hemolymph, and much of the apical surface of the epidermis faces perimicrovillar rather than cuticular fluid space.

Vesicular Membrane Barriers

In the caterpillar rectum the cuticle enlarges into a separate fluid-filled compartment. The rectum itself lies with Malpighian tubules in a hemolymph pool enclosed by a complex of 1-cell-thick membranous extensions of the epidermis (35). These separate the (posterior) tokus compartment from the main (anterior) hemolymph pool. The cellular membranes facing the anterior hemolymph contain little but vesicles, most connected to one surface or the other, resembling some vertebrate capillaries where there is a *trans*-cellular vesicular flow. These structures are clearly related to the way that the perirectal compartment takes up fluid from the rectum.

PRETRANSITION REGION COMPARTMENTS

Transport Across the Nuclear Envelope

Most larval insect activities follow the intermolt/molt cycle, preparation for which involves phases of RNA synthesis and cell growth controlled by ecdysone (24). The exact timing of this rhythm in the epidermis depends on hormonal signals (11). At the beginning of the intermolt and again at the beginning of molting, elevated titers of hemolymph ecdysteroid precede nucleolar activity, ribosome and rough endoplasmic reticulum (RER) formation, and the movement of much material through the nuclear pores (52). The route between the cytosol and the nucleus is usually thought to be restricted to the nuclear pores, but at stages of epidermal development when the pores are occupied in this way, the nuclear envelope contains microvesicles (Figure 3). Microvesicles occur with their lumina connected to the cytoplasmic face and the nuclear face, as well as free in the interior of the envelope. In freeze-fracture preparations, the vesicle connections appear as tiny cone-shaped tubercles on both inner and outer envelope membranes. The discovery of these intra-envelope vesicles shows the probability of *trans*-nuclear envelope transport. Nucleocytoplasmic communication of this kind may be a general mechanism for the specific routing of signaling molecules made necessary by the preoccupation of the pores with RNA transport.

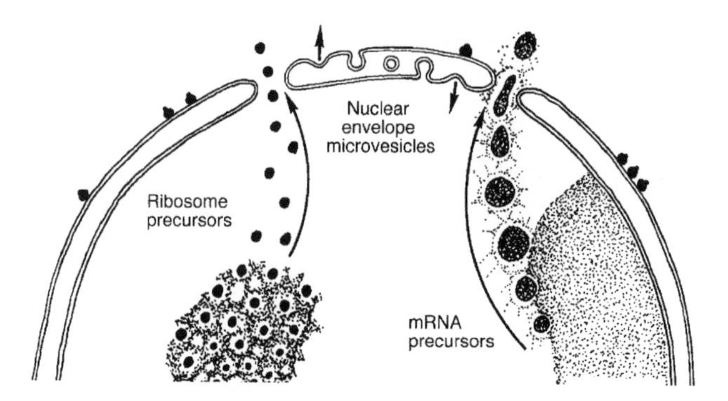

Figure 3 *Trans*-nuclear envelope transport by microvesicles. The nuclear envelope of epidermal cells contains microvesicles at the time that the transport of bulk RNA would be expected to prevent nucleocytoplasmic communication through the nuclear pores.

Smooth Endoplasmic Reticulum

Smooth endoplasmic reticulum (SER) is defined as a membrane compartment of the vacuolar system between the nuclear envelope and Golgi complex (GC) that does not bear ribosomes because it is primarily devoted to activities other than protein metabolism. The two main activities are related either to detoxification, as in liver SER, or to lipids, as in steroid and wax metabolism. In liver the SER is reticulate with many small irregularly flattened cisternae. The fat body of insects is often likened to the liver but it almost never has SER: Detoxification is largely a function of the midgut where it is abundant. Lipid metabolism is associated with tubular SER in insects, as in oenocytes, prothoracic glands, or specialized epidermal wax-secreting cells. Midgut SER also tends to be tubular, although probably functioning in detoxification.

Rough Endoplasmic Reticulum

Rough endoplasmic reticulum (RER) is defined as a membrane compartment of the vacuolar system between the nuclear envelope and GC that bears ribosomes because it is primarily devoted to protein metabolism. It tends to be in stacks of flattened cisternae. It is abundant in the fat body, the primary protein-synthesizing tissue. It may become distended with protein at the beginning of the intermolt, when synthesis is out of synchrony with its exit gate through the GC (39). The nuclear envelope of larval *Tenebrio* epidermis may fill with protein, forming storage granules superficially appearing as though the nucleus itself had become a protein store. The RER may also become distended as a storage vacuole for metals and their salts (iron, copper, calcium, magnesium).

Ferritin Distribution in RER and SER

Vertebrate holoferritin (i.e., apoferritin loaded with iron) is largely cytosolic, as in the gut and liver, although apoferritin in liver RER is presumably the origin of serum ferritin (73). Homoptera store holoferritin in the cytosol in response to the iron overload from feeding on plant sap (13, 66) or nectar (39, 67), but in most insects ferritin is largely vacuolar (59, 69). Most fat body contains apoferritin in the RER (55, 61), as in *C. ethlius* where it becomes holoferritin only after secretion into the hemolymph (68). Cell iron does not equilibrate in the RER lumen of live fat body. Even adding enough iron to the hemolymph for holoferritin to appear abundantly in athrocytes (62) does not load fat body ferritin with iron. However, holoferritin does appear in fat body RER cisternae of hemocytes (33) or fat body (55) if the membranes are permeabilized in glycerol before exposure to iron.

The midgut, unlike the fat body, contains holoferritin in the RER and SER (Figure 4). The SER is intimately associated with a reticular system on the hemolymph face and with the apical surface apposing the lumen. We suppose that ferritin in the SER lumen facilitates iron transport across the cell in either direction depending upon availability and demand. An iron overload causes excess iron to be secreted through the GC into the lumen as hemosiderin.

Vacuoles for Iron Storage in Derivatives of the RER

In some insects, especially those that may be subject to an iron overload through their diet, both fat body cytosol and RER cisternae may contain holoferritin (66). Iron concretions of honey bees form from RER or the nuclear envelope (67). In the early stages, holoferritin builds up in small RER vacuoles. The protein shells are lost to make the iron-rich hemosiderin cores of the concretions. Concretions similar to those accumulating iron in fat body concentrate other metal ions in midgut and Malpighian tubule cells (70, 75).

Peroxisomes

Peroxisomes are pretransition vacuoles occurring in most fat body cells (Figure 5), as well as oenocytes (23), Malpighian tubules, lenticles (38), and fire fly lanterns (12). They often occur in small groups, sometimes next to lipid droplets. Because their thin boundary membranes must be permeable to solutes, they are relatively unaffected by the osmolarity of fixatives and tend to have uniformly dense contents even when nearby organelles are swollen. They can usually be distinguished from secretory vesicles by a flattened core on one side, making them kidney shaped.

Peroxisomes contain a variable assortment of oxidases and usually catalase, and work by oxidizing soluble substrates that have diffused into their interior. Those in the fat body contain urate oxidase, which converts relatively insoluble urate to allantoate and other soluble products that can move out of the cell during the intermolt. In the absence of peroxisomal oxidases, urate is stored rather than metabolized. At metamorphosis the loss of peroxisomes correlates with a

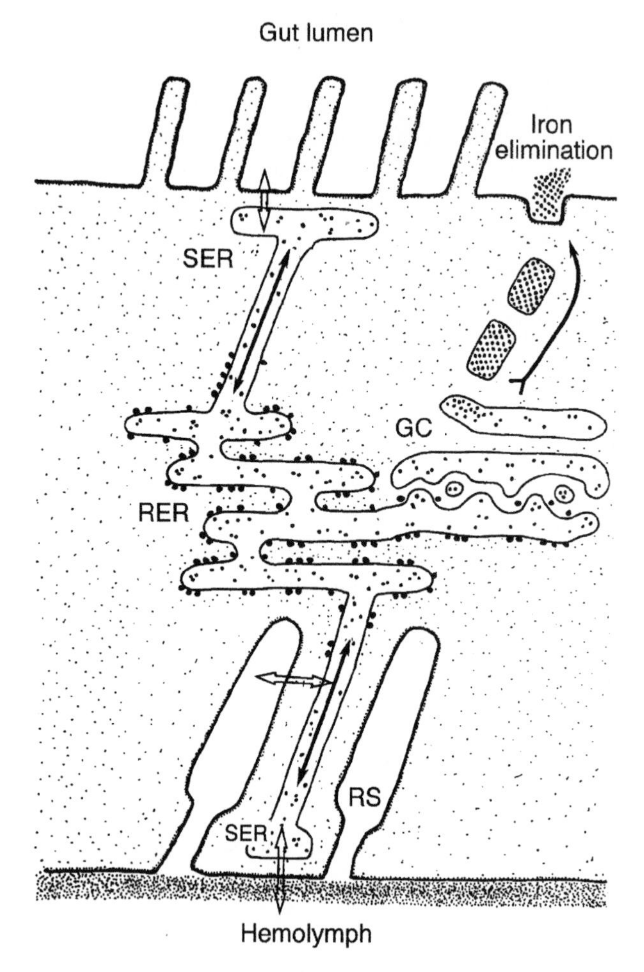

Figure 4 Ferritin occurs in the vacuolar system but not in the cytosol of midgut cells. Iron may be equilibrated between the hemolymph, lymph, and midgut lumen using ferritin in the smooth (SER) and rough endoplasmic reticulum (RER) as a transporter. Iron overload results in the excretion of insoluble iron to the midgut lumen through Golgi complex (GC) secretory vesicles.

switch in urate metabolism that allows granules containing urate to accumulate (9, 11).

Peroxisomes develop for use in the intermolt, with each stadium having a cycle of formation and destruction (Figure 5). In the fourth larval stadium of *C. ethlius*, they decrease in size before ecdysis to the fifth stage. Old peroxisomes atrophy after ecdysis and are replaced by a new fifth stage population. Peroxisome primordia form from diverticula of the RER and grow by the addition of material from confronting RER cisternae (60). In vertebrates the origin of peroxisomes as

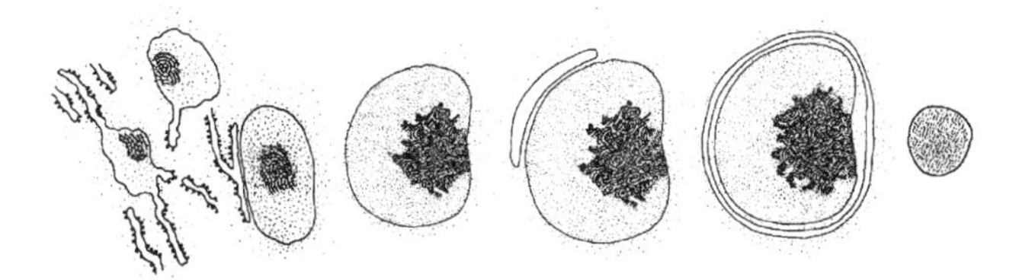

Figure 5 The life and death of peroxisomes in a fifth stage caterpillar. Peroxisome primordia separate from the RER early in the stadium in preparation for their function in the intermolt. Peroxisomes are isolated by envelopes for destruction in autophagic vacuoles late in the stadium, allowing the pupal fat body to store urate.

membranous vesicles is an embryological event. This has led to the erroneous statement that vertebrate peroxisomes do not arise from the RER because replication and growth occur in adult tissues without RER continuity. Fifth stage peroxisomes are the first organelles to be lost in a specific sequence of autophagy at the beginning of pupation (9, 27, 39, 45). A new generation of peroxisomes develops during the formation of the adult at the end of the pupal stadium (19).

POSTTRANSITION REGION COMPARTMENTS

The Golgi Complex

Membranes become thicker as they mature through the Golgi complex (GC). Increases in thickness begin in the transition region where transfer vesicles ferry membranes and their contents between the RER and the first GC saccule (28). Pretransition membranes, such as those of the ER, are 6–8 nm thick after the usual procedures for electron microscopy. Posttransition membranes, such as the plasma membrane, those lining most vacuoles and secretory vesicles, are in the range of 8–10 nm. In general, posttransition elements have membranes about 20% thicker than pretransition membranes (65).

Insect GCs are structurally simpler than those of most vertebrate or plant cells. They begin with a cloud of transition vesicles that have budded through rings of beads on a smooth face of RER (46). These fuse to one or two outer saccules, followed by two or three inner saccules. Depending on the stage of development, the inner face gives rise to condensing vacuoles, secretory vesicles, primary lysosomes, and isolation envelope precursor vesicles. The complete complex lies within a zone of exclusion, i.e., a region of microfibers not penetrated by organelles and even most ribosomes. Cationic ferritin does penetrate and has been injected experimentally into fat body cells to show that GC membranes are not negatively charged on their cytosolic face (3).

In the fat body early in a stadium, secretion is “constitutive”; there is no build up of secretory vesicles waiting for a signal to allow them to fuse with the plasma membrane. Condensing vacuoles bud off from the inner saccule and become smaller with a more spherical form as the contents become denser, sometimes crystalline.

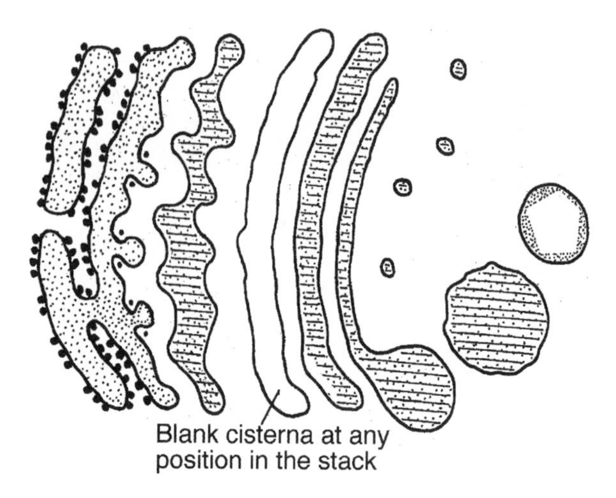

Figure 6 All components in a Golgi complex can react with osmium or lead, but in each stack one compartment is unstainable. The unstained compartment may be at any level in the sequence from transition vesicles to forming vacuoles. If osmiophilia indicates maturation of the cisternal environment, then the modification of GC cisternal contents through the stack may be a batch process.

Secretion may be regulated later when vesicles tend to accumulate, suggesting that hemolymph protein accumulation switches off first at the level of secretion and exocytosis. The ER then switches to metamorphic functions as the GC turns to autophagy (9, 28, 39).

Batch Processing by the Golgi Complex

It is usually assumed that membrane and cisternal contents move through GCs in a continuous assembly line of protein modification. However, the modification of GC cisternal contents through the stack may be discontinuous (Figure 6). All GC components can react with osmium or lead, but in each individual GC one compartment in the stack always remains unstained (54). Each GC is out of phase with others in the same cell. If osmiophilia is an indicator for the cisternal environment needed for a particular step in protein maturation (such as the reducing condition needed to allow disulfide bonds to break and reform), then maturation through the GC is a batch process rather than one of continuous assembly. This means that packages for different destinations might vary their contents in a temporal sequence rather than by having concurrent multiple addresses.

Golgi Complex Beads at the RER-GC Boundary

The ER face where the transition region begins lies within the GC zone of exclusion and lacks ribosomes. It also differs from other ER in having rings of GC beads attached to it (Figure 7). Transfer vesicles arise on the smooth face of ER next

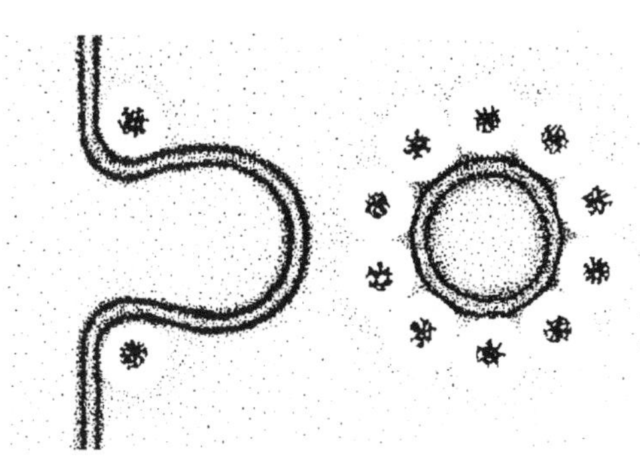

Figure 7 Golgi complex beads are arranged in rings on the smooth face of RER facing the transition region. The beads form rings through which transfer vesicles arise (32, 47).

to the outer or *cis* saccule of the GC. In the cytosol above this smooth face, particles about half the size of ribosomes are arranged in rings (46, 47). Beads stain characteristically with bismuth in all arthropods except Onychophora but not in GCs from other organisms (50). Vertebrates have particles in the same location but they lack distinctive staining, making them difficult to distinguish from other cell components (48). Insects therefore offer a special advantage for studies aimed at determining the role of beads (or their vertebrate equivalent) in the function of GCs (32, 40).

The Location and Arrangement of GC Beads

The beads are particles 10–12 nm in diameter lying 14 nm from the ER membrane surface. They occur only in the transition region, which is usually just wide enough to contain transfer vesicles separated from each surface. They are not present on the *cis* face of the first GC cisterna or on vesicles on other membrane surfaces. In normal, active GCs, the beads are in rings of 8–12 particles with a center-to-center separation of 27 nm, that is, they are equidistant from the membrane and one another. Some rings are located near flat membrane surfaces; others lie around the necks of attached transfer vesicles. After a vesicle has separated, the ring remains behind, keeping its constant 14-nm separation from the membrane surface. The beads are specific in their association with transfer vesicles and are therefore presumed to be concerned with movement from the RER.

The Structure and Chemistry of the Beads

Beads can only be identified easily through the specificity of their bismuth staining for electron microscopy. The reaction is extraordinarily specific. After glutaraldehyde fixation has blocked amino groups, bismuth reacts with few other cell components (49).

The staining of beads in unfixed cell fractions suggests that stable groups are present in live tissue (32). In vitro tests show that only exposed phosphate and guanidyl groups continue to bind bismuth after glutaraldehyde blocking (5, 49). Electron spectroscopic imaging detects enough phosphorus in the beads to account for their bismuth staining, suggesting that they are phosphorylated (6). A phosphorylated structure might be predicted for the beads if they have a role in the energy-dependent step in transport between the RER and the GC.

Bead Arrangement in Relation to Function

Low intracellular ATP blocks transfer vesicle formation and changes the configuration of the beads from rings to sheets (4). Beads attached to their ER membranes also collapse in sheets in cell fractions, except in the presence of ATP when they tend to survive in rings (32, 40). These effects of cellular ATP suggest that beads are part of an energy-transducing mechanism, the structural correlate for the energy-dependent step in the movement of transfer vesicles from the ER to the GC. They may be comparable to the GTP-bound conformation of the Rab9 guanosine triphosphatases in vertebrate GC cargo vesicle interactions (7, 80). If the rings are concerned only with departing vesicles, they could prevent returning smooth vesicles from having access to the central area of the saccule. They may determine the location where transfer vesicles return to the ER, directing traffic in a one-way central flow into the GC with a peripheral return.

Beads and the Formation of New GCs

At the beginning of the fifth larval stage of *C. ethlius*, the fat body has few GCs and RER is mainly in large distended cisternae (Figure 8). Some ER membrane surfaces lack polysomes but have one or a few forming transfer vesicles. At this stage of incipient GC formation, the beads form single or small numbers of rings on the smooth ER face. New rings appear as more abundant transfer vesicles fuse to make *cis* face saccules, which later mature into small stacks and complete GCs. Thus new GCs arise where there is an interaction between the ER and beads in the active ring form. The beads are organizers for GC development in the sense that they are the first recognizable elements to appear during the formation of new GCs (32).

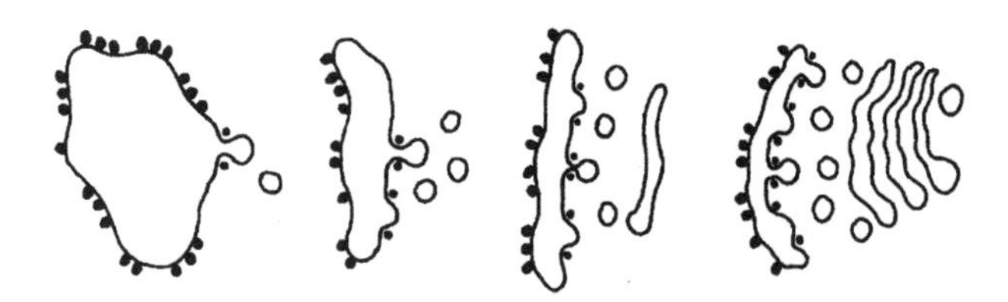

Figure 8 Golgi complex beads are organizers for Golgi complexes in development. Bead rings and Golgi complexes develop concurrently in the fat body from 5 to 40 h into the fifth stadium. The beads are organizers in the sense that they are the first recognizable elements to appear in the formation of new GCs.

Golgi Complex Survival During Mitoses

Vertebrate tissue culture cells close down many of their activities during mitosis. Endocytosis, phagocytosis, secretion, recycling, and the intracellular transport of newly synthesized proteins may all be inhibited when GCs and RER fragment to microvesicles (82). The accompanying vesiculation may be the way that these cells accomplish an equal partitioning of organelles between daughters at division. Such studies gave rise to the dogma that cell activities other than those related to cell division shut down during mitosis. However, unlike vertebrate tissues where differentiated cells no longer divide, insect epidermal cells have overlapping phases of division and secretion (24, 83). Their GCs are uninfluenced by mitosis, remaining structurally indistinguishable from those in interphase. If insects were like vertebrates, the intermolt would have to be given over to cell division rather than having division concurrent with the main phase of lamellate cuticle deposition. Although a contractile ring of cortical microfilaments cuts dividing epidermal cells in two, the apical cortex with microvilli remains functional for cuticle secretion. Even in cells with chromosomes in metaphase, chitin/protein microfibers continue to appear and orient into lamellae. GCs are structurally the same in dividing cells as in those only secreting cuticle precursors. GC beads, for example, continue to be in rings (83). Insect epidermal cells divide and secrete cuticle as easily as people walk and chew gum, contradicting narrow ideas arising from work restricted to vertebrate cells.

AUTOPHAGY

In contrast to vertebrate cells where time's arrow tends to lead irreversibly to differentiation, death, and whole-tissue regeneration, insect cells follow a path of "make do and mend," with cells regenerating and changing their function through life. Insect cells lend themselves to studies on remodeling. Except in the Diptera, insects reuse most of their larval cells at metamorphosis to make the adult. Cells even change their function and structure from molt to molt. Autophagy and remodeling are therefore common and important parts of the life cycle of most insect cells (45). Several mechanisms are used to externalize a whole or part of a cell for digestion in preparation for replacement.

Isolation Envelopes and Organelle-Specific Turnover in Phagic Vacuoles

At the end of the fifth larval intermolt, the fat body of a caterpillar switches from larval synthesis and secretion to storage for re-utilization during pupal and adult development (9, 27, 28, 39, 42, 45). The switch involves the selective removal of organelles and is induced by β-ecdysone in vitro (10). The most general mechanism for organelle turnover begins with the separation of the components to be digested in isolation envelopes. Isolation envelopes create convenient mitochondrion-sized isolation bodies by a mechanism for curving through the RER like an ice cream scoop. Primary lysosomes fuse with these isolation bodies, many of which may fuse together and lose their inner envelope membrane as they become autophagic vacuoles. All peroxisomes, many mitochondria, and most RER are destroyed in this way.

Organelle-Specific Autophagy

The structures involved in autophagy displayed in metamorphosing fat body suggest that the cytosolic membrane face of an isolation envelope recognizes different organelles and attaches to them at the appropriate time. Organelle turnover is not random. There is an organelle-specific sequence of isolation and fusion of the isolation bodies into autophagic vacuoles. Peroxisomes are the first to be destroyed followed by mitochondria and later the RER. The RER is cut up by the isolation envelopes into pieces about the same size as mitochondria. Autophagic vacuoles, especially those in late stages containing RER, often fuse with endocytosed hemolymph protein to create mixed auto- and heterophagic vacuoles that go on to become storage granules. At the completion of autophagy, the fat body has been turned into a store for iron, protein, lipid, glycogen, and urate, everything needed for adult development (9, 39, 42, 44).

Basal Lamina Turnover in Phagocytic Vacuoles

On the outside of some tissues, such as nerves, the basal lamina is in layers, as though it survives from molt to molt with a new layer being added at each molt. In other tissues, such as the fat body, the basal lamina is a single thin layer. The reason that it does not record successive laminae laid down in earlier stages is that the old basal lamina is phagocytosed and a new one is secreted at molting (28). For example, in third-to-fourth stage *C. ethlius* caterpillars, hemidesmosome-like adhesions between basal lamina and fat body plasma membrane concentrate in patches. These involute to become vacuoles, dragging the lamina into the vacuole lumen with them. Later vacuoles show stages of degradation, presumably after fusing with primary lysosomes. Basal lamina turnover in endocytic vacuoles is probably a specific example of a general primitive property of fat body and other cells, i.e., being able to phagocytose foreign material such as histolysed tissues. For example, in *Rhodnius* epidermis at metamorphosis, some cells undergo a phase of autophagy before being ingested by their neighbors (26, 29).

Multivesicular Bodies and Membrane Turnover in the Molt Cycle

At the cell level, a corollary of the metamorphic change in the form of an insect is the turnover of plasma membranes. The organelle by which this is accomplished is the multivesicular body (MVB). MVBs contain material from two sources, endocytic vesicles from the plasma membrane surface and primary lysosomes from the GC.

They lie in a cage of microfilaments that excludes ribosomes but not pinocytic vesicles or tubular membrane connections.

Epidermal microvilli disappear at the end of fifth stage lamellate cuticle formation, leaving a flat surface from which larval plaques (the tips of microvilli structured for both adhesion to and secretion of the lamellate cuticle) are endocytosed for digestion in MVBs (25, 51, 58). New pupal plaques then develop. Plaques are only absent from the apical membrane for a brief period prior to the formation of the new cuticle (51). This time marks the beginning of separation of the old cuticle from the epidermis, Hinton's (18) old fashioned and imprecise word for which was apolysis. The brief time when epidermis can be freed from the old cuticle continues into the subsequent phase, also apolysis in Hinton's terminology. The old cuticle is then easily detached because it is floating above molting fluid. It is not only the epidermis that separates at this time, but also its new protective envelope of epicuticle. Lamellate cuticle deposition follows envelope assembly on the same plaque surface. The reverse sequence does not occur. Plaques that have been used for fiber deposition are not then involved in envelope deposition. The clusters of enzymes needed for glucosamine chain elongation are different from those for envelope deposition. Envelope deposition signals a new molt and requires a new membrane with properties suitable for the next stage.

The turnover of apical plaques in MVBs is part of the more general process of replacement of the larval epidermal surface by the new pupal one. The feet disappear from the basal surface and their large membrane area involutes into lamellar bodies. Membrane turnover also occurs in the fat body at this time.

Multivesicular Bodies and Cuticle Uptake

Epidermal MVBs take up both cuticle and hemolymph components, endocytosing from both basal and apical surfaces. MVBs accumulate endocytosed tracers from the hemolymph (22) and fluorescently tagged hemolymph proteins. At molting they receive cuticle proteins. Phenolases secreted into the cuticle for the stabilization of the epicuticle envelope are afterward endocytosed into MVBs (21, 22, 57, 58). The cuticular environment is thus controlled dynamically by secretion and uptake.

Multivesicular Bodies in the Fat Body

MVBs are continually present in fat body cells, receiving plasma membrane, proteins bound to surface receptors (specific uptake), and proteins trapped in the center of the vesicle (nonspecific bulk flow). The proportion of membrane, specific and nonspecific protein uptake, varies with development. Pinocytosis occurs from all plasma membrane surfaces. Excess membrane brought to the MVB surface is compensated by inwardly budding microvesicles from which the MVB gets its name.

During the intermolt, fat body pinocytic vesicles carry little protein to MVBs and the turnover is mainly of membrane (27, 44). Prior to pupation many more microvesicles carry much more protein and the MVBs become granules storing concentrated hemolymph (9, 39). The switch to massive heterophagy is sudden, taking only a few hours. If the membrane needed to carry the mass of protein accumulating in the fat body is turned over rather than recycled, rough calculations show that it would be enough to replace the fat body plasma membrane every 10 minutes. Even allowing for a large margin of error, the larval fat body plasma membrane is probably completely replaced at metamorphosis. Membrane turnover would be required to remove larval receptors and introduce the new ones needed for internal signals in an imaginal environment.

Lamellar Bodies and Fat Body Membrane Turnover

When tyrosine storage vacuoles release their contents to the hemolymph just before ecdysis, there is a sudden return of membrane to the cell surface (65). Excess membrane folds into the cell and collapses to form lamellar bodies whose contents degrade after receiving primary lysosomes. Plasma membranes labeled with cationic ferritin may also pass to MVBs before turnover in lamellar bodies (2). Membranes from old fifth stage reticular systems are also disposed of in lamellar bodies prior to pupation. The loss of feet in the epidermis results in surplus membrane that also ends up in lamellar bodies. Lamellar bodies are probably of general importance in massive membrane turnover (28).

STORAGE VACUOLES, GRANULES, AND CONCRETIONS

The variety of insect secretions is legendary. Their cells have evolved vacuoles to serve as reaction and storage vessels for a wide range of molecules, another consequence of insects exploiting the efficiency of small-scale structures (Figure 9).

Vacuoles and Granules for Storing Urate

The storage of urate may be incompatible with the presence of the urate oxidase in peroxisomes needed for normal metabolism in trophocytes. Cockroaches and locusts have solved the problem by evolving separate cells specialized for urate storage that lack peroxisomes. Caterpillars have temporal separation, with urate storage confined to stages of development when peroxisomes are absent.

Urate storage vacuoles occur in urocytes, cells specialized for the accumulation and release of urate in cockroach fat body (8, 9). Urocytes have all cell components reduced except for vacuoles filled with urate. They lie adjacent to mycetocytes, packed between the more numerous trophocytes. The uniform structure and complexity of urocyte vacuoles suggest that they are long-lived structures, not constantly being formed from plasma membrane and recycled like the vacuoles for phenolic storage.

Locust urate cells lie between trophocytes on the outside of the fat body. They contain little except nucleus and vacuoles, which are smaller and more numerous

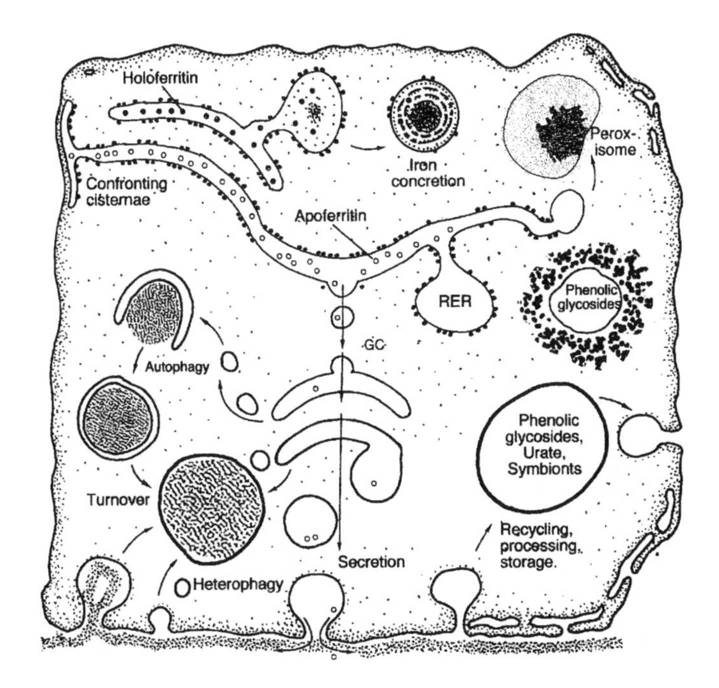

Figure 9 Vacuolar system compartments in a fat body cell (see Table 1). Nuclear envelope omitted.

than those of cockroaches. The vacuoles have a fibrous cortex, a core, and occasional microvesicles. In many ways they are intermediate in structure between cockroach urate vacuoles and lepidopteran urate granules.

Lepidoptera begin to store urate only when peroxisomes have been endocytosed prior to formation of the pupa. At this sharply defined stage a kind of MVB arises distinguished by its dense microvesicles and fibrous contents. They contain 75% urate and 25% protein in *Hyalophora cecropia* (81). The urate may come from the turnover of nucleic acids accompanying prepupal autophagy. The granules and their urate contents disappear during development of the adult (19).

Vacuoles for Storing Phenolics

Large fluid-filled vacuoles have been described in the fat body of few insects [*Chironomus*, *Aedes*, *Rhodnius*, *Leptinotarsa*, *Calliphora* (28)], but they are probably often overlooked, mistaken for dissolved lipid droplets. They are vacuoles for storing phenolics, probably in solution as the soluble b-glucosyl-o-tyrosine. In *C. ethlius* larvae the vacuoles occupy 40% of the cell volume at ecdysis to the fourth stage (65). They disappear during the next 6–12 h and a new population of fourth instar vacuoles arises. Provacuoles form from surface infolds and fuse to form vacuoles. By further fusion and growth, each cell comes to contain a few large vacuoles or even a single giant one 20–40 mm in diameter. Tyrosine increases in the fat body in step with vacuole growth and declines as it rises in the hemolymph prior to ecdysis. It disappears from the hemolymph as the new cuticle forms. A similar cycle of vacuole formation and release occurs at the fifth-to-pupal ecdysis.

Vacuoles Associated with Glycogen

Small vacuoles appear in the heart of the masses of glycogen accumulating in both fat body and epidermis during the fifth stage intermolt. They react like tyrosine glucoside storage vacuoles. The association of tyrosine vacuoles with glycogen could be functional, making carbohydrates readily available to transform phenolics into a safe and conveniently storable form. Glycogen-associated vacuoles could be general examples of a class of vacuole for storing phenolic glucosides, tyrosine vacuoles being an exaggerated form, specialized for the cyclical need of phenolics in cuticle formation (28, 39).

Heterophagy and Protein Granules

Hemolymph protein is endocytosed by all cells, especially at metamorphosis, but most conspicuously by the fat body. Through the early part of the fifth stadium, MVBs contain little but membrane fragments. They are largely autophagic, concerned with membrane turnover. Later in the stadium, when elevated ecdysteroids have signaled the beginning of pupation, proteins may make up 15% of the hemolymph. MVBs then switch to the heterophagy of hemolymph protein. Later still the MVB environment becomes suitable for either urate or protein storage.

Protein storage granules are composite structures. They contain endocytosed hemolymph proteins and lytic enzymes from primary lysosomes. They also fuse with autophagic vacuoles containing RER (9, 28, 43). Proteins crystallize out of this mixture and are stored until needed for adult development in the late pupa and early adult (19). Ferritin, derived from both endocytosed hemolymph and autophagocytosed RER, forms an iron store for adult development. It may occur throughout the crystal but often crystallizes out separately. Protein storage granules are therefore reaction vessels with two phases of activity. In the late larva they maintain conditions for the autophagy of RER and the crystallization of proteins for storage. In the late pupa and young adult they control conditions for the hydrolysis of their contents to the building blocks needed for adult development (19). Each phase is separate and probably needs primary lysosomes with different appropriate activities.

Vacuoles for Symbionts

Many insects live in symbiotic relationships with microorganisms lodged within some of their cells that form specialized mycetocytes. Fat body cells are commonly differentiated as mycetocytes in orthopteroids and hemipteroids (9, 27, 28). Most vacuoles have separate phases of filling, processing, and emptying, but symbiont vacuoles only rarely become digestive vacuoles. Symbiont cells deliver nutrients and receive bacterially synthesized molecules from the symbiont vacuoles concurrently. The problem for the fat body is how to control the environment in which symbionts are reared without compromising other functions. The mycetocytes of *Periplaneta americana* containing gram-negative bacteria are probably representative of the way that this is achieved. Cells specialized for symbiont care are surrounded by urocytes in the center of fat body lobes of trophocytes. Their association with urocytes must surely be meaningful. Symbiont cells contain glycogen, some RER and GCs, hundreds of bacteria housed in permanent vacuoles, and little else. The vacuoles have a thick membrane often ruffled, appearing superficially like SER, presumably derived from the cell surface, since bacteria infect the cell from outside. Growth of vacuole membranes keeps pace with that of the bacteria, dividing with them, ensuring that each bacterium has its own compartment.

Pigment Vacuoles

Some insect coloring depends on the contents of different kinds of epidermal vacuole. Most commonly, reddish pigments (16) occur in thin-walled vacuoles of unknown origin. Blue colors can be due to cyanoproteins endocytosed from the hemolymph.

APICAL-POLARIZED VACUOLES

The Creation of Gradients and the Coordination of Growth

The lateral face in the apical region of epidermal cells involutes on one side in microvesicles. These apical-polarized vesicles are location-specific caveolae perhaps identical to the argosomes described in *Drosophila* development (14). Growth varies in different regions of the epidermis. A general question for a cell in a tissue such as the epidermis is how it knows how much to grow, how many times to divide. Tracheae in culture are stimulated to molt by ecdysone but they do not increase in diameter; they do not grow (76). In a whole insect the diameter of main tracheae increases in proportion to growth of terminal tracheae, which varies with local respiratory demand. It is adaptive. Information about terminal growth is relayed to the main branch in a polarized fashion (20). The gradient that repeats in body and appendage segments may also function to control growth, since surface integument grafts that are excluded from the host gradient fail to thrive (41). Malpighian tubules also require epithelial continuity for normal growth from molt to molt. We might expect to find a polarized structure at the cell level correlated with such polarized transport of information. The subapical microvesicles found on one side of epidermal cells might be the structural correlate needed to account for gradient and polarized epithelial behavior. It is of especial interest that argosomes, membrane fragments carrying the wingless morphogen in *Drosophila* imaginal discs, are found predominantly in endosomes (14). Uptake of molecules by microendocytosis on one side of a cell in an epithelium could explain much of the polarized distribution of morphogens. Polarized subapical microvesicles may occur widely and have general significance for morphogenesis.

TRANSEPIDERMAL TRANSPORT AND CUTICLE SECRETION

The Epidermis

The epidermis of *C. ethlius* is involved with peptides from four main routing categories (56, 77, 79). Some cuticle peptides are secreted apically. Others follow a basal route into the hemolymph. A third class is routed both apically and basally while a fourth class is transported across the epidermis from the hemolymph (Figure 10). The transported class includes peptides synthesized by hemocytes (78) or the epidermis (63, 64) that are taken up later for transport to the cuticle. The problem is to determine how the structures in an epidermal cell can bring this about (Figure 11). Peptides in the first class have been detected in both secretory vesicles and the cuticle. Basal secretion could be by specific GC routing, but more probably it involves apical secretion and return because basally appearing peptides are detectable in the perimicrovillar compartment. Apical MVBs receive peptides from both apical and basal surfaces. They are presumably sorting compartments, directing peptides from the perimicrovillar compartment basally and hemolymph peptides apically to the cuticle. This could explain transepithelial transport as well as basal and bidirectional peptide routing.

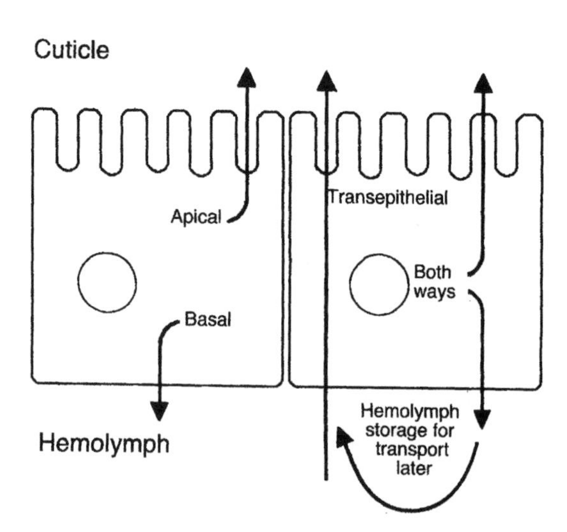

Figure 10 The routing categories for particular peptide classes in an epidermal cell (see Figure 11).

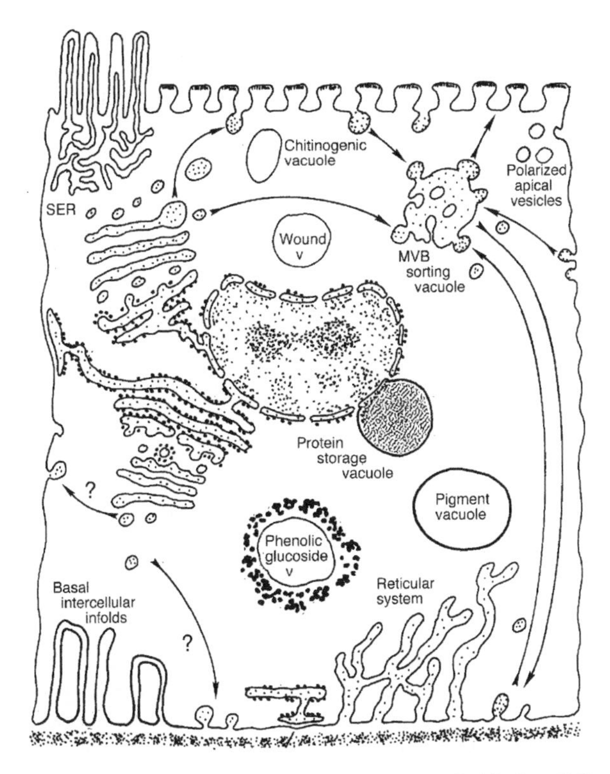

Figure 11 The vacuolar system of an epidermal cell (see Table 1 and Figure 10). Peptides are secreted apically and basally, in both directions, and transported across the cell from the hemolymph. The apical MVB is probably a sorting vacuole. Basally secreted peptides are probably secreted apically and endocytosed for basal return through the MVB. Some hemolymph peptides transported to the cuticle may have been synthesized earlier by the epidermis and other tissues.

CONCLUSIONS

Differences from Vertebrates

Insect cells differ from those of vertebrates. They often function within their own compartments, in a lymph space interfacing with the hemolymph or a perimicrovillar space separate from that of the cuticle. Inside the cell their GCs have a batch processing system and the release of transfer vesicles involves GC beads. Ferritin occurs in their vacuolar system rather than in the cytosol.

Insects differ from vertebrates in the mechanism for change in development. Vertebrates change by differentiation to create a functionally mature tissue followed by cell death. There is no redifferentiation. Insects, on the other hand, are continually changing through the autophagy and redifferentiation of cell components. They are continually changing their structure in the molt/intermolt/metamorphic sequence. Cell division coincides with other activities. Rather than turn over whole cells, they divide and retool organelles and surface membranes. Insect cells can therefore seem enormously complex. For example, a liver cell has a structure, but a fat body cell has many structures depending on the stage of development.

ACKNOWLEDGMENT

This work was supported by NSERC grant A6607.

The *Annual Review of Entomology* is online at http://ento.annualreviews.org

LITERATURE CITED

1. Brac T. 1983. Charged sieving by the basal lamina and the distribution of anionic sites on the external surfaces of fat body cells. *Tissue Cell* 15:489–98
2. Brac T. 1984. Lamellar bodies are the intracellular site of membrane turnover in the fat body. *Tissue Cell* 15:873–84
3. Brac T. 1984. The charge distribution in the rough endoplasmic reticulum Golgi complex transitional area injected by the injection of charged tracers. *Tissue Cell* 16:859–71
4. Brodie DA. 1981. Bead rings at the endoplasmic reticulum/Golgi complex boundary: morphological changes accompanying inhibition of intracellular transport of secretory proteins in arthropod fat body tissue. *J. Cell Biol.* 90:92–100
5. Brodie DA. 1982. Bismuth chemistry and the glutaraldehyde insensitive staining reaction. *Tissue Cell* 14:39–46
6. Brodie DA, Locke M, Ottensmeyer FP. 1982. High resolution microanalysis for phosphorus in golgi complex beads of insect fat body tissue by electron spectroscopic imaging. *Tissue Cell* 14:1–11
7. Carroll KS, Hanna J, Simon I, Krise J, Barbero P, et al. 2001. Role of Rab9 GTPase in facilitating receptor recruitment by TIP 47. *Science* 292:1373–78
8. Cochran DG, Mullins DE, Mullins KJ. 1979. Cytological changes in the fat body of the American cockroach, *Periplaneta americana*, in relation to dietary nitrogen levels. *Ann. Ent. Soc. Am.* 72:197–207
9. Dean R, Collins JV, Locke M. 1985. See Ref. 18a, pp. 155–210
10. Dean RL. 1978. The induction of autophagy in isolated insect fat body by β-ecdysone. *J. Insect Physiol.* 24:439–47
11. Dean RL, Bollenbacher WE, Locke M, Smith SL, Gilbert LI. 1980. Hemolymph ecdysteroid levels and cellular events in the intermoult/moult sequence of *Calpodes ethlius*. *J. Insect Physiol.* 26:267–80
12. Ghiradella H. 1998. The anatomy of light production: the fine structure of the firefly lantern. In *Microscopic Anatomy of Invertebrates, Insecta*, ed. FW Harrison, M Locke, pp. 363–81. New York:Wiley
13. Gouranton J, Folliot R. 1968. Presence de cristaux de ferritine de grande taille dans les cellules de l'Intestin moyen de *Campylenchia latipes* Say. (Homoptera, Membracidae). *Rev. Can. Biol.* 27:77–81
14. Greco V, Hannus M, Eaton S. 2001. Argosomes: a potential vehicle for the spread of morphogens through epithelia. *Cell* 106:633–45
15. Heckman DS, Geiser DM, Eidell BR, Stauffer R, Kardos NL, et al. 2001. Molecular evidence for the early colonization of land by fungi and plants. *Science* 293:1129–33

16. Hori M, Riddiford LM. 1981. Isolation of ommochromes and 3-hydroxykynurenine from the tobacco hornworm, *Manduca sexta*. *Insect Biochem.* 11:507–13
17. Jackson A, Locke M. 1989. The formation of the plasma membrane reticular systems in the oenocytes of an insect. *Tissue Cell* 21:463–73
18. Jenkin PM, Hinton HE. 1966. Apolysis in arthropod moulting cycles. *Nature* 211: 871
18a. Kerkut GA, Gilbert LI. 1985. *Comprehensive Insect Physiology, Biochemistry and Pharmacology*. Oxford: Pergamon
19. Larsen WJ. 1976. Cell remodelling in the fat body of an insect. *Tissue Cell* 8:73–92
20. Locke M. 1958. The coordination of growth in the tracheal system of insects. *Q. J. Microbiol. Sci.* 99:373–91
21. Locke M. 1969. The localization of a peroxidase associated with hard cuticle formation in an insect, *Calpodes ethlius*, Stoll, Lepidoptera, Hesperiidae. *Tissue Cell* 1:555–74
22. Locke M. 1969. The structure of an epidermal cell during the formation of the protein epicuticle and the uptake of molting fluid in an insect. *J. Morphol.* 127:7–40
23. Locke M. 1969. The ultrastructure of the oenocytes in the moult/intermoult cycle of an insect. *Tissue Cell* 1:103–54
24. Locke M. 1970. The moult/intermoult cycle in the epidermis and other tissues of an insect *Calpodes ethlius* (Lepidoptera, Hesperiidae). *Tissue Cell* 2:197–223
25. Locke M. 1976. The role of plasma membrane plaques and golgi complex vesicles in cuticle deposition during the molt/intermolt cycle. In *The Insect Integument*, ed. HR Hepburn, pp. 237–58. Amsterdam: Elsevier
26. Locke M. 1980. Cell structure during insect metamorphosis. In *Metamorphosis: A Problem in Developmental Biology*, ed. E Frieden LI Gilbert, pp. 75–103. New York: Plenum
27. Locke M. 1980. The cell biology of fat body development. In *V. B. W. 80, Insect Biology in the Future*, ed. M Locke, DS Smith, pp. 227–52. New York: Academic
28. Locke M. 1984. The structure and development of the vacuolar system in the fat body of insects. In *Insect Ultrastructure*, ed. H Akai, R King, pp. 151–97. New York: Plenum
29. Locke M. 1985. A structural analysis of postembryonic development. See Ref. 18a, pp. 87–149
30. Locke M. 1986. The development of the plasma membrane reticular system in the fat body of an insect. *Tissue Cell* 18:853–67
31. Locke M. 1988. Insect cells for the study of general problems in biology-somatic inheritance. *Int. J. Insect Morphol. Embryol.* 1:419–36
32. Locke M. 1990. Golgi complex beads and the transition region. Problems and paradigms. *BioEssays* 12:495–501
33. Locke M. 1991. Apoferritin in the vacuolar system of insect hemocytes. *J. Insect Physiol.* 23:367–75
34. Locke M. 1991. Insect epidermal cells. In *Physiology of the Insect Epidermis*, ed. A Retnakaran, K Binnington, pp. 1–22. Melbourne: CSIRO
35. Locke M. 1998. Caterpillars have evolved lungs for hemocyte gas exchange. *J. Insect Physiol.* 44:1–20
36. Locke M. 1998. Reticular systems and intercellular lymph spaces. In *Microscopic Anatomy of Invertebrates, Insecta*, ed. FW Harrison, M Locke, pp. 51–71. New York: Wiley
37. Locke M. 1998. Epidermis. In *Microscopic Anatomy of Invertebrates, Insecta*, ed. FW Harrison, M Locke, pp. 75–138. New York: Wiley
38. Locke M. 1998. Lenticles. In *Microscopic Anatomy of Invertebrates, Insecta*, ed. FW Harrison, M Locke, pp. 209–17. New York: Wiley
39. Locke M. 1998. The fat body. In *Microscopic Anatomy of Invertebrates, Insecta*, ed. FW Harrison, M Locke, pp. 641–86. New York: Wiley

40. Locke M. 1998. Golgi complexes and GC beads. In *Microscopic Anatomy of Invertebrates, Insecta*, ed. FW Harrison, M Locke, pp. 1177–83. New York: Wiley
41. Locke M. 2001. The Wigglesworth lecture: insects for studying fundamental problems in biology. *J. Insect Physiol.* 47:495–507
42. Locke M, Collins JV. 1965. The structure and formation of protein granules in the fat body of an insect. *J. Cell Biol.* 26:857–85
43. Locke M, Collins JV. 1966. The sequestration of protein by the fat body of an insect. *Nature* 210:552–53
44. Locke M, Collins JV. 1968. Protein uptake into multivesicular bodies and storage granules in the fat body of an insect. *J. Cell Biol.* 36:453–83
45. Locke M, Collins JV. 1980. Organelle turnover in insect metamorphosis. In *Pathological Aspects of Cell Membranes*, ed. BF Trump, A Arstila, pp. 223–48. New York: Academic
46. Locke M, Huie P. 1975. The golgi complex/endoplasmic reticulum transition region has rings of beads. *Science* 188: 1219–18
47. Locke M, Huie P. 1976. The beads in the golgi complex/endoplasmic reticulum region. *J. Cell Biol.* 70:384–94
48. Locke M, Huie P. 1976. Vertebrate golgi complexes have beads in a similar position to those found in arthropods. *Tissue Cell* 8:739–43
49. Locke M, Huie P. 1977. Bismuth staining for light and electron microscopy. *Tissue Cell* 9:347–71
50. Locke M, Huie P. 1977. Bismuth staining of the golgi complex is a characteristic arthropod feature lacking in *Peripatus*. *Nature* 270:341–43
51. Locke M, Huie P. 1979. Apolysis and the turnover of plasma membrane plaques during cuticle formation in an insect. *Tissue Cell* 11:277–91
52. Locke M, Huie P. 1980. The nucleolus during epidermal development in an insect. *Tissue Cell* 12:175–94
53. Locke M, Huie P. 1983. A function for plasma membrane reticular systems. *Tissue Cell* 15:885–902
54. Locke M, Huie P. 1983. The mystery of the unstained golgi complex cisternae. *J. Histochem. Cytochem.* 31:1019–32
55. Locke M, Ketola-Pirie C, Leung H, Nichol H. 1991. Vacuolar apoferritin synthesis by the fat body of an insect (*Calpodes ethlius*). *J. Insect Physiol.* 37:297–309
56. Locke M, Kiss A, Sass M. 1994. The cuticular localization of integument peptides from particular routing categories. *Tissue Cell* 26:707–34
57. Locke M, Krishnan N. 1971. Distribution of phenoloxidases and polyphenols during cuticle formation. *Tissue Cell* 3:103–26
58. Locke M, Krishnan N. 1973. The formation of the ecdysial droplets and ecdysial membrane in an insect. *Tissue Cell* 5:441–50
59. Locke M, Leung H. 1984. The induction and distribution of an insect ferritin—a new function for the endoplasmic reticulum. *Tissue Cell* 16:739–66
60. Locke M, McMahon JT. 1971. The origin and fate of microbodies in the fat body of an insect. *J. Cell Biol.* 48:61–78
61. Locke M, Nichol H. 1992. Iron economy in insects: transport, metabolism and storage. *Annu. Rev. Entomol.* 37:195–215
62. Locke M, Russell VW. 1998. Pericardial cells or athrocytes. In *Microscopic Anatomy of Invertebrates, Insecta*, ed. FW Harrison, M Locke, pp. 687–709. New York: Wiley
63. Marcu O, Locke M. 1998. A cuticular protein from the moulting stage of an insect. *Insect Biochem. Mol. Biol.* 28:659–69
64. Marcu O, Locke M. 1999. The origin, transport and cleavage of the molt-associated cuticular protein CECP22 from *Calpodes ethlius* (Lepidoptera, Hesperiidae). *J. Insect Physiol.* 45:861–70
65. McDermid H, Locke M. 1983. Tyrosine storage vacuoles in insect fat body. *Tissue Cell* 15:137–58

66. Nichol H, Locke M. 1990. The localization of ferritin in insects. *Tissue Cell* 22:767–77
67. Nichol H, Locke M. 1995. Honeybees and magnetoreception. Technical comments. *Science* 269:1888–89
68. Nichol H, Locke M. 1999. Secreted ferritin subunits are of two kinds in insects: molecular cloning of cDNAs encoding two major subunits of secreted ferritin from *Calpodes ethlius*. *Insect Biochem. Mol. Biol.* 29:999–1013
69. Nichol H, Locke M. 1989. The characterization of ferritin in an insect. *Insect Biochem.* 19:587–602
70. Raes H, Bohyn W, De Rycke PH, Jacobs F. 1989. Membrane bound iron rich granules in fat cells and midgut cells of the adult honey bee (*Apis mellifera* L.). *Apidologie* 20:327–37
71. Rebers JE, Willis JH. 2001. A conserved domain in arthropod cuticular proteins binds chitin. *Insect Biochem. Mol. Biol.* 31:1083–93
72. Reddy JT, Locke M. 1989. The size limited penetration of gold particles through insect basal laminae. *J. Insect Physiol.* 36:397–407
73. Renaud DL, Nichol H, Locke M. 1991. The visualization of apoferritin in the secretory pathway of vertebrate liver cells. *J. Submicrosc. Cytol. Pathol.* 23: 501–7
74. Rothman JE, Wieland FT. 1996. Protein sorting by transport vesicles. *Science* 272: 227–34
75. Ryerse JS. 1979. Developmental changes in Malpighian tubule cell structure. *Tissue Cell* 11:533–51
76. Ryerse JS, Locke M. 1978. Ecdysterone-mediated cuticle deposition and the control of growth in insect tracheae. *J. Insect Physiol.* 24:541–50
77. Sass M, Kiss A, Locke M. 1993. Classes of integument peptides. *Insect Biochem. Mol. Biol.* 23:845–57
78. Sass M, Kiss A, Locke M. 1994. Integument and hemocyte peptides. *J. Insect Physiol.* 40:407–21
79. Sass M, Kiss A, Locke M. 1994. The localization of surface integument peptides in tracheae and tracheoles. *J. Insect Physiol.* 40:5621–75
80. Segev N. 2001. A TIP about Rabs. *Science* 292:1313–14
81. Tojo S, Betchaku T, Ziccardi VJ, Wyatt GR. 1978. Fat body protein granules and storage proteins in the silk moth *Hyalophora cecropia*. *J. Cell Biol.* 78:823–38
82. Warren G. 1985. Membrane traffic and organelle division. *TIBS* 10:439–43
83. Zeng W, Locke M. 1993. The persistence of Golgi complexes during cell division in an insect epidermis. *Tissue Cell* 25: 709–23

Annu. Rev. Entomol. 2003. 48:29–50
doi: 10.1146/annurev.ento.48.091801.112605

First published online as a Review in Advance on October 8, 2002

COMMUNICATION WITH SUBSTRATE-BORNE SIGNALS IN SMALL PLANT-DWELLING INSECTS[1]

Andrej Čokl and Meta Virant-Doberlet
Department of Invertebrate Physiology, National Institute of Biology, Večna pot 111, P.O.Box 141, SI-1001 Ljubljana, Slovenia; e-mail: andrej.cokl@uni-lj.si; meta.virant-doberlet@uni-lj.si

Key Words Hemiptera, Heteroptera, Auchenorrhyncha, substrate vibration, behavior, orientation

■ **Abstract** Vibratory signals of plant-dwelling insects, such as land bugs of the families Cydnidae and Pentatomidae, are produced mainly by stridulation and/or vibration of some body part. Signals emitted by the vibratory mechanisms have low-frequency characteristics with a relatively narrow frequency peak dominant around 100 Hz and differently expressed frequency modulation and higher harmonics. Such spectral characteristics are well tuned to the transmission properties of plants, and the low attenuation enables long-range communication on the same plant under standing wave conditions. Frequencies of stridulatory signals extend up to 10 kHz. In some groups, vibratory and stridulatory mechanisms may be used simultaneously to produce broadband signals. The subgenual organ, joint chordotonal organs, campaniform sensilla and mechanoreceptors, such as the Johnston's organ in antennae, are used to detect these vibratory signals. Species-specific songs facilitate mate location and recognition, and less species-specific signals provide information about enemies or rival mates.

CONTENTS

[1]Definitions and descriptions for this article: (*a*) *Bending waves*: For bending (or flexural) waves it is characteristic that particles move in a plane perpendicular to the direction of wave propagation and to the surface with stresses and strains in the longitudinal direction. Their propagation is dispersive (frequency-dependent). In short pulses of sinusoidal carrier wave, the propagation velocity of the pulse envelope (the *group velocity*) is twice that of the *phase velocity* (propagation velocity of the carrier wave, the velocity of movement necessary

0066-4170/03/0107-0029$14.00

INTRODUCTION

Although vibratory communication was recognized long ago in many insect lineages (109), intensive investigations of its basic phenomena started only in the past few decades, enabled by the development of advanced methods for signal recording from natural substrates. The basic principles of plants as transmission media for vibratory signals have been described (5, 86), and in many respects these characteristics determine signal production and the mode of reception. Many songs and signals are species specific and facilitate mate location, recognition, and pair formation. Many insects cannot communicate over long distances using airborne signals, as bush crickets do, owing to their small size. In the latter, bimodal processing of auditory and vibratory components significantly improves signal resolution (74). Vibratory signals are also produced by potential host, prey, or enemy. The resolution of vibratory signals can be influenced by abiotic factors such as the vibratory noise produced by wind and raindrops.

Substrate-borne communication has been described in a variety of small insects such as bees (85), termites (61), ants (81), lacewings (58), Auchenorrhyncha (15),

---

to remain at the same phase of a sinusoidal wave motion). (*b*) *Frequency*: the number of complete oscillations (or cycles) made per unit time by a vibrating or oscillating medium or object. Frequency modulation is modulation in which the instantaneous frequency of the modulated wave or carrier is made to vary in proportion to the instantaneous amplitude of the modulating wave. Frequency spectrum is the range of frequencies covered by the frequency components of a signal. When a periodic quantity (an oscillatory quantity whose values recur for certain equal increments of time) is not a pure sine wave, it can be resolved by Fourier's Theorem into sinusoidal components (which, in a note, represent the constituent pure tones). The first of which has the same frequency as the quantity itself, called the *fundamental frequency*; and the remainder of which, called *harmonics*, are whole multiples of this. The expressions *dominant* or *principal frequency* are often used synonymously with fundamental frequency, although the dominant frequency should be reserved for the frequency of the harmonic that has the greatest amplitude. (*c*) *Resonance*: condition of peak vibratory response where a small change in excitation frequency causes a decrease in system response. (*d*) *Spike phase locked response*: action potentials (spikes) of a neuron appear constantly as a response to a certain phase of stimulatory sine waves. (*e*) *Standing wave*: a periodic wave having a fixed distribution in space that is the result of interference of progressive waves of the same frequency and kind and characterized by the existence of maxima and minimal amplitudes that are fixed in space. (*f*) *Vibration*: the oscillatory motion of a body upon which a force (whose magnitude is proportional to the displacement) is acting. *Intensity* of vibrations may be expressed as displacement, velocity or acceleration. *Displacement* is a vector quantity that specifies the change of position of a body, usually measured from the rest position. *Velocity* is a vector quantity that specifies time rate of change of displacement. *Acceleration* is a vector quantity that specifies rate of change of velocity. *Threshold acceleration* (*displacement, velocity*) *sensitivity* is the lowest intensity (expressed in acceleration, displacement or velocity units) of a vibratory stimulus that elicits responses of an animal, sensory organ, or sensory cell. (*g*) *Wavelength*: the distance along a periodic wavefront between points of comparable amplitude with a phase difference of one period.

stone flies (109), and beetles (60). In 1984 Gogala provided a brief overview of vibration-producing mechanisms and song characteristics as a taxonomic character of land bugs (48). However, a review including different aspects of substrate-borne communication in plant-dwelling land bugs is lacking. To fill this gap we summarized the data on vibrational communication in land bugs, focusing mainly on Pentatomidae and the southern green stink bug *Nezara viridula*. *N. viridula* is an economically important pest on different plants (94) and has been the focus of the scientific community (115) as a model species within the whole family. This view compares the vibratory signal production, transmission and recognition between different groups of small plant-dwelling insects. Here the term Auchenorrhyncha refers to auchenorrhynchus insects other than the cicada (leaf hoppers, treehoppers, frog hoppers, spittlebugs, and plant hoppers).

SIGNAL PRODUCTION

Insects produce signals by stridulation, percussion, vibration, click mechanisms, and air expulsion (43). Stridulation, the process whereby sound or vibration is produced by friction of two body parts moving across one another, is widespread among Heteroptera (48). Many land bugs also emit low-frequency vibratory signals, produced by vibration of body parts by direct muscular action, without the intervention of some method of frequency multiplication. Production of low-frequency signals is accompanied in land bugs by movement of the abdominal tergal plate, previously termed the tymbal. When movement of this plate was prevented, by waxing it to the scutellum, vibratory signals were not produced. Morphologically different abdominal tergal plates have been described in Reduviidae, Piesmatidae, Lygaeidae, Coreidae, Cydnidae, Scutelleridae, Pentatomidae, and Acanthosomatidae (48). In addition, low-frequency vibratory signals recorded in the Plataspidae and Rhopalidae (49) indicate that the same vibration-producing mechanism emits them.

In the southern green stink bug *N. viridula*, the first and second abdominal tergites are fused into a forward-backward movable tymbal-like plate that is loosely fixed, anterior and posterior, to the thorax and to the third abdominal tergum, by a chitinous membrane, and more firmly, laterally, to the pleurites. Recording the electrical activity from the tergal plate longitudinal and lateral compressor muscles in freely moving and singing animals demonstrated that most of them are directly involved in producing vibratory signals (77). They contract synchronously and in phase with these vibratory waves. The first and second pair of tergal longitudinal muscles (TLI and TLII) move the plate forward and backward, and contraction of compressor muscles moves it in a latero-ventral direction. Multiple peaks of electromyogram potentials indicate that motor units within a muscle are not contracted simultaneously but one after another with short time delays. Anterior downward movement of the plate accompanies posterior downward movement of the rest of the abdomen by lifted wings. Waxing the suture line between the two fused tergites does not prevent signal production in the southern green stink bug but stopped their emission in Cydnidae (41). This indicates that signals are produced by movement of the plate as a unit, in contrast to Cydnidae, in which opening and closing the fold at the suture line between the two abdominal tergites was described as essential for signal production (48). Although the tymbal-like, abdominal tergal plate is present in most land bugs investigated, some of them produce vibratory signals without its movement. In the bean bug *Riptortus clavatus* (Alydidae), male songs were also recorded after the plate was fixed to the thorax and other abdominal terga (91). Exactly how these vibratory signals are produced by this mechanism is not known.

In many respects principles of vibratory signal production in land bugs can be compared with those of Auchenorrhyncha (89, 92, 106, 108). In the latter, specialized thin and often striated areas of cuticle, which are located dorsolaterally on either side of the first abdominal segment, constitute tymbal-like, sound-producing organs that are morphologically similar to the cicada tymbal but lack associated resonant structures such as air sacs. The specialized muscles attach to internal apodemes within the metathorax and the first two abdominal segments. The tymbal-like, vibratory sound-producing apparatus is less developed in females, and in some species no obvious tymbal apparatus is found. In some species signal production is associated with dorsoventral vibrations of the whole abdomen, without striking the substrate (89). Detailed physiological investigations are still lacking owing to the small size of these insects, and the mechanism of sound production is not yet understood. The vibrational songs of Auchenorrhyncha also contain components produced by abdominal percussion and wing flicks (64). Such signals probably carry relevant information for the receiving insects (86).

A sound source can efficiently transfer power to the surrounding medium if its radius is 1/6 (monopole) or 1/4 (dipole) of the sound wavelength (7). The intensity of the airborne component of 100–300 Hz vibratory signals emitted by bugs, 1 cm or less in length, is low. According to Markl (80) a 1-cm animal radiates poorly below 10 kHz in air. Sound pressure values for the airborne component of land bugs' vibratory and stridulatory emissions were not measured, but comparative data were available for drosophilid fruit flies, which are some of the smallest insects known to communicate through air at an effective distance of a few mm (7). At this range, the estimated sound pressure of 200–450 Hz signals was just 35 dB SPL (re 2×10^{-5} Pa) (6). The intensity of the vibratory component of the emitted signal measured on the back of the singing *N. viridula* reached values between 16 and 32 mm s^{-1} (26), and an approximately 10-times-lower velocity was measured on the surface of a *Hedera helix* plant immediately below the legs of the singing animals (2). These values lie in the range of those measured normal to the surface of the plants for a variety of cydnid bugs and plant hopper and leaf hopper songs (86).

The dominant frequency of signals produced by the vibratory mechanism lies below 200 Hz and above 50 Hz in most Heteroptera and Auchenorrhyncha species (48, 86). Spectra characterized by relatively narrow frequency peaks are, in many land bug species, enlarged by the presence of distinct higher harmonics, as in *Platyplax salviae* (Lygaeidae), or by extensive frequency modulation, as for example in

Piezodorus lituratus (Pentatomidae) (53). Nevertheless, intensities of frequency components above 500 Hz lie below the threshold of most sensitive vibrational receptors, as determined in the southern green stink bug (25).

Spectra of stridulatory signals cover a wider range of frequencies. In Reduviidae, the carrier frequency ranges from 700–800 Hz, as is characteristic for the male-deterring stridulations (99), to 1600–3200 Hz, as determined for disturbance stridulations (104). The differences in carrier frequencies (1500 and 2000 Hz) between two different stridulatory signals of *Rhodnius prolixus* were attributed to the different velocities of rubbing the proboscis against the prosternal stridulatory organ (79). Frequency modulation is also present in stridulatory signals; in *Enoplops scapha* (Coreidae), Gogala (48) registered fluctuation of the dominant frequency between 1.5 and 11 kHz.

In many land bug families such as Cydnidae and Reduviidae, the presence of the abdominal tergal plate and stridulatory apparatus indicates that vibratory signals may be produced by both mechanisms. In *Rhinocoris iracundus* (Reduviidae), low-frequency components of carrier frequency below 200 Hz are exchanged with frequency-modulated stridulatory components whose dominant frequency lies between 1 and 2 kHz (48). The repertoire of Cydnidae contains signals produced by stridulatory or vibratory mechanisms, as well as by mixed signals (48, 52). Although stridulatory and vibratory signals cover a broad range of frequencies, the low-frequency components are more suitable for longer-range communication through plants. Spectral analyses of signals recorded simultaneously as substrate-borne and airborne sound have shown that the frequency range of both components is almost identical but that the main energy is emitted at lower frequencies of the vibrational signals (86). The energy of stridulatory components of the cydnid bug songs, measured as weak airborne sounds, extends to at least 12 kHz, with the main energy emitted between 3 and 4 kHz (51). Plant vibrations induced by their singing are below 2–3 kHz, with the main energy below 1.5 kHz and in most cases below 0.5 kHz.

TRANSMISSION THROUGH PLANTS

The pioneer work of Michelsen et al. (86) described properties of plants that act as transmission channels for insect vibrational songs. They proved that bending or flexural waves are used for substrate-borne sound communication. The group propagation velocity depends on both the vibratory signal frequency and the plant's mechanical properties. Higher-frequency signals propagate faster than those of lower-frequency signals. The measured-group propagation velocity for 200 Hz signals in bean is 36 m s^{-1} and for 2000 Hz it is 120 m s^{-1} (86), and in an *Agave* leaf, values of 8 m s^{-1} for 30 Hz, 10 m s^{-1} for 59 Hz, and 18 m s^{-1} for 80 Hz vibrations were measured (5). The 30-Hz vibration group velocity was 4.4 m s^{-1}, measured at the apical third in an *Agave* leaf, and increased to 35.7 m s^{-1} on the leaf's basal region. Finally 200-Hz signals are transmitted through *Phragmites communis* with a group velocity of 75 m s^{-1} and through *Acer pseudoplatanus* at 95 m s^{-1} (86).

Low attenuation of vibratory signals transmitted through plants enables communication over longer distances. Behavioral tests demonstrated that the male southern green stink bug *N. viridula* responded to female calling when they were on top of a different cyperus (*Cyperus alternifolius*) stem 2 m apart that were coupled mechanically only by roots and surrounding earth (A. Čokl, unpublished data). The intensity of signals recorded on a plant below the leg of the singing bug was about 4 mm s^{-1}, on the bottom of the same stem (distance 80 cm) it decreased to around 3 mm s^{-1}, and at the top of the neighboring stem 200 cm away it was approximately 0.5 mm s^{-1} (N. Stritih & A. Čokl, unpublished data). This results in about 0.1 dB/cm calculated attenuation between apical points on neighboring stems and just about 0.04 dB/cm on the same stem (80-cm distance). Low-average attenuation (0.3 dB/cm) was demonstrated also on a banana leaf for 75-Hz signals (4). Magal et al. (78) showed that, in ~6-cm-long apple leaves, the loss of signal energy for the midvein decreased from 80% at the leaf base to 40% at the apex, for minor veins from 70% to 31%, and for homogeneous regions about 40%. The authors demonstrated that the midvein acts as a low pass filter for signal frequencies below 1.7 kHz.

Because of vibratory signals, low internal damping, and reflections, standing waves occur in stems and other rod-like parts of a plant. In such conditions the intensity of the pure tone vibratory signals does not decrease monotonically with distance. The distance between nodes and internodes decreases with increasing frequency of the signal. In a cyperus stem, the intensity of a 124-Hz pure tone signal varied about 20 dB when measured on nodes and internodes, and the intensity values measured 17 and 32 cm from the source (positions of the first and second internode) were almost the same (26). Together with the 124-Hz component, a new 84-Hz component appeared whose velocity remained above the input value throughout the entire length of the stem. The velocity value at the first two internodes (20 cm apart) was almost the same. The dominant frequency peak of the vibratory signal decreased from around 141 Hz, when measured on the back of the singing animal, to about 91 Hz at the surface of a cyperus stem 29 cm away, which confirms that spectral changes of a signal occur during transmission through a plant (26). Michelsen et al. (86) measured filtering properties of *Thesium bavarum*, which is a characteristic host plant for many cydnid bug species. They found that the vibration amplitude might change 10–30 dB when the frequency of vibration changes by less than 10%. In the frequency range below 1 kHz, they found that pure tones around 100 Hz are amplified at a distance of 3 cm from the source. Low-frequency vibratory signals with the dominant frequency around 100 Hz are thus well tuned with the transmission properties of plants.

Although broadbanded-mixed stridulatory and vibratory signals are supposed to be more convenient for substrate-borne communication through plants (86), there are many land bug species that emit only narrow-band vibratory signals with dominant frequencies around 100 Hz. Even in mixed signals, the main emitted energy is concentrated below 500 Hz, and the low-frequency vibratory component does not always appear simultaneously with the higher-frequency stridulatory one.

Higher attenuation of higher-frequency signals (86) makes stridulatory components less relevant for long-distance communication through a plant. Attention has to be paid to higher harmonics and discrete frequency-modulated components of narrow-banded natural signals, both of which may minimize extensive amplitude differences characteristic of pure tones in standing wave conditions. Frequency sweeps well above 10%, in the frequency range of ~100 Hz, are present in signals of the southern green stink bug and many other land bug species. According to high-amplitude differences that occur by frequency changes of 10% (86), we may expect that some components of the signal will come through at any given point of a plant, so that the position of the singing and listening animal will not critically influence communication through the plant.

A vibrational pulse reflects both at the root and top of the plant, and because of small internal damping, the reflected waves travel up and down the stem several times. Reflections change the pattern of the input signal. A short sine wave pulse (2–3 cycles of a 2 kHz) repeated several times caused 20-ms or longer movement of a plant, with a frequency-dependent pattern (86). Variation in duration of 100-ms signals transmitted through a cyperus stem depended on the intensity of the induced 84-Hz component, and at internodes a signal duration of more than 200 ms could be measured (26).

Several abiotic factors may mask insect vibratory signals in the environment and decrease the signal-to-noise ratio. Casas et al. (12) demonstrated that a drop falling on a leaf produces a vibratory signal, whose 9- to 29-ms-long irregular phase of maximal velocity between 76.1 and 137 mm s^{-1} had broadbanded-frequency characteristics spanning up to 25 kHz. The basic frequency of the regular phase ranged between 5.7 and 10.5 Hz and its amplitude decreased exponentially. Wind-induced basic oscillations of a leaf below 15 Hz with maximal intensity range from 30–60 mm s^{-1} for low to 70–130 mm s^{-1} for high wind speed. Although the basic oscillation of 15 Hz did not depend on wind speed, the irregular vibrations induced by air movement contained frequencies up to 25 kHz.

SENSORY ORGANS

Legs are the site of sensory organs that detect vibratory signals with highest sensitivity. Studies of leg transmission properties and the biophysics of sensory organs comparable with those performed in bees (75, 101) are lacking in land bugs. The anatomical (84) and functional (25) properties of the leg vibrational receptors have been studied in detail in the southern green stink bug *N. viridula*. At the dorsal side of each leg, four scolopidial organs are anchored distally to different leg structures. The femoral chordotonal organ, composed of the proximal and distal scoloparium, is distally fixed, partly to the muscle (musculus levator tibiae) and partly to the tibial apodeme. The tibial chordotonal organ containing two scolopidia is distally anchored to the joint membrane between tibia and tarsus, and the tarsopretarsal chordotonal organ is fixed at the tendon of the unguitractor plate (proximal scoloparium) and to the anterior and posterior claw. A similar tarsal

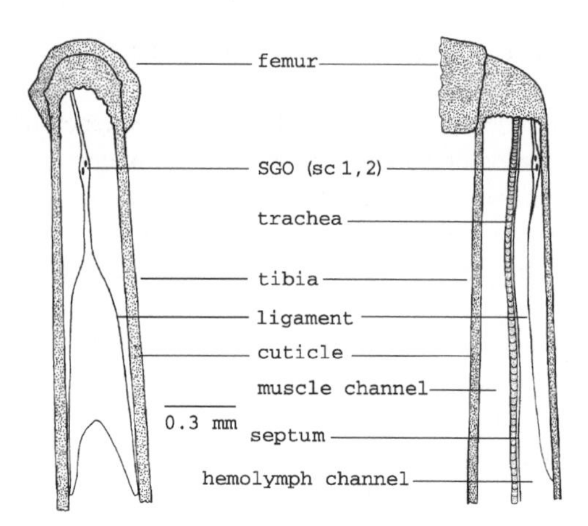

Figure 1 Schematic drawing of subgenual organ (SGO) of *N. viridula* with two scolopidia (sc 1, 2) [after Michel et al. (84)].

chordotonal organ was described in a water bug species *Notonecta* sp. (116). The subgenual organ lies in the hemolymph channel in the tibial subgenual region and contains only two scolopidia, each with one sensory cell (Figure 1). Proximally, both sensory cells are attached to the epithelium of the tibial wall, and distally, scolopals with cilia and the ligament (formed by cap cells of two scolopidia) are stretched out in the hemolymph. The ligament runs longitudinally in the tibial hemolymph channel, flattens distally to a thin membrane, and is fixed by two strands, partly to the epidermis of the tibial wall and partly to the two main tibial nerves. Similar anatomy and low number of sensory cells of the subgenual and joint chordotonal organs were described in *Pyrrhocoris apterus* (Pyrrhocoridae) (34).

Responses of the southern green stink bug leg vibratory receptor neurons were divided into low- and high-frequency groups (25). The low-frequency receptor (LFR) neurons, attributed to joint chordotonal organs and/or campaniform sensilla, are most sensitive between 50 and 100 Hz. Their threshold velocity sensitivity lies around 0.05 mm s^{-1}, with threshold curves generally following the line of equal displacement values (around 10^{-7} m) in the frequency range above 100 Hz. The higher frequency receptor neurons are of two types: The middle frequency receptor cell (MFR) shows highest velocity sensitivity around 200 Hz (0.01 mm s^{-1}) and the higher frequency receptor (HFR) between 700 and 1000 Hz (around 0.002 mm s^{-1}). In both, threshold curves follow the line of equal acceleration values (around 10^{-2} m s^{-2}) in the frequency range below the highest velocity sensitivity. Above the highest velocity sensitivity, they follow the line of equal displacement values (between 10^{-9} and 10^{-10} m for HFR and between 10^{-7} and 10^{-8} m for MFR). Spike phase locked response was characteristic for the LFR neurons in the frequency range below 120 Hz and for MFR and HFR neurons below 200 Hz.

The latter are characterized also by prolonged responses at frequencies around 200 Hz. Although direct proof is lacking, we may suppose that responses of the MFR and HFR neurons belong to both subgenual organ sensory cells. The highest sensitivity of the subgenual organ receptor cells is not tuned to the range of the dominant frequencies of *N. viridula* songs. The intensity of ~100-Hz vibratory signals measured on a plant below the singing animal lies between 14 dB and 20 dB above threshold sensitivity of the leg low- and high-frequency receptor organs. This relatively narrow range above the threshold intensity does not critically limit communication because of the low attenuation of 100 Hz signals transmitted through plants.

Fewer data on the architecture and function of leg vibratory receptor organs are available in other land bug species, and detailed comparative investigations are lacking. Devetak et al. (38) investigated vibrational receptors in three cydnid species and found receptors in the subgenual region of tibiae, with threshold curves following the line of equal acceleration values ($5–10 \times 10^{-3}$ m s^{-2}), without obvious peak acceleration sensitivity in the broad frequency range between 100 and 2000 Hz. Receptors located in the femur showed sensitivity below 10^{-2} m s^{-2} at frequencies between 100 and 200 Hz. Although detailed morphological investigations are lacking, the high-frequency responses at least may be attributed to stimulation of the subgenual organ.

In land bugs a subgenual organ acting as a velocity receiver, similar to that in honey bees (75), has not been found, and there are no data about scolopidia fixed to some accessory structures that would enlarge sensitivity in the higher-frequency range. In the green lacewing *Chrysoperla carnea* (36), for example, the best velocity sensitivity lies between 1 and 2 kHz (37). Scolopidia of the subgenual organ are fixed to the velum, a lens-like part of the organ that divides the hemolymph channel into two separate parts (39). Among land bugs, no sensory organs for airborne sound, sensitive enough to detect weak airborne components of the emitted signals, have been found. Behavioral experiments with Cydnidae demonstrated that the airborne component of the emitted mixed vibratory and stridulatory signals is not effective, even at distances of a few millimeters (51).

In *N. viridula*, Jeram & Pabst (73) found that 45 scolopidia of the Johnston's organ and 7 of the central chordotonal organ are most sensitive in the frequency range between 30 and 100 Hz (72). A threshold velocity sensitivity of both organs around 1 mm s^{-1} does not allow them to be relevant for long-distance communication, but antennae may be used as an additional vibratory input in close vicinity, or at leaf edges or plant tips, where low-frequency components of the emitted signals reach intensities above the input value.

Almost nothing is known about the vibroreceptors in Auchenorrhyncha. Howse & Claridge (62) advanced anatomical evidence that a Johnston's organ of the leaf hopper *Oncopsis flavicollis* may be used as an efficient sound receptor. The antennal flagellum is inserted into a radial membrane in the pedicel, and its articulation suggests that the flagellum is lightly damped and highly resonant. Small auchenorrhynchus insects probably possess chordotonal and subgenual organs in their legs, as do other Hemiptera, although there is no information about their anatomy and function.

In contact vibrational communication (80), receptors other than those located in and on the legs may be involved because signal characteristics do not change and the ratio between signal power and receptor sensitivity is of less concern. Hematophagous reduviid species such as *R. prolixus* and *Triatoma infestans* emit stridulatory male-deterring and defensive (disturbance) signals (79). As in other Heteroptera, no auditory organs or behavioral evidence for airborne communication have been found. Autrum & Schneider (3) recorded no responses to vibratory stimuli from tibial distal scoloparia of *R. prolixus* in the frequency range above 400 Hz. Because leg vibrational organs tuned below 400 Hz probably cannot detect low intensity, deterrent stridulatory signals of fundamental frequencies above 1500 Hz, we can expect that antennal receptors such as trichobotria or Johnston's organ are involved during contact communication.

Males, females, and nymphs of the phymatid species *Phymata crassipes* emit signals produced by locomotory, stridulatory, and/or vibratory mechanisms (50). Airborne stimuli of frequencies between 200 and 4000 Hz triggered the stereotyped vibrational response with peak average sensitivity of 65–70 dB SPL between 700 and 2000 Hz when bugs were standing on the protecting grid of the microphone, and between 50 and 60 dB SPL when standing on the membrane of a condenser microphone. Between 200 and 2000 Hz, the average vibratory threshold sensitivity ranged from 0.01 to 0.02 mm s^{-1}. The similar shapes of auditory and vibratory threshold curves in the frequency range below 1500 Hz indicate that airborne signals directly or indirectly stimulate vibrational receptors. This hypothesis is supported by the different thresholds observed when bugs were standing on the membrane or on the protection grid of the condenser microphone, and by the fact that bugs within a group respond to each other only via substrate, even in close proximity.

LONG-RANGE COMMUNICATION AND ORIENTATION

Vibrational communication is a single but important step in the whole process leading to male-female meeting in the field. Host-plant finding is probably the primary cause of interplant movement by homopteran insects within a habitat (97), and in many land bugs the male pheromone attracts females over long distances, to the same plant(s) (1, 9, 115). Hunt & Nault (65) demonstrated that a "call-fly" strategy, suggested first for species of the chloropid genus *Lipara* (13), three plant hopper species (66, 67), and the tick-tock cicada (54), may also be applied to pair formation in leaf hoppers and represents an adaptation of small insects to problems such as how to locate a singing partner by the use of substrate-borne signals. Virgin females perch on the upper half of a plant above mated ones and express less movement than males, which fly from one plant to another calling on each of them. When they detect a response from a virgin female (mated females do not respond) they move toward her. The movement of virgin females is to the upper part of the plant, and male search is oriented toward the light.

The general pattern of singing during pre-mating behavior is similar for all Pentatomidae (76) (Figure 2). Calling starts by the emission of the female calling song, which (*a*) triggers males to respond with the calling and courtship songs, (*b*) activates them to walk on a plant, and (*c*) enables directional movement toward the female, with search behavior on plant crossings. The female calling song of pentatomids is composed of a characteristic sequence of readily repeated pulse

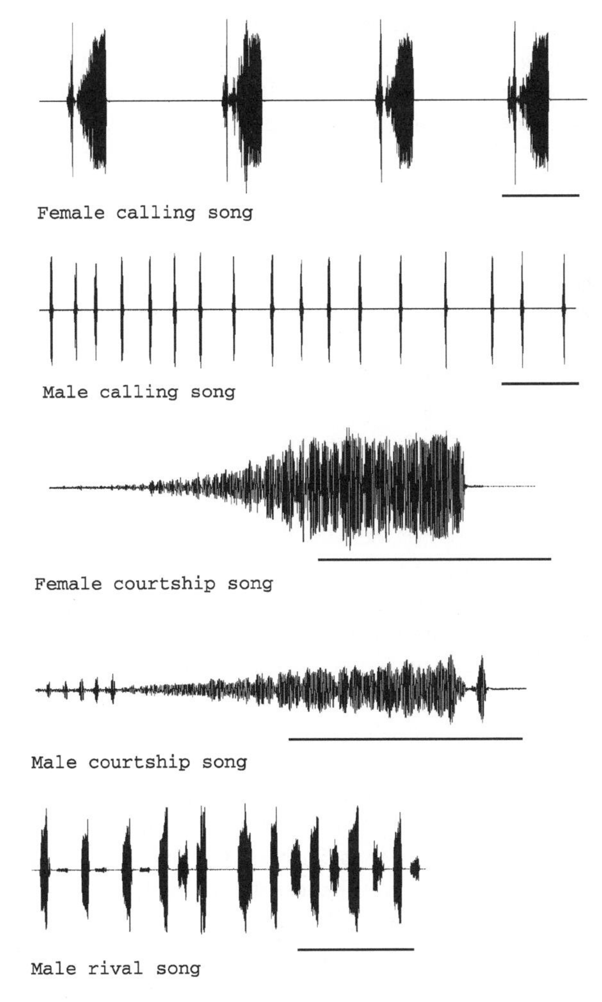

Figure 2 Oscillograms of female and male vibratory signals of the southern green stink bug *Nezara viridula*. Scale bars: 2 s.

trains. In the southern green stink bug, two temporally and spectrally different types of pulse trains are exchanged in the same sequence (30, 31, 102). The female calling songs of related and sympatric species *Acrosternum hilare* (28), *Palomena prasina* and *P. viridissima* (27), and *Thyanta pallidovirens* and *T. custator accerra* (H. L. McBrien, A. Čokl & J. G. Millar, unpublished data) are less species specific and differ mainly in the pulse train duration and repetition rate. In all of them, the basic and dominant frequencies lie in the frequency range between 80 and 150 Hz.

In the pentatomid species *N. viridula* (30, 31), *P. prasina*, and *P. viridissima* (27), a male calling song was recorded. The male calling song (MS1), characterized by a sequence of single and/or grouped pulses of different spectral characteristics, is emitted during male approach to the female and always as a response to her calling. The role of the song is not clear because males, in many cases at longer distances from the female, respond to the calling female immediately with the courtship song, or respond with the calling song as a short transition to the courtship or rival song.

In the species *A. hilare*, sympatric with *N. viridula*, (28) the male song, of comparable time and spectral characteristics to the *N. viridula* calling song, was emitted in the courtship phase when partners were close together, and led directly to copulatory attempts. A reverse situation was found in the pentatomid species *Holcostethus strictus* (95) where the male calling song pulse train triggered the emission of the female calling song phrase. Vibrational directionality could not be observed in the species, which may explain the lack of the typical female calling song, as emitted by larger pentatomid land bugs. The spectral characteristics of male songs show all the properties of signals produced by the vibratory mechanism in land bugs: The dominant frequency ranges between 80 and 150 Hz with different extents of frequency modulation.

The existence in the southern green stink bug of temporally and spectrally different units within the same sequence of the female calling song raises the question about its recognition by the male (N. Miklas, A. Čokl, M. Renou & M. Virant-Doberlet, unpublished data). In experiments carried out on a loudspeaker membrane, on which the emitted signal has similar characteristics at every point, males responded significantly less to pulse trains with several short and quickly repeated pulses than to equally long-pulse trains of another type, containing a single long pulse preceded by a shorter one (88). In natural conditions on a bean plant this differentiation was not seen. We can explain this by the fact that short, quickly repeated pulses in standing wave conditions on a plant are fused into a signal that is recognized by a male as one unit. It is not clear why temporally and spectrally different units appear within the same song sequence in both the female and male calling songs. One explanation is that the different positions of internodes and nodes on a plant ensure that a listening partner gets some above-threshold component of the signal at every point.

Since the classic study of Ossiannilsson (92), many authors have described the signals and acoustic repertoire of different species of Auchenorrhyncha (8, 17–21, 33, 35, 40, 42, 44, 45, 55–57, 63, 64, 69–71, 90, 98, 103, 105, 107, 110–114).

Vibrational signals are usually complex and consist of amplitude-modulated trains of pulses with distinctive sex- and species-specific temporal patterns. The dominant frequency of emitted signals is limited to the range below 1 kHz, and in some species, signals are frequency modulated (57, 63, 64, 86, 107). The male signals are structurally more complicated than those of females of the same species and usually consist of repeated sequences, each composed of several different sound elements (14, 15, 33, 35, 64). The female calls, on the other hand, consist of simple trains of regularly repeated pulses (14, 33, 35, 45, 63). The acoustic repertoire differs among species. Vibrational signals described in the treehopper *Enchenopa binotata* (64) are far more complex and diverse than those used during male-female interactions in treehopper *Spissistilus festinus* (63).

Acoustic signals emitted by Auchenorrhyncha are usually defined in terms of their behavioral context. The most common are calling signals (8, 14, 19, 20, 32, 33, 35, 42, 44, 45, 55–57, 63, 64, 71, 105, 107). The male usually initiates calling. Females only rarely call spontaneously, but they respond readily with their own calling signals to male calls or to playback of a conspecific male calling signal (15, 16, 32).

Vibrational directionality was demonstrated during search behavior in *N. viridula* (29, 93). It was also demonstrated in other species: in the context of recruitment in leaf-cutting ants (100), and in host or prey searching in a stink bug predator *Podisus maculiventris* (96) and in parasitic wasps (83). The mechanism underlying resolution of direction by vibratory cues in such small insects is still under discussion. Differences in intensity, time (the time delay between arrival of a substrate-borne signal at spatially separated receptors) or phase of stimulation of spatially separated sense organs, are small when compared with those of large arthropods, such as nocturnal scorpions and wandering spiders. For scorpions the threshold time-of-arrival differences for triggering directional response was determined to be 0.2 ms (10), and 4 ms for the wandering spider *Cupiennius salei* (59). Maximal distances between the legs of the southern green stink bug do not exceed 1 cm, and only in some natural situations when a bug is standing on close, parallel running branches of a crossing, can they be extended. Propagation velocities between 40 and 80 m s^{-1}, as expected for songs in the region of 100 Hz, create a time-of-arrival difference between 0.125 and 0.250 ms at a distance of 1 cm. When an individual is standing on different branches, at 1 cm from the crossing, the distance between the legs may be extended to 2 cm, which causes the difference in the time of arrival to increase from 0.25 to 0.50 ms. These values are close to the behaviorally determined threshold for scorpions, and their use as a directionality cue demands precise timing of response latencies of different receptor neurons located in different legs. Attenuation of less than 1 dB/cm by itself cannot create an intensity difference large enough to be recognized by the male as a signal to turn to the vibrated branch. Amplitude differences cannot be excluded because, at a distance of a few cm, the amplitude may change by more than 10 dB if the frequency changes by less than 10% (86). Frequency sweeps of more than 10 Hz around a 100-Hz dominant frequency are common for the land bug calling songs. Stimulation of the vibratory interneurons with simultaneously recorded calling signals from both branches of a crossing may provide information about the relevance of discrete amplitude, spectral, and/or time-of-arrival differences for resolving the direction of the vibratory source.

Although the exchange of vibratory songs while a male moves around and searches for a sedentary calling female on the same plant is characteristic for auchenorrhynchus insects, it is still unclear how insects as small as plant hoppers, leaf hoppers, and treehoppers extract directional information from conspecific vibrational signals. Recently Cocroft et al. (24) proposed for the treehopper *Umbonia crassicornis* another mechanism that facilitates vibrational directionality on a small spatial scale. The response of the insect body to the substrate shows resonance at lower frequencies and attenuation at higher frequencies. Transfer functions on the body differ significantly depending on the direction of the vibratory stimulus. The mechanical response of the body to substrate vibration is thus available to provide directional information.

SHORT-RANGE COMMUNICATION, COURTSHIP, AND RIVALRY

Significant changes of temporal, amplitude, and spectral characteristics of vibratory signals, which occur during transmission through a plant, suggest that precise mate recognition can be performed at shorter distances and that singing at longer distances is devoted mainly to responding to the calling partner. Thus, it is not surprising that in many species courtship songs exchanged at short distances express the highest level of species specificity.

Gogala (46, 47) described two types of species-specific male courtship songs produced by the stridulatory and vibratory mechanisms in some cydnid land bug species. Females respond only to the first type, suggesting that it serves as the identification code and that the second type stimulates pair formation. Females respond with the species-specific agreement song, and this alternation is essential for pair formation and copulation.

In the southern green stink bug and other pentatomid species that have been investigated, emission of the male courtship song is not always limited to the presence of a female in the vicinity; it can be triggered by a female calling from a longer distance (29). In a fully developed duet, a male adapts the duration of a single pulse train to the pause between two consecutive female calling song pulse trains, and an a-b-a-b-... male/female alternation follows. In most of the pentatomid species investigated, the male courtship song pulse train originates from single male calling song pulses, whose repetition rate increases, leading finally to their fusion in a pulse train of species-specific temporal characteristics. In *N. viridula*, single male calling song pulses may be distinguished only at the beginning of a male courtship song pulse train of only a few seconds duration (31). In *P. prasina* single pulses are joined in the male courtship song pulse train without fusion, and in *P. viridissima* four pulses of characteristically decreasing duration form the male courtship song pulse train (27). On the other hand, the MS1 pulse train of the sympatric species *A. hilare* exhibits species-specific and complex temporal characteristics similar to those described above. It is, however, emitted mainly in response to the calling female during his approach to her; when he reaches her, MS1 is changed to a sequence of short pulses (28) resembling the male calling song of *N. viridula*, *P. prasina*, and *P. viridissima*. In another pentatomid species, *H. strictus* (95), a male courtship song phrase (sequence) consists of three types of pulses and pulse trains.

Among the Pentatomidae, the female courtship song emitted when a male is in the close vicinity has been described only in *N. viridula* (31), and its characteristics are similar to those of the male courtship song. Females of this species also emit a song that rejects copulatory attempts of males and stops their courting (31). This song is comparable to the deterrent stridulatory signals in a reduviid bug species *R. prolixus* (79).

In Auchenorrhyncha, the distinction between calling and courtship songs is not always sharp and both terms are sometimes used for signals with the same function, especially when songs emitted during courtship do not differ greatly from those emitted during the initial stage of pair formation (14–16, 63, 64). Distinctive vibrational signals associated with courtship and copulation have been described in genus *Oncopsis* (20), *Amrasca* (103), *Dalbulus* (57), *Enchenopa* (64), *Agallia* (105), *Agalliopsis* (105), and *Empoasca* (107).

Pentatomid (31) and cydnid (47) bugs engage in rival singing. The rival songs of *N. viridula* (31) and *Rhaphigaster nebulosa* (R. Razpotnik, unpublished data) show a lower level of species specificity, and in a well-established duet, prolonged single pulses alternate in an a-b-a-b-... fashion until one or both stop singing. In another pentatomid species, *P. lituratus* (53), the rival song of males shows extreme modulation of the dominant frequency between 90 and 190 Hz, and its recognition depends mainly on its frequency characteristics (11). A similar pattern of male rival singing has been described in Auchenorrhyncha (19, 57, 68, 92). The rival song produced by males during aggressive interactions and intrasexual competition disrupts calling between another male and the female (19). When both males emit rivalry songs, one of them eventually becomes quiet and the other will begin to court the female (57).

DISTURBANCE SIGNALS

Distress (disturbance or alarm) signals produced by larvae and adults are generally associated with predator-repelling behavior. Among the Heteroptera, most examples have been described in the Reduviidae and Cydnidae. In the Cydnidae (46, 48), signals show low species specificity and are emitted by males and females with an irregular repetition rate. In several species of Auchenorrhyncha, distinct signals are produced when insects are disturbed (18, 57, 92). In Heteroptera, distress signals are usually emitted by a stridulatory mechanism, the phymatid species *P. crassipes* being an exception, in which Gogala & Čokl (50) recorded stridulatory signals whose typical, broad banded stridulatory spectra also contained low-frequency components between 100 and 500 Hz, indicating that a vibratory mechanism may also be involved.

Disturbance signals are also produced in the context of parental care of offspring. Cocroft (22, 23) described coordinated group vibrational signals emitted by a number of aggregated nymphs of the subsocial treehopper *U. crassicornis* in the form of synchronized bursts that elicit anti-predator behavior in the mother. The nymphs responded by signaling in response to the presence of a predator and also to playback of vibratory signals produced by their siblings.

In the blood-sucking bug *T. infestans* (Reduviidae), stridulatory disturbance signals are produced, as in other reduviid species, by rubbing of the tip of the proboscis against the groove (99). The main vibration energy is concentrated about 1 kHz above the values of stridulatory deterring signals that have peak frequency ranges between 700 and 800 Hz. A similar spectral distinction between the two types of stridulatory signals was also described in another reduviid species, *R. prolixus* (79). Recent comparison of reduviid disturbance signals demonstrated similar syllable repetition rates and a carrier frequency reaching about 2000 Hz despite differences in animal size, inter-ridge distances of their stridulatory grooves, and the temporal pattern of their vibratory signals (104).

ACKNOWLEDGMENTS

This review is dedicated to our teacher and mentor Prof. Dr. Matija Gogala. We thank all our friends and colleagues, present and past, who have shared with us their interest in sound and vibration research. We are grateful to Dr. Roger H. Pain for critical reading of the manuscript and language advice. We would like to thank the anonymous reviewers for their helpful comments. The Slovenian Ministry of Education, Science and Sports (Program 0503-0105) provides financial support to projects on vibrational communication in insects.

The *Annual Review of Entomology* is online at http://ento.annualreviews.org

LITERATURE CITED

1. Aldrich JR. 1995. Chemical communication in the true bugs and parasitoid exploitation. In *Chemical Ecology of Insects 2*, ed. RT Cardé, WJ Bell, 9:318–63. New York: Chapman & Hall
2. Amon T, Čokl A. 1990. Transmission of the vibratory song of the bug *Nezara viridula* (Pentatomidae, Heteroptera) on the *Hedera helix* plant. *Scopolia* (Suppl. 1): 133–41
3. Autrum H, Schneider W. 1948. Vergleichende Untersuchungen über den Erschütterungssinn der Insekten. *Z. Vergl. Physiol.* 31:77–88
4. Barth FG. 1985. Neuroethology of the spider vibration sense. In *Neurobiology of Arachnids*, ed. FG Barth, pp. 203–29. Berlin: Springer
5. Barth FG. 1998. The vibrational sense of spiders. In *Comparative Hearing: Insects*, ed. RR Hoy, AN Popper, RR Fay, 7:228–78. New York: Springer

6. Bennet-Clark HC. 1971. Acoustics of insect song. *Nature* 234:255–59
7. Bennet-Clark HC. 1998. Size and scale effects as constraints in insect sound communication. *Philos. Trans. R. Soc. London Ser. B* 353:407–19
8. Booij CJH. 1982. Biosystematics of the *Muellerianella* complex (Homoptera, Delphacidae), interspecific and geographic variation in acoustic behavior. *Z. Tierpsychol.* 58:31–52
9. Brézot P, Malosse C, Mori K, Renou M. 1994. Bisabolene epoxides in sex pheromone in *Nezara viridula* (L.) (Heteroptera:Pentatomidae): role of *cis* isomer and relation to specificity of pheromone. *J. Chem. Ecol.* 20:3133–47
10. Brownell P, Farley RD. 1979. Orientation to vibrations in sand by the nocturnal scorpion *Paruroctonus mesaensis*: mechanism of target localization. *J. Comp. Physiol. A* 131:31–38
11. Brvar V. 1987. *Vibratory signal parameters important for communication in Piezodorus lituratus (Heteroptera).* BSc thesis. Univ. Ljubljana, Ljubljana. 80 pp.
12. Casas J, Bacher S, Tautz J, Meyhöfer R, Pierre D. 1998. Leaf vibrations and air movements in a leafminer-parasitoid system. *Biol. Control* 11:147–53
13. Chvala M, Doskocil J, Mook H, Pokorny V. 1974. The genus *Lipara* Meigen (Diptera, Chloropidae), systematics, morphology, behavior, and ecology. *Tijdschr. Entomol.* 117:1–25
14. Claridge MF. 1985. Acoustic behavior of leafhoppers and planthoppers: species problems and speciation. In *The Leafhoppers and Planthoppers*, ed. LR Nault, JG Rodriquez, pp. 103–25. New York: Wiley
15. Claridge MF. 1985. Acoustic signals in the Homoptera: behavior, taxonomy and evolution. *Annu. Rev. Entomol.* 30:297–317
16. Claridge MF, de Vrijer PWF. 1994. Reproductive behavior: the role of acoustic signals in species recognition and speciation. In *Planthoppers: Their Ecology and Management*, ed. RF Denno, TJ Perfect, pp. 216–33. New York: Champan & Hall
17. Claridge MF, den Hollander J, Morgan JC. 1985. Variation in courtship signals and hybridization between geographically definable populations of the rice brown planthopper *Nilaparvata lugens* (Stål). *Biol. J. Linn. Soc.* 24:35–49
18. Claridge MF, Howse PE. 1968. Songs of some British *Oncopsis* species (Hemiptera: Cicadellidae). *Proc. R. Entomol. Soc. London Ser. A* 43:57–61
19. Claridge MF, Morgan JC. 1993. Geographical variation in acoustic signals of the planthopper *Nilaparvata bakeri* (Muir) in Asia: species recognition and sexual selection. *Biol. J. Linn. Soc.* 48:267–81
20. Claridge MF, Nixon GA. 1986. *Oncopsis flavicollis* (L.) associated with tree birches (*Betula*): a complex of biological species or a host plant utilization polymorphism? *Biol. J. Linn. Soc.* 27:381–97
21. Claridge MF, Reynolds WJ. 1973. Male courtship songs and siblings species in the *Oncopsis flavicollis* species group (Hemiptera: Cicadellidae). *J. Entomol. Ser. B* 42:29–39
22. Cocroft RB. 1999. Offspring-parent communication in a subsocial treehopper (Hemiptera: Membracidae: *Umbonia crassicornis*). *Behavior* 136:1–21
23. Cocroft RB. 1999. Parent-offspring communication in response to predators in a subsocial treehopper (Hemiptera: Membracidae: *Umbonia crassicornis*). *Ethology* 105:553–68
24. Cocroft RB, Tieu TD, Hoy RR, Miles RN. 2000. Directionality in the mechanical response to substrate vibration in a treehopper (Hemiptera:Membracidae: *Umbonia crassicornis*). *J. Comp. Physiol. A* 186:695–705
25. Čokl A. 1983. Functional properties of vibroreceptors in the legs of *Nezara viridula* (L.) (Heteroptera: Pentatomidae). *J. Comp. Physiol. A* 150:261–69

26. Čokl A. 1988. Vibratory signal transmission in plants as measured by laser vibrometer. *Period. Biol.* 90:193–96
27. Čokl A, Gogala M, Blaževič A. 1978. Principles of sound recognition in three pentatomide bug species (Heteroptera). *Biol. Vestn.* 26:81–94
28. Čokl A, McBrien HL, Millar JG. 2001. Comparison of substrate-borne vibrational signals of two stink bug species *Acrosternum hilare* and *Nezara viridula* (Heteroptera: Pentatomidae). *Ann. Entomol. Soc. Am.* 94:471–79
29. Čokl A, Virant-Doberlet M, McDowell A. 1999. Vibrational directionality in the southern green stink bug *Nezara viridula* is mediated by female song. *Anim. Behav.* 58:1277–83
30. Čokl A, Virant-Doberlet M, Stritih N. 2000. Temporal and spectral properties of the songs of the southern green stink bug *Nezara viridula* (L.) from Slovenia. *Pflügers Arch. Eur. J. Physiol.* 439 (Suppl.):R168–70
31. Čokl A, Virant-Doberlet M, Stritih N. 2000. The structure and function of songs emitted by southern green stink bugs from Brazil, Florida, Italy and Slovenia. *Physiol. Entomol.* 25:196–205
32. de Vrijer PWF. 1984. Variability in calling signals of the planthopper *Javesella pellucida* (F.) (Homoptera: Delphacidae) in relation to temperature and consequences for species recognition during distant communication. *Neth. J. Zool.* 34: 388–406
33. de Vrijer PWF. 1986. Species distinctiveness and variability of acoustic calling signals in the planthopper genus *Javesella* (Homoptera: Delphacidae). *Neth. J. Zool.* 36:162–75
34. Debasieux P. 1938. Organes scolopidiaux des pattes d'insectes. *Cellule* 47:77–202
35. den Bieman CFM. 1986. Acoustic differentiation and variation in planthoppers of the genus *Ribautodelphax* (Homoptera, Delphacidae). *Neth. J. Zool.* 36:461–80
36. Devetak D. 1998. Detection of substrate vibration in Neuropteroidea: a review. *Acta Zool. Fenn.* 209:87–94
37. Devetak D, Amon T. 1997. Substrate-vibration sensitivity of the leg scolopidial organs in the green lacewing *Chrysoperla carnea*. *J. Insect Physiol.* 43:433–37
38. Devetak D, Gogala M, Čokl A. 1978. A contribution to the physiology of the vibration receptors in bugs of the family Cydnidae (Heteroptera). *Biol. Vestn.* 26:131–39
39. Devetak D, Pabst MA. 1994. Structure of the subgenual organ in the green lacewing *Chrysoperla carnea*. *Tissue Cell* 26:249–57
40. Downham MCA, Claridge MF, Morgan JC. 1997. Preliminary observations on the mating behavior of *Cicadulina storeyi* China and *C. mbila* Naudé (Homoptera: Cicadellidae). *J. Insect Behav.* 10:753–60
41. Drašlar K, Gogala M. 1976. Structure of stridulatory organs of insects from family Cydnidae (Heteroptera). *Biol. Vestn.* 24:175–200
42. Drosopoulos S. 1985. Acoustic communication and mating behavior in the *Muellerianella* complex (Homoptera-Delphacidae). *Behavior* 94:183–201
43. Ewing AW. 1989. Mechanisms of sound production. In *Arthropod Bioacoustics. Neurobiology and Behavior*, ed. AW Ewing, 2:16–57. Edinburgh: Edinburgh Univ. Press
44. Gillham MC. 1992. Variation in acoustic signals within and among leafhopper species of the genus *Alebra* (Homoptera, Cicadellidae). *Biol. J. Linn. Soc.* 45:1–15
45. Gillham MC, de Wrijer PWF. 1995. Patterns of variation in the acoustic calling signals of *Chloriona* planthoppers (Homoptera: Delphacidae) coexisting on the common reed *Phragmites australis*. *Biol. J. Linn. Soc.* 54:245–69
46. Gogala M. 1970. Artspezifität der Lautäußerungen bei Erdwanzen (Heteroptera, Cydnidae). *Z. Vergl. Physiol.* 70:20–28
47. Gogala M. 1978. Acoustic signals of four

bug species of the family Cydnidae (Heteroptera). *Biol. Vestn.* 26:153–68

48. Gogala M. 1984. Vibration producing structures and songs of terrestrial Heteroptera as systematic character. *Biol. Vestn.* 32:19–36
49. Gogala M. 1990. Distribution of low frequency vibrational songs in local Heteroptera. *Scopolia* (Suppl. 1):125–32
50. Gogala M, Čokl A. 1983. The acoustic behavior of the bug *Phymata crassipes* (F.) (Heteroptera). *Rev. Can. Biol. Exptl.* 42:249–56
51. Gogala M, Čokl A, Drašlar K, Blaževič A. 1974. Substrate-borne sound communication in Cydnidae (Heteroptera). *J. Comp. Physiol.* 94:25–31
52. Gogala M, Hočevar I. 1990. Vibrational songs in three sympatric species of *Tritomegas*. *Scopolia* (Suppl. 1):117–23
53. Gogala M, Razpotnik R. 1974. An oscillographic-sonagraphic method in bioacoustical research. *Biol. Vestn.* 22:209–16
54. Gwynne DT. 1987. Sex-biased predation and the risky mate-locating behavior of male tick-tock cicadas (Homoptera: Cicadidae). *Anim. Behav.* 35:571–76
55. Heady SE, Denno RF. 1991. Reproductive isolation in *Prokelisia* planthoppers (Homoptera:Delphacidae): acoustic differentiation and hybridization failure. *J. Insect Behav.* 4:367–90
56. Heady SE, Nault LR. 1991. Acoustic signals of *Graminella nigrifrons* (Homoptera: Cicadellidae). *Great Lakes Entomol.* 24:9–16
57. Heady SE, Nault LR, Shambaugh GF, Fairchild L. 1986. Acoustic and mating behavior of *Dalbulus* leafhoppers (Homoptera: Cicadellidae). *Ann. Entomol. Soc. Am.* 79:727–36
58. Henry CS. 1980. The importance of low-frequency, substrate-borne sounds in lacewing communication (Neuroptera: Chrysopidae). *Ann. Entomol. Soc. Am.* 73: 617–21
59. Hergenröder R, Barth FG. 1983. Vibratory signals and spider behavior: How do the sensory inputs from the eight legs interact in orientation? *J. Comp. Physiol. A* 152:361–72
60. Hirschberger P. 2001. Stridulation in *Aphodius* dung beetles: behavioral context and intraspecific variability of song patterns in *Aphodius ater* (Scarabaeidae). *J. Insect Behav.* 14:69–88
61. Howse PE. 1964. The significance of the sound produced by the termite *Zootermopsis augusticollis* (Hagen). *Anim. Behav.* 12:284–300
62. Howse PE, Claridge MF. 1970. The fine structure of Johnston's organ of the leafhopper *Oncopsis flavicollis*. *J. Insect. Physiol.* 16:1665–75
63. Hunt RE. 1993. Role of vibrational signals in mating behavior of *Spissistilus festinus* (Homoptera: Membracidae). *Ann. Entomol. Soc. Am.* 86:356–61
64. Hunt RE. 1994. Vibrational signals associated with mating behavior in the treehopper *Enchenopa binotata* Say (Hemiptera: Membracidae). *J. NY Entomol. Soc.* 102:266–70
65. Hunt RE, Nault LR. 1991. Roles of interplant movement, acoustic communication and phototaxis in mate-location behavior of the leafhopper *Graminella nigrifrons*. *Behav. Ecol. Sociobiol.* 28:315–20
66. Ichikawa T. 1976. Mutual communication by substrate vibrations in the mating behavior of planthoppers (Homoptera: Delphacidae). *Appl. Entomol. Zool.* 11:8–21
67. Ichikawa T. 1979. Studies on the mating behavior of the four species of Auchenorrhynchous Homoptera which attack the rice plant. *Mem. Fac. Agric. Kagawa Univ.* 34:1–60
68. Ichikawa T. 1982. Density-related changes in male-male competative behavior in the rice brown planthopper *Nilaparvata lugens* (Stål) (Homoptera: Delphacidae). *Appl. Entomol. Zool.* 17:439–52
69. Ichikawa T, Ishii S. 1974. Mating signal of the brown planthopper *Nilaparvata*

lugens (Stål) (Homoptera: Delphacidae): vibration of the substrate. *Appl. Entomol. Zool.* 9:196–98
70. Ichikawa T, Sakuma M, Ishii S. 1975. Substrate vibrations: mating signal of three species of planthoppers which attack the rice plant (Homoptera:Delphacidae). *Appl. Entomol. Zool.* 10:162–71
71. Inoue H. 1982. Species-specific calling sounds as a reproductive isolating mechanism in *Nephotettix* spp. (Hemiptera: Cicadellidae). *Appl. Entomol. Zool.* 17:253–62
72. Jeram S, Čokl A. 1996. Mechanoreceptors in insects: Johnston's organ in *Nezara viridula* (L.) (Pentatomidae, Heteroptera). *Pflügers Arch. Eur. J. Physiol.* 431 (Suppl.):R281–82
73. Jeram S, Pabst MA. 1996. Johnston's organ and central organ in *Nezara viridula* (L.) (Heteroptera, Pentatomidae). *Tissue Cell* 28:227–35
74. Kalmring K. 1985. Vibrational communication in insects (reception and integration of vibratory information). In *Acoustic and Vibrational Communication in Insects*, ed. K Kalmring, N Elsner, pp. 127–34. Berlin: Paul Parey
75. Kilpinen O, Storm J. 1997. Biophysics of the subgenual organ of the honeybee, *Apis mellifera*. *J. Comp. Physiol A* 181:309–18
76. Kon M, Oe A, Numata H, Hidaka T. 1988. Comparison of the mating behavior between two sympatric species *Nezara antennata* and *N. viridula* (Heteroptera: Pentatomidae) with special reference to sound emission. *J. Ethol.* 6:91–98
77. Kuštor V. 1989. *Activity of vibratory organ muscles in the bug Nezara viridula (L.).* MsD thesis. Univ. Ljubljana, Ljubljana. 68 pp.
78. Magal C, Schöller M, Tautz J, Casas J. 2000. The role of the leaf structure in vibration propagation. *J. Acoust. Soc. Am.* 108(5):2412–18
79. Manrique G, Schilman PE. 2000. Two different vibratory signals in *Rhodnius prolixus* (Hemiptera: Reduviidae). *Acta Trop.* 77:271–78
80. Markl H. 1983. Vibrational communication. In *Neuroethology and Behavioral Physiology. Roots and Growing Points*, ed. F Huber, H Markl, V.2:332–53. Berlin/Heidelberg: Springer
81. Markl H, Fuchs S. 1972. Klopfsignale mit Alarmfunktion bei Roßameisen (*Camponotus*, Formicidae, Hymenoptera). *Z. Vergl. Physiol.* 76:204–25
82. Deleted in proof
83. Meyhöfer R, Casas J. 1999. Vibratory stimuli in host location by parasitic wasps. *J. Insect. Physiol.* 45:967–71
84. Michel K, Amon T, Čokl A. 1983. The morphology of the leg scolopidial organs in *Nezara viridula* (L.) (Heteroptera, Pentatomidae). *Rev. Can. Biol. Exptl.* 42:139–50
85. Michelsen A. 1999. The dance language of honey bees: recent findings and problems. In *The Design of Animal Communication*, ed. MD Hauser, M Konishi, 4:111–31. Cambridge, MA/London: Bradford Book, MIT Press
86. Michelsen A, Fink F, Gogala M, Traue D. 1982. Plants as transmission channels for insect vibrational songs. *Behav. Ecol. Sociobiol.* 11:269–81
87. Deleted in proof
88. Miklas N, Stritih N, Čokl A, Virant-Doberlet M, Renou M. 2001. The influence of substrate on male responsiveness to the female calling song in *Nezara viridula*. *J. Insect. Behav.* 14:313–32
89. Mitomi M, Ichikawa T, Okamoto H. 1984. Morphology of the vibration-producing organ in adult rice brown planthopper *Nilaparvata lugens* (Stål) (Homoptera: Delphacidae). *Appl. Entomol. Zool.* 19:407–17
90. Moore TE. 1961. Audiospectrographic analysis of sounds of Hemiptera and Homoptera. *Ann. Entomol. Soc. Am.* 54:273–91
91. Numata H, Kon M, Fuji H, Hidaka T. 1989. Sound production in the bean bug

Riptortus clavatus Thunberg (Heteroptera: Alydidae). *Appl. Entomol. Zool.* 24: 169–73

92. Ossiannilsson F. 1949. Insect drummers. A study on the morphology and function of the sound-producing organ of Swedish Homoptera Auchenorrhyncha with notes on their sound production. *Opusc. Entomol. Suppl.* 10:1–145
93. Ota D, Čokl A. 1991. Mate location in the southern green stink bug *Nezara viridula* (Heteroptera: Pentatomidae) mediated through substrate-borne signals on ivy. *J. Insect. Behav.* 4:441–47
94. Panizzi AR. 1997. Wild hosts of pentatomids: ecological significance and role in their pest status on crops. *Annu. Rev. Entomol.* 42:99–122
95. Pavlovčič P, Čokl A. 2001. Songs of *Holcostethus strictus* (Fabricius): a different repertoire among land bugs (Heteroptera: Pentatomidae). *Behav. Proc.* 53:65–73
96. Pfannenstiel RS, Hunt RE, Yeargan KV. 1995. Orientation of a hemipteran predator to vibrations produced by feeding caterpilar. *J. Insect Behav.* 8:1–9
97. Power AG. 1987. Plant community diversity, herbivore movement, and an insect-transmitted disease of maize. *Ecology* 68:1658–69
98. Purcell AH, Loher W. 1976. Acoustical and mating behavior of two taxa in the *Macrosteles fascifrons* species complex. *Ann. Entomol. Soc. Am.* 69:513–18
99. Roces F, Manrique G. 1996. Different stridulatory vibrations during sexual behaviour and disturbance in the blood-sucking bug *Triatoma infestans* (Hemiptera: Reduviidae). *J. Insect. Physiol.* 42:231–38
100. Roces F, Tautz J, Hölldobler B. 1993. Stridulation in leaf-cutting ants. Short-range recruitment through plant-borne vibrations. *Naturwissenschaften* 80:521–24
101. Rohrseitz K, Kilpinen O. 1997. Vibration transmission characteristics of the legs of freely standing honeybees. *Zoology* 100:80–84
102. Ryan MA, Čokl A, Walter GH. 1996. Differences in vibratory sound communication between a Slovenian and an Australian population of *Nezara viridula* (L.) (Heteroptera: Pentatomidae). *Behav. Proc.* 36:183–93
103. Saxena KN, Kumar H. 1984. Acoustic communication in the sexual behavior of the leafhopper *Amrasca devastans*. *Physiol. Entomol.* 9:77–86
104. Schilman PE, Lazzari CR, Manrique G. 2001. Comparison of disturbance stridulations in five species of triatomine bugs. *Acta Trop.* 79:171–78
105. Shaw KC. 1976. Sounds and associated behavior of *Agallia constricta* and *Agalliopsis novella* (Homoptera: Auchenorrhyncha: Cicadellidae). *J. Kans. Entomol. Soc.* 49:1–17
106. Shaw KC, Carlson OV. 1979. Morphology of the tymbal organ of the potato leafhopper *Empoasca fabae* Harris (Homoptera: Cicadellidae). *J. Kans. Entomol. Soc.* 52:701–11
107. Shaw KC, Vargo A, Carlson OV. 1974. Sounds and associated behavior of some species of *Empoasca*. *J. Kans. Entomol. Soc.* 47:284–307
108. Smith JW, Georghiou GP. 1972. Morphology of the tymbal organ of the beet leafhopper *Circulifer tenellus*. *Ann. Entomol. Soc. Am.* 65:221–26
109. Stewart KW. 1997. Vibrational communication in insects. Epitome in the language of stoneflies? *Am. Entomol.* Summer 1997: 81–91
110. Strübing H. 1970. Zur Artberechtigung von *Euscelis alsius* Ribaut gegenüber *Euscelis plebejus* Fall (Homoptera-Cicadina). Ein Beitrag zur neuen Systematik. *Zool. Beitr.* 16:441–78
111. Strübing H. 1976. *Euscelis ormaderensis* Remane 1968. I. Saisonformenbildung und akustische Signalgebung. *Sitzungsber. Ges. Naturforsch. Freunde Berlin* 16:151–60
112. Strübing H. 1978. *Euscelis lineolatus* Brullé 1832 und *Euscelis ononidis*

Remane 1967. 1. Ein ökologischer, morphologischer und bioakusticher Vergleich. *Zool. Beitr.* 24:123–54

113. Tischeckin DJ. 1992. Acoustic signalization in Paralimnini leafhoppers (Homoptera, Cicadellidae, Deltacephalinae). *Zool. Zh.* 71:58–65
114. Tischeckin DJ. 1996. Acoustic communication and classification of higher taxa in Typhlocybinae leafhoppers (Homoptera, Cicadellidae) and some related groups. *Zool. Zh.* 75:1007–20
115. Todd JW. 1989. Ecology and behavior of *Nezara viridula*. *Annu. Rev. Entomol.* 34:273–92
116. Wiese K, Schmidt K. 1974. Mechanorezeptoren im Insektentarsus. Die Konstruktion des tarsalen Scolopidialsorgan bei *Notonecta* (Hemiptera, Heteroptera). *Z. Morph. Tiere* 79:47–63

Annu. Rev. Entomol. 2003. 48:51–72
doi: 10.1146/annurev.ento.48.091801.112733

First published online as a Review in Advance on August 19, 2002

Tomato, Pests, Parasitoids, and Predators: Tritrophic Interactions Involving the Genus *Lycopersicon*

George G. Kennedy
Department of Entomology, North Carolina State University, Raleigh, North Carolina 27695-7630; e-mail: george_kennedy@ncsu.edu

Key Words trichomes, host plant resistance, insect-plant interactions, constitutive plant defenses, induced plant defenses

■ **Abstract** Insect-plant interactions involving the cultivated tomato and its relatives in the genus *Lycopersicon* have been intensively studied for several decades, resulting in one of the best documented and in-depth examples of the mechanistic complexities of insect-plant interactions, which encompass both herbivores and their natural enemies. Trichome-mediated defenses are particularly significant in *L. hirsutum f. glabratum* and have been extensively implicated in negative tritrophic effects mediated by direct contact of parasitoids and predators with trichomes, as well as indirect effects mediated through their hosts or prey. Both constitutive and inducible defense traits of *L. esculentum* exert effects on selected parasitoids and predators. The effects of any particular plant defense trait on parasitoids and predators depend on the specific attributes of the plant trait and the details of the physical, biochemical, and behavioral interaction between the natural enemy, its host (prey), and the plant.

CONTENTS

INTRODUCTION

The cultivated tomato, *Lycopersicon esculentum*, is attacked by a number of serious arthropod pests (81). During the past 25 years, considerable research has been directed toward identifying and developing a mechanistic understanding of traits that confer pest resistance within the genus *Lycopersicon*. That body of research reveals a level of detail and complexity in plant-herbivore-natural enemy interactions that is unique among well-studied systems involving crop plants and their close relatives. This review focuses primarily on defensive traits in *L. esculentum* and *L. hirsutum f. glabratum* and their interactions with several pests of tomato and some of the parasitoids and predators that attack them. General reviews of relevant aspects of tritrophic interactions are presented in (9b, 121a). Farrar & Kennedy (32) provide a more complete overview of resistance traits in *Lycopersicon*. For information on genetics, breeding, biology, and production of tomato and its relatives see (65, 89).

The genus *Lycopersicon* is characterized by great diversity within and among its nine species (65, 87). Arthropod resistance has been studied most intensively in *L. esculentum*, *L. hirsutum*, and *L. pennellii*. Of these, the highest levels of resistance to the greatest number of arthropod species are found in *L. hirsutum f. typicum* (= *hirsutum*) and *L. hirsutum f. glabratum* (32). There is extensive variation in the spectrum and level of arthropod resistance among accessions of these species. Arthropod resistance has been associated with a diverse array of traits, including the physical and chemical properties of glandular trichomes, and constitutively expressed and wound-induced chemical defenses associated with the leaf lamella (33, 39, 113). Resistance traits in each of these categories exert effects on the third trophic level. Although the genetics and mechanisms of arthropod resistance in *L. pennellii* and *L. hirsutum f. typicum* have been studied extensively, tritrophic effects of resistance traits in these species have received little attention. Hence, neither species is treated in detail in this review. It must be pointed out, however, that the potential exists for effects such as those described below for *L. hirsutum f. glabratum* to occur in other highly defended *Lycopersicon* species.

TRICHOMES AND DEFENSE IN *LYCOPERSICON*

Trichomes, both glandular and nonglandular, are prominent features of the foliage and stems of *Lycopersicon* spp. Four types of glandular trichomes occur most commonly (85). Of these, type IV and type VI trichomes have been associated with high levels of arthropod resistance. Type IV trichomes have a short, multicellular stalk on a monocellular base and produce a droplet of exudate at their tip. High densities of type IV glandular trichomes and the presence of high levels of toxic acylsugars in their exudate play a major role in the resistance of *L. pennellii* to a number of arthropods, including aphids (*Macrosiphum euphorbiae* and *Myzus persicae*), whiteflies (*Bemisia argentifolii*, *Trialeurodes vaporariorum*), tomato fruitworm (*Helicoverpa zea*), beet armyworm (*Spodoptera exigua*), and the agromyzid leafminer *Liriomyza trifolii* (8, 41–43, 47–50, 61, 82, 94, 99). Genes from *L. pennellii* conferring resistance to arthropods can be introgressed into *L. esculentum* for the development of resistant cultivars. However, glandular trichome-mediated whitefly resistance is not considered suitable for use in tomato cultivars for greenhouse production because the sticky trichomes entrap *Encarsia formosa* wasps used in the biological control of whitefly (99). High densities of type IV trichomes have been implicated in resistance of *L. hirsutum* f. *typicum* to two-spotted spider mite (*Tetranychus urticae*) (15a, 15b, 123).

Type VI trichomes, which have a four-celled glandular head on a short multicellular stalk and a monocellular base (85), have been implicated in resistance of several *Lycopersicon* species to a number of arthropod pests. Type VI trichomes have been specifically implicated in the entrapment of aphids and other small arthropods (20, 86). They correspond to the well-studied type A trichomes of the wild potato species *Solanum berthaultii*, which play an important role in resistance to aphids by fouling their legs and mouthparts and entangling them in exudate (44). The chemical mechanisms responsible for entanglement are the same in *Lycopersicon* and *S. berthaultii*. The trichome tips contain several phenolics [primarily rutin (80%–90%)] but also chlorogenic acid and conjugates of caffeic acid and polyphenol oxidase and peroxidase. When an insect discharges the trichome tip, the contents are mixed. The ensuing enzymatic reaction results in oxidation of the phenolic substrates to quinones, which polymerize (browning reaction) or react with proteins, reducing or eliminating their nutritive value. In addition, the quinones may be directly toxic to the insect. The product of the browning reaction collects on and entangles appendages and mouthparts of small arthropods (20, 23, 44).

There is extensive variation both within and among *Lycopersicon* species in the level of browning reaction associated with type VI trichomes, which reflects differences in polyphenol oxidase/peroxidase activity as well as density of type VI trichomes (20). Even in the absence of a significant browning reaction, the phenolics contained in the trichome tips of some *Lycopersicon* species may alkylate plant proteins, thereby reducing their nutritional value to leaf-chewing insects (20, 37, 57). Type VI trichomes contain only about 15% of the total foliar catecholic phenolics. The remainder is contained within the leaf lamella where it contributes to plant defense. The glandular tips of type VI trichomes of *L. hirsutum f. typicum* contain several sequiterpenes, including zingiberene γ-elemene, δ-elemene, α-curcumene, and α-humulene, which are acutely toxic to *Spodoptera exigua*. As a group, these compounds do not fully account for the toxicity of the crude trichome tip content, indicating the presence of additional, unidentified toxins (24, 24a). Zingiberene, the predominant sesquiterpene in the tips of type VI trichomes of at least some accessions, is also toxic to Colorado potato beetle larvae and has been implicated in resistance to that important pest (14, 15).

Trichome-Mediated Effects in *Lycopersicon hirsutum f. glabratum*

The effects of glandular and nonglandular trichomes on parasitoids and predators attacking phytophagous insects have been extensively investigated in a number of plant systems (95). Nonglandular trichomes may impede searching behavior of parasitoids and predators (54, 80, 95, 100, 102, 103). Glandular trichomes entrap small hymenopterous parasitoids in sticky exudates (66, 96, 97, 101) and reduce predator mobility. In addition, exudates of glandular trichomes can be directly toxic to natural enemies of pests (69, 70). Virtually all of these effects have been documented in *L. hirsutum f. glabratum*.

BASIS OF RESISTANCE *L. hirsutum f. glabratum* expresses traits conferring resistance to at least 19 arthropod pest species of tomato (32, 83, 88). Glandular trichomes have been implicated in the resistance to most species, although factors not associated with trichomes contribute to the overall resistance to at least some species. One accession of *L. hirsutum f. glabratum*, PI134417, has been investigated extensively and possesses multiple defenses against a number of phytophagous arthropods. Foliage of PI134417 is lethal to a number of phytophagous species because of the presence of toxic methyl ketones, 2-tridecanone and 2-undecanone, in the tips of the type VI glandular trichomes, which abound on the foliage and stems (32, 125). These ketones comprise 90% of the tip contents of type VI trichomes of PI134417, but only trace amounts are present in the type VI trichomes of *L. esculentum* (19, 84). 2-tridecanone is acutely toxic to a number of phytophagous species when assayed in contact bioassays (19, 76, 84, 125). For example, in bioassays involving neonates confined on treated filter paper, LC_{50} values of 2-tridecanone for *Manduca sexta*, *Helicoverpa* (=*Heliothis*) *zea*, and *Leptinotarsa decemlineata* are 17.0, 17.1, and 26.5 ug/cm^2 treated surface (19, 72). Published concentrations of 2-tridecanone associated with PI134417 foliage range from 6.3 ng to 146 ng per trichome tip (72, 84). Because 2-tridecanone is contained in the trichome tips, the amount per cm^2 of leaflet surface is determined by both the amount produced per trichome tip and by the number of type VI trichomes per cm^2. These traits vary independently with plant age and with leaf age and position on the plant. Consequently, there is substantial variation in resistance expression and in the potential for tritrophic level effects among plants and among leaves on individual plants (84, 105). 2-tridecanone concentration averaged 44.6 μg/cm^2 of leaflet surface on young fully expanded foliage of PI134417 (72).

Both 2-tridecanone level and type VI trichome density are under separate genetic control but there are epistatic effects (38, 90). Expression of both traits is highly influenced by environmental conditions, including day length, light intensity, and plant nutrient status (2, 78, 104, 123). High levels of 2-tridecanone and resistance to *M. sexta* and *L. decemlineata* are conditioned by at least three major genes inherited in a recessive manner (38, 106). Restriction fragment length polymorphism (RFLP) analyses of quantitative trait loci (QTLs) associated with 2-tridecanone levels in F_2 progeny of crosses between *L. esculentum* and PI134417 identified three different linkage groups associated with the expression of 2-tridecanone. One of the RFLP loci having the highest correlation with 2-tridecanone levels is primarily associated with expression of type VI trichome density (75, 90).

2-undecanone, a second ketone, in the tips of type VI trichomes of PI134417 is less abundant than 2-tridecanone [range 1.1 to 47 ng per tip and less acutely toxic to *H. zea* in contact bioassays (73, 84)]. In contrast, 2-undecanone and 2-tridecanone do not differ in acute contact toxicity to *Keiferia lycopersicella* and to *S. exigua* (84). At the ratio that occurs in type VI trichome tips, 2-undecanone synergizes the toxicity of 2-tridecanone to neonate *H. zea*, *K. lycopersicella*, and *S. exigua* (28, 84). At levels present in foliage, 2-undecanone has no effect on survival, growth, or development of *M. sexta* (28).

Resistance of PI134417 to *M. sexta* and *L. decemlineata* results from exposure of neonates to lethal concentrations of 2-tridecanone, although other factors associated with the leaf lamella contribute to resistance to *L. decemlineata* (38, 72, 77). Despite the presence of 2-tridecanone at levels potentially lethal to *K. lycopersicella* and *H. zea*, on plants both species avoid significant mortality from exposure to the ketone. *K. lycopersicella* larvae avoid lethal exposure because they are leafminers (84). The situation is more complex with *H. zea* (19, 28–30, 33, 36, 71, 74). Sublethal exposure of eggs and neonates on the foliage of PI134417 to 2-tridecanone vapors from type VI trichomes induces elevated levels of detoxifying enzymes (cytochrome P-450 isozymes) in neonates. Induced *H. zea* larvae tolerate subsequent exposure to 2-tridecanone with no apparent effects; they are also more tolerant to the insecticide carbaryl, which is metabolized by inducible P-450 isozymes. Approximately 15% of *H. zea* neonates are killed by their initial exposure to 2-tridecanone. The remaining larvae quickly recover and begin to feed on the foliage. Most are subsequently killed prior to pupation by resistance factors contained in the leaf lamellae. Unlike 2-tridecanone/trichome-mediated resistance, which is inherited in a recessive fashion, the lamella-based resistance is inherited as a dominant trait. The few *H. zea* that complete larval development on the foliage ingest 2-undecanone during the last instar and die in the pupal stage. 2-undecanone ingested during the last larval instar appears to interfere with lipid metabolism essential for successful pupation and pupal survival.

A series of laboratory and field experiments, involving F_1 and F_1 backcross progeny from crosses between *L. esculentum* and PI134417, which expressed a range of densities of type VI trichomes and methyl ketone levels, demonstrated that glandular trichomes of PI134417 exert dramatic effects on egg and larval parasitoids and predators of *H. zea* and *M. sexta*. F_1 plants express intermediate densities of type VI trichomes but only trace amounts of the methyl ketones; backcross plants are intermediate between F_1 hybrids and PI134417 in both type VI density and methyl ketone levels (27, 69).

EFFECTS ON EGG PARASITOIDS *Trichogramma* spp. are a major source of mortality of *H. zea* eggs on tomato (27, 69). Parasitism rates of *H. zea* eggs by *Trichogramma* (primarily *T. pretiosum* and *T. exiguum*) over a four-year period averaged 43% on *L. esculentum*, 14% on F_1 hybrids, and <2% on a backcross line and PI134417. The latter two plant lines expressed comparable, high trichome densities and methyl ketone levels (27). Because plant lines with the highest trichome densities also had the highest concentrations of 2-tridecanone, it was not possible to separate effects of trichome density from those of 2-tridecanone (69). Similar reductions in parasitism of *Manduca* eggs were observed on *L. esculentum*, F_1, and PI134417 plants by *Trichogramma* and by the somewhat larger *Telenomus sphingis* (27).

Both physical effects of elevated trichome densities and toxic effects of the methyl ketones contribute to the near elimination of egg parasitism on PI134417. Walking speed of adult *T. pretiosum* was reduced on foliage having high type VI trichome densities and moderate-to-high levels of 2-tridecanone (67, 68). *T. pretiosum* adults contacting PI134417 foliage or exposed to the 2-tridecanone-rich volatiles produced by the foliage also suffered higher mortality (36% and 10%) within 4 h than adults exposed to foliage or volatiles from *L. esculentum* (5% mortality for both). In addition, emergence of adult *T. pretiosum* from parasitized *H. zea* eggs was lower when eggs were incubated on PI134417 foliage than when incubated on *L. esculentum* foliage (emergence 42%–56% and 80%–86%). Mortality of *T. pretiosum* within host eggs is apparently due to exposure of the developing parasitoid to 2-undecanone. Exposure within the egg to 2-tridecanone has no effect on adult emergence.

Telenomus sphingis, which parasitizes *Manduca* eggs, is larger than *T. pretiosum* (adult lengths about 2.0 and 0.5 mm). Although both species are adversely affected by the trichome/methyl ketone-mediated defenses of *L. hirsutum f. glabratum*, the interactions that result in those effects are different (31, 67). Parasitism of host eggs by *T. sphingis* in the field is significantly reduced on PI134417 and F_1 plants compared to *L. esculentum* (parasitism rates 6, 6, and 23%). Fewer adults land on PI134417 than on *L. esculentum* or F_1 foliage because they are repelled by the 2-tridecanone vapors produced by PI134417 foliage. In addition, searching efficiency by adult *T. sphingis* is greatly reduced on both PI134417 and F_1 foliage. Adults placed on *L. esculentum* foliage spent 78% of their time searching (walking). In contrast, adults on PI134417 and F_1 foliage spent 57% and 41% of the time grooming or resting. Excessive grooming, which occurs on PI134417 and F_1 foliage, cannot be attributed to the methyl ketones because they are not produced by F_1 foliage. Although parasitism is greatly reduced on PI134417 plants, it is not completely prevented. Unlike *T. pretiosum*, *T. sphingis* larvae and pupae within eggs incubated on PI134417 foliage develop normally, unaffected by either 2-tridecanone or 2-undecanone. *T. sphingis* is not entrapped and killed by the glandular trichomes of *L. hirsutum f. glabratum*, an observation consistent with findings in other systems in which large-bodied parasitoid species were less affected by glandular trichomes than were small-bodied parasitoids (7, 96). However, despite its larger size, *T. sphingis* does not effectively utilize host eggs on PI134417 or F_1 plants because the trichome constituents elicit behavioral responses that severely impede effective searching for host eggs.

EFFECTS ON LARVAL PARASITOIDS OF *H. zea* Larval parasitoids interact with their host and the food plant of their host in a variety of ways depending on their specific biological attributes and host-finding behaviors. The details of the interactions between parasitoids and plant defensive traits are important determinants of the type and magnitude of potential tritrophic effects. A comparison of the tritrophic interactions involving PI134417, *H. zea*, and three species of larval parasitoids illustrates this point and some of the subtleties involved.

In the field, larval parasitism of *H. zea* and *Heliothis virescens* by *Campoletis sonorensis* is lower on PI134417 than *L. esculentum* and F_1 (27). Adult *C. sonorensis* are less efficient at locating host larvae on PI134417 plants, although the specific behavioral causes are not known (69). However, the effects of PI134417 on *C. sonorensis* extend beyond behavioral modification. *C. sonorensis* oviposits in second and early-third instar hosts. The parasitoid larvae kill their host when it is in the late-third or early-fourth instar. They then emerge from the dead host onto their host's food plant, where they spin a cocoon and pupate. On PI134417, this cocoon-spinning behavior results in disruption of glandular trichomes and direct contact with the methyl ketones contained in their tips. Both in the field and in the laboratory, mortality of *C. sonorensis* larvae during cocoon spinning is high on PI134417 (90% versus 10% on *L. esculentum*). 2-tridecanone is lethal to *C. sonorensis* larvae at the concentrations expressed on PI134417 foliage, and mortality of *C. sonorensis* larvae on foliage from parental and backcross plant lines expressing different levels of 2-tridecanone is directly related to the 2-tridecanone concentration of the foliage (69, 70). Although 2-tridecanone is similarly toxic to both *C. sonorensis* and *H. zea*, the parasitoid is much more affected by PI134417 foliage than its host is. *C. sonorensis* larvae are killed because they experience a massive dose of the toxin as they spin their cocoons. In contrast, neonate *H. zea* rarely discharge trichome tips before exposure to 2-tridecanone vapors induces elevated levels of cytochrome P-450, which then allow the larvae to tolerate subsequent exposures to 2-tridecanone (70, 74).

The tachinids *Archytas marmoratus* and *Eucelatoria bryani* both parasitize *H. zea* but interact with the food plants of their host in distinctly different ways. The methyl ketone/glandular trichome-based resistance of PI134417 differentially affects both *A. marmoratus* and *E. bryani*. At least a portion of the difference between the two species can be attributed to differences in their interactions with the host plant (34, 35). *A. marmoratus* larviposits minute planidia on its host's food plant. The planidia attach to passing hosts and penetrate the cuticle. The parasitoid undergoes limited development within its host until the host pupates. It then develops rapidly and emerges from the host pupa as an adult (53). In contrast, *E. bryani* larviposits directly into larvae of its host. The parasitoid larvae develop rapidly and kill the host within a few days. Mature larvae emerge from the dead host larva and pupate. A single host often yields several parasitoids (12, 59). Because the parasitized host typically drops to the ground before it is killed by the parasitoid, *E. bryani*, unlike *A. marmoratus*, has minimal direct contact with its host food plant.

In field cage studies, parasitism of *H. zea* larvae by *E. bryani* was not affected by plant line. In contrast, parasitism by *A. marmoratus* was lower on PI134417 and on a backcross line expressing high levels of 2-tridecanone and high trichome densities (mean parasitism about 55%) than on *L. esculentum* and F_1 lines (mean parasitism 74% on both). More detailed studies revealed that the difference in parasitism was due to elevated mortality of *A. marmoratus* planidia on PI134417 and backcross plants due to a combination of toxic effects of 2-tridecanone and physical effects of high trichome densities (34, 35). In addition, there is a host-mediated effect of 2-undecanone but not 2-tridecanone on survival of *A. marmoratus* within parasitized *H. zea* pupae. 2-undecanone ingested during the last larval instar causes premature death of *H. zea* pupae. At high RH, this does not prevent parasitoid larvae from completing development; but at low RH, host pupae desiccate, killing the parasitoid larvae before they complete development. 2-undecanone in the diet of *H. zea* does not affect *E. bryani*, but 2-tridecanone in the host diet reduces by about 45% the number of *E. bryani* puparia produced per host. It is not known if this effect is due to reduced larviposition by *E. bryani* in hosts reared on a diet containing 2-tridecanone or to mortality of parasitoid larvae within their host.

FIELD OBSERVATIONS In a four-year field study, both egg and larval parasitism of *Manduca* spp. were lower on PI134417 plants (27). Egg parasitism of *M. sexta* and *M. quinquemaculata* by *T. sphingis* and *T. pretiosum* was significantly lower on PI134417 and F_1 than on *L. esculentum* plants. In addition, larval parasitism of *Manduca* spp. by *Cotesia congregata* was consistently lower on PI134417 than on *L. esculentum*. However, because *M. quinquemaculata* larvae are less affected than *M. sexta* by the resistance and because *C. congregata* prefers *M. sexta* to *M. quinquemaculata*, reduced larval parasitism on PI134417 could reflect the greater relative abundance of *M. quinquemaculata* on PI134417 in this study rather than an effect of PI134417 on the parasitoid.

Egg populations of *H. zea* and *H. virescens* on PI134417 averaged 3.2 and 16 times greater than on *L. esculentum* over a three-year period. Oviposition was also higher on F_1 plants than on *L. esculentum*. This difference is likely due to the presence of high levels of an oviposition stimulant associated with the foliage of PI134417 (60). Egg parasitism by *Trichogramma* on *L. esculentum*, F_1, and PI134417 averaged 43, 14, and 1% over four years. *H. zea* larvae were parasitized by *Campoletis sonorensis* and *Cotesia marginiventris*. *H. virescens* larvae were parasitized by *C. sonorensis*, *C. marginiventris*, and *Cardiochiles nigriceps*. The incidence of parasitism in both *H. zea* and *H. virescens* by *C. sonorensis* was lower on PI134417 than on *L. esculentum* and F_1. Parasitism by *C. marginiventris* was low and variable, revealing no consistent differences among plant lines. No differences in parasitism rates of *H. virescens* by *C. nigriceps* were observed among plant lines. Given that parasitism rates by *C. nigriceps* on PI134417 plants exceeded 75% in each of two years, it is unlikely that *C. nigriceps* is adversely affected by high densities of methyl ketone-rich, glandular trichomes on the foliage of its host's food plants.

Despite higher egg densities and greatly reduced levels of egg parasitism by *Trichogramma* spp. on PI134417 and F_1, and a 70% lower incidence of larval parasitism by *C. sonorensis* on PI134417 than on *L. esculentum*, densities of large *H. zea* larvae were similar on each of the plant lines. Overall, for *H. zea* higher oviposition and reduced parasitism and predation rates (3) on PI134417 were offset by higher larval mortality owing to the combined effects of trichome- and nontrichome-mediated resistance traits. For *H. virescens*, which is less affected by the resistance traits, the combined effects of trichome- and nontrichome-mediated resistance traits did not compensate for the higher oviposition and lower parasitism and predation rates. As a result, populations of large *H. virescens* larvae tended to be higher on PI134417 than on *L. esculentum*.

NONTRICHOME DEFENSES IN *LYCOPERSICON*

Constitutive and induced defenses other than those associated with glandular trichomes have been well documented in *Lycopersicon* spp. However, only the glycoalkaloid α-tomatine, the growth-inhibiting phenolics rutin and chlorogenic acid, and the jasmonic acid-inducible defenses have been studied from the perspective of tritrophic level effects.

Constitutive Defenses

α-tomatine has been hypothesized as a possible contributor to the lamella-based resistance of *Lycopersicon* spp. to *H. zea* and *S. exigua*. It is present in both tomato foliage and unripe fruit, but not in ripe fruits or in the tips of type VI trichomes (23). Foliar α-tomatine levels vary both within and among *Lycopersicon* species, with the highest levels occurring in *L. esculentum* var. *cerasiforme* and *L. pimpinellifolium* (64). Variation in α-tomatine levels among progeny of crosses between high- (*L. esculentum* var. *cerasiforme* and *L. pimpinellifolium*) and low- (*L. esculentum*) expressing plants is controlled by segregation of two codominant alleles at a single locus (62). α-tomatine content of tomato fruit was positively correlated with development time and mortality of *H. zea* larvae fed fruit of different ages, and it was negatively correlated with larval growth rate and adult weight. In contrast, there were no correlations between α-tomatine content of fruit and growth or survival of *S. exigua* (63). When incorporated into artificial diets at concentrations representative of those found in commercial tomato varieties, α-tomatine is a potent growth inhibitor of both *H. zea* and *S. exigua* (9, 20, 25). However, the toxic effects of α-tomatine are dependent on concentrations of 3-beta-hydroxy-sterols present in the diet. In the presence of equimolar concentrations of dietary sterols, α-tomatine is nontoxic to *H. zea* and only minimally toxic to young *S. exigua* larvae. In addition, the sensitivity of *S. exigua* but not *H. zea* larvae to α-tomatine declines with age to the point that later stage *S. exigua* larvae are insensitive to its effects (9).

Hyposoter exiguae, a larval parasitoid of *H. zea* and *S. exigua*, is adversely affected when its hosts feed on diet containing α-tomatine (13). The toxicity of α-tomatine is mediated through the host larva and is manifested as prolonged larval development, disruption or prevention of pupal eclosion, deformation of antennal, abdominal and genital structures, and a reduction in adult weight and longevity. *H. exiguae* is affected by α-tomatine in its host diet at lower concentrations when parasitizing *S. exigua* than when parasitizing *H. zea*. α-tomatine in the host's diet at a concentration of 0.9 μmol/g.fr.wt. is lethal to *H. exiguae* when parasitizing

S. exigua, but it merely slows larval and pupal development and reduces adult longevity when parasitizing *H. zea* (21). As with *H. zea* and *S. exigua*, the toxic effects of α-tomatine on *H. exiguae* are eliminated by the presence of equimolar concentrations of sterols in the host diet (13a). The occurrence of adverse effects of α-tomatine in *Lycopersicon* foliage on *H. exiguae* or other parasitoids has not been documented.

The actual significance of α-tomatine as a mechanism of resistance in *Lycopersicon* to *H. zea* and *S. exigua* is not clear, nor is the occurrence, in the field, of α-tomatine-mediated tritrophic effects. Many *L. esculentum* cultivars contain levels of phytosterols sufficient to negate the toxic effects of α-tomatine, but the levels of antidotal phytosterols and α-tomatine in tomato foliage are not correlated (13a, 21, 22). The presence of high levels of phytosterols in the foliage may be related to the role of α-tomatine as a defense against fungi, which requires high levels of phytosterols (22). There is some evidence that α-tomatine in *Lycopersicon* foliage could inhibit infection of phytophagous insects by the entomopathogenic fungi *Beauveria bassiana* and *Nomuraea rileyi* (18, 40, 46). In addition, there is some evidence that α-tomatine may influence predation on *M. sexta* larvae by *Podisus maculiventris* (120). *P. maculiventris*, which had previously preyed upon tomatine-fed larvae, subsequently rejected *M. sexta* larvae as prey whether or not the larvae had fed upon a diet containing tomatine.

Phenolics have been implicated as possible factors in the growth-inhibitory effects of tomato foliage on *H. zea* and *S. exigua* larvae. The catecholic phenols rutin and chlorogenic acid in the leaf lamella and the tips of type VI trichomes account for over 60% of the total phenolic content of *L. esculentum* foliage (57, 58). These compounds are common in green plants (45, 79). When incorporated into artificial diet they act in an additive fashion to produce a dose-dependent growth inhibition of *H. zea* and *S. exigua* larvae (21, 25, 57, 58). However, the growth-inhibitory effects are dependent on the type and level of protein in the diet (21). Levels of both phenolics and proteins in tomato foliage are influenced by nitrogen fertilization, water stress, and light intensity (26, 116, 124). They also vary widely and independently both within and among plants of the same and different *L. esculentum* cultivars, although the concentrations of soluble phenolics are consistently higher in young than mature foliage (110). The interaction between phenolics and protein may explain a lack of correlation between total foliar phenolic content and growth of *H. zea* larvae on different *L. esculentum* cultivars, when phenolic levels in the foliage are within a range that retards larvae growth on artificial diet (57).

In artificial diet studies, rutin exerts mild effects on a parasitoid and a predator of *M. sexta*. Dietary rutin prolongs development of *M. sexta* larvae at cool but not at warm temperatures (52, 107–110, 122). This effect involves more than a reduction in availability of dietary protein; it includes interference with physiological processes relating to the initiation of molting. Development time of the parasitoid *C. congregata* is also increased when the host larvae are raised on artificial diet containing rutin at concentrations found in plants. Because *C. congregata* does not develop beyond the first instar until *M. sexta* reaches the last larval stage, prolonged development of the parasitoid is thought to result from delayed development of its host rather than from an effect of rutin on the parasitoid (1, 6). Presumably the effect of rutin on *C. congregata* is greater at cool than warm temperatures. When nymphs of the predatory stink bug *P. maculiventris* were fed *M. sexta* larvae whose diet contained rutin, they experienced a reduction in weight gain at warm but not cool temperatures, a temperature effect opposite that seen for *M. sexta* larvae. Effects of rutin were observed when prey was scarce but not when abundant. In addition, *P. maculiventris*, which had previously preyed upon chlorogenic acid–fed larvae, subsequently rejected chlorogenic acid–fed *M. sexta* larvae as prey but not *M. sexta* larvae that had fed on diet that contained no chlorogenic acid (120).

Induced Defenses

Even minor injury to *L. esculentum* leaves by insects, pathogens, or mechanical wounding can elicit a systemic response that results in synthesis of more than 20 defense-related proteins, including antinutritional proteins, signaling pathway compounds, and proteinases (101a). These responses alter the suitability of foliage for some insect species and plant pathogens. Thus, feeding by *H. zea* larvae on *L. esculentum* induces elevated resistance to *H. zea* and *S. exigua*, as well as *M. euphorbiae*, *T. urticae*, and the phytopathogen *Pseudomonas syringae* (114).

Stout et al. (113) suggest that inducible defenses of *L. esculentum* consist of an array of sets of defensive compounds, with different sets being differentially inducible by different insects or pathogens or combinations thereof. Because individual insect and pathogen species vary in their sensitivity to each set of defensive compounds, injury to the plant by different agents may result in different patterns of resistance to any given array of insects and pathogens. For example, feeding by *H. zea* induces a systemic resistance to *H. zea*, mediated primarily by the induction of proteinase inhibitors, and resistance to the phytopathogen *P. syringae*, mediated primarily by induction of pathogenesis-related protein P4. The induced resistance to *P. syringae* is less than that induced by a localized *P. syringae* infection, which elicits a production of higher levels of pathogenesis-related protein P4. Because local infections of *P. syringae* induce levels of proteinase inhibitors comparable to those induced by *H. zea* feeding, *P. syringae* infections and *H. zea* feeding induce comparable levels of resistance to *H. zea* (39, 55, 113). Detailed descriptions of these induced defenses can be found elsewhere (10, 11, 39, 55, 56, 111, 112, 114, 115, 117, 119).

Caterpillar feeding on *L. esculentum* induces proteinase inhibitors, peroxidase, polyphenol oxidase, and lipoxygenase. The production of elevated levels of proteinase inhibitors, and to a lesser extent polyphenol oxidase, appears to be the primary mechanism for induced resistance to caterpillars (11, 111, 115, 117). Feeding on foliage from induced plants causes reduced growth and delayed development in *H. zea*, *M. sexta*, and *S. exigua*, as well as elevated mortality in *S. exigua* (11, 111, 112). Growth of *M. sexta* larvae on transgenic tobacco plants expressing the tomato gene coding for proteinase inhibitor II, which inhibits both trypsin and chymotrypsin, was reduced from 50% to 67% relative to controls, depending on expression level in the plant. In contrast, larval growth was only slightly reduced on transgenic tobacco plants expressing tomato proteinase inhibitor I, which is a strong inhibitor of chymotrypsin but only weakly inhibits trypsin (59a) The induction of proteinase inhibitors and polyphenol oxidase is systemic, occurs within hours, and is long lasting (≥21 days) (111). The systemic induction of defensive compounds reduces the ability of phytophagous arthropods to avoid locally induced compounds by moving to uninjured leaves (17a). However, induction is not uniform throughout the plant (98, 111, 112), and therefore some portions of the plant are more defended than others. Consequently, the distribution of insect feeding or damage on the plant may change following induction (5). In addition, levels of induction of proteinase inhibitors and polyphenol oxidase in response to feeding decline with plant age (16, 111). On plants of a given age, younger leaves have higher constitutive levels of polyphenol oxidase than older leaves but inducible levels of polyphenol oxidase are higher in older leaves (111). They also vary among *L. esculentum* cultivars (11).

The production of jasmonic acid via the octadecanoid pathway serves as a signal for expression of proteinase inhibitors, polyphenol oxidase, and peroxidase in tomato foliage in response to caterpillar feeding (39, 83a, 101a, 119). Thaler (117, 118) used foliar applications of jasmonic acid to induce elevated levels of these compounds in field-grown *L. esculentum* plants. In her experiment, the number of leaves damaged by insect feeding was 38% lower on the induced plants than on control plants (117). In a separate experiment, twice as many *S. exigua* larvae were parasitized by *H. exiguae* on the jasmonic acid–treated tomato plants as on untreated control plants (118). This difference was not due to an attraction of the wasps to jasmonic acid, nor was it due to a greater period of susceptibility to attack by *H. exiguae* resulting from slower growth of *S. exigua* larvae on induced plants. Thaler hypothesized that higher parasitism resulted from the attraction of *H. exiguae* adults to the blend of volatiles produced by the induced plants. Parasitoid survival was comparable when host larvae fed on induced or noninduced foliage; however, *H. exiguae* developing in *S. exigua* fed induced foliage developed more slowly than those developing in hosts fed foliage from noninduced plants. It seems unlikely that this delayed development would reduce the impact of elevated levels of parasitism on induced plants.

Effects on Predators

Although most research on tritrophic effects involving *Lycopersicon* has focused on parasitoids, effects involving predaceous arthropods have been documented as well. The predatory mite *Phytoseiulus persimilis* is used in the biological control of the two-spotted spider mite, *T. urticae*. Feeding on *L. esculentum* foliage by *T. urticae* induces leaves to produce volatiles that are attractive to the *P. persimilis*. The response of *P. persimilis* to *T. urticae*-induced plant volatiles is affected by starvation, specific hunger, and experience of the predator, as well as the presence of competitors and pathogen infestations (18a, 116a).

Type VI trichomes on the foliage and stems of *L. esculentum* entrap *P. persimilis*. Entrapment rates range from 2.5 to 20% (91) but vary among cultivars, with higher entrapment rates associated with high trichome densities (121). Rapid increases in temperature appear to temporarily increase entrapment rates by reducing the rupture threshold of the trichome tips and may contribute to variation in the effectiveness of mite control by *P. persimilis* (91–93).

Oviposition by the whitefly *B. argentifolii* on *L. esculentum* is positively correlated with trichome density (type not specified) (50). Although lifetime fecundity and walking speed of its coleopterous predator *Delphastus pusillus* are significantly reduced on cultivars with high trichome densities, the ability of the predator to suppress whitefly populations is not reduced in the presence of high trichome densities. It seems that in the presence of high trichome densities, significantly longer residence time by beetles compensates for their reduced fecundity and walking speed (51).

In a three-year field study, populations of predaceous arthropods (*Coleomegilla maculata*, *Geocoris punctipes*, *Jalysus wickhami*, and spiders) were similar on PI134417, *L. esculentum*, F_1, and backcross plantings (4). Laboratory results confirmed that survival of adult *C. maculata* and *G. punctipes* was similar on these plant lines but demonstrated that consumption of *H. zea* eggs by adults and immatures of these predators was lower on PI134417 and F_1 than on *L. esculentum* foliage. *H. zea* egg consumption by *C. maculata* and *G. punctipes* adults on *L. esculentum* foliage was about 13 and 10 times greater than on PI134417 and about 2 times greater (both species) than on F_1 foliage. Reduced egg consumption on F_1 foliage appears to be due to the direct physical effects of the glandular trichomes on the predators, whereas the even greater reduction on PI134417 foliage appears to be due to the combined effects of high trichome densities and 2-tridecanone (3). The effects of tomato trichomes on prey consumption are consistent with observations of reduced searching efficiency of *Orius insidiosus* on tomato relative to bean and maize (17).

GENERAL CONCLUSIONS

The most intensively studied *Lycopersicon* species, *L. esculentum*, *L. pennellii*, *L. hirsutum f. typicum*, and *L hirsutum f. glabratum*, express a diverse array of defense traits. These include glandular trichomes that physically impede, entrap, and in some cases intoxicate phytophagous arthropods. They also include constitutively expressed and induced defenses associated with the leaf lamellae that act as growth inhibitors or toxins. A number of these traits affect parasitoids and predators as well as phytophagous species. Trichome-mediated defenses are particularly significant in *L. pennellii*, *L. hirsutum f typicum*, and *L hirsutum f. glabratum* and have been most extensively implicated in negative tritrophic effects in both the field and laboratory. These effects range from reduced searching efficiency due to direct physical interference with movement to elevated mortality due to entrapment or lethal intoxication of parasitoids and predators that discharge and contact sticky or toxic contents of trichome tips. Effects also include those that are indirectly mediated through the host or prey, such as those of 2-undecanone on *Trichogramma* or *A. marmoratus*. Even seemingly simple traits, such as high concentrations of 2-tridecanone or 2-undecanone, can have profound and complex effects on the second and third trophic levels. The effects of any particular plant defensive trait on parasitoids and predators depend on the specific attributes of the plant trait and the details of the interaction between the natural enemy, its host (prey), and the plant, as well as the physiological ability of the natural enemy to de-toxify or sequester plant toxins. The great diversity that exists in parasitoid and predator biology with respect to the details of these interactions suggests that predictions linking types of plant defenses to tritrophic effects that are likely to be important ecologically or of value in deploying host plant resistance for crop protection may be limited and general. Clearly, one such generalization that emerges from work on *Lycopersicon*, as well as other species, is that glandular trichome-mediated resistance traits are more likely to exert general adverse tritrophic effects than nontrichome-mediated traits.

One striking feature of defense traits discussed in this review is the extent to which their expression is strongly influenced by environmental conditions, as well as plant age, leaf age, and leaf position on the plant. The implications of this variation for herbivore/parasitoid/predator interactions and adaptation to resistance traits represent an inviting avenue for future research. Only a limited portion of the research on tritrophic interactions in *Lycopersicon* has involved field studies. Virtually all these have been conducted in areas far removed from the plants' native habitats and have focused on phytophagous species that are pests on *L. esculentum* cultivars. Additional field studies of arthropod population and community dynamics involving a greater spectrum of *Lycopersicon* species, resistance traits, and locations, including native habitats of the plants under study, have the potential to provide new insights and understanding of the ecological impacts of plant defense traits.

Lycopersicon provides an excellent model system for research on the mechanisms, ecology, and evolution of plant defenses, and on tritrophic interactions. Recent progress in understanding the biochemical basis and molecular genetics of induced defenses in tomato, combined with the discovery in tomato of a gene that blocks biosynthesis of jasmonic acid and a gene that blocks the ability of the plant to respond to jasmonic acid (52a), provide the tools necessary to greatly advance our understanding of the mechanistic basis for and ecological consequences of tritrophic effects of induced defenses. *Lycopesicon* is also a rich source of resistance traits of potential use in breeding pest-resistant tomato cultivars. From a crop protection perspective, it is the net effects of plant resistance on the targeted pest(s) and their natural enemies that determine the final level of crop damage and the potential utility of a particular resistance trait or set of traits for crop protection.

ACKNOWLEDGMENTS

I would like to thank James D. Barbour, Michael B. Dimock, Robert R. Farrar Jr., and William C. Kauffman for their thought provoking ideas and research on *L. hirsutum f. glabratum*, and Molly Puente and Fred Gould and his lab group for helpful comments on the manuscript. This review is dedicated to Sean S. Duffey whose genius and friendship have served as an inspiration to me. Research support from the USDA-NRI Program, USDA-ARS, and North Carolina Agricultural Research Service is gratefully acknowledged.

The *Annual Review of Entomology* is online at http://ento.annualreviews.org

LITERATURE CITED

1. Barbosa P, Gross P, Kemper J. 1991. Influence of plant allelochemicals on the tobacco hornworm and its parasitoid *Cotesia congregata*. *Ecology* 72:1567–75
2. Barbour JD, Farrar RR Jr, Kennedy GG. 1991. Interaction of fertilizer regime with host-plant resistance in tomato. *Entomol. Exp. Appl.* 60:289–300
3. Barbour JD, Farrar RR Jr, Kennedy GG. 1993. Interaction of *Manduca sexta* resistance in tomato with insect predators of *Helicoverpa zea*. *Entomol. Exp. Appl.* 68:143–55
4. Barbour JD, Farrar RR Jr, Kennedy GG. 1997. Populations of predaceous natural enemies developing on insect-resistant and susceptible tomato in North Carolina. *Biol. Control* 9:173–84
5. Barker AM, Wratten SD, Edwards PJ. 1995. Wound-induced changes in tomato leaves and their effects on the feeding patterns of larval Lepidoptera. *Oecologia* 10:251–57
6. Beckage NE, Riddiford LM. 1983. Growth and development of the endoparasitic wasp *Apanteles congregata*: dependence on host nutritional status and parasite load. *Physiol. Entomol.* 8:231–41
7. Belcher DW, Thurston R. 1982. Inhibition of movement of larvae of the convergent ladybeetle *Hippodamia convergens* by leaf trichomes of tobacco. *Environ. Entomol.* 11:91–94
8. Blauth SL, Churchill GA, Mutschler MA. 1998. Identification of quantitative trait loci associated with acylsugar accumulation using intraspecific populations of the wild tomato, *Lycopersicon pennellii*. *Theor. Appl. Genet.* 967:458–67
9. Bloem K, Kelley KC, Duffey SS. 1989. Differential effect of tomatine and its alleviation by cholesterol on larval growth and efficiency of food utilization in *Heliothis zea* and *Spodoptera exigua*. *J. Chem. Ecol.* 15:387–98

9a. Boethel DJ, Eikenbary RD, eds. 1986. *Interactions of Plant Resistance and Parasitoids and Predators of Insects*. Chichester: Horwood. 224 pp.

9b. Bottrell DG, Barbosa P, Gould F. 1998. Manipulating natural enemies by plant variety selection and modification: a realistic strategy? *Annu. Rev. Entomol.* 43:347–67

10. Broadway RM, Duffey SS. 1988. The effect of plant protein quality on insect digestive physiology and the toxicity of plant proteinase inhibitors. *J. Insect Physiol.* 34:1111–17
11. Broadway RM, Duffey SS, Pearce G, Ryan CA. 1986. Plant proteinase inhibitors: a defense against herbivorous insects? *Entomol. Exp. Appl.* 41:33–38
12. Bryan DE, Jackson CG, Stoner A. 1969. Rearing cotton insect parasites in the laboratory. *USDA Agric. Prod. Res. Rep. 109*. 13 pp.
13. Campbell BC, Duffey SS. 1979. Tomatine and parasitic wasps: potential incompatibility of plant antibiosis with biological control. *Science* 205:700–2

13a. Campbell BC, Duffey SS. 1981. Alleviation of α-tomatine-induced toxicity to the parasitoid *Hyposoter exiguae* by phytosterols in the diet of the host *Heliothis zea*. *J. Chem. Ecol.* 7:927–46

14. Carter CD, Gianfagna TJ, Sacalis JN. 1989. Sesquiterpenes in glandular trichomes of a wild tomato species and toxicity to the Colorado potato beetle. *J. Agric. Food Chem.* 37:1425–28
15. Carter CD, Sacalis JN, Gianfagna TJ. 1989. Zingiberene and resistance to Colorado potato beetle in *Lycopersicon hirsutum f. hirsutum*. *J. Agric. Food Chem.* 37:206–10

15a. Carter CD, Snyder JC. 1985. Mite response in relation to trichomes of *Lycopersicon esculentum* x *L. hirsutum* F_2 hybrids. *Euphytica* 34:177–85

15b. Carter CD, Snyder JC. 1986. Mite responses and trichome characters in a full-sib F_2 family of *Lycopersicon esculentum* x *L. hirsutum*. *J. Am. Soc. Hort. Sci.* 111:130–33

16. Cipollini DF Jr, Redman AM. 1999. Age-dependent effects of jasmonic acid treatment and wind exposure on foliar oxidase activity and insect resistance in tomato. *J. Chem. Ecol.* 25:271–81
17. Coll M, Smith LA, Ridgway RL. 1997. Effect of plants on the searching efficiency of a generalist predator: the importance of predator-prey spatial associations. *Entomol. Exp. Appl.* 83:1–10

17a. Constabel CP, Ryan CA. 1998. A survey of wound and methyl jasmonate-induced leaf polyphenol oxidase in crop plants. *Phytochemistry* 44:507–11

18. Costa SD, Gaugler RR. 1989. Sensitivity of *Beauveria bassiana* to solanine and tomatine: plant defensive chemicals inhibit an insect pathogen. *J. Chem. Ecol.* 15:697–706

18a. Dicke M, Takabayashi J, Posthumus MA, Schutte C. 1998. Plant-phytoseiid interactions mediated by herbivore-induced plant volatiles: variation in production of cues and in responses of predatory mites. *Expt. Appl. Acarol.* 22: 311–33

19. Dimock MB, Kennedy GG. 1983. The role of glandular trichomes in the resistance of *Lycopersicon hirsutum f. glabratum* to *Heliothis zea*. *Entomol. Exp. Appl.* 33:263–68
20. Duffey SS. 1986. Plant glandular trichomes: their potential role in defense against insects. In *Insects and the Plant Surface*, ed. B Juniper, R Southwood, pp. 151–72. London: Arnold. 360 pp.
21. Duffey SS, Bloem KA. 1986. Plant defense-herbivore-parasite interactions and biological control. In *Ecological Theory and Integrated Pest Management Practice*, ed. M Kogan, pp. 135–83. New York: Wiley. 361 pp.
22. Duffey SS, Bloem KA, Campbell BC. 1986. Consequences of sequestration of plant natural products in plant-insect-parasitoid interactions. See Ref. 9a, pp. 151–72
23. Duffey SS, Isman MB. 1981. Inhibition of insect larval growth by phenolics in glandular trichomes of tomato leaves. *Experientia* 37:574–76
24. Eigenbrode SD, Trumble JT, Millar JG, White KK. 1994. Topical toxicity of tomato sesquiterpenes to the beet armyworm and the role of these compounds in resistance derived from an accession of *Lycopersicon hirsutum f typicum*. *J. Agric. Food Chem.* 42:807–10

24a. Eigenbrode SD, Trumble JT, White KK. 1996. Trichome exudates and resistance to beet armyworm (Lepidoptera: Noctuidae) in *Lycopersicon hirsutum f. typicum* accessions. *Environ. Entomol.* 25: 90–95

25. Elliger CA, Wong Y, Chan BG, Waiss AC Jr. 1981. Growth inhibitors in tomato (*Lycopersicon*) to tomato fruitworm (*Heliothis zea*). *J. Chem. Ecol.* 7:753–58
26. English-Loeb G, Stout MJ, Duffey SS. 1997. Drought stress in tomatoes: changes in plant chemistry and potential nonlinear consequences for insect herbivores. *Oikos* 79:456–68
27. Farrar RR, Barbour JD, Kennedy GG.

1994. Field evaluation of insect resistance in a wild tomato and its effect on insect parasitoids. *Entomol. Exp. Appl.* 71:211–26
28. Farrar RR Jr, Kennedy GG. 1987. 2-undecanone, a constituent of the glandular trichomes of *Lycopersicon hirsutum f. glabratum*: effects on *Heliothis zea* and *Manduca sexta* growth and survival. *Entomol. Exp. Appl.* 43:17–23
29. Farrar RR Jr, Kennedy GG. 1987. Growth, food consumption and mortality of *Heliothis zea* larvae on the foliage of the wild tomato *Lycopersicon hirsutum f. glabratum* and the cultivated tomato, *L. esculentum. Entomol. Exp. Appl.* 44:213–19
30. Farrar RR Jr, Kennedy GG. 1988. 2-undecanone, a pupal mortality factor in *Heliothis zea*: sensitive larval stage and *in planta* activity in *Lycopersicon hirsutum f. glabratum. Entomol. Exp. Appl.* 47:205–10
31. Farrar RR Jr, Kennedy GG. 1991. Inhibition of *Telenomus sphingis* an egg parasitoid of *Manduca* spp. by trichome/2-tridecanone-based host plant resistance in tomato. *Entomol. Exp. Appl.* 60:157–66
32. Farrar RR Jr, Kennedy GG. 1991. Insect and mite resistance in tomato. See Ref. 65, pp. 121–42
33. Farrar RR Jr, Kennedy GG. 1991. Relationship of lamellar-based resistance to *Leptinotarsa decemlineata* and *Heliothis zea* in a wild tomato, *Lycopersicon hirsutum f. glabratum. Entomol. Exp. Appl.* 58:61–67
34. Farrar RR Jr, Kennedy GG. 1993. Field cage performance of two tachinid parasitoids of the tomato fruitworm on insect resistant and susceptible tomato lines. *Entomol. Exp. Appl.* 67:73–78
35. Farrar RR Jr, Kennedy GG, Kashyap RK. 1992. Influence of life history differences of two tachinid parasitoids of *Helicoverpa zea* (Boddie) (Lepidoptera: Noctuidae) on their interactions with glandular trichome/methyl ketone-based insect resistance in tomato. *J. Chem. Ecol.* 18:499–515
36. Farrar RR Jr, Kennedy GG, Roe RM. 1992. The protective role of dietary unsaturated fatty acids against 2-undecanone-induced pupal mortality and deformity in *Helicoverpa zea. Entomo. Exp. Appl.* 62:191–200
37. Felton GW, Duffey SS. 1991. Reassessment of the role of gut alkalinity and detergency in insect herbivory. *J. Chem. Ecol.* 1821–36
38. Fery RL, Kennedy GG. 1987. Genetic analysis of 2-tridecanone concentration, leaf trichome characteristics, and tobacco hornworm resistance in tomato. *J. Am. Soc. Hort. Sci.* 112:886–91
39. Fidantsef AL, Stout MJ, Thaler, Duffey SS, Bostock RM. 1999. Signal interactions in pathogen and insect attack: expression of lipoxygenase, proteinase inhibitor II, and pathogenesis-related protein P4 in the tomato, *Lycopersicon esculentum. Physiol. Mol. Plant Pathol.* 54:97–114
40. Gallardo F, Boethel DJ, Fuxa JR, Richter A. 1990. Susceptibility of *Heliothis zea* (Boddie) larvae to *Nomuraea rileyi* (Farlow) Samson: effects of alpha-tomatine at the third trophic level. *J. Chem. Ecol.* 16:1751–59
41. Gentile AG, Stoner AK. 1968. Resistance in *Lycopersicon* and *Solanum* species to potato aphid. *J. Econ. Entomol.* 61:1152–54
42. Goffreda JC, Mutschler MA. 1989. Inheritance of potato aphid resistance in hybrids between *Lycopersicon esculentum* and *L. pennellii. Theor. Appl. Genet.* 78:210–16
43. Goffreda JC, Steffens JC, Mutschler MA. 1990. Association of epicuticular sugars with aphid resistance in hybrids with wild tomato. *J. Am. Soc. Hort. Sci.* 115:161–65
44. Gregory P, Ave DA, Bouthyette PY, Tingey WM. 1986. Insect-defensive

chemistry of potato glandular trichomes. See Ref. 59a, pp. 173–83

45. Harborne JB. 1979. Variation in and functional significance of phenolic conjugation in plants. *Rec. Adv. Phytochem.* 12:457–74
46. Hare JD, Andreadis TG. 1983. Variation in the susceptibility of *Leptinotarsa decemlineata* (Coleoptera: Chrysomelidae) when reared on different host plants to the fungal pathogen *Beauveria bassiana* in the field and laboratory. *Environ. Entomol.* 12:1892–97
47. Hartman JB, St. Clair DA. 1999. Combining ability for beet armyworm, *Spodoptera exigua*, resistance and horticultural traits of selected *Lycopersicon pennellii*-derived inbred backcross lines of tomato. *Plant Breed.* 118:523–30
48. Hartman JB, St. Clair DA. 1999. Variation for aphid resistance and insecticidal acyl sugar expression among and within *Lycopersicon pennellii*-derived inbred backcross lines of tomato and their F_1 progeny. *Plant Breed.* 118:531–36
49. Hawthorne DM, Shapiro JA, Tingey WM, Mutschler MA. 1992. Trichome-borne and artificially applied acylsugars of wild tomato deter feeding and oviposition of the leafminer, *Liriomyza trifolii*. *Entomol. Exp. Appl.* 65:65–73
50. Heinz KM, Zalom FG. 1995. Variation in trichome-based resistance to *Bemisia argentifolii* (Homoptera: Aleyrodidae) oviposition on tomato. *J. Econ. Entomol.* 88:1494–1502
51. Heinz KM, Zalom FG. 1996. Performance of the predator *Delphastus ausillus* on *Bemisia* resistant and susceptible tomato lines. *Entomol. Exp. Appl.* 81: 345–52
52. Horwath KL, Stamp NE. 1993. Use of dietary rutin to study molt initiation in *Manduca sexta* larvae. *J. Insect Physiol.* 39:987–1000

52a. Howe GA, Schilmiller L. 2002. Oxylipin metabolism in response to stress. *Curr. Opinions Plant Biol.* 5:230–36

53. Hughes PS. 1975. The biology of *Archytas marmoratus* (Townsend). *Ann. Entomol. Soc. Am.* 68:557–67
54. Hulspas-Jordaan PM, van Lenteren JC. 1978. The relationship between host-plant leaf structure and parasitization efficiency of the parasitic wasp *Encarsia formosa* Graham (Hymenoptera: Aphelinidae). *Med. Fac. Landbouww. Rijksuniv. Gent.* 43:431–40
55. Inbar M, Doostdar H, Leibee GL, Mayer RT. 1999. The role of plant rapidly induced responses in assymetric interspecific interactions among insect herbivores. *J. Chem. Ecol.* 25:1961–79
56. Inbar M, Doostdar H, Sonada RM, Leibee GL, Mayer RT. 1998. Elicitors of plant defense systems reduce insect densities and disease incidence. *J. Chem. Ecol.* 24:135–48
57. Isman MB, Duffey SS. 1982. Phenolic compounds in foliage of commercial tomato cultivars as growth inhibitors to the fruitworms, *Heliothis zea*. *J. Am. Soc. Hort. Sci.* 107:167–70
58. Isman MB, Duffey SS. 1982. Toxicity of tomato phenolic compounds to the fruitworm, *Heliothis zea*. *Entomol. Exp. Appl.* 31:370–76
59. Jackson CG, Bryan DE, Patana R. 1969. Laboratory studies of *Eucelatoria armigera* (Coquillett), a tachinid parasite of *Heliothis* spp. *J. Econ. Entomol.* 62:907–10

59a. Johnson R, Narvaez J, An G, Ryan CA. 1989. Expression of proteinase inhibitor I and II in transgenic tobacco plants: effects on natural defense against *Manduca sexta* larvae. *Proc. Natl. Acad. Sci. USA* 86:9871–75

60. Juvik JA, Babka BA, Timmerman EA. 1988. Influence of trichome exudates from species of *Lycopersicon* on oviposition behavior of *Heliothis zea* (Boddie). *J. Chem. Ecol.* 12:1261–78
61. Juvik JA, Shapiro JA, Young TE, Mutschler MA. 1994. Acylglucoses from wilt tomatoes alter behavior and

reduce growth and survival of *Helicoverpa zea* and *Spodoptera exigua* (Lepidoptera: Noctuidae). *J. Econ. Entomol.* 87:482–9262.

62. Juvik JA, Stevens MA. 1982. Inheritance of foliar alpha tomatine content in tomato. *J. Am. Soc. Hort. Sci.* 107:1061–65
63. Juvik JA, Stevens MA. 1982. Physiological mechanisms of host-plant resistance in the genus *Lycopersicon* to *Heliothis zea* and *Spodoptera exigua*, two insect pests of the cultivated tomato. *J. Am. Soc. Hort. Sci.* 107:1065–69
64. Juvik JA, Stevens MA, Rick CM. 1982. Survey of the genus *Lycopersicon* for variability in alpha-tomatine content. *HortScience* 17:764–66
65. Kalloo G, ed. 1991. Genetic improvement of tomato. Monogr. Theor. Appl. Genet., Vol. 14. Berlin: Springer-Verlag. 358 pp.
66. Kantanyukul W, Thurston R. 1973. Seasonal parasitism and predation of eggs of the tobacco hornworm on various host plants in Kentucky. *Environ. Entomol.* 2:939–45
67. Kashyap RK, Kennedy GG, Farrar RR Jr. 1991. Behavioral response of *Trichogramma pretiosum* Riley and *Telenomus sphingis* (Ashmead) to trichmone/methyl ketone mediated resistance in tomato. *J. Chem. Ecol.* 17:543–56
68. Kashyap RK, Kennedy GG, Farrar RR Jr. 1991. Mortality and inhibition of *Helicoverpa zea* egg parasitism rates by *Trichogramma* in relation to trichome/methyl ketone-mediated insect resistance of *Lycopersicon hirsutum f. glabratum* accession PI134417. *J. Chem. Ecol.* 17:2381–95
69. Kauffman WC, Kennedy GG. 1989. Inhibition of *Campoletis sonorensis* parasitism of *Heliothis zea* and of parasitoid development by 2-tridecanone mediated insect resistance of wild tomato. *J. Chem. Ecol.* 15:1919–30
70. Kauffman WC, Kennedy GG. 1989. Toxicity of allelochemicals from the wild insect-resistant tomato *Lycopersicon hirsutum f. glabratum* to *Campoletis sonorensis*, a parasitoid of *Helicoverpa zea*. *J. Chem. Ecol.* 15:2051–60
71. Kennedy GG. 1984. 2-tridecanone, tomatoes and *Heliothis zea*: potential incompatibility of plant antibiosis with insecticidal control. *Entomol. Exp. Appl.* 35:305–11
72. Kennedy GG. 1986. Consequences of modifying biochemically mediated insect resistance in *Lycopersicon* species ACS Symp. Ser. 296, pp. 130–41
73. Kennedy GG, Farrar RR Jr, Kashyap RK. 1991. 2-tridecanone-glandular trichome-mediated insect resistance in tomato: effect on parasitoids and predators of *Heliothis zea*. In *Naturally Occurring Pest Bioregulators*, ed. PA Hedin, pp. 150–65. Washington, DC: ACS Symp. Ser. 449. Am. Chem. Soc. 456 pp.
74. Kennedy GG, Farrar RR, Riskallah MR. 1987. Induced tolerance in *Heliothis zea* neonates to host plant allelochemicals and carbaryl following incubation of eggs on foliage of *Lycopersicon hirsutum f. glabratum*. *Oecologia* 73:615–20
75. Kennedy GG, Nienhuis J, Helentjaris T. 1987. Mechanisms of arthropod resistance in tomatoes. See Ref. 89, pp. 145–54
76. Kennedy GG, Sorenson CE. 1985. Role of glandular trichomes in the resistance of *Lycopersicon hirsutum f. glabratum* to Colorado potato beetle (Coleoptera: Chrysomelidae). *J. Econ. Entomol.* 547–51
77. Kennedy GG, Sorenson CE, Fery RL. 1985. Mechanisms of resistance to Colorado potato beetle in tomato. *Mass. Agric. Expt Stn. Res. Bull. No. 704*, pp. 107–16
78. Kennedy GG, Yamamoto RT, Dimock MB, Williams WG, Bordner J. 1981. Effect of daylength and light intensity on 2-tridecanone levels and resistance in *Lycopersicon hirsutum f. glabratum* to

Manduca sexta. *J. Chem. Ecol.* 7:707–16
79. Krewson CF, Naghaski J. 1953. Occurrence of rutin in plants. *Am. J. Pharmacol.* 117:190–200
80. Kumar A, Triphanti CPM, Singh R, Pandey RK. 1983. Bionomics of *Trioxys* (*Binodosys*) *indicas*, an aphid parasitoid of *Aphis craccivora*. 17. Effect of host plants on the activities of the parasitoid. *Z. Agnew. Entomol.* 96:304–7
81. Lange WH, Bronson L. 1981. Insect pests of tomato. *Annu. Rev. Entomol.* 26:345–371
82. Leidl BE, Lawson DM, White KK, Shapiro JA, Cohen DE, et al. 1995. Acylglucoses of the wild tomato *Lycopersicon pennellii* (Corr.) D'Arcy alters settling and reduces oviposition of *Bemisia argentifolii* (Homoptera: Aleyrodidae). *J. Econ. Entomol.* 88:742–48
83. Leite GLD, Picanco M, Guedes RNC, Zanuncio JC. 2001. Role of plant age in the resistance of *Lycopersicon hirsutum.f. glabratum* to the tomato leafminer *Tuta absoluta* (Lepidoptera: Gelechiidae). *Sci. Hortic.* 89:103–13
83a. Li L, Li C, Lee GI, Howe GA. 2002. Distinct roles for jasmonate synthesis and action in systemic wound response of tomato. *PNAS* 99:6416–21
84. Lin, SYH, Trumble JT, Kumamoto J. 1987. Activity of volatile compounds in glandular trichomes of *Lycopersicon* species against two insect herbivores. *J. Chem. Ecol.* 13:837–50
85. Luckwill LC. 1943. *The Genus* Lycopersicon*: A Historical, Biological and Taxonomic Survey of the Wild and Cultivated Tomato*. Aberdeen, Scotland: Aberdeen Univ. Studies, 120. Aberdeen Univ. Press. 44 pp.
86. McKinney KB. 1938. Physical characteristics on the foliage of beans and tomatoes that tend to control some small insect pests. *J. Econ. Entomol.* 31:360–61
87. Miller JC, Tanksley SD. 1990. RFLP analysis of phylogenetic relationships and genetic variation in the genus *Lycopersicon*. *Theor. Appl. Genet.* 80:437–48
88. Musetti L, Neal JJ. 1997. Resistance to the pink potato aphid, *Macrosiphum euphorbiae*, in two accessions of *Lycopersicon hirsutum f. glabratum*. *Entomol. Exp. Appl.* 84:137–45
89. Nevins DJ, Jones RA, eds. 1987. *Tomato Biotechnology*. New York: Wiley-Liss. 339 pp.
90. Nienhuis J, Helentjaris T, Slocum M, Ruggero B, Schaefer A. 1987. Restriction fragment length polymorphism analysis of loci associated with insect resistance in tomato. *Crop Sci.* 27:797–803
91. Nihoul P. 1993. Controlling glasshouse climate influences the interaction between tomato glandular trichome, spider mite and predatory mite. *Crop Prot.* 12:443–47
92. Nihoul P. 1993. Do light intensity, temperature and photoperiod affect the entrapment of mites on glandular hairs of cultivated tomatoes? *Exp. Appl. Acarol.* 17:709–18
93. Nihoul P. 1994. Phenology of glandular trichomes related to entrapment of *Phytoseiulus persimilis* A.-H. in the greenhouse tomato. *J. Hortic. Sci.* 69:783–89
94. Nombela F, Beitia F, Muniz M. 2000. Variation in tomato host response to *Bemisia tabaci* (Hemiptera: Aleyrodidae) in relation to acyl sugar content and presence of the nematode and potato aphid resistance gene Mi. *Bull. Entomol. Res.* 90:161–67
95. Obrycki JJ. 1986. The influence of foliar pubescence on entomophagous species. See Ref. 9a, pp. 61–83
96. Obrycki JJ, Tauber MJ. 1984. Natural enemy activity on glandular pubescent potato plants in the greenhouse: an unreliable predictor of effects in the field. *Environ. Entomol.* 13:679–83
97. Obrycki JJ, Tauber MJ, Tauber CA, Gollands B. 1985. *Edovum puttleri* (Hymenoptera: Eulophidae), an exotic egg parasitoid of the Colorado potato beetle

(Coleoptera: Chrysomelidae): responses to temperate zone conditions and resistant potato plants. *Environ. Entomol.* 14:48–54

98. Orians CM, Pomerleau J, Ricco R. 2000. Vascular architecture generates fine scale variation in systemic induction of proteinase inhibitors in tomato. *J. Chem. Ecol.* 26:471–83
99. Ponti OMB de, Steenhuis MM, Elzinga P. 1983. Partial resistance of tomato to the green house whitefly (*Trialeurodes vaporariorum* Westw.) to promote its biological control. *Med. Fac. Landbouww. Rijksuniv. Gent.* 48:195–98
100. Putman WL. 1965. Bionomics of *Stethorus punctillum* Weise (Coleoptera: Coccinellidae) in Ontario. *Can. Entomol.* 87: 9–33
101. Rabb RL, Bradley JR Jr. 1968. The influence of host plants on parasitism of eggs of the tobacco hornworm. *J. Econ. Entomol.* 61:1249–52

101a. Ryan CA. 2000. The systemin signaling pathway: differential activation of plant defensive genes. *Biochem. Biophys. Acta* 1477:112–21

102. Schuster MF, Calderon M. 1986. Interactions of host plant resistant genotypes and beneficial insects in cotton ecosystem. See Ref. 9a, pp. 84–97
103. Sengonca C, Gerlach S. 1984. Einfluss der Blattoberflache auf die Wirksamkeit des rauberischen Thrips, *Scolothrips longicornis* (Thysanoptera: Thripidae). *Entomophaga* 29:55–61
104. Snyder JC, Hyatt JP. 1984. Influence of daylength on trichome densities and leaf volatiles of *Lycopersicon* species. *Pl. Sci. Lett.* 37:177–81
105. Sorenson CE. 1984. *Resistance of Lycopersicon hirsutum f. glabratum (C. H. Mull) to the Colorado potato beetle, Leptinotarsa decemlineata (Say)*. MS thesis. N. C. State Univ., Raleigh. 70 pp.
106. Sorenson CE, Fery RL, Kennedy GG. 1989. Relationship between Colorado potato beetle (Coleoptera: Chrysomelidae) and tobacco hornworm (Lepidoptera: Sphingidae) resistance in *Lycopersicon hirsutum f. glabratum*. *J. Econ. Entomol.* 82:1743–48
107. Stamp NE. 1990. Growth versus molting time of caterpillars as a function of temperature, nutrient concentration and the phenolic rutin. *Oecologia* 82:107–13
108. Stamp NE. 1994. Interactive effects of rutin and constant versus alternating temperatures on performance of *Manduca sexta* caterpillars. *Entomol. Exp. Appl.* 72:125–33
109. Stamp NE. 1994. Simultaneous effects of potassium, rutin and temperature on performance of *Manduca sexta* caterpillars. *Entomol. Exp. Appl.* 72:135–43
110. Stamp NE, Horwath KL. 1992. Interactive effects of temperature and concentration of the flavonol rutin on growth, molt, and food utilization of *Manduca sexta* caterpillars. *Entomol. Exp. Appl.* 64:135–50
111. Stout MJ, Brevont RA, Duffey SS. 1998. Effect of nitrogen availability on expression of constitutive and inducible chemical defenses in tomato. *J. Chem. Ecol.* 24:945–63
112. Stout MJ, Duffey SS. 1996. Characterization of induced resistance in tomato plants. *Entomol. Exp. Appl.* 79:273–83
113. Stout MJ, Fidantsef AL, Duffey SS, Bostock RM. 1999. Signal interaction in pathogen and insect attack: systemic plant-mediated interactions between pathogens and herbivores of the tomato: *Lycopersicon esculentum*. *Physiol. Mol. Plant Pathol.* 54:115–30
114. Stout MJ, Workman KV, Bostock RM, Duffey SS. 1998. Specificity of induced resistance in the tomato *Lycopersicon esculentum*. *Oecologia* 113:74–81
115. Stout MJ, Workman KV, Bostock RM, Duffey SS. 1998. Stimulation and attenuation of induced resistance by elicitors and inhibitors of chemical induction in tomato (*Lycopersicon esculentum*) foliage. *Entomol. Exp. Appl.* 86:267–79

116. Stout MJ, Workman KV, Workman JS, Duffey SS. 1996. Temporal and ontogenetic aspects of protein induction in foliage of the tomato, *Lycopersicon esculentum*. *Biochem. Syst. Ecol.* 24:611–25
116a. Takabayashi J, Shimoda T, Dicke M, Ashihara W, Takafuji A. 2000. Induced response of tomato plants to injury by green and red strains of *Tetranychus urticae*. *Expt. Appl. Acarol.* 24: 377–83
117. Thaler JS. 1999. Induced resistance in agricultural crops: effects of jasmonic acid on herbivory and yield in tomato plants. *Environ. Entomol.* 28:30–37
118. Thaler JS. 1999. Jasmonate-inducible plant defenses cause increased parasitism of herbivores. *Nature* 399:686–88
119. Thaler JS, Fidantsef AL, Duffey SS, Bostock RM. 1999. Trade-offs in plant defense against pathogens and herbivores: a field demonstration of chemical elicitors of induced resistance. *J. Chem. Ecol.* 25:1597–1609
120. Traugott MS, Stamp NE. 1996. Effects of chlorogenic acid- and tomatine-fed caterpillars on the behavior of an insect predator. *J. Insect Behav.* 9:461–76
121. VanHaren RJF, Steenhuis MM, Sabelis MW, de Ponti OMB. 1987. Tomato stem trichomes and dispersal success of *Phytoseiulus persimilis* relative to its prey *Tetranychus urticae*. *Exp. Appl. Acarol.* 3:115–21
121a. Vet LEM, Dicke M. 1992. Ecology of infochemical use by natural enemies in a tritrophic context. *Annu. Rev. Entomol.* 37:141–72
122. Weiser LA, Stamp NE. 1998. Combined effects of allelochemicals, prey availability, and supplemental plant material on growth of a generalist insect predator. *Entomol. Exp. Appl.* 87:181–89
123. Weston PA, Johnson DA, Burton HT, Snyder JC. 1989. Trichome secretion, composition, trichome densities and spider mite resistance of ten accessions of *Lycopersicon hirsutum*. *J. Am. Soc. Hort. Sci.* 114:492–98
124. Wilkens RT, Spoerke JM, Stamp NE. 1996. Differential responses of growth and two soluble phenolics of tomato to resource availability. *Ecology* 77:247–58
125. Williams WG, Kennedy GG, Yamamoto RT, Thacker JD, Bordner J. 1980. 2-tridecanone: a naturally occurring insecticide from the wild tomato species *Lycopersicon hirsutum f. glabratum*. *Science* 207:888–89

Annu. Rev. Entomol. 2003. 48:73–88
doi: 10.1146/annurev.ento.48.060402.102812
First published online as a Review in Advance on August 19, 2002

ROLE OF ARTHROPOD SALIVA IN BLOOD FEEDING: Sialome and Post-Sialome Perspectives*

José M. C. Ribeiro and Ivo M. B. Francischetti
Medical Entomology Section, Laboratory of Malaria and Vector Research, National Institute of Allergy and Infectious Diseases, National Institutes of Health, 4 Center Drive, Bethesda, Maryland, 20892-0425; e-mail: Jribeiro@nih.gov; Ifrancischetti@niaid.nih.gov

Key Words hemostasis, inflammation, nociception, mast cells, salivary glands

■ **Abstract** This review addresses the problems insects and ticks face to feed on blood and the solutions these invertebrates engender to overcome these obstacles, including a sophisticated salivary cocktail of potent pharmacologic compounds. Recent advances in transcriptome and proteome research allow an unprecedented insight into the complexity of these compounds, indicating that their molecular diversity as well as the diversity of their targets is still larger than previously thought.

CONTENTS

INTRODUCTION

Fifteen years ago it was proposed that (71), based on enzymatic and bioassay data, saliva of bloodsucking arthropods served mainly an antihemostatic role and, additionally, that hard tick saliva served to evade their hosts' inflammation and immunity [a concept more detailed in (72)]. Except for tick salivary prostaglandins (25, 34), no other hematophagous arthropod salivary compound had been molecularly characterized at that time. Thanks to the revolutionary techniques of molecular biology as well as the miniaturization of high-performance liquid chromatography, mass spectrometry, and Edman degradation techniques, detailed knowledge of such molecules and their role in blood feeding has increased at a fast pace. Accordingly, it was confirmed that almost all bloodsucking arthropods studied (a small sample consisting of only a few species from over 500 genera and ~19,000 known species) (73) have at least one anticlotting, one vasodilator, and one antiplatelet compound. However, the molecular diversity of the nature of such compounds was very large. We have also learned that hematophagous insects and ticks have many different and sometimes apparently conflicting salivary strategies, which were not predicted before. In addition, at least one half of the message expressed in the salivary glands of such animals, leading to apparently secreted proteins, has no known function, thus challenging the researchers' knowledge and imagination. The reader unfamiliar with saliva of vector arthropods is encouraged to read previous reviews (11, 73). Several excellent reviews have been written on the role of saliva in host immunity and parasite transmission (31, 87, 100, 101).

WHY IT IS DIFFICULT TO STEAL BLOOD FROM A VERTEBRATE

To a bloodsucking animal, paradise is a place where the host blood does not clot, the blood flow at the feeding site is intense, and the host will not bother (or kill) the guest. Real life is different. Vertebrates have three efficient systems that make life potentially difficult for hematophagous animals: hemostasis, inflammation, and immunity. These three complex physiological responses interact with each other and, at times, are in opposition.

Hemostasis

Hemostasis is the host response that controls the loss of blood following injury to a blood vessel. It consists of platelet aggregation, blood coagulation, and vasoconstriction. All these phenomena are redundant. There are several independent agonists of platelet aggregation [ADP, collagen, thrombin, platelet-activating factor, thromboxane A_2 (TXA_2)], at least two vasoconstrictors are released by platelets (TXA_2 and serotonin), and the clotting cascade is a complex system with many potential points of amplification and control. For more details on hemostasis see References 11, 73. Hemostasis thus takes care of the host blood loss following injury and places a major barrier to any blood-feeding arthropod.

Inflammation

Inflammation is the host response following tissue injury. Classically, it consists of the triple response of Lewis: pain, redness, and heat; the last two last being the result of tissue vasodilatation. Although vasodilatation is favorable to blood feeding, pain triggers the host's awareness to the blood sucker. To ticks, tissue repair may lead to encapsulation and isolation of the feeding mouthparts from live tissue. An aseptic tissue injury produces a series of events leading to tissue repair. ATP, released by injured cells, is also responsive to the immediate and acute pain following tissue injury (16). Serotonin and histamine, released by platelets and mast cells, are also inducers of pain and increased vascular permeability, as is bradykinin, which is produced following activation of factor XII by tissue-exposed collagen (38). Activated factor XII converts prekallikrein to kallikrein, which hydrolyzes blood kininogen to produce the vasodilatory peptide, bradykinin. Note that many of the hemostasis mediators are linked to pain production in inflammation.

Polymorphonuclear cells and monocytes are important mediators of inflammation. ATP, released by injured cells, activates neutrophils that accumulate and degranulate at the injury site (46, 58). Thrombin from the blood-coagulation cascade and other proinflammatory molecules, such as platelet-activating factor, also activate neutrophils that produce prostaglandins and platelet-activating factor itself (8, 44). Neutrophil activation is also accompanied by the release of several proteases modulating platelet function, such as cathepsin G (82), or enzymes that act on the tissue matrix, such as elastase (33, 80, 85). Importantly, thrombin also has inflammatory properties including causing fibroblast proliferation and increasing neutrophil adhesion, whereas clotting factor Xa functions as mediator of acute inflammation by binding to effector cell protease receptor-1, inducing vascular permeability and leukocyte exudation (18, 32, 52). Activation of neutrophil and other cell types is also accompanied by generation of pain-inducing prostaglandins (38). Pain is also induced by chemokines (inflammatory proteins) such as interleukin-1 (IL-1) generated by neutrophils, as well as bradykinin produced by the intrinsic pathway of blood coagulation. Therefore, several molecules work in concerted manner to generate pain (55). In this regard, bradykinin induces TNF-α release from neutrophils (27, 61), which in turn stimulates the release of IL-1β and IL-6 from various cell types including those of the phagocyte mononuclear system. These cytokines contribute to the phenomenon of increased sensitivity to pain, or hyperalgesia, that accompanies inflammation. Cytokine-mediated inflammatory hyperalgesia is accompanied by production of cyclo-oxygenase products and IL-8 released by monocytes, macrophages, and endothelial cells, which stimulate the production of sympathomimetic mediators also involved in increased pain reception (19, 20).

In the case of septic injury to the tissue, the inflammatory response is amplified. Activation of the alternative or colectin complement pathway by bacterial or fungal surfaces leads to the production of anaphylatoxins, which are potent molecules attracting granulocytes and monocytes to the injury site (40, 99). Bacterial lipopolysaccharide also induces activation of various leukocytes, leading, for example, monocytes, neutrophils, and eosinophils to produce vasodilatory prostaglandins, various cytokines, superoxide, and nitric oxide, as well as releasing their granules (39, 50, 62, 88, 103). Both macrophages and neutrophils actively phagocytose bacteria, and their products influence each other (80).

Following this acute phase, the resultant inflammatory, complement, and hemostatic reactions produce a unique environment within wounds that promotes repair. Repair involves four components: angiogenesis (formation of new blood vessels), migration and proliferation of fibroblasts (fibroplasia), deposition of extracellular matrix, and remodeling. Repair begins early in inflammation, sometimes as early as 24 h after injury. Remarkably, endothelial cell proliferation is an indispensable early process in the formation of new blood vessels, and it is fundamental to tissue repair because blood vessels carry oxygen and nutrients necessary to sustain cell metabolism (104). Other cell types are also involved in repair mechanisms. Macrophages provide a continuing source of cytokines necessary to stimulate fibroplasia and angiogenesis, whereas fibroblasts construct new extracellular matrix necessary to support cell growth. In this process, proliferative responses are triggered by a number of growth factors released by endothelial cells, platelets, macrophages, and fibroblasts (89). Because some of these cell types are activated by enzymes of the blood-coagulation cascade (e.g., thrombin), angiogenesis and hemostasis are interrelated (7). Accordingly, platelet aggregation and blood-coagulation inhibitors found in the salivary gland of blood feeders may also negatively modulate angiogenesis in vivo. However, for most fast feeders such as mosquitoes, sand flies, triatomines, and fleas, tissue repair is not a major barrier as the events are on a timescale of hours and days, whereas feeding takes minutes. On the other hand, hard ticks, which feed continuously for 3–10 days with their mouthparts imbedded in their hosts, face host-repair mechanisms that may get in the way of a satisfactory meal.

Immunity

Exposure of foreign antigens to a vertebrate in an inflammatory context leads the immune system to further recognize these molecules and react accordingly. The human immune response to mosquito bites was described by Mellanby (54) as progressing from no reaction, to delayed-type hypersensitivity (DTH), to immediate-type hypersensitivity, and to desensitization. Exposure of skin antigens to naïve animals leads dendritic cells to initially process the antigen, and then to activate and clonally expand the so-called DTH–T cells. After this T cell expansion, and following new antigen deposition in the skin, these circulating cells congregate in the vicinity of the antigen and start producing various inflammatory cytokines, including γ-interferon, which activates local macrophages to produce TNF-α and IL-1, two mediators of inflammatory pain (26, 27). Because some of these cells are not in the skin and have to accumulate from the blood circulation, it takes 6–12 h for the monocytic infiltrate to be barely noticeable, and they achieve peak infiltration at 24-h postdermal antigen exposure, thus the name DTH. We may speculate that the vertebrate perception of the DTH induces the vertebrates to avoid behaviorally the re-exposure to the source of the nuisance. The DTH may actually be of advantage to some sand flies, and to nest-associated blood feeders such as *Cimex* and fleas, as proposed before (4). To ticks, a special type of DTH, the basophil infiltrate characterizing the Motte Jones response (2), is associated with tick-rejection reactions (1, 6), especially in guinea pig models. The rejection occurs by a behaviorally defensive action of the host (scratching), as well as by disturbance of feeding, wherein blood is substituted by a purulent infiltrate. These leukocytes, particularly eosinophils, may also prove noxious by acting on the tick gut (101, 102).

Following continuous exposure of the antigen, a new subset of lymphocytes is activated, leading to the production of IgE by a B-cell subtype. Mast cells, loaded with histamine (or histamine and serotonin, depending on the vertebrate species) and residing on the connective tissue such as the dermis, have high-affinity receptors to IgE (35). When divalent antigens cross-link two IgE bound to the mast cell IgE receptor, the cell degranulates, releasing their vasoactive amines. After activation, mast cells also produce and release several arachidonic acid metabolites and a diversity of cytokines, including IL-4, which stimulates the immune response to progress toward a Th2- or antibody-mediated response. Mast cells also produce nerve-growth factor, which acts on nociceptive neurons to decrease their threshold to pain-inducing molecules such as bradykinin, serotonin, and histamine (38). Histamine promotes more vasodilatation in the arteriolar side than in the venular side of the skin circulation, thus creating an increase in the hydrostatic pressure of the capillaries. Histamine (as well as the aforementioned inflammatory substances serotonin and bradykinin) also increases the spacing of the endothelial cells, which, together with the larger capillary hydrostatic pressure, leads to extravasation of plasma into the interstitial tissue, creating edema. These vasoactive substances are quick acting, producing a visible response within a minute or two of their release, but they do not last long because they are washed out of the tissue or metabolized. After 20–30 min, most of the reaction is gone. To the vertebrate host, this reaction leads to a behavioral response leading to the identification and removal of the annoying bloodsucker and, if possible, avoidance of the area of exposure. The host may also display life-threatening anaphylaxis response at this stage of its immune response to the arthropod bite. To the bloodsucker, this response may result in hunger, if it is lucky enough to escape, or death.

Continuation of the antigenic exposure to the vertebrate leads to maturation of the immune response into different subtypes of IgG. Although in humans IgG_4 binds to mast cell receptors (with much less affinity than IgE), other immunoglobulin types recognize the antigen, and the complex will likely be taken by macrophages that then digest these molecules—and possibly present them to lymphocytes—further maturing (i.e., increasing antibody specificity and affinity) the IgG response. The tissue response following antigen encounter within this immune response state is minimal, causing this stage of the reaction to an arthropod bite to be somewhat misleadingly named the desensitization stage. However, if the antigen happens to bind to host skin cells, it may lead to complement fixation and tissue necrosis [the result of an Arthus reaction (15)]—a rare outcome to insect bites described in some types of allergy to triatomine bites (17). To the vertebrate host, this stage of the immune response (without complement fixation) leads to minimal annoyance. To the insect or tick, this stage may result in the neutralization of several molecules it uses in the feeding process.

In reality, these stages of the immune response may not be distinguished in a clear-cut way. It is common to have both a DTH and an immediate response, and to have a DTH and specific IgG antibodies simultaneously (a mixed-type response). Even in the presence of a mature response, a less intense but noticeable, immediate response may occur due to IgG4 (of human) or IgG1 (in guinea pigs) binding to mast cells. With all these mechanisms operating against them, how do bloodsuckers succeed?

THE SALIVARY PHARMACOLOGIC COMPLEXITY OF HEMATOPHAGOUS ARTHROPODS

Perhaps because disarming a complex and redundant system such as hemostasis with a magic bullet is impossible, saliva of bloodsucking animals evolved a "magic potion," allowing them to succeed against all the complex barriers imposed by their hosts. As a rule, these animals' saliva contains at least one anticlotting, one antiplatelet, and one vasodilatory substance. In many cases, more than one molecule exists in each category. In some, compounds such as adenosine and nitric oxide that are at once antiplatelet and vasodilatory are found in saliva. It also became clear that, although only a few of the >15,000 species of >500 genera were studied (73), an enormous diversity of concoctions exist in these magic potions. The few vasodilators known thus far offer a good example of this diversity: The triatomine bug *Rhodnius prolixus* makes use of nitric oxide, as does the cimicid bug *Cimex lectularius* (76, 98). Because nitric oxide is an unstable gas, each bug developed a different heme protein that stabilizes and carries this gas to the host. *Rhodnius* nitrophorin is a member of the lipocalin family (9) and *Cimex* nitrophorin is a member of the inositol phosphatase family (97). Ticks have salivary prostaglandins in large amounts, including PGE_2 and PGF_2 (25, 34, 75). Old World sand flies of the genus *Phlebotomus* have adenosine as vasodilators (66, 68), but New World sand flies of the genus *Lutzomyia* have a 6.5-kDa peptide, maxadilan, the most potent vasodilator known, acting on PACAP receptors (48, 56). Note that *Phlebotomus* flies do not have maxadilan and *Lutzomyia* do not have adenosine in their saliva. The black fly *Simulium vittatum* has a 15-kDa vasodilator that acts on ATP-dependent K-channels (22, 23). This vasodilatory protein has no similarity to other known proteins. Finally, *Aedes* mosquitoes have a vasodilatory tachykinin decapeptide named sialokinin (10), while *Anopheles* mosquitoes have a vasodilatory peroxidase, of ~65 kDa, which destroys skin-vasoconstricting norepinephrine and serotonin (78, 95). Accordingly, the salivary vasodilatory catalog goes from NO and prostaglandins to peptides up to a 65-kDa protein. Note also, members of the same insect family but not the same genus, such as the *Lutzomyia* and *Phlebotomus* or *Aedes* and *Anopheles*, have completely different vasodilators. The same diversity can be found for salivary anticlotting peptides. For example, *Aedes* has a salivary inhibitor of factor Xa that is a member of the serpin family (84), while *Anopheles* has a smaller antithrombin peptide unrelated to other known peptides (28, 95). A ubiquitous salivary enzyme is apyrase, which hydrolyzes both ATP and ADP to AMP, thus having an anti-pain, anti-inflammatory and antihemostatic activity (74). However, at least two different families of this enzyme exist. Mosquito apyrases are members of the 5′ nucleotidase family (12). In bacteria, these enzymes hydrolyze ATP, ADP, and AMP to adenosine and orthophosphate (105), while sand fly and the bed bug *Cimex* have an enzyme of a novel protein family (14, 91, 92). Other different anticlotting peptides exist and diversity in antiplatelet compounds was also found, which will not be discussed here. The picture of bloodsucker saliva that emerges is one of very diverse composition. To help understand the origins of this diversity, it is useful to consider that *Aedes* and *Anopheles* diverged more than 150 million years ago (63), or 100 million years before the radiation of mammals, while *Lutzomyia* and *Phlebotomus* diverged before the last tectonic plate separation, thus at a time coinciding with or prior to mammal radiation. We accordingly postulate that all genera that diverged before mammal radiation have considerable variation in their salivary composition.

It is becoming apparent that the salivary cocktail of hematophagous animals contains many substances that counteract host pain. True anesthetic substances inhibit nerve conduction, while substances that inhibit the action of pain agonists (nociceptive agents) have analgesic effects. The saliva of the bug *Triatoma infestans* inhibits sodium channel activity in nerves by an unspecified molecule (24), and this report contains the only account of such activity in arthropod saliva. Salivary components with potential antinociceptive effects are varied, including apyrase (by destroying ATP), histamine, and serotonin-binding proteins, thus far found in ticks and triatomine bugs (59, 70), and kininase, which destroys bradykinin (67). Active search of salivary components acting on nerve conduction and on mast cells should yield many more activities and the possible discovery of novel compounds.

Because ticks stay attached for several days on their hosts, they have potent anti-inflammatory and immunomodullatory components that may prevent or retard deleterious host responses or counteract pharmacologically their host's immunopharmacologic mediators. These include anticomplement (94), anti-IL-2 (30), protease inhibitors (47), antineutrophil activity (79), and other immunomodullatory molecules with unclear mechanism of action (5). Salivary gland extracts, or tick feeding, modify host cytokine expression in several ways (3, 41–43, 45, 51, 53). Considering the redundancy and complexity of the vertebrate immune system, individual tick species must have a complex immunosupressant cocktail, as is the case with their salivary antihemostatic cocktails. A challenge for the next few years will be to determine the extent and redundancy of such cocktails.

Ticks also contain salivary compounds that affect tissue repair. Notably, calreticulin, a molecule with potent antiangiogenesis properties and recently identified in the supernatants of Epstein-Barr virus–immortalized cells (60), has also been characterized in the salivary gland of the tick *Amblyomma americanum* (36). We have also reported the presence of the N terminus of calreticulin in the SDS/PAGE of *Ixodes scapularis* saliva (94a), and a potent inhibitory activity of both endothelial cell proliferation and chick aorta sprouting formation has been demonstrated in *I. scapularis* saliva (I.M.B. Francischetti & J.M.C. Ribeiro, unpublished observation).

The face fly *Haematobia irritans*, which is a cattle feeder, provides an interesting exception to the rule of a magic potion in saliva. Only a salivary anticlotting substance was found in this fly. We failed to find vasodilatory or antiplatelet activity (21); however, this fly feeds sparingly several times a day, taking a few microliters or less of blood at a time. Most of the time it stays attached to the cattle's face, where it feeds. We propose that this fly and possibly also the stable fly *Stomoxys* are in their evolutionary infancy of adaptation to blood feeding. These flies have an ancestor in common with house flies, leading to preadaptations such as long and robust mouthparts and a robust body. Their feeding apparatus is coarse when compared with the fine tools of a mosquito or a tsetse. They inflict quite a severe injury, with consequent pain, while feeding. Against this negative effect of feeding, they behaviorally found a place to feed where the host cannot do much about it. The cattle's tail or tongue cannot reach the flies, despite severe annoyance to the animal. Nor can movements of the dermal musculature have an impact. One day, millions of years from now, these flies may feed as well as their tsetse relatives, assuming their hosts will continue to be around.

SALIVARY TRANSCRIPTOMES AND PROTEOMES (SIALOMES)

In the past three years it has become practically feasible to sequence full-length cDNA libraries obtained from the salivary glands of bloodsucking arthropods. Initially, we attempted a PCR-based subtractive hybridization protocol to enrich salivary-specific molecules (14). However, these protocols usually led to obtaining fragmentary, not full-length sequences, owing to the use of restriction enzymes in the manufacture of the library. By mass sequencing nonsubtracted full-length sequence libraries, we obtained the same information as in the subtracted library, with the bonus of being much easier to obtain full-length sequence information by RACE (rapid amplification of cDNA ends) protocols (96). By using PCR-based techniques to construct these libraries, as few as 20–50 pairs of glands from mosquitoes or sand flies yield complex libraries. By sequencing 500–1000 clones of each library, and clustering the cDNA sequences by their similarity, we identified almost all or most of the previously reported sequences of proteins or peptides from salivary glands of hematophagous arthropods deposited in GenBank, and many more. For example, in a cDNA library of the mosquito *Aedes aegypti*, we found all the six previously deposited sequences and described 30 new ones (96). In the sand fly *Phlebotomus papatasi*, the salivary gland library information led to identification of a vaccine candidate against *Leishmania major* (90) based on sand fly salivary allergens. Although we can identify with some certainty the possible role of some of the proteins synthesized with the obtained DNA information, the majority of the sequences are either of an unknown or unexpected nature.

UNEXPECTED SEQUENCES OR CONFLICTING STRATEGIES IN THE BLOODSUCKING ARENA

From their salivary gland cDNA library information, we have identified several unexpected activities in the blood-feeding strategy of arthropods that expanded our view of the salivary complexity of these animals. For example, in the sand fly *Lutzomyia longipalpis* we found, in addition to the ubiquitous apyrase activity that hydrolyzes ATP and ADP to AMP (14), a secreted 5′ nucleotidase hydrolyzing AMP to adenosine (69) and an adenosine deaminase that further hydrolyzes adenosine to inosine (13). While the 5′ nucleotidase may play a role in transforming AMP to the vasodilatory and antiplatelet compound adenosine, the primary vasodilator of Old World *Phlebotomus* sand flies (66), the conversion of adenosine to inosine does not fit with the salivary antihemostatic paradigm because inosine lacks vasodilatory and antiplatelet activities. We postulated that *L. longipalpis* might have evolved to produce these two salivary enzymes for two reasons. First, adenosine is a potent inducer of mast cell degranulation (86); second, because *Lutzomyia* has the potent peptidic vasodilator maxadilan, which does not exist in *Phlebotomus*, to produce vasodilatation, it has no need for the antihemostatic action of adenosine. If preventing mast cell degranulation by destroying adenosine is important for *Lutzomyia*, the unanswered question arises as to how *Phlebotomus* deals with mast cells, inasmuch as they secrete copious amounts of adenosine. It thus appears that preventing mast cell degranulation by destroying adenosine is important for *Lutzomyia*; like *Lutzomyia*, the mosquito *Ae. aegypti* appears to have the same dislike for adenosine. From mosquito salivary cDNA sequences, we found not only adenosine deaminase (65) but also purine nucleosidase sequences (96), which code for enzymes hydrolyzing inosine to hypoxanthine and ribose. The enzyme was indeed found in mosquito saliva, where it is one of the richest natural sources (78a). Finding this enzyme in any metazoan animal is unusual, as purine hydrolases of the enzyme class were thought to be exclusive of unicellular organisms and plants, not of animals. Perhaps *Aedes* destroys inosine because it still induces mast cell degranulation, albeit 10 times less potently than adenosine (86). Anopheline mosquitoes, however, do not have these adenosine/inosine-catabolizing enzymes, and how they deal with mast cells, if they do, remains to be discovered. Other enzymes found in both cDNA sequences (14) and insect saliva include hyaluronidase in sand flies and black flies (64), an enzyme that might help to spread the salivary pharmacologic agents through the skin matrix, and metalloprotease in ticks (I.M.B. Francischetti, T.N. Mather & J.M.C. Ribeiro, manuscript submitted), which may disrupt tissue repair. These enzymatic activity discoveries increase our understanding of the complex salivary potions of bloodsucking arthropods in general as well as show us the love/hate relationship of bloodsucking Diptera with adenosine, which raises the question of whether saliva evolution was influenced by host mast cells.

Many cDNA sequences coding for nonenzymatic peptides were also identified, together with a good presumption of their possible action. For example, the tick *I. scapularis* has a variety of specific protease inhibitors that have been confirmed by biochemical activity [described in (29, 94a)]. The variety of antiprotease activities in ticks points to the diverse cocktail they produce against host enzymes of the clotting cascade and inflammation.

UNKNOWN SEQUENCES, OR THE PROBLEM OF ORPHAN MOLECULES

About 40% of the cDNA clusters we find from salivary gland libraries are of unknown function. In many cases, these sequences have a clear signal peptide indicating secretion, and their amino terminal or internal sequence, can be located by Edman degradation of protein bands transferred to PVDF membranes from saliva or salivary gland homogenates submitted to SDS-polyacrylamide gel electrophoresis (SDS-PAGE). Among the several classes of these unknown novel proteins, the D7 family exists in several bloodsucking dipteran species (93). The D7 family belongs to the odorant-binding superfamily but constitutes a distinct subset found in bloodsucking mosquitoes and sand flies. They are among the most abundant salivary proteins in these insects, have known allergenic properties (83), and may function by binding to host hemostatic/inflammatory agonists or serve to deliver some low-molecular-weight pharmacologic compound from the insect salivary gland to the vertebrate skin. After writing this review, a D7 protein from *Anopheles stephensi* was reported to inhibit activation of factor XII and prekallikrein (35a), thus having an effect in preventing bradykinin formation and activation of the intrinsic clotting pathway. Another ubiquitous and abundantly expressed family of salivary proteins in bloodsucking Diptera [from mosquitoes (96), sand flies (14) and tsetse (49)] is the antigen-5 family, a group of extracellular proteins found in wasp venom, seminal fluid, and plant defense proteins (81). Their function remains unknown. Sand fly (14) and mosquito salivary glands (96) also possess members of the *Drosophila* yellow protein family, which may have a function in oxidation of norepinephrine or DOPA agonists (37). Many putative peptide sequences (from 1 kDa to 15 kDa) are found in most bloodsucking arthropods studied thus far—often over 6–10 different peptides per species studied. Their biologic role is not evident and identifying their function remains a challenge. Because expression of large amounts of correctly folded proteins will be necessary for the many bioassays needed to identify their function, a network of specialized laboratories would be advantageous. Periodic comparison of such sequences to public databases will also help both to identify sequence similarities that may indicate function.

Obtaining crystal structure information of such proteins may also help elucidate their function.

POSSIBLE APPLICATIONS

It is clear that saliva of bloodsucking arthropods represents a vast range of novel molecules affecting hemostasis, inflammation, and immunity. With over 15,000 species in ~500 genera (73), we have barely scratched the surface of such biochemical and pharmacologic wealth. In addition, because the saliva of such arthropods affects the immune response of the host, it may become a novel vaccine target against various vector-borne diseases, as recently described for leishmaniasis (57, 90). The advanced salivary evolution of these bloodsucking insects may one day be the basis for alleviating not only the diseases they transmit but also other diseases of vascular origin. It is also appealing that these insects could be transformed with transposable elements (77) to synthesize and deliver vaccines—any vaccine—to their hosts.

ACKNOWLEDGMENTS

We are grateful to Drs. Robert Gwadz, Thomas Kindt, and Louis Miller for encouragement and support, and to Ms. Nancy Schulman for editorial assistance.

The *Annual Review of Entomology* is online at http://ento.annualreviews.org

LITERATURE CITED

1. Allen JR. 1973. Tick resistance: basophil in skin reactions of resistant guinea pigs. *Int. J. Parasitol.* 3:195–200
2. Askenase PW, Atwood JE. 1976. Basophils in tuberculin and "Jones-Mote" delayed reactions of humans. *J. Clin. Invest.* 58:1145–54
3. Barriga OO. 1999. Evidence and mechanisms of immunosuppression in tick infestations. *Genet. Anal.* 15:139–42
4. Belkaid Y, Valenzuela JG, Kamhawi S, Rowton E, Sacks DL, Ribeiro JM. 2000. Delayed-type hypersensitivity to *Phlebotomus papatasi* sand fly bite: an adaptive response induced by the fly? *Proc. Natl. Acad. Sci. USA* 97:6704–9
5. Bergman DK, Palmer MJ, Caimano MJ, Radolf JD, Wikel SK. 2000. Isolation and molecular cloning of a secreted immunosuppressant protein from *Dermacentor andersoni* salivary gland. *J. Parasitol.* 86: 516–25
6. Brossard M, Wikel SK. 1997. Immunology of interactions between ticks and hosts. *Med. Vet. Entomol.* 11:270–76
7. Browder T, Folkman J, Pirie-Shepherd S. 2000. The hemostatic system as a regulator of angiogenesis. *J. Biol. Chem.* 275:1521–24
8. Camussi G, Tetta C, Bussolino F, Baglioni C. 1989. Tumor necrosis factor stimulates human neutrophils to release leukotriene B4 and platelet activating factor. *Eur. J. Biochem.* 182:661–66
9. Champagne D, Nussenzveig RH, Ribeiro JMC. 1995. Purification, characterization, and cloning of nitric oxide-carrying heme proteins (nitrophorins) from

salivary glands of the blood sucking insect *Rhodnius prolixus*. *J. Biol. Chem.* 270:8691–95
10. Champagne D, Ribeiro JMC. 1994. Sialokinins I and II: two salivary tachykinins from the Yellow Fever mosquito, *Aedes aegypti*. *Proc. Natl. Acad. Sci. USA* 91:138–42
11. Champagne DE. 1994. The role of salivary vasodilators in bloodfeeding and parasite transmission. *Parasitol. Today* 10:430–33
12. Champagne DE, Smartt CT, Ribeiro JM, James AA. 1995. The salivary gland-specific apyrase of the mosquito *Aedes aegypti* is a member of the 5′-nucleotidase family. *Proc. Natl. Acad. Sci. USA* 92: 694–98
13. Charlab R, Rowton ED, Ribeiro JM. 2000. The salivary adenosine deaminase from the sand fly *Lutzomyia longipalpis*. *Exp. Parasitol.* 95:45–53
14. Charlab R, Valenzuela JG, Rowton ED, Ribeiro JM. 1999. Toward an understanding of the biochemical and pharmacological complexity of the saliva of a hematophagous sand fly *Lutzomyia longipalpis*. *Proc. Natl. Acad. Sci. USA* 96: 15155–60
15. Cochrane CG. 1967. Mediators of the Arthus and related reactions. *Prog. Allergy* 11:1–35
16. Cook SP, McCleskey EW. 2002. Cell damage excites nociceptors through release of cytosolic ATP. *Pain* 95:41–47
17. Costa CH, Costa MT, Weber JN, Gilks GF, Castro C, Marsden PD. 1981. Skin reactions to bug bites as a result of xenodiagnosis. *Trans. R. Soc. Trop. Med. Hyg.* 75: 405–8
18. Coughlin SR. 2001. Protease-activated receptors in vascular biology. *Thromb. Haemost.* 86:298–307
19. Cunha FQ, Lorenzetti BB, Poole S, Ferreira SH. 1991. Interleukin-8 as a mediator of sympathetic pain. *Br. J. Pharmacol.* 104:765–67
20. Cunha FQ, Poole S, Lorenzetti BB, Ferreira SH. 1992. The pivotal role of tumour necrosis factor alpha in the development of inflammatory hyperalgesia. *Br. J. Pharmacol.* 107:660–64
21. Cupp EW, Cupp MS, Ribeiro JM, Kunz SE. 1998. Blood-feeding strategy of *Haematobia irritans* (Diptera: Muscidae). *J. Med. Entomol.* 35:591–95
22. Cupp M, Ribeiro J, Champagne D, Cupp E. 1998. Analyses of cDNA and recombinant protein for a potent vasoactive protein in saliva of a blood-feeding black fly, *Simulium vittatum*. *J. Exp. Biol.* 201: 1553–61
23. Cupp MS, Ribeiro JMC, Cupp EW. 1994. Vasodilative activity in black fly salivary glands. *Am. J. Trop. Med. Hyg.* 50: 241–46
24. Dan A, Pereira MH, Pesquero JL, Diotaiuti L, Beirao PS. 1999. Action of the saliva of *Triatoma infestans* (Heteroptera: Reduviidae) on sodium channels. *J. Med. Entomol.* 36:875–79
25. Dickinson RG, O'Hagan JE, Shotz M, Binnington KC, Hegarty MP. 1976. Prostaglandin in the saliva of the cattle tick *Boophilus microplus*. *Aust. J. Exp. Biol. Med. Sci.* 54:475–86
26. Ferreira SH, Lorenzetti BB, Bristow AF, Poole S. 1988. Interleukin-1 beta as a potent hyperalgesic agent antagonized by a tripeptide analogue. *Nature* 334:698–700
27. Ferreira SH, Lorenzetti BB, Poole S. 1993. Bradykinin initiates cytokine-mediated inflammatory hyperalgesia. *Br. J. Pharmacol.* 110:1227–31
28. Francischetti IM, Valenzuela JG, Ribeiro JM. 1999. Anophelin: kinetics and mechanism of thrombin inhibition. *Biochemistry* 38:16678–85
29. Francischetti IM, Valenzuela JG, Andersen JF, Mather TN, Ribeiro JMC. 2002. Ixolaris, a novel recombinant tissue factor pathway inhibitor (TFPI) from the salivary gland of the tick, *Ixodes scapularis*: identification of factor X and factor Xa as scaffolds for the inhibition of factor VIIa/tissue factor complex. *Blood* 99:3602–12

30. Gillespie RD, Dolan MC, Piesman J, Titus RG. 2001. Identification of an IL-2 binding protein in the saliva of the Lyme disease vector tick, *Ixodes scapularis*. *J. Immunol.* 166:4319–26
31. Gillespie RD, Mbow ML, Titus RG. 2000. The immunomodulatory factors of blood-feeding arthropod saliva. *Parasite Immunol.* 22:319–31
32. Gillis S, Furie BC, Furie B. 1997. Interactions of neutrophils and coagulation proteins. *Semin. Hematol.* 34:336–42
33. Gompertz S, Stockley RA. 2000. Inflammation—role of the neutrophil and the eosinophil. *Semin. Respir. Infect.* 15:14–23
34. Higgs GA, Vane JR, Hart RJ, Porter C, Wilson RG. 1976. Prostaglandins in the saliva of the cattle tick, *Boophilus microplus* (Canestrini) (Acarina, Ixodidae). *Bull. Entomol. Res.* 66:665–70
35. Huntley JF. 1993. Mast cells and basophils: a review of their heterogeneity and function. *J. Comp. Path.* 107:349–72
35a. Isawa H, Yuda M, Orito Y, Chinzei Y. 2002. A mosquito salivary protein inhibits activation of the plasma contact system by binding to factor XII and high molecular weight kininogen. *J. Biol. Chem.* 277(31):27651–58
36. Jaworski DCSFA, Lamoreaux W, Coons LB, Muller MT, Needham GR. 1995. A secreted calreticulin protein in ixodid tick (*Amblyomma americanum*) saliva. *J. Insect Physiol.* 41:369–75
37. Johnson JK, Li J, Christensen BM. 2001. Cloning and characterization of a dopachrome conversion enzyme from the yellow fever mosquito, *Aedes aegypti*. *Insect Biochem. Mol. Biol.* 31:1125–35
38. Julius D, Basbaum AI. 2001. Molecular mechanisms of nociception. *Nature* 413:203–10
39. Kalmar JR, Van Dyke TE. 1994. Effect of bacterial products on neutrophil chemotaxis. *Methods Enzymol.* 236:58–87
40. Kirschfink M. 1997. Controlling the complement system in inflammation. *Immunopharmacology* 38:51–62
41. Kopecky J, Kuthejlova M. 1998. Suppressive effect of *Ixodes ricinus* salivary gland extract on mechanisms of natural immunity in vitro. *Parasite Immunol.* 20:169–74
42. Kopecky J, Kuthejlova M, Pechova J. 1999. Salivary gland extract from *Ixodes ricinus* ticks inhibits production of interferon-gamma by the upregulation of interleukin-10. *Parasite Immunol.* 21:351–56
43. Kovar L, Kopecky J, Rihova B. 2001. Salivary gland extract from *Ixodes ricinus* tick polarizes the cytokine profile toward Th2 and suppresses proliferation of T lymphocytes in human PBMC culture. *J. Parasitol.* 87:1342–48
44. Kroegel C. 1988. The potential pathophysiological role of platelet-activating factor in human diseases. *Klin. Wochenschr.* 66:373–78
45. Kubes M, Fuchsberger N, Labuda M, Zuffova E, Nuttall PA. 1994. Salivary gland extracts of partially fed *Dermacentor reticulatus* ticks decrease natural killer cell activity in vitro. *Immunology* 82:113–16
46. Kuroki M, Minakami S. 1989. Extracellular ATP triggers superoxide production in human neutrophils. *Biochem. Biophys. Res. Comm.* 162:377–80
47. Leboulle G, Crippa M, Decrem Y, Mejri N, Brossard M, et al. 2002. Characterization of a novel salivary immunosuppressive protein from *Ixodes ricinus* ticks. *J. Biol. Chem.* 277:10083–89
48. Lerner EA, Ribeiro JMC, Nelson RJ, Lerner MR. 1991. Isolation of maxadilan, a potent vasodilatory peptide from the salivary glands of the sand fly *Lutzomyia longipalpis*. *J. Biol. Chem.* 266:11234–36
49. Li S, Kwon J, Aksoy S. 2001. Characterization of genes expressed in the salivary glands of the tsetse fly, *Glossina morsitans morsitans*. *Insect Mol. Biol.* 10:69–76

50. Liew FY. 1993. The role of nitric oxide in parasitic diseases. *Ann. Trop. Med. Parasitol.* 87:637–42
51. Macaluso KR, Wikel SK. 2001. *Dermacentor andersoni*: effects of repeated infestations on lymphocyte proliferation, cytokine production, and adhesion-molecule expression by BALB/c mice. *Ann. Trop. Med. Parasitol.* 95:413–27
52. McEver RP. 2001. Adhesive interactions of leukocytes, platelets, and the vessel wall during hemostasis and inflammation. *Thromb. Haemost.* 86:746–56
53. Mejri N, Franscini N, Rutti B, Brossard M. 2001. Th2 polarization of the immune response of BALB/c mice to *Ixodes ricinus* instars, importance of several antigens in activation of specific Th2 subpopulations. *Parasite Immunol.* 23:61–69
54. Mellanby K. 1946. Man's reaction to mosquito bites. *Nature* 158:554751–53
55. Millan MJ. 1999. The induction of pain: an integrative review. *Prog. Neurobiol.* 57:1–164
56. Moro O, Lerner EA. 1997. Maxadilan, the vasodilator from sand flies, is a specific pituitary adenylate cyclase activating peptide type I receptor agonist. *J. Biol. Chem.* 272:966–70
57. Morris RV, Shoemaker CB, David JR, Lanzaro GC, Titus RG. 2001. Sandfly maxadilan exacerbates infection with *Leishmania major* and vaccinating against it protects against *L. major* infection. *J. Immunol.* 167:5226–30
58. O'Flaherty J, Cordes JF. 1994. Human neutrophil degranulation responses to nucleotides. *Lab. Invest.* 70:816–21
59. Paesen GC, Adams PL, Harlos K, Nuttall PA, Stuart DI. 1999. Tick histamine-binding proteins: isolation, cloning, and three-dimensional structure. *Mol. Cell* 3: 661–71
60. Pike SE, Yao L, Jones KD, Cherney B, Appella E, et al. 1998. Vasostatin, a calreticulin fragment, inhibits angiogenesis and suppresses tumor growth. *J. Exp. Med.* 188:2349–56
61. Poole S, Lorenzetti BB, Cunha JM, Cunha FQ, Ferreira SH. 1999. Bradykinin B1 and B2 receptors, tumour necrosis factor alpha and inflammatory hyperalgesia. *Br. J. Pharmacol.* 126:649–56
62. Qureshi ST, Gros P, Malo D. 1999. The Lps locus: genetic regulation of host responses to bacterial lipopolysaccharide. *Inflamm. Res.* 48:613–20
63. Rai KS, Black WC IV. 1999. Mosquito genomes: structure, organization, and evolution. *Adv. Genet.* 41:1–33
64. Ribeiro JM, Charlab R, Rowton ED, Cupp EW. 2000. *Simulium vittatum* (Diptera: Simuliidae) and *Lutzomyia longipalpis* (Diptera: Psychodidae) salivary gland hyaluronidase activity. *J. Med. Entomol.* 37:743–47
65. Ribeiro JM, Charlab R, Valenzuela JG. 2001. The salivary adenosine deaminase activity of the mosquitoes *Culex quinquefasciatus* and *Aedes aegypti*. *J. Exp. Biol.* 204:2001–10
66. Ribeiro JM, Katz O, Pannell LK, Waitumbi J, Warburg A. 1999. Salivary glands of the sand fly *Phlebotomus papatasi* contain pharmacologically active amounts of adenosine and 5′-AMP. *J. Exp. Biol.* 202:1551–59
67. Ribeiro JM, Mather TN. 1998. *Ixodes scapularis*: salivary kininase activity is a metallo dipeptidyl carboxypeptidase. *Exp. Parasitol.* 89:213–21
68. Ribeiro JM, Modi G. 2001. The salivary adenosine/AMP content of *Phlebotomus argentipes* Annandale and Brunetti, the main vector of human kala-azar. *J. Parasitol.* 87:915–17
69. Ribeiro JM, Rowton ED, Charlab R. 2000. The salivary 5′-nucleotidase/phosphodiesterase of the hematophagus sand fly, *Lutzomyia longipalpis*. *Insect Biochem. Mol. Biol.* 30:279–85
70. Ribeiro JMC. 1982. The antiserotonin and antihistamine activities of salivary secretion of *Rhodnius prolixus*. *J. Insect Physiol.* 28:69–75
71. Ribeiro JMC. 1987. Role of arthropod

saliva in blood feeding. *Annu. Rev. Entomol.* 32:463–78
72. Ribeiro JMC. 1989. Role of saliva in tick/host associations. *Exp. Appl. Acarol.* 7:15–20
73. Ribeiro JMC. 1995. Blood-feeding arthropods: live syringes or invertebrate pharmacologists? *Infect. Agents Dis.* 4:143–52
74. Ribeiro JMC, Endris TM, Endris R. 1991. Saliva of the tick, *Ornithodorus moubata*, contains anti-platelet and apyrase activities. *Comp. Biochem. Physiol.* 100A:109–12
75. Ribeiro JMC, Evans PM, MacSwain JL, Sauer J. 1992. *Amblyomma americanum*: characterization of salivary prostaglandins E_2 and F_2 alpha by RP-HPLC/bioassay and gas chromatography-mass spectrometry. *Exp. Parasitol.* 74:112–16
76. Ribeiro JMC, Hazzard JMH, Nussenzveig RH, Champagne D, Walker FA. 1993. Reversible binding of nitric oxide by a salivary nitrosylhemeprotein from the blood sucking bug, *Rhodnius prolixus*. *Science* 260:539–41
77. Ribeiro JMC, Kidwell MG. 1994. Transposable elements as population drive mechanisms: specification of critical parameter values. *J. Med. Entomol.* 31:10–16
78. Ribeiro JMC, Nussenzveig RH. 1993. The salivary catechol oxidase/peroxidase activities of the mosquito, *Anopheles albimanus*. *J. Exp. Biol.* 179:273–87
78a. Ribeiro JMC, Valenzuela JG. 2002. The salivary purine nucleosidase of the mosquito *Aedes aegypti*. *Insect Biochem. Mol. Biol.* In press
79. Ribeiro JMC, Weis JJ, Telford SR III. 1990. Saliva of the tick *Ixodes dammini* inhibits neutrophil function. *Exp. Parasitol.* 70:382–88
80. Sampson AP. 2000. The role of eosinophils and neutrophils in inflammation. *Clin. Exp. Allergy* 30(Suppl.)1:22–27
81. Schreiber MC, Karlo JC, Kovalick GE. 1997. A novel cDNA from *Drosophila* encoding a protein with similarity to mammalian cysteine-rich secretory proteins, wasp venom antigen 5, and plant group 1 pathogenesis-related proteins. *Gene* 191:135–41
82. Selak MA, Chignard M, Smith JB. 1988. Cathepsin G is a strong platelet agonist released by neutrophils. *Biochem. J.* 251:293–99
83. Simons FE, Peng Z. 2001. Mosquito allergy: recombinant mosquito salivary antigens for new diagnostic tests. *Int. Arch. Allergy Immunol.* 124:403–5
84. Stark KR, James AA. 1998. Isolation and characterization of the gene encoding a novel factor Xa-directed anticoagulant from the yellow fever mosquito, *Aedes aegypti*. *J. Biol. Chem.* 273:20802–9
85. Stockley RA. 1999. Neutrophils and protease/antiprotease imbalance. *Am. J. Respir. Crit. Care Med.* 160:S49–52
86. Tilley SL, Wagoner VA, Salvatore CA, Jacobson MA, Koller BH. 2000. Adenosine and inosine increase cutaneous vasopermeability by activating A(3) receptors on mast cells. *J. Clin. Invest.* 105: 361–67
87. Titus RG, Ribeiro JMC. 1990. The role of vector saliva in transmission of arthropod-borne diseases. *Parasitol. Today* 6:157–60
88. Tonnesen MG. 1989. Neutrophil-endothelial cell interactions: mechanisms of neutrophil adherence to vascular endothelium. *J. Invest. Dermatol.* 93:53S–58S
89. Tonnesen MG, Feng X, Clark RA. 2000. Angiogenesis in wound healing. *J. Invest. Dermatol. Symp. Proc.* 5:40–46
90. Valenzuela JG, Belkaid Y, Garfield MK, Mendez S, Kamhawi S, et al. 2001. Toward a defined anti-*Leishmania* vaccine targeting vector antigens: characterization of a protective salivary protein. *J. Exp. Med.* 194:331–42
91. Valenzuela JG, Belkaid Y, Rowton E, Ribeiro JM. 2001. The salivary apyrase of the blood-sucking sand fly *Phlebotomus papatasi* belongs to the novel *Cimex*

family of apyrases. *J. Exp. Biol.* 204:229–37

92. Valenzuela JG, Charlab R, Galperin MY, Ribeiro JM. 1998. Purification, cloning, and expression of an apyrase from the bed bug *Cimex lectularius.* A new type of nucleotide-binding enzyme. *J. Biol. Chem.* 273:30583–90
93. Valenzuela JG, Charlab R, Gonzalez EC, Miranda-Santos IKF, Marinotti O, et al. 2002. The D7 family of salivary proteins in blood sucking Diptera. *Insect Mol. Biol.* 11:149–55
94. Valenzuela JG, Charlab R, Mather TN, Ribeiro JM. 2000. Purification, cloning, and expression of a novel salivary anticomplement protein from the tick, *Ixodes scapularis. J. Biol. Chem.* 275:18717–23

94a. Valenzuela JG, Francischetti IM, Pham VM, Garfield MK, Mather TN, Ribeiro JM. 2002. Exploring the sialome of the tick, *Ixodes scapularis. J. Exp. Biol.* 205(Pt. 18):2843–64

95. Valenzuela JG, Francischetti IM, Ribeiro JM. 1999. Purification, cloning, and synthesis of a novel salivary anti-thrombin from the mosquito *Anopheles albimanus. Biochemistry* 38:11209–15
96. Valenzuela JG, Pham VM, Garfield MK, Francischetti IM, Ribeiro JMC. 2002. Toward a description of the sialome of the adult female mosquito *Aedes aegypti. Insect Biochem. Mol. Biol.* 32(9):1101
97. Valenzuela JG, Ribeiro JM. 1998. Purification and cloning of the salivary nitrophorin from the hemipteran *Cimex lectularius. J. Exp. Biol.* 201:2659–64
98. Valenzuela JG, Walker FA, Ribeiro JM. 1995. A salivary nitrophorin (nitric-oxide-carrying hemoprotein) in the bedbug *Cimex lectularius. J. Exp. Biol.* 198:1519–26
99. Vogt W. 1974. Activation, activities and pharmacologically active products of complement. *Pharmacol. Rev.* 26:125–69
100. Wikel S, Ramachandra RN, Bergman DK. 1994. Tick-induced modulation of the host immune response. *Int. J. Parasitol.* 24:59–66
101. Wikel SK. 1996. Host immunity to ticks. *Ann. Rev. Entomol.* 41:1–22
102. Willadsen P. 1980. Immunity to ticks. *Adv. Parasitol.* 18:293–311
103. Wilson ME. 1985. Effects of bacterial endotoxins on neutrophil function. *Rev. Infect. Dis.* 7:404–18
104. Yancopoulos GD, Klagsbrun M, Folkman J. 1998. Vasculogenesis, angiogenesis, and growth factors: ephrins enter the fray at the border. *Cell* 93:661–64
105. Zimmermann H. 1992. 5′-Nucleotidase: molecular structure and functional aspects. *Biochem. J.* 285:345–65

Annu. Rev. Entomol. 2003. 48:89–110
doi: 10.1146/annurev.ento.48.091801.112654

First published online as a Review in Advance on August 19, 2002

KEY INTERACTIONS BETWEEN NEURONS AND GLIAL CELLS DURING NEURAL DEVELOPMENT IN INSECTS

Lynne A. Oland and Leslie P. Tolbert
Arizona Research Laboratories Division of Neurobiology, University of Arizona, Tucson, Arizona 85721; e-mail: lao@neurobio.arizona.edu; tolbert@neurobio.arizona.edu

Key Words axon guidance, embryogenesis, metamorphosis, intercellular interactions

■ **Abstract** Nervous system function is entirely dependent on the intricate and precise pattern of connections made by individual neurons. Much of the insightful research into mechanisms underlying the development of this pattern of connections has been done in insect nervous systems. Studies of developmental mechanisms have revealed critical interactions between neurons and glia, the non-neuronal cells of the nervous system. Glial cells provide trophic support for neurons, act as struts for migrating neurons and growing axons, form boundaries that restrict neuritic growth, and have reciprocal interactions with neurons that govern specification of cell fate and axonal pathfinding. The molecular mechanisms underlying these interactions are beginning to be understood. Because many of the cellular and molecular mechanisms underlying neural development appear to be common across disparate insect species, and even between insects and vertebrates, studies in developing insect nervous systems are elucidating mechanisms likely to be of broad significance.

CONTENTS

INTRODUCTION

Complex nervous systems, such as those of insects, are characterized by a degree of cellular specificity that is extraordinary. Neurons, present by the thousands to trillions, form functional connections with each other with dizzying precision, generating specific behavior in appropriate response to the environment.

The challenges faced by developing neurons in acquiring their particular cellular identities, differentiating appropriate ion channels, neurotransmitter machinery, and other molecular specializations, extending axons toward target fields and connecting with their specific cellular targets, and elaborating dendrites of the correct shape and in the correct areas are almost too overwhelming to ponder; however, investigators have made significant progress in understanding the cellular and molecular mechanisms that underlie these processes. A key insight in recent years has been that intercellular interactions play key roles in development. Neuron-neuron interactions as well as two-way interactions between neurons and glial cells are essential for normal development of the nervous system.

Elucidation of specific neuron-glia interactions has been especially rapid in insect nervous systems, where these interactions are readily identified and readily probed with modern imaging and molecular genetic tools. Insects offer particular advantages for study: Sensory neurons reside in the periphery, and therefore can be experimentally manipulated independent of their targets in the central nervous system (CNS); certain neurons are uniquely identifiable from animal to animal, so that one can know with unsurpassed precision the normal developmental history of a neuron to compare with that in experimentally perturbed specimens; postembryonic development is protracted, offering large, accessible systems for study; and the genetics of one species, *Drosophila melanogaster*, are especially amenable to characterization and manipulation.

Glial Cells in Insect Nervous Systems

Glial cells serve many essential functions in the adult insect nervous system (24, 60, 112). Glial cells of the blood-brain barrier as well as those within the parenchyma of the CNS and those in close association with peripheral neurons support homeostasis in the extracellular environment of neurons. Glia are involved in the trophic support of neurons, and glia provide electrical insulation for axons. Microglia-like cells may participate in wound-healing. In insects, as in vertebrates, glial cells outnumber neurons (30). Insect brains have a lower glia:neuron ratio than vertebrates (75); however, insect glial cells provide types and levels of support similar to those of vertebrate glial cells.

Glial cells in insects are a diverse group of cells. Investigators using different criteria have categorized them according to many classification schemes (23, 24, 44, 50, 92, 106, 115), no two of which are alike and none of which directly match the vertebrate astrocytes, oligodendrocytes, and Schwann cells. Rather than choose one scheme for this article, we refer to glial cells by the names given to them in the studies we review.

In the *D. melanogaster* CNS, most glial cells arise from common progenitors, the neuroglioblasts (25, 33, 44, 50, 114). Determination of progeny to become glial cells is under the control of the transcription factor encoded by *glial cells missing*. In the peripheral nervous system (PNS), glia arise from stem cells often called sensory organ precursor cells (73, 82). Glial cells are present as early cellular components of the insect nervous system, and therefore are in a position to play important roles in neural development.

Study of Neuron-Glia Interactions

A rich literature has been written about neuron-glia interactions in the developing insect nervous system. A number of excellent recent reviews focus on roles of glial cells in embryonic insect development (1, 25, 29, 57, 83, 95) and on roles of glia in metamorphic adult development (24, 25, 60, 76). In addition, some more general reviews on glial cells and neuron-glia interactions include special emphasis on insect neural development (64). Our goal here is not to be encyclopedic, but rather to focus on providing examples of the breadth of types of interactions between neurons and glial cells in embryonic and metamorphic neural development in insects, as an impetus to readers to explore further the full range of interactions available for study.

COMMISSURES AND LONGITUDINAL TRACTS

The axonal tracts of the embryonic ventral nerve cord of insects form a ladder-like array, the sidepieces comprising the longitudinal tracts and the crosspieces comprising in each segment the anterior and posterior commissures. These tracts are pioneered at a time when glial cells are prominent features in the cellular terrain and potentially the source of diffusible or surface-based signals for growing axons. Evidence from *D. melanogaster* and grasshopper suggests, however, that pioneer axon pathfinding is essentially cell-autonomous. The glia instead are far more important in influencing the growth and survival of neurons whose axons are considered followers of the pioneers, a finding that supports the intriguing possibility (98) that the development of these later-extending neurons is much more susceptible to extrinsic influences, including those of glia, than previously thought.

Midline Glia of the Embryonic Ventral Nerve Cord

In *D. melanogaster*, midline glia arise from mesectodermal precursors (44), unlike all the other glial types we mention in this chapter. With the exception of a gene that suppresses neuronal identity [*tramtrack* (27)], they express markers different from those of the "true" glia (45). They do not express either *glial cells missing* or *reversed polarity* (*repo*, which encodes a transcription factor regulated by *gcm*) (31, 118), both of which are considered glia-specific and hallmarks of glial identity elsewhere in the nervous system (38). Most authors, however, grant them glial status based on functional roles that include axon ensheathment (46), axon guidance (4, 66), and participation in organization of the midline structure (103).

The cells destined to give rise to the midline lineage, which includes both neurons and glia, first appear in the blastoderm at the boundary between the presumptive mesoderm and neurectoderm. During gastrulation, the mesectodermal cells move to the midline. Their midline identity depends upon expression of *single-minded* (*sim*), a transcription factor, and *rhomboid*, a cell surface protein involved in the EGF receptor–mediated signaling pathway (45). Those that follow the glial lineage continue to express both *single-minded* and *rhomboid*, while those entering the neuronal path cease expressing *rhomboid* and either cease expression of *single-minded* or reduce Sim protein production (6, 19). Within several hours of the onset of gastrulation several of the neurons pioneer the axonal tracts in the ventral nerve cord, and two to four pairs of glia (45) appear in the anterior part of the segment where their glial fate is determined by *wingless* (a segment polarity gene) function (42). Of the six midline glia, only three survive to the end of embryogenesis (21). Early in their differentiation as glia, the cells express *slit* [encodes a secreted glycoprotein (87)], *netrins* [encode small secreted proteins (66)], and *wrapper* [encodes an Ig superfamily protein (69)], all of which are important in axon guidance or ensheathment.

The glial cells and the pioneer midline neurons participate in an intricate choreography that leads to the formation and separation of the two commissures in each segment (55). The first growth cones in each segment extend toward the most anterior ventral unpaired median (VUM) neuron and pioneer the posterior commissure. The VUM neurons then send their leading processes anteriorly, pioneering the anterior commissure and providing a substrate along which the most anterior pair of midline glia will migrate. The glial cells extend processes to contact the VUM axons and then migrate along the axons to a position over the cluster of VUM cell bodies and between the now-separated commissures. In the absence of VUM neurons, the midline glia fail to migrate and the commissures fail to separate (56). Studies in *spitz* mutants (56, 102), in which glia also fail to migrate, indicate that it is the migration of these glia that is required for commissure separation. During the late stages of embryogenesis, the two forward pairs of glia enwrap the anterior commissure, and the posterior pair, which migrates forward from the more posterior segment, enwraps some of the posterior commissure.

Because the midline glia have not migrated into place at the time the commissures initially form, they are unlikely to have a role in the initial formation of the commissures. Once the basic commissure scaffold is in place, however, the midline glial cells do have a direct role in axon guidance. The genetically and biochemically well-documented midline signaling array active during axon outgrowth includes *roundabout* (*robo*), *commissureless*, and *slit*, as well as *netrin* and *frazzled*. Interactions among these players provide a clear example of the richness of the signaling environment and the capability of growth cones to detect and integrate numerous and often antagonistic signals. Growth cones of axons that are to cross the midline express Robo, a receptor for Slit (9, 52), and Frazzled, a receptor for Netrin (58). Netrins A and B are expressed in the midline before the midline glial lineage emerges as well as later just in the differentiated glia (32, 66). Slit expression requires midline glial differentiation (10). Growth cones attracted to the midline by the presence of the Netrins would nevertheless be prevented from crossing by the repellant Slit were it not for the presence of Commissureless on the surfaces of the midline glia. By an unknown mechanism, Commissureless downregulates Robo, in essence allowing the axons to ignore Slit and proceed across the midline. Robo expression is strongly upregulated and the growth cones lose their sensitivity to Netrin as they move away from the glia, the combination thus preventing axonal re-entry into the midline (52, 93, 108). *netrin A* and *netrin B* mutants show a phenotype in which axons never approach the midline (43, 93), indicating the loss of an attractive cue. In *robo* mutants, axons cross and recross the midline, while in *commissureless* or *netrin* mutants, the commissures are thin or absent. Axons in *slit* mutants project ectopically, resulting in a fused midline phenotype (4, 87), indicating loss of a repulsive cue.

Slit released from the midline glia also prevents ipsilaterally projecting axons from crossing the midline and further acts as a long-range signal to control which of the three parallel subdivisions of the longitudinal tract the axons will follow. *D. melanogaster* axons may express any of three Robo receptors: robo, robo 2, and/or robo 3. They project in successively more lateral zones of the longitudinal pathway depending on whether they express no Robo or one or two Robo receptors (Robo 2 or 3) (78, 79).

During formation of the commissures, axonal growth cones intensively explore the surfaces of the midline glial cells, which robustly express Gliolectin, a carbohydrate-binding protein (97). When the *gliolectin* locus is deleted, the commissural axons bundle tightly together and arch across the midline, effectively reducing their contact with the glia. Many axons actually stall and fail to reach the contralateral longitudinal tract. The interaction between the n-acetylglucosamine-terminated molecules on the axonal surfaces and the glia-based lectin may ensure sufficient surface contact to permit a high-fidelity readout of local guidance cues.

Longitudinal Glia of the Embryonic Ventral Nerve Cord

Development of the longitudinal tracts has been described in grasshopper (3) and in *D. melanogaster* (34, 47). Development of these tracts begins with the formation of two small lateral clusters of cells in each neuromere, all transiently expressing *glial cells missing* (51). One of the cells, which maintains *glial cells missing* expression, is a glioblast that divides symmetrically to produce two glial precursors, the more laterally located one giving rise to glia associated with the nerve roots and most of the peripheral glia, and the more medially located one giving rise to the longitudinal glia. The longitudinal precursor migrates medially (47, 51) and divides to form an elongated cluster of 6–8 cells near the cell bodies of the neurons that pioneer the longitudinal tracts. The glia migrate dorsally with the neurons. Interestingly, transient expression of *robo* by longitudinal glia limits the medial extent of their migration and prevents them from establishing positions in the midline (53). As the pioneers begin to extend their axons, the glia also are extending processes, eventually forming a continuous glial canopy linking the longitudinal glia within and between segments (33). Growth cones have been observed extending in regions lacking glial processes, so the glial pattern must not prefigure the longitudinal tracts (34). The longitudinal glial cells gradually form sheet-like processes that enwrap the axon fascicles and the developing neuropil.

Several observations suggest that contact between particular glial cells and pioneer axons may influence axon direction: Pioneer growth cones insert filopodia into the glial cell directly above their cell body, before they have turned onto the longitudinal pathway but only rarely afterward (46), and several enhancer trap lines directed to the longitudinal glia reveal only subsets of the group, which suggests that they are molecularly heterogeneous (55). Contacts between the axons and glia often are extensive, with growth cones flattened over much of the glial surfaces, and the ablation of particular glial cells results in many pathway defects, particularly at choice points marked by the position of particular longitudinal glia (34).

Experimental manipulation of longitudinal glia, however, suggests that the primary role for the longitudinal glia is ensuring high fidelity of axon pathfinding, not of the pioneer axons, but of those that follow the pioneers. In mutations affecting the differentiation of the midline glia (54), such as spitz group members *single-minded*, *slit*, *faint little ball*, *rhomboid*, and *Star*, the longitudinal glia are unaffected and the tracts form normally, although they are displaced toward the midline by the loss of midline-specific cells. When mutations, such as *orthodenticle*, *hindsight*, *prospero* (all transcription factors), or *repo*, affect the differentiation of the longitudinal glia, some, but not all, of the longitudinal tracts are interrupted. In *glial cells missing* mutants, longitudinal cells take on a neuronal identity, but many of the transformed cells retain some glial characteristics. In these mutants, the longitudinal tracts often have numerous gaps and thick commissures, suggesting that the axons take abnormal paths, sometimes even veering into the commissures, but these defects appear long after the longitudinal tracts are established by the pioneer axons (14, 20, 31, 38, 51). When the longitudinal glia are eliminated via expression of ricin toxin in the lateral glioblast (35), the longitudinal tracts have gaps, or pathway or fasciculation defects in some of the segments (34), but the defects seen were in the follower axons, not in the pioneers. The apparent normalcy of some tracts in all these experiments indicates that the longitudinal glia do not guide initial formation of the tracts, but instead play a supportive role.

Studies of development of the axon tracts also have revealed glial dependence on axons. In the *commissureless* mutant, in which axons do not enter the commissures (93), more than the usual 50% of midline glia die, and those remaining typically migrate laterally to contact longitudinal tract axons (103). One model, based on analysis of *rhomboid* mutants, suggests that axons normally provide EGF to the glia. Activation of the EGF receptor–mediated signaling pathway, in which Rhomboid is an element, is required to suppress apoptosis (62). In the absence of crossing axons, the midline glia EGF receptor would fail to be activated and the cell would die. Similarly, when the ricin toxin is expressed in longitudinal axons, killing many of them, many longitudinal glial cells also die, presumably because they lack trophic factors from the axons (53).

Midline and Longitudinal Glia in the Embryonic Brain

The process of forming the major commissures and longitudinal tracts is far less well understood in the brain than it is in the ventral nerve cord, in large part for the obvious reason that the brain is a much more complicated structure. Nevertheless, many of the underlying patterns of development are similar. Brain formation in grasshopper (8) and *D. melanogaster* (33, 68, 109) begins with the appearance of bilaterally symmetrical groups of neuroblasts in the neurogenic regions of the future brain hemispheres. They and their progeny form proliferative clusters that become the centers for the various regions of the brain. The clusters become surrounded and infiltrated by the processes of glial cells. Pioneer neurons extend axons whose growth cones navigate along the cluster surfaces, probing the cell surfaces with their filopodia. The axons arising from cells in the more peripheral clusters extend toward the midline, fasciculating as they proceed with those of more medially located pioneer neurons. In the developing *D. melanogaster* brain, *repo*-expressing cells have been detected in a small cluster of glial progenitor cells found near the boundary between the deuto- and the trito-cerebrum. The progenitors migrate away from the cluster, mainly in association with the developing fiber tracts, and eventually give rise to most, if not all, of the interface glia associated with the neuropil of the brain (33).

Connection between the two brain hemispheres is established in *D. melanogaster* as a column of cells that extends from the medial edge of each developing hemisphere and connects to form a bridge of neuronal cell bodies and glia, each in separate layers. Axons extend toward the midline from a pair of pioneer neurons at the edge of each hemisphere, and their growth cones meet in the midline, then fasciculate with their contralateral partner's axon to complete the traverse and thus establish the first brain commissure. Their growth across the midline occurs in close association with the bridge midline glial cells, an arrangement reminiscent of the glial sling that prefigures the corpus callosum in the vertebrate brain (101) but not present during commissure formation in the ventral nerve cord. As in the ventral nerve cord, the pioneer axons in *D. melanogaster* brains are fully capable of establishing the first commissure in *commissureless* mutants, but in the absence of Commissureless expression by the glial cells to repress Robo and block detection of the repellant Slit, follower axons cannot cross (109).

Longitudinal glia are involved in the formation of axonal tracts connecting the brain and the ventral ganglia (8, 109). Two bilaterally symmetrical arrays of glia form at the medial edge of the procephalic and subesophageal neurogenic regions and extend around the developing gut toward the ventral neurogenic regions. Again the pioneer axons, in this case of both ascending and descending neurons, extend in close association with the glial array, but it is not yet clear whether the glia actually prefigure the pathway. In mutants that have lost or show gaps in the glial array, the axonal pathway has gaps, fails to form, or forms in aberrant positions (109), a phenotype like that seen when the longitudinal tracts of the embryonic ventral nerve cord develop in the absence of longitudinal glial cells.

SEGMENTAL AND INTERSEGMENTAL NERVES OF THE PERIPHERAL NERVOUS SYSTEM

The pathways established to and from the periphery include glial cells, almost all of which arise in early embryogenesis from glioblasts in each CNS neuromere (33). They migrate to the lateral edge of the neuromere, proliferate, and initially form a cone-shaped array at the CNS/PNS border where they may have a transient role as intermediate targets for pioneer axons (3, 92, 95). Pioneering motoneuron growth cones make extensive contacts with these glia and interact selectively with them (2, 3). In the grasshopper, for example, the growth cones of the U and aCC axons pioneering the intersegmental nerve normally turn laterally at the segment boundary cell, which is a homolog of the glial cells that populate the CNS/PNS transition region in *D. melanogaster*. The growth cones of other pioneer axons do not change direction despite similarly extensive contact.

After the motoneuron pioneer growth cones pass the array of glia at the CNS/PNS border, the array breaks up as the glial cells begin to migrate along the axon pathway, never extending processes ahead of the leading tip of the pioneer's growth cone. By the end of embryogenesis, the glia maintain a presence at the CNS/PNS transition region and have reached and enwrapped the axons of the ingrowing sensory neurons; ensheathment of the motoneuron axons is not completed until the third larval instar (95).

When the segment boundary cell in the grasshopper ventral ganglion is ablated (3), intersegmental nerve pioneers fail to turn and instead continue along the longitudinal tract. An intersegmental nerve nevertheless forms later in development, perhaps via ingrowing sensory neurons whose path is not dependent on interaction with the segment boundary cell. In *D. melanogaster*, deletion of peripheral glia in both *glial cells missing* and *repo* mutants reveals formation of peripheral axon tracts at unusual positions and with incorrect fasciculation (14, 31, 38, 51, 118); the defects, however, appear to be partially corrected by late in embryogenesis, indicating that the glia at the CNS/PNS boundary are supportive of, but not essential to, motor axon pathfinding to the periphery.

glial cells missing mutants, of course, can affect CNS as well as CNS-derived peripheral glia, making it impossible to determine whether any defects seen were a direct or indirect consequence of the loss of the peripheral glia. When peripheral glia were selectively killed by targeted expression of apoptotic genes, including *grim* (119) and *ced-3* (99), the initial routes of both motor and sensory axons often were abnormal in trajectory and pattern of fasciculation, and sensory axons sometimes stalled as they progressed toward the CNS. While the misrouting of sensory axons often resulted in their entry into the CNS at unusual positions, the motoneurons typically reached their correct targets (96). Taken together, the experimental data to date indicate that the peripheral glia, like their central counterparts, serve to enhance the fidelity of axon pathfinding but are not necessary to the process.

Mutations in *pointed* (a glial transcription factor present in differentiated glia) have revealed a possible role in controlling neuronal pathfinding capability through induction or regulation of particular axonal proteins (54). Axons in *pointed* mutants lose expression of a microtubule-associated protein (Futsch) required for axonal and dendritic growth (41), and neurons directly adjacent to transplanted glia that express *pointed* do show Futsch expression (54), which implies an interaction mediated either by a local diffusible factor or by contact. As expected if Futsch production is induced or regulated by glia, Futsch staining also was reduced in *glial cells missing* and glia-ablated embryos (7).

VISUAL SYSTEMS

Larval Visual System

During embryologic development of the larval visual system in *D. melanogaster*, larval photoreceptor cells originate in or near the optic lobe placode to populate the light-sensitive Bolwig's organ of the pharynx (15). On their way to brain targets, their growing axons contact several different cell types, including glial cells and the optic lobe pioneer neurons (15, 111). As larval optic axons grow toward the brain, they grow rather diffusely along a path of glial cells derived from optic placode. The axons then fasciculate to form the larval optic nerve, which elongates as the photoreceptor organ and developing brain move apart. Optic lobe pioneer neurons arise from an early CNS neuroblast and from the optic lobe primordium. During fusion of the optic lobe primordium with the adjacent brain hemisphere, the corner optic lobe pioneer cells differentiate and extend axons that bundle with the larval optic nerve. As the axons project through the optic lobe primordium, they turn toward the central optic lobe pioneer before entering the presumptive medulla area of the optic lobe.

The larval optic nerve requires the presence of the pioneer neurons to form correct connections (15). In *disconnected* (*disco*) mutants, optic lobe pioneer neurons, which normally express Disco, are absent, and the larval optic nerve either fails to find the brain or finds the brain but fails to form appropriate connections. In the abnormally projecting optic nerve are glial cells, which normally would express *disco*, that now have abnormal morphologies. Although the corner optic lobe pioneers are thought to be required intermediate targets for optic nerve axons, it may also be that the altered development of glial cells underlies the misprojection of axons.

Adult Visual System

In holometabolous insects, the compound eyes of the adult arise from imaginal discs. Each compound eye of *D. melanogaster* comprises 750–800 ommatidia, each containing 8 photoreceptor cells that send their axons along the larval optic nerve through the optic stalk to terminate in retinotopic patterns in the optic lobe of the brain. The axons of photoreceptors R1–R6 terminate in a retinotopic pattern in the optic lamina irrespective of whether they are accompanied by neighboring axons (61), and R7 and R8 axons both grow through the lamina to terminate in the medulla even though R8 axons are the first and R7 the last growth cones to reach the brain. These findings indicate that interactions among photoreceptor growth cones are unlikely to play key roles in pathfinding and targeting, and that other cues—potentially including those from non-neuronal cells—must dominate (65).

Subretinal glial cells originate in the optic stalk during the third larval instar and migrate into the developing eye disc (16, 80, 81). Two diffusible factors, Decapentaplegic (Dpp) and Hedgehog (Hh), play important roles in photoreceptor differentiation and also in glial precursor development (39, 81). Whereas Dpp regulates glial motility and proliferation, Hh has a strong effect only on proliferation. Both work through transcriptional regulation. In flies with mutations in the *gilgamesh* (*gish*) locus (which encodes a casein kinase), glial cells migrate precociously into the eye disc, even before photoreceptor cells have differentiated, indicating that the glia do not require axons to migrate (40). *gish* acts together with the eye specification genes *eyeless* and *sine oculis* to prevent precocious entry of glial cells into the eye disc; Hh expression can suppress the precocious migration in *gish* mutants. When glial cells are caused to migrate ectopically in either *Hh* or *gish* mutants, photoreceptor axons also grow ectopically, indicating that the axons follow the glia. Once in the disc, they are essential for instructing or allowing growing photoreceptor axons to enter the optic stalk; without the glial cells present, axons do extend and grow in the correct direction, but they fail to enter the stalk. Evidence suggests that the interaction between axons and retinal basal glia is contact dependent.

Experiments on several mutants have revealed that glial cells in the larval optic lobes are essential for photoreceptor axons to find their cellular targets. Photoreceptor axons normally grow past subretinal glial cells at the base of the optic stalk, the giant glial cells of the outer chiasm, and satellite glial cells scattered among lamina neurons (73, 111, 116). Axons from R1 to R6 terminate in the lamina, between the epithelial and the marginal glial cells that demarcate layers of the lamina, whereas axons from R7 and R8 grow deeper, past the medulla glial cells that lie between lamina and medulla, to terminate in the medulla. Unlike other glial cells, the epithelial and marginal glial cells extend many fine, short processes in close association with the R axons, suggesting that they may serve as intermediate targets (77). Loss-of-function mutations in *nonstop*, which encodes a ubiquitin-dependent protease, produce a disruption in migration of epithelial, marginal, and medulla glial cells into a laminar organization and a mistargeting of the R1–R6 axons. The mistargeting of axons may be a result of the decreased number of glial cells there, or, conversely, the laminar arrangement of glia may be dependent on the presence of axons (65). In *hedgehog* mutants, with severely decreased numbers of lamina neurons, R1–R6 axons terminate normally (77). Thus, R1–R6 photoreceptor axons may require contact with glial cells, but not with their target neurons, to select the correct neuropil layer for termination. In *argos* (encoding a secreted protein in the EGF pathway) and *repo* mutants, glial cells of the larval optic lobe fail to differentiate and, in both, photoreceptor axons fail to find their targets (11, 90, 118). Furthermore, glial cells are present deep to the medulla in the region of the inner optic chiasm, where axons of medulla neurons cross to form an inverted retinotopic map as they project to the lobula/lobula plate neuropil (111). If *optomotor-blind* (a T-box transcription factor) expression is blocked in inner optic chiasm glial cells, the axons fail to form a proper chiasm (74). Clearly, abundant evidence points to key roles for glial cells in guiding photoreceptor axons to their targets.

Nonstop mutations raised the possibility that although ingrowth of photoreceptor axons is not necessary for glial proliferation, it may be essential for differentiation of glial cells in the lamina. In enhancer trap lines that mark these glial cells, although glial cells are produced normally, the glial markers do not develop when photoreceptor axons are not present (116). A particular gene, the JAB1/CSN5 subunit of the COP9 signalosome complex, must be expressed in photoreceptor cells for them to have the ability to induce lamina glial cell migration (107).

ANTENNAL SYSTEMS

Antennal Systems in Hemimetabolous Insects

In hemimetabolous insects such as cockroaches, the antenna develops during embryogenesis and is progressively elaborated upon with each nymphal molt (84). The earliest receptor axons are guided toward the brain by special cells in the lumen of the antenna that express neuronal properties, but they do not appear to be chemosensory neurons (94). At each molt, new segments are added at the base of the antennal flagellum and, with this, the numbers of different types of antennal-receptor neurons grow dramatically (91). It is not known whether non-neuronal, including glial, cells influence the growth of the later-growing axons toward the brain.

The antennal lobes continue to grow in hemimetabolous species, but the neurons of the antennal lobes are born during embryogenesis. In addition, in *Blaberus craniifer* glomeruli have been counted and found to be constant, at 109, throughout the life of the animal (84). Early in development, a glial border develops between the cortex of neuronal cell bodies and the neuropil of the developing antenna lobe of *Periplaneta americana* (88). Olfactory receptor axons grow into the neuropil and then arborize, never extending beyond the glial border. Once the axon terminals coalesce into glomerular clusters, the border glial cells extend processes to surround them. In the absence of axons, glial proliferation is mildly reduced, but the glia that are present extend processes into the existing neuropil (89), suggesting that the envelopment of glomeruli may not be specifically in response to axons.

Antennal Systems in Holometabolous Insects

More is known about the development of the antennal systems in holometabolous species, such as moths and flies. The rudimentary antennae and other chemosensory structures of larvae are shed and replaced by large, complex adult antennae that arise from imaginal discs during the pupal stage of metamorphosis. In *D. melanogaster*, the axons of 1200–1300 adult olfactory receptor neurons residing in the maxillary palps and in the third segment of the antenna, the funiculus, project to the antennal lobes where they terminate in 43 glomeruli; the receptor neurons of different types of sensilla project to specific subsets of glomeruli ipsilaterally or bilaterally, and many glomeruli are uniquely identifiable from animal to animal [reviewed in (105)]. The adult antennal nerve is pioneered by axons from a nonchemosensory antennal segment; the axons use the larval chemosensory nerve for guidance (110). As they grow, peripheral glial cells of the *atonal* sensory-neuron lineage direct them into three fascicles based roughly on location of axon origin (49). In *Manduca sexta*, approximately 300,000 adult olfactory receptor neurons reside in the long, segmented flagellum of each antenna [reviewed in (71)]. The sensory neurons of the antenna are born soon after the imaginal discs evert, and immediately begin to extend axons down the antenna toward the brain. The axons grow along a structure called the pupal nerve, which connects the tip of the antenna to the brain. This pupal nerve serves some function as a guidance strut, but is not essential for axons to reach the brain (70, 86). Glial cells populate the antennal nerve only relatively late in the formation of the antennal nerve (85) and therefore are unlikely to play roles in guidance.

While small, simple antennal lobes are generated during embryogenesis to serve the rudimentary larval antennae, the adult antennal lobes arise during metamorphosis. In *D. melanogaster*, the neurons of the adult lobe are born during larval life and gradually substitute for cells of the larval center; some of them may function as differentiated neurons in the larva (49, 110). In *M. sexta* [reviewed in (71)], antennal-lobe neurons arise almost completely de novo from a small set of neuroblasts that divide throughout larval life but whose progeny do not differentiate until metamorphosis (104). They send processes into a neuropil that is surrounded by a continuous border of glial cells. As the first olfactory receptor axons begin to arrive in the brain, they pierce the glial border and terminate in a fringe around the edge of the existing neuropil. Eventually, the terminals segregate into nodular protoglomeruli (the precursors of glomeruli), which become enwrapped by glia. The neurites of uniglomerular projection neurons reach outward to overlap with the axon terminals almost as soon as protoglomeruli form, whereas the neurites of local interneurons of the antennal lobe extend outward to meet the axons only after glia form boundaries. Considerable similarity has been shown in the development of antennal lobes in *D. melanogaster* (48, 49).

Glomeruli are common to the olfactory neuropils of virtually all animal species with highly developed olfactory systems. Glomeruli generally are discrete structures, often surrounded by an incomplete envelope of glia. Each glomerulus contains the axon terminals of olfactory receptor neurons that express a particular species of olfactory receptor protein [reviewed in (37)], which is thought to bind a particular "odotope" or molecular feature of molecules that act as odorous stimuli. In *D. melanogaster*, olfactory receptor neurons expressing different putative receptor genes occur in overlapping clusters on the antennae and maxillary palps (17). In *M. sexta*, the olfactory receptor genes have not been identified yet, but each of the 80 segments of the antenna is known to house receptor neurons of many sensitivities [reviewed in (37)]. Because receptor neurons expressing different olfactory receptor proteins or sensitivities to odors are generally intermingled, the projections of receptor axons onto the glomerular array is not a simple spatial one; instead, most spatial topography is lost as axons grow to particular glomeruli based on their odor specificity.

How do olfactory receptor axons extending from the receptor epithelium find their correct target glomeruli within the glomerular array? Work from several laboratories suggests differential axonal adhesion as one likely mechanism. Fasciclin II, a member of the immunoglobulin superfamily, is expressed by a subset of developing olfactory receptor neurons in each antennal segment in *M. sexta* (36). Their axons are scattered throughout the width of the antennal nerve until they reach a glia-rich "sorting zone" at the entrance to the antennal lobe, where they abruptly shift their associations and fasciculate with other fasciclin II–positive axons targeted for specific glomeruli. If the number of glial cells in the sorting zone is severely reduced using γ-radiation, fasciclin II–positive axons do not fasciculate with each other, and many axons grow right past their normal targets in the antennal lobe (85), indicating that an interaction with glia is necessary for fasciclin II–based axonal sorting. In cell culture, growing olfactory receptor axons pause their extension and form elaborate growth cones when they contact isolated sorting zone glia (113), suggesting that contact with glial cells has a direct influence on growth cone dynamics, perhaps via changes in adhesive properties.

A number of investigators (67) have adduced evidence that the olfactory receptor proteins themselves, expressed on receptor axons, might act as adhesive guidance molecules. However, in *D. melanogaster*, olfactory receptor neurons begin to express olfactory receptor proteins only after glomeruli have formed (17), suggesting that receptor protein-based sorting of the type seen in mice is not likely to occur. Thus, changes in growth cone properties initiated by contact with glia, as demonstrated in *M. sexta*, may have prime importance.

In *M. sexta*, the larval antennal lobes are replaced by large adult antennal lobes during metamorphosis, even if the antennal anlage are removed [reviewed in (71)]. Similarly, in the hemimetabolous *P. americana*, no interaction between olfactory receptor axons and the brain is required during embryogenesis to produce the primary olfactory centers (88). Even in these species, however, and in fact in every species so far examined, formation of glomeruli requires the presence of olfactory receptor axons. Olfactory receptor axons have the special property of forming protoglomerulus-like "knots" even when they grow into ectopic regions of the brain or form a blind neuroma (86). Nevertheless, studies of intercellular interactions in developing *M. sexta* indicate that olfactory axons alone are not sufficient and that glial cells play key roles in glomerulus development.

In antennal lobes developing without antennal-axon input, neuropil-associated glial cells remain restricted to a rim surrounding the neuropil, multiglomerular local interneurons of the lobe develop diffuse rather than tufted branching patterns, and uniglomerular projection neurons develop arbors that are restricted in extent but larger than the normal glomerular size [reviewed in (71)]. Taking advantage of the fact that glia proliferate later than neurons in the antennal system, we used γ-radiation or hydroxyurea to reduce the number of glia present in the developing antennal lobe (72). These experiments produced strong evidence that glia confine growing receptor axons to branching within their protoglomerulus (71). A separate line of experiments revealed that tenascin-like molecules on the surfaces of glial cells might have a direct influence on the branching patterns of antennal-lobe neurons. Neuropil-associated glial cells express molecules similar in size and antigenically to mammalian tenascin, and if antennal-lobe neurons are plated on dishes painted with stripes of purified tenascin, many avoid growing on the tenascin (71).

Recent studies suggest a role for nitric oxide (NO) in neuron-glia interactions in the developing antennal lobe in *M. sexta*. In situ hybridization and immunocytochemistry revealed that NO synthase is expressed in the axons of olfactory receptor neurons throughout development. Pharmacological inhibition of NO synthase during development of the antennal lobe causes the glomeruli to be small and misshapen, and some areas of the neuropil are devoid of glomerular organization; most strikingly, glial cells do not migrate to form normal glomerular borders (26). It is possible that the abnormal appearance of glomeruli results from the abnormal migration of glial cells in response to the ingrowth of receptor axons.

Taken together, these studies suggest that antennal axons in *M. sexta* induce glial cells to surround developing glomeruli, perhaps through the action of NO that they release, and that these glial cells subsequently restrict the growth of the axons and the dendrites they target to the confines of individual glomeruli, perhaps through the action of tenascin-like molecules. In *D. melanogaster*, the processes of neuropil-associated glia extend into the antennal-lobe neuropil along with receptor axons, but, similarly to *M. sexta*, the protoglomeruli formed by axon terminals become surrounded by glial processes (48, 49).

NEURONAL SURVIVAL

In the vertebrate nervous system, cell death is a critical and normal developmental process that contributes to an optimal matching of input to target, rids the system of superfluous neurons and glial cells, and aids in pattern formation (13). Initiation of cell death by apoptosis is believed to depend on the ability to maintain an adequate supply of trophic factors to stave off activation of an intrinsic cell death pathway.

Vertebrate glial cells long have been known to provide signals that enhance the likelihood of survival in certain neurons, and the reciprocal also appears to occur in that the survival of glial cells can be influenced by neuronal factors.

In *D. melanogaster*, glia secrete a glycoprotein, Anachronism, that regulates the timing of postembryonic neuronal proliferation (22). The first evidence in the developing insect nervous system that glial cells might also contribute to the survival of neurons surfaced in a study of the *drop-dead* mutant (12). In these mutants, which behaviorally are normal at the time of eclosion to the adult but die within about a week, structural abnormalities such as stunted processes, incomplete wrapping of neuronal cell bodies, and signs of degeneration were found in most glial cells at the time of eclosion, well before the onset of neuronal degeneration in the adult brain. Loss of the glia followed by loss of the neurons suggested neuronal dependence on the glia. Wild-type tissue in genetically mosaic brains could rescue the mutant portions of the brain, suggesting that the *drop-dead* gene encodes a diffusible factor, but neither the cellular source nor the function of the putative factor is as yet known. A similar pattern was seen in early studies examining the effects of *repo* on the developing visual system. The frequency of apoptosis was noted to be higher than normal in lamina neurons of *repo* mutant optic lobes (117), and nearly all the lamina glia showed signs of apoptosis.

Glia in the developing ventral nerve cord also could have a role in regulating neuron survival, in particular survival of the follower neurons but not of the pioneer neurons (7). When interface glia were ablated via targeted expression of a ricin toxin or of *reaper* (a cell death gene) at a time after the longitudinal tracts had been established, the tracts comprised fewer axons and many neurons were apoptotic. Similar increases in neuronal apoptosis were seen when *glial cells missing* was directed to the longitudinal glia (51), moving them toward a neuronal identity and clearly disrupting the mechanism by which these glial cells normally promote neuronal survival. Again, the underlying mechanism is unknown.

Glial Ensheathment of Axons

Although insect axons are not enwrapped by complex myelinated sheaths as are many vertebrate axons, glia in insects do wrap single large axons or bundles of axons in a manner that has a morphological appearance similar to the ensheathment of unmyelinated vertebrate axons. Failure to ensheathe axons adequately due to glial ablation, transformation of cell identity, failure of glial migration, or disruption of pathways required for glial ensheathment affect fasciculation, axonal conduction, and, in some instances, the survival of neurons, presumably because the protective blood-brain barrier has been lost (5, 63, 69, 95).

Investigators have begun to examine the role of axon-derived FGF in signaling glial cells to initiate ensheathment. In *heartless* mutants, which lack the glia-expressed FGF receptor, glia migrate to and distribute themselves along axon tracts but then fail to develop flattened cytoplasmic sheets that normally would enwrap the axons (28, 100). In vitro, grasshopper glia were attracted to FGF2-coated beads and further showed strong MAPK (mitogen-activated protein kinase) activation localized to the axon-glia interface (18), suggesting a model in which FGF from axons is bound by glial FGF receptors at the surface adjacent to the axons. The resulting activation of the MAPK pathway may generate a change in glial shape that would permit the cell to surround the axon.

swiss cheese encodes a protein expressed in neurons that is similar to the regulatory subunit of protein kinase A. In mutants, the neurons become wrapped by many more glial layers than usual, beginning late in metamorphic development, and then become vacuolated and die (59). Mosaic flies showed the damage to be limited to mutant tissue, suggesting that the protein does not act as a diffusible signal. The defect may lie in failure of the neurons to provide a signal, whose effect is possibly cAMP-mediated, to the glia that establishes the correct amount of glial wrapping. The mechanism by which hyperwrapping would lead to neuronal apoptosis is not yet understood.

SUMMARY AND PROSPECTS FOR THE FUTURE

Our appreciation of the repertoire of neuron-glia interactions in the developing insect nervous system is still rudimentary. Already, undeniable evidence implicates glial cells in processes related to axon growth and guidance. Early indications also point to the importance of two-way conversations between neurons and glial cells in the determination of cell fate, migration, and survival, and many other aspects of differentiation of both cell types.

As research on neuron-glia interactions continues, undoubtedly, a network of interactions richer and more varied than ever imagined at both the cellular and molecular levels will be uncovered to help explain the daunting specificity of neuronal connections. Insects, with their many particular experimental advantages, will continue to contribute essential aspects to the basic foundation of our understanding.

ACKNOWLEDGMENTS

Our work was supported by NIH grants NS28495 and DC04598.

The *Annual Review of Entomology* is online at http://ento.annualreviews.org

LITERATURE CITED

1. Auld V. 1999. Glia as mediators of growth cone guidance: studies from insect nervous systems. *Cell Mol. Life Sci.* 55:1377–85
2. Bastiani MJ, du Lac S, Goodman CS. 1986. Guidance of neuronal growth cones in the grasshopper embryo. I. Recognition of a specific axonal pathway by the pCC neuron. *J. Neurosci.* 6:3518–31
3. Bastiani MJ, Goodman CS. 1986. Guidance of neuronal growth cones in the grasshopper embryo III. Recognition of specific glial pathways. *J. Neurosci.* 6: 3542–51

4. Battye R, Adrienne S, Jacobs JR. 1999. Axon repulsion from the midline of the *Drosophila* CNS requires *slit* function. *Development* 126:2475–81
5. Baumgartner S, Littleton J, Broadie K, Bhat M, Harbecke R, et al. 1996. A *Drosophila* neurexin is required for septate junction and blood-nerve barrier formation and function. *Cell* 87:1059–68
6. Bier E, Jan L, Jan Y. 1990. *rhomboid*, a gene required for dorsoventral axis establishment and peripheral nervous system development in *Drosophila melanogaster*. *Genes Dev.* 4:190–203
7. Booth G, Kinrade E, Hidalgo A. 2000. Glia maintain follower neuron survival during *Drosophila* CNS development. *Development* 127:237–44
8. Boyan G, Therianos S, Williams J, Reichert H. 1995. Axogenesis in the embryonic brain of the grasshopper *Schistocerca gregaria*: an identified cell analysis of early brain development. *Development* 121:75–86
9. Brose D, Bland K, Wang K, Arnott D, Henzel W, et al. 1999. Slit proteins bind robo receptors and have an evolutionarily conserved role in repulsive axon guidance. *Cell* 96:795–806
10. Brose K, Tessier-Lavigne M. 2000. Slit proteins: key regulators of axon guidance, axonal branching, and cell migration. *Curr. Opin. Neurobiol.* 10:95–102
11. Brunner A, Twardzik T, Schneuwly S. 1994. The *Drosophila giant lens* gene plays a dual role in eye and optic lobe development: inhibition of differentiation of ommatidial cells and interference in photoreceptor axon guidance. *Mech. Dev.* 48:175–85
12. Buchanan R, Benzer S. 1993. Defective glia in the *Drosophila* brain degeneration mutant *drop-dead*. *Neuron* 10:839–50
13. Burek M, Oppenheim R. 1999. Cellular interactions that regulate programmed cell death in the developing vertebrate nervous system. *Cell Death Dis. Nerv. Syst.* 1:145–80
14. Campbell G, Göring H, Lin T, Spana E, Andersson S, et al. 1994. RK2, a glial-specific homeodomain protein required for embryonic nerve cord condensation and viability in *Drosophila*. *Development* 120:2957–66
15. Campos A, Lee K, Steller H. 1995. Establishment of neuronal connectivity during development of the *Drosophila* larval visual system. *J. Neurobiol.* 28:313–29
16. Choi K-W, Benzer S. 1994. Migration of glia along photoreceptor axons in the developing *Drosophila* eye. *Neuron* 12:423–31
17. Clyne P, Warr C, Freeman M, Lessing D, Kim J, Carlson J. 1999. A novel family of divergent seven-transmembrane proteins: candidate odorant receptors in *Drosophila*. *Neuron* 22:327–38
18. Condron B. 1999. Spatially discrete FGF-mediated signalling directs glial morphogenesis. *Development* 126:4635–41
19. Crews S, Thomas J, Goodman C. 1988. The *Drosophila single-minded* gene encodes a nuclear protein with sequence similarity to the *per* gene product. *Cell* 52:143–51
20. Doe C, Chu-Lagraff Q, Wright D, Scott M. 1991. The *prospero* gene specifies cell fates in the *Drosophila* central nervous system. *Cell* 65:451–64
21. Dong R, Jacobs J. 1997. Origin and differentiation of supernumerary midline glia in *Drosophila* embryos deficient for apoptosis. *Dev. Biol.* 190:165–77
22. Ebens A, Garren H, Cheyette B, Zipursky S. 1993. The *Drosophila anachronism* locus: A glycoprotein secreted by glia inhibits neuroblast proliferation. *Cell* 74:15–27
23. Edwards JS. 1969. Postembryonic development and regeneration of the insect nervous system. *Adv. Insect Physiol.* 6:97–137
24. Edwards JS, Tolbert LP. 1998. Insect neuroglia. In *Microscopic Anatomy of the Invertebrates*, ed. M Locke, 11B:449–66. New York:Wiley

25. Giangrande A. 1996. Development and organization of glial cells in *Drosophila melanogaster*. *Int. J. Dev. Biol.* 40:917–27
26. Gibson NJ, Rossler W, Nighorn AJ, Oland LA, Hildebrand JG, Tolbert LP. 2001. Neuron-glia communication via nitric oxide is essential in establishing antennal-lobe structure in *Manduca sexta*. *Dev. Biol.* 240:326–39
27. Giesen K, Hummel T, Stollewerk A, Harrison S, Travers A, Klambt C. 1997. Glial development in the *Drosophila* CNS requires concomitant activation of glial and repression of neuronal differentiation genes. *Development* 124:2307–11
28. Gisselbrecht S, Skeath J, Doe C, Michelson A. 1996. *heartless* encodes a fibroblast growth factor receptor (DFR1/DFGF-R2) involved in the directional migration of early mesodermal cells in the *Drosophila* embryo. *Genes Dev.* 10:3003–17
29. Granderath S, Klambt C. 1999. Glia development in the embryonic CNS of *Drosophila*. *Curr. Opin. Neurobiol.* 9:531–36
30. Gymer A, Edwards J. 1967. The development of insect nervous system. I. An analysis of postembryonic growth in the terminal ganglion of *Acheta domesticus*. *J. Morphol.* 123:191–97
31. Halter D, Urban J, Rickert C. 1995. The homeobox gene *repo* is required for the differentiation and maintenance of glia function in the embryonic nervous system of *Drosophila melanogaster*. *Development* 121:317–32
32. Harris R, Sabatelli L, Seeger M. 1996. Guidance cues at the *Drosophila* CNS midline: identification and characterization of two *Drosophila* netrin/UNC-6 homologs. *Neuron* 17:217–28
33. Hartenstein V, Nassif C, Lekven A. 1998. Embryonic development of the *Drosophila* brain. II. Pattern of glial cells. *J. Comp. Neurol.* 402:32–47
34. Hidalgo A, Booth G. 2000. Glia dictate pioneer axon trajectories in the *Drosophila* embryonic CNS. *Development* 127:393–402
35. Hidalgo A, Urban J, Brand A. 1995. Targeted ablation of glia disrupts axon tract formation in the *Drosophila* CNS. *Development* 121:3703–12
36. Higgins MR, Gibson NJ, Eckholdt PA, Nighorn A, Copenhaver P, et al. 2002. Different isoforms of fasciclin II are expressed by a subset of developing olfactory receptor neurons and by olfactory-nerve glial cells during formation of glomeruli in the moth *Manduca sexta*. *Dev. Biol.* 244:134–54
37. Hildebrand JG, Shepherd GM. 1997. Mechanisms of olfactory discrimination: converging evidence for common principles across phyla. *Annu. Rev. Neurosci.* 20:595–631
38. Hosoya T, Takiwaza K, Nitta K, Hotta Y. 1995. *glial cells missing*: a binary switch between neuronal and glial differentiation in *Drosophila*. *Cell* 82:1025–36
39. Huang Z, Kunes S. 1998. Signals transmitted along retinal axons in *Drosophila*: hedgehog signal reception and the cell circuitry of lamina cartridge assembly. *Development* 125:3753–64
40. Hummel T, Attix S, Gunning D, Zipursky S. 2002. Temporal control of glial cell migration in the *Drosophila* eye requires *gilgamesh*, *hedgehog*, and eye specification genes. *Neuron* 33:193–203
41. Hummel T, Krukkert K, Roos J, Davis G, Klambt C. 2000. *Drosophila* Futsch/22C10 is a MAP1B-like protein required for dendritic and axonal development. *Neuron* 26:357–70
42. Hummel T, Schimmelpfeng K, Klämbt C. 1999. Commissure formation in the embryonic CNS of *Drosophila*. I. Identification of gene function. *Dev. Biol.* 208:381–98
43. Hummel T, Schimmelpfeng K, Klämbt C. 1999. Commissure formation in the embryonic CNS of *Drosophila*. II. Function

of the different midline cells. *Development* 126:771–79
44. Ito D, Urban J, Technau G. 1995. Distribution, classification, and development of *Drosophila* glial cells in the late embryonic and early larval ventral nerve cord. *Roux's Arch. Dev. Biol.* 204:284–307
45. Jacobs JR. 2000. The midline glia of *Drosophila*: a molecular genetic model for the developmental functions of glia. *Prog. Neurobiol.* 62:475–508
46. Jacobs JR, Goodman CS. 1989. Embryonic development of axon pathways in the *Drosophila* CNS. I. A glial scaffold appears before the first growth cones. *J. Neurosci.* 9:2402–11
47. Jacobs JR, Hiromi Y, Patel M, Goodman CS. 1989. Lineage, migration, and morphogenesis of longitudinal glia in the *Drosophila* CNS as revealed by a molecular lineage marker. *Neuron* 2:1625–31
48. Jhaveri D, Rodrigues V. 2002. Sensory neurons of the Atonal lineage pioneer the formation of glomeruli within the adult *Drosophila* olfactory lobe. *Development* 129:1251–60
49. Jhaveri D, Sen A, Rodrigues V. 2000. Mechanism underlying olfactory neuronal connectivity in *Drosophila*—the Atonal lineage organizes the periphery while sensory neurons and glia pattern the olfactory lobe. *Dev. Biol.* 226:73–87
50. Jones B. 2001. Glial cell development in the *Drosophila* embryo. *BioEssays* 23: 877–87
51. Jones B, Fetter R, Tear G, Goodman CS. 1995. *glial cells missing*: a genetic switch that controls glial versus neuronal fate. *Cell* 82:1013–23
52. Kidd T, Bland KS, Goodman CS. 1999. Slit is the midline repellent for the Robo receptor in *Drosophila*. *Cell* 96:785–94
53. Kinrade E, Brates T, Tear G, Hidalgo A. 2001. Roundabout signaling, cell contact and trophic support confine longitudinal glia and axons in the *Drosophila* CNS. *Development* 128:207–16
54. Klaes A, Menne T, Stollewerk A, Scholz H, Klämbt C. 1994. The ets transcription factors encoded by the *Drosophila* gene *pointed* direct glial cell differentiation in the embryonic CNS. *Cell* 78:149–60
55. Klämbt C, Goodman CS. 1991. The diversity and pattern of glia during axon pathway formation in the *Drosophila* embryo. *Glia* 4:205–13
56. Klämbt C, Jacobs J, Goodman C. 1991. The midline of the *Drosophila* central nervous system: a model for the genetic analysis of cell fate, cell migration, and growth cone guidance. *Cell* 64:801–15
57. Klämbt D, Hummel T, Menne T, Sadlowski E, Scholz H, Stollewerk A. 1996. Development and function of embryonic central nervous system glial cells in *Drosophila*. *Dev. Genet.* 18:40–49
58. Kolodziej P, Timpe L, Mitchell K, Fried S, Goodman C, et al. 1996. *frazzled* encodes a *Drosophila* member of the DCC immunoglobulin subfamily and is required for CNS and motor axon guidance. *Cell* 87:197–204
59. Kretzschmar D, Hasan G, Sharma S, Heisenberg M, Benzer S. 1997. The *swiss cheese* mutant causes glial hyperwrapping and brain degeneration in *Drosophila*. *J. Neurosci.* 17:7425–32
60. Kretzschmar D, Pflugfelder G. 2002. Glia in development, function, and neurodegeneration of the adult insect brain. *Brain Res. Bull.* 57(1):121–31
61. Kunes S, Wilson C, Steller H. 1993. Independent guidance of retinal axons in the developing visual system of *Drosophila*. *J. Neurosci.* 13:752–67
62. Lanoue B, Jacobs J. 1999. *rhomboid* function in the midline of the *Drosophila* CNS. *Dev. Genet.* 25:321–30
63. Leiserson W, Harkins E, Keshishian H. 2000. Fray, a *Drosophila* serine/threonine kinase homologous to mammalian PASK, is required for axonal ensheathment. *Neuron* 28:793–806
64. Lemke G. 2001. Glial control of neural development. *Annu. Rev. Neurosci.* 24:87–105

65. Martin KA, Poeck B, Roth H, Ebens AJ, Ballard LC, Zipursky SL. 1995. Mutations disrupting neuronal connectivity in the *Drosophila* visual system. *Neuron* 14:229–40
66. Mitchell K, Doyle J, Serafini T, Kennedy T, Tessier-Lavigne M, et al. 1996. Genetic analysis of *netrin* genes in *Drosophila*: Netrins guide CNS commissural axons and peripheral motor axons. *Neuron* 17: 203–15
67. Mombaerts P. 1996. Targeting olfaction. *Curr. Opin. Neurobiol.* 6:481–86
68. Nassif C, Noveen A, Hartenstein V. 1998. Embryonic development of the *Drosophila* brain. I. Pattern of pioneer tracts. *J. Comp. Neurol.* 402:10–31
69. Noordermeer J, Kopczynski C, Fetter R, Bland K, Chen W, Goodman C. 1998. Wrapper, a novel member of the Ig superfamily, is expressed by midline glia and is required for them to ensheath commissural axons in *Drosophila. Neuron* 21: 991–1001
70. Oland LA, Pott WM, Higgins MR, Tolbert LP. 1998. Targeted ingrowth and axon-glial relationships of olfactory receptor axons in the primary olfactory pathway of an insect. *J. Comp.Neurol.* 398:119–38
71. Oland LA, Tolbert LP. 1996. Multiple factors shape the development of olfactory glomeruli: insights from an insect model system. *J. Neurobiol.* 30:92–109
72. Oland LA, Tolbert LP, Mossman KL. 1988. Radiation-induced reduction of the glial population during development disrupts the formation of olfactory glomeruli in an insect. *J. Neurosci.* 8:353–67
73. Perez SE, Steller H. 1996. Migration of glial cells into retinal axon target field in *Drosophila melanogaster. J. Neurobiol.* 30:359–73
74. Pflugfelder GO, Heisenberg M. 1995. *Optomotor-blind* of *Drosophila melanogaster*: a neurogenetic approach to optic lobe development and optomotor behaviour. *Comp. Biochem. Physiol. A* 110: 185–202
75. Pfreiger FW, Barres B. 1995. What the fly's glia tell the fly's brain. *Cell* 83:671–74
76. Pielage J, Klämbt C. 2001. Glial cells aid axonal target selection. *Trends Neurosci.* 24:432–33
77. Poeck B, Fischer S, Gunning D, Zipursky L, Salecker I. 2001. Glial cells mediate target layer selection of retinal axons in the developing visual system of *Drosophila. Neuron* 29:99–113
78. Rajagopalan S, Nicolas E, Vivancos V, Berger J, Dickson B. 2000. Crossing the midline: roles and regulation of robo receptors. *Neuron* 28:767–77
79. Rajagopalan S, Vivancos V, Nicolas E, Dickson B. 2000. Selecting a longitudinal pathway: Robo receptors specify the lateral position of axons in the *Drosophila* CNS. *Cell* 103:1033–45
80. Rangarajan R, Courvoisier H, Gaul U. 2001. Dpp and Hedgehog mediate neuron-glia interactions in *Drosophila* eye development by promoting the proliferation and motility of subretinal glia. *Mech. Dev.* 108:93–103
81. Rangarajan R, Gong Q, Gaul U. 1999. Migration and function of glia in the developing *Drosophila* eye. *Development* 126:3285–92
82. Reddy GV, Rodrigues V. 1999. A glial cell arises from an additional division within the mechanosensory lineage during development of the microchaete on the *Drosophila* notum. *Development* 126: 4617–22
83. Reichert H, Boyan G. 1997. Building a brain: developmental insights in insects. *Trends Neurosci.* 20:258
84. Rospars J. 1988. Structure and development of the insect antennodeutocerebral system. *Int. J. Insect Morphol. Embryol.* 17:243–94
85. Rössler W, Oland LA, Higgins MR, Hildebrand JG, Tolbert LP. 1999. Development of a glia-rich axon-sorting zone in the olfactory pathway of the moth *Manduca sexta. J. Neurosci.* 19:9865–77

86. Rössler W, Randolph PW, Tolbert LP, Hildebrand JG. 1999. Axons of olfactory receptor cells of trans-sexually grafted antennae induce development of sexually dimorphic glomeruli in *Manduca sexta*. *J. Neurobiol.* 38:521–41
87. Rothberg J, Jacobs J, Goodman C, Artavanis-Tsakonas S. 1990. *slit*: An extracellular protein necessary for the development of midline glia and commissural axon pathways contains both EGF and LRR domains. *Genes Dev.* 4:2169–87
88. Salecker I, Boeckh J. 1995. Embryonic development of the antennal lobes of a hemimetabolous insect, the cockroach *Periplaneta americana*: light and electron microscopic observations. *J. Comp. Neurol.* 352:33–54
89. Salecker I, Boeckh J. 1996. Influence of receptor axons on the formation of olfactory glomeruli in a hemimetabolous insect, the cockroach *Periplaneta americana*. *J. Comp. Neurol.* 370:262–79
90. Sawamoto K, Okabe M, Tanimura T, Hayashi S, Mikoshiba K, Okano H. 1996. *argos* is required for projection of photoreceptor axons during optic lobe development in *Drosophila*. *Dev. Dyn.* 205: 162–71
91. Schafer R. 1973. Postembryonic development in the antenna of the cockroach, *Leucophaea maderae*: growth, regeneration, and the development of the adult pattern of sensory organs. *J. Exp. Zool.* 183:353–64
92. Schmidt H, Rickert C, Bossing T, Vef O, Urban J, Technau GM. 1997. The embryonic central nervous system lineages of *Drosophila melanogaster*. II. Neuroblast lineages derived from the dorsal part of the neuroectoderm. *Dev. Biol.* 189:186–204
93. Seeger M, Tear G, Ferres-Marco D, Goodman C. 1993. Mutations affecting growth cone guidance in *Drosophila*: genes necessary for guidance toward or away from the midline. *Neuron* 10:409–26
94. Selzer R, Schaller-Selzer L. 1987. Structure and function of luminal neurons in the early embryonic antenna of the American cockroach, *Periplaneta americana*. *Dev. Biol.* 122:363–72
95. Sepp K, Schulte J, Auld V. 2000. Developmental dynamics of peripheral glia in *Drosophila melanogaster*. *Glia* 30:122–33
96. Sepp K, Schulte J, Auld V. 2001. Peripheral glia direct axon guidance across the CNS/PNS transition zone. *Dev. Biol.* 238:47–63
97. Sharrow M, Tiemeyer M. 2001. Gliolectin-mediated carbohydrate binding at the *Drosophila* midline ensures the fidelity of axon pathfinding. *Development* 128:4585–95
98. Shepherd D. 2000. Glial dependent survival of neurons in *Drosophila*. *BioEssays* 22:407–9
99. Shigenaga A, Kimura K, Kobayakawa Y, Tsujimoto Y, Tanimura T. 1997. Cell ablation by ectopic expression of cell death genes, *ced-3* and *Ice*, in *Drosophila*. *Dev. Growth Differ.* 39:429–36
100. Shishido E, Ono N, Kojima T, Saigo K. 1997. Requirements of DFR1/Heartless, a mesoderm-specific *Drosophila* FGF-receptor, for the formation of heart, visceral and somatic muscles, and ensheathing of longitudinal axon tracts in CNS. *Development* 124:2119–28
101. Silver J, Lorenz S, Wahlsten D, Coughlin J. 1982. Axonal guidance during development of the great cerebral commissures: descriptive and experimental studies, in vivo, on the role of preformed glial pathways. *J. Comp. Neurol.* 210:10–29
102. Sonnenfeld MJ, Jacobs JR. 1994. Mesectodermal cell fate analysis in *Drosophila* midline mutants. *Mech. Dev.* 46:3–13
103. Sonnenfeld MJ, Jacobs JR. 1995. Apoptosis of the midline glia during *Drosophila* embryogenesis: a correlation with axon contact. *Development* 121:569–78
104. Sorensen KA, Davis NT, Hildebrand JG. 1990. Postembryonic neurogenesis in the

brain of *Manduca sexta*. *Soc. Neurosci.* 16:1147 (Abstr.)

105. Stocker R. 1994. The organization of the chemosensory system in *Drosophila melanogaster*: a review. *Cell Tissue Res.* 275:3–26
106. Strausfeld NJ. 1976. *Atlas of the Fly Brain*. Berlin: Springer
107. Suh GS, Poeck B, Chouard T, Oron E, Segal D, et al. 2002. *Drosophila* JAB1/CSN5 acts in photoreceptor cells to induce glial cells. *Neuron* 33:35–46
108. Tear G, Harris R, Sutaria S, Kilomanski K, Goodman C, Seeger M. 1996. *commissureless* controls growth cone guidance across the CNS midline in *Drosophila* and encodes a novel membrane protein. *Neuron* 16:501–14
109. Therianos S, Leuzinger S, Hirth F, Goodman C, Reichert H. 1995. Embryonic development of the *Drosophila* brain: formation of commissural and descending pathways. *Development* 121:3849–60
110. Tissot M, Gendre N, Hawken A, Störtkuhl KF, Stocker RF. 1997. Larval chemosensory projections and invasion of adult afferents in the antennal lobe of *Drosophila*. *J. Neurobiol.* 32:281–97
111. Tix S, Eckhart E, Fischbach K-F, Benzer S. 1997. Glia in the chiasms and medulla of the *Drosophila melanogaster* optic lobes. *Cell Tissue Res.* 289:397–409
112. Treherne JE, Schofield PK. 1981. Mechanisms of ionic homeostasis in the central nervous system of an insect. *J. Exp. Biol.* 95:61–73
113. Tucker ES, Oland LA, Tolbert LP. 2000. *In vitro* study of interactions between olfactory receptor growth cones and glial cells of axonal sorting zone. *Soc. Neurosci.* 26:1611 (Abstr.)
114. Udolph G, Prokop A, Bossing T, Technau GM. 1993. A common precursor for glia and neurons in the embryonic CNS of *Drosophila* gives rise to segment-specific lineage variants. *Development* 118:765–75
115. Wigglesworth VB. 1959. The histology of the nervous system of an insect, *Rhodnius prolixus* (Hemiptera). II. The central ganglia. *Q. J. Microsc. Sci.* 100:299–13
116. Winberg ML, Perez SE, Steller H. 1992. Generation and early differentiation of glial cells in the first optic ganglion of *Drosophila melanogaster*. *Development* 115:903–11
117. Xiong W, Montell C. 1995. Defective glia induce neuronal apoptosis in the *repo* visual system of *Drosophila*. *Neuron* 14:581–90
118. Xiong WC, Okano H, Patel NH, Blendy JA, Montell C. 1994. *repo* encodes a glial-specific homeo domain protein required in the *Drosophila* nervous system. *Genes Dev.* 8:981–94
119. Zhou L, Schnitzler A, Agapite J, Schwartz L, Steller H, Nambu J. 1997. Cooperative functions of the reaper and head involution defective genes in the programmed cell death of *Drosophila* central nervous system midline cells. *Proc. Natl. Acad. Sci. USA* 94:5131–36

Annu. Rev. Entomol. 2003. 48:111–39
doi: 10.1146/annurev.ento.48.091801.112647

First published online as a Review in Advance on August 27, 2002

MOLECULAR SYSTEMATICS OF *ANOPHELES*: From Subgenera to Subpopulations

Jaroslaw Krzywinski and Nora J. Besansky
Department of Biological Sciences, Center for Tropical Disease Research and Training, University of Notre Dame, Notre Dame, Indiana 46556; e-mail: krzywinski.1@nd.edu; Besansky.1@nd.edu

Key Words chromosomal inversions, gene flow, introgression, molecular phylogenetics, sibling species

■ **Abstract** The century-old discovery of the role of *Anopheles* in human malaria transmission precipitated intense study of this genus at the alpha taxonomy level, but until recently little attention was focused on the systematics of this group. The application of molecular approaches to systematic problems ranging from subgeneric relationships to relationships at and below the species level is helping to address questions such as anopheline phylogenetics and biogeography, the nature of species boundaries, and the forces that have structured genetic variation within species. Current knowledge in these areas is reviewed, with an emphasis on the *Anopheles gambiae* model. The recent publication of the genome of this anopheline mosquito will have a profound impact on inquiries at all taxonomic levels, supplying better tools for estimating phylogeny and population structure in the short term, and ultimately allowing the identification of genes and/or regulatory networks underlying ecological differentiation, speciation, and vectorial capacity.

CONTENTS

0066-4170/03/0107-0111$14.00

INTRODUCTION

> *The stream of heredity makes phylogeny; in a sense it is phylogeny. Complete genetic analysis would provide the most priceless data for the mapping of this stream.*
>
> *G. G. Simpson, 1945*

> *All species are members of groups of more or less closely related species, and each species can only be understood in the historical context of belonging to groups united in the past into single lineages.*
>
> *J. R. Powell, 1997*

Genus *Anopheles* is by far the largest of the three genera comprising the most basal mosquito (Diptera: Culicidae) lineage, subfamily Anophelinae. Representatives of its nearly 500 recognized species can be found on every continent except Antarctica. At the alpha taxonomy level, anophelines are among the insects most thoroughly studied owing to their medical importance as vectors of malaria, filariasis, and arboviruses. However, despite efforts spanning a century, the present system of *Anopheles* classification remains largely untested for its accurate representation of *Anopheles* phylogeny. Critical examinations of evolutionary relationships at all levels of divergence, intraspecific as well as interspecific, have focused on the minor subset of *Anopheles* that are major vectors of human disease. This review, like previous ones (106), reflects that bias. However, the fullest appreciation of the evolutionary changes that led to disease vector status can be gained only through study of phylogenetic relationships at several hierarchical levels within and between diverse anopheline nonvectors and vectors alike. At present, the genus *Anopheles* is not represented in the Tree of Life Project (http://tolweb.org/tree/phylogeny.html). An ambitious but not unrealistic goal would be to rectify that omission within the next decade.

The main theme uniting the breadth of *Anopheles* research covered by this review is the notion that systematics is a discipline that interprets genetic diversity in terms of hierarchical evolutionary relationships, beginning at the level of gene flow within populations. It was with this in mind that the burgeoning field of *Anopheles* population genetics was included here (48, 92, 95). The word "molecular" in the title of this review reflects the huge stimulus to the field provided by the development of a variety of DNA-based markers and technology for their analysis. However, it is the comparison of multiple classes of markers—morphological, chromosomal, and molecular—that has proved most revealing.

SPECIES IDENTIFICATION

Correct species identification is essential for both basic and applied research. This is especially true for anopheline disease vectors, whose correct identification has far-reaching practical consequences, as illustrated by a malaria control campaign in Vietnam that mistakenly targeted a nonvector species misidentified as the vector *An. minimus* (147). Problems in classical *Anopheles* taxonomy include not only strong morphological similarity between species, but also pronounced morphological variation within species, and accurate species identification usually requires rearing to correlate adult with immature morphology. However, identifications for molecular studies are sometimes based on individual adults, and expert taxonomists are seldom consulted to verify their accuracy. As a result sequences deposited in public databases may be attributed to the wrong species; several such instances are already known (90a; R. Harbach, personal communication). Resolving specific taxonomic questions requires implementation of a multidisciplinary approach that includes morphological, molecular, distribution, and bionomic data (93, 108). Retaining voucher specimens and providing details on collection localities, long practiced by morphological systematists, should also become standard practice in molecular systematics. A recent study in which the neotype for *An. sundaicus* was designated has set the standard (90).

The recognition that members of cryptic species complexes often differ in their capacity to transmit malaria has been, and continues to be, a driving force in the development of identification methods alternative to morphological ones. An ideal method should be fast, cost-effective, easy to implement, and applicable to both sexes and to all developmental stages. Above all, however, the method should be consistent, meaning that it produces the same result species-wide. This requires that the marker(s) on which the method is based is fixed within, and exclusive to, the species of interest. The difficulty in finding such markers increases with the recency of lineage splitting, larger descendant population sizes, and evolutionary conservation of the marker. Here we provide a brief critical account of methods and markers most widely used in *Anopheles*. Additional coverage may be found in (5, 14, 27, 28, 154).

Cytogenetics

Chromosomal inversions have been implicated in the speciation process (30, 109). Analysis of the banding pattern of polytene chromosomes often reveals fixed differences between sibling species due to chromosomal inversions. Unfortunately, polytene chromosomes are stage- and/or sex-limited, and interpretation of the banding pattern requires considerable time and expertise. Not all anophelines have polytene chromosomes with discernible banding patterns and, more importantly, not all anopheline sibling species differ in banding patterns (73, 135).

Biochemical Methods

Protein electrophoresis has been used extensively to discover and diagnose cryptic species through fixed allozyme differences; recent examples include *An. dirus* and *An. minimus* taxa in Thailand, the *An. sundaicus* complex in Indonesia, and the Australasian *An. punctulatus* group (59, 67, 137, 148). The need to preserve enzyme activity dictates that specimens be fresh or stored frozen, a requirement difficult to meet under field conditions. Moreover, for species that have not been reproductively isolated for long, this method may not offer sufficient resolution, given that it only detects the small fraction of variation causing amino acid replacements that affect net charge.

DNA-Based Methods

Because DNA is easily preserved by desiccation or immersion in alcohol and offers virtually unlimited polymorphic markers in coding and noncoding regions, there has been a shift toward DNA-based methods of species identification, which rely on probe hybridization or the polymerase chain reaction (PCR).

The hybridization assays employ species-specific probes complementary to families of highly repetitive DNA that differ in copy number or nucleotide sequence among related species (39). Potentially high throughput, this approach can be unreliable due to its sensitivity both to unequal amounts of target DNA loaded on a membrane and to variation in copy number across the species range (7, 36).

A significant advantage of PCR over other approaches is the minute amount of template DNA usually required. Because a few scales or a leg segment may suffice (113), specimens can be kept alive for crossing experiments, submitted to other analyses, or preserved as morphological vouchers. Two main PCR strategies for species identification amplify anonymous regions of the genome or target specific sequences such as ribosomal DNA (rDNA).

As random amplified polymorphic DNA PCR (RAPD-PCR) relies on the analysis of banding patterns generated with arbitrary decamer primers, it requires no prior knowledge of the genome and tends to target repetitive and rapidly evolving genome regions. Some success has been recorded (159), but this technique is notoriously unreliable. For example, markers identified for the differentiation of *An. minimus* species A and C (136) subsequently appeared inconsistent (77). Nevertheless, starting with species-diagnostic RAPD bands, more robust allele-specific primers can be developed for species identification (77).

The most widely applied PCR assays target nuclear rDNA, present at a single X-linked locus in anophelines (121). It is organized as a tandem-repeated array of conserved genes (*18S*, *5.8S*, and *28S*) punctuated by rapidly evolving noncoding spacers: internal transcribed spacers 1 and 2 (ITS1, ITS2) and the intergenic spacer (IGS). Within interbreeding populations the arrays undergo rapid homogenization through concerted evolution, which drives new sequence variants to fixation and gives rise to species-specific differences. Most recently developed species identification assays are based on differences within the ITS2 (8, 9, 70, 79, 98, 120, 122, 142, 149, 152), although the IGS and a variable region within the *28S* gene also have been exploited successfully (56, 80, 129, 131). Species-specific differences among PCR products similar in size may be detected by restriction fragment length polymorphism (RFLP), single-strand conformation polymorphism (SSCP), or heteroduplex analysis (HDA). The most time-efficient species identification assays rely on a cocktail including both universal (conserved) and allele-specific primers that amplify products sufficiently different in length for each species to be unequivocally discriminated on an agarose gel.

The different types of markers used for species identification are direct or indirect estimators of DNA sequence variation, and each evolves under different constraints and at different rates. Furthermore, alternative assay methods are not equally efficient at revealing extant variation. For example, allele-specific PCR, unless followed by RFLP, SSCP, or HDA, is unable to detect sequence variation between the primer binding regions other than major insertions or deletions. To be maximally useful, the marker(s) employed must evolve at a high enough rate to be informative for the taxa under study, and adequate inter- and intraspecific sampling must validate the method. Application of an inadequately tested assay to field-collected specimens can lead to lack of signal (36) or to erroneous species identification if overlap occurs between markers thought to be diagnostic (70).

As stressed elsewhere (14), the development and application of methods alternative to morphological identification is neither cheap nor trivial. Morphology remains the simplest, fastest, and least expensive means of identification for many of the *Anopheles* species encountered during epidemiological surveys (R. Harbach, personal communication). Moreover, to quote Harbach, "... morphology is a prerequisite for identifying species to complex or group prior to the application of molecular and other methods. You need to get into the ball park before you can play the game."

HIGHER LEVEL SYSTEMATICS

Traditionally, the subfamily Anophelinae is subdivided into three genera, *Anopheles*, *Bironella*, and *Chagasia* (Figure 1). *Anopheles*, most numerous in species, is further subdivided into subgenera, *Anopheles*, *Cellia*, *Kerteszia*, *Lophopodomyia*, *Nyssorhynchus*, and *Stethomyia*, and various informal infrasubgeneric groupings (71). Although Ross (124) presented the first hypothesis about evolutionary relationships within the family Culicidae 50 years ago, surprisingly few phylogenetic studies involving the genus *Anopheles* followed until relatively recently.

Modern morphological and molecular studies using rDNA, mitochondrial DNA (mtDNA), and single-copy nuclear genes strongly support most deeper relationships, including monophyly of the subfamily Anophelinae, the basal position of *Chagasia*, monophyletic origins of the subgenera *Cellia*, *Kerteszia*, and

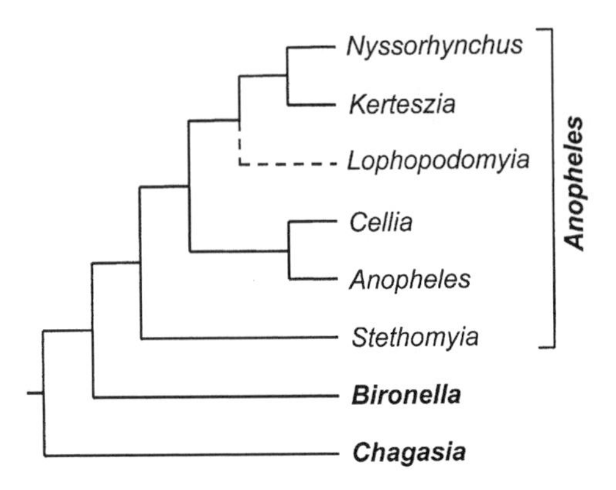

Figure 1 Hypothetical phylogenetic relationships within the subfamily Anophelinae. Only genera (in bold) and subgenera of *Anopheles* are shown. Note that the position of the subgenus *Lophopodomyia* remains unsupported.

Nyssorhynchus, and the sister taxon relationship of *Kerteszia* + *Nyssorhynchus* (11, 58, 72, 81, 82, 125, 126). A recent proposal to sink genus *Bironella* and subgenera *Lophopodomyia* and *Stethomyia* into synonymy with subgenus *Anopheles* (126) seems premature, as it is supported by homoplasious characters and contradicted by independent molecular and morphological evidence (11, 58, 72, 81, 82, 125). However, it is noteworthy that basal relationships within the *Bironella* + *Anopheles* lineage were poorly resolved in all anopheline phylogenetic studies to date. The most likely explanation for this phenomenon is rapid radiation of *Bironella* and basal clades within *Anopheles* (82). If this is the case, a better understanding of early anopheline evolution may be elusive, although acquisition of sequence data from more genes and their combined analysis holds some promise. In this regard, single-copy nuclear genes such as *white* offer maximum information across a breadth of taxonomic levels while avoiding the strong A+T bias in base composition typical of mtDNA and alignment difficulties for which rDNA is notorious.

Notwithstanding some remaining discrepancies, a working hypothesis of anopheline diversification emerges from independent lines of evidence (Figure 1). All studies suggest that *Chagasia* was the first lineage to radiate from the Anophelinae tree. The *white* gene predicts that the next split was between *Bironella* and genus *Anopheles* (82). Within the *Anopheles* lineage the basal split gave rise to *Stethomyia*, and although branching order in the remaining groups is less certain, evidence suggests the radiation of two sister clades: *Cellia* + subgenus *Anopheles*, and (*Lophopodomyia* + (*Kerteszia* + *Nyssorhynchus*)). This hypothetical phylogeny of anophelines has important biogeographical implications.

Phylogenetic studies at the lower level have encompassed several species groups and complexes of *Anopheles*. Relationships among the Nearctic members of the *An. maculipennis* complex were poorly resolved using the D2 region of 28S rDNA (116). In contrast, the phylogeny of the Palaearctic representatives of this group was well supported based on the more variable ITS2 region (99). Members of the Australasian *An. punctulatus* group were studied using several markers. Allozyme analyses (59) were incongruent with DNA sequence data, of which *18S* rDNA provided better resolution than mtDNA genes *COII* and *12S* (6, 35, 58). All the above studies of the *An. punctulatus* group consistently clustered the coastal malaria vector *An. farauti s.s.*, which breeds in brackish water, with a nonvector *An. farauti* 7. Interestingly, these two species shared saltwater tolerance (57), even though larvae of *An. farauti* 7 have been collected exclusively from freshwater (4). This example shows the value of an established phylogeny as a foundation from which the distribution of various traits can be predicted.

Biogeography

Among the most interesting issues in Anophelinae evolutionary history are causes of their peculiar geographical distribution. Of the six *Anopheles* subgenera, *Kerteszia*, *Lophopodomyia*, *Nyssorhynchus*, and *Stethomyia* inhabit South America, *Cellia* is found in the Old World, and subgenus *Anopheles* is cosmopolitan. The phylogenetic framework for Anophelinae yields some of the first insights into their historical biogeography. The basal placement of Neotropical *Chagasia* relative to other anophelines and the Neotropical distribution of four out of six subgenera of *Anopheles* led to the conclusion that Anophelinae originated in the New World (72, 82), in accord with earlier speculation (10). No attempts have yet been made to estimate the divergence times based on molecular data. However, the inference of a South American origin of Anophelinae, a likely monophyletic origin of subgenera *Anopheles* + *Cellia* and their derived position in the phylogeny, together with the geographic distribution data, place the first branching events within the subfamily well before the breakup of Gondwana (82). The first radiations within subgenus *Anopheles* might have taken place before the loss of the land connection between Africa and South America, ~95 million years ago, and the breakup of the continents might have sundered the Neotropical *An. pseudopunctipennis* group (one of the basal lineages of the subgenus *Anopheles*) from the other lineages in Africa. The land bridges between Africa and Europe (created in the Paleocene) and the connection from Europe to North America (existing until the end of the Eocene) allowed further dispersal of the subgenus *Anopheles* into Laurasia, and some of its lineages may have reentered South America from the north. The absence of *Cellia* in the New World suggests that the radiation of this subgenus might not have been triggered until the late Eocene. According to this sequence of events, Australasian *Anopheles* emigrated from the Oriental region as early as the late Miocene, 11 million years ago (6, 10, 58). Further study with denser sampling of subgenera *Anopheles* and *Cellia* and timing of the divergence events is needed to test this hypothesis and to assess the likelihood of alternative routes of *Anopheles* dispersal from Gondwana. Unfortunately, the mosquito fossil record is too young and rare to be informative, but finding *Anopheles* (*Nyssorhynchus*) in Dominican amber (15–20 to 30–45 million years ago) is consistent with a long history of anophelines in the New World (162). A cautionary note is raised by the possibility that multiple instances of secondary dispersal and speciation in the derivative clades might blur ancient vicariant separation (2).

There are some striking parallels between the hypothesis presented above and the proposed phylogeny of placental mammals, which likely originated in Gondwana. Like anophelines, their basal lineages experienced rapid diversification, probably coincidental with the separation of Africa and South America (107). Remarkably, like subgenera *Cellia* and *Anopheles* in the anopheline phylogeny, the placental mammal lineage Boreoeutheria not only occupies a derived position in the mammal phylogeny but also is the most diverse and cosmopolitan. It is conceivable that radiation of anophelines was a by-product of the radiation of their hosts.

SPECIES BOUNDARIES

The splitting of a polymorphic ancestral population into two independent lineages does not immediately result in taxa diagnosable at loci that are evolving under selective neutrality. Coalescent theory, "the name now applied to formal mathematical and statistical treatments of gene genealogies within and among related species" (2), predicts that initially both descendant lineages will share ancestral polymorphism and thus appear polyphyletic in gene genealogies. Over time, stochastic loss of variation in these lineages will lead to a period of paraphyly (one lineage nested within the other) before reciprocal monophyly (mutual exclusivity) is finally achieved. The time to reciprocal monophyly increases with effective population size (N_e), requiring $\sim 4N_e$ generations for mtDNA and even longer for nuclear genes (2). This is usually not a problem for species separated for long time spans, unless the duration of the speciation event was short, as suggested for *Bironella* and basal clades of *Anopheles*. For recently diverged taxa, the expectation of shared ancestral variation at a large proportion of loci greatly complicates inferences about reproductive isolation and historical relationships, and poses practical problems for accurate species identification. It is likely that young species whose speciation durations were relatively short have fixed differences only at genes directly responsible for speciation.

Chromosomal Forms

In West Africa, the species *An. gambiae* is composed of sympatric taxa that are at least partially reproductively isolated (15, 31, 143). They were initially described based on characteristic combinations of paracentric inversions of chromosome 2, between which the expected karyotype frequencies assuming random mating were not found. This observation led to the splitting of *An. gambiae* into "chromosomal forms," given the non-Linnean names Mopti, Bamako, Savanna, Bissau, and Forest. In addition to characteristic combinations of inversions, these taxa also display different ecological tolerances and behaviors. Mopti assumes particular ecological and epidemiological importance in its unique capability of breeding throughout the dry season in irrigated areas.

The ecological and chromosomal differences were suggestive of reproductive isolation between forms, but they also could be explained by differential selection on inversions conferring adaptations to alternative ecological niches exploited by a polymorphic but panmictic *An. gambiae*. Laboratory crossing experiments failed to show evidence of any premating or postmating reproductive isolation between the chromosomal forms (143). On the other hand, Milligan et al. (103) found subtle differences in the concentrations of four cuticular hydrocarbons between sympatric populations of Savanna, Bamako, and Mopti in Mali. Cuticular hydrocarbon differences may be an important mechanism underlying mate recognition (127), although their role in the *An. gambiae* complex remains speculative.

Indirect genetic evidence for premating reproductive isolation between the chromosomal forms of *An. gambiae* has been sought in regions of the genome unlinked to chromosome 2 inversions. Fixed differences, or even significant frequency differences, measured at loci outside of inversions would support the hypothesis of reproductive isolation. An initial indication of fixed differences using anonymous DNA-based markers proved premature when additional field samples were tested (55, 105). However, fixed differences in the IGS and ITS of X-linked rDNA were discovered that correspond to chromosomal forms in parts of Africa (42, 54, 56, 62, 105). Taxonomic units defined by ITS- or IGS-based sequence differences were originally referred to as "Types" and "molecular forms" (42, 62). As both these rDNA definitions are coincident with the exception of the island of São Tomé (61, 62), hereafter they are referred to collectively as molecular forms for simplicity.

However, the molecular and chromosomal definitions do not always coincide. Within Mali and Burkina Faso, the M and S molecular forms correspond to the Mopti and Savanna + Bamako chromosomal forms; outside this region, chromosomal and molecular definitions diverge. For example, in Cameroon the S and M molecular forms both have the monomorphic standard karyotype associated with the Forest chromosomal form of *An. gambiae* (160). The discrepancy between chromosomal and molecular definitions can be explained if it is assumed that inversion arrangements are not directly involved in any reproductive isolating mechanism, and therefore do not actually specify different taxonomic units. Rather, arrangements are shared across units but may differ significantly in relative frequencies between units codistributed at a given locale. Thus in Burkina Faso and Mali, inversions are accidental markers of partially or completely reproductively isolated gene pools; whereas in Cameroon they are uninformative. This highlights the importance of careful and consistent use of nomenclature ("chromosomal" or "molecular" form, as appropriate) when reporting and interpreting results pertaining to *An. gambiae*.

The discovery of fixed differences in the rDNA between M and S molecular forms, and the relative paucity of M/S hybrids in areas of sympatry (42, 138, 144; T. Lehmann, M. Licht, N. Elissa, B.T.A. Maega, J.M. Chimumbwa, F.T.

Watsenga, C.S. Wondji, F. Simard & W.A. Hawley, manuscript submitted), favors the interpretation of restricted gene flow. However, because the rDNA multigene family is subject to concerted evolution, this result does not necessarily support the complete absence of gene flow. Unfortunately, but perhaps predictably, extensive DNA sequencing of additional nuclear and mtDNA loci was inconclusive despite an abundance of nucleotide variation (62, 105). The inability to detect differentiation was consistent with recent cessation of gene flow according to coalescent theory, but current gene flow could not be ruled out.

The theoretically higher level of resolution afforded by microsatellites over limited time scales, and their high throughput, makes microsatellites a tool of choice for this application. Using 21 microsatellite loci representing all chromosome arms, Lanzaro et al. (83) compared Mopti and Bamako chromosomal forms of *An. gambiae* from Mali. They found significant differentiation at all loci on chromosome 2, and fewer differences elsewhere. This was interpreted as selection against gene flow on chromosome 2 due to inversion effects rather than reproductive isolation, although the latter possibility was not ruled out. However, Lanzaro et al. (83) emphasized estimates of gene flow based on Nm, the biological interpretation of which is ambiguous in nonequilibrium situations (see Remembrance of Things Past, below). Direct inspection of the F_{ST} values shows that three of four interform comparisons of chromosome 3 loci were significantly different from zero. Subsequent microsatellite studies aimed at genome-wide patterns (but employing different sets of loci) detected differentiation at loci outside chromosome 2. Although sympatric samples of S and M from Mali were largely indistinguishable across the genome, Wang et al. (156) found two loci at the base of the X chromosome near the rDNA that showed unambiguous differences between molecular forms. Sympatric populations of S and M sampled in Cameroon and the Democratic Republic of Congo showed statistically significant levels of differentiation throughout the genome, although the highest levels of differentiation again were measured on the X chromosome near the rDNA (160; T. Lehmann, M. Licht, N. Elissa, B.T.A. Maega, J.M. Chimumbwa, F.T. Watsenga, C.S. Wondji, F. Simard & W.A. Hawley, manuscript submitted).

One of the most elegant and incisive contributions to the debate about reproductive isolation was a study aimed at direct measurement of gene flow between sympatric molecular forms in Mali. Analyzing wild-caught females and the sperm stored in their reproductive tracts, Tripet et al. (144) found strong evidence of positive assortative mating, with nearly 99% intraform matings among the 250 samples analyzed. As one of the females involved in cross-form mating was a hybrid, these data argue against complete reproductive isolation. Indeed, the frequency of adult interform hybrids (0.4%) is roughly 20-fold higher than the frequency of interspecific hybrids (0.02%) between *An. gambiae* and *An. arabiensis* in Mali (143).

Consistent with strong but incomplete barriers to gene flow between molecular forms are data on the distribution of a mutant allele of the *para* sodium channel gene that confers a phenotype called knock-down resistance (*kdr*) to pyrethroid insecticides (100). The *kdr* mutation, present in the S molecular form in Senegal,

Ivory Coast, Burkina Faso, Mali, Benin, and the Central African Republic, is absent from the M molecular form at most locales even where it is sympatric and synchronous with the S form, despite presumably uniform insecticide pressures on both taxa (23, 42, 53, 53a, 158). Moreover, where specimens were characterized both chromosomally and molecularly in Mali, the *kdr* mutation was not only exclusive to the S molecular form but also was only found within the Savanna and not the Bamako chromosomal form (53a), suggesting further gene flow barriers. On the other hand, in Benin the *kdr* mutation introgressed from the S into the M molecular form (158).

Evidence suggests that an analogous situation exists within *An. funestus*. Populations sampled from the same sites in Burkina Faso showed a significant deficit of chromosomal inversion heterokaryotypes (38). The data are consistent with the coexistence of two taxonomic units, named Kiribina and Folonzo, that differ not only in the degree of chromosomal polymorphism, but also in host preference and resting habits (38). It is possible that similar heterogeneities exist in Senegal (45, 91). Attempts to find molecular differences between Kiribina and Folonzo have been unsuccessful using mtDNA and rDNA sequences (105); efforts using microsatellites are underway.

Evidence for strong, if incomplete, barriers to gene flow is irrefutable for the molecular forms of *An. gambiae*, despite the unsuccessful search for fixed differences except in the rDNA. The issue of their taxonomic status is deferred, following a review of hybridization and introgression among "good" biological species.

Species Complexes: Hybridization and Introgression

When it comes to anophelines, it may be said that their creator had an inordinate fondness for sibling species. At least half of the important vectors of malaria belong to sibling (or cryptic) species complexes whose members are isomorphic or similar (28). Their close genetic relationship complicates the recovery of phylogenetic relationships among sibling species for two reasons. First, the species are typically recently diverged and therefore may share ancestral polymorphism. Second, postmating reproductive barriers are incomplete, in that F1 hybrid sterility usually applies to males but not to females, allowing the latter to serve as a potential bridge for introgression (gene flow between species). In high–gene flow species such as *An. gambiae* it can be difficult to distinguish the effects of these two processes.

The case for genetic introgression is best developed for the sibling species *An. gambiae* and *An. arabiensis* within the *An. gambiae* complex (119), although more examples within and outside this complex are known or suspected (141, 151; R. Wang, A. della Torre, C. Schwager, C. Blass, G. Dolo, Y.T. Touré, F.H. Collins, G.C. Lanzaro, F.C. Kafatos & L. Zheng, manuscript submitted) and others undoubtedly await discovery. Besides the sympatric species just named, the *An. gambiae* complex contains five other allopatric species, two that breed in freshwater (*An. quadriannulatus* and *An. quadriannulatus* B) and three that breed in brackish/mineral water (*An. bwambae*, *An. melas*, and *An. merus*). In the first attempt to infer phylogenetic relationships among members of the *An. gambiae* complex, Coluzzi et al. (32) applied fixed paracentric chromosomal inversions and the principle of parsimony in concluding that *An. gambiae* and *An. merus* were most closely related (sister taxa) through sharing of the sex-linked Xag inversion. To explain the conflicting distribution of inversions on chromosome 2 that were apparently shared between *An. gambiae* and *An. arabiensis*, secondary contact and genetic introgression was invoked at a time when it was much less fashionable. Coluzzi's hypothesis predicts that DNA sequences from *An. gambiae* should cluster with those of *An. merus* if derived from loci within the Xag inversion, or with those of *An. arabiensis* if derived from loci within the shared chromosome 2 inversions. These predictions were met (20, 60, 105; N.J. Besansky, J. Krzywinski, M. Kern, O. Mukabayire, D. Fontenille, Y. Touré & N'F. Sagnon, manuscript submitted). The hypothesis that it was chromosome 2 inversions rather than Xag that had been exchanged between species pairs was reinforced by laboratory crossing experiments showing that heteromorphic X chromosomes could not be introgressed, generally suggesting that inversions not already shared between species in nature could not be introgressed (43).

Given that *An. gambiae* and *An. merus* are indeed sister taxa, the evidence for broadscale genetic introgression is overwhelming. That the cytoplasmic mtDNA has crossed species boundaries (12, 13, 19) is perhaps not too surprising assuming little or no selection against mtDNA introgression (117). That nuclear loci across the genome have experienced interspecific gene flow through introgressive hybridization is more challenging to traditional species concepts. However, in addition to DNA sequence data from inside inversions 2La and 2Rb, *An. gambiae* and *An. arabiensis* sequences form elsewhere in the genome cluster or even intertwine in gene trees. This is true of nine loci to date, from all chromosomes (29; R. Wang, A. della Torre, C. Schwager, C. Blass, G. Dolo, Y.T. Touré, F.H. Collins, G.C. Lanzaro, F.C. Kafatos & L. Zheng, manuscript submitted; N.J. Besansky, J. Krzywinski, M. Kern, O. Mukabayire, D. Fontenille, Y. Touré & N'F. Sagnon, manuscript submitted). These data are compatible with "a speciation model in which species continue to exchange genes at some loci and not at others" (109, 157, 161). This "mosaic" model was developed based on observations of flies other than anophelines, but all available data from the *An. gambiae* complex are consistent with this speciation model.

Returning to the question of taxonomic status posed by the chromosomal/molecular forms within *An. gambiae*, it can now be seen in a larger perspective. Current or at least recent gene flow has occurred between "good" biological species in the *An. gambiae* complex without resulting in their fusion. The fact that gene flow also occurs between taxa within *An. gambiae* at some low level is not inconsistent with the mosaic model, as long as variation at genes directly responsible for speciation is not shared. Fixed differences have been found in the rDNA between the M and S forms; considering all molecular evidence together, there is evidence for genome-wide differentiation. However, unless speciation loci can be identified (see Prospects in the Postgenomic Era, below), this debate has no easy resolution (61). From a practical perspective, regardless of taxonomic status, it seems advisable to treat these taxa up front as independent evolutionary units within the context of any future population genetic study.

GENE FLOW WITHIN SPECIES

The amount and pattern of genetic differentiation detected within a species—its population structure—is influenced by a complex array of factors, including population sizes, dispersal rates, and historical perturbations such as population bottlenecks or expansions. If population structure can be envisioned as the geographic pattern of anastomosing lineages within a species, its connection with the traditionally separate discipline of phylogenetics becomes evident (2).

Markers and Natural Selection

Both the detection and interpretation of population structure critically depend on the class of marker and the location of marker loci in the genome. Results obtained from the same species using different types of marker or different sets of loci will not necessarily agree, particularly if based on summary statistics that encompass loci that strongly deviate from genome-wide trends (T. Lehmann, M. Licht, N. Elissa, B.T.A. Maega, J.M. Chimumbwa, F.T. Watsenga, C.S. Wondji, F. Simard & W.A. Hawley, manuscript submitted). Polytene chromosomes have revealed dramatically different levels of paracentric inversion polymorphism, from virtually nil in *A. albimanus* to high in *An. arabiensis* (78). Because recombination in inversion heterozygotes is suppressed both within and outside the breakpoints for considerable distances (as inferred by asynapsis; A. della Torre, personal communication), and because inversions may be subject to selection (32, 143), neither they nor DNA-based markers located within or nearby can be assumed to behave neutrally. Protein electrophoresis also has revealed sharp contrasts in the level of variation among loci within species, related to their different metabolic roles and the effects of selection. Far higher rates of evolution are typical of DNA-based markers such as microsatellites [$\sim 10^{-5}$ (128)] as compared to chromosomal inversions or enzymes (25, 84). As is the case for chromosomal inversions and allozymes, DNA markers can exhibit dramatic variations in level of polymorphism within populations. This is due to locus-specific rates of mutation and to physical location with respect to genes or inversions subject to selection (101, 111).

Population Size

The basic unit of population structure, N_e (roughly, the number of reproductively active adults), determines the rate of genetic drift and is strongly influenced by the smallest population size between inter-generational fluctuations. It is also influenced by the rate of gene flow between populations. For *An. gambiae* and *An.*

arabiensis, several independent estimates were made using both direct (ecological) and indirect (genetic) approaches [see (48, 138) for discussion and contrasting views on reconciling these approaches]. Indirect estimates of N_e based on temporal variation of microsatellites or chromosomal inversions were on the order of 10^3 (87, 133, 139). All estimates of N_e for both species were inconsistent with severe population bottlenecks predicted across the dry season.

If bottlenecks are ruled out, how can a minimum population size of $\sim 10^3$ be reconciled with the difficulty of collecting these species during the dry season? Attempts to sample aestivating adults and to discover embryo dormancy in moist soil (24, 104, 110) have met with limited success. Instead, evidence favors an alternative hypothesis of relatively isolated dry season refugia populations that breed continuously throughout the year and re-invade during the wet season (24, 50, 104). Although flight range of anophelines is traditionally thought to be limited (perhaps 3 km), there are records of 19–32-km flights in *An. albimanus*, and where there are no environmental features to deter it, such long-distance dispersal would not be surprising, particularly in arid zones (24, 64, 130). In fact, Lehmann et al. (87) coined the phrase "diffused deme" to describe a view of *An. gambiae* population structure in which large populations are "maintained by extensive mobility of adults, resulting in a low-density deme which is spread over a large geographical area."

Geographic Population Structure

In East Africa west of the Rift Valley, chromosomal inversion studies revealed reduced levels of polymorphism and apparent panmixia (114, 115). Although McLain et al. (102) concluded to the contrary, reinterpretation of the rDNA-based results following removal of an apparent outlier site (Kisian) also suggests little or no detectable population structure in Kenya west of the Rift. By contrast, rDNA differentiation across the Rift Valley was maximal (102). Microsatellite and mtDNA studies reinforced the conclusion of no apparent microgeographic structure in western Kenya while identifying the Rift Valley as a barrier to gene flow for *An. gambiae* (75, 76, 85, 86, 88; T. Lehmann, M. Licht, N. Elissa, B.T.A. Maega, J.M. Chimumbwa, F.T. Watsenga, C.S. Wondji, F. Simard & W.A. Hawley, manuscript submitted). The role of the Rift Valley as a barrier to gene flow in *An. arabiensis* is less certain (76, 115); unlike *An. gambiae*, *An. arabiensis* occurs throughout this series of high, dry, and relatively uninhabited plateaus (104a). Significant chromosomal and microsatellite differentiation across the Rift Valley have been reported in *An. funestus* (74; O.P. Braginets, N. Minakawa, C.N. Mbogo, & G. Yan, unpublished data), a species that has been recorded in the Rift Valley, albeit at relatively low densities compared to *An. arabiensis* (104a).

Is the barrier to gene flow posed by the Rift Valley due to the relatively large distance ($\sim$700 km) spanning the western and coastal Kenyan sites? A number of studies using allozymes, microsatellites, and mtDNA have suggested that distance alone plays a relatively minor role in shaping population structure in *An. gambiae*

(12, 21, 76, 89, 111). Assuming mutation-migration-drift equilibrium in many subpopulations, estimates of gene flow (Nm) can be derived from F_{ST} (134), and estimates from the studies cited above [except (21)] were interpreted as exceeding the threshold of genetic differentiation by drift. However, evidence suggests that neither *An. arabiensis* nor *An. gambiae* have achieved equilibrium (47), casting doubts on the biological interpretation of Nm values in these species. The mtDNA-based F_{ST} values, however low, were significantly different from zero across Africa (12), and microsatellite-based F_{ST} values increased by distance in nine *An. gambiae* populations spanning western Kenya and Senegal, with a significant though shallow slope (T. Lehmann, M. Licht, N. Elissa, B.T.A. Maega, J.M. Chimumbwa, F.T. Watsenga, C.S. Wondji, F. Simard & W.A. Hawley, manuscript submitted). "Genetic differentiation for microsatellite DNA loci is consistent with traditional models of isolation by distance" species-wide in *An. gambiae* (21), a conclusion that applies equally to *An. arabiensis* from mainland Africa (46, 49, 112, 132).

Aside from *An. gambiae* and sibling species, few other anophelines have received much attention. The population structure of a handful of neotropical anopheline species (including *An. darlingi*, *An. nuneztovari*, *An. albimanus*, *An. pseudopunctipennis*) has been examined using allozymes, mtDNA, rDNA, or randomly amplified polymorphic DNA (RAPDs) in various combinations (92); microsatellite data have not yet been reported. Significant population structure was found within each species, although its depth varied considerably, being most shallow in the widespread vector *An. albimanus* and deepest in *An. nuneztovari*, a vector whose populations are disjunct in Venezuela/Colombia and the Amazon Basin (92). In most cases, genetic and geographic distances were significantly correlated, suggesting isolation by distance. The exception, found in *An. nuneztovari*, could be an artifact of pooling genetically independent lineages (33, 92).

Despite their extensive distributions over heterogeneous environments, there is no evidence for cryptic species within either *An. darlingi* or *An. albimanus* (34, 40, 41, 97, 94). However, in *An. pseudopunctipennis*, evidence from crossing experiments, cytogenetics, and allozymes suggests barriers to gene flow between northern, southern, and Caribbean population samples (26, 51, 52, 96). The absence of present-day geographic barriers suggests the possibility of more ancient vicariance phenomena involving sea level changes. This hypothesis predicts that a similar pattern should be found in species whose ranges coincide with *An. pseudopunctipennis*, notably *An. albimanus* and *An. darlingi*. In this regard, the observation by De Merida et al. (41) of abrupt changes in *An. albimanus* mtDNA haplotype frequencies between northern and southern samples merits follow-up study. Moreover, careful inspection of Manguin et al. (97) reveals a pattern in *An. darlingi* consistent with the vicariance hypothesis. Allozyme data indicated that the population sampled from Belize was most distant from the South American population samples. Stronger support was derived from rDNA sequence data that uncovered a fixed insertion/deletion difference between specimens from Belize and all other specimens (97). Application of microsatellite markers to *An. darlingi* populations with careful attention to sampling across Central America in addition to populations to the north and south will confirm and better define this phenomenon.

Another example of concordant patterns across taxa may be found in Thailand. Microsatellite data from *An. maculatus* populations in Thailand indicated restricted gene flow between northern and southern populations, possibly due to intervening mountain ranges (123). It is suggestive that in *An. dirus* C, differentiation at mtDNA and microsatellite loci was much greater across this same region, but other explanations for this pattern cannot be ruled out without additional sampling (151, 153).

Remembrance of Things Past

Many of the classical tools available for the analysis of population structure depend upon the simplifying assumption of mutation-migration-drift equilibrium. Unfortunately, for many species, particularly synanthropic malaria vectors, this assumption appears to have been violated due to a population expansion that may have coincided with the expansion of human populations (12, 47, 48, 87, 112, 153). Even for less anthropophilic anophelines, it is clear that recent deforestation, while devastating to some species, actually creates new habitat for others (e.g., *An. darlingi* in Brazil) and has allowed dramatic population expansions (150) that greatly reduce the rate of lineage sorting. In practical terms, this means that traditional F_{ST}-based estimates of "current" gene flow will be inflated by "historical" gene flow, registered by the sharing of alleles through recent common ancestry. Thus, a low level of differentiation and a correspondingly high level of estimated gene flow do not necessarily imply meaningful levels of current exchange between populations.

In cases where microsatellites are used to estimate genetic differentiation, their high levels of polymorphism compound the problem. Balloux & Lugon-Moulin (3) present a hypothetical example in which two highly diverse subpopulations that do not share any alleles (i.e., they are maximally differentiated) nevertheless have an F_{ST} value of only 0.053, a result normally interpreted as indicating considerable gene flow in an empirical study.

This is not an untenable situation. Templeton [(140) and refs therein; see also (2, pp.79–90)] proposed a new approach called nested clade phylogeographic analysis. Evolutionary relationships among alleles or haplotypes, as inferred from a haplotype tree, can be used to define a nested series of branches, called clades. These are used to examine the geographic as well as historical pattern of genetic variation in a nested statistical framework. The analysis begins by testing the null hypothesis of no association between geography and the haplotype tree. Given that the sampling has been adequate, "such nested analyses can discriminate between phylogeographical associations due to recurrent but restricted gene flow vs. historical events operating at the population level (e.g., past fragmentation, colonization, or range expansion events)" (140). Programs to perform this analysis (TCS1.12, GEODIS2.0) are available on the website of K. Crandall (http://zoology.byu.edu/crandall_lab/programs.htm).

Apparent Conflict or Real Consensus?

Apparent conflicts between studies are inevitable and potentially informative. However, to arrive at biologically meaningful conclusions, certain aspects of the study design over which the investigator has control should be standardized. A problem encountered in the study of *An. gambiae* population genetics is that different authors use nonoverlapping—or only partially overlapping—sets of loci, which introduces uncertainty into comparisons among different regions in Africa. It is unfortunate that the suggestion to employ a large universal set of loci for studies of *An. gambiae* (155) has been so widely ignored. For other anopheline species yet to be studied so intensively, similar standards should be adopted early on.

A subtler variable concerns the definition of a "population sample" and how it is collected in a given study. Not infrequently, the terms "sample" and "population" are used interchangeably in reporting results, which can be misleading; the population is the biological reality, the sample is the investigator's window on that reality. Was the sample collected from a single (compact) village on a single day, a series of collections over time, or a series of collections over space? Was the adult population sampled indoors, outdoors, or both [given that there can be genetic differences between indoor- and outdoor-resting components (38)]? Perhaps larvae were sampled, which may represent siblings, or which may show important differences in genetic variation relative to adults depending upon as yet poorly understood patterns of dispersal and/or selection. It is not always possible to control these variables within the confines of a given study, but an attempt should be made to test whether different sampling strategies yield statistically indistinguishable results before pooling or comparing across studies (3).

Even if a canonical set of loci were employed and sampling was standardized, comparison across studies might still be problematic if different analytical approaches were adopted, as is common. One possible solution to this problem would be the adoption by the scientific community of a protocol similar to the one already implemented for DNA sequence data, in which the raw data in a standard format are deposited into a public database as a condition of publication (e.g., AnoDB, konops.imbb.forth.gr/AnoDB/). In the absence of this solution, we have an unresolved debate over the appropriateness of different distance measures (3, 156), with the consequence that different authors may apply different measures to different effect. Because microsatellite datasets can be especially large, space considerations require the summary of much information in the printed version of publications, which may present only mean distance measures across all loci and all population samples. This practice can mask effects specific to particular samples and/or to particular loci. It is often assumed (and not always tested) that either of these effects are "averaged out," given that enough loci or enough samples are used; this assumption is not necessarily justified (3). Not only should the contribution of individual localities to a significant result of isolation by distance be investigated (17), but locus-specific contributions to summary statistics also should be routinely investigated (3; T. Lehmann, M. Licht, N. Elissa, B.T.A. Maega, J.M. Chimumbwa, F.T. Watsenga, C.S. Wondji, F. Simard & W.A. Hawley, manuscript submitted).

The final issue concerns the accuracy of allele sizing across different laboratories. Obviously, discrepancies can result in significant differences in estimates of differentiation, gene flow, and population size. Studies have demonstrated that there can be strong discrepancies in allele sizing among automated slab gel, automated capillary, and manual methods (44, 69), possibly due to compression or stretching of allele sizes on the automated sequencer. Unfortunately, the magnitude and direction of size calling errors appear to be locus-specific and follow no size-related trend, precluding the use of a normalizing factor (44). The problem of inaccurate allele sizing emphasizes the necessity for reference standard DNA genotypes that should be shared among labs (44, 155).

PROSPECTS IN THE POSTGENOMIC ERA

The publication of the complete *An. gambiae* genome (The Anopheles Genome Sequencing Consortium; ftp://ftp.ncbi.nih.gov/genbank/genomes/Anopheles_gam biae/) will undoubtedly exert a vast influence on systematic studies in mosquitoes, and Diptera in general. Here we preview some major impacts and opportunities that can be anticipated.

Phylogenetic and Population Genetic Markers

In their recent review of insect molecular systematics, Caterino et al. (22) advocate the use of a common set of phylogenetic markers in all insects to allow datasets from independent studies of different taxa to be combined into supertrees. Although the idea has merit, identifying universal markers informative at various divergence levels has been problematic across a group as old and diverse as insects, particularly in the absence of comparative genome data. Indeed, two of the proposed markers, mtDNA *COI* and nuclear *18S* rDNA, provided limited resolution within *Anopheles* (104b, 125). Prior studies have shown that phylogenetic inference in anophelines and other dipteran groups can be difficult, more so because of an insufficient number of reliable and informative markers to resolve deeper relationships. The availability of the genome sequence of a higher fly, *Drosophila melanogaster* (1), and a lower fly, *An. gambiae*, opens enormous possibilities to explore novel phylogenetic markers in a directed manner, with potential application in all groups of Diptera, if not all insects. Single-copy nuclear genes may be the most promising candidates; comparison of such genes from *Drosophila* and *Anopheles* may reveal phylogenetically informative variable regions flanked by more conserved regions allowing for PCR primer design.

At much shallower taxonomic depths, the *An. gambiae* genome sequence will provide a wealth of new markers for population genetic inference within this species and for phylogenetic inference at the infraspecific and species complex level. Obvious candidates include SNPs and Y chromosome sequences. Other promising candidates are short interspersed transposable elements such as MITES and SINES (145), whose polymorphic pattern of presence/absence at specific loci across the genome may represent recent integration (or excision) events exclusive to certain lineages.

Evolutionary Genomics

As alluded to previously, rearrangements of gene order may play an important role in anopheline speciation. To what extent has gene order (and content) been conserved between *An. gambiae* and *D. melanogaster* or other anophelines? Three studies have begun the task of comparing chromosomal location of orthologous genes between *An. gambiae* (subgenus *Cellia*) and *An. funestus* (subgenus *Cellia*), *An. albimanus* (subgenus *Nyssorhynchus*), or *D. melanogaster*. Initial results revealed rearrangements between chromosomes by whole-arm translocations, and within chromosome arms by paracentric inversions in each pairwise comparison (16, 37; I. Sharakhov, A. Serazin, O. Grushko, A. Dana, N. Lobo, R. Westerman, P. San-Miguel, J. Romero-Severson, C. Costantini, N'F. Sagnon, F.H. Collins & N.J. Besansky, manuscript submitted), the extent of conservation negatively correlating with phylogenetic distance. It is possible, although perhaps not yet practical, that differences in gene order could be used for phylogenetic inference among groups of anophelines as envisioned by the late C.A. Green (65, 66).

An. gambiae as a Model System

As a model organism *par excellence*, *D. melanogaster* illuminates biological problems in genetics, development, molecular biology, and molecular evolution (118), but is wanting when it comes to the ectoparasitic lifestyle of mosquitoes and the transmission of human malaria parasites by anophelines. With the complete genome sequence and a system for germline transformation (68), *An. gambiae* is a viable model system in which evolutionary adaptations associated with parasitism and disease transmission can be studied. These include host-seeking behavior and host selection, in particular the preference for human blood (anthropophily) and the tendency toward domesticity, as well as blood-feeding physiology, blood meal digestion, and susceptibility to human malaria parasites. It is premature to predict whether lessons learned from this model will apply broadly to other anopheline species that may have evolved anthropophilic tendencies independently, or have been evolutionarily isolated for at least 100 million years.

Epidemiological Implications

As discussed above, a problem of both practical and academic importance known for nearly a century is that within a single cryptic species complex are members whose anthropophilic, catholic, or zoophilic behavior makes them major, minor, or nonvectors of malaria. These are typically quite closely related and recently diverged species, which often cannot be distinguished morphologically or even molecularly without great difficulty. What underlies differences in their vectorial capacity, such as attraction to alternative hosts? What underlies ecological differences, such as the saltwater tolerance of *An. merus*, the ability of the *An. gambiae* M molecular form to breed during the dry season, or the ability of the *An. gambiae* Bamako chromosomal form to breed in flowing water of the Niger River? What underlies premating isolation among species and forms, and what drives such rapid diversification? Can the answers to these questions provide a window into the speciation process as well as a set of practical tools to differentiate the vector from nonvector species? To begin to address these questions, what is needed is a tool for surveying genome-wide patterns of gene expression underlying particular phenotypic traits. The genome project makes possible the construction of such a tool: microarrays of cDNAs or gene-specific oligonucleotides that can be used to delineate physiological pathways, study mechanisms of adaptation and divergence, and characterize genetic differences among isolates and closely related species (63). It is not at the ultimate level of reduction, the genomic DNA with its endless string of nucleotides "as featureless as a plate of tofu" [(18) as cited by (146)], where we are likely to find the answers, but at the transcriptional level, among interconnected regulatory pathways best revealed with a global perspective.

ACKNOWLEDGMENTS

T. Lehmann, G. Yan, G. Lanzaro, C. Taylor, A. della Torre, G. Caccone, R. Wang, F. Simard, S. Manguin, and W. Black generously provided prepublications and personal communications. Thanks to T. Lehmann, M. Donnelly, Y. Linton, R. Harbach, S. Manguin, C. Louis, J. Powell, and F. Collins for constructive comments on this manuscript. Research support to our laboratory from the NIH (AI44003, AI48842) and the State of Indiana (CRTF 042700-0207) is gratefully acknowledged.

The *Annual Review of Entomology* is online at http://ento.annualreviews.org

LITERATURE CITED

1. Adams MD, Celniker SE, Holt RA, Evans CA, Gocayne JD, et al. 2000. The genome sequence of *Drosophila melanogaster*. *Science* 287:2185–95
2. Avise JC. 2000. *Phylogeography: The History and Formation of Species*. Cambridge, MA: Harvard Univ. Press
3. Balloux F, Lugon-Moulin N. 2002. The estimation of population differentiation with microsatellite markers. *Mol. Ecol.* 11:155–65
4. Beebe NW, Bakote'e B, Ellis JT, Cooper RD. 2000. Differential ecology of *Anopheles punctulatus* and three members of the *Anopheles farauti* complex of mosquitoes on Guadalcanal, Solomon Islands, identified by PCR-RFLP analysis. *Med. Vet. Entomol.* 14:308–12
5. Beebe NW, Cooper RD. 2000. Systematics of malaria vectors with particular reference to the *Anopheles punctulatus* group. *Int. J. Parasitol.* 30:1–17
6. Beebe NW, Cooper RD, Morrison DA, Ellis JT. 2000. A phylogenetic study of the *Anopheles punctulatus* group of malaria vectors comparing rDNA sequence

alignments derived from the mitochondrial and nuclear small ribosomal subunits. *Mol. Phylogenet. Evol.* 17:430–36

7. Beebe NW, Foley DH, Cooper RD, Bryan JH, Saul A. 1996. DNA probes for the *Anopheles punctulatus* complex. *Am. J. Trop. Med. Hyg.* 54:395–98
8. Beebe NW, Maung J, van den Hurk AF, Ellis JT, Cooper RD. 2001. Ribosomal DNA spacer genotypes of the *Anopheles bancroftii* group (Diptera: Culicidae) from Australia and Papua New Guinea. *Insect Mol. Biol.* 10:407–13
9. Beebe NW, Saul A. 1995. Discrimination of all members of the *Anopheles punctulatus* complex by polymerase chain reaction-restriction fragment length polymorphism analysis. *Am. J. Trop. Med. Hyg.* 53:478–81
10. Belkin JN. 1962. *Mosquitoes of the South Pacific*, Vol. 1. Berkeley: Univ. Calif. Press
11. Besansky NJ, Fahey GT. 1997. Utility of the *white* gene in estimating phylogenetic relationships among mosquitoes (Diptera: Culicidae). *Mol. Biol. Evol.* 14: 442–54
12. Besansky NJ, Lehmann T, Fahey GT, Fontenille D, Braack LE, et al. 1997. Patterns of mitochondrial variation within and between African malaria vectors, *Anopheles gambiae* and *An. arabiensis*, suggest extensive gene flow. *Genetics* 147:1817–28
13. Besansky NJ, Powell JR, Caccone A, Hamm DM, Scott JA, Collins FH. 1994. Molecular phylogeny of the *Anopheles gambiae* complex suggests genetic introgression between principal malaria vectors. *Proc. Natl. Acad. Sci. USA* 91: 6885–88
14. Black WC 4th, Munstermann LE. 2002. Molecular taxonomy of arthropod vectors. In *The Biology of Disease Vectors, 2nd Edition*, ed. WC Marquardt, WC Black, JR Freir, B Kondratieff, HJ Hagedorn, et al. Harcourt Academic Press. In press
15. Black WC 4th, Lanzaro GC. 2001. Distribution of genetic variation among chromosomal forms of *Anopheles gambiae s.s*: introgressive hybridization, adaptive inversions, or recent reproductive isolation? *Insect Mol. Biol.* 10:3–7
16. Bolshakov VN, Topalis P, Blass C, Kokoza E, della Torre A, et al. 2002. A comparative genomic analysis of two distant diptera, the fruit fly, *Drosophila melanogaster*, and the malaria mosquito, *Anopheles gambiae*. *Genome Res.* 12:57–66
17. Bossart JL, Prowell DP. 1998. Genetic estimates of population structure and gene flow: limitations, lessons and new directions. *Trends Ecol. Evol.* 13:202–6
18. Bray D. 2001. Reasoning for results. *Nature* 412:863
19. Caccone A, Garcia BA, Powell JR. 1996. Evolution of the mitochondrial DNA control region in the *Anopheles gambiae* complex. *Insect Mol. Biol.* 5:51–59
20. Caccone A, Min GS, Powell JR. 1998. Multiple origins of cytologically identical chromosome inversions in the *Anopheles gambiae* complex. *Genetics* 150: 807–14
21. Carnahan J, Zheng L, Taylor CE, Toure YT, Norris DE, et al. 2002. Genetic differentiation of *Anopheles gambiae s.s.* populations in Mali, West Africa using microsatellite loci. *J. Hered.* In press
22. Caterino MS, Cho S, Sperling FA. 2000. The current state of insect molecular systematics: a thriving Tower of Babel. *Annu. Rev. Entomol.* 45:1–54
23. Chandre F, Manguin S, Brengues C, Dossou Yovo J, Darriet F, et al. 1999. Current distribution of a pyrethroid resistance gene (*kdr*) in *Anopheles gambiae* complex from west Africa and further evidence for reproductive isolation of the Mopti form. *Parassitologia* 41:319–22
24. Charlwood JD, Vij R, Billingsley PF. 2000. Dry season refugia of malaria-transmitting mosquitoes in a dry

savannah zone of east Africa. *Am. J. Trop. Med. Hyg.* 62:726–32
25. Cianchi R, Urbanelli S, Villani F, Sabatini A, Bullini L. 1985. Electrophoretic studies in mosquitoes: recent advances. *Parassitologia* 27:157–67
26. Coetzee M, Estrada-Franco JG, Wunderlich CA, Hunt RH. 1999. Cytogenetic evidence for a species complex within *Anopheles pseudopunctipennis* Theobald (Diptera: Culicidae). *Am. J. Trop. Med. Hyg.* 60:649–53
27. Collins FH, Kamau L, Ranson HA, Vulule JM. 2000. Molecular entomology and prospects for malaria control. *Bull. WHO* 78:1412–23
28. Collins FH, Paskewitz SM. 1996. A review of the use of ribosomal DNA (rDNA) to differentiate among cryptic *Anopheles* species. *Insect Mol. Biol.* 5:1–9
29. Collins FH, Paskewitz SM, Finnerty V. 1989. Ribosomal RNA genes of the *Anopheles gambiae* species complex. In *Advances in Disease Vector Research*, 6:1–28. New York: Springer
30. Coluzzi M. 1982. Spatial distribution of chromosomal inversions and speciation in anopheline mosquitoes. In *Mechanisms of Speciation*, pp. 143–53. New York: Liss
31. Coluzzi M, Petrarca V, DiDeco MA. 1985. Chromosomal inversion intergradation and incipient speciation in *Anopheles gambiae*. *Boll. Zool.* 52:45–63
32. Coluzzi M, Sabatini A, Petrarca V, Di Deco MA. 1979. Chromosomal differentiation and adaptation to human environments in the *Anopheles gambiae* complex. *Trans. R. Soc. Trop. Med. Hyg.* 73:483–97
33. Conn JE, Mitchell SE, Cockburn AF. 1998. Mitochondrial DNA analysis of the neotropical malaria vector *Anopheles nuneztovari*. *Genome* 41:313–27
34. Conn JE, Rosa-Freitas MG, Luz SL, Momen H. 1999. Molecular population genetics of the primary neotropical malaria vector *Anopheles darlingi* using mtDNA. *J. Am. Mosq. Control. Assoc.* 15:468–74
35. Cooper RD, Waterson DG, Bangs MJ, Beebe NW. 2000. Rediscovery of *Anopheles (Cellia) clowi* (Diptera: Culicidae), a rarely recorded member of the *Anopheles punctulatus* group. *J. Med. Entomol.* 37:840–45
36. Cooper RD, Waterson DGE, Frances SP, Beebe NW, Sweeney AW. 2002. Speciation and distribution of the members of the *Anopheles punctulatus* (Diptera: Culicidae) group in Papua New Guinea. *J. Med. Entomol.* 39:16–27
37. Cornel AJ, Collins FH. 2000. Maintenance of chromosome arm integrity between two *Anopheles* mosquito subgenera. *J. Hered.* 91:364–70
38. Costantini C, Sagnon NF, Ilboudo-Sanogo E, Coluzzi M, Boccolini D. 1999. Chromosomal and bionomic heterogeneities suggest incipient speciation in *Anopheles funestus* from Burkina Faso. *Parassitologia* 41:595–611
39. Crampton JM, Hill SM. 1997. Generation and use of species-specific DNA probes for insect vector identification. In *The Molecular Biology of Insect Disease Vectors: A Methods Manual*, ed. JM Crampton, CB Beard, C Louis. pp. 384–98. London: Chapman & Hall
40. De Merida AM, De Mata MP, Molina E, Porter CH, Black WC 4th. 1995. Variation in ribosomal DNA intergenic spacers among populations of *Anopheles albimanus* in South and Central America. *Am. J. Trop. Med. Hyg.* 53:469–77
41. De Merida AM, Palmieri M, Yurrita M, Molina A, Molina E, Black WC 4th. 1999. Mitochondrial DNA variation among *Anopheles albimanus* populations. *Am. J. Trop. Med. Hyg.* 61:230–39
42. della Torre A, Fanello C, Akogbeto M, Dossou-Yovo J, Favia G, et al. 2001. Molecular evidence of incipient speciation within *Anopheles gambiae s.s.* in West Africa. *Insect Mol. Biol.* 10:9–18

43. della Torre A, Merzagora L, Powell JR, Coluzzi M. 1997. Selective introgression of paracentric inversions between two sibling species of the *Anopheles gambiae* complex. *Genetics* 146:239–44
44. Delmotte F, Leterme N, Simon JC. 2001. Microsatellite allele sizing: difference between automated capillary electrophoresis and manual technique. *Biotechniques* 31:810–18
45. Dia I, Lochouarn L, Boccolini D, Costantini C, Fontenille D. 2000. Spatial and temporal variations of the chromosomal inversion polymorphism of *Anopheles funestus* in Senegal. *Parasite* 7:179–84
46. Donnelly MJ, Cuamba N, Charlwood JD, Collins FH, Townson H. 1999. Population structure in the malaria vector, *Anopheles arabiensis* Patton, in East Africa. *Heredity* 83:408–17
47. Donnelly MJ, Licht MC, Lehmann T. 2001. Evidence for recent population expansion in the evolutionary history of the malaria vectors *Anopheles arabiensis* and *Anopheles gambiae*. *Mol. Biol. Evol.* 18:1353–64
48. Donnelly MJ, Simard F, Lehmann T. 2002. Evolutionary studies of malaria vectors. *Trends Parasitol.* 18:75–80
49. Donnelly MJ, Townson H. 2000. Evidence for extensive genetic differentiation among populations of the malaria vector *Anopheles arabiensis* in Eastern Africa. *Insect Mol. Biol.* 9:357–67
50. Dukeen MYH, Omer SM. 1986. Ecology of the malaria vector *Anopheles arabiensis* Patton (Diptera: Culicidae) by the Nile in northern Sudan. *Bull. Entomol. Res.* 76:451–67
51. Estrada-Franco JG, Lanzaro GC, Ma MC, Walker-Abbey A, Romans P, et al. 1993. Characterization of *Anopheles pseudopunctipennis sensu lato* from three countries of neotropical America from variation in allozymes and ribosomal DNA. *Am. J. Trop. Med. Hyg.* 49: 735–45
52. Estrada-Franco JG, Ma MC, Gwadz RW, Sakai R, Lanzaro GC, et al. 1993. Evidence through crossmating experiments of a species complex in *Anopheles pseudopunctipennis sensu lato*: a primary malaria vector of the American continent. *Am. J. Trop. Med. Hyg.* 49:746–55
53. Fanello C, Akogbeto M, della Torre A. 2000. Distribution of the pyrethroid knockdown resistance gene (*kdr*) in *Anopheles gambiae s.l.* from Benin. *Trans. R. Soc. Trop. Med. Hyg.* 94:132

53a. Fanello C, Petrarca V, della Torre A, Santolamazza F, Alloueche A, et al. 2002. The pyrethroid *knock-down resistance* gene in the *Anopheles gambiae* complex in Mali and further evidence of reproductive isolation within *An. gambiae* s.s. *Insect Mol. Biol.* In press

54. Favia G, della Torre A, Bagayoko M, Lanfrancotti A, Sagnon N, et al. 1997. Molecular identification of sympatric chromosomal forms of *Anopheles gambiae* and further evidence of their reproductive isolation. *Insect Mol. Biol.* 6: 377–83
55. Favia G, Dimopoulos G, della Torre A, Toure YT, Coluzzi M, Louis C. 1994. Polymorphisms detected by random PCR distinguish between different chromosomal forms of *Anopheles gambiae*. *Proc. Natl. Acad. Sci. USA* 91: 10315–19
56. Favia G, Lanfrancotti A, Spanos L, Siden-Kiamos I, Louis C. 2001. Molecular characterization of ribosomal DNA polymorphisms discriminating among chromosomal forms of *Anopheles gambiae s.s. Insect Mol. Biol.* 10:19–23
57. Foley DH, Bryan JH. 2000. Shared salinity tolerance invalidates a test for the malaria vector *Anopheles farauti s.s* on Guadalcanal, Solomon Islands. *Med. Vet. Entomol.* 14:450–52
58. Foley DH, Bryan JH, Yeates D, Saul A. 1998. Evolution and systematics of *Anopheles*: insights from a molecular phylogeny of Australasian mosquitoes. *Mol. Phylogenet. Evol.* 9:262–75

59. Foley DH, Cooper RD, Bryan JH. 1995. A new species within the *Anopheles punctulatus* complex in Western Province, Papua New Guinea. *J. Am. Mosq. Control. Assoc.* 11:122–27
60. Garcia BA, Caccone A, Mathiopoulos KD, Powell JR. 1996. Inversion monophyly in African anopheline malaria vectors. *Genetics* 143:1313–20
61. Gentile G, della Torre A, Maegga B, Powell JR, Caccone A. 2002. Genetic differentiation in the African malaria vector, *Anopheles gambiae s.s.*, and the problem of taxonomic status. *Genetics* 161:1561–78
62. Gentile G, Slotman M, Ketmaier V, Powell JR, Caccone A. 2001. Attempts to molecularly distinguish cryptic taxa in *Anopheles gambiae s.s. Insect Mol. Biol.* 10:25–32
63. Gibson G. 2002. Microarrays in ecology and evolution: a preview. *Mol. Ecol.* 11:17–24
64. Gillies MT, De Meillon B. 1968. *The Anophelinae of Africa South of the Sahara*. Johannesburg: S. Afr. Inst. Med. Res. 2nd ed.
65. Green CA. 1982. Cladistic analysis of mosquito chromosome data (*Anopheles* (Cellia) Myzomyia). *J. Hered.* 73:2–11
66. Green CA, Harrison BA, Klein TA, Baimai V. 1985. Cladistic analysis of polytene chromosome rearrangements in anopheline mosquitoes, subgenus *Cellia*, series Neocellia. *Can. J. Genet. Cytol.* 27:123–33
67. Green CA, Munstermann LE, Tan SG, Panyim S, Baimai V. 1992. Population genetic evidence for species A, B, C and D of the *Anopheles dirus* complex in Thailand and enzyme electromorphs for their identification. *Med. Vet. Entomol.* 6:29–36
68. Grossman GL, Rafferty CS, Clayton JR, Stevens TK, Mukabayire O, Benedict MQ. 2001. Germline transformation of the malaria vector, *Anopheles gambiae*, with the piggyBac transposable element. *Insect Mol. Biol.* 10:597–604
69. Haberl M, Tautz D. 1999. Comparative allele sizing can produce inaccurate allele size differences for microsatellites. *Mol. Ecol.* 8:1347–49
70. Hackett BJ, Gimnig J, Guelbeogo W, Costantini C, Koekemoer LL, et al. 2000. Ribosomal DNA internal transcribed spacer (ITS2) sequences differentiate *Anopheles funestus* and *An. rivulorum*, and uncover a cryptic taxon. *Insect Mol. Biol.* 9:369–74
71. Harbach RE. 1994. Review of the internal classification of the genus *Anopheles* (Diptera: Culicidae): the foundation for comparative systematics and phylogenetic research. *Bull. Entomol. Res.* 84:331–42
72. Harbach RE. 1998. Phylogeny and classification of the Culicidae (Diptera). *Syst. Entomol.* 23:327–70
73. Hunt RH, Coetzee M, Fettene M. 1998. The *Anopheles gambiae* complex: a new species from Ethiopia. *Trans. R. Soc. Trop. Med. Hyg.* 92:231–35
74. Kamau L, Hunt R, Coetzee M. 2002. Analysis of the population structure of *Anopheles funestus* (Diptera: Culicidae) from western and coastal Kenya using paracentric chromosomal inversion frequencies. *J. Med. Entomol.* 39:78–83
75. Kamau L, Lehmann T, Hawley WA, Orago AS, Collins FH. 1998. Microgeographic genetic differentiation of *Anopheles gambiae* mosquitoes from Asembo Bay, western Kenya: a comparison with Kilifi in coastal Kenya. *Am. J. Trop. Med. Hyg.* 58:64–69
76. Kamau L, Mukabana WR, Hawley WA, Lehmann T, Irungu LW, et al. 1999. Analysis of genetic variability in *Anopheles arabiensis* and *Anopheles gambiae* using microsatellite loci. *Insect Mol. Biol.* 8:287–97
77. Kengne P, Trung HD, Baimai V, Coosemans M, Manguin S. 2001. A multiplex PCR-based method derived from

random amplified polymorphic DNA (RAPD) markers for the identification of species of the *Anopheles minimus* group in Southeast Asia. *Insect Mol. Biol.* 10: 427–35

78. Kitzmiller JB. 1976. Genetics, cytogenetics, and evolution of mosquitoes. *Adv. Genet.* 18:315–433
79. Koekemoer L, Weeto MM, Kamau L, Hunt RH, Coetzee M. 2002. A cocktail polymerase chain reaction (PCR) assay to identify members of the *Anopheles funestus* (Diptera: Culicidae) group. *Am. J. Trop. Med. Hyg.* 66:804–11
80. Koekemoer LL, Lochouarn L, Hunt RH, Coetzee M. 1999. Single-strand conformation polymorphism analysis for identification of four members of the *Anopheles funestus* (Diptera: Culicidae) group. *J. Med. Entomol.* 36:125–30
81. Krzywinski J, Wilkerson RC, Besansky NJ. 2001. Evolution of mitochondrial and ribosomal gene sequences in Anophelinae (Diptera: Culicidae): implications for phylogeny reconstruction. *Mol. Phylogenet. Evol.* 18:479–87
82. Krzywinski J, Wilkerson RC, Besansky NJ. 2001. Toward understanding Anophelinae (Diptera, Culicidae) phylogeny: insights from nuclear single-copy genes and the weight of evidence. *Syst. Biol.* 50:540–56
83. Lanzaro GC, Toure YT, Carnahan J, Zheng L, Dolo G, et al. 1998. Complexities in the genetic structure of *Anopheles gambiae* populations in west Africa as revealed by microsatellite DNA analysis. *Proc. Natl. Acad. Sci. USA* 95:14260–65
84. Lanzaro GC, Zheng L, Toure YT, Traore SF, Kafatos FC, Vernick KD. 1995. Microsatellite DNA and isozyme variability in a west African population of *Anopheles gambiae*. *Insect Mol. Biol.* 4:105–12
85. Lehmann T, Besansky NJ, Hawley WA, Fahey TG, Kamau L, Collins FH. 1997. Microgeographic structure of *Anopheles gambiae* in western Kenya based on mtDNA and microsatellite loci. *Mol. Ecol.* 6:243–53
86. Lehmann T, Blackston CR, Besansky NJ, Escalante AA, Collins FH, Hawley WA. 2000. The Rift Valley complex as a barrier to gene flow for *Anopheles gambiae* in Kenya: the mtDNA perspective. *J. Hered.* 91:165–68
87. Lehmann T, Hawley WA, Grebert H, Collins FH. 1998. The effective population size of *Anopheles gambiae* in Kenya: implications for population structure. *Mol. Biol. Evol.* 15:264–76
88. Lehmann T, Hawley WA, Grebert H, Danga M, Atieli F, Collins FH. 1999. The Rift Valley complex as a barrier to gene flow for *Anopheles gambiae* in Kenya. *J. Hered.* 90:613–21
89. Lehmann T, Hawley WA, Kamau L, Fontenille D, Simard F, Collins FH. 1996. Genetic differentiation of *Anopheles gambiae* populations from East and West Africa: comparison of microsatellite and allozyme loci. *Heredity* 77:192–200
90. Linton YM, Harbach RE, Seng CM, Anthony TG, Matusop A. 2001. Morphological and molecular identity of *Anopheles (Cellia) sundaicus* (Diptera: Culicidae), the nominotypical member of a malaria vector species complex in Southeast Asia. *Syst. Entomol.* 26:357–66

90a. Linton YM, Samanidou-Voyadjoglou A, Harbach RE. 2002. Ribosomal ITS2 sequence data for *Anopheles maculipennis* and *An. messeae* in northern Greece, with a critical assessment of previously published sequences. *Insect Mol. Biol.* 11:379–83

91. Lochouarn L, Dia I, Boccolini D, Coluzzi M, Fontenille D. 1998. Bionomical and cytogenetic heterogeneities of *Anopheles funestus* in Senegal. *Trans. R. Soc. Trop. Med. Hyg.* 92:607–12
92. Lounibos LP, Conn JE. 2000. Malaria vector heterogeneity in South America. *Am. Entomol.* 46:238–49
93. Lounibos LP, Wilkerson RC, Conn JE,

Hribar LJ, Fritz GN, Danoff-Burg JA. 1998. Morphological, molecular, and chromosomal discrimination of cryptic *Anopheles* (*Nyssorhynchus*) (Diptera: Culicidae) from South America. *J. Med. Entomol.* 35:830–38
94. Malafronte RS, Marrelli MT, Marinotti O. 1999. Analysis of ITS2 DNA sequences from Brazilian *Anopheles darlingi* (Diptera: Culicidae). *J. Med. Entomol.* 36:631–34
95. Manguin S, Fontenille D, Chandre F, Lochouarn L, Mouchet J, et al. 1999. Anopheline population genetics. *Bull. Soc. Pathol. Exot.* 92:229–35
96. Manguin S, Roberts DR, Peyton EL, Fernandez-Salas I, Barreto M, et al. 1995. Biochemical systematics and population genetic structure of *Anopheles pseudopunctipennis*, vector of malaria in Central and South America. *Am. J. Trop. Med. Hyg.* 53:362–77
97. Manguin S, Wilkerson RC, Conn JE, Rubio-Palis Y, Danoff-Burg JA, Roberts DR. 1999. Population structure of the primary malaria vector in South America, *Anopheles darlingi*, using isozyme, random amplified polymorphic DNA, internal transcribed spacer 2, and morphologic markers. *Am. J. Trop. Med. Hyg.* 60:364–76
98. Manonmani A, Townson H, Adeniran T, Jambulingam P, Sahu S, Vijayakumar T. 2001. rDNA-ITS2 polymerase chain reaction assay for the sibling species of *Anopheles fluviatilis*. *Acta Trop.* 78:3–9
99. Marinucci M, Romi R, Mancini P, Di Luca M, Severini C. 1999. Phylogenetic relationships of seven palearctic members of the maculipennis complex inferred from ITS2 sequence analysis. *Insect Mol. Biol.* 8:469–80
100. Martinez-Torres D, Chandre F, Williamson MS, Darriet F, Berge JB, et al. 1998. Molecular characterization of pyrethroid knockdown resistance (*kdr*) in the major malaria vector *Anopheles gambiae s.s.* *Insect Mol. Biol.* 7:179–84
101. Mathiopoulos KD, Lanzaro GC. 1995. Distribution of genetic diversity in relation to chromosomal inversions in the malaria mosquito *Anopheles gambiae*. *J. Mol. Evol.* 40:578–84
102. McLain DK, Collins FH, Brandling-Bennett AD, Were JB. 1989. Microgeographic variation in rDNA intergenic spacers of *Anopheles gambiae* in western Kenya. *Heredity* 2:257–64
103. Milligan PJM, Phillips A, Broomfield G, Molyneux DH, Toure Y, Coluzzi M. 1993. A study of the use of gas chromatography of cuticular hydrocarbons for identifying members of the *Anopheles gambiae* (Diptera: Culicidae) complex. *Bull. Entomol. Res.* 83:613–24
104. Minakawa N, Githure JI, Beier JC, Yan G. 2001. Anopheline mosquito survival strategies during the dry period in western Kenya. *J. Med. Entomol.* 38:388–92
104a. Minakawa N, Sonye G, Mogi N, Githeko AK, Yan G. 2002. The effects of climatic factors on the distribution and abundance of malaria vectors in Kenya. *J. Med. Entomol.* In press
104b. Mitchell A, Sperling FAH, Hickey DA. 2002. Higher level phylogeny of mosquitoes (Diptera: Culicidae): mtDNA data support a derived placement for *Toxorhynchites*. *Insect Syst. Evol.* 33:163–74
105. Mukabayire O, Caridi J, Wang X, Toure YT, Coluzzi M, Besansky NJ. 2001. Patterns of DNA sequence variation in chromosomally recognized taxa of *Anopheles gambiae*: evidence from rDNA and single-copy loci. *Insect Mol. Biol.* 10:33–46
106. Munstermann LE, Conn JE. 1997. Systematics of mosquito disease vectors (Diptera: Culicidae): impact of molecular biology and cladistic analysis. *Annu. Rev. Entomol.* 42:351–69
107. Murphy WJ, Eizirik E, O'Brien SJ, Madsen O, Scally M, et al. 2001. Resolution of the early placental mammal

radiation using Bayesian phylogenetics. *Science* 294:2348–51

108. Nguyen DM, Tran DH, Harbach RE, Elphick J, Linton YM. 2000. A new species of the Hyrcanus Group of *Anopheles*, subgenus *Anopheles*, a secondary vector of malaria in coastal areas of southern Vietnam. *J. Am. Mosq. Control. Assoc.* 16:189–98
109. Noor MA, Grams KL, Bertucci LA, Reiland J. 2001. Chromosomal inversions and the reproductive isolation of species. *Proc. Natl. Acad. Sci. USA* 98:12084–88
110. Omer SM, Cloudsley-Thompson JL. 1970. Survival of female *Anopheles gambiae* Giles through a 9-month dry season in Sudan. *Bull. WHO* 42:319–30
111. Onyabe DY, Conn JE. 2001. Genetic differentiation of the malaria vector *Anopheles gambiae* across Nigeria suggests that selection limits gene flow. *Heredity* 87:647–58
112. Onyabe DY, Conn JE. 2001. Population genetic structure of the malaria mosquito *Anopheles arabiensis* across Nigeria suggests range expansion. *Mol. Ecol.* 10:2577–91
113. Paskewitz SM, Collins FH. 1997. PCR amplification of insect ribosomal DNA. In *The Molecular Biology of Disease Vectors: A Methods Manual*, ed. JC Crampton, CB Beard, C Louis. pp. 374–83. London: Chapman & Hall
114. Petrarca V, Beier JC. 1992. Intraspecific chromosomal polymorphism in the *Anopheles gambiae* complex as a factor affecting malaria transmission in the Kisumu area of Kenya. *Am. J. Trop. Med. Hyg.* 46:229–37
115. Petrarca V, Nugud AD, Ahmed MA, Haridi AM, Di Deco MA, Coluzzi M. 2000. Cytogenetics of the *Anopheles gambiae* complex in Sudan, with special reference to *An. arabiensis*: relationships with East and West African populations. *Med. Vet. Entomol.* 14:149–64
116. Porter CH, Collins FH. 1996. Phylogeny of nearctic members of the *Anopheles maculipennis* species group derived from the D2 variable region of 28S ribosomal RNA. *Mol. Phylogenet. Evol.* 6:178–88
117. Powell JR. 1983. Interspecific cytoplasmic gene flow in the absence of nuclear gene flow: evidence from *Drosophila*. *Proc. Natl. Acad. Sci. USA* 80:492–95
118. Powell JR. 1997. *Progress and Prospects in Evolutionary Biology: The Drosophila Model*. Oxford: Oxford Univ. Press
119. Powell JR, Petrarca V, della Torre A, Caccone A, Coluzzi M. 1999. Population structure, speciation, and introgression in the *Anopheles gambiae* complex. *Parassitologia* 41:101–13
120. Proft J, Maier WA, Kampen H. 1999. Identification of six sibling species of the *Anopheles maculipennis* complex (Diptera: Culicidae) by a polymerase chain reaction assay. *Parasitol. Res.* 85: 837–43
121. Rai KS, Black WC 4th. 1999. Mosquito genomes: structure, organization, and evolution. *Adv. Genet.* 41:1–33
122. Romi R, Boccolini D, Di Luca M, La Rosa G, Marinucci M. 2000. Identification of the sibling species of the *Anopheles maculipennis* complex by heteroduplex analysis. *Insect Mol. Biol.* 9:509–13
123. Rongnoparut P, Sirichotpakorn N, Rattanarithikul R, Yaicharoen S, Linthicum KJ. 1999. Estimates of gene flow among *Anopheles maculatus* populations in Thailand using microsatellite analysis. *Am. J. Trop. Med. Hyg.* 60:508–15
124. Ross HH. 1952. Conflict with *Culex*. *Mosq. News* 11:128–32
125. Sallum MAM, Schultz TR, Foster PG, Aronstein K, Wirtz RA, Wilkerson RC. 2002. Phylogeny of Anophelinae (Diptera: Culicidae) based on nuclear ribosomal and mitochondrial DNA sequences. *Syst. Entomol.* 27:361–82
126. Sallum MAM, Schultz TR, Wilkerson RC. 2000. Phylogeny of Anophelinae (Diptera: Culicidae) based on morphological characters. *Ann. Entomol. Soc. Am.* 93:745–75

127. Savarit F, Sureau G, Cobb M, Ferveur JF. 1999. Genetic elimination of known pheromones reveals the fundamental chemical bases of mating and isolation in Drosophila. *Proc. Natl. Acad. Sci. USA* 96:9015–20
128. Schug MD, Hutter CM, Wetterstrand KA, Gaudette MS, Mackay TF, Aquadro CF. 1998. The mutation rates of di-, tri- and tetranucleotide repeats in *Drosophila melanogaster*. *Mol. Biol. Evol.* 15: 1751–60
129. Scott JA, Brogdon WG, Collins FH. 1993. Identification of single specimens of the *Anopheles gambiae* complex by the polymerase chain reaction. *Am. J. Trop. Med. Hyg.* 49:520–29
130. Service MW. 1997. Mosquito (Diptera: Culicidae) dispersal—the long and short of it. *J. Med. Entomol.* 34:579–88
131. Sharpe RG, Hims MM, Harbach RE, Butlin RK. 1999. PCR-based methods for identification of species of the *Anopheles minimus* group: allele-specific amplification and single-strand conformation polymorphism. *Med. Vet. Entomol.* 13:265–73
132. Simard F, Fontenille D, Lehmann T, Girod R, Brutus L, et al. 1999. High amounts of genetic differentiation between populations of the malaria vector *Anopheles arabiensis* from West Africa and eastern outer islands. *Am. J. Trop. Med. Hyg.* 60:1000–9
133. Simard F, Lehmann T, Lemasson JJ, Diatta M, Fontenille D. 2000. Persistence of *Anopheles arabiensis* during the severe dry season conditions in Senegal: an indirect approach using microsatellite loci. *Insect Mol. Biol.* 9:467–79
134. Slatkin M. 1995. A measure of population subdivision based on microsatellite allele frequencies. *Genetics* 139:457–62
135. Somboon P, Walton C, Sharpe RG, Higa Y, Tuno N, et al. 2001. Evidence for a new sibling species of *Anopheles minimus* from the Ryukyu Archipelago, Japan. *J. Am. Mosq. Control. Assoc.* 17: 98–113
136. Sucharit S, Komalamisra N. 1997. Differentiation of *Anopheles minimus* species complex by RAPD-PCR technique. *J. Med. Assoc. Thai* 80:598–602
137. Sukowati S, Baimai V, Harun S, Dasuki Y, Andris H, Efriwati M. 1999. Isozyme evidence for three sibling species in the *Anopheles sundaicus* complex from Indonesia. *Med. Vet. Entomol.* 13:408–14
138. Taylor C, Toure YT, Carnahan J, Norris DE, Dolo G, et al. 2001. Gene flow among populations of the malaria vector, *Anopheles gambiae*, in Mali, West Africa. *Genetics* 157:743–50
139. Taylor CE, Toure YT, Coluzzi M, Petrarca V. 1993. Effective population size and persistence of *Anopheles arabiensis* during the dry season in west Africa. *Med. Vet. Entomol.* 7:351–57
140. Templeton AR. 1998. Nested clade analyses of phylogeographic data: testing hypotheses about gene flow and population history. *Mol. Ecol.* 7:381–97
141. Thelwell NJ, Huisman RA, Harbach RE, Butlin RK. 2000. Evidence for mitochondrial introgression between *Anopheles bwambae* and *Anopheles gambiae*. *Insect Mol. Biol.* 9:203–10
142. Torres EP, Foley DH, Saul A. 2000. Ribosomal DNA sequence markers differentiate two species of the *Anopheles maculatus* (Diptera: Culicidae) complex in the Philippines. *J. Med. Entomol.* 37:933–37
143. Toure YT, Petrarca V, Traore SF, Coulibaly A, Maiga HM, et al. 1998. The distribution and inversion polymorphism of chromosomally recognized taxa of the *Anopheles gambiae* complex in Mali, West Africa. *Parassitologia* 40:477–511
144. Tripet F, Toure YT, Taylor CE, Norris DE, Dolo G, Lanzaro GC. 2001. DNA analysis of transferred sperm reveals significant levels of gene flow between molecular forms of *Anopheles gambiae*. *Mol. Ecol.* 10:1725–32

145. Tu Z. 2001. Eight novel families of miniature inverted repeat transposable elements in the African malaria mosquito, *Anopheles gambiae*. *Proc. Natl. Acad. Sci. USA* 98:1699–704
146. Turner M, Blackwell J, Newbold C, Vickerman K. 2002. Introduction. *Philos. Trans. R. Soc. London B Biol. Sci.* 357:3–4
147. Van Bortel W, Harbach RE, Trung HD, Roelants P, Backeljau T, Coosemans M. 2001. Confirmation of *Anopheles varuna* in Vietnam, previously misidentified and mistargeted as the malaria vector *Anopheles minimus*. *Am. J. Trop. Med. Hyg.* 65:729–32
148. Van Bortel W, Trung HD, Manh ND, Roelants P, Verle P, Coosemans M. 1999. Identification of two species within the *Anopheles minimus* complex in northern Vietnam and their behavioural divergences. *Trop. Med. Int. Health* 4:257–65
149. Van Bortel W, Trung HD, Roelants P, Harbach RE, Backeljau T, Coosemans M. 2000. Molecular identification of *Anopheles minimus s.l.* beyond distinguishing the members of the species complex. *Insect Mol. Biol.* 9:335–40
150. Walsh JF, Molyneux DH, Birley MH. 1993. Deforestation: effects on vector-borne disease. *Parasitology* 106(Suppl.): S55–75
151. Walton C, Handley JM, Collins FH, Baimai V, Harbach RE, et al. 2001. Genetic population structure and introgression in *Anopheles dirus* mosquitoes in South-East Asia. *Mol. Ecol.* 10:569–80
152. Walton C, Handley JM, Kuvangkadilok C, Collins FH, Harbach RE, et al. 1999. Identification of five species of the *Anopheles dirus* complex from Thailand, using allele-specific polymerase chain reaction. *Med. Vet. Entomol.* 13:24–32
153. Walton C, Handley JM, Tun-Lin W, Collins FH, Harbach RE, et al. 2000. Population structure and population history of *Anopheles dirus* mosquitoes in Southeast Asia. *Mol. Biol. Evol.* 17:962–74
154. Walton C, Sharpe RG, Pritchard SJ, Thelwell NJ, Butlin RK. 1999. Molecular identification of mosquito species. *Biol. J. Linn. Soc.* 68:241–56
155. Wang R, Kafatos FC, Zheng L. 1999. Microsatellite markers and genotyping procedures for *Anopheles gambiae*. *Parasitol. Today* 15:33–37
156. Wang R, Zheng L, Toure YT, Dandekar T, Kafatos FC. 2001. When genetic distance matters: measuring genetic differentiation at microsatellite loci in whole-genome scans of recent and incipient mosquito species. *Proc. Natl. Acad. Sci. USA* 98:10769–74
157. Wang RL, Wakeley J, Hey J. 1997. Gene flow and natural selection in the origin of *Drosophila pseudoobscura* and close relatives. *Genetics* 147:1091–106
158. Weill M, Chandre F, Brengues C, Manguin S, Akogbeto M, et al. 2000. The *kdr* mutation occurs in the Mopti form of *Anopheles gambiae s.s.* through introgression. *Insect Mol. Biol.* 9:451–55
159. Wilkerson RC, Parsons TJ, Klein TA, Gaffigan TV, Bergo E, Consolim J. 1995. Diagnosis by random amplified polymorphic DNA polymerase chain reaction of four cryptic species related to *Anopheles (Nyssorhynchus) albitarsis* (Diptera: Culicidae) from Paraguay, Argentina, and Brazil. *J. Med. Entomol.* 32:697–704
160. Wondji C, Simard F, Fontenille D. 2002. Evidence for genetic differentiation between the molecular forms M and S within the Forest chromosomal form of *Anopheles gambiae* in an area of sympatry. *Insect Mol. Biol.* 11:11–19
161. Wu C-I. 2001. The genic view of the process of speciation. *J. Evol. Biol.* 14:851–65
162. Zavortink TJ, Poinar GO Jr. 2000. *Anopheles (Nyssorhynchus) dominicanus* sp. n. (Diptera: Culicidae) from Dominican amber. *Ann. Entomol. Soc. Am.* 93:1230–35

Annu. Rev. Entomol. 2003. 48:141–61
doi: 10.1146/annurev.ento.48.091801.112722

First published online as a Review in Advance on October 17, 2002

MANIPULATION OF MEDICALLY IMPORTANT INSECT VECTORS BY THEIR PARASITES

Hilary Hurd
Centre for Applied Entomology and Parasitology, School of Life Sciences, Keele University, Keele, Staffordshire ST5 5BG, United Kingdom; e-mail: h.hurd@keele.ac.uk

Key Words vector blood feeding, vector fecundity, vector longevity, malaria, *Leishmania*

■ **Abstract** Many of the most harmful parasitic diseases are transmitted by blood-feeding insect vectors. During this stage of their life cycles, selection pressures favor parasites that can manipulate their vectors to enhance transmission. Strategies may include increasing the amount of contact between vector and host, reducing vector reproductive output and consequently altering vector resource management to increase available nutrient reserves, and increasing vector longevity. Manipulation of these life-history traits may be more beneficial at some phase of the parasite's developmental process than at others. This review examines empirical, experimental, and field-based evidence to evaluate examples of changes in vector behavior and physiology that might be construed to be manipulative. Examples are mainly drawn from malaria-infected mosquitoes, *Leishmania*-infected sandflies, and *Trypanosoma*-infected tsetse flies.

CONTENTS

INTRODUCTION

Ten diseases that disproportionately affect poor and marginalized populations form the focus of the UNDP/World Bank/WHO special program for research and training in tropical diseases (TDR). Of these, seven are transmitted by insects (Figure 1, see color insert), the exceptions being tuberculosis, schistosomiasis, and leprosy. These seven diseases account for untold misery and death, their distribution affecting population dynamics, economic achievements, and land use. The causative agents of these and other vector-transmitted disease are diverse, ranging from rickettsia-like organisms to helminths. Yet they have in common a crucial aspect of their biology, namely a reliance on insect hematophagy for the continuation of their life cycle. As a consequence, there are shared elements in their route of entry, initial experience in the insect host, and final transmission pathway to a new human host (Figure 1).

The majority of vector-borne diseases do not rely on mechanical passage but undergo a period of growth, development, and sometimes reproduction within their vector that can, relative to the life span of the respective insect, be protracted. As a consequence many, and in some cases, the majority of infected vectors die before they take a blood meal that transmits infective parasites. Parasite fitness is thus intrinsically linked to vector fitness.

Evolutionary biology suggests that a parasite's success would be improved if it could develop faster or if its vector lived longer. One or both of these traits should thus be under selective pressure (50). Vector longevity is determined in part by size or physiological factors, such as nutrient reserves, and partly by behavior traits. Blood-feeding behavior is particularly risky due to the defensive action of hosts, and feeding strategies have a large influence on longevity. Parasites that influence these crucial factors to their advantage are likely to be selected.

Once extrinsic incubation periods are concluded and parasites are infective, alternative parasite strategies should be adopted such that vector/host contact is favored and the probability of transmission during hematophagy is enhanced. Various physiological and mechanical changes occur that would enhance parasite fitness, often now at the expense of the vector.

MANIPULATION AS A CONCEPT

Much has been written about host manipulation, especially with respect to changes that enhance trophic transmission or passage from host to host via the food chain. By their mere invasion and colonization, parasites are bound to alter their host in some way. However, whether this is true manipulation is difficult to substantiate (67). Parasites by definition cause harm; if the consequence of this harm increases parasite fitness (transmission success), then parasite virulence will probably be selected (22). Is this then adaptive manipulation or the side effect of infection?

The debate concerning pathological versus adaptive changes thus revolves around the elements of cause and effect. Exactly what causes the changes that occurred in the vector and is it a specific action of the parasite, having no other benefit (19)? Additionally, and equally important, did the effect enhance transmission (67, 68)? Answers to the first questions can be sought by examining the mechanisms underlying parasite-induced changes in their hosts. The ultimate substantiating evidence is the identification of manipulator molecules produced by the parasite, serving no other function than to instigate changes in the host that will benefit the parasite (41). Clues that are easier to obtain come from reviewing the diversity of these changes and the phylogenetic patterns induced in related hosts. Examples of seemingly convergent evolution may be supportive of host manipulation (62, 68). Although parasite-induced changes in host behavior or physiology are easier to demonstrate in the laboratory, it is usually difficult to obtain evidence that these changes directly enhance parasite transmission. This effect is usually inferred. Ideally, confirmation that parasite-induced changes in vector insects do enhance transmission must come from studies of natural infections in the wild. However, unfortunately the literature contains few examples of these [but see (52)].

The issue of manipulation is further muddied if both sides of the symbiosis are considered, for host defenses against the parasite are just as likely to evolve in these "arms race" situations. Thus, we also need to consider whether we are actually observing a host's adaptive response to infection that benefits the host. Or indeed the consequence of adaptation by both organisms in the symbiosis may be of mutual benefit once an infection has been established (37).

Interactions between parasites and their vector insects can occur at various points in their development, starting with the arrival of an infected blood meal in the insect midgut (Figure 1). Some parasites, such as *Trypanosoma cruzi*, spend all vector phases of their life cycle in the gut, others penetrate into the hemocoel and invade other tissues early in the infection, such as *Onchocerca volvulus*, or later, such as *Plasmodium* spp. Any changes in vector physiology or behavior could thus take place at a local tissue level or at a distance from the infection site and at any time during the infection.

There are many reports of vector manipulation by parasites of medical importance [see reviews in (60, 61)]. However, examples of parasite-induced changes in vector insects that have been explored in detail are sparse, and few have been demonstrated unequivocally to be examples of manipulation rather than pathology. Few manipulator molecules have been identified, and enhanced transmission as a direct result of manipulation has been rarely demonstrated in the field. Nevertheless, parasites do interact with their vectors at all stages of their development and many of the outcomes are intuitively suggestive of increased parasite fitness. This review examines selected examples that have been studied in some depth and have been chosen to illustrate changes in vector physiology and behavior that are likely to have been induced by parasites at different stages of infection. Where possible, examples are assessed in terms of potential transmission enhancement and evaluated using criteria that define manipulation.

Points of contact between vector and parasites begin with ingestion of the infective blood meal. Once infected, the immune system may be manipulated, circulating metabolite titers changed, vector reproductive success and longevity altered, and the process of further blood feeding affected. The three aspects that have been most studied, namely blood feeding, egg production, and longevity, are reviewed.

HEMATOPHAGY

Many authors have reported that infection results in altered vector feeding behavior such that more host contact occurs [see the many examples cited in comprehensive reviews in (60, 61)]. This thus appears to be a strategy common to many parasite/vector associations. However, whether it is the result of manipulation is, in most cases, unclear. First, to benefit the parasite, impaired vector feeding must increase transmission prospects and thus only occur when the parasite has reached the infective stage. Second, the cause of changes in feeding behavior must arise directly as a result of some action by the parasite. Before the transmission stage has been reached, increased contact may actually be disadvantageous because it increases the risk of vector death with no chance of associated transmission (6, 78). Poor feeding success during parasite development also affects vector fitness. It decreases fecundity and possibly also decrease survivorship. The latter occurs if the vector tries to compensate for a small meal by multiple feeding attempts and is thereby made more vulnerable to host defenses; this may impact negatively upon the parasite. Thus, for both organisms, parasite trade-offs exist between the costs and benefits derived from increasing host contact and at times these may be in conflict. In addition, for the parasite, optimal trade-off positions may alter during development.

Leishmania-Infected Sandflies

Early in the twentieth century it was recognized that the feeding behavior of sandflies infected with *Leishmania* was modified such that parasite transmission appeared to be enhanced (80). Difficulties in ingesting a blood meal result in increased probing, producing a recorded case of 11 biting attempts by a single fly (7, 8, 47). This phenomenon is common to a variety of *Leishmania*/sandfly associations and has been dubbed the "blocked fly hypothesis" (45) due to the observation that occlusion of the stomodeal valve causes swelling and may prevent blood from flowing into the midgut even though it produces a permanently open valve. The occlusion is caused by a filamentous gel-like matrix, which embeds masses of infectious metacyclic promastigotes and fills the cardia and stomodeal valve, extending backward through the thoracic midgut. This promastigote secretory gel plug exerts mechanical pressure on the surrounding gut wall (74). Parasites located anterior to this plug are deposited in the bite wound when material from the plug is refluxed during a feeding attempt. Although the biochemical nature and origin of the gel-like plug is not known, a parasite secretion, filamentous promastigote proteophosphoglycan (fPPG), is the major component and forms a three-dimensional meshwork within the gel. The mucin-like fPPG is secreted in vitro by *L. major* and *L. mexicana* and has been observed to be present in 10 *Leishmania*/sandfly associations. In addition to the effect it has on sandfly probing, fPPG is thought to prevent the flushing of unattached promastigotes from the cardia back into the abdominal midgut with the blood meal. This parasite product thus appears to be a modulator molecule that alters the parasite environment and thereby increases parasite fitness by enhancing transmission [(82) and references therein]. Promastigotes also secrete an enzyme, a polymeric phosphoglycoprotein, from the flagella pocket but this does not form part of the filamentous gel plug and its function in vivo is currently unknown (42).

In an elegant study of feeding behavior, Rogers and coworkers (74) produced definitive evidence in support of the "blocked fly hypothesis." They blood-fed sandflies exposed to *Leishmania* infection through chick skin membranes, or on anesthetized mice, and immediately compared blood meal size with infection intensity. In both feeding regimes, full blood meals were taken by lightly infected or uninfected sandflies, whereas of the flies that had only taken partial meals, over 80% were heavily infected. Parasites were transmitted during probing, even if feeding attempts were unsuccessful, leading to the conclusion that heavily infected flies were likely to make further probing attempts and transmit more parasites.

The *Leishmania* amastigotes ingested with an infective blood meal develop via rapidly dividing procyclic and nectomonad forms to form nondividing, infective promastigotes within 3 to 5 days. Vector probing behavior is affected during the first feed postinfection.

Malaria-Infected Mosquitoes

Parasite-induced changes in probing behavior have also been associated with malaria-infected mosquitoes, but several blood meals could be taken postinfection before this is apparent. This is because changes to mosquito probing behavior are only caused when infective sporozoites have invaded the salivary glands, an event that takes 10 or more days depending on species and environmental conditions. As with infected sandflies, the female mosquito makes many attempts to feed, each time depositing parasites at the feeding site. However, the mechanism underlying increased probing is different. Malaria sporozoites inhabit the distal and intermediate areas of the lateral and median lobes of salivary glands and are deposited intradermally with saliva during the probing phase, before blood ingestion begins (11).

As in other hematophagous insects (53, 70, 71), saliva of female mosquitoes contains apyrase, an ADP-degrading enzyme that inhibits the stimulatory activity of ADP upon platelet recruitment at a wound site, and thereby counters host hemostasis and promotes easier and longer blood feeding (55). Salivary apyrase levels are inversely correlated with probing duration for three anopheline mosquito species (72). Apyrase is produced in the distal regions of the salivary gland lobes, the same location as sporozoites (81). Apyrase activity is reduced to one fourth in the salivary glands of *Plasmodium gallinaceum*-infected *Aedes aegypti*, resulting in median blood location time being increased threefold, although saliva volume remains constant (73, 75). Wekesa et al. (84) collected *Anopheles gambiae* with natural infections of *P. falciparum* and found that 65% probed on a guinea pig at least three times compared with 27% of uninfected females. Laboratory experiments eliminated mosquito age as a factor and attributed increased probing to sporozoite infection. Rossignol et al. (76) also reported that *P. gallinaceum* infection increases vector response to host odor, and they conclude, but did not demonstrate, that these two effects increase vector biting rate and therefore increase parasite transmission. Rossignol & Rossignol (77) produced a model of the effect of sporozoite-induced salivary gland pathology that predicted that the relationship between number of hosts contacted and salivary gland pathology should be exponential, probing failure being positively related to parasite-induced lesions (72). This assumed link is reasonable, given the relationship between apyrase activity and probing. However, if the findings of a field study of *P. falciparum* infected *An. gambiae* are common to other malaria/mosquito associations, no link exists between sporozoite intensity and probing rate (84). A parasite-associated increase in biting rate may thus, in reality, be an all-or-nothing effect, rather that infection-intensity related. Studies of several natural malaria/mosquito associations need to be performed to clarify this. Caution must also be observed when extrapolating from observations of one or two malaria/mosquito associations to all others, as Li and coworkers (56) were unable to detect any effect of *Plasmodium berghei* sporozoites on the ability of *Anopheles stephensi* to locate blood, possibly due to constitutively low apyrase levels in this species.

Just two studies [both demonstrating that sporozoite infection cause impaired blood location and increased probing (75, 84)] have given rise to a perceived view in the literature that vector feeding behavior is manipulated by malaria parasites. How true is this interpretation? Currently we do not know the mechanism that underlies sporozoite-induced reduction in apyrase activity. It could result from an inhibitor produced directly by the parasite (adaptive manipulation) or mechanical damage resulting from tissue invasion (by-product of infection). To qualify as true manipulation, host changes must be advantageous to the parasite. So, increased biting must also be demonstrated to result in increased parasite transmission. This occurs if infected mosquitoes desist feeding and seek an alternative host, thereby transmitting the parasite to more than one new host during a single gonotrophic cycle. This behavior greatly affects the epidemiology of malaria because biting rate is a major factor determining vectorial capacity and hence the basic case reproduction number, R_0 (the number of new infections that arise from a single current infection) (29, 58). There is one study that elegantly demonstrates increased host contact by sporozoite-infected mosquitoes. Koella and coworkers (52) used microsatellite loci as markers to identify the source of mosquito blood meals and demonstrated that 22% of wild *P. falciparum*-infected *An. gambiae* had bitten more than one person in a single night compared with only 10% of uninfected females. Larger infected mosquitoes were most likely to take multiple meals and to become fully engorged. An earlier field study supported this finding by showing that the genotypes of *P. falciparum* infecting pairs of children sleeping in the same house were more likely to be identical, and thus originating from the same mosquito, than the genotypes of pairs of children chosen at random (15). Koella et al. (52) concluded that infected *An. gambiae* may feed less efficiently because of apyrase deficiency or because infection had increased the threshold blood volume at which blood-seeking behavior was inhibited.

Additional observations made during a field study of feeding behavior of *Anopheles punctulatus* in Papua New Guinea indicated that infected females became fully engorged earlier in the night (51), although opposite results were obtained by Bockarie et al. (10). Koella & Packer (51) suggested that their findings resulted either from infected females obtaining more blood from one feed or from multiple feeding and further suggested that infection may increase the feeding tenacity of mosquitoes. Both probing and persistence are crucial aspects of feeding that could enhance transmission. In a laboratory study using a rodent malaria, feeding persistence was enhanced by malaria infection; only 20% of sporozoite-infected *An. stephensi* desist in feeding attempts upon a human when continually interrupted in contrast to 33% of uninfected females (6). Increased biting does not, however, benefit the parasite during phases in the life cycle before it is infectious and indeed may be detrimental because increased host contact is risky for the vector and could result in death before transmission. As predicted, during the oocyst stage, *Plasmodium yoelii nigeriensis* infection significantly decreased feeding persistence, 53% having given up feeding attempts during a 10-min period compared with the figures for uninfected and sporozoite-infected mosquitoes quoted above (6) (Figure 2). Feeding behavior thus changes in completely opposite, but intuitively advantageous, ways depending on the stage of parasite development.

Examples from Other Infected Vectors

Although malaria and *Leishmania* infections represent the most-studied examples of parasites affecting vector feeding behavior, reduced blood meal location efficiency has also been attributed to other parasite infections of vectors, e.g., *Rhodnius prolixus* infected with *Trypanosoma rangeli*, tsetse flies infected with *Trypanosoma* spp., and the rat flea *Xenopsylla cheopis* infected with the plague bacterium *Yersinia pestis*. Explanations for changes in feeding behavior include physical blockage of the foregut by parasites, as in plague-infected fleas, obscured phagoreceptors in tsetse flies infected with trypanosomes, and reduced apyrase activity in the salivary glands of *Rhodnius prolixus* infected with *T. rangeli* [reviewed in (54, 60, 61, 72)]. Behavioral changes purely associated with obstruction are more likely to be due to pathology than direct manipulation but could nevertheless

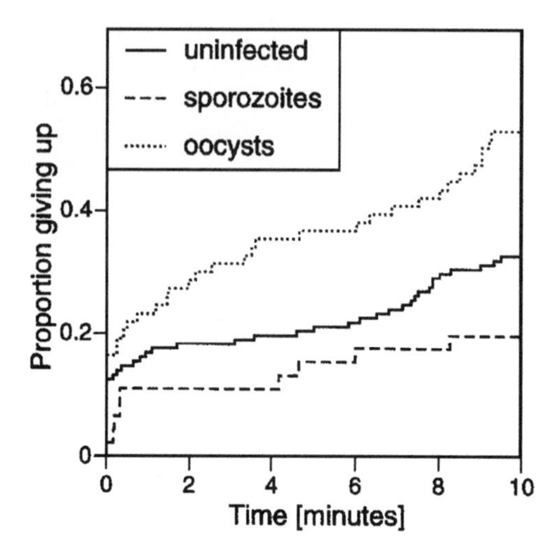

Figure 2 The effect of *P. yoelii nigeriensis* infection on the feeding persistence of *An. stephensi*. Feeding behavior was interrupted as soon as a mosquito landed on the arm of a volunteer, and the time taken to give up feeding attempts was recorded up to a maximum of 10 min. Mosquitoes were uninfected, infected with oocysts, or with sporozoites. Infection status was determined by dissection at the end of the trial period. Taken with permission from (6).

give rise to enhanced transmission success. In these cases selective pressure may favor parasite-induced pathology (virulence) and more virulent parasites will then be selected.

Transmission from the Vertebrate Host

Although the remit of this article is to review parasite manipulation of vector insects, it should be noted that parasitemia in the vertebrate hosts also affects insect hematophagous behavior. In some cases host attractiveness is enhanced, possibly due to modified host odor plumes (65), and vectors such as tsetse flies feed more frequently on infected hosts [reviewed in (54)]. In addition, host defensive behavior may be reduced, making feeding on infected hosts less risky [reviewed in (29)]. Pathology associated with infection could also assist the blood-feeding efforts of vectors. For example, on initial infection, thrombocytopenia induced by malaria (64), dengue, African trypanosomiasis, and babesiosis decrease hemostasis [discussed in (72)]. In addition anemia enhances blood flow (18) and, in the case of malaria, at certain parasitemias (i.e., slight anemia) increases erythrocyte intake (83). Young vectors contribute most eggs for the next generation because they outnumber the old, and mosquitoes thus gain by being aided in blood feeding on infected hosts (71). In the 1980s this argument was persuasive. However, we now know that infection-induced loss of reproductive fitness occurs in young and old irrespective of blood intake and that in several cases blood intake is not affected by sporozoite infection (32, 84). Thus the odds appear to be stacked in favor of the parasite.

FECUNDITY, BLOOD FEEDING, AND INFECTION

Blood feeding is as essential for the vector as it is for the parasite. This is because, in most cases, hematophagous female insects require the amino acids acquired from erythrocyte and plasma protein digestion to synthesize yolk proteins for egg production. Egg production is influenced by a variety of factors in hematophagous insects, many of them being interlinked, but with blood meal quantity and quality playing major roles (39) (Figure 3).

In many insects the normal process of oogenesis is disrupted by parasites of various taxa [reviewed in (35, 36)], resulting in loss of reproductive fitness. Observations on parasite-induced fecundity reduction include several vector/parasite associations (39), but here attention focuses on malaria-infected mosquitoes with a view to evaluating the manipulative nature of what appears to be a strategy so widespread as to constitute an example of convergent evolution.

Fecundity Reduction in *Plasmodium*-Infected Mosquitoes

Originally, studies of *Ae. aegypti* infected with the avian malaria *P. gallinaceum* attributed loss of fecundity to reduced blood meal intake or the intake of impoverished blood when mosquitoes fed on infected hosts (26). In addition problems associated with sporozoite-induced changes in probing behavior were thought to affect fecundity (76).

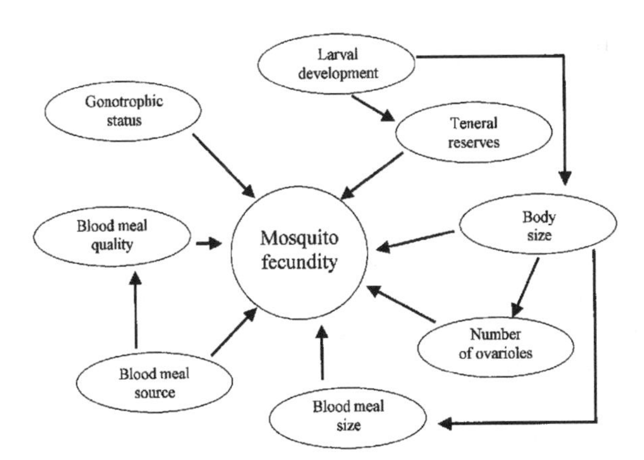

Figure 3 Factors affecting mosquito fecundity. Adapted with permission from (39).

Egg Production and the Blood Meal

The mechanical process of prediuresis makes blood meal measurement more complicated in anopheline than in aedine mosquitoes. Prediuresis occurs during feeding and enables the mosquitoes to concentrate the erythrocyte component of the meal by expelling plasma via a sieve made from spines located in the hindgut. Gravimetric quantification is thus unsuitable as a measure of blood meals in these mosquitoes. Protein intake should be assessed by measuring hemoglobin intake or by measuring its digested product, hematin (12). As expected, a highly significant positive correlation between egg production and hematin production was observed when *An. stephensi* fed on mice (31, 32). This relationship disappeared when females fed on highly anemic mice with heavy *P. y. nigeriensis* asexual parasitemias, but no gametocytes. Nor does a relationship exist when females fed on much less anemic mice with low parasitemia, but containing infective gametocytes. Surprisingly, fecundity reduction was most severe in mosquitoes that developed oocyst infections (fed on gametocytemic mice), although the protein content of their blood meal was less reduced than those feeding on noninfective mice with high parasitemias (31). Taylor & Hurd (83) designed laboratory experiments that closely monitored mouse parasitemia and exposed mosquitoes to blood meals containing early infections that were highly infectious to mosquitoes but caused only slightly lowered packed cell volume in the mice (42%–45%). They demonstrated that lower blood viscosity significantly enhanced hemoglobin intake at this stage and thus impoverished blood meals cannot explain the parasite-induced fecundity reduction seen in this system. Furthermore, blood meal protein content, enzyme activity, and digestion were not affected by *P. y. nigeriensis* infection of *An. stephensi* when mosquitoes fed on gametocytemic mice with little evidence of anemia (43). *P. y. nigeriensis* infection also causes significant reproductive curtailment during later stages of infection when oocysts are present on the midgut or sporozoites in the salivary glands, even though females feed on the same uninfected mice as the experimental, uninfected control mosquitoes and ingested similar amounts of hemoglobin (32). Similarly fecundity reduction was not caused by a reduction in blood meal size in wild-caught *An. gambiae* infected with the human malaria parasite *P. falciparum* (33). These data gave rise to the hypothesis that some aspect of malaria infection in the mosquito reduces reproductive output in a manner divorced from, or synergistic to, any affect caused by feeding on an infected host. Substantial anemia in the chicken hosts used for *Ae. aegypti* infections (26) may thus have masked malaria/mosquito interactions in the avian malaria/aedine mosquito model. Alternatively this mosquito may respond to malaria infection in a different manner to that of anopheline mosquitoes. This issue will not be resolved until the results from experiments designed to eliminate all factors apart from malaria infection from the assessment of *Ae. aegypti* fecundity are assessed.

Parasite-induced fecundity reduction is often attributed to competition between the two symbionts for limited nutrients. Production of a batch of mosquito eggs requires a large supply of nutrients that are largely derived from the blood meal.

However, only 19% of the blood meal protein taken by *An. gambiae* is incorporated into the ovaries during normal vitellogenesis, and anophelines as a whole utilize less than one third of the energetic input from a blood meal for oogenesis (13). Although isoleucine was identified as a possible limiting factor for oogenesis when mosquitoes fed on human blood (13), this could not explain the fecundity reduction observed when mosquitoes fed on mice infected with rodent malaria because rodent blood has a high isoleucine content compared with human blood. It is therefore unlikely that amino acids are a limiting factor to oogenesis and indeed, as described below, vitellogenin actually accumulates in the hemolymph of infected mosquitoes. Furthermore, in a resource competition situation, intensity of infection should be related to the degree of reproductive loss. This is not the case in malaria infections of anopheline mosquitoes where mean infections of ~4 oocysts per midgut caused a mean cumulative reduction in fecundity (21%) that was not significantly different from that seen in mosquitoes infected with more than 75 oocysts (28%) (32). Field infection, where oocyst densities are usually only one or two per midgut, also causes fecundity reduction (33). If we are not observing a resource-led, density-dependent mechanism, is this reduction in vector reproductive fitness a result of adaptive manipulation?

As with changes in vector feeding behavior, parasite-induced fecundity reduction is a strategy that is evident in many vector parasite associations [e.g., *Leishmania*-infected sandflies (21), filarial worm–infected mosquitoes (17,44), filarial nematode–infected blackflies (28, 69)]. What is not clear is whether fecundity reduction is a parasite strategy, a host strategy, or a by-product of infection (37). A greater understanding of the biochemical and molecular events that alter egg development in infected vectors may help answer these questions. Some headway has recently been made but many questions still remain.

Mechanisms Underlying Fecundity Reduction

Using *P. y. nigeriensis* infection of *An. stephensi* or *An. gambiae* as model systems, we have shown that all stages of vitellogenesis are disrupted by the parasite. A blood meal immediately initiates a gonotrophic cycle that results in the production of a batch of eggs in 3 days. If the blood meal is infected with malaria gametocytes, egg production initially proceeds normally. The first effects of infection coincides with the time at which ookinetes begin to penetrate the midgut epithelium. In the ovary a significant proportion of developing oocytes begin to resorb (14, 34) (Figure 4), and the total ovarian protein content (30) and, more specifically, the yolk protein or vitellin content (3) are significantly reduced. During the second gonotrophic cycle postinfection, when malaria oocysts are developing on the midgut wall, *An. gambiae* ovaries have significantly reduced vitellin content throughout the egg production cycle. The resultant egg batch is significantly reduced, as is the hatch rate of eggs produced by infected females (4). Some reduction in the protein content of vitellogenic ovaries may be due to parasite-induced retardation in development (14) but the majority is due to the resorption of developing follicles.

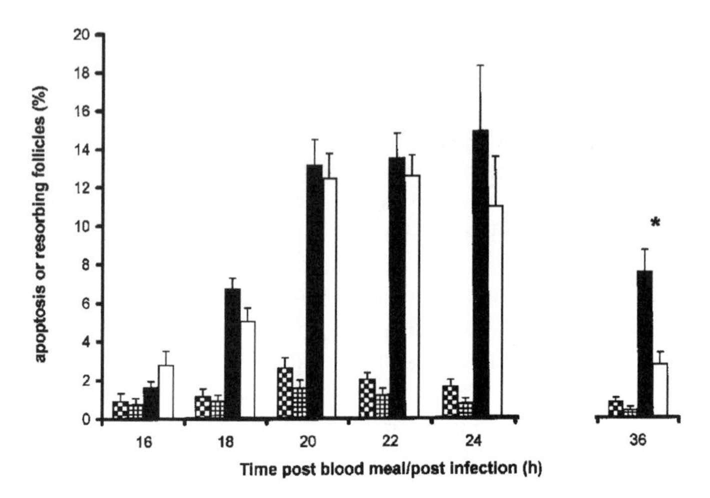

Figure 4 The effect of malaria infection on follicle resorption and the occurrence of apoptosis in the follicles of ovaries of *An. stephensi*. Mean percentage resorbing follicles per ovary, *bars with checkered squares* = uninfected mosquitoes; *solid black bars* = infected mosquitoes. Mean percentage follicles containing apoptotic cells, *bars with white squares* = uninfected mosquitoes; *white bars* = infected mosquitoes. Sample size 20 in each case, error bars = S.E.M.; * = significant difference between infected and control values at that time point. Taken with permission from (34).

IN THE OVARY Follicle resorption rarely occurs in well-fed laboratory mosquitoes, and the signaling mechanisms and cellular processes are poorly understood. Hopwood et al. (34) demonstrated that patches of follicular epithelial cells surrounding the resorbing follicles of infected mosquitoes were dying by programmed cell death or apoptosis. The nuclei of these cells were condensed, stained intensely with acridine orange, and DNA fragmentation could be detected using in situ terminal-deoxynucleotidyl-transferase-mediated dUPT-biotin nick end labeling (TUNEL) (Figure 5, see color insert). Caspases are site-specific proteases at the core of the biochemical mechanisms initiating apoptosis. Egg production was restored to control levels in infected mosquitoes injected with caspase inhibitors, thus establishing that malaria-induced follicle resorption is caused by follicle cell apoptosis (34). Although this is the first report that establishes apoptosis as a mechanism underlying parasite-induced fecundity reduction, the follicle cells surrounding the ovaries of *Brugia pahangi*–infected *Ae. aegypti*, pictured in (23), exhibit similar signs of the condensed nuclei typical of apoptotic cells (Figure 5).

IN THE FAT BODY Malaria infection also affects the synthesis of vitellogenin (Vg) in the mosquito fat body. The abundance of vitellogenin mRNA was significantly reduced during the second gonotrophic cycle, when oocysts were developing, and a slight reduction also occurred 36 h post initial infection (3). Vg is secreted into the hemolymph immediately after synthesis. *Plasmodium*-induced changes in Vg mRMA abundance were thus reflected in the titer of circulating Vg, significant reductions occurring 30 h post infection. During the first 12 h of the second gonotrophic cycle (when oocysts were developing on the midgut), the titer of circulating Vg in transit from fat body to ovaries was also significantly depleted. However, by 24 h post blood feeding the yolk protein was accumulating in the hemolymph to a significantly greater degree than in uninfected females, despite the significant reduction in production by the fat body. This additional Vg is likely to result from events occurring in the ovary, as Vg uptake is retarded and many follicles are resorbing (3). Decreased sequestration into the ovary must be more pronounced than decreased fat body synthesis, resulting in an increase in hemolymph titer. A similar temporal change in Vg profile was observed in black flies infected with the cattle filarial nematode *Onchocerca lienalis*. Here too infection causes reproductive curtailment associated with decreased fat body synthesis of Vg and decreased ovarian protein content (69).

It is possible that events occurring in the ovarian follicular epithelium of infected mosquitoes are also responsible for the reduction in Vg mRNA abundance. In response to a blood meal, these cells produce the hormone ecdysone. Ecdysone is converted to 20-hydroxyecdysone in the fat body and regulates the transcription of Vg genes (20). Destruction of developing follicles is thus likely to reduce the ecdysone titer, which could in turn affect Vg gene transcription.

Although the mechanisms underlying parasite-induced vector fecundity reduction have not been investigated in other associations, the process of vitellogenesis is affected in a similar manner in mermithid infections of locusts and black flies [reviewed in (35)] and *Tenebrio molitor* infected with metacestodes of the rat tapeworm *Hymenolepis diminuta* [reviewed in (38)]. In this latter system, the parasite produces a manipulator molecule that downregulates Vg synthesis in the fat body. However, events occurring in the ovary of infected beetles may be caused by a host-derived competitive inhibitor of juvenile hormone binding to receptors on the follicular epithelium (38).

Is Egg Production Manipulated?

Reproductive curtailment in malaria-infected anopheline mosquitoes is clearly the result of a complex series of molecular and biochemical events. But experimental studies so far have drawn us no nearer to determining whether they result from direct host manipulation or whether they represent a response on the part of the vector to the stress imposed by infection. No inhibitory molecule of parasite origin has been identified, or indeed has been reported to have been sought. If present, these manipulator molecules may act directly to initiate apoptosis in the follicular epithelium or to inhibit Vg gene transcription. Alternatively, parasite manipulators may act indirectly via pre-existing control mechanisms, such as the endocrine system. If fecundity reduction is a host response to infection, regulation mechanisms that control reproduction are likely to be linked to the insect surveillance mechanism such that egg production is reduced when an infection is detected. This could be a host strategy to conserve resources that may be required for defense. There is a growing field of literature linking immune response to changes in life-history traits such as reproductive success (63, 79). It is interesting to note that direct stimulation of the mosquito humoral defense system, by injection of lipopolysaccharide, results in both a reduction in the protein content of developing eggs and a reduction in the number of eggs produced in the cycle immediately following treatment (2).

An alternative hypothesis postulates that, in host populations where parasite prevalence is low or parasites are not highly virulent, the evolution and deployment of an efficient surveillance and immune response may involve more costs than benefits (79). Response to infection that results in shifting resources away from processes that are not immediately essential, such as reproduction, may be a more efficient strategy on behalf of these hosts because costs are not fixed and only paid by the host upon infection.

If fecundity reduction is an adaptive strategy on the part of parasite or host, long-term advantages must accrue. Life-history theory predicts that a trade-off exists between survival and reproduction whereby increased reproductive effort decreases lifespan and vice versa. Thus, Forbes (25) suggested that fecundity reduction may stop the parasite adversely affecting host longevity, and Perrin (66) extended this concept to encompass total host reproductive effort. This is clearly demonstrated in the *H. diminuta*/*T. molitor* association where survival time to 50% mortality is increased by 40% in infected female beetles that exhibit reduced fecundity (40).

Similarly, if fecundity reduction results in vector survival until parasite transmission can occur, the parasite obviously increases its chances of successful transmission. However, the vector may also gain from reduced reproductive output per gonotrophic cycle if it results in more efficient resource management, an increase in life span, and more time in which to produce further, albeit reduced, egg batches. Thus both parasite and host may gain from host fecundity reduction during the intrinsic period of parasite growth before parasites invade the salivary glands (1, 37). As discussed above, evolutionary pressures change once a parasite can be transmitted by blood feeding. At this stage virulence and transmission may be positively correlated such that biting rate (and risk of mortality) is higher than the optimal rate for the mosquito (50).

INFECTION AND SURVIVORSHIP

These theoretical considerations suggest that a parasite manipulates its vector such that survivorship is increased during the parasite's developmental period but that, as a result of changes in host contact and biting behavior, the opposite may occur once a patent infection has developed. What evidence is there that this actually occurs? Many studies of the effect of parasites on vector survivorship have been made, most of them focused on malaria/mosquito associations.

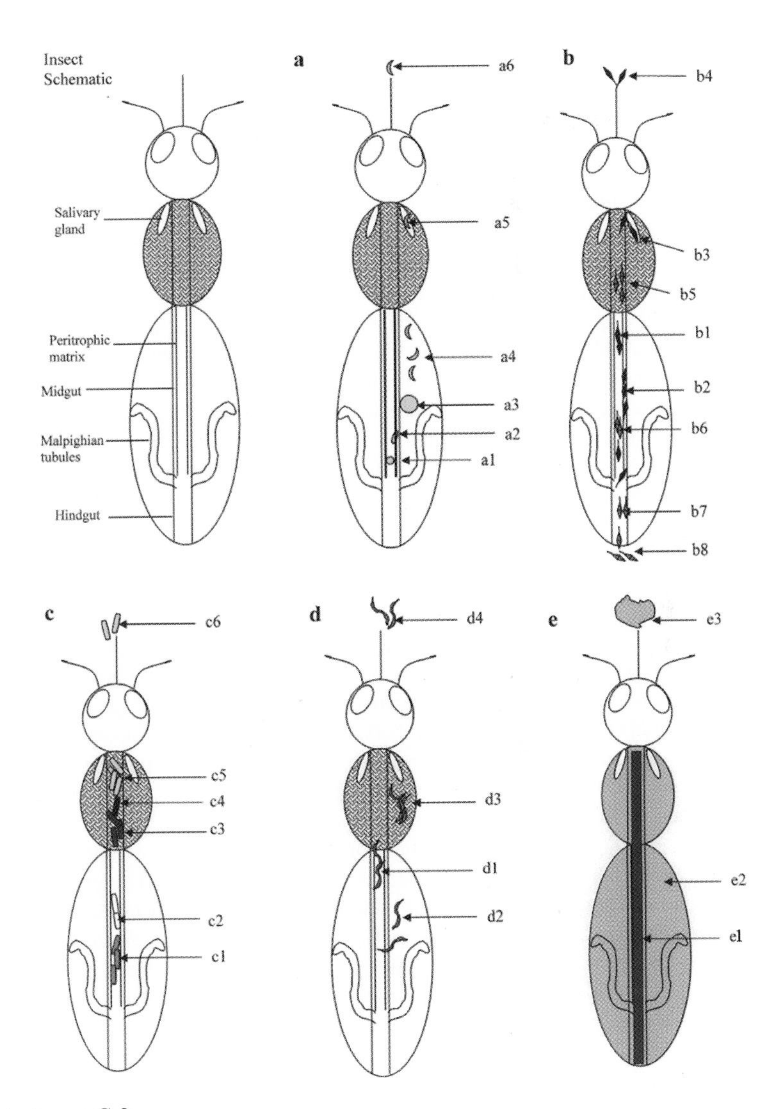

See text page C-2

Figure 1 (page C-1) Parasites transmitted by medically important insects. (*a*1–6) Malaria. *Plasmodium* spp. are the causative agents of malaria and are transmitted by anopheline mosquitoes. (*a*1) Gametes are fertilized in the midgut lumen and zygotes develop into motile ookinetes. (*a*2) Ookinetes invade the midgut epithelium and transform into oocysts. (*a*3) Oocysts develop beneath the basal lamina and produce sporozoites. (*a*4) Sporozoites are released into the hemocoel. (*a*5) Sporozoites invade the salivary glands. (*a*6) Sporozoites pass into the host with the saliva (9). (*b*1–4) Sleeping sickness. *Trypanosoma brucei gambiense* or *T. brucei rhodesiense*, which cause African sleeping sickness infection, are transmitted by the tsetse flies of the genus *Glossina.* (*b*1) Ingested stumpy bloodstream forms rapidly differentiate into procyclic forms in the midgut lumen. (*b*2) Procyclic forms form a persistent, actively dividing population in the ectoperitrophic space, but their route of entry to this space is controversial. (*b*3) A proportion of this population migrates to the salivary gland and transforms to epimastigotes and nondividing metacyclic stages. (*b*4) Metacyclic forms are deposited in the dermal connective tissue during biting (85). (*b*5–8) *Trypanosoma cruzi* causes Chagas' disease and is transmitted by the triatomine bug *Triatoma infestans*. (*b*5) Trypomastigote forms enter the gut with the blood meal and within hours transform into epimastigotes. (*b*6) They multiply in the crop and midgut. (*b*7) Epimastigote forms spread to the rectum and mature into nondividing metacyclic trypomastigotes. (*b*8) Metacyclic forms in the rectum are deposited on the host skin with the feces (46). (*c*1–6) Visceral, cutaneous, and mucocutaneous leishmaniasis are caused by flagellates of the genus *Leishmania.* Phlebotamine sandflies act as vectors. (*c*1) *Leishmania mexicana* amastigotes are ingested and transform to procyclic promastigotes on day 1, these multiply in the abdominal midgut. (*c*2) Nectomonad promastigotes become the predominant form by day 3 then persist in low numbers for the rest of the infection. (*c*3) Leptomonad promastigotes form approximately 50% of the population by day 5. (*c*4) Haptomonad promastigotes appear on day 4 and persist at low prevalence for the entire infection. (*c*5) Metacyclic promastigotes appear on day 4 and increase to represent half the population by day 10. (*c*6) Metacyclic promastigotes are deposited into the wound when part of the gel-like plug from the cardia is refluxed (74). (*d*1–4) Lymphatic filariasis and river blindness are caused by filarial nematodes. Important vectors of *Wuchereria bancrofti* include *Culex quinquefasciatus* and species of *Anopheles* and *Aedes. Onchocerca volvulus*, the causative agent of river blindness, is transmitted by black flies of the genus *Simulium.* (*d*1) Microfilaria are taken up by the vector. (*d*2) Microfilariae migrate through the midgut wall to the thoracic muscles. (*d*3) Worms molt to L2 and L3 larvae in the thoracic muscles. (*d*4) Microfilaria are released from the mouthparts during blood feeding. (*e*1–3) Dengue fever is a disease caused by an arbovirus that is transmitted by mosquitoes such as *Ae. aegypti* and *Ae. albopictus.* (*e*1) The virus is taken in with an infected blood meal. (*e*2) It invades the gut, spreads to all tissues, and survives as long as the mosquito. (*e*3) It is passed on to another host in mosquito saliva and can also be transmitted transovarially.

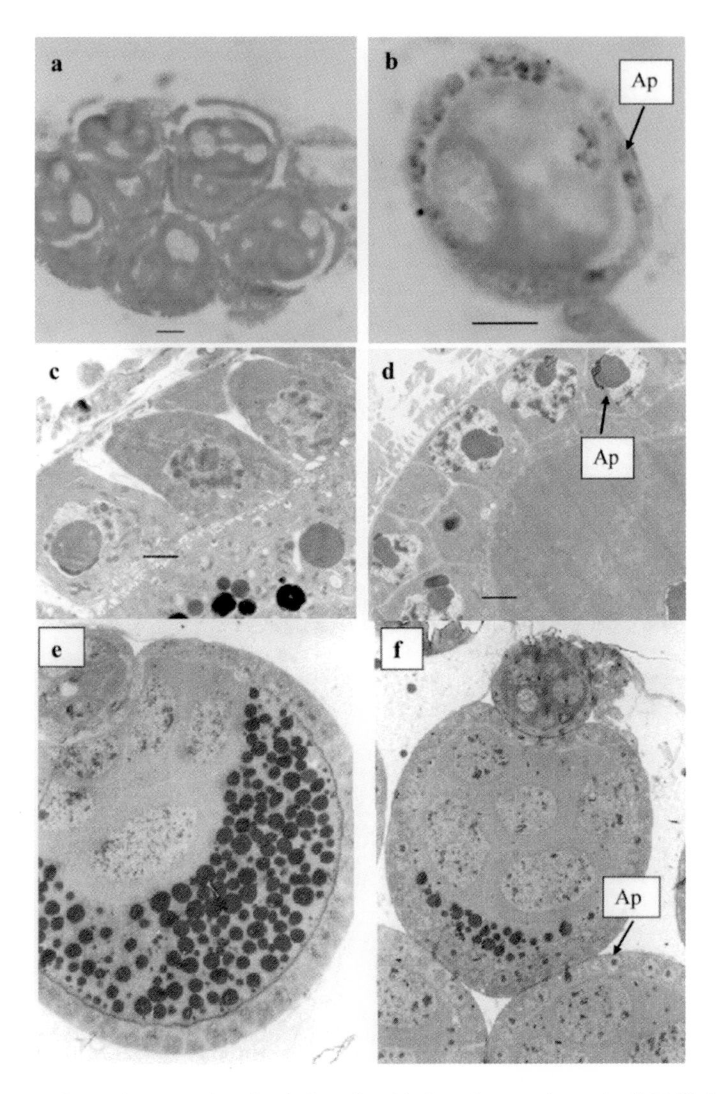

Figure 5 Follicles from ovaries of uninfected and infected mosquitoes. (*a*, *b*) Follicles from ovaries of *An. stephensi* stained with methyl green and treated by in situ terminal-deoxynucleotidyl-transferase-mediated dUTP-biotin nick end labeling (TUNEL). (*a*) Uninfected; (*b*) *P. y. nigeriensis* infected. (*c*, *d*) Electron micrographs of follicles from ovaries of *An. stephensi*. (*c*) Uninfected; (*d*) *P. y. nigeriensis* infected. (*e*, *f*) Toluidine blue-stained thick sections of follicles from ovaries of *Armigeres subalbatus*. (*e*) Uninfected mosquito; (*f*) mosquito infected with the filarial nematode, *Brugia malayi*. Ap, follicular epithelial cell with apoptotic nucleus. Scale bars: A and B = 10 μm; C and D = 2 μm. (*a–d*) Adapted with permission from Reference 34. (*e*) Produced with permission from Reference 23.

Malaria and Mosquito Longevity

The picture that emerges from a review of these studies is far from conclusive and the source of this confusion has been analyzed and discussed recently by Ferguson et al. (24). What is clear is that neither laboratory nor field studies have demonstrated that infected mosquitoes exhibit enhanced survivorship in comparison with uninfected mosquitoes at any time during the development of the parasite. Klein et al. (49) found that survival of heavily infected *Anopheles dirus* was only reduced after the first 8 days post infection and that survival decreased with oocyst count. However, several studies have concluded that mosquito mortality, during the first day or two post infection, is related to parasite density (27, 48, 57). This has been our experience where laboratory infection of mosquitoes feeding on mice with heavy gametocytemia (32) result in high mortality, whereas no substantial mortality occurs if mouse gametocytemia is low. Furthermore, the low parasite densities encountered in naturally infected mosquitoes collected in the field do not appear to be detrimental to survivorship (16, 33). Mosquito mortality that occurs early in infection is likely to be the result of the pathology associated with ookinete midgut invasion. Experiments designed to compare mortality when mosquitoes feed on hosts with differing gametocytemias and producing differing ookinete densities in the mosquito midgut should be conducted to enable this hypothesis to be fully tested. If heavy ookinete invasion causes death, then selection is likely to favor parasite/mosquito associations in which ookinete density is restricted below the level that could cause vector death. We have recently demonstrated that a large proportion of ookinetes undergo apoptosis in the mosquito midgut, suggesting that this may be a mechanism to reduce parasite density (5).

Of the 22 studies of mosquito survival analyzed by Ferguson et al. (24) none of the natural combinations of *Plasmodium* and mosquito species exhibited changes in survivorship, whereas 7 out of 10 of the unnatural combinations did so. It is not clear whether this is the due to lower gametocyte/ookinete density in natural parasite/mosquito combinations.

Mosquito mortality is likely to rise once salivary gland infections occur if mosquitoes have access to hosts and are exposed to their defensive behavior. This decrease in survivorship is advantageous for the parasite if it enhances transmission chances and thus could be manipulative. Meaningful assessments of the effect of infection on survivorship during the period of oocyst development are difficult to make in the laboratory and impossible to undertake in the field. The real effect of infection and concomitant changes in resource management associated with reproductive curtailment may only be evident when normal environmental and biotic stresses associated with host seeking, feeding, and oviposition are imposed on infected females. To my knowledge, these factors have never been incorporated in laboratory measurement of survivorship and, although present in a field situation, cannot be investigated due to the ethical considerations associated with releasing and monitoring infected females.

Survival of Infected Tsetse Flies

Similar equivocal conclusions exist from studies of the effect of trypanosome infections upon tsetse flies (59, 54). They demonstrated that the hazard of the tsetse fly *Glossina mortisans mortisans* dying increased with age more rapidly in flies exposed to infection with *Trypanosoma brucei rhodesiense* and was more severe in males than in females. In contrast to these salivary gland infections, no significant reduction in mortality was observed in flies with the midgut-exclusive infections associated with *Trypanosoma congolense*. Maudlin et al. (59) discuss the possible implication of these findings on our understanding of the vectorial capacity of wild tsetse flies and highlight the difficulty of assessing the impact of infection on natural populations.

Is Survival Manipulated?

Despite the obvious importance of vector survivorship to the likelihood of parasite transmission, there appears to be no conclusive evidence that this life-history trait is manipulated by parasites such that infected insects have a greater chance of survival during the parasite's prepatent phase than their uninfected counterparts. Any benefits that accrue from the depression of egg production in parasitized hematophagous insects may protect the vector from an early death, at least in natural vector/parasite associations. However, the experimental regimes used to date have not demonstrated the predicted inverse relationship between reproductive success and longevity in infected vectors. Perhaps this relationship only becomes evident when additional environmental and biotic stressors are present, but this hypothesis awaits investigation. The only evidence in its favor is the demonstration that mosquitoes infected with *P. y. nigeriensis* oocysts exhibit decreased persistence in returning to feed in the face of host defensive behavior and are thus less likely to be killed than uninfected females (6). From the parasite's perspective, the optimum trade-off between risky host contact and blood feeding is likely to change once infective stages can be transmitted and vector survivorship prospects may be reduced by parasites that manipulate feeding behavior and increase host contact (6, 50).

SUMMARY

From the moment a parasite is imbibed with a blood meal, reciprocal interactions between vector and parasite are initiated. Some of these occur at the interface between symbiont cells or tissues, but equally, parasite secretory/excretory products may affect the vector, and vector responses may operate at some distance from the invader. Many of these interactions have no long-term effect on vector physiology or life-history traits, but some, as illustrated above, cause profound changes that may be detrimental. If these interactions are of an adaptive nature they will, at least in the field, lead to outcomes that result from trade-off positions between the interests of the parasite and the interests of the vector. These may alter as biotic and environmental factors change. This is particularly likely in the case of parasites that cause disease or economic loss. In these cases intervention strategies rapidly impose severe selection pressures on the vector/parasite association. If we do not understand these interactions, we cannot predict the likely, long-term, outcomes of programs for vector or parasite control. It is therefore crucial that we rapidly use both empirical and experimental means to gain a broader understanding of all facets of both parasite manipulation of medically important insects and the response these insects make to infection.

The *Annual Review of Entomology* is online at http://ento.annualreviews.org

LITERATURE CITED

1. Agnew P, Koella JC, Michalakis Y. 2000. Host life history responses to parasitism. *Microbes Infect.* 2:891–96
2. Ahmed AM, Baggott SL, Maingon R, Hurd H. 2002. The costs of mounting an immune response are reflected in the reproductive fitness of the mosquito *Anopheles gambiae*. *Oikos* 97:371–77
3. Ahmed AM, Maingon R, Romans P, Hurd H. 2001. Effects of malaria infection on vitellogenesis in *Anopheles gambiae*. *Insect Mol. Biol.* 10:347–56
4. Ahmed AM, Maingon RD, Taylor PJ, Hurd H. 1999. The effects of infection with *Plasmodium yoelii nigeriensis* on the reproductive fitness of the mosquito *Anopheles gambiae*. *Invertebr. Reprod. Dev.* 36:217–22
5. Al-Olayan EM, Williams GT, Hurd H. 2002. Apoptosis dependent on a caspase-like enzyme in the protozoan malaria parasite, *Plasmodium berghei*. *Int. J. Parasitol.* In press
6. Anderson RA, Koella JC, Hurd H. 1999. The effect of *Plasmodium yoelii nigeriensis* infection on the feeding persistence of *Anopheles stephensi* Liston throughout the sporogonic cycle. *Proc. R. Soc. London Sci. Ser. B* 266:729–33
7. Beach R, Kilu G, Hendricks L, Oster C, Leewenburg J. 1984. Cutaneous leishmaniasis in Kenya: transmission of *Leishmania major* to man by the bite of a naturally infected *Phlebotomus duboscqi*. *Trans. R. Soc. Trop. Med. Hyg.* 78:747–51
8. Beach R, Kilu G, Leewenburg J. 1985. Modification of sand fly biting behaviour by *Leishmania* leads to increased transmission. *Am. J. Trop. Med. Hyg.* 34:278–82
9. Beier JC. 1998. Malaria parasite development in mosquitoes. *Annu. Rev. Entomol.* 43:519–43
10. Bockarie MJ, Alexander N, Bockarie F, Ibam E, Barnish G, Alpers M. 1996. The late biting habit of parous *Anopheles* mosquitoes and pre-bedtime exposure of humans to infective female mosquitoes. *Trans. R. Soc. Trop. Med. Hyg.* 90:23–25
11. Boyd MF, Stratman-Thomas WK. 1937. Studies on benign tertian malaria. 7. Some observations on inoculation and onset. *Am. J. Hyg.* 20:488–95
12. Briegel H. 1980. Determination of uric acid and hematin in a single sample of excretia from blood fed insects. *Experientia* 36:142
13. Briegel H. 1990. Fecundity, metabolism and body size in *Anopheles* (Diptera: Culicidae), vectors of malaria. *J. Med. Entomol.* 27:839–50
14. Carwardine SL, Hurd H. 1997. *Anopheles stephensi*: the effects of infection with *Plasmodium yoelii nigeriensis* on egg development and resorption. *J. Med. Vet. Entomol.* 11:265–69
15. Conway DJ, McBride JS. 1991. Genetic evidence for the importance of interrupted

feeding by mosquitoes in the transmission of malaria. *Trans. R. Soc. Trop. Med. Hyg.* 85:454–56
16. Cheng GMM, Beier JC. 1990. Effect of *Plasmodium falciparum* on the survival of naturally infected Afrotropical *Anopheles* (Diptera: Culicidae). *J. Med. Entomol.* 27:454–58
17. Christensen BM. 1981. Effect of *Dirofilaria immitis* on the fecundity of *Aedes trivittatus*. *Mosq. News* 41:78–81
18. Daniel TL, Kingsolver JG. 1983. Feeding strategy and the mechanics of blood sucking insects. *J. Theor. Biol.* 105:661–72
19. Dawkins R. 1982. *The Extended Phenotype*. Oxford, UK: Oxford Univ. Press
20. Dhadialla TS, Raikhel AS. 1994. Endocrinology of mosquito vitellogenesis. In *Perspectives in Comparative Endocrinology*, ed. KG Davey, RE Peter, SS Tobe, pp. 275–81. Canada: Natl. Res. Counc.
21. El Sawaf BM, El Sattar SA, Shehata MG, Lane RP, Morsy TA. 1994. Reduced longevity and fecundity in *Leishmania* infected sand flies. *Ann. J. Trop. Med. Hyg.* 51:767–70
22. Ewald PW. 1995. The evolution of virulence: a unifying link between parasitology and ecology. *J. Parasitol.* 81:659–69
23. Ferdig MT, Beernsten BT, Spray FJ, Li J, Christensen BM. 1993. Reproductive costs associated with resistance in a mosquito-filarial worm system. *Am. J. Trop. Med. Hyg.* 49:756–62
24. Ferguson HM, Rivero A, Read AF. 2002. The influence of malaria parasite genetic diversity on mosquito feeding and fecundity. *Trends Parasitol.* 18:256–62
25. Forbes MRL. 1993. Parasitism and host reproductive effort. *Oikos* 67:444–50
26. Freier JE, Friedman S. 1976. Effect of host infection with *Plasmodium gallinaceum* on the productive capacity of *Aedes aegypti*. *J. Invertebr. Pathol.* 28:61–66
27. Gad AM, Maier WA, Piekarski G. 1979. Pathology of *Anopheles stephensi* after infection with *Plasmodium berghei berghei*. *Z. Parasitenkd.* 60:249–61
28. Ham PJ, Banya AJ. 1984. The effect of experimental *Onchocerca* infection on the fecundity and oviposition of laboratory reared *Simulium* spp. (Diptera: Simuliidae). *Trop. Parasitol.* 35:62–66
29. Hamilton JGC, Hurd H. 2002. Parasite manipulation of vector behaviour. In *The Behavioural Ecology of Parasites*, ed. EE Lewis, JF Cambell, MVK Sukhdeo, pp. 259–81. London: CAB Int. 384 pp.
30. Hogg JC, Carwardine S, Hurd H. 1997. The effect of *Plasmodium yoelii nigeriensis* infection on ovarian protein accumulation by *Anopheles stephensi*. *Parasitol. Res.* 83: 374–79
31. Hogg JC, Hurd H. 1995. Malaria induced reduction of fecundity during the first gonotrophic cycle of *Anopheles stephensi* mosquitoes. *Med. Vet. Entomol.* 9:176–80
32. Hogg JC, Hurd H. 1995. *Plasmodium yoelii nigeriensis*: the effect of high and low intensity of infection upon the egg production and blood meal size of *Anopheles stephensi* during three gonotrophic cycles. *Parasitology* 111:555–62
33. Hogg JC, Hurd H. 1997. The effect of natural *Plasmodium falciparum* infection on the fecundity and mortality of *Anopheles gambiae s.l.* in northeast Tanzania. *Parasitology* 114:325–31
34. Hopwood JA, Ahmed AM, Polwart A, Williams GT, Hurd H. 2001. Malaria-induced apoptosis in mosquito ovaries: a mechanism to control vector egg production. *J. Exp. Biol.* 204:2773–80
35. Hurd H. 1990. Physiological and behavioural interactions between parasites and invertebrate-hosts. *Adv. Parasitol.* 29:271–317
36. Hurd H. 1993. Reproductive disturbances induced by parasites and pathogens of insects. In *Parasites and Pathogens of Insects*, ed. NE Beckage, SN Thompson, BA Federici, 1:87–105. San Diego: Academic. 364 pp.
37. Hurd H. 1998. Parasite manipulation of insect reproduction: who benefits? *Parasitology* 116:S13–221

38. Hurd H. 2001. Parasite regulation of insect reproduction: similar strategies, different mechanisms? In *Endocrine Interactions of Parasites and Pathogens*, ed. JP Edwards, RJ Weaver, pp. 207–19. Oxford: Bios Sci. Publ. Ltd. 314 pp.
39. Hurd H, Hogg JC, Renshaw M. 1995. Interactions between blood-feeding, fecundity and infection in mosquitoes. *Parasitol. Today* 11:411–16
40. Hurd H, Warr E, Polwart A. 2001. A parasite that increases host lifespan. *Proc. R. Soc. London Sci. Ser. B* 68:749–53
41. Hurd H, Webb TJ. 1997. The role of endocrinologically active substances in mediating changes in insect hosts and insect vectors. In *Parasites and Pathogens: Effects on Host Hormones and Behaviour*, ed. NE Beckage, pp. 179–97. New York: Chapman & Hall. 338 pp.
42. Ilg T. 2001. Lipophosphoglycan of the protozoan parasite *Leishmania:* stage- and species-specific importance for colonization of the sandfly vector, transmission and virulence to mammals. *Med. Microbiol. Immunol.* 190:13–17
43. Jahan N, Docherty PT, Billingsley PF, Hurd H. 1999. Blood digestion in the mosquito, *Anopheles stephensi* the effects of *Plasmodium yoelii nigeriensis* on midgut enzyme activities. *Parasitology* 119:535–41
44. Javadian E, Macdonald WW. 1974. The effect of infection with *Brugia pahangi* and *Dirofilaria repens* on the egg-production of *Aedes aegypti*. *Ann. Trop. Med. Parasitol.* 68:477–81
45. Jeffries D, Livesey JL, Molyneux DH. 1986. Fluid mechanics of bloodmeal uptake by *Leishmania*-infected sandflies. *Acta Trop.* 43:43–53
46. Kaslow DC, Welburn S. 1996. Insect transmitted pathogens in the insect midgut. In *Biology of the Insect Midgut*, ed. MJ Lehane, PF Billingsley, pp. 433–62. London: Chapman & Hall. 486 pp.
47. Killick-Kendrick R, Bryson ADM, Peters W, Evans DA, Leaney AJ, Rioux J-A. 1985. Zoonotic cutaneous leishmaniasis in Saudi Arabia: lesions healing naturally in man followed by a second infection with the same zymodeme of *Leishmania major*. *Trans. R. Soc. Trop. Med. Hyg.* 79:363–65
48. Klein TA, Harrison BA, Andre RG, Whitmire RE, Inlao I. 1982. Detrimental effects of *Plasmodium cynomolgi* infections on the longevity of *Anopheles dirus*. *Mosq. News* 42:265–71
49. Klein TA, Harrison BA, Grove JS, Dixon SV, Andre RG. 1986. Correlation of survival rates of *Anopheles dirus* A (Diptera: Culicidae) with different infection densities of *Plasmodium cynomolgi*. *Bull. WHO Org.* 64:901–7
50. Koella JC. 1999. An evolutionary view of the interactions between anopheline mosquitoes and malaria parasites. *Microb. Infect.* 1:303–8
51. Koella JC, Packer MJ. 1996. Malaria parasites enhance blood-feeding of their naturally infected vector *Anopheles punctulatus*. *Parasitology* 113:105–9
52. Koella JC, Sørensen FL, Anderson RA. 1998. The malaria parasite, *Plasmodium falciparum,* increases the frequency of multiple feeding of its mosquito vector, *Anopheles gambiae*. *Proc. R. Soc. London Sci. Ser. B.* 265:763–68
53. Law J, Ribeiro JMC, Wells M. 1992. Biochemical insights derived from diversity in insects. *Annu. Rev. Biochem.* 61:87–112
54. Leak SGA. 1998. *Tsetse Biology and Ecology: Their Role in the Epidemiology and Control of Trypanosomosis*. New York: CABI. 568 pp.
55. Lehane MJ. 1991. *The Biology of Blood-Sucking Insects*. London: Harper Collins Academic. 288 pp.
56. Li X, Sina B, Rossignol PA. 1992. Probing behaviour and sporozoite delivery by *Anopheles stephensi* infected with *Plasmodium berghei*. *Med. Vet. Entomol.* 6:57–61
57. Lyimo EO, Koella JC. 1992. Relationship between body size of adult *Anopheles gambiae s.l.* and infection with the malaria

parasite *Plasmodium falciparum*. *Parasitology* 104:233–37

58. Macdonald G. 1957. *The Epidemiology and Control of Malaria*. Oxford, UK: Oxford Univ. Press
59. Maudlin I, Welburn SC, Millligan PJM. 1998. Trypanosome infections and survival in tsetse. *Parasitology* 116:S23–28
60. Molyneux DH, Jefferies D. 1986. Feeding behaviour of pathogen-infected vectors. *Parasitology* 92:721–36
61. Moore J. 1993. Parasites and the behaviour of biting flies. *J. Parasitol.* 79:1–16
62. Moore J, Gotelli NJ. 1990. A phylogenetic perspective on the evolution of altered host behaviour: a critical look at the manipulation hypothesis. In *Parasitism and Host Behaviour*, ed. CJ Barnard, JM Behnke, pp. 193–233. London: Taylor & Francis. 331 pp.
63. Moret Y, Schmid-Hempel P. 2000. Survival for immunity: the price of immune system activation for bumblebee workers. *Science* 290:1166–68
64. Neva FA, Sheagreb JN, Shulman NR, Canfield CJ. 1970. Malaria, host-defences mechanisms and complications. *Ann. Intern. Med.* 73:295–306
65. O'Shea B, Rebollar-Tellez E, Ward RD, Hamilton JGC, El Naiem D, Polward A. 2002. Enhanced sandfly attraction to Leishmania-infected hosts. *Trans. R. Soc. Trop. Med. Hyg.* 96:1–2
66. Perrin N. 1996. On host life-history response to parasitism. *Oikos* 75:317–20
67. Poulin R. 1995. 'Adaptive' changes in the behaviour of parasitized animals: a critical review. *Int. J. Parasitol.* 25:1371–83
68. Poulin R. 1998. Evolution and phylogeny of behavioural manipulation of insects by parasites. *Parasitology* 116:S3–9
69. Renshaw M, Hurd H. 1994. The effect of *Onchocerca* infection on the reproductive physiology of the British blackfly, *Simulium ornatum*. *Parasitology* 109:337–45
70. Ribeiro JMC. 1995. Blood-feeding arthropods: live syringes or invertebrate pharmacologists? *Infect. Agents Dis.* 4:143–52
71. Ribeiro JMC. 2000. Blood-feeding in mosquitoes: probing time and salivary gland anti-haemostatic activities in representatives of three genera (*Aedes, Anopheles, Culex*). *Med. Vet. Entomol.* 14:42–48
72. Ribeiro JMC, Rossignol PA, Spielman A. 1985. *Aedes aegypti:* model for blood finding strategy and prediction of parasite manipulation. *Exp. Parasitol.* 60:118–32
73. Ribeiro JMC, Rossignol PA, Spielman A. 1985. Salivary gland apyrase determines probing time in anopheline mosquitoes? *J. Insect Physiol.* 31:689–92
74. Rogers ME, Chance ML, Bates PA. 2002. The role of promastigote secretory gel in the origin and transmission of the infective stage of *Leishmania mexicana* by the sandfly *Lutzomyia longipalpis*. *Parasitology* 124:495–507
75. Rossignol PA, Ribeiro JMC, Spielman A. 1984. Increased intradermal probing time in sporozoite-infected mosquitoes. *Am. J. Trop. Med. Hyg.* 33:17–20
76. Rossignol PA, Ribeiro JMC, Spielman A. 1986. Increased biting rate and reduced fertility in sporozoite-infected mosquitoes. *Am. J. Trop. Med. Hyg.* 35:277–79
77. Rossignol PA, Rossignol AM. 1988. Simulations of enhanced malaria transmission and host bias induced by modified vector blood location behaviour. *Parasitology* 97:363–72
78. Schwartz A, Koella JC. 2001. Trade-offs, conflicts of interest and manipulation in *Plasmodium*-mosquito interactions. *Trends Parasitol.* 17:189–94
79. Sheldon BS, Verhulst S. 1996. Ecological immunology: costly parasite defenses and trade-offs in evolutionary ecology. *Trends Ecol. Evol.* 11:317–21
80. Shortt HE, Swaminath CS. 1927. The method of feeding of *Phlebotomus argentipes* with relation to its bearing on the transmission of Kala-azar. *Indian J. Med. Res.* 15:827–836
81. Sterling CR, Aikawa M, Vanderberg JP. 1973. The passage of *Plasmodium berghei*

sporozoites through the salivary glands of *Anopheles stephensi*: an electron microscope study. *J. Parasitol.* 59:593–605
82. Stierhof Y-D, Bates PA, Jacobson RL, Rogers MS, Schlein Y, et al. 1999. Filamentous proteophosphoglycan secreted by *Leishmania* promastigotes forms gel-like three-dimensional networks that obstruct the digestive tract of infected sandfly vectors. *Eur. J. Cell Biol.* 78:675–89
83. Taylor P, Hurd H. 2001. The influence of host haematocrit on the blood feeding success of *Anopheles stephensi*: implications for enhanced malaria transmission. *Parasitology* 122:491–96
84. Wekesa JW, Copeland RS, Mwangi RW. 1992. Effect of *Plasmodium falciparum* on blood-feeding behaviour of naturally infected *Anopheles* mosquitoes in western Kenya. *Am. J. Trop. Med. Hyg.* 47:484–88
85. Welburn SC, Maudlin I. 1997. Current trends in parasite vector interactions. In *Trypanosomiasis and Leishmaniasis*, ed. G Hide, JC Mottram, GH Coombs, PH Holmes, pp. 315–34. London: CAB Int. 366 pp.

Annu. Rev. Entomol. 2003. 48:163–84
doi: 10.1146/annurev.ento.48.091801.112657

First published online as a Review in Advance on August 27, 2002

MALE ACCESSORY GLAND SECRETIONS: Modulators of Female Reproductive Physiology and Behavior

Cedric Gillott
Department of Biology, University of Saskatchewan, 112 Science Place, Saskatoon, SK S7N 5E2, Canada; e-mail: gillott@duke.usask.ca

Key Words egg laying, fecundity, mating, receptivity, sperm

■ **Abstract** Secretions of male accessory glands contain a variety of bioactive molecules. When transferred during mating, these molecules exert wide-ranging effects on female reproductive activity and they improve the male's chances of siring a significant proportion of the female's offspring. The accessory gland secretions may affect virtually all aspects of the female's reproductive activity. The secretions may render her unwilling or unable to remate for some time, facilitating sperm storage and ensuring that any eggs laid will be fertilized by that male's sperm. They may stimulate an increase in the number and rate of development of eggs and modulate ovulation and/or oviposition. Antimicrobial agents in the secretions ensure that the female reproductive tract is a hospitable environment during sperm transfer. In a few species the secretions include noxious chemicals. These are sequestered by developing eggs that are thereby protected from predators and pathogens when laid.

CONTENTS

INTRODUCTION

The accessory reproductive glands (ARG) of male insects produce secretions essential for the transfer of sperm to the female. In many insect species, especially those in more primitive orders, components of the secretion combine (in a way that remains largely a biochemical mystery) to form one or more sac-like structures, spermatophores, in which the sperm are conveyed to the female reproductive tract. In other taxa spermatophores are not formed, and sperm are transferred in a liquid medium. This seminal fluid is also a product of the ARG, though components secreted by other regions of the male reproductive tract, notably the ejaculatory duct, may be added.

Facilitation of sperm transfer, however, is not the only role of ARG secretions. Specific components have a variety of functions that collectively serve to improve the likelihood that the male will sire a significant proportion of a female's offspring. The ARG components exert their effects at all phases of the reproductive biology of the mated female, from the moment sperm is deposited in the reproductive tract to egg deposition. Processes affected may include sperm protection, storage and activation, sperm competition (effects on rival males' sperm), female behavior (notably induction of refractoriness and reduction in attractiveness), fecundity, ovulation, oviposition, and protection of laid eggs. Components with sperm-related functions seem likely to act mainly within the female reproductive tract, whereas those molecules that influence female behavior and physiology typically appear to pass through the wall of the tract into the hemocoel to reach their site of action. These statements are not intended to imply that males alone, through their ARG secretion, regulate these processes. As is already clear in regard to sperm-related processes (24, 33), female insects are far from being "silent partners." Thus, it may be anticipated that females strongly influence the manner and extent to which male ARG materials function. However, as discussed below, there is as yet only slight evidence to support this contention.

The past two decades have seen the publication of two comprehensive reviews of male ARG (14, 33), as well as several that deal with specific aspects, namely, structure and development of ARG (44, 45), endocrine regulation of ARG secretory activity (34, 37), and the role of ARG secretions in egg production (including nuptial gifts) (35, 42, 43, 103) and protection (6, 25, 35). In addition, the ARG products of male *Drosophila melanogaster* (15, 16, 62, 63, 105, 106) and aedine mosquitoes (59) have been extensively reviewed. The well-known genetics of *D. melanogaster*, combined with its amenability for biochemical, physiological, and genomic analysis, has made this species by far the most extensively studied with respect to ARG function.

The objective of this review is to provide a synthesis of the nature and functions of male ARG secretions that affect the reproductive physiology and behavior of the mated female. Constraints of space and the author's particular interests and expertise dictate that coverage is selective rather than all-inclusive. Thus, the roles of ARG in the provision of nuptial gifts and in egg protection have been omitted because these topics have been recently reviewed by other authors (see above).

BIOCHEMISTRY OF ACCESSORY GLAND SECRETION

ARG secretion generally includes carbohydrates (both free and complexed with protein), some lipid (normally bound to protein), and small amounts of amino acids and amines [references in (33)]. Unexpected materials are found in the ARG of some species, for example, uric acid (cockroaches) (88), prostaglandins (Lepidoptera) [references in (33)], juvenile hormones (JH) (Lepidoptera and mosquitoes) (8, 84), and various toxic materials that serve as egg protectants following their transfer to the female [see (6, 25)]. However, the major components of ARG secretions, in both quantity and importance as modulators of female reproductive activity, are proteins.

The proteinaceous components of ARG secretion can be arranged in three classes based principally on molecular size, though there are also correlations with functions and sites of action (105). In the first group (simple peptides) are molecules containing fewer than 100 amino acids. These include about 75% of *Drosophila* ARG proteins (Acps). For example, the sex peptide (SP) (Acp70A) of *D. melanogaster* is synthesized as a 55–amino acid precursor that includes a 19–amino acid signal sequence (15). The signal peptide is removed as the SP is released into the ARG lumen (63); thus, the active form of the molecule is transferred to the female. In *Drosophila funebris* two ARG polypeptides have been characterized; the first, PS-1, comprises two polypeptide chains, each with 27 amino acids, that differ in a single amino acid at position 2, whereas the second, with sequence homologies to serpins (serine protease inhibitors), is composed of 63 amino acids (92). Perhaps also in this group are "substance A" (mol wt 500–1000 kDa) from the ARG of the bean weevil *Acanthoscelides obtectus* (21) and the small receptivity-inhibiting peptides of *Musca domestica* (mol wt 750 kDa) (produced by the ejaculatory duct) and *Cochliomyia hominivorax* (mol wt 3000 kDa), whose source is unknown (77). Molecules in this first group are released from the site of production in their functional form and, when their function requires it, are easily able to move through the wall of the reproductive tract.

The second group includes larger molecules (200–400 amino acids), commonly glycosylated, that may have to be cleaved before becoming biologically active. An example is Acp26Aa (264 amino acids) from *D. melanogaster*, which is a prohormone-like molecule (75). Acp26Aa contains a sequence of amino acids similar to that found in the egg-laying hormone of the mollusk *Aplysia californica*, whose active form is arrived at by cleavage of the 271–amino acid precursor. It is possible that some structural and physiological proteins from the ARG of other species are also in this category. For example, several spermatophore-forming proteins in *Tenebrio molitor* undergo proteolysis to their active form (38). The oviposition-stimulating proteins of *Melanoplus sanguinipes* (mol wt 30 kDa) and *Locusta migratoria* (mol wt 13 kDa) also fall within the size range of this group (see 110). These are presumed to act in the hemocoel, as their injection directly into the body cavity induces their effect. However, it must be stressed that the necessary studies have not been carried out to examine whether either protein is structurally similar to prohormones or in what form they move across the wall of the genital tract.

Requiring special consideration as a possible member of this category is matrone, the purported multifunctional molecule of aedine [but not, apparently, anopheline (60)] mosquitoes. More than 30 years ago, Fuchs and collaborators (30–32), using whole-body extracts of *Aedes aegypti*, identified and partially characterized matrone as a protein, separable into α and β fractions with molecular weights of 60 kDa and 30 kDa, respectively. Both fractions were necessary to induce refractoriness when injected into virgin females but α matrone alone would stimulate oviposition. However, more recent analyses, based on extracts of isolated ARG, indicate that the active fractions in mosquito ARG have molecular weights in the low kilodalton range and are comparable with those of other Diptera. Thus, for the receptivity-inhibiting substances (RIS) of *Culex tarsalis*, a molecular weight of about 2 kDa has been suggested (112), whereas for *A. aegypti*, a 7.6 kDa peptide that both induces refractoriness and modulates host-seeking behavior has been reported (64). Also relevant is the early report that implantation of the SP-containing ARG of *D. melanogaster* also renders virgin *Aedes* refractory (19). The high-molecular-weight values for the mosquito RIS may have resulted from binding of the bioactive molecule with other nonspecific proteins (112). In view of these conflicting reports and the variety of effects claimed for matrone (sometimes solely on the basis of injection of whole ARG extracts) (see following sections), a thorough reexamination of the biochemical nature of mosquito ARG proteins is of utmost importance.

Large, often glycosylated, proteins form the third proteinaceous component of ARG secretions. Included are the larger structural proteins, for example, the spermatophore protein SP-62 (mol wt 85 kDa), of *M. sanguinipes* (13) and the sperm-storage protein Acp36DE (mol wt 122 kDa) of *D. melanogaster* (5). SP-62 is cleaved to a 62-kDa fraction during spermatophore formation (13), whereas Acp36DE is processed to a 68-kDa molecule starting about 20 min after transfer to the female genital tract (5, 78). Also likely to belong to this group are many of the enzymes identified in ARG secretions. These include the proteinases responsible for production of the active forms of spermatophore proteins in *T. molitor* (38) and *M. sanguinipes* (13) and the trehalases in the ARG of *Periplaneta americana* (98) and *T. molitor* (109).

Though the majority of the ARG compounds noted above are synthesized by the ARG per se, there are a few reports that proteins synthesized in the fat body are then sequestered by the ARG [see (33)]. In addition, the mass of uric acid found in cockroach ARG is probably synthesized in the fat body and Malpighian tubules and then accumulated by the glands. In contrast, some male beetles are able to ingest noxious chemicals and store them in the ARG ready for transfer to the female during mating. For example, males of the spotted cucumber beetle

Diabrotica undecimpunctata acquire cucurbitacins from food plants, and a fire beetle (*Neopyrochroa flabellata*) avidly ingests cantharidin under experimental conditions (though how it acquires this naturally is not known) [see (25)].

INDUCTION OF FEMALE REFRACTORINESS AND REDUCTION OF ATTRACTIVENESS

After a successful mating, females of a range of species (but especially Diptera) do not remate for a varied period of time. This nonmating status may be seen as an active rejection of would-be mates (refractoriness) or as a passive decline in attractiveness. Ablation experiments, injection of tissue homogenates, or injection of purified components showed that the male ARG generally produce RIS (see 36). However, in some species testis or ejaculatory duct material is important, and rarely, mating provides a physical rather than a chemical cue [references in (33)]. Biochemical analyses of the secretion and characterization of the RIS have shown that these are peptides (e.g., *Drosophila* SP and the RIS of *Musca* and *Cochliomyia*) or perhaps proteins (matrone).

Discussion of the site and mode of action of RIS must include consideration of the fate of the RIS after its deposition in the female's genital tract. In the case of peptidic RIS, it is likely that, based on their size, they will be able to pass through the wall of the genital tract into the hemocoel to be transported to their site of action. Such ability has not yet been demonstrated for any RIS, though three decades ago it was shown that in *Musca* radiolabeled male peptides transferred during mating move rapidly across the thin-walled vaginal pouches into all regions of the body (66, 99). In *Drosophila* the peptides Acp26Aa and Acp62F move readily through the thin, nonmuscular posterior wall of the vagina (71).

Early experiments showed that injection of matrone into the hemocoel could induce refractory behavior in virgin *A. aegypti* (30–32). Disregarding the questionable nature of matrone, it seems unlikely that an intact protein of this size could penetrate the wall of the female genital tract after a normal mating. Rather, the protein may stimulate the genital tract itself to release a hormone that then moves to the target, or the protein may be cleaved to release the active fraction. Though conclusive evidence is available for neither of these options, it has been elegantly demonstrated that the RIS in the grasshopper *Gomphocerus rufus* exerts its effect only from within the spermatheca (46, 47). Injection of ARG extract into a female's body cavity does not render her refractory; however, microinjection of this material into the spermatheca rapidly induces the characteristic "secondary defense" (the female rejects mating attempts by vigorous kicking of the hindlegs). Further, the spermathecal cells appear to take up and process ARG material, which it is suggested may then stimulate contact chemoreceptors in the spermathecal canal (47).

The rather limited evidence supports the assumption that the site of action of RIS is the nervous system. For example, in *Gomphocerus*, nerve severance experiments indicate that the information from the putative spermathecal canal chemoreceptors travels to the terminal abdominal ganglion en route to a "higher center," likely the brain (67). In *Musca*, radiolabeled male peptides become concentrated and, relative to other body regions, are more persistent in the head (66). Further, decapitated, decerebrated, or cervically ligated virgin females remain receptive, mating up to five times in an 8-h period (65). Using ^{125}I-labeled synthetic SP and synthetic DUP99B [an ejaculatory duct peptide with structural and functional homology with SP (89)] injected into the body cavity, Ottiger et al. (81) showed that the RIS bind to multiple sites in the nervous system and to the genital tract of female *D. melanogaster* (but not those of *D. funebris*, an SP-insensitive species). Binding to afferent nerves is especially strong, whereas binding to the brain is weak. However, owing to a lack of resolution, it was not possible to discriminate between binding to neurons and binding to the surrounding glial cells. The latter may indeed play an active role in the transmission of humoral signals to, and thus modify the activity of, the nervous system (4). The observation that virgin females of the learning- and memory-deficient *dunce* mutant do not become refractory after injection of SP (9b) led Fleischmann et al. (29) to examine whether the mushroom bodies within the brain of *D. melanogaster* had a role in regulation of refractory behavior. However, flies with chemically ablated mushroom bodies continued to show reduced receptivity after injection of synthetic SP, suggesting that these centers are not implicated (29). Ottiger et al. (81) also observed binding of the peptides to the male nervous system in a pattern broadly similar to that seen in females, though whether this has any functional significance is questionable. As an exception, in *Aedes* the terminal abdominal ganglion may be the site of action of the RIS (41).

For species from a range of taxa, the duration of the refractory period is correlated with the "quality" of the mating, that is, how much ARG material and sperm are transferred and, in some instances, how long the spermatophore remains in the female tract. In the extreme, as for example in *A. aegypti* and *Lucilia cuprina* (Australian sheep blowfly), a sexually mature male mating for the first time may switch off a female for life [hence the term "monocoitic substance" for the RIS (77)]. More typically, receptivity returns after a period of time, especially in females that have had the opportunity to lay several batches of eggs or had previously mated to ARG-depleted (multiply mated) males [references in (33, 60, 94)]. Only rarely, however, has the relationship between mating quality and the duration of refractoriness been examined in detail. Thus, in contrast to normally mated *Drosophila* females, which remain refractory for about 7–9 days, females injected with SP reject males for only about 24 h.

Though more than 40 years ago Manning (73) proposed that sperm may play a role in the long-term refractoriness of mated females, the nature of the so-called "sperm effect" is only now becoming clear. Kubli (62) has argued that SP is required for both short- and long-term refractoriness, with sperm potentiating SP's effect. Hihara (50) noted that repeatedly mated males, which continue to transfer sperm but transfer decreasing amounts of ARG fluid, are progressively less effective at inducing long-term refractoriness. Further, transgenic virgin females that continuously express SP ectopically show persistent refractoriness (1). Support for this proposal has been obtained using male *D. melanogaster* mutants that lack the ability to produce either sperm or ARG fluid (56, 108). Females mated to *prd* males (which generate sperm but have severely reduced ARG) remain as receptive as virgins. Females mated to *tud* males (which have normal ARG but produce no sperm) were refractory for 2 days or less, after which they were again receptive to males (56). In other words, though both SP and sperm are required for full expression of refractoriness, the sperm act only indirectly. Thus, Kubli (62) suggested that sperm might stabilize the SP. However, it is equally possible that the sperm signal, via neural triggers, that the sperm-storage organs are full. It must also be noted that the experiments performed to date do not rule out the participation of ARG proteins other than SP in the induction and persistence of refractoriness.

In addition to inducing refractoriness (when females actively reject males), in *Drosophila* and some Lepidoptera mating also lowers the attractiveness of the female. In Lepidoptera reduced attractiveness results from the arrest of pheromone synthesis and release triggered by chemicals in the seminal fluid. Thus, in the corn earworm moth, *Helicoverpa zea*, injection of combined extracts of ARG and ejaculatory duct stops the synthesis of sex pheromone (58). In *Helicoverpa assulta* the pheromonostatic factors are in the ARG of males in scotophase but not in those in photophase (93). In other Lepidoptera, for example the cecropia moth (*Hyalophora cecropia*) and the gypsy moth (*Lymantria dispar*), sperm or testicular secretions bring about pheromonostasis [references in (58)]. Kingan et al. (57) isolated a 57–amino acid peptide (mol wt 6.6 kDa) from the ARG of *H. zea* that partially suppressed pheromone synthesis. For complete pheromonostasis, however, an intact ventral nerve cord is necessary, indicating that neuronal signals are also involved. Remarkably, synthetic *D. melanogaster* SP depresses pheromone biosynthesis in *Helicoverpa armigera* (27), though the amino acid sequence of SP bears no similarity to that of the *H. zea* peptide reported by Kingan et al. Further, there is no indication that ARG products have pheromonostatic effects in *Drosophila* (101). Pheromone synthesis in Lepidoptera is regulated by pheromone biosynthesis–activating neuropeptide (PBAN), released from the brain during scotophase [references in (87)]. However, the steps between deposition of male materials in the bursa and the arrest of pheromone synthesis remain largely a mystery. Because they used intact insects, Kingan et al. were unable to localize the site of action of the *H. zea* pheromonostatic peptide. Using an in vitro preparation, Fan et al. (27) demonstrated that synthetic *Drosophila* SP significantly inhibits PBAN-induced pheromone synthesis in a dose-dependent manner, suggesting that the male factor acts directly on the pheromone gland in *H. armigera*. Synthetic *Drosophila* SP also stimulates in vitro JH biosynthesis in the corpora allata (CA) of *H. armigera* (27). These two observations may seem contradictory given that in *H. armigera* (26) and other Noctuidae (20) there is a correlation between CA activity and pheromone synthesis and release. In vivo, however, the effects of an RIS on the CA and on the pheromone glands may be age related and modulated by other factors (notably PBAN).

In contrast to Lepidoptera, virgin and 24-h postmating female *D. melanogaster* have the same amount of pheromone (101), and in this species reduced attractiveness may be due to the relative immobility of mated females compared to virgins (100). Further, the initial change is due to neither ARG products nor sperm, though the latter are required for maintenance of reduced attractiveness (101).

SPERM-RELATED FUNCTIONS

Rendering a female refractory or less attractive is an important means by which males of many species ensure paternity of at least some eggs laid by that female. However, other strategies have evolved to ensure that the sperm of a successfully mating male receive their "just desserts." For the first successful male, these strategies include the use of physical "plugs" to prevent further seminal fluid transfer, facilitation of sperm storage, and the acceleration of egg production (which is considered in subsequent sections). For the last mating male, destruction or displacement of previously deposited sperm is paramount. Protection and activation of sperm in preparation for fertilization may also be important.

The earliest means of preventing further mating may have been the presence of a spermatophore, though this structure generally is ejected or digested in situ within a few days. In many species that do not use spermatophores the seminal fluid coagulates within the female reproductive tract to form a mating plug. Mating plugs have been reported in Orthoptera, Lepidoptera, mosquitoes, *Apis mellifera*, and *Drosophila* spp. [references in (33)], though except for the plug of *D. melanogaster* there is little information on the means by which they are formed. In *Drosophila* the mating plug is divisible into anterior and posterior regions (72). The larger posterior portion, which forms within 3 min after the start of mating, is primarily a product of the ejaculatory duct (3) and has a 38-kDa protein (PEB-me) as a major component (68, 72). PEB-me is especially rich in proline and glycine (which make up >40% of the total amino acids), and in one region of the polypeptide chain there are 12 proline-glycine-glycine repeats. It is also probable that PEB-me is highly glycosylated (72). These two features are common to two other invertebrate proteins, spider silk and mussel byssus threads, both of which undergo homopolymerization (self-conversion from the secreted liquid form to the final solid state) (72). Posterior mating plug formation begins before sperm transfer and occurs in the absence of ARG products and sperm (3, 72), suggesting that the posterior plug may have functions in addition to simply preventing further mating and sperm loss. In contrast, formation of the anterior region of the mating plug occurs some 20 min after the start of mating, that is, after sperm transfer, and involves ARG proteins. In particular, the sperm-storage protein, Acp36DE, appears to be an important constituent of the anterior plug (72).

Like those of other animals, the genital tracts of insects are potential avenues of entry for microorganisms, some of which may negatively impact reproductive success [see (9a, 74)]. Though information is limited, insects of both sexes produce antibacterial proteins within the reproductive system, including the male ARG.

Comparison of genital tract extracts from virgin and just-mated *D. melanogaster* females revealed that three prominent antibacterial proteins are transferred during mating (69). One of these proteins (mol wt 28 kDa) is a product of the ARG, whereas the others are from the ejaculatory duct. It has been speculated that these proteins protect unstored sperm, though this effect would only be temporary, as any unstored seminal fluid is expelled with the laying of the first egg (69). Antimicrobial proteins synthesized within the female reproductive system likely afford longer-term sperm protection.

Sperm in the female storage organs may be physiologically quite different from those found in the male. During or shortly after their transfer the sperm may become more motile and develop the ability to fertilize eggs. In the seminal vesicles of Orthoptera, for example, groups of sperm are held by their heads within a gelatinous "cap" (a testicular product); however, within the spermatheca the sperm are free. Apparently, disintegration of the cap occurs within the spermatophore and involves ARG material (104). Using gel filtration, Viscuso et al. (104) identified a fraction of the ARG secretion of the tettigoniid *Rhacocleis annulata* with a prominent 29-kDa protein. In in vitro incubations this fraction can destroy the cap to liberate the sperm. Electrophoresis showed that the fraction includes a number of proteins, and further study is necessary to characterize the 29-kDa protein and to clarify whether the other proteins in the fraction are also necessary for cap breakdown.

The nature of capacitation of insect sperm and the factors that regulate it remain largely a mystery. However, there are signs that ARG components may be involved. That ARG material is necessary for sperm capacitation in *Drosophila* has been claimed following experiments with *prd* and *tud* mutants (108). *prd* males are infertile, suggesting that sperm cannot fertilize eggs in the absence of ARG material. Further, when a *prd* mating is followed (within 2 days) by a *tud* mating, eggs of a small proportion of females tested can be fertilized. Though Xue & Noll (108) suggested an effect on capacitation, it seems equally possible that these observations reflect improper storage of the sperm from *prd* males in the absence of ARG materials, as discussed in the next paragraph. In a study of sperm competition in *Drosophila*, using doubly mated females, Price et al. (86) noted that the second male could both physically displace and incapacitate sperm remaining from the first mating (incapacitation is the inhibition of the use of sperm without removing them). However, whereas displacement occurred only if the second male transferred sperm, incapacitation required only the presence of seminal fluid. Interestingly, incapacitation was seen only when the mating interval was long (7 days). This may result from a gradual reduction of viability and/or female-induced changes in the quality of the first male's sperm (86). Clearly, much work remains to be done with respect to ARG involvement in sperm capacitation.

Uncommonly, as in some Orthoptera, the spermatophore opens directly into the spermatheca or its duct, so that sperm storage is readily accomplished. Generally, however, seminal fluid is deposited in the female's bursa or uterus, and sperm must move (or be moved) to their site of storage. Failure to reach this site results in poor fertilization when eggs move down the oviduct during oviposition. Largely circumstantial evidence indicates that in some species male ARG secretion stimulates peristalsis of the female tract to move sperm. The clearest work, done over 40 years ago (22), indicates that the secretion of the opaque ARG in *Rhodnius prolixus* acts on a small area of the dorsal bursa wall. When stimulated, nerves running from this region to the common oviduct appear to induce rhythmic muscular contractions that move sperm into the spermatheca. A similar mechanism has been proposed for some Lepidoptera and *Periplaneta americana*, though experimental proof is lacking [see (33)]. An alternative is for the sperm to migrate under their own power into the storage organ. Although there is no significant evidence for the involvement of ARG components in inducing sperm motility, a few authors have reported the existence of motility-inducing factors in the secretion produced by the noncuticular region of the ejaculatory duct of Lepidoptera [references in (33)]. Neither peristalsis of the female tract nor sperm motility ensures that sperm reach their storage sites. For this to occur, such activity must have a directional component. For example, migrating sperm may be attracted into the spermatheca by secretion released within this organ, as suggested for the boll weevil, *Anthonomus grandis* (39). In *Drosophila* sperm are initially deposited in the uterus, from which they are transferred into the seminal vesicle (80–90% of the total) and the paired spermathecae. Using transgenic male flies, Tram & Wolfner (102) demonstrated that in matings with males deficient in ARG products, 90% of the sperm was not properly stored. Simultaneously, it was determined that Acp36DE is essential for sperm storage (78). Acp36DE binds to the oviduct wall just anterior to the openings of the sperm storage organs (5) and is present in the anterior mating plug (72). It is also tightly associated with the sperm mass in the uterus (5) and subsequently enters the storage organs (78).

The precise role(s) of Acp36DE in sperm storage remains to be determined. Its binding to the oviduct wall suggests a role similar to that of the opaque gland secretion in *Rhodnius* (22). Its strong binding to sperm and its concurrent movement into the storage organs may relate to a role in the orderly storage and prevention of loss of the sperm, both of which imply an indirect role in sperm competition. The latter suggestion is supported by experiments in which sperm-storage ability was compared following double-matings with normal, Acp36DE-deficient, and truncated-Acp36DE male *Drosophila* (12). The experiments showed that Acp36DE does not itself displace sperm; rather, its role is to facilitate sperm storage and use.

REGULATION OF EGG DEVELOPMENT

Egg production and eventual deposition occur in virgin females of most insect species, though at much reduced rates compared with those in mated insects. The increased fecundity that results from mating is due in many species to the presence in male ARG secretion of fecundity-enhancing substances (FES) (36). Typically, the FES stimulate egg laying (i.e., ovulation and/or oviposition), as discussed in the following section. However, a few reports demonstrate that egg development is accelerated in the presence of ARG material. In the bean weevil, *Acanthoscelides obtectus*, early oogenesis is promoted by substance A, a putative peptide (52). For *Aedes* spp. there are reports that matrone promotes egg development, though the effect may be indirect and it is vitellogenesis (not oogenesis) that is stimulated. Thus, injection of purified matrone into the abdomen of virgin or mated *A. aegypti* accelerates digestion of the blood meal, an effect that can be prevented by neck ligation (23). It is not clear from this study whether matrone stimulates digestion per se or vitellogenesis, which might enhance food processing through a "sink" effect. However, work on an autogenous population of *Aedes taeniorhynchus* showed that injection of male ARG extract stimulates egg development in more than half the experimental (virgin) females (79), indicating that the effect of matrone is on vitellogenesis. Borovsky (7), also using *A. taeniorhynchus*, stimulated vitellogenin synthesis by injection of ARG extract into intact virgins, but not by injection into decapitated or ovariectomized insects. Incubation of fat body with ARG extract and ovary ecdysteroidogenic hormone (produced in the brain) does not stimulate vitellogenin synthesis. This suggests that matrone may act on the ovaries that produce a hormone to trigger ovary ecdysteroidogenic hormone release (7). Further support for an effect of matrone on vitellogenesis came from the work of Klowden & Chambers (61), who stimulated egg development in nutritionally stressed virgin *A. aegypti* by injecting ARG homogenate or by topically applying JH III. The similar effect of these treatments is of interest because the role of JH in mosquito egg development remains unclear (61) and may vary among species. Also, it has not yet been demonstrated that mosquito ARG material can trigger JH synthesis by the CA (compare with *Drosophila* below).

In those Lepidoptera in which vitellogenesis is regulated by JH, there appears to be a dual role for ARG material. First, male-produced JH received from the ARG during mating may directly stimulate both yolk production and its uptake into the maturing oocytes. Thus, *Heliothis virescens* females decapitated immediately after mating nevertheless produce three times as many eggs as decapitated virgins (83). Second, the CA of mated female *H. virescens* synthesize and release JH at 2.5-fold the rate seen in CA of same-age virgins, suggesting that ARG material also stimulates endogenous JH synthesis (85). Though it remains to be shown that conspecific ARG products can act directly on the CA in Lepidoptera, it has been reported that synthetic *D. melanogaster* SP stimulates in vitro JH synthesis by the CA of virgin female *H. armigera* (27, 28).

The clearest demonstration of involvement of ARG products in stimulating egg development is in *D. melanogaster*. In this species during sexual maturation, oocytes gradually develop, but unless mating occurs, a large proportion of the oocytes never mature. Rather, their development is arrested in early vitellogenesis. The path followed by the oocytes depends on the interplay between SP, JH, and 20-HE. In vitro incubation of CA with SP significantly increases synthesis of JHB_3 (which comprises about 95% of the JH in this species), but not JH III (28, 76). However, ^{125}I-labelled SP does not bind to CA after injection into virgin females (81). Synthetic SP injected into sexually mature virgin females also stimulates the progression of developing oocytes through early vitellogenesis, leading to increased yolk protein uptake from the hemolymph and increased production of yolk protein by the follicle cells. These effects are mimicked by topical treatment of such females with methoprene (a JH analog) (95, 96). In contrast, 20-HE inhibits oocyte progression through early vitellogenesis and is responsible for the oocyte resorption that occurs in mature virgin females (96). Thus, it appears that SP may influence egg development indirectly, by regulating the synthetic activity of the CA. Whether SP also acts directly on the ovary to affect egg development was not resolved by Soller et al. (95, 96), who used whole insects. Nor does the possible effect of SP on 20-HE production by ovarian tissue appear to have been considered. It must also be noted that, in addition to SP, sperm play a role in regulating oocyte development. Comparison of females mated to spermless males with those mated to normal males showed that oocyte progression through early vitellogenesis is significantly lower in the absence of sperm (49), though the mechanism of the sperm effect remains unclear. Further, as Heifetz et al. (49) stress, vitellogenesis, ovulation, and oviposition must be considered as parts of a continuous process, so that as eggs leave the ovary and are laid, the development of additional eggs will be stimulated. Thus, factors (including ARG materials) that affect these later stages must also be considered to indirectly influence egg development.

INDUCTION OF OVULATION AND OVIPOSITION

By far the most commonly demonstrated effect of FES is the rapid induction of egg laying (33). In early studies, however, including those in which purified FES was used, the experimental design usually did not permit a distinction to be made between an effect on ovulation (release of oocytes from the ovary) and an effect on oviposition (peristalsis of the lateral and common oviducts that moves eggs to the exterior). This difficulty is compounded by the fact that ovulation is quickly followed by oviposition in many insects. A few species, for example *M. sanguinipes* and other acridid grasshoppers, can retain ovulated eggs in the lateral oviducts for a short time until egg-laying conditions are suitable. In these insects are ARG components that specifically trigger oviposition [see (111)]. When injected into virgin females, these molecules stimulate egg deposition only if ovulation has occurred. In in vitro preparations using isolated lateral oviducts, they induce increases in the frequency, amplitude, and tonus, respectively, in a dose-dependent manner (111). Extracts (not purified components) from female tissues, for example the spermatheca and brain, can also modulate oviduct peristalsis; the nature of the modulation differs depending on whether the tissue is taken from virgin or mated insects (111). Clearly, in an intact female the decision of whether or not to lay eggs is reliant on input from a variety of both exogenous and endogenous sources.

In *D. melanogaster* the availability first of synthetic SP and, more recently, of genetically altered males and females has provided the opportunity to separate control of ovulation from stimulation of oviposition. Interestingly, however, even after purified and synthetic SP became available, it was not initially considered whether the peptide was modulating ovulation or oviposition (the bioassay simply measured the number of eggs laid over time). Later, a so-called ovulation bioassay was developed that measured the presence of eggs in the uterus (62). It should be noted, however, that the assay does not measure ovulation alone, but is the result of ovulation plus egg movements less oviposition. In *D. melanogaster* females able to express SP ectopically when heat shocked, the proportion of females with uterine eggs increases from 5% to 80% within 1.5–2 h (1). Likewise, injection of purified or synthetic SP into the body cavity of virgins induces an increase in the number of uterine eggs within 3–24 h (17, 90, 91).

In view of the rapidity with which this occurs, it seems unlikely that this effect of SP is the result of an increased rate of egg maturation (see previous section). Using this bioassay, so-called ovulation-stimulating substances (OSS) have been identified from other *Drosophila* species [*D. suzukii* (80, 91), *D. biarmipes* (53)]. Curiously, in *D. suzukii* (80, 91) and *D. biarmipes* (53), there are indications that more than one ovulation-inducing molecule may exist. Thus, in *D. suzukii* the OSS (80) has a different amino acid sequence from those of the two peptides characterized by Schmidt et al. (91). One of these peptides has strong homology with *D. melanogaster* SP, whereas the other is suggested to be a contaminant (91). In *D. suzukii* the three reported molecules are produced in the ARG. In contrast, in *D. biarmipes* the OSS (which could not be purified) is from the ARG, whereas the second molecule (ED-OSS), which has some homology with *D. melanogaster* SP, was isolated from the ejaculatory duct and furthermore is strain specific. The occurrence of several components of apparently similar function suggests that there is redundancy in the reproductive system. However, in the bioassays used for *D. suzukii* and *D. biarmipes* the occurrence of uterine eggs was not checked for 24 h and 14–16 h after injection, respectively. Thus, it may be that one of the molecules is not directly concerned with ovulation but, perhaps, with egg development [vitellogenesis is completed in about 16 h at 25°C in *D. melanogaster* (2)].

A different scheme has emerged from Wolfner's laboratory, where Heifetz et al. (48) have shown that Acp26Aa ("ovulin") triggers ovulation in *Drosophila*. In females mated to Acp26Aa-deficient male mutants there is a ninefold lower chance of finding an egg in the lateral oviducts compared with females mated to normal males. The action of Acp26Aa appears to be mediated rapidly and locally, as immunostaining of the genital tract of females mated to wild-type males localized the Acp26Aa antibody on cells at the base of the ovary by 10 min after the start of mating. As noted above, Acp26Aa is a prohormone-like molecule and undergoes cleavage presumed to release bioactive peptides. Normal cleavage occurs only in the presence of products of the main cells of the ARG and in the mated female genital tract, suggesting a role for the female system in ensuring that Acp26Aa acts at the proper time and place (82). About one half of the Acp26Aa remains in the genital tract; the remainder passes through the nonmuscular posterior wall of the tract and enters the hemocoel (71), though no extra-genital targets have yet been identified (48). It is proposed that by moving eggs into the lateral oviducts,

Acp26Aa stimulates egg laying (48). Further, by clearing the ovary of mature oocytes, Acp26Aa terminates feedback inhibition, permitting development of new oocytes. Indirectly, therefore, Acp26Aa may be considered to coordinate egg development with proper storage (and hence use) of sperm (10), which requires up to 9 h to complete (40). Acp26Aa appears not to affect oviposition, as females mated to Acp26Aa-deficient males did not lay significantly fewer eggs than females mated to normal males at 6 h postmating. At 24 h postmating, females mated to the Acp26Aa-deficient males had laid only about two thirds the number of eggs laid by females mated to normal males, but this is likely due to the lack of ovulated eggs in the former group. In the scheme proposed by Heifetz et al. (48), SP promotes egg development (by its action on JHB_3 synthesis). However, the rapid induction of egg laying that follows its injection suggests that SP also may have a more proximate effect. Is it possible that SP is also an oviposition stimulant that enhances peristalsis of the musculature of the oviducts and uterus? This idea is supported by the prominent and specific binding of ^{125}I-labelled synthetic SP to the uterus and oviducts of virgins (81).

OTHER EFFECTS OF ACCESSORY REPRODUCTIVE GLAND SECRETIONS

Additional effects of ARG secretions have been reported that indirectly affect female reproductive physiology and behavior. These include modulation of host-seeking behavior and circadian rhythmicity, alteration of metabolism, and decrease in life span of the mated female. It should be stressed, however, that for the most part these effects have been demonstrated using entire homogenates or partially purified extracts of ARG; that is, the identity of the causal agent remains unknown. Further, in at least some cases, the effect is likely to be simply a consequence of ARG material acting directly on a reproductive process.

A number of these actions have been attributed to matrone in aedine mosquitoes. For example, in *A. aegypti*, injection of purified matrone promotes feeding and digestion of the blood meal (23). Injection of ARG extract (by implication, matrone) causes a switch in behavior from swarming (mating) to host-seeking behavior in *Culex quinquefasciatus* (55). In this species these activities occur in different locations. By contrast, in *A. aegypti*, which commonly mates near its host, male ARG components transferred during insemination precipitate a decrease in host-seeking and biting behavior. However, this effect may not be due to matrone but rather to a peptide (mol wt 7600 kDa) (64). A switch from attraction to male pheromone to attraction to host-fruit odor following injection of ARG extract has been reported for virgin Mediterranean fruit fly, *Ceratitis capitata* (54).

In addition to those discussed, a number of other *Drosophila* genes and their products have been characterized, though their roles remain speculative (97, 107). For example, Acp62F, which is toxic when expressed ectopically in *Drosophila*, is highly similar in its amino acid sequence and other features to extracellular protease inhibitors from the roundworm *Ascaris* (70). After mating, about 90% of the Acp62F remains in the female reproductive tract (including the sperm-storage organs), while the rest enters the hemocoel (71). The localization of Acp62F in the reproductive tract suggests a role in the protection of sperm and even other seminal fluid proteins, perhaps by preventing attack by proteases (70). However, the small amount of Acp62F that enters the hemocoel may contribute to the decreased life span of mated females, perhaps by interfering with the action of proteases that regulate essential extracellular processes, for example, the immune response (11, 70). Other ARG proteins [e.g., Acp76A in *D. melanogaster* (18, 107) and the protease inhibitor in *D. funebris* (92)] also have sequence similarities to, or activities of, serpins. Such molecules, which prevent untimely activation of proteolytic enzymes, could have diverse pre- and postcopulatory functions.

CONCLUSIONS

It will be readily evident that a veritable explosion of knowledge has occurred since the last review of male ARG function for the *Annual Review of Entomology* was prepared by Professor Chen (14). Great strides have been made in our understanding of both the nature of the molecules involved and their specific functions. To a significant degree these advances have been achieved in a single species, *D. melanogaster*, in which both traditional genetics and current genome technology facilitate manipulation and experimentation. To date, more than 75 *Drosophila* Acp genes and their products have been identified and characterized [representing ~90% of the total (97, 105)], and a clear role for some of these has been established. It is only a matter of time before the role(s) of additional *Drosophila* Acps becomes known. The advantage of working with a single species is that new observations are easily fitted into the existing picture. The downside is the difficulty of extrapolating ideas, hypotheses, and experimental methods to other insects, especially those whose reproductive strategies are different from that of *Drosophila*. Not wishing to sound critical, work on the ARG of other species lags significantly behind that of *Drosophila*. Too often, methods in place two or three decades ago (e.g., the use of homogenates or partially purified extracts of ARG) are still in use, when it should be relatively easy to obtain pure fractions and synthetic forms of the active principle(s). Most conspicuous is the continued lack of attention paid to which specific components of mosquito ARG secretion are responsible for the range of functions attributed to it. Reexamination of the nature of matrone is urgently required in light of its reported existence in α and β forms, its purported multifunctional nature (59), and the observation that *Drosophila* ARG extracts mimic the effect of matrone when injected into *A. aegypti* (51).

As our knowledge of the nature and roles of male insect ARG products grows, so will the interest of a wide range of biological scientists. Biochemists will be able to compare similarly constructed molecules from other animal taxa in their search for structure-function relationships at the molecular level. Comparative physiologists, especially those curious to learn the "ins and outs" of mating, will see parallels between insect and vertebrate mechanisms. Likewise, evolutionary biologists fascinated by animal mating strategies, sexual selection, and speciation will identify principles that are broadly applicable to other taxa, and applied entomologists may evolve new ideas based on ARG functions as they strive to develop novel means of managing insect pests.

ACKNOWLEDGMENTS

I am most grateful to Marc Klowden, Margaret Bloch Qazi, and Mariana Wolfner for their invaluable comments on a draft of this review. The Natural Sciences and Engineering Research Council of Canada and the University of Saskatchewan supported original work of the author cited in the review.

The *Annual Review of Entomology* is online at http://ento.annualreviews.org

LITERATURE CITED

1. Aigaki T, Fleischmann I, Chen PS, Kubli E. 1991. Ectopic expression of sex peptide alters reproductive behavior of female *D. melanogaster*. *Neuron* 7:557–63
2. Ashburner M. 1989. *Drosophila,Vol. 1*. Cold Spring Harbor, NY: Cold Spring Harbor Lab. Press. 1331 pp.
3. Bairati A. 1968. Structure and ultrastructure of the male reproductive system in *Drosophila melanogaster* Meig. 2. The genital duct and accessory glands. *Monit. Zool. Ital.* 2:105–82
4. Barres BA. 1991. New roles for glia. *J. Neurosci.* 11:3685–94
5. Bertram MJ, Neubaum DM, Wolfner MF. 1996. Localization of the *Drosophila* male accessory gland protein Acp36DE in the mated female suggests a role in sperm storage. *Insect Biochem. Mol. Biol.* 26: 971–80
6. Blum MS, Hilker M. 2002. Chemical protection of insect eggs. See Ref. 50a, pp. 61–90
7. Borovsky D. 1985. The role of the male accessory gland fluid in stimulating vitellogenesis in *Aedes taeniorhynchus*. *Arch. Insect Biochem. Physiol.* 2:405–13
8. Borovsky, D, Carlson DA, Hancock RG, Rembold H, Van Handel E. 1994. *De novo* biosynthesis of juvenile hormone III and I by the accessory glands of the male mosquito. *Insect Biochem. Mol. Biol.* 24: 437–44

9a. Bulet P, Hetru C, Dimarcq JL, Hoffman D. 1999. Antimicrobial peptides in insects: structure and function. *Dev. Comp. Immunol.* 23:329–44

9b. Chapman T, Choffat Y, Lucas WE, Kubli E, Partridge L. 1996. Lack of response to sex-peptide results in increased cost of mating in *dunce Drosophila melanogaster* females. *J. Insect Physiol.* 42:1007–15

10. Chapman T, Herndon LA, Heifetz Y, Partridge L, Wolfner MF. 2001. The Acp-26Aa seminal fluid protein is a modulator of early egg hatchability in *Drosophila melanogaster*. *Proc. R. Soc. London Ser. B* 268:1647–54
11. Chapman T, Liddle LF, Kalb JM, Wolfner MF, Partridge L. 1995. Cost of mating in *Drosophila melanogaster* females is mediated by male accessory gland products. *Nature* 373:241–44
12. Chapman T, Neubaum DM, Wolfner MF, Partridge L. 2000. The role of male accessory gland protein Acp36DE in sperm competition in *Drosophila melanogaster*. *Proc. R. Soc. London Ser. B* 267:1097–105

13. Cheeseman MT, Gillott C, Ahmed I. 1990. Structural spermatophore proteins and a trypsin-like enzyme from the accessory reproductive glands of the male grasshopper, *Melanoplus sanguinipes*. *J. Exp. Zool.* 255:193–204
14. Chen PS. 1984. The functional morphology and biochemistry of insect male accessory glands and their secretions. *Annu. Rev. Entomol.* 29:233–55
15. Chen PS. 1991. Biochemistry and molecular regulation of the male accessory gland secretions in *Drosophila* (Diptera). *Ann. Soc. Entomol. Fr.* 27:231–44
16. Chen PS. 1996. The accessory gland proteins in male *Drosophila*: structural, reproductive, and evolutionary aspects. *Experientia* 52:503–10
17. Chen PS, Stumm-Zollinger E, Aigaki T, Balmer J, Bienz M, Böhlen P. 1988. A male accessory gland peptide that regulates reproductive behavior of female *D. melanogaster*. *Cell* 54:291–98
18. Coleman S, Drähn B, Petersen G, Stolorov J, Kraus K. 1995. A *Drosophila* male accessory gland protein that is a member of the serpin superfamily of proteinase inhibitors is transferred to females during mating. *Insect Biochem. Mol. Biol.* 25: 203–7
19. Craig GB Jr. 1967. Mosquitoes: female monogamy induced by male accessory gland substance. *Science* 156:1499–501
20. Cusson M, McNeil JN. 1989. Involvement of juvenile hormone in the regulation of pheromone release activities in a moth. *Science* 243:210–12
21. Das AK, Huignard J, Barbier M, Quesneau-Thierry A. 1980. Isolation of the two paragonial substances deposited into the spermatophores of *Acanthoscelides obtectus* (Coleoptera, Bruchidae). *Experientia* 36:918–19
22. Davey KG. 1958. The migration of spermatozoa in the female of *Rhodnius prolixus* Stål. *J. Exp. Biol.* 35:694–701
23. Downe AER. 1975. Internal regulation of rate of digestion of blood meals in the mosquito, *Aedes aegypti*. *J. Insect Physiol.* 21:1835–39
24. Eberhard WG. 1996. *Female Control: Sexual Selection by Cryptic Female Choice*. Princeton, NJ: Princeton Univ. Press. 501 pp.
25. Eisner T, Rossini C, González A, Iyengar VK, Siegler MVS, Smedley SR. 2002. Paternal investment in egg defence. See Ref. 50a, pp. 91–116
26. Fan Y, Rafaeli A, Gileadi C, Applebaum SW. 1999. Juvenile hormone induction of pheromone gland PBAN-responsiveness in *Helicoverpa armigera* females. *Insect Biochem. Mol. Biol.* 29:635–41
27. Fan Y, Rafaeli A, Gileadi C, Kubli E, Applebaum SW. 1999. *Drosophila melanogaster* sex peptide stimulates juvenile hormone synthesis and depresses sex pheromone production in *Helicoverpa armigera*. *J. Insect Physiol.* 45:127–33
28. Fan Y, Rafaeli A, Moshitzky P, Kubli E, Choffat Y, Applebaum SW. 2000. Common functional elements of *Drosophila melanogaster* seminal peptides involved in reproduction of *Drosophila melanogaster* and *Helicoverpa armigera* females. *Insect Biochem. Mol. Biol.* 30:805–12
29. Fleischmann I, Cotton B, Choffat Y, Spengler M, Kubli E. 2001. Mushroom bodies and post mating behaviors of *Drosophila melanogaster* females. *J. Neurogenet.* 15: 1–27
30. Fuchs MS, Craig GB Jr, Despommier DD. 1969. The protein nature of the substance inducing female monogamy in *Aedes aegypti*. *J. Insect Physiol.* 15:701–9
31. Fuchs MS, Craig GB Jr, Hiss EA. 1968. The biochemical basis of monogamy in mosquitoes. I. Extraction of the active principle from *Aedes aegypti*. *Life Sci.* 7: 835–39
32. Fuchs MS, Hiss EA. 1970. The partial purification and separation of the protein components of matrone from *Aedes aegypti*. *J. Insect Physiol.* 16:931–39

33. Gillott C. 1988. Arthropoda-Insecta. In *Reproductive Biology of Invertebrates, Vol. III. Accessory Sex Glands*, ed. KG Adiyodi, RG Adiyodi, pp. 319–471. New York: Wiley
34. Gillott C. 1996. Male insect accessory glands: functions and control of secretory activity. *Invertebr. Reprod. Dev.* 30:199–205
35. Gillott C. 2002. Insect accessory reproductive glands: key players in production and protection of eggs. See Ref. 50a, pp. 37–59
36. Gillott C, Friedel T. 1977. Fecundity-enhancing and receptivity-inhibiting substances produced by male insects: a review. *Adv. Invertebr. Reprod.* 1:199–218
37. Gillott C, Gaines SB. 1992. Endocrine regulation of male accessory gland development and activity. *Can. Entomol.* 124: 871–86
38. Grimnes KA, Bricker CS, Happ GM. 1986. Ordered flow of secretion from accessory glands to specific layers of the spermatophore of mealworm beetles: demonstration with a monoclonal antibody. *J. Exp. Zool.* 240:275–86
39. Grodner ML, Steffens WL. 1978. Evidence of a chemotactic substance in the spermathecal gland of the female boll weevil (Coleoptera: Curculionidae). *Trans. Am. Microsc. Soc.* 97:116–20
40. Gromko MH, Gilbert DG, Richmond RC. 1984. Sperm transfer and use in the multiple mating system of *Drosophila*. In *Sperm Competition and the Evolution of Animal Mating Systems*, ed. RL Smith, pp. 371–426. New York: Academic
41. Gwadz RW. 1972. Neuro-hormonal regulation of sexual receptivity in female *Aedes aegypti*. *J. Insect Physiol.* 18:259–66
42. Gwynne DT. 1997. The evolution of edible 'sperm sacs' and other forms of courtship feeding in crickets, katydids and their kin (Orthoptera: Ensifera). In *The Evolution of Mating Systems in Insects and Arachnids*, ed. JC Choe, BJ Crespi, 1:110–29. New York: Cambridge Univ. Press
43. Gwynne DT. 2001. *Katydids and Bush-Crickets: Reproductive Behavior and Evolution of the Tettigoniidae*. Ithaca, NY: Cornell Univ. Press. 317 pp.
44. Happ GM. 1984. Structure and development of male accessory glands in insects. In *Insect Ultrastructure*, ed. RC King, H Akai, 2:365–96. New York: Plenum. 624 pp.
45. Happ GM. 1992. Maturation of the male reproductive system and its endocrine regulation. *Annu. Rev. Entomol.* 37:303–20
46. Hartmann R, Loher W. 1996. Control mechanisms of the behavior 'secondary defense' in the grasshopper *Gomphocerus rufus L.* (Gomphocerinae: Orthoptera). *J. Comp. Physiol. A* 178:329–36
47. Hartmann R, Loher W. 1999. Post-mating effects in the grasshopper, *Gomphocerus rufus* L. mediated by the spermatheca. *J. Comp. Physiol. A* 184:325–32
48. Heifetz Y, Lung O, Frongillo EA Jr, Wolfner MF. 2000. The *Drosophila* seminal fluid protein Acp26Aa stimulates release of oocytes by the ovary. *Curr. Biol.* 10:99–102
49. Heifetz Y, Tram U, Wolfner MF. 2001. Male contributions to egg production: the role of accessory gland products and sperm in *Drosophila melanogaster*. *Proc. R. Soc. London Ser. B* 268:175–80
50. Hihara F. 1981. Effects of the male accessory gland secretion on oviposition and remating in females of *Drosophila melanogaster*. *Zool. Mag.* 90:307–16
50a. Hilker M, Meiners T, eds. 2002. *Chemoecology of Insect Eggs and Egg Deposition*. Berlin: Blackwell. 416 pp.
51. Hiss EA, Fuchs MS. 1972. The effect of matrone on oviposition in the mosquito, *Aedes aegypti*. *J. Insect Physiol.* 18:2217–27
52. Huignard J, Quesneau-Thierry A, Barbier M. 1977. Isolement, action biologique

et evolution des substances paragoniales contenues dans le spermatophore d' *Acanthoscelides obtectus* (Coleoptère). *J. Insect Physiol.* 23:351–57

53. Imamura M, Haino-Fukushima K, Aigaki T, Fuyama Y. 1998. Ovulation stimulating substances in *Drosophila biarmipes* males: their origin, genetic variation in the response of females, and molecular characterization. *Insect Biochem. Mol. Biol.* 28:365–72
54. Jang EB. 1995. Effects of mating and accessory gland injections on olfactory-mediated behavior in the female Mediterranean fruit fly, *Ceratitis capitata. J. Insect Physiol.* 41:705–10
55. Jones MDR, Gubbins SJ. 1979. Modification of female flight-activity by a male accessory gland pheromone in the mosquito, *Culex pipiens quinquefasciatus. Physiol. Entomol.* 4:345–51
56. Kalb JM, DiBenedetto AJ, Wolfner MF. 1993. Probing the function of *Drosophila melanogaster* accessory glands by directed cell ablation. *Proc. Natl. Acad. Sci. USA* 90:8093–97
57. Kingan TG, Bodnar WM, Raina AK, Shabanowitz J, Hunt DF. 1995. The loss of female sex pheromone after mating in the corn earworm moth *Helicoverpa zea*: identification of a male pheromonostatic peptide. *Proc. Natl. Acad. Sci. USA* 92:5082–86
58. Kingan TG, Thomas-Laemont PA, Raina AK. 1993. Male accessory gland factors elicit change from 'virgin' to 'mated' behaviour in the female corn earworm moth *Helicoverpa zea. J. Exp. Biol.* 183:61–76
59. Klowden MJ. 1999. The check is in the male: male mosquitoes affect female physiology and behavior. *J. Am. Mosq. Control Assoc.* 15:213–20
60. Klowden MJ. 2001. Sexual receptivity in *Anopheles gambiae* mosquitoes: absence of control by male accessory gland substances. *J. Insect Physiol.* 47:661–66
61. Klowden MJ, Chambers GM. 1991. Male accessory gland substances activate egg development in nutritionally stressed *Aedes aegypti* mosquitoes. *J. Insect Physiol.* 37:721–26
62. Kubli E. 1992. The sex-peptide. *BioEssays* 14:779–84
63. Kubli E. 1996. The *Drosophila* sex-peptide: a peptide pheromone involved in reproduction. *Adv. Dev. Biochem.* 4:99–128
64. Lee J-J, Klowden MJ. 1999. A male accessory gland protein that modulates female mosquito (Diptera: Culicidae) host-seeking behavior. *J. Am. Mosq. Control Assoc.* 15:4–7
65. Leopold RA, Terranova AC, Swilley EM. 1971. Mating refusal in *Musca domestica*: effects of repeated mating and decerebration upon frequency and duration of copulation. *J. Exp. Zool.* 176:353–60
66. Leopold RA, Terranova AC, Thorson BJ, Degrugillier ME. 1971. The biosynthesis of the male housefly accessory secretion and its fate in the mated female. *J. Insect Physiol.* 17:987–1003
67. Loher W. 1966. Die Steuerung sexueller Verhaltensweisen und der Oocytenentwicklung bei *Gomphocerus rufus* L. *Z. Vgl. Physiol.* 53:277–316
68. Ludwig MZ, Uspensky II, Ivanov AI, Kopantseva MR, Dianov CM, et al. 1991. Genetic control and expression of the major ejaculatory bulb protein PEB-me in *Drosophila melanogaster. Biochem. Genet.* 29:215–40
69. Lung O, Kuo L, Wolfner MF. 2001. Drosophila males transfer antibacterial proteins from their accessory gland and ejaculatory duct to their mates. *J. Insect Physiol.* 47:617–22
70. Lung O, Tram U, Finnerty CM, Eipper-Mains MA, Kalb JM, Wolfner MF. 2002. The *Drosophila melanogaster* seminal fluid protein Acp62F is a protease inhibitor that is toxic upon ectopic expression. *Genetics* 160:211–24

71. Lung O, Wolfner MF. 1999. *Drosophila* seminal fluid proteins enter the circulatory system of the mated female fly by crossing the posterior vaginal wall. *Insect Biochem. Mol. Biol.* 29:1043–52
72. Lung O, Wolfner MF. 2001. Identification and characterization of the major *Drosophila melanogaster* mating plug protein. *Insect Biochem. Mol. Biol.* 31: 543–51
73. Manning A. 1962. A sperm factor affecting the receptivity of *Drosophila melanogaster*. *Nature* 194:252–53
74. Meister M, Lemaitre B, Hoffmann JA. 1997. Antimicrobial peptide defense in Drosophila. *BioEssays* 19:1019–26
75. Monsma SA, Wolfner MF. 1988. Structure and expression of a Drosophila male accessory gland gene whose product resembles a peptide pheromone precursor. *Genes Dev.* 2:1063–73
76. Moshitzky P, Fleischmann I, Chaimov N, Saudan P, Klauser S, et al. 1996. Sex-peptide activates juvenile hormone biosynthesis in the *Drosophila melanogaster* corpus allatum. *Arch. Insect Physiol. Biochem.* 32:363–74
77. Nelson DR, Adams TS, Pomonis JG. 1969. Initial studies on the extraction of the active substance inducing monocoitic behavior in house flies, black blow flies, and screw-worm flies. *J. Econ. Entomol.* 62:634–39
78. Neubaum DM, Wolfner MF. 1999. Mated *Drosophila melanogaster* females require a seminal fluid protein, Acp36DE, to store sperm efficiently. *Genetics* 153:845–57
79. O'Meara GF, Evans DG. 1977. Autogeny in saltmarsh mosquitoes induced by a substance from the male accessory gland. *Nature* 267:342–44
80. Ohashi YY, Haino-Fukushima K, Fuyama Y. 1991. Purification and characterization of an ovulation stimulating substance from the male accessory glands of *Drosophila suzukii*. *Insect Biochem.* 4: 413–19
81. Ottiger M, Soller M, Stocker, RF, Kubli E. 2000. Binding sites of *Drosophila melanogaster* sex peptide pheromones. *J. Neurobiol.* 44:57–71
82. Park M, Wolfner MF. 1995. Male and female cooperate in the prohormone-like processing of a *Drosophila melanogaster* seminal fluid protein. *Dev. Biol.* 171:694–702
83. Park YI, Ramaswamy SB. 1998. Role of brain, ventral nerve cord, and corpora cardiaca-corpora allata complex in the reproductive behavior of female tobacco budworm (Lepidoptera: Noctuidae). *Ann. Entomol. Soc. Am.* 91:329–34
84. Park YI, Ramaswamy SB, Srinivasan A. 1998. Spermatophore formation and regulation of egg maturation and oviposition in female *Heliothis virescens* by the male. *J. Insect Physiol.* 44:903–8
85. Park YI, Shu S, Ramaswamy SB, Srinivasan A. 1998. Mating in *Heliothis virescens*: transfer of juvenile hormone during copulation by male to female and stimulation of biosynthesis of endogenous juvenile hormone. *Arch. Insect Physiol. Biochem.* 38:100–7
86. Price CSC, Dyer KA, Coyne JA. 1999. Sperm competition between *Drosophila* males involves both displacement and incapacitation. *Nature* 400:449–52
87. Raina AK. 1993. Neuroendocrine control of sex pheromone biosynthesis in Lepidoptera. *Annu. Rev. Entomol.* 38:320–49
88. Roth LM. 1967. Uricose glands in the accessory sex gland complex of male Blattaria. *Ann. Entomol. Soc. Am.* 60:1203–11
89. Saudan P, Hauck K, Soller M, Choffat Y, Ottinger M, et al. 2002. Ductus ejaculatorius peptide 99B (DUP99B), a novel *Drosophila melanogaster* sex-peptide pheromone. *Eur. J. Biochem.* 269: 989–97
90. Schmidt T, Choffat Y, Klauser S, Kubli E. 1993. The *Drosophila melanogaster* sex-peptide: a molecular analysis of

structure-function relationships. *J. Insect Physiol.* 39:361–68

91. Schmidt T, Choffat Y, Schneider M, Hunziker P, Fuyama Y, Kubli E. 1993. *Drosophila suzukii* contains a peptide homologous to the *Drosophila melanogaster* sex-peptide and functional in both species. *Insect Biochem. Mol. Biol.* 23: 571–79
92. Schmidt T, Stumm-Zollinger E, Chen PS. 1989. A male accessory gland peptide with protease inhibitory activity in *Drosophila funebris*. *J. Biol. Chem.* 264: 9745–49
93. Si S, Xu S, Du J. 2000. Pheromonostatic activity of male accessory gland factors in female *Helicoverpa assulta*. *Acta Entomol. Sinica* 43:120–26
94. Smith PH, Gillott C, Barton Brown L, Van Gerwen ACM. 1990. The mating-induced refractoriness of *Lucilia cuprina* females: manipulating the male contribution. *Physiol. Entomol.* 15:469–81
95. Soller M, Bownes M, Kubli E. 1997. Mating and sex peptide stimulate the accumulation of yolk in oocytes of *Drosophila melanogaster*. *Eur. J. Biochem.* 243:732–38
96. Soller M, Bownes M, Kubli E. 1999. Control of oocyte maturation in sexually mature *Drosophila* females. *Dev. Biol.* 208: 337–51
97. Swanson WJ, Clark AG, Waldrip-Dail HM, Wolfner MF, Aquadro CF. 2001. Evolutionary EST analysis identifies rapidly evolving male reproductive proteins in *Drosophila*. *Proc. Natl. Acad. Sci. USA* 98:7375–79
98. Takahashi SY, Higashi S, Minoshima S, Ogiso M, Hanaoka K. 1980. Trehalases from the American cockroach, *Periplaneta americana*: multiple occurrence of the enzymes and partial purification of enzymes from male accessory glands. *Int. J. Invertebr. Reprod.* 2:373–81
99. Terranova AC, Leopold RA, Degrugillier ME, Johnson JR. 1972. Electrophoresis of the male accessory secretion and its fate in the mated female. *J. Insect Physiol.* 18:1573–91
100. Tompkins L, Gross AC, Hall JC, Gailey DA, Siegel RW. 1982. The role of female movement in the sexual behavior of *D. melanogaster*. *Behav. Genet.* 12:295–307
101. Tram U, Wolfner MF. 1998. Seminal fluid regulation of female attractiveness in *Drosophila melanogaster*. *Proc. Natl. Acad. Sci. USA* 95:4051–54
102. Tram U, Wolfner MF. 1999. Male seminal fluid proteins are essential for sperm storage in *Drosophila melanogaster*. *Genetics* 153:837–44
103. Vahed K. 1998. The function of nuptial feeding in insects: a review of empirical studies. *Biol. Rev.* 73:43–78
104. Viscuso R, Narcisi L, Sottile L, Violetta Brundo M. 2001. Role of male accessory glands in spermatodesm reorganization in Orthoptera Tettigonioidea. *Tissue Cell* 33:33–39
105. Wolfner MF. 1997. Tokens of love: functions and regulation of Drosophila male accessory gland products. *Insect Biochem. Mol. Biol.* 27:179–92
106. Wolfner MF. 2002. The gifts that keep on giving: physiological functions and evolutionary dynamics of male seminal proteins in *Drosophila*. *Heredity* 88:85–93
107. Wolfner MF, Harada HA, Bertram MJ, Stelick TJ, Kraus KW, et al. 1997. New genes for male accessory gland proteins in *Drosophila melanogaster*. *Insect Biochem. Mol. Biol.* 27:825–34
108. Xue L, Noll M. 2000. *Drosophila* female sexual behavior induced by sterile males showing copulation complementation. *Proc. Natl. Acad. Sci. USA* 97:3272–75
109. Yaginuma T, Happ GM. 1988. Trehalase from the male accessory gland and the spermatophore of the mealworm beetle, *Tenebrio molitor*. *J. Comp. Physiol. B* 157:765–70
110. Yi S-X, Gillott C. 1999. Purification

and characterization of an oviposition-stimulating protein of the long hyaline tubules in the male migratory grasshopper, *Melanoplus sanguinipes*. *J. Insect Physiol.* 45:143–50

111. Yi S-X, Gillott C. 2000. Effects of tissue extracts on oviduct contraction in the migratory grasshopper, *Melanoplus sanguinipes*. *J. Insect Physiol.* 46:519–25

112. Young ADM, Downe AER. 1987. Male accessory gland substances and the control of sexual receptivity in female *Culex tarsalis*. *Physiol. Entomol.* 12:233–39

Annu. Rev. Entomol. 2003. 48:185–209
doi: 10.1146/annurev.ento.48.091801.112725

First published online as a Review in Advance on August 27, 2002

FEATHER MITES (ACARI: ASTIGMATA): Ecology, Behavior, and Evolution

Heather C. Proctor[1]
Australian School of Environmental Studies, Griffith University, Nathan 4111 Queensland, Australia; e-mail: hproctor@ualberta.ca

Key Words functional morphology, diet, coevolution, Aves

■ **Abstract** Birds host many lineages of symbiotic mites, but the greatest diversity is shown by the three superfamilies of astigmatan feather mites: Analgoidea, Pterolichoidea, and Freyanoidea. Members of this diphyletic grouping have colonized all parts of the avian integument from their ancestral nidicolous habitat. Whereas some clearly feed on feather pith or skin, acting as parasites, other feather mites are paraphages and consume feather oils without causing structural damage. Sexual dimorphism in feather mites is often extreme, and little is known of the function of many elaborate male structures. Abundance and location of vane-dwelling mites is affected by season, temperature, light, humidity, and host body condition. Because transmission between hosts usually depends on host body contact, it is unsurprising that feather mite phylogeny often parallels host phylogeny; however, recent cladistic analyses have also found evidence of host-jumping and "missing the boat" in several mite lineages.

CONTENTS

[1]address after July 1, 2002: Department of Biological Sciences, University of Alberta, Edmonton, Alberta T6G 2E9; e-mail: hproctor@ualberta.ca

0066-4170/03/0107-0185$14.00

INTRODUCTION

Ever since Hamilton & Zuk (52) proposed that gaudy male birds might be advertising their parasite resistance, interest in avian parasites has flowed beyond taxonomic backwaters into mainstream evolutionary theory (22, 70). Arthropods such as lice, fleas, and mites have been among the most frequently investigated avian symbiotes (17, 23, 46, 109, 119). Although the biology of some blood-feeding mites is discussed in medical-veterinary texts (64), the ecology and behavior of feather mites are recorded in a smaller and more obscure body of literature. This is unfortunate because feather mites display a plethora of intriguing forms, behaviors, and relationships. The recent publication of keys to the world genera by Gaud & Atyeo (48, 49) has made identification of feather mites possible for nonspecialists and should herald a new surge of interest in the ecology of these now readily identifiable symbiotes. In this review I assemble the scattered research on feather mite biology with the aim of enlightening ecologists and evolutionary biologists of all taxonomic persuasions.

MITES ASSOCIATED WITH BIRDS

Birds host a rich diversity of acarine symbiotes, some dwelling on the surface of feathers but others inhabiting skin, nostrils, and respiratory passages. Most major mite taxa are represented in this menagerie. Among the ticks (suborder Ixodida), both hard (Ixodidae) and especially soft (Argasidae) ticks feed on bird blood (11, 20, 106, 113). The most common blood-feeding nest mites are members of the suborder Mesostigmata, especially Dermanyssidae (e.g., *Dermanyssus gallinae*) and Macronyssidae (e.g., *Ornithonyssus sylviarum*). Ecological relationships between these hematophages and domestic fowl (62, 74, 85) and wild birds (33, 79–81) have been thoroughly investigated. Another family of Mesostigmata, the Rhinonyssidae, feed within the respiratory passages of their hosts (94). These three families damage host health and reproduction [see lists in (105, 118)], with the rhinonyssid *Sternostoma tracheacolum* being implicated in the decline of the Gouldian finch (*Erythrura gouldiae*) (9, 115). Another group of mesostigmatan mites associated with avian respiratory systems, the hummingbird flower mites (Ascidae: *Rhinoseius*, *Tropicoseius*, and certain *Proctolaelaps*), are more benign, using hosts as transport between flowers rather than as meals (25, 26). Some members of the Ameroseiidae also use bird beaks to get from flower to flower (111).

Many members of the acarine suborder Prostigmata view birds as habitat. Some are temporary ectoparasites, such as larval chiggers (Trombiculidae, Leeuwenhoekiidae) (66), whereas others are permanent associates. Most members of the family Cheyletidae are free-living predators, but some prowl through nests and feathers, searching for other mites to eat (37, 99). Cheyletiellidae, in contrast, are all obligatory skin-dwelling parasites of birds and mammals (66). Their relatives, the Harpyrhynchidae, live on or in skin and feathers (83, 84). Another cheyletoid group, the Syringophilidae, occupy quills where they feed by piercing the calamus wall to feed on the tissue fluids of the feather follicle (16, 63). Although the cheyletoid family Cloacaridae is best known for mites that inhabit the cloacae of turtles, one species has been collected from the lungs of owls (45). Finally, there are many species in the tydeoid family Ereynetidae that inhabit nasal passages of birds (55).

Although this list of avian symbiotes seems impressive, the greatest diversity of bird-associated mites is in the suborder Astigmata. This group of mites is probably derived from within the suborder Oribatida (86, 89), a large group of free-living detritivores. The astigmatid superfamilies Glycyphagoidea and Acaroidea [sensu OConnor (87)] have many bird and mammal associates, most of them nidicoles or skin parasites. Within the first superfamily, the Aeroglyphidae occasionally inhabits nests of birds. The monotypic Euglycyphagidae is known only from nests of raptors. Among the Acaroidea, members of the Acaridae, Suidasiidae, and Lardoglyphidae are often found in bird nests, whereas the monotypic Glycacaridae is known only from nests of the white-chinned petrel (*Procellaria aequinoctialis*) (66, 87). These nest dwellers probably feed on sloughed skin and feathers, fungi associated with nesting material, and other detritus (66).

The Hypoderoidea [sensu OConnor (87)] consists of the single family Hypoderidae (=Hypoderatidae = Hypodectidae). In *Hypodectes propus*, nest-dwelling females produce eggs that develop directly into deutonymphs that invade the fatty tissue of their hosts (42). Following tissue invasion and growth, the enlarged deutonymphs emerge and transform directly to the adult stage. Deutonymphs have nonfunctional mouthparts, and food intake apparently occurs via trans-integumental absorption (43, 66). Adults either feed on nest detritus (88) or rely on nutrition garnered by the deutonymph.

The cohort Psoroptidia is composed almost entirely of mites closely associated with the bodies of birds and mammals. The Pyroglyphidae are exceptions, being primarily inhabitants of nests (66). Knemidocoptidae are significant skin parasites of wild and domestic birds, inducing "scaly-leg disease" (67, 95); however, at least one species makes its home inside quills (91). The Laminosioptidae are another group of subcutaneous parasites, and the Cytoditidae and Turbinoptidae are parasites of nasal cavities, tracheae, bronchi, air sacs, and lungs (55, 66, 87). Finally, and most importantly for this review, are the three groups of psoroptoids that make up the "true" feather mites, the Analgoidea, Freyanoidea, and Pterolichoidea (32, 48, 66). The approximately 2000 named species of feather mites are almost all obligatory symbiotes of birds, living in and on the skin (dermicoles), inside the quills (syringicoles), and on the surface of feathers (plumicoles) (48). The term feather mites is used in the remainder of this review as a convenient umbrella for members of these three groups rather than an ecological term for mites restricted to the feather habitat.

FEATHER MITE ORIGINS AND RELATIONSHIPS

Fossil Record

There are no authenticated fossils of any member of the three feather mite superfamilies (32). The only mite associated with birds in the fossil record is a larval argasid tick found with feathers in Cretaceous amber from 90–94 million years ago (65), and the only known fossils of Astigmata are of *Amphicalvolia hurdi* (Winterschmidtiidae) (88). Most extant winterschmidtiids are associated with insect nests or fungi (87). One recent claim of fossil feather mites published in *Nature* (73) is likely to be the result of misinterpretation. Martill & Davis (73) argue that small, spherical objects associated with a fossilized feather from 120-million-year-old strata are most likely mite eggs. They describe these objects as hollow, subspherical structures, 68–75 μm in diameter, with circular apertures 35–40 μm wide. The authors dismiss the possibility that the objects could be louse eggs on the basis that lice have much larger, elongate eggs, and state that the morphology of the structures resembles that of "the eggs of ectoparasitic feather mites." However, these objects are unlikely to be feather mite eggs. Eggs of modern feather mites are larger (150–400 μm long), sausage-shaped, and have longitudinal seams rather than circular openings (28, 37, 97, 98, 103). If the objects are indeed eggs, they may be those of ostracods rather than mites. The feather was found in a fossil bed well known for fish, and hence was likely preserved under water. Ostracods readily deposit their small, round eggs on submerged detritus (H.C. Proctor, personal observation).

Phylogenetic Relationships

The lack of fossil records and higher-level cladistic analyses of the Astigmata renders interpretation of feather mite origins and phylogenetic relationships problematic. Phylogenetic analyses of family-level relationships among feather mite taxa are rare, with Dabert & Mironov (32) and Ehrnsberger et al. (40) as the only recent examples. Despite this, there is general agreement that feather mites represent a diphyletic grouping, with Pterolichoidea and Freyanoidea derived from one ancestral astigmatan lineage and Analgoidea from another (3, 32, 48, 88).

The familial constituency of these superfamilies has undergone much change since Krantz (66) listed 20 families in 1978. Gaud & Atyeo (48) listed 33 families, the increase occurring mainly from the raising of subfamilial taxa to familial status, but also from the discovery of new forms [e.g., Oconnoriidae (50)]. In this review I use the taxonomy of Gaud & Atyeo (48) (Table 1); however, this is not to say that their scheme is undebated. There is particular disagreement over the placement of Pyroglyphidae and the three freyanoid families. Gaud & Atyeo (48) consider pyroglyphids to belong among the Analgoidea, but this is contended by OConnor (87) and Dabert & Mironov (32). OConnor (87) proposed that Freyanoidea should be placed within Pterolichoidea. Although Gaud & Atyeo (48) expressed some

TABLE 1 Host breadth* of feather mite taxa based on host records compiled from Gaud & Atyeo (48) and Proctor (104)

Analgoidea	Freyanoidea	Pterolichoidea
Alloptidae	Caudiferidae (i)	Ascouracaridae (klptuyz)
Alloptinae (eg-jmn)	Freyanidae	Cheylabididae (ko)
Echinacarinae (g)	Burhinacarinae (mn)	Crypturoptidae (d)
Microspalacinae (g)	Diomedeacarinae (g)	Eustathiidae (u)
Oxyalginae (g-j)	Freyaninae (j)	Falculiferidae (op)
Analgidae	Michaeliinae (h)	Gabuciniidae (ktx-z)
Analginae (z)	Vexillariidae	Kiwilichidae (c)
Ancyralginae (k)	Calaobiinae (xz)	Kramerellidae (hikms)
Anomalginae (nxz)	Vexillariinae (x)	Ochrolichidae (z)
Kiwialginae (c)		Oconnoriidae (s)
Megniniinae (bdlmoqrvx)		Pterolichidae
Protalginae (uy)		Ardeacarinae (i)
Tillacarinae (l)		Ardeialginae (i)
Apionacaridae (klnp)		Magimeliinae (n)
Avenzoariidae		Pterolichinae (aik-mo-rtvx)
Avenzoariinae (n)		Xoloptoidinae (i)
Bonnetellinae (ghjkn)		Ptiloxenidae (fin)
Hemifreyaninae (n)		Rectijanuuidae (j)
Pteronyssinae (uw-z)		Syringobiidae (n)
Dermationidae (ghjsxz)		Thoracosathesidae (l)
Dermoglyphidae (adhkln-pux-z)		
Epidermoptidae (gi-nsuyz)		
Gaudoglyphidae (l)		
Proctophyllodidae		
Proctophyllodinae (mnpuz)		
Pterodectinae (uwyz)		
Rhamphocaulinae (u)		
Psoroptoididae		
Pandalurinae (mnpsuyz)		
Psoroptoidinae (rx)		
Ptyssalgidae (u)		
Pyroglyphidae		
Dermatophagodinae (ouyz)		
Paralgopsinae (p)		
Pyroglyphinae (vyz)		
Thysanocercidae (u)		
Trouessartiidae (rtx-z)		
Xolalgidae		
Ingrassiinae (f-kmnpqstx-z)		
Xolalginae (rx-z)		
Zumptiinae (m)		

*Host ordinal codes: a, Struthioniformes; b, Casuariiformes; c, Apterygiformes; d, Tinamiformes; e, Gaviiformes; f, Podicipediformes; g, Procellariformes; h, Pelecaniformes; i, Ciconiiformes; j, Anseriformes; k, Falconiformes; l, Galliformes; m, Gruiformes; n, Charadriiformes; o, Columbiformes; p, Psittaciformes; q, Musophagiformes; r, Cuculiformes; s, Strigiformes; t, Caprimulgiformes; u, Apodiformes; v, Coliiformes; w, Trogoniformes; x, Coraciiformes; y, Piciformes; z, Passeriformes concern about giving this small group superfamilial rank, they considered the freyanoids to have distinct synapomorphies that separate them from both other superfamilies. Ehrnsberger et al. (40) addressed this conflict in their cladistic analysis of the family Freyanidae. The monophyly of the three freyanoid families was supported, but they were placed within the superfamily Pterolichoidea, as OConnor (87) had argued. For convenience, I continue to call the clade "Freyanoidea" but acknowledge that there is strong evidence against superfamilial status.

Evolution of Bird Association

Despite arguments about where Pyroglyphidae belongs, most authors agree that pyroglyphid-like nidicoles were the ancestors of feather mites (3, 41, 44, 88). Extant Pyroglyphidae exhibit a range of behaviors that suggest a possible evolutionary route from living in nests to the colonization of feathers (48). This family includes free-living detritivores commonly found in nests of birds and mammals (e.g., *Dermatophagoides* spp., dust mites), nidicoles that occasionally wander among the feathers of their host (e.g., *Sturnophagoides*), those that are found regularly in feathers of a range of host species that share nesting sites (e.g., *Hirstia*), those that occur throughout their life history on the feathers of a particular host genus (e.g., *Onychalges*), and one genus of obligate quill dwellers (*Paralgopsis*). Atyeo & Gaud (3) suggest that feather mites originated from ancestors that inhabited bird nests in the Cretaceous period, 65–130 million years ago. They argue that the more primitive mites should be associated with nonpasserine orders of birds, which are known to predate the Passeriformes [but see (76)]. The Pterolichoidea, and to a lesser extent the Freyanoidea, are characteristic of nonpasserines, whereas feather mites of passerines are almost exclusively Analgoidea. Thus Atyeo & Gaud (3) hypothesize that the ancestors of pterolichoids and freyanoids arose and diversified with their avian hosts in the late Cretaceous, and analgoids were latecomers that colonized Passeriformes when the order arose in the Eocene. However, this does not explain why pterolichoids and freyanoids are rare on passerines or why analgoids are relatively well represented on nonpasserine as well as passerine hosts.

Dabert & Mironov (32) present an alternative hypothesis. They argue that because analgoids are more widely distributed across host taxa, and across habitat types within hosts (occupying skin, down, vanes, and quills), they and not the pterolichoid/freyanoid families were the original colonizers of birds. The analgoids' generally smaller size gave them an advantage when the Passeriformes arose because analgoids better fit the niche space of these relatively small-bodied hosts and they outcompeted the Pterolichoidea and Freyanoidea. A third, and possibly more likely, scenario is that analgoid and nonanalgoid lineages were present on nonpasserines, but when the explosive radiation of passerines took place, by chance only analgoids were present on the passerine ancestor. There is growing evidence that such lineage sorting, or "missing the boat," is important in host-symbiote evolution (see Cospeciation, below).

MORPHOLOGICAL AND PHYSIOLOGICAL ADAPTATIONS TO THE AVIAN HABITAT

A brief digression on feather morphology is useful at this point [definitions from (71)]. The feather shaft, or quill, is the longitudinal axis and is composed of two segments. The calamus is the short, hollow base. The rachis is the long solid section of the shaft above the calamus. Where the calamus becomes the solid rachis is a small opening, the superior umbilicus. The rachis has a thick inner medulla composed of dense spongy pith. On each side of the rachis is a row of closely set branches termed barbs that collectively form a vane. Each barb has a solid axis called the ramus, which supports two combs of fine barbules. Viewed dorsally, a flight feather vane is a flat expanse because the feet of the L-shaped barbules are presented, but ventrally the long arms are arranged like a series of parallel hedges.

Although the skin's surface and the interior of quills would be micro-climatically similar to the feather mite's ancestral nidicolous habitat, physical stresses on the surface of feathers are different (48). Especially on the flight feathers, humidity is much lower and aerodynamic forces more powerful than what free-living astigmatans experience. Temperature is also different, being warmer and more constant in the plumage than in the nest. Relative humidity of flight feathers is usually less than 50%, whereas ancestral habitats of litter and stored product substrates usually range from 60 to 100% relative humidity (37). Dubinin (37) found that plumicolous feather mites have a much lower percentage of moisture in their bodies (15%–50%) than do free-living Astigmata (70%–90%). Birds have an average body temperature of 42–45°C (122); feather mites show the greatest activity at 40°C and become immobile when temperatures drop below 15°C (37). In response to daily or seasonal fluctuations in ambient humidity or temperature, feather mites seek their preferred climate by moving closer to or further from the bodies of their hosts (37, 75, 120) (see also Host Colonization and Movements Within the Host, below).

Adaptations to substrate structure and aerodynamic stress are linked. Feather mites that live in quills or downy feathers near the host's body surface experience a habitat in which they are relatively free to move with little or no air disturbance. Body setae in these mites tend to be long and fine, as in free-living Astigmata (32, 37). Long setae help the mite orient in their habitat and detect substrate vibrations, which may denote the presence of conspecifics or predators [e.g., lice (99)]. In contrast, the dorsal body setae of vane-dwelling mites are greatly reduced or absent, whereas lateral and terminal setae are often thickened and bladelike (32, 37). Setal reduction may be an adaptive response to the danger of being blown off the wing during flight, i.e., smaller setae reduce surface area (37). Long dorsal setae might also extend above the level of the barbs and be snagged by flight feathers sliding over each other. However, it is also possible that the reduction in setal length is a response to excessive stimulation from air currents. This is one hypothesis for why oribatid mites that dwell on exposed leaf surfaces exhibit shorter, thicker sensory setae than those of soil-dwelling relatives (24).

Vane-dwelling mites typically orient themselves so that their longitudinal axis is parallel to the rami of adjacent barbs. Their strengthened lateral and terminal setae help wedge the mites between rami (laterally) and possibly barbules (terminally) and allow them to resist displacement when the host flies. Dubinin (37) calculated that for mites living in interbarb chutes of similar widths, the sum of body width plus length of outer humeral setae tends to exceed the width of the chute. These setae, and other body structures, sometimes show bizarre asymmetry in males (e.g., Kramerellidae: *Freyanella*) (48). Dubinin (37) and Krantz (66) state that males use these enlarged setae to wedge themselves between adjacent barbs; however, nymphal and female mites never display unilaterally enlarged structures, implying that asymmetry functions in a sexual role specific to adult males (see "Mating and Oviposition").

In addition to modified setae, feather mites have diverse leg and body modifications. Syringicolous mites are the least modified relative to free-living astigmatans, having ovoid or cylindrical bodies with poorly sclerotized opisthosomas and legs with ventral insertions that give them a relatively upright stance (32, 48). Mites that clamber about in down also exhibit fairly weak body sclerotization. They have long legs with apophyses or clasping structures on legs I and II to help hold on to plumaceous barbules (e.g., Analgidae: Analginae) (32). Vane-dwelling mites show the greatest departure from the ancestral free-living form (32, 37). They are usually strongly dorso-ventrally flattened, with short, stocky, laterally inserted legs. Large tarsal ambulacra may help grip the flat plates of the pennaceous barbs. Vane mites are typically well sclerotized dorsally with both pro- and opisthonotal plates. The internal leg apodemes are thickened and often anastomose to form an internal framework (37). Body flattening of vane mites is a form of aerodynamic streamlining, and heavy sclerotization could both protect mites when flight feathers rub together and/or prevent water loss on the exposed vane surface (32). Finally, dermicolous mites have a morphology reminiscent of psoroptoid skin parasites of mammals (37). They are round bodied, dorso-ventrally flattened, poorly sclerotized, and have short hooked legs for grasping skin (32). Dermicolous mites fall at the lower end of feather mite body-length range at 0.25–0.3 mm (Epidermoptidae: *Microlichus*; Dermationidae: *Dermation*), whereas vane dwellers from seabirds are among the largest of feather mites (Freyanidae: *Diomedacarus gigas* 1.0–1.3 mm; Alloptidae: *Alloptes phaethornis* 1.0–2.2 mm) (37). Most feather mites are 0.3–0.7 mm long (48).

ECOLOGY AND BEHAVIOR OF FEATHER MITES

Life Cycle

The ancestral life history of Astigmata includes four motile feeding stages (larva, protonymph, tritonymph, and adult) and one nonfeeding stage, the deutonymph (a.k.a. hypopus). Deutonymphs have nonfunctional mouthparts and morphological modifications for phoretic attachment to hosts (e.g., ventral setae modified into a sucker-plate). Feather mites and other members of the Psoroptidia lack the deutonymphal stage. This may reflect the reduced need for phoretic transport in obligatorily symbiotic mites. Some non-Psoroptidia that live in permanent habitats have also lost the phoretic deutonymph (90). The only group of Psoroptidia with nonsymbiotic members, the dust mites (Pyroglyphidae), also lacks deutonymphs, but in this case it is harder to postulate why. Although the great Russian acarologist V.B. Dubinin (37–39) was usually an accurate observer of feather mite ecology, he mistakenly believed feather mites of the families Falculiferidae and Dermoglyphidae had a deutonymphal stage that passed as a parasite beneath the host's skin (37). He had confused the subdermal hypopodes of a member of the Acaridia (the family Hypoderidae) with the "missing" stage of feather mites (see Mites Associated With Birds, below).

Feather mites often molt entirely or partially inside the cast skins of other members of their species (99; H.C. Proctor, personal observation). This can result in long chains of nested exuviae. Feather mites also molt inside the empty eggshells of feather lice. Pérez & Atyeo (99) termed this use of nonliving parts of other organisms "thanatochresis" (*thanatos* = death; *chresis* = utility). Thanatochresis may help mites avoid predation, reduce loss of moisture during molt, or provide a frictional surface against which to rub off the old skin. Cast skins often remain attached to feathers at the junctions of barbs and rachis, potentially posing problems for rapid estimation of feather mite number by the wing-against-the-light method (105). Behnke et al. (7) tested this commonly used method of estimating mite load and found that it worked for many, but not all, small passerines.

Sexual dimorphism is typically strong in feather mites (see "Mating and Oviposition"), but it is rarely expressed until the final, adult stage. An exception is the kiwi mite *Kiwialges palametrichus* (Analgidae), whose nymphs are sexually dimorphic as well (51). Gaud & Atyeo (48) suggest that this may be a reflection of sexual habitat partitioning, with male nymphs living in feathers and females living in cutaneous pores. There are also two nymphal morphs in the megapode mite *Gallilichus jonesi* (103); however, because pharate adults have not been observed, the spiny and smooth morphs could not be assigned to sexes.

Diet

Waage (117) described two evolutionary routes toward occupation of and feeding from a host. Adaptations for close association between the two species may precede adaptations for feeding from the host, or, adaptations that allow feeding precede those for association. There is general agreement that feather mites followed the first route, colonizing bird bodies from an original nidicolous habitat. However, it also appears that their mouthparts were also preadapted to dining on avian integument. There is remarkably little difference in the chelicerae and palps of free-living pyroglyphids and those of most feather mites, in strong contrast to the change from chelate to stylettiform chelicerae observed in transitions from free-living to hematophagous Mesostigmata (107). Feather mite chelicerae have the same robust, chelate appearance as those of free-living nidicoles but frequently exhibit reduced numbers of cheliceral teeth (37). Exceptions to this are males in strongly sexually dimorphic taxa (e.g., Avenzoariidae: *Bdellorhynchus*; Falculiferidae: *Falculifer*), in which the enormous chelicerae are probably used for reproductive rather than nutritive purposes (see Mating and Oviposition, below).

Skin-dwelling mites such as those in the families Dermationidae and Epidermoptidae are likely to feed on skin flakes, although to my knowledge their diets have not been examined. Some epidermoptids have turned their normally phoretic relationship with lice and hippoboscid flies into a hyperparasitic one, feeding on the hemolymph of these insects (38, 66). The keratinous medulla of the rachis is eaten by many syringicolous feather mites, including the Syringobiidae (88), Ascouracaridae (100, 103), and the aberrant knemidocoptid *Lukoschoscoptes asiaticus* (91). In contrast, syringicolous Dermoglyphidae consume fluids from the papilla at the base of the feather (88).

Plumicolous mites feed primarily on uropygial gland oil that they lick from the surface of the barbules (12, 37, 66, 88, 105). Consumption of secretions or cast-off parts of a host is termed paraphagy (32). Chelicerae of plumicolous mites are usually smaller and more poorly musculated than those of syringicolous pith-eating mites, and there is little evidence that they actively gnaw on intact feathers (15, 37). Dubinin (37) examined the stomach contents of 26 species of vane-dwelling mites and observed finely grained material containing fatty droplets rather than the bolus of keratin typical to feather chewers. Dubinin hypothesized that mite loads are determined in part by the volume of oil production, which would explain the high loads of mites on waterfowl and sandpipers (which have well developed uropygial glands), low loads on small passerines (where the gland is poorly developed), and aberrantly high loads—for a raptor—on the osprey *Pandion haliaetus*, which also has a large uropygial gland. Dubinin (37) also noted that the weight of oil per feather varied within species and seemed to be associated with season and proximity of molt. Such changes in oiliness may affect population dynamics of plumicolous mites (12).

Although Dubinin (37) observed occasional skin flakes and tiny feather fragments in numerous mites, they were common only in the guts of *Ardeacarus ardeae* (Pterolichidae). These mites inhabit the plumage of herons, which have special feathers without a rachis that break up to produce an almost greasy powder down (122); however, Dubinin (37) considered down fragments in the gut to be too few to constitute a major part of the diet of *A. ardeae*. Besides these tiny keratinous fragments, he found that almost all feather mite species have fungal spores in their guts (basidiomycetes, ascomycetes, and fungi imperfecti). Most abundant were spores of *Cladosporium*, *Alternaria*, and rust fungi (Uredinales). Blanco et al. (15) examined the guts of two species of feather mites. Fungal mycelia and spores were found in 53% and 38% of *Pterodectes rutilus* (Proctophyllodidae) and *Pteronyssoides nuntiaeveris* (Avenzoariidae). Keratinolytic fungi grow in bird plumage (18), but rust spores are more likely to be filtered from foliage by bird feathers. Plumicolous mites also swallow other organic particles. Dubinin (37) examined the guts of 18,735 specimens of *Freyana* (Freyanidae) from waterfowl and found diatoms in 135 mites. Other authors have observed pollen and algae inside feather mites (15). Pollen is frequently present in the guts of mites from nectar-feeding birds (H.C. Proctor, personal observation). Bacteria are also common on bird feathers (18). Uropygial gland secretions are composed mainly of waxes and fatty acids (56) and hence are likely to be deficient in nitrogen. It is possible that microorganisms in rancid feather oil (bacteria, yeasts), plus the occasional spore or pollen grain, may provide feather mites with an important source of nitrogen.

Mating and Oviposition

Feather mites share many reproductive traits with other Astigmata. These include males with a sclerotized intromittent organ (the aedeagus) and females with a secondary copulatory opening (copulatory pore, sperm pore) that opens on the posterior dorsum (48, 66, 118). Aedeagus length is quite variable, even within a genus (e.g., *Proctophyllodes*) (37). Ancestrally, the aedeagus was a short finger-like projection, and this form is still manifested by most (if not all) syringicolous and dermicolous feather mites. In many plumicolous mites found on the vane of the feather, the aedeagus is much longer, sometimes one third to half the length of the male's body (e.g., Proctophyllodidae: *Proctophyllodes*, *Anisodiscus*; Alloptidae: *Microspalix*) (48). This may be an adaptive response to the elongation of the body common to vane-dwelling mites. Some female feather mites have an extension of the primary spermaduct (the usually internal, sclerotized tube of the copulatory pore) that extends beyond the body terminus (e.g., Caudiferidae: *Caudifera*; Crypturoptidae: all genera). In such species the male's aedeagus is usually short and in some has moved from the typical position between the third or fourth coxae to between the first coxae (the throat) of the male (Crypturoptidae: *Sternosathes*, *Tinamolichus*; Thoracosathidae: *Thoracosathes*) (48). Oviposition occurs through a ventrally located gonopore. Male feather mites, and those of most other Astigmata, also possess a pair of ventral suckers (adanal discs) used for clinging to the female during copulation (66).

Other common forms of dimorphism involve bifurcation of the male's posterior idiosoma and enlargement of the third pair of legs (48) (Figure 1). Posterior lobes of males can be simple and rounded or elaborately rayed like the tail of a peacock (104). Likewise, legs may be mildly enlarged or modified into hypertrophied pincers (Analgidae: *Analges*). Other types of male modifications include enlarged and flattened terminal setae (e.g., many Pterolichidae), enlarged chelicerae (e.g., Avenzoariidae: *Bdellorhynchus*; Falculiferidae: *Falculifer*), and enlarged and strangely branched palps (e.g., the tellingly named falculiferid *Cheiloceras cervus*). One of the oddest phenomena is the skewed body symmetry that has evolved independently in males of several lineages (e.g., Alloptidae: *Dinalloptes*; Freyanidae: *Sulanyssus*; Kramerellidae: *Freyanella*, *Michaelichus*) (37, 48).

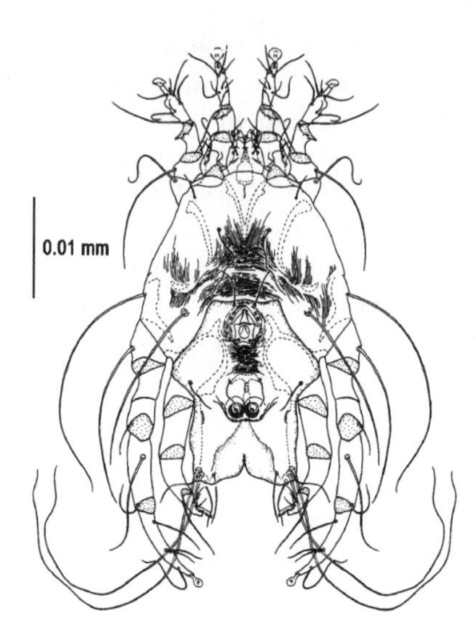

Figure 1 Ventral view of a male *Dubininia melopsittaci* (Analgoidea: Xolalgidae) showing enlarged third legs, adanal discs, and terminal lobes (copyright D.E. Walter).

Sperm-transfer behavior has rarely been documented in feather mites, and so the functions of many male modifications are mysterious. Feather mites, like many other Astigmata, mate with male and female facing in opposite directions (118), the male's venter apposed to the female's dorsum. The aedeagus is inserted into the dorsally opening copulatory pore. In species in which the female has an elongate external spermaduct, the male receives the tip of the spermaduct in his aedeagus or modified genital bursa (48); hence, the female's spermaduct acts as the intromittent device. Atyeo & Gaud (3) hypothesize that much of the elaborate modifications of the male hind body and hind legs is a response to the problem of alignment of aedeagus to the female spermaduct opening. These flanges, lobes, lamelliform setae, leg clamps, and adanal discs help line up the bodies so successful intromission can occur. Although this explains many aspects of sexual dimorphism, it fails to account for some of the most elaborate male modifications. The enlarged chelicerae of *Bdellorhynchus*, the antler-like palps of *Cheiloceras*, and many instances of body asymmetry are not expressed by all males of a species. Rather, they are present only in heteromorphic males, whereas homeomorphic males show at most slight enlargement of these structures. Mesomorphic intermediates have been postulated for some feather mite taxa (78). Other forms of polymorphism are discussed by Gaud & Atyeo (48).

Male polymorphism is found in many other Astigmata and has been particularly well studied in stored-product mites in the family Acaridae, in which up to four morphs have been observed (118). In *Caloglyphus berlesei*, the most strongly heteromorphic males are aggressive fighters that have high reproductive success in small colonies because they kill off most of their rivals (108). In large colonies, however, heteromorphs have lower reproductive success because they spend too much time fighting and too little time mating. Morph type in this species is environmentally determined, with a higher proportion of males developing into fighters at low population densities, probably mediated by pheromonal concentrations in the colonies (116). Are the strange elaborations of heteromorphic male feather mites also used for fighting rather than mating? There is no literature on intermale aggression in feather mites, nor are there studies of how morph ratios vary with population density. To further amplify the mystery surrounding male polymorphism, heteromorphic males often display reduced size of adanal discs and rutella (scrapers located at the base of the mouthparts) (28). Reductions in disc size may be explicable if heteromorphs use some other parts of their bodies to hold on to the female, but no explanation has emerged to explain reduced rutella. Dubinin (37) argued that strong body asymmetry in male feather mites helps wedge the male between the rami of adjacent barbs, and stated that it was necessary because during precopula, the male was doubly laden with the aerodynamic burden of holding on to the feather himself and clasping the inert (premolting) tritonymph. When asymmetry is expressed by all males in a species (48), this may be a plausible explanation, but when it only manifests in heteromorphic males it is unsatisfying.

Precopulatory guarding of immobile, pharate female nymphs is widespread across the Acari (118) as well as within the Astigmata (121). Dubinin (37) mistook precopulatory guarding for actual copulation and dismissed such obvious copulatory structures as the adult female's spermaduct and the absence of a functional copulatory pore in tritonymphs. It is possible that many of the sexually dimorphic structures displayed by male feather mites, including highly modified heteromorphs, are adaptations for preventing females from being stolen by other males. If true, one would predict a bias toward heteromorphic males in situations of high male-to-female ratios, when a willing female would be a rare commodity. In contrast to the situation in most feather mites, male quill mites in the family Ascouracaridae display no obvious modifications for holding on to the female, and even lack adanal discs [based on illustrations in (49)]. It would be informative to determine whether sex ratios of ascouracarids are strongly female biased as they are in the quill-dwelling Syringophilidae (Prostigmata) (<5% male) (63).

Some Astigmata produce offspring asexually (118), but there is no evidence for parthenogenesis in feather mites (37). Feather mites are primarily oviparous; however, Dubinin (37) observed evidence of seasonal ovoviviparity in some species. In spring and summer females produced eggs with thin shells that hatched upon oviposition, or sometimes within the mother. In autumn females produced thick-shelled eggs that hatched the following spring. Female feather mites usually mature only one large, sausage-shaped egg at a time (37). In plumicolous feather mites the eggs are firmly glued to feathers in often species-specific spots (97). In the syringicolous mite *Gallilichus jonesi* each egg is contained within a large membranous envelope that is plastered to the inner wall of the calamus (103). Exceptions to the single-egg rule are members of the epidermoptid genus *Myialges*, in which females are ectoparasites of feather lice and hippoboscid flies (37). In these species females become physogastric and may mature more than a dozen eggs simultaneously. Eggs are laid on stalks attached to the integument of the host insect.

Host Colonization and Movements Within the Host

Most feather mites are so highly modified for dwelling on or in bird feathers that they cannot walk well when removed from them (37). This obviously limits the potential for infestation of hosts that are not in direct body contact. The best opportunity for transfer occurs at fledging, when young birds have developed their adult plumage but are still cared for by parents. Dubinin (37) noted that mites that transfer to fledglings are chiefly nymphs and that they leave the parent birds in such numbers that the adults may become almost defaunated. Initially, feather mites are distributed over the body of a fledgling in a relatively haphazard manner, but once the young bird begins flying they are redistributed to match their typical stations on adult birds (37). This may be via a process as simple as those mites on the wrong part of the plumage falling off (3). Syringicolous mites either chew their way out of and into feathers at the time of transfer, or move through the existing opening of the superior umbilicus (48, 91, 100).

Courtship, mating, and communal roosting are other situations in which close body contact occurs between hosts. For some birds these are the only opportunities for acquiring host-specific species of mites. Nest-parasitic species such as cuckoos and cowbirds oviposit in the nests of other bird species, leaving the cuckolded parents to raise foreign young. Although some cuckoos are reported not to acquire the feather mites of their host parents (3, 48), both nestling and adult diederik cuckoos (*Chrysococcyx caprius*) harbor mites and lice typical of their ploceine weaver hosts (68). Persistence of host symbiotes on cuckoo and cowbird fledglings may allow recognition of which host species did the rearing (53, 68). Mound-building birds (Galliformes: Megapodiidae) rear their own young but do it without body contact. Eggs are placed in mounds of decomposing litter or buried in warm volcanic sands that do the incubation in place of the parent. When the young hatch they depart the nest site immediately and live a solitary life until they reach juvenile status (58). Megapodes have an extremely diverse array of feather mites (1, 103), many genera of which are specific to this family. Young brush turkeys (Megapodiidae: *Alectura lathami*) lack feather mites (H.C. Proctor, personal observation), so it is unlikely that mites found on adults are descendents of mites that lay in wait on the mound. The species-specific acarofauna of many cuckoos and probably all megapodes is maintained through horizontal rather than vertical transmission.

Horizontal transfer can occur between as well as within species. The most important form of transfer involves phoresy of dermicolous Epidermoptidae on winged hippoboscid flies (48, 61, 72, 88), a behavior also exhibited by the prostigmatan Cheyletiellidae (61) and bird lice (57). Plumicolous feather mites have been sporadically reported from hippoboscids (101), but a survey of flies taken from swifts and pigeons found no vane-dwelling mites on the hippoboscids despite an almost 100% prevalence of these mites on the birds (61). Philips & Fain (101) suggest that plumicolous mites avoid hippoboscids for transmission because flies tend to be more host generalist than plumicolous mites. In contrast, skin mites are less host specific than plumicoles, and thus the generalized behavior of hippoboscids would not be problematic. Because the flies feed on blood by burrowing through feathers to reach the skin, dermicolous mites may have more opportunities to attach to hippoboscids than would plumicolous mites (61). However, Johnson et al. (57) observed that body lice of doves were reported less frequently from hippoboscid flies than were lice from wing feathers. Body lice are more host specific than wing lice (57), supporting the hypothesis of Philips & Fain (101) that host specificity is the likely reason why plumicolous mites are rare on hippoboscid flies.

Dubinin (37) demonstrated another form of horizontal transmission through an ingenious experiment. He had observed that during mass molting of waterfowl, huge rafts of feathers would be blown into reed beds, and feather mites could be observed climbing among the floating feathers and even onto the vegetation. Dubinin covered a recently molted duck with a gauze wrapper and induced it to swim through a raft of feathers. After each of 10 passes he examined the gauze for mites and removed a total of 39 feather mites from three species. Dubinin also observed that mites in these rafts were attracted to heat, be it from a passing duck, a human arm, or a warmed metal plate.

Once on the host, mites assume their typical positions in the plumage. When more than one species of mite lives on a bird, habitat partitioning is often apparent. This has been relatively well studied in vane-dwelling mites (5, 21, 37, 100), and to a lesser extent in syringicoles (100). Passerines show little variation in feather morphology among the flight feathers, which is reflected in the low site specificity in mites of these birds (37). Partitioning can occur by species-specific settling of certain flight feathers (100) or by segregation within individual feathers. One species may inhabit the more basal or medial section of the feather, and another the apical or distal section. Despite this, there are some generalities to distributions of vane mites (37). The ventral surfaces of flight feathers, with their hedges of barb rami, are home to more species of mites than the smoother, more aerodynamically disturbed dorsal surface. *Trouessartia* spp. (Trouessartiidae) are rare exceptions, as they are more common on the dorsal surface of feathers. Their strong coxosternal skeletons and dark pigmentation may be adaptations to the stresses of this exposed condition (37). The medial part of the feathers where the rami are deep and there is relatively less rubbing of feather on feather during flight is preferred to the distal sections. Mites avoid certain feather features including the shiny mirror spots on some duck feathers and white spots on otherwise dark feathers (37).

Site specificity, although often apparent, is not invariable. Many environmental and host life-history factors induce movement in plumicolous mites. When the host is still, either roosting or brooding, mites leave their normal stations and crawl about the plumage (37). When examining birds captured at midday and released at sunset, McClure (75) observed a "crepuscular rush" of feather mites from near the rachis of flight feathers out along the vane. Birds noted at the time of capture as having few mites were suddenly re-graded as having medium loads.

Perhaps for some species of wing-dwelling mites, normal locations are actually refuges occupied when the host is active during the day, and at night the mites range widely to graze on feather oils. Changes in ambient temperature, whether seasonal or diurnal, also induce movement of wing mites (14, 37, 120), with mites moving closer to the body during winter or cold snaps. Exposure of feathers to strong light, e.g., when cormorants spread their wings to dry, causes movement toward feather bases (37). Vibration of loosening feathers prior to molt seems to signal feather mites to move to more stable surroundings (37, 60). Feather mites find certain substances unpleasant or toxic, and loading of wings with salt (35, 37), petroleum (37), and to a lesser extent formic acid introduced into feathers by anting birds (37) results in reduced mite loads.

Other aspects of host ecology can affect feather mite prevalence and abundance. Figuerola (46) compared the degree of sexual dichromatism in 70 species of passerines with prevalence of their wing-dwelling mites, testing the Hamilton & Zuk (52) prediction that mites would be more commonly observed on more dimorphic species. Results ran counter to the prediction, with the degree of plumage dichromatism being negatively correlated with feather mite prevalence. Figuerola (46) also compared mite prevalence with host body mass, breeding coloniality, migratory behavior, and winter sociality. Only the last factor was associated with increased prevalence. Likewise, Poulin (102) observed no relationship between body size and mite prevalence. Rózsa (110) found a weak positive relationship between host mass and feather mite numbers, suggesting that abundance rather than prevalence may be more responsive to size of host habitat.

COEVOLUTION OF MITES AND HOSTS

Cospeciation

Feather mites are known from all families of birds except penguins (Spheniscidae). The greatly modified feathers and subaquatic lifestyle of these diving birds may render them inappropriate habitat for feather mites. There is one published report of an *Ingrassia* sp. (Xolalgidae) from the blue penguin (*Eudyptula minor*) in New Zealand (10); however, the purported collector, Bob Pilgrim (University of Canterbury) actually did not collect this specimen (B. Pilgrim, personal communication), and so the host assignation may be spurious. Until recently, feather mites were also unknown from cassowaries (Casuariidae), but this has been rectified with the discovery of the analgid *Megninia casuaricola* (104). Another group of ratites, the rheas (Rheidae) are not known conclusively to host feather mites. Dubinin (39) collected two species of mites also known from ostriches, *Struthiopterolichus bicaudatus* (Pterolichidae) and *Paralges pachycnemis* (Dermoglyphidae), from captive rheas (*Rhea americana*); however, he was concerned that they were contaminants from ostriches held in the same enclosure. The host orders with the greatest recorded number of mite families and subfamilies are the Passeriformes (N = 18), Charadriiformes (N = 15), Coraciiformes,

Piciformes, and Apodiformes (N = 13) (Table 1). Given the disparity in attention paid to different orders of birds, this is as likely to reflect intensity of study as it is true mite richness. Regardless, feather mites occupy both ancient and modern birds and hence may have cospeciated with them.

One of the prerequisites (but not necessarily a predictor) of cospeciation is host specificity (57). For feather mites, specificity is likely tied to feather morphology and the nature of the uropygial secretions, as well as to the mode of transmission (typically between parents and offspring or between mates) (105). There are numerous claims of high host specificity among mites of birds [e.g., Turbinoptidae (55), Rhinonyssidae (94), Harpyrhynchidae (83)]. How host specific are feather mites? Unfortunately, there is no species-level compendium of host-symbiote relationships available for birds and feather mites. Scattered analyses hint at high specificity, such as 38% of 123 species of *Proctophyllodes* (Proctophyllodidae) being restricted to single host species (2). Černý (19) provided a nonexhaustive list of feather mite specificity; however, this was published 30 years ago and intervening collections have expanded host records. In more recent literature Gaud & Atyeo (48) tabulated the number of mite genera restricted to each avian order [using a pre-Sibley & Ahlquist (112) classification of birds; see footnote for Table 1]. Some mite taxa appear to have diversified mainly within particular host taxa. For example, the Pterolichinae (Pterolichidae) has 40 genera specific to the Galliformes, 26 to the Psittaciformes, and 12 to the Gruiformes. All 18 genera of Eustathiidae are restricted to the Apodiformes. Other bird mites are more widely distributed. Dubinin (37) lists *Freyana anatina* (Freyanidae) on 15 species of ducks, and *Proctophyllodes ampelidis* (Proctophyllodidae) and *Analges passerinus* (Analgidae) on 27 species of small passerines from different genera and families. Although there is little evidence that species of mites can cross host ordinal boundaries, genera often do. For example, species of *Brephosceles* (Alloptidae) are found on Gaviiformes, Procellariiformes, Anseriformes, Gruiformes, and Charadriiformes (19). Based on lists in Gaud & Atyeo (48) the Xolalginae are the most widely distributed subfamily of feather mites, being known from 15 orders of birds (Table 1).

Clearly, some taxa of feather mites are good candidates for having cospeciated with their hosts (19), whereas others are not. Is there any evidence of parallel host-symbiote phylogenies? At the microevolutionary scale Lombert (69) observed that races of the silvereye *Zosterops lateralis* possessed morphologically distinct populations of the plumicole *Trouessartia megadisca* (Trouessartiidae). At the macroevolutionary level several recent morphological and molecular analyses have tested coevolutionary hypotheses (29, 40, 77). All demonstrate a considerable degree of cospeciation but also many exceptions to the ideal of identical bird-mite cladograms. There are host-jumps, where a feather mite taxon appears to have colonized a distantly related host. There are also examples of duplication, where feather mites speciate without the host doing so. The greatest number of congeners on a single species of host are seven species of *Fainalges* (Xolalgidae) on the green conure *Aratinga holochlora* (96); however, it is not known whether mites diversified on this host or if some originated on other conures and invaded *A. holochlora* secondarily. Finally, there are many instances of lineage sorting, in which a feather mite lineage disappears from a host lineage. Paterson et al. (93) also observed evidence of lineage sorting, or missing the boat, in lice of naturally and artificially introduced birds in New Zealand. The number of ways in which symbiotes can fail to diversify with their hosts is problematic for those interested in cophylogenetic studies (92). Dabert (27) attempted an apparently modest reconstruction of the origin of feather mites of one species of sandpiper, *Actitis hypoleucos*, and discovered five different sources of colonization, of which cospeciation was only one.

Are Feather Mites Parasites?

The relationship between feather mites and birds has often been termed parasitic (37, 54, 102, 114). If we use an ecological definition of parasitism, i.e., that a parasite causes harm by taking resources garnered by the host (105), is there any evidence that feather mites are parasites? In some cases it seems possible. Syringicoles that feed on feather pith or gnaw through the calamus may weaken the feather and possibly cause premature breakage (34, 100), and some dermicolous mites (Epidermoptidae) cause skin lesions through their feeding (29). Vane-dwelling mites, however, do not obviously cause structural damage (see "Diet"). Pathology caused by plumicolous mites is usually restricted to situations in which birds maintained in captivity are unable to fly and build up unusual densities of mites that cause itching and feather-pulling (e.g., in budgerigars, *Melopsittacus undulatus*) (4, 6) and can lead to production losses (e.g., in poultry) (48). In wild birds there is only slight correlative evidence for negative effects of feather mites. Thompson et al. (114) found that house finches (*Carpodacus mexicanus*) with high loads of *Proctophyllodes* also had pox lesions and poor feather quality after molt. Harper (54) found a negative relationship between proctophyllodid numbers and plumage brightness in several passerines. Recently, however, there have been numerous studies showing no relationship between mite load and avian condition (13), or even a positive one in which high loads are correlated with good condition (12, 14, 36, 59). Blanco et al. (15) suggest that feather mites may be mutualists, removing excess body oils and potentially harmful fungus. However, mite numbers may simply respond positively to improved host condition, if condition happens to be correlated with a greater abundance of uropygial oil (12).

CONCLUSIONS AND RESEARCH NEEDS

The study of feather mites is experiencing a renaissance. Rigorous phylogenetic methods are being applied to test systematic and coevolutionary hypotheses (e.g., 29–32, 40). Feather mites even have the potential to help resolve controversial aspects of host phylogenies (1, 103) or to test biodiversity theories. Bird species differ in the richness of feather mite fauna they host, with some small passerines bearing only one or two species (37) and some parrots having over 20 (98). The literature on other host-symbiote systems suggests that host body size may be the determinant of symbiote richness (82), but Dubinin (37) argued that diversity of feather morphology was more important. This straightforward hypothesis could readily be tested using modern comparative methods.

The surge in ecological studies of feather mites and their hosts is also an encouraging trend (8, 12–14, 35, 36, 59–61). A logical extension to this correlational research—and the only way to truly test whether feather mites are parasites, commensals, or mutualists—is experimental defaunation of birds, as has been done for lice and hematophagous mites (23, 79, 105). Missing from the current trends are investigations of the basic biology of feather mites themselves. Despite some flaws, the 50-year-old works of Dubinin (37, 38, 39) still stand as the most thorough investigations of feather mite behavior and autecology. They are filled with astonishing observations that should inspire young behavioral ecologists and evolutionary biologists. Without understanding fundamental aspects of diet and reproduction, we will be unable to understand how feather mites have adapted to the host habitat or unravel the mysteries of their fascinating sexual polymorphisms.

ACKNOWLEDGMENTS

Bruce Halliday (CSIRO) generously allowed me access to his English translation of Dubinin's works. Serge Mironov (Zoological Institute, Russian Academy of Sciences) and Jacek Dabert (A. Mickiewicz University) provided many references, and Dave Walter (University of Queensland) kindly allowed me to use his *Dubininia* image.

The *Annual Review of Entomology* is online at http://ento.annualreviews.org

LITERATURE CITED

1. Atyeo WT. 1992. The pterolichoid feather mites (Acarina, Astigmata) of the Megapodiidae (Aves, Galliformes). *Zool. Scr.* 21:265–305
2. Atyeo WT, Braasch NL. 1966. The feather mite genus *Proctophyllodes* (Sarcoptiformes: Proctophyllodidae). *Bull. Univ. Nebr. State Mus.* 5:1–354
3. Atyeo WT, Gaud J. 1979. Feather mites and their hosts. See Ref. 109a, pp. 355–61
4. Atyeo WT, Gaud J. 1987. Feather mites (Acarina) of the parakeet, *Melopsittacus undulatus* (Shaw) (Aves: Psittacidae). *J. Parasitol.* 73:203–6
5. Atyeo WT, Pérez TM. 1988. Species in the genus *Rhytidelasma* Gaud (Acarina: Pterolichidae) from the green conure, *Aratinga holochlora* (Sclater) (Aves: Psittacidae). *Syst. Parasitol.* 11:85–96
6. Baker JR. 1996. Survey of feather diseases of exhibition budgerigars in the United Kingdom. *Vet. Rec.* 139:590–94
7. Behnke J, McGregor P, Cameron J, Hartley I, Shepherd M, et al. 1999. Semi-quantitative assessment of wing feather mite (Acarina) infestations on passerine birds from Portugal: evaluation of the criteria for accurate quantification of mite burdens. *J. Zool.* 248:337–47

8. Behnke JM, McGregor PK, Shepherd M, Wiles R, Barnard C, et al. 1995. Identity, prevalence and intensity of infestation with wing feather mites on birds (Passeriformes) from the Setubal Peninsula of Portugal. *Exp. Appl. Acarol.* 19:443–58
9. Bell PJ. 1996. The life history and transmission biology of *Sternostoma tracheacolum* Lawrence (Acari: Rhinonyssidae) associated with the Gouldian finch *Erythrura gouldiae*. *Exp. Appl. Acarol.* 20:323–41
10. Bishop DM, Heath ACG. 1998. Checklist of ectoparasites of birds in New Zealand. *Surveillance* 25:13–31
11. Black WC, Piesman J. 1994. Phylogeny of hard-tick and soft-tick taxa (Acari, Ixodida) based on mitochondrial 16s rDNA sequences. *Proc. Natl. Acad. Sci. USA* 91:10034–38
12. Blanco G, Frias O. 2001. Symbiotic feather mites synchronize dispersal and population growth with host sociality and migratory disposition. *Ecography* 24:113–20
13. Blanco G, Seoane J, de la Puente J. 1999. Showiness, non-parasitic symbionts, and nutritional condition in a passerine bird. *Ann. Zool. Fenn.* 36:83–91
14. Blanco G, Tella JL, Potti J. 1997. Feather mites on group-living red-billed choughs: a non-parasitic interaction? *J. Avian Biol.* 28:197–206
15. Blanco G, Tella JL, Potti J, Baz A. 2001. Feather mites on birds: costs of parasitism or conditional outcomes? *J. Avian Biol.* 32:271–74
16. Bochkov A, Galloway T. 2001. Parasitic cheyletoid mites (Acari: Cheyletoidea) associated with passeriform birds (Aves: Passeriformes) in Canada. *Can. J. Zool.* 79:2014–28
17. Burley N, Tidemann SC, Halupka K. 1991. Bill colour and parasite levels of zebra finches. See Ref. 70, pp. 359–76
18. Burtt EH Jr, Ichida JM. 1999. Occurrence of feather-degrading bacilli in the plumage of birds. *Auk* 116:364–72
19. Černý V. 1971. Parasite-host relationships in feather mites. In *Proc. 3rd Int. Congr. Acarol.*, ed. M Daniel, B Rosický, pp. 761–64. The Hague: Junk
20. Chapman BR, George JE. 1991. The effects of ectoparasites on cliff swallow growth and survival. See Ref. 70, pp. 69–92
21. Choe JC, Kim KC. 1989. Microhabitat selection and coexistence in feather mites (Acari: Analgoidea) on Alaskan seabirds. *Oecologia* 79:10–14
22. Clayton DH, Moore J. 1997. *Host-Parasite Evolution: General Principles and Avian Models*. Oxford: Oxford Univ. Press
23. Clayton DH, Tompkins DM. 1995. Comparative effects of mites and lice on the reproductive success of rock doves (*Columba livia*). *Parasitology* 110:195–206
24. Colloff MJ, Niedbala W. 1996. Arboreal and terrestrial habitats of phthiracaroid mites (Oribatida) in Tasmanian rainforests. See Ref. 78a, pp. 607–11
25. Colwell RK. 1995. Effects of nectar consumption by the hummingbird flower mite *Proctolaelaps kirmsei* on nectar availability in Hamelia patens. *Biotropica* 27:206–17
26. Colwell RK, Naeem S. 1999. Sexual sorting in hummingbird flower mites (Mesostigmata: Ascidae). *Ann. Entomol. Soc. Am.* 92:952–59
27. Dabert J. 1992. Feather mites of the common sandpiper *Actitis hypoleucos* (Charadriiformes, Scolopaci): an attempt at a reconstruction of acarofauna origin (Acari: Astigmata). *Genus* 3:1–11
28. Dabert J, Atyeo WT. 1997. The feather mite genus *Grenieria* Gaud & Mouchet, 1959 (Acarina, Syringobiidae). I. Systematics and descriptions of species. *Mitt. Hambg. Zool. Mus. Inst.* 94:125–44
29. Dabert J, Dabert M, Mironov SV. 2001. Phylogeny of feather mite subfamily

Avenzoariinae (Acari: Analgoidea: Avenzoariidae) inferred from combined analyses of molecular and morphological data. *Mol. Phylogenet. Evol.* 20: 124–35

30. Dabert J, Ehrnsberger R. 1995. Zur Systematik und Phylogenie der Gattung *Thecarthra* Trouessart, 1896 (Astigmata, Pterolichoidea, Syringobiidae) mit Beschreibung zweier neuer Arten. *Mitt. Hamb. Zool. Mus. Inst.* 92:87–116
31. Dabert J, Ehrnsberger R. 1998. Phylogeny of the feather mite family Ptiloxenidae GAUD, 1982 (Acari: Pterolichoidea). In *Arthropod Biology: Contributions to Morphology, Ecology and Systematics*, ed. E Eberman, Biosyst. Ecol. Ser., 14:145–78. Vienna: Austrian Acad. Sci.
32. Dabert J, Mironov SV. 1999. Origin and evolution of feather mites (Astigmata). *Exp. Appl. Acarol.* 23:437–54
33. Darolová A, Hoi H, Schleicher B. 1997. The effect of ectoparasite nest load on the breeding biology of the Penduline Tit *Remiz pendulinus*. *Ibis* 139:115–20
34. Dorrestein GM, VanderHorst HHA, Cremers HJWM, VanderHage M. 1997. Quill mite (*Dermoglyphus passerinus*) infestation of canaries (*Serinus canaria*): diagnosis and treatment. *Avian Pathol.* 26:195–99
35. Dowling DK, Richardson DS, Blaakmeer K, Komdeur J. 2001. Feather mite loads influenced by salt exposure, age and reproductive stage in the Seychelles Warbler *Acrocephalus sechellensis*. *J. Avian Biol.* 32:364–69
36. Dowling DK, Richardson DS, Komdeur J. 2001. No effects of a feather mite on body condition, survivorship, or grooming behavior in the Seychelles warbler, *Acrocephalus sechellensis*. *Behav. Ecol. Sociobiol.* 50:257–62
37. Dubinin VB. 1951. Feather mites (Analgesoidea). Part 1. Introduction to their study. *Fauna USSR* 6(5):1–363
38. Dubinin VB. 1953. Feather mites (Analgesoidea). Part II. Families Epidermoptidae and Freyanidae. *Fauna USSR* 6(6):1–411
39. Dubinin VB. 1956. Feather mites (Analgesoidea). Part III. Family Pterolichidae. *Fauna USSR* 6(7):1–813
40. Ehrnsberger R, Mironov SV, Dabert J. 2001. A preliminary analysis of phylogenetic relationships of the feather mite family Freyanidae Dubinin, 1953 (Acari: Astigmata). *Biol. Bull. Poznan* 38:181–201
41. Fain A. 1979. Specificity, adaptation and parallel host-parasite evolution in acarines, especially Myobiidae, with a tentative explanation for the regressive evolution caused by the immunological reactions of the host. See Ref. 109a, pp. 321–28
42. Fain A, Bafort J. 1967. Cycle évolutif et morphologie de *Hypodectes* (*Hypodectoides*) *propus* (Nitzsch) acarien nidicole à deutonymphe parasite tissulaire des pigeons. *Bull. Acad. R. Belg.* 53:501–33
43. Fain A, Clark JM. 1994. Description and life cycle of *Suladectes hughesae antipodus* subspec. nov. (Acari: Hypoderatidae) associated with *Sula bassana serrator* Gray (Aves: Pelecaniformes) in New Zealand. *Acarologia* 35:361–71
44. Fain A, Hyland KW Jr. 1985. Evolution of astigmatid mites on mammals. In *Coevolution of Parasitic Arthropods and Mammals*, ed. KC Kim, pp. 641–58. New York: Wiley-Intersci.
45. Fain A, Smiley RL. 1989. A new cloacarid mite (Acari: Cloacaridae) from the lungs of the great horned owl, *Bubo virginanus*, from the U.S.A. *Int. J. Acarol.* 15:111–15
46. Figuerola J. 2000. Ecological correlates of feather mite prevalence in passerines. *J. Avian Biol.* 31:489–94
47. Deleted in proof
48. Gaud J, Atyeo WT. 1996. Feather mites of the world (Acarina, Astigmata): the supraspecific taxa. Part I. Text. *Ann. Zool. Wet.* 277:1–193

49. Gaud J, Atyeo WT. 1996. Feather mites of the world (Acarina, Astigmata): the supraspecific taxa. Part II. Illustrations. *Ann. Zool. Wet.* 277:1–436
50. Gaud J, Atyeo WT, Klompen JSH. 1989. Oconnoriidae, a new family of feather mites (Acarina, Pterolichoidea). *J. Entomol. Sci.* 24:417–21
51. Gaud J, Laurence BR. 1981. Surprenant polymorphisme des formes immatures chez lacarien plumicole *Kiwialges palametrichus* (Astigmata, Analgidae). *Acarologia* 22:209–15
52. Hamilton WD, Zuk M. 1982. Heritable true fitness and bright birds: a role for parasites? *Science* 218:384–87
53. Hahn DC, Price RD, Osenton PC. 2000. Use of lice to identify cowbird hosts. *Auk* 117:943–51
54. Harper DGC. 1999. Feather mites, pectoral muscle condition, wing length and plumage coloration of passerines. *Anim. Behav.* 58:553–62
54a. Houck MA, ed. 1994. *Mites: Ecological and Evolutionary Analyses of Life-History Patterns*. New York: Chapman & Hall. 357 pp.
55. Hyland KE. 1979. Specificity and parallel host-parasite evolution in the Turbinoptidae, Cytoditidae and Ereynetidae living in the respiratory passages of birds. See Ref. 109a, pp. 363–69
56. Jacob J, Ziswiler V. 1982. The uropygial gland. In *Avian Biology VI*, ed. DS Farner, JR King, pp. 199–324. New York: Academic
57. Johnson KP, Williams BL, Drown DM, Adams RJ, Clayton DH. 2002. The population genetics of host specificity: genetic differentiation in dove lice (Insecta: Pthiraptera). *Mol. Ecol.* 11:25–38
58. Jones DN, Dekker RWRJ, Roselaar CS. 1995. *The Megapodes*. Oxford: Oxford Univ. Press. 262 pp.
59. Jovani R, Blanco G. 2000. Resemblance within flocks and individual differences in feather mite abundance on long-tailed tits, *Aegithalos caudatus* (L.). *Écoscience* 7:428–32
60. Jovani R, Serrano D. 2001. Feather mites (Astigmata) avoid moulting wing feathers of passerine birds. *Anim. Behav.* 62: 723–27
61. Jovani R, Tella JL, Sol D, Ventura D. 2001. Are hippoboscid files a major mode of transmission of feather mites? *J. Parasitol.* 87:1187–89
62. Kells SA, Surgeoner GA. 1997. Sources of northern fowl mite (*Ornithonyssus sylviarum*) infestation in Ontario egg production facilities. *J. Appl. Poultry Res.* 6:221–28
63. Kethley J. 1971. Population regulation in quill mites (Acarina: Syringophilidae). *Ecology* 52:113–18
64. Kettle DS. 1995. *Medical and Veterinary Entomology*. Wallingford, UK: CAB Int. 725 pp. 2nd ed.
65. Klompen H, Grimaldi D. 2001. First mesozoic record of a parasitiform mite: a larval argasid tick in Cretaceous amber (Acari: Ixodida: Argasidae). *Ann. Entomol. Soc. Am.* 94:10–15
66. Krantz GW. 1978. *A Manual of Acarology*. Corvallis: Ore. State Univ. Book Stores, Inc. 509 pp. 2nd ed.
67. Latta SC, OConnor BM. 2001. Patterns of *Knemidokoptes jamaicensis* (Acari: Knemidokoptidae) infestations among eight new avian hosts in the Dominican Republic. *J. Med. Entomol.* 38:437–40
68. Lindholm AK, Venter GJ, Ueckermann EA. 1998. Persistence of passerine ectoparasites on the diederik cuckoo *Chrysococcyx caprius*. *J. Zool.* 244:145–53
69. Lombert HAMP. 1988. *Co-evolution of three eastern Australian subspecies of the passerine bird* Zosterops lateralis *(Latham, 1801) and their ectoparasitic feather mite* Trouessartia megadisca *Gaud, 1962*. PhD thesis. Griffith Univ., Qld., Aust.
70. Loye JE, Zuk M. 1991. *Bird-Parasite Interactions: Ecology, Evolution and Behaviour*. Oxford: Oxford Univ. Press

71. Lucas AM, Stettenheim PR. 1972. *Avian Anatomy*, Part 1. *Integument. Agriculture Handbook 362*. Washington, DC: US GPO
72. Madden D, Harmon WM. 1998. First record and morphology of *Myialges caulotoon* (Acari: Epidermoptidae) from Galapagos hosts. *J. Parasitol.* 84:186–89
73. Martill DM, Davis PG. 1998. Did dinosaurs come up to scratch? *Nature* 396: 528–29
74. Maurer V, Baumgärtner J, Fölsch DW. 1995. The impact of *Dermanyssus gallinae*-infestations on performance and on physiological and behavioural parameters of laying hens. In *The Acari: Physiological and Ecological Aspects of Acari-Host Relationships*, ed. D Kropczynska, J Boczek, A Tomczyk, pp. 551–56. Warsaw: Dabor
75. McClure HE. 1989. Occurrence of feather mites (Proctophyllodidae) among birds of Ventura County lowlands, California. *J. Field Ornithol.* 60:431–50
76. Mindell DP, Sorenson MD, Huddleston CJ, Miranda HC Jr, Kinght A, et al. 1997. Phylogenetic relationships among and within select avian orders based on mitochondrial DNA. In *Avian Molecular Evolution and Systematics*, ed. DP Mindell, pp. 213–47. San Diego, CA: Academic
77. Mironov SV, Dabert J. 1999. Phylogeny and co-speciation in feather mites of the subfamily Avenzoariinae (Analgoidea : Avenzoariidae). *Exp. Appl. Acarol.* 23:525–49
78. Mironov SV, Dabert J, Atyeo WT. 1993. A new species of the feather mite genus Bregetovia (Analgoidea, Avenzoariidae) with notes on the systematics of the genus. *Entomol. Mitt. Zool. Mus. Hambg.* 11:75–87
78a. Mitchell R, Horn DJ, Needham GR, Welbourn WC, eds. 1996. *Acarology IX: Proceedings*. Columbus: Ohio Biol. Survey
79. Møller AP. 1990. Effects of parasitism by a haematophagous mite on reproduction in the barn swallow. *Ecology* 71:2345–57
80. Møller AP. 2000. Survival and reproductive rate of mites in relation to resistance of their barn swallow hosts. *Oecologia* 124:351–57
81. Møller AP, De Lope F. 1999. Senescence in a short-lived migratory bird: age-dependent morphology, migration, reproduction and parasitism. *J. Anim. Ecol.* 68:163–71
82. Morand S, Simková A, Matejusová I, Plaisance L, Verneau O, Desdevises Y. 2002. Investigating patterns may reveal processes: evolutionary ecology of ectoparasite monogeneans. *Int. J. Parasitol.* 32:111–19
83. Moss WW. 1979. Patterns of host-specificity and co-evolution in the Harpyrhynchidae. See Ref. 109a, pp. 379–84
84. Moss WW, Oliver JH Jr, Nelson BC. 1968. Karyotypes and developmental stages of *Harpyrhynchus novoplumaris* sp. n. (Acari: Cheyletoidea: Harpyrhynchidae), a parasite of North American birds. *J. Parasitol.* 54:377–92
85. Nordenfors H, Hoglund J, Uggla A. 1999. Effects of temperature and humidity on oviposition, molting, and lonevity of *Dermanyssus gallinae* (Acari: Dermanyssidae). *J. Med. Entomol.* 36: 68–72
86. Norton RA. 1998. Morphological evidence for the evolutionary origin of the Astigmata (Acari: Acariformes). *Exp. Appl. Acarol.* 22:559–94
87. OConnor BM. 1982. Acari: Astigmata. In *Synopsis and Classification of Living Organisms*, ed. SP Parker, pp. 146–69. New York: McGraw-Hill. 1232 pp.
88. OConnor BM. 1982. Evolutionary ecology of astigmatid mites. *Annu. Rev. Entomol.* 27:385–409
89. OConnor BM. 1984. Phylogenetic relationships among higher taxa in the Acariformes, with particular reference to the Astigmata. In *Acarology VI*, ed. DA

Griffiths, CE Bowman, pp. 19–27. Chichester, UK: Ellis-Horwood

90. OConnor BM. 1994. Life-history modifications in astigmatid mites. See Ref. 54a, pp. 136–59
91. OConnor BM, Klompen JSH, Lombert HAPM. 1987. A new subfamily of quill-inhabiting mites (Acari: Knemidokoptidae) from Asian swifts (Aves: Apodidae) with observations on phylogenetic relationships in the Knemidokotidae. *Int. J. Acarol.* 14:261–66
92. Paterson AM, Banks J. 2001. Analytical approaches to measuring cospeciation of host and parasites: through a glass, darkly. *Int. J. Parasitol.* 31:1012–22
93. Paterson AM, Palma RL, Gray RD. 1999. How frequently do avian lice miss the boat? Implications for coevolutionary studies. *Syst. Biol.* 48:214–23
94. Pence DB. 1979. Congruent interrelationships of the Rhinonyssinae (Dermanyssidae) with their avian hosts. See Ref. 109a, pp. 371–77
95. Pence DB, Cole RA, Brugger KE, Fischer JR. 1999. Epizootic podoknemidokoptiasis in American robins. *J. Wildlife Dis.* 35:1–7
96. Pérez TM. 1995. Seven species of *Fainalges* Gaud and Berla (Analgoidea, Xolalgidae) from *Aratinga holochlora* (Sclater) (Aves, Psittacidae). *Zool. Scr.* 24:203–23
97. Pérez TM. 1996. The eggs of seven species of *Fainalges* Gaud and Berla (Xolalgidae) from the green conure (Aves, Psittacidae). See Ref. 78a, pp. 297–300
98. Pérez TM. 1997. Eggs of feather mite congeners (Acarina: Pterolichidae, Xolalgidae) from different species of new world parrots (Aves, Psittaciformes). *Int. J. Acarol.* 23:103–6
99. Pérez TM, Atyeo WT. 1984. Feather mites, feather lice, and thanatochresis. *J. Parasitol.* 70:807–12
100. Pérez TM, Atyeo WT. 1984. Site selection of the feather and quill mites of Mexican parrots. In *Acarology 6*, ed. DA Griffiths, CE Bowman, pp. 1:563–70. Chichester, UK: Ellis-Horwood
101. Philips JR, Fain A. 1991. Acarine symbionts of louseflies (Diptera: Hippoboscidae). *Acarologia* 32:377–84
102. Poulin R. 1991. Group-living and infestation by ectoparasites in passerines. *Condor* 93:418–23
103. Proctor HC. 1999. *Gallilichus jonesi* sp n. (Acari: Ascouracaridae): a new species of feather mite from the quills of the Australian brush-turkey (Aves: Megapodiidae). *Aust. J. Entomol.* 38:77–84
104. Proctor HC. 2001. *Megninia casuaricola* sp. n. (Acari: Analgidae), the first feather mite from a cassowary (Aves: Struthioniformes: Casuariidae). *Aust. J. Entomol.* 40:335–41
105. Proctor H, Owens I. 2000. Mites and birds: diversity, parasitism and coevolution. *Trends Ecol. Evol.* 15:358–64
106. Pruett-Jones M, Pruett-Jones S. 1991. Analysis and ecological correlates of tick burdens in a New Guinea avifauna. See Ref. 70, pp. 154–76
107. Radovsky FJ. 1994. The evolution of parasitism and the distribution of some dermanyssoid mites (Mesostigmata) on vertebrate hosts. See Ref. 54a, pp. 186–217
108. Radwan J. 1993. The adaptive significance of male polymorphism in the acarid mite *Caloglyphus berlesei*. *Behav. Ecol. Sociobiol.* 33:291–96
109. Rendell WB, Verbeek NAM. 1996. Old nest material in nestboxes of tree swallows: effects on reproductive success. *Condor* 98:142–52

109a. Rodriguez JG, ed. 1979. *Recent Advances in Acarology*, Vol. II. New York: Academic

110. Rózsa L. 1997. Wing feather mite (Acari: Proctophyllodidae) abundance correlates with body mass of passerine hosts: a comparative study. *Can. J. Zool.* 75:1535–39

111. Seeman OD. 1996. Flower mites and phoresy: the biology of *Hattena panopla* Domrow and *Hattena cometis* Domrow (Acari: Mesostigmata: Ameroseiidae). *Aust. J. Zool.* 44:193–203
112. Sibley CG, Ahlquist JE. 1990. *Phylogeny and Classification of Birds: A Study in Molecular Evolution*. New Haven, CT: Yale Univ. Press
113. Teel PD, Hopkins SW, Donahue WA, Strey OF. 1998. Population dynamics of immature *Amblyomma maculatum* (Acari: Ixodidae) and other ectoparasites on meadowlarks and northern bobwhite quail resident to the coastal prairie of Texas. *J. Med. Entomol.* 35:483–88
114. Thompson CW, Hillgarth N, Leu M, McClure HE. 1997. High parasite load in house finches (*Carpodacus mexicanus*) is correlated with reduced expression of a sexually selected trait. *Am. Nat.* 149:270–94
115. Tidemann SC. 1996. Causes of the decline of the Gouldian finch *Erythrura gouldiae*. *Bird Conserv. Int.* 6:49–61
116. Timms S, Ferro DN, Emberson RM. 1981. Andropolymorphism and its heritability in *Sancassania berlesei* (Michael) (Acari: Acaridae). *Acarologia* 22:391–98
117. Waage JK. 1979. The evolution of insect/vertebrate associations. *Biol. J. Linn. Soc.* 12:187–224
118. Walter DE, Proctor HC. 1999. *Mites: Ecology, Evolution and Behaviour*. Wallingford, UK: CAB Int. 322 pp.
119. Weatherhead PJ, Dufour KW, Lougheed SC, Eckert CG. 1999. A test of the good-genes-as-heterozygosity hypothesis using red-winged blackbirds. *Behav. Ecol.* 10:619–25
120. Wiles PR, Cameron J, Behnke JM, Hartley IR, Gilbert FS, McGregor PK. 2000. Season and ambient air temperature influence the distribution of mites (*Proctophyllodes stylifer*) across the wings of blue tits (*Parus caeruleus*). *Can. J. Zool.* 78:1397–407
121. Witalinski W, Dabert J, Walzl MG. 1992. Morphological adaptation for precopulatory guarding in astigmatic mites (Acari: Acaridida). *Int. J. Acarol.* 18:49–54
122. Young JZ. 1981. *The Life of Vertebrates*. Oxford: Clarendon. 645 pp. 3rd ed.

Annu. Rev. Entomol. 2003. 48:211–34
doi: 10.1146/annurev.ento.48.091801.112756

First published online as a Review in Advance on October 16, 2002

THE GENOME SEQUENCE AND EVOLUTION OF BACULOVIRUSES

Elisabeth A. Herniou,[1,2] Julie A. Olszewski,[1] Jennifer S. Cory,[2] and David R. O'Reilly[1,3]
[1]*Department of Biological Sciences, Imperial College of Science, Technology and Medicine, London SW7 2AZ, United Kingdom; e-mail: elisabeth.herniou@ic.ac.uk; j.olszewski@ic.ac.uk; david.oreilly@syngenta.com*
[2]*Ecology and Biocontrol group, Centre for Ecology and Hydrology, Oxford, Mansfield Road, Oxford OX1 3SR, United Kingdom; e-mail: jsc@ceh.ac.uk*
[3]*Syngenta, Jealotts Hill International Research Station, Bracknell, Berkshire RG42 6EY, United Kingdom*

Key Words Baculoviridae, phylogeny, gene order, gene composition, gene function

■ **Abstract** Comparative analysis of the complete genome sequences of 13 baculoviruses revealed a core set of 30 genes, 20 of which have known functions. Phylogenetic analyses of these 30 genes yielded a tree with 4 major groups: the genus *Granulovirus* (GVs), the group I and II lepidopteran nucleopolyhedroviruses (NPVs), and the dipteran NPV, CuniNPV. These major divisions within the family Baculoviridae were also supported by phylogenies based on gene content and gene order. Gene content mapping has revealed the patterns of gene acquisitions and losses that have taken place during baculovirus evolution, and it has highlighted the fluid nature of baculovirus genomes. The identification of shared protein phylogenetic profiles provided evidence for two putative DNA repair systems and for viral proteins specific for infection of lymantrid hosts. Examination of gene order conservation revealed a core gene cluster of four genes, *helicase*, *lef-5*, *ac96*, and *38K*(*ac98*), whose relative positions are conserved in all baculovirus genomes.

CONTENTS

INTRODUCTION

Interest in insect diseases can be traced to the sixteenth century when a "wilting disease" of silkworms was first formally described (8). Early in the twentieth century this disease was attributed to a virus infection, and in 1947, visualization of rod-shaped virions, which are now known to be characteristic of the virus family Baculoviridae, was reported (81). Insect-infecting viruses have now been described from at least 12 distinct virus families (113). For most of these, we only have a rudimentary understanding of how the disease is transmitted, the role that the virus plays in insect population dynamics, the molecular details of viral infection, or how these viruses have arisen and evolved.

This last issue, determining a family tree describing the evolutionary relatedness of individual members of a virus family, is particularly challenging. Traditional approaches for classification of organisms, such as using morphological features or using molecular sequences such as ribosomal DNA (120), do not easily transfer to viruses. Reasons for this include a lack of fossil records for viruses, their streamlined gene complements, and the plasticity of their genomes compared with cellular organisms. Nevertheless, viral gene sequences useful for reconstructing phylogenies have been identified. For example, DNA polymerase gene homologs have been used to infer phylogenetic relationships among DNA viruses (24, 58, 85, 98). The wealth of genome sequence information available in this postgenomic era facilitates new approaches to determining how groups of viruses are related to one another, and hence, how they might have evolved. Here, we focus on the family Baculoviridae and examine what genomic sequence data have allowed us to deduce about the evolution of these viruses.

The family Baculoviridae comprises a diverse group of arthropod-specific viruses. They have been reported worldwide from over 600 host species (78), mostly from the order Lepidoptera but also from the orders Diptera, Hymenoptera, and the crustacean order Decapoda (11). Baculoviruses are relatively host specific and have thus been used for biocontrol applications (84). A desire to improve baculoviruses as pesticides, as well as the development of baculovirus-based gene expression systems, has led to many studies characterizing the infection process at the molecular level (82).

Baculoviruses are rod-shaped, enveloped viruses with circular double-stranded DNA genomes ranging in size from 90 kb to 180 kb. The family Baculoviridae is currently subdivided into two genera based largely on the morphology of the occlusion bodies (OBs) they form in infected cells (11). The genus *Nucleopolyhedrovirus* (NPV) is characterized by viruses forming polyhedral OBs, each containing many virions (100), whereas viruses of the genus *Granulovirus* (GV) typically produce ovoid OBs with a single virion (119). GVs have been described solely from lepidopteran hosts, whereas NPVs have been isolated from a wider range of insects. However, the taxonomical status of nonlepidopteran baculoviruses is still uncertain (11).

The baculovirus life cycle typically involves the production of two virus forms (14, 82). The occluded form mentioned above comprises one or more virions embedded in a crystalline protein matrix. Occlusion is thought to protect virions from environmental degradation until a host caterpillar ingests the OB. After ingestion, the virions are released in the insect midgut and infect midgut epithelial cells. This initial round of infection produces budded virus (BV), which disseminates the infection throughout susceptible insect tissues. These infected tissues eventually produce more OBs, which are liberated upon disintegration of the caterpillar cadaver.

The first baculovirus to be completely sequenced was *Autographa californica* NPV (AcMNPV) (5). The number of complete sequences has since grown rapidly, reaching 13 genomes at the time of writing this review, including 3 GVs (50, 51, 75), 9 lepidopteran NPVs (2, 5, 26, 27, 47, 55, 56, 69, 91), and 1 dipteran NPV (1). Each complete sequence has provided a wealth of data about the complement of genes present in each individual virus. However, whole genome sequences, when compared between viruses, can also suggest how virus family members are related to one another and how they have evolved since their last common ancestor. This chapter reviews the insights into baculovirus biology and evolution that can be gained by taking comparative genomic and phylogenetic approaches.

We have previously presented comparative genomic analyses of 9 (53) or 10 (55) complete baculovirus genomes. So that this review is as current as possible, we have now included three additional sequences that have become available in the interim: CuniNPV (1), HzSNPV (27), and SpltNPV (91). The methods used are the same as those described previously (53) and all trees presented here are consistent with previously published phylogenies. In agreement with others, we find that the CuniNPV baculovirus is highly divergent from other sequenced baculoviruses and represents an additional baculovirus group.

CORE BACULOVIRUS GENES

The 13 completely sequenced baculoviruses have 30 identified genes in common. A putative function can be assigned to 20 of these (Table 1). They include genes that act at various stages of the baculovirus infection cycle and can be grouped into five functional categories: RNA transcription, DNA replication, structural proteins, auxiliary proteins, and proteins of unknown function.

Transcription

Baculovirus gene expression occurs in a temporally regulated fashion, which is primarily regulated at the level of transcription (42). Baculovirus transcription

TABLE 1 Genes present in all baculovirus genomes

Gene function	Genes present in all baculoviruses	Additional genes present in all lepidopteran baculoviruses
Replication	*lef-1 (ac14), lef-2 (ac6), dnapol (ac65), helicase (ac95)*	*dbp1 (ac25), lef-3 (ac67), ie-1 (ac147), me53 (ac139)*
Transcription	*p47 (ac40), lef-8 (ac50), lef-9 (ac62), vlf-1 (ac77), lef-4 (ac90), lef-5 (ac99)*	*Pp31/39K (ac36), lef-6 (ac28), lef-11 (ac37)*
Structural proteins	*ac23 (ld130), gp41 (ac80). odv-ec27 (ac144), odv-e56 (ac148), p6.9 (ac100), p74 (ac138), vp91/p95 (ac83), vp39 (ac89), vp1054 (ac54)*	*fp25K (ac61), odv-e18 (ac143), odv-e25 (ac94), odv-e66 (ac46), pk1 (ac10), polh (ac8),*
Auxiliary*	*alk-exo (ac133)*	*fgf (ac32), ubiquitin (ac35)*
Unknown	*38K (ac98), ac22, ac68, ac81, ac92, ac96, ac109, ac115, ac119, ac142*	*38.7K (ac13), ac29, ac38, ac53, ac66, ac75, ac76, ac78, ac82, ac93, ac106, ac110, ac145, ac146, p40 (ac101), p12 (ac102), p45 (ac103)*

*Auxiliary genes are genes that are not directly involved in viral gene expression or genome replication, or in the formation of progeny virus particles.

is typically categorized into three classes: early, late, and very late. Host RNA polymerase II transcribes early genes (42), but a viral RNA polymerase transcribes late and very late genes (44, 48). The AcMNPV late RNA polymerase is composed of four subunits, all of them viral proteins (48). The genes that encode these, *lef-4*, *lef-8*, *lef-9*, and *p47*, are present in all sequenced baculoviruses. *lef-8* and *lef-9* (74) display homology to subunits of the DNA-dependent RNA polymerases of prokaryotes and eukaryotes, with *lef-8* containing a motif conserved in β- and β'-subunits of RNA polymerases that is thought to form part of the catalytic site of the enzyme (92). The lef-4 protein of AcMNPV, which has guanylyltransferase (48), RNA 5′-triphosphatase, and ATPase (61) activities, is proposed to provide a capping function for RNA transcribed by the baculovirus polymerase (61). The function of *p47* is unknown.

Two other genes implicated in transcription, *lef-5* and *vlf-1*, are in the core set. The function of *lef-5* is not clear. VLF-1 is required for very late gene transcription in AcMNPV and has sequence homology to a family of integrases/resolvases (80). Studies indicate that VLF-1 acts by physically interacting with promoter sequences, which allows for the burst of expression observed from very late promoter-driven genes (121).

DNA Replication

Although baculovirus DNA replicates in the nucleus, baculoviruses carry their own complement of genes encoding DNA replication proteins. Four of these, *DNA polymerase*, *DNA helicase* (*p143*), *lef-1*, and *lef-2*, are found in all sequenced genomes.

lef-1 encodes a putative DNA primase (6). The Lef-1 and Lef-2 proteins are thought to form hetero-oligomeric complexes during baculovirus replication (38).

Structural Proteins

Newly replicated baculovirus genomes are packaged with viral proteins to form nucleocapsids, which 'bud' from infected cells, acquiring a virally modified plasma membrane envelope as they do so to form BV. At very late times of infection, nucleocapsids instead become enveloped within the nucleus prior to being occluded. The source of the envelope is not clear. However, it contains several viral proteins (45). Within the NPVs, some viruses have many individually enveloped nucleocapsids per OB (SNPV, singly embedded NPV), whereas others have bundles of enveloped nucleocapsids within each OB (MNPV, multiply embedded NPV) (117). This trait has no phylogenetic relevance (122) and its biological basis has not been determined. Four genes implicated in nucleocapsid structure or assembly are conserved among the 13 baculovirus genomes: *p6.9*, *vp39*, *vp1054*, and *vp91* (Table 1). The p6.9 protein is a small basic protein thought to facilitate condensation of the DNA associated with packaging within the nucleocapsid (112, 118). Vp39 is the major capsid protein (12, 96, 110). Vp1054 and Vp91 are both associated with BV and occluded virus (88, 103).

Ld130, now termed baculovirus F protein (94a, 114b), is an envelope fusion protein found in infected cell membranes and in BV envelopes (57, 93). It is required for fusion of the BV envelope with the host endosomal membranes to release nucleocapsids for initiation of infection. Interestingly, an unrelated BV envelope protein, Gp64, provides this fusion activity for viruses such as AcMNPV, BmNPV, and OpMNPV, even though they possess an *ld130* gene (13, 15, 116). Gp64 homologs are more closely related to one another than the less-conserved Ld130 homologs (101). It has therefore been suggested that *ld130* represents an ancestral envelope fusion protein gene whose activity has been displaced in some baculoviruses by *gp64* (95), a process known as non-orthologous gene displacement (46). Presumably, *ld130* homologs in *gp64*-containing genomes have evolved some other essential viral function (95).

Odv-ec27, Odv-ec56, p74, and Gp41 are all conserved proteins found in association with occluded virus (20, 21, 68, 115). Odv-ec27 has homology to cellular cyclins and interacts with cyclin-dependent kinases and other cell cycle–related host proteins in in vitro assays, leading to the suggestion that this protein is a multifunctional viral cyclin (7). p74 is important for the oral infectivity of occluded viruses (39), while the Gp41 glycoprotein is required for egress of nucleocapsids out of the infected cell nucleus to form BV (87).

Auxiliary Proteins/Unknown Function

The only auxiliary protein conserved among all 13 sequenced baculoviruses is a putative alkaline exonuclease (alk-exo). In addition, another 10 genes of unknown function are conserved among all 13 genomes. The broad conservation of these genes strongly suggests that they provide an essential function for virus replication. They are prime candidates for future molecular and biochemical studies.

Lepidopteran versus Dipteran Baculoviruses

It is worth noting that a much greater number of genes are conserved among all the lepidopteran baculovirus genomes. An additional 32 genes are found in all lepidopteran viruses sequenced to date that are apparently absent from the mosquito CuniNPV genome (Table 1). Some of these differences may reflect a different spectrum of genes required to infect a lepidopteran host compared with a dipteran host. Alternatively, they may reflect the different biology of the CuniNPV compared with the lepidopteran viruses that have been sequenced. Unlike the lepidopteran baculoviruses, CuniNPV is restricted to replication in a subset of midgut cells (85); therefore, these "additional" lepidopteran genes may be involved in utilization of various host tissues. However, it is important to bear in mind that the biology of these viruses might not be as distinct as the sequence data appear to indicate. Phylogenetically unrelated genes may provide similar functions in different virus groups. One example of this, the *ld130/gp64* genes, has been discussed already. Another example is the genes baculoviruses carry to block apoptosis in the infected cell. Two unrelated families of anti-apoptotic proteins are found in baculoviruses: p35 homologs and IAPs (30). All lepidopteran baculoviruses have at least one *iap* gene, whereas AcMNPV, BmNPV, SpltNPV, and CuniNPV all have putative *p35* genes. Finally, it is possible that homologs of some of the lepidopteran core virus genes are present in CuniNPV, but they are too highly diverged to be identified by BLAST searches (1).

TOWARD A GENOME PHYLOGENY OF BACULOVIRUSES

Traditionally, baculovirus phylogenies have been based on individual gene sequences. The polyhedrin/granulin (*polh*) gene has been the most widely used (9, 22, 32, 73, 122), but other genes such as DNA polymerase (*dpol*), *egt*, *gp41*, *chitinase*, *cathepsin*, *lef2*, and *gp37* have also been used (22, 25, 28, 29, 62, 63, 71, 73). In general, these studies agree that the lepidopteran NPVs and GVs constitute distinct well defined groups and that the lepidopteran NPVs can be subdivided into two subgroups: the group I NPVs including AcMNPV and OpMNPV, and the group II NPVs typically including HaSNPV and SeMNPV (22, 32, 122). However, these analyses have produced some conflicting results. In particular, *polh* phylogenies often disagree with other gene phylogenies, especially in the relationship between AcMNPV and BmNPV. Phylogenies based on *polh* always have AcMNPV as a sister group to the rest of the group I NPVs, whereas other phylogenetic analyses consistently cluster AcMNPV and BmNPV together (29). These disagreements between different single-gene trees may reflect inaccurate phylogenetic inferences due to unequal rates of evolution or to lack of a robust phylogenetic signal. Alternatively, they could indicate genuine differences in the phylogeny of individual genes as a result of recombination events, duplications, and gene losses (90). Exchanges of genetic material occur among baculoviruses. Heterologous recombination can happen between genomes of different baculoviruses coinfecting the same cell, most likely between viruses belonging to different strains of a viral population infecting a common host (33), or between closely related viruses, such as AcMNPV and BmNPV, that normally infect different host species but can be forced to coinfect the same cell (65). Illegitimate recombination between host genomes and baculoviruses has also been observed in the form of diverse classes of transposable elements. The TED *gypsy* retrotransposon and the IFP2 transposon from *Trichoplusia ni* can spontaneously insert themselves in the genome of AcMNPV when this virus is grown in *T. ni* cells (41, 43). Similarly, lepidopteran Tc1-like transposable elements can jump into the genome of CpGV (59, 60). Furthermore, phylogenetic analysis of the envelope gene of *gypsy* retrotransposons showed that these errantiviruses from the Metaviridae family have probably acquired this gene from baculoviruses, thus permitting viral infection (76, 101). However, the degree to which baculovirus evolution has been influenced by recombination events is unclear. A fundamental question therefore is whether the representation of the evolution of baculovirus species by a phylogenetic tree is appropriate or possible, or whether frequent horizontal transfers obscure such a "backbone" tree.

To circumvent the problem of conflicting single-gene trees, entire genomes have been used to reconstruct baculovirus phylogenies (53). Several different approaches have been explored to take advantage of complete genome sequences. Genomes contain multiple levels of phylogenetic information. In addition to the nucleotide and amino acid sequences, complete genomes also contain structural information such as the order of genes or gene composition (66). The rationale for using such characters for phylogenetic inference is that two genomes sharing similar organization are more likely to have inherited it from a common ancestor than by chance. An advantage of evaluating evolution on a genomic scale is that inferences are based on the complete genetic makeup of species.

The inference of phylogeny based on either nucleotide or amino acid sequences is greatly extended by the ability to use all genes common between the genomes of interest. There are two potential approaches: (*a*) to analyze each gene separately and derive a consensus from the resulting phylogenies or (*b*) to concatenate the genes and analyze them together. In principle, the second approach gives better results because each gene contributes to the overall phylogenetic signal and a synergistic effect is produced by combination of all the signals. This approach has proved useful for the phylogenetic analysis of metazoans based on mitochondrial genomes (86) and for the phylogeny of herpesviruses and baculoviruses (3, 53, 79). It has also been successful when employed on a larger scale for the analysis of prokaryotic genomes (104).

Gene order was first used in phylogenetics when the sequences of some mitochondrial genomes became available (17). Since then several approaches have been developed to reconstruct phylogenies based on gene order, using either distance methods or parsimony (10, 16). Mitochondrial genomes, however, constitute a serious challenge because their small size limits the number of possible rearrangements (70). The availability of sequences from larger, closely related genomes, such as herpesviruses, baculoviruses, and plant chloroplasts, has provided new opportunities for the application of such approaches (19, 49, 53).

Evolution of genome content is one of biology's most fundamental questions. How genes are inherited, acquired, lost, exchanged, duplicated, and evolve are questions that have become predominant in recent years (36, 67). It is not surprising therefore that the completion of recent genome projects has provoked renewed interest in these questions, and phylogenetic analyses based on the presence or absence of genes, gene families, or protein folds have been carried out on a range of microorganisms (3, 40, 53, 72, 83, 105, 106, 108).

Genomic Characters for Baculovirus Phylogenies

We have previously described baculovirus phylogenies based on 9 (53) or 10 (55) complete genome sequences. For this review, we have repeated these analyses using all 13 complete genome sequences available at the time of writing [for methodology, see (53)]. A key step in understanding relationships between genomes is to determine which gene homologs represent orthologs (descended from the same gene copy in the last common ancestor of two species) or paralogs (descended from different genes in the last common ancestor due, for example, to gene duplication). For a gene phylogeny to reflect the species phylogeny, orthologs must be used (111). Evaluating gene orthology is therefore crucially important, as this determines which genes are used for all subsequent studies. In our analyses, we have deliberately omitted the *bro* gene family. This gene family is highly repeated in some baculovirus genomes, making it difficult to establish orthology relationships (i.e., it is difficult to judge which *bro* gene is the ortholog of which).

The different analysis methods use different subsets of baculovirus genes (Figure 1, see color insert). The data set for phylogenies based on gene sequences (nucleotides or amino acids) is the set of genes common to all genomes. Neighbor pair analysis is based on the order of this same set of genes. Breakpoint analysis is also based on gene order but considers all genes shared by at least two genomes. Gene content analysis is based on all genes except those that are present in all genomes. Thus it is based on a non-overlapping gene set to that used for the gene sequence analyses.

GENE SEQUENCES Phylogenetic trees, based on the protein sequences of the 30 common genes, were reconstructed either by analyzing each gene separately and summarizing the single-gene phylogenies with a consensus tree, or by concatenating the alignments and performing one analysis. By first analyzing each gene separately, the heterogeneity of the phylogenetic signal contained within each gene can be evaluated. If all the individual genes give the same phylogenetic tree, then the signal contained in each gene is as good as that of all of them combined. However, if each gene gives a different tree, suggesting weak individual data sets, it is likely that combining data sets would reinforce the signal. Strong disagreements between individual gene phylogenies could indicate that those genes have had different evolutionary histories.

The results of each of the 30 single-gene analyses were extremely heterogeneous. No two single-gene analyses gave the same tree (data not shown). However, the majority rule consensus tree (Figure 2*A*) demonstrates that there is a single tree underlying the different topologies. The figures above each branch show the percentage of gene trees supporting that branch. The majority of genes support the separation of CuniNPV from the lepidopteran viruses, the division between GVs and NPVs, and the separation and relationships among the group I NPVs. However, the low percentage values of branches leading to the group II NPVs, of branches within the group II NPVs and of branches within the GVs, show that there is little agreement between genes on these relationships. Similar heterogeneity has been observed in previous phylogenetic analyses of some of these 30 genes (22, 25, 52, 73, 85).

As expected in the case of weak single-gene data sets, the concatenation of all 30 gene alignments gives a much stronger phylogenetic signal. The combined gene tree (Figure 2*B–C*) has the same topology as the majority rule consensus tree (Figure 2*A*), but it has high bootstrap values and shows that four main groups can be distinguished within the baculoviruses. The group I NPVs, which includes AcMNPV, BmNPV, OpMNPV, and EppoNPV, is the best-supported and most highly resolved group. There is strong support for the existence of the group II NPVs, comprising HaSNPV, HzSNPV, SeMNPV, SpltNPV, and LdMNPV. However, relationships within the group are not well supported. Both groups of lepidopteran NPVs are more closely related to each other than they are to either the GVs or to CuniNPV. The lepidopteran NPVs and GVs also group strongly together to the exclusion of the dipteran virus. This corroborates previous results where CuniNPV was identified as a divergent baculovirus based on morphological data and a DNA polymerase phylogeny (85).

GENE ORDER Gene parity plots, which compare the positions of orthologous genes on different genomes, have been widely used to assess synteny conservation between baculovirus genomes (26, 50, 54, 56, 75). They have shown that the gene order of GVs is fairly well conserved, but quite different to that of the NPVs (50, 75), that the group I NPVs also share similar gene arrangements, and that gene order is far less conserved among the group II NPVs (26, 54, 56). Gene parity plots demonstrate the utility of synteny comparison for assessing the relationships of baculoviruses. However, they only provide a qualitative estimate and their use is limited for phylogenetic studies due to lack of quantitative data. Other methods are necessary to interpret gene order within an evolutionary context.

Two approaches have been used to infer phylogenies based on gene order. The first, neighbor pair analysis, is based on the arrangements of the 30 genes common to all baculovirus genomes (53, 107), whereas the second, relative breakpoint

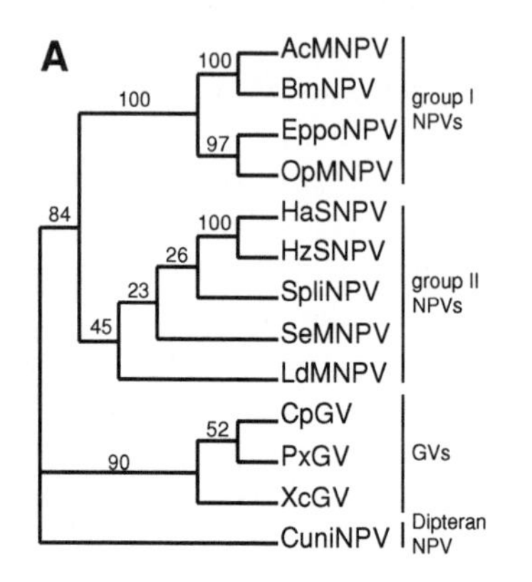
A
AcMNPV
BmNPV
EppoNPV
OpMNPV
group I NPVs
HaSNPV
HzSNPV
SpliNPV
SeMNPV
LdMNPV
group II NPVs
CpGV
PxGV
XcGV
GVs
CuniNPV
Dipteran NPV
100
100
97
84
100
26
23
45
52
90

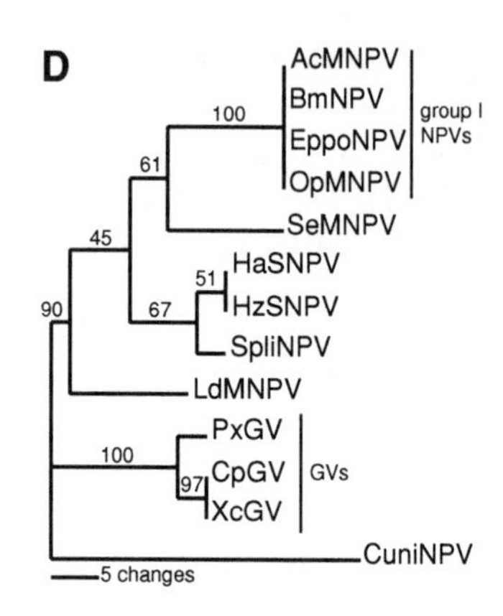
D
AcMNPV
BmNPV
EppoNPV
OpMNPV
group I NPVs
SeMNPV
HaSNPV
HzSNPV
SpliNPV
LdMNPV
PxGV
CpGV
XcGV
GVs
CuniNPV
5 changes
100
61
45
51
67
90
100
97

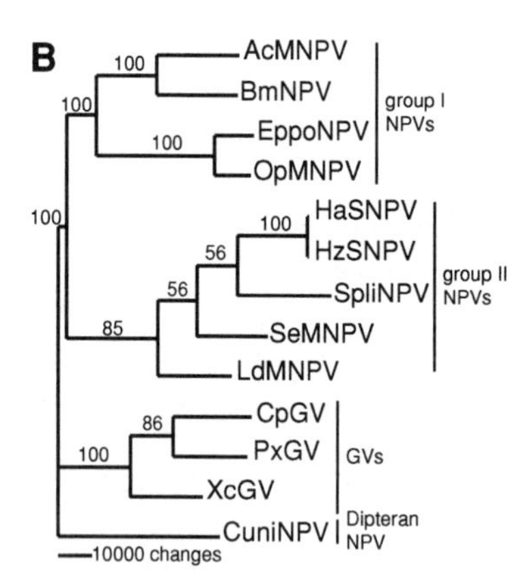
B
AcMNPV
BmNPV
EppoNPV
OpMNPV
group I NPVs
HaSNPV
HzSNPV
SpliNPV
SeMNPV
LdMNPV
group II NPVs
CpGV
PxGV
XcGV
GVs
CuniNPV
Dipteran NPV
10000 changes
100
100
100
100
100
56
56
85
86
100

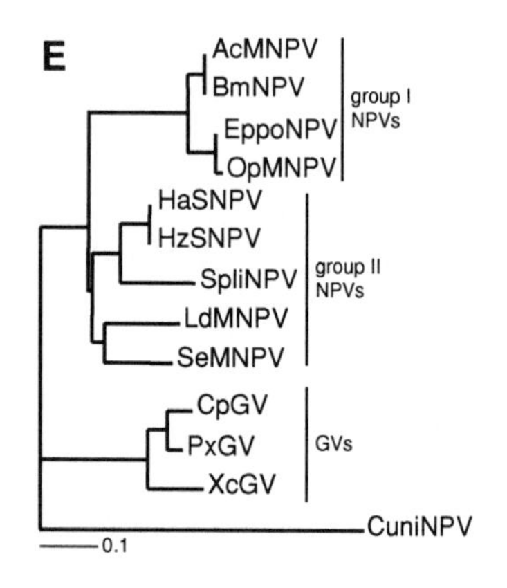
E
AcMNPV
BmNPV
EppoNPV
OpMNPV
group I NPVs
HaSNPV
HzSNPV
SpliNPV
LdMNPV
SeMNPV
group II NPVs
CpGV
PxGV
XcGV
GVs
CuniNPV
0.1

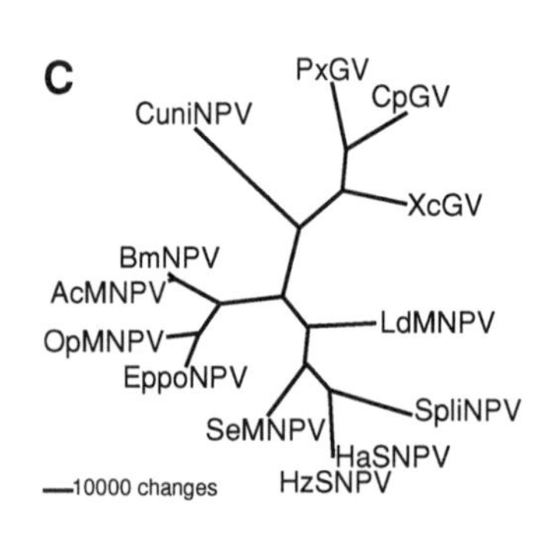
C
CuniNPV
PxGV
CpGV
XcGV
BmNPV
AcMNPV
OpMNPV
EppoNPV
LdMNPV
SpliNPV
SeMNPV
HaSNPV
HzSNPV
10000 changes

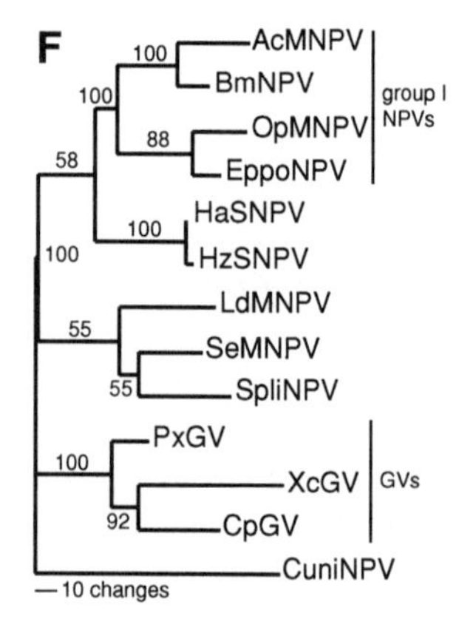
F
AcMNPV
BmNPV
OpMNPV
EppoNPV
group I NPVs
HaSNPV
HzSNPV
LdMNPV
SeMNPV
SpliNPV
PxGV
XcGV
CpGV
GVs
CuniNPV
10 changes
100
100
88
58
100
100
55
55
100
92

analysis, is based on a comparison of gene order between pairs of genomes and is therefore based on all genes present in more than one virus (10, 53) (Figure 1). In neighbor pair analysis, the order of the 30 common genes is recorded by scoring the presence or absence of contiguous pairs of genes. The binary matrix created is used to reconstruct phylogenies using parsimony. The advantage of this method is that the matrix can be used in statistical analyses to assess the robustness of the data and the phylogenetic signal. The disadvantage is that the matrix comprises a small number of characters because it is based on the order of only 30 genes. In relative breakpoint analysis, the number of breakpoints between two genomes (abstract points on one genome that need to be drawn to represent the breaks necessary to recreate the gene order of another genome) are counted and subsequently divided by the number of genes common to the pair of genomes. These distances, calculated for all pairs of genomes, are used to reconstruct a tree by neighbor joining. The advantage of this method is that it is based on a large data set—all genes present in at least two viruses. The main drawback of using distances in phylogenetic analyses is that statistical methods to test the robustness of the resulting tree are not available.

The neighbor pair analysis tree (Figure 2*D*) shows clear divisions between CuniNPV and the lepidopteran baculoviruses and also between the NPVs and the GVs. However, it is poorly resolved for and fails to establish relationships within the group I NPVs. This is mostly due to the inclusion of the dipteran virus, which reduced the common gene pool from 62 genes between all lepidopteran baculoviruses (55) to 30 genes. In fact, there are no differences in the order of these 30 common genes among the group I NPVs, between HaSNPV and HzSNPV, or between CpGV and XcGV, as shown by the null branch lengths on the tree.

←

Figure 2 Baculovirus phylogenies based on complete genome data. (*A*) Majority rule consensus tree of the most parsimonious trees obtained from the separate phylogenetic analysis of each of the 30 common genes; numbers above each branch indicate the percentage of individual gene trees possessing that branch. (*B*) Most parsimonious tree based on an analysis of the combined sequences of the 30 common genes. (*C*) Unrooted version of *B*. (*D*) One of two most parsiminious trees generated by neighbor pair phylogenetic analysis. (*E*) Neighbor-joining tree based on relative breakpoint distances. (*F*) One of three most parsimonious trees generated by the gene content phylogenetic analysis. With the exception of *C*, trees are rooted using CuniNPV. With the exception of *A*, the numbers indicate the percentage bootstrap support for each node from 1000 replicates. AcMNPV (*Autographa californica* MNPV), BmNPV (*Bombyx mori* NPV), CpGV (*Cydia pomonella* GV), CuniNPV (*Culex nigripalpus* NPV), EppoMNPV (*Epiphyas postvittana* MNPV), HaSNPV (*Helicoverpa armigera* SNPV), HzSNPV (*Helicoverpa zea* SNPV), LdMNPV (*Lymantria dispar* MNPV), OpMNPV (*Orgyia pseudotsugata* MNPV), PxGV (*Plutella xylostella* GV), SeMNPV (*Spodoptera exigua* MNPV), SpltMNPV (*S. litura* MNPV), XcGV (*Xestia c-nigrum* GV).

Furthermore, the neighbor pair analysis fails to recover the group II NPVs, and the low bootstrap values indicate that branching in this area of the tree is weakly supported. In contrast, the relative breakpoint distance tree (Figure 2*E*) is well resolved. It recovers the four main groups of baculoviruses, and its topology only differs from the 30 gene tree (Figure 2*B*) in the grouping of SeMNPV and LdMNPV within the group II NPVs.

GENE CONTENT When all the gene homology and orthology relationships have been ascertained, each genome can be scored for the presence or absence of each baculovirus gene (53). This generates a binary gene content matrix that can be utilized to reconstruct a phylogenetic tree using parsimony. Genes present in only one species provide no information to infer phylogenetic relationships between the viruses, but are used to calculate the branch lengths of the tree (Figure 1). The gene content tree (Figure 2*F*) again shows a clear division between CuniNPV, the GVs, and the lepidopteran NPVs. High bootstrap values show that relationships within the group I NPVs and the GVs are strongly supported. However, compared with the 30 gene trees and the relative breakpoint tree (Figure 2*A–C,E*), this analysis does not support the group II NPVs. This is due to the fact that only a few genes are specific to the group II NPVs. As observed for bacterial genome analysis (105), a limitation of gene content analysis is that, regardless of their relatedness, larger genomes are more likely to share more genes and thus to appear more closely related, i.e., they are prone to long branch attraction artifacts. An example is the apparent relationships within the GVs. The gene content analysis places PxGV, which has a much smaller genome than the other two GVs, basal to the group. However the 30 gene tree and other observations show that in fact it is more closely related to CpGV and that XcGV should be at the base of the group. Nonetheless, the gene content analysis managed to recover the main relationships between the baculoviruses.

A Phylogeny of the Baculoviridae

The different methods to reconstruct baculovirus phylogenies based on complete genomes each have positive points in their own right. In gene order analyses, as well as resolving the basic relationships of the baculoviruses, the branch lengths of the trees provide an approach to visualizing divergent genome arrangements. As noted, they are limited in their resolving power when genome orders are too similar or too divergent. However, they could prove extremely useful in intermediate cases. The gene content analyses are conceptually simple but also have limited power. However, they are extremely useful to address questions of genome evolution. The combined analysis of all genes that are common between all the genomes is the most robust approach to infer baculovirus phylogenies (53). The proposed phylogeny of these 13 baculoviruses has 4 distinct groups: CuniNPV, the GVs, group I NPVs, and group II NPVs (Figure 2*B*). It is worth noting that the underlying branching pattern of the phylogeny presented here based on all 13 sequenced baculoviruses is the same as the phylogenies we presented previously based on 9 or 10 genomes (53, 55), apart from the addition of the extra genomes. Thus, we are confident that this represents a robust and plausible phylogeny of these viruses. Based on this phylogeny, the mosquito virus clearly belongs to a different group to the lepidopteran NPVs. Thus, we believe that the taxonomy of the Baculoviridae should be reexamined. It is no longer appropriate to include all NPVs within a single genus *Nucleopolyhedrovirus*.

COMPARATIVE GENOMICS

Following the many genome projects that have been completed recently, extensive efforts have been made to develop methods to analyze those sequences in a comparative and evolutionary context, to improve our understanding of genome evolution (36, 66). For instance, the study of gene content has shown the extent of variation between bacterial genomes (4). Protein phylogenetic profiles have been used to infer the functions and interactions of proteins (18, 34, 35, 37). Comparative genomics has the potential to provide valuable insights into baculovirus evolution and biology. The comparison of closely related viruses can indicate which genes are involved in fine-tuning their biology. Conversely, the study of more distantly related genomes can highlight the molecular innovations characterizing each group. In this section, using the phylogeny determined above, we review some comparative analyses of these genomes and place them in an evolutionary framework.

The Evolution of Genome Content

The evolution of genome content can be visualized by mapping gene content onto the phylogenetic tree (53). Character changes, i.e., gene acquisition or loss, are retraced on the tree using parsimony algorithms. The tree in Figure 3 (see color insert) therefore summarizes the most parsimonious hypothesis of changes in gene composition that have occurred during the evolution of these 13 baculovirus genomes (for a complete listing of the hypothesized gene acquisitions and losses, see the Supplemental Material link in the online version of this chapter or at http://www.annualreviews.org/). The ancestor of the lepidopteran viruses (Figure 3) acquired 45 genes, 32 of which are still present in all lepidopteran virus genomes sequenced so far (Table 1), 12 of which have been subsequently lost in at least one lineage, and 1 of which (ac111; Figure 3) seems to have been lost in some lineages and subsequently reacquired (Figure 3). One can speculate that the acquisition of these 45 genes might have been crucial to the colonization of lepidopteran hosts. However, it is important to remember that because only one nonlepidopteran virus has been sequenced to date, the evidence that these genes are specific to lepidopteran viruses is weak. Interestingly, several of the 12 genes that were lost in some lineages had already been classified as auxiliary genes: *egt*, *chitinase*, *cathepsin*, *sod*, and *p10* (89). This supports the idea that these are beneficial but not essential genes.

The gene content tree indicates that 27 genes specific to GVs and 14 to lepidopteran NPVs (Figure 3) could underlie the biological differences of these two baculovirus groups (53). Unfortunately, potential functions have only been ascribed to six of the NPV-specific genes and two of the GV-specific genes. Some of the structural differences between NPVs and GVs could be associated with three NPV-specific genes (*vp80*, *pp34/calyx*, and *orf1629*) that code for structural proteins (45, 52). The presence in the NPVs of *arif 1*, which is implicated in rearrangement of the cytoskeleton (102), may be linked to differences in subcellular architecture during NPV and GV infections (119). The *iap* genes are implicated in the inhibition of apoptosis (31) in baculovirus-infected cells. It is intriguing that individual members of this gene family appear to be unique to both GVs (*iap5*) and NPVs (*iap2*). The only other GV-unique gene with a potential function encodes a metalloproteinase thought to contribute to the proteolysis of infected tissue (64). Further correlation between differences in gene content and biological differences of NPVs and GVs awaits the characterization of genes of unknown function.

Seventeen genes are only found in group I NPVs, 14 in all of them, and 3 (*ets*, *etm*, and *vp80a*) in some of them (Figure 3). The best studied of these is probably *gp64*, which has the same function as the *ld130* homologs from group II NPVs, as discussed earlier (57, 93). This represents the best example of non-orthologous gene displacement in baculoviruses. Acquisition of this gene is thought to have promoted the diversification of group I NPVs because of its unique structural role (93). However, another 16 genes are specific to this group of viruses, 3 of which also encode structural proteins (*odve26*, *ptp1*, *vp80a*) (53).

Mapping of baculovirus genes onto the phylogeny (Figure 3) also revealed that a number of genes appear to have been gained independently by several lineages. Based on these 13 genomes, 33 genes have been acquired separately at least twice and 9 of them (*ac63*, *ctl1*, *ctl2*, *he65*, *hlx2*, *p35*, *ptp2*, *rr1*, *rr2a*) three times. Note that "independent acquisition" here could reflect either two lineages separately acquiring a gene from the genome of another organism, or one lineage acquiring the gene that is then transferred horizontally to other baculovirus lineages. In either case, the picture that emerges is one of baculoviruses continuously "sampling" their genomic environment for beneficial genes during the course of their evolution (53).

Discovering Functional Linkage

Several computational approaches have been developed to predict the function and physical interactions of proteins from genomic sequence data (37, 46). Protein phylogenetic profiles have been used in prokaryotes to infer protein function (97). Genes that share the same phylogenetic profile are defined as those genes that are always present in the same genomes. When they cannot be correlated with single-gene acquisitions in a common ancestor, protein phylogenetic profiles can provide clues toward the understanding of protein function. The Rosetta Stone method, which deduces protein interactions from instances of genes fusion and fission (77), can also be used to infer or reinforce functional hypotheses. Finally, the examination of gene neighborhood can be useful to detect operon-style gene clusters (35).

When these approaches are applied to baculovirus genomes a number of interesting features come to light. The first is the presence of two putative DNA repair systems in a number of distantly related genomes. The second is the presence of host-related factors in nonrelated viruses infecting larvae from the family Lymantriidae. The final observation is of a cluster of four genes that appear to be a physically linked core baculovirus feature.

TWO PUTATIVE DNA REPAIR SYSTEMS IN BACULOVIRUSES Gene mapping reveals only one example of a conserved protein phylogenetic profile that cannot be explained by the baculovirus species phylogeny. The genes ribonucleotide reductase 1 (*rr1*) and *ld138* are always present together in OpMNPV, SpltNPV, SeMNPV, LdMNPV, and CpGV (Figure 4*A*, see color insert). In addition, *rr1* is itself always linked to a ribonucleotide reductase 2 gene (*rr2*). Ribonucleotide reductase is normally composed of two subunits encoded by two different genes (99). In baculoviruses, there are two kinds of *rr2* genes, *rr2a* and *rr2b*, that are both contained within the genome of LdMNPV but otherwise have different distributions on the tree. One or both of these *rr2* genes is always found with *rr1*. A further association can be seen within the lepidopteran NPVs, where the *dUTPase* gene has a phylogenetic distribution clearly linked with the ribonucleotide reductase genes and *ld138*. It is present in OpMNPV, SpltNPV, SeMNPV, and LdMNPV. The idea that these four genes share a functional linkage is further reinforced by the fact that the *ld138* and *dUTPase* genes are fused in OpMNPV, forming a Rosetta Stone gene (75). Ribonucleotide reductases catalyze the production of deoxynucleotides (dNTPs) required for DNA replication and repair (99). The function of the dUTPase is to prevent the mutagenic incorporation of uracyl into DNA (114). These homologies suggest that these genes constitute a DNA repair system in baculoviruses. Protein phylogenetic profiling and Rosetta Stone analysis show that *ld138* is clearly linked to this system, strongly suggesting that the protein encoded by this gene is also involved in DNA repair.

Another pair of genes involved in DNA metabolism has related phylogenetic profiles unrelated to baculovirus phylogeny. The DNA ligase gene is only present in genomes containing a helicase 2 gene: CpGV, PxGV, XcGV, and LdMNPV (Figure 4*A*, see color insert). So far, only the SpltNPV genome has a helicase 2 gene on its own. However this gene only has low homology to the other helicase 2 genes, and it is not clear that it would have the same activity. There are currently two types of helicase genes identified in baculoviruses. The first is part of the core set of baculovirus genes (Table 1) and has low homology to helicase genes of other organisms. The second, helicase 2, is part of the helicase superfamily I, including PIF1 from eukaryotes and RecD from *Escherichia coli*. These enzymes are typically involved in DNA metabolism, such as replication, recombination, or repair. The baculovirus DNA ligase is homologous to the DNA ligase III gene family and is therefore most probably involved in DNA metabolism. Pearson &

Rohrmann (94) have characterized the DNA ligase from LdMNPV and have shown that it is an active ligase. They also showed that it does not enhance DNA replication and that helicase 2 is not a functional analog of the core helicase gene. Therefore, it is most likely that both genes are involved together in a DNA recombination or repair system.

LYMANTRIID HOST-RELATED FACTORS The *hrf1* gene from LdMNPV expands the host range of AcMNPV both in vitro and in vivo, allowing it to infect nonpermissive hosts (23, 109). The only baculovirus homolog of this gene is found in OpMNPV. Both of these viruses infect larvae of the Lymantriidae family, namely *Lymantria dispar* and *Orgyia pseudotsugata*. Two conotoxin-like genes, *ctl1* and *ctl2*, are also found in both of these genomes (Figure 4*B*). Other baculoviruses encode one or the other, but only OpMNPV and LdMNPV encode both. The mode of action of these genes remains unclear, but there is a clear phylogenetic link between *hrf1* and the presence of both conotoxin genes. The fact that this linkage is associated with viruses infecting a particular family of Lepidoptera suggests the proteins may function as part of a viral response to host resistance.

BACULOVIRIDAE CORE GENE CLUSTER Examination of synteny conservation across the sequenced baculoviruses shows that gene order is poorly conserved. Even the relative arrangement of the 30 conserved genes is not maintained, showing that extensive genome rearrangements have occurred during the course of baculovirus evolution. It is thus particularly notable that a set of four genes, present in all baculoviruses, has remained in the same relative position in all the sequenced genomes, including the mosquito virus CuniNPV (Figure 5). Apart from this cluster of four genes, there is not another single pair of genes that is kept together in all the genomes. It is therefore highly unlikely that this feature is random. This cluster of four genes includes *helicase* and *lef5*, which are essential genes for baculovirus replication and transcription. It also comprises two genes of unknown function, *ac96* and *38K* (*ac98*) (Figure 5). The fact that these genes are found in the same relative position in all genomes indicates that there is some physical constraint preventing them, or at least these DNA sequences, from being separated. One possibility is that the genes are part of a single transcriptional unit, in an operon-type arrangement or as part of a spliced transcript(s). This in turn might suggest that there is some functional linkage between all four genes. However the fact that opposite strands of viral DNA code for them makes a potential transcriptional/splicing

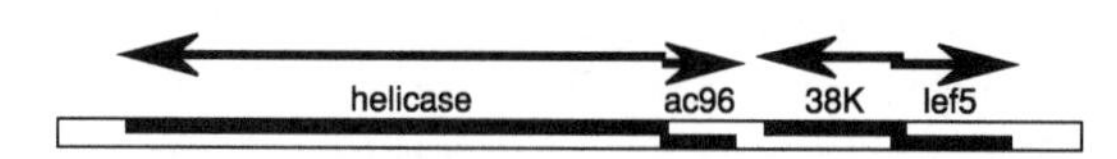

Figure 5 Schematic map of the Baculoviridae core gene cluster. The relative positions of the *helicase*, *ac96*, *38K*, and *lef5* genes are conserved in all 13 sequenced baculovirus genomes.

A. Homology distribution of genes from 13 baculovirus genomes

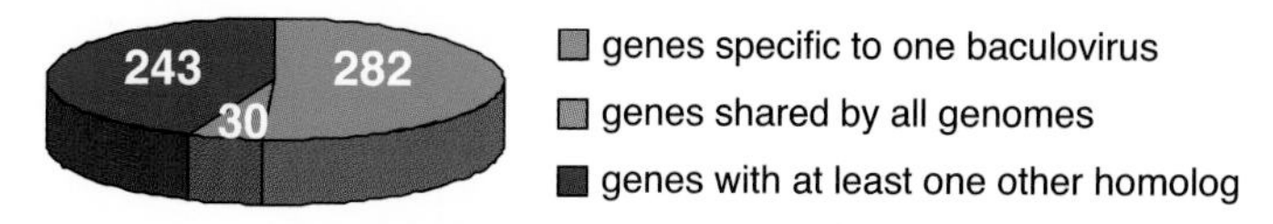

B. Homologs used for phylogenetic analyses

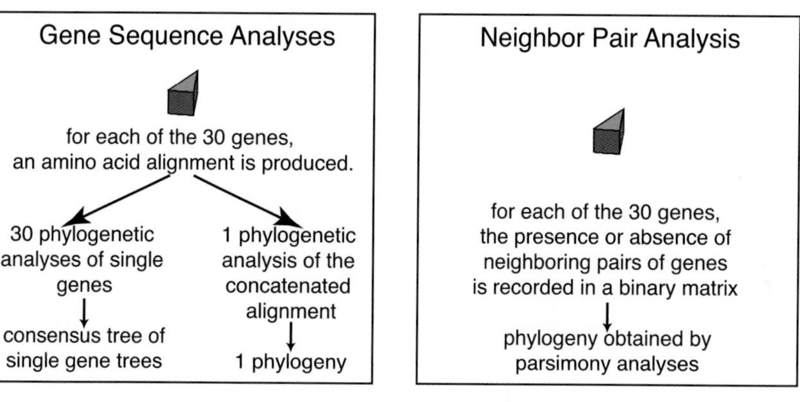

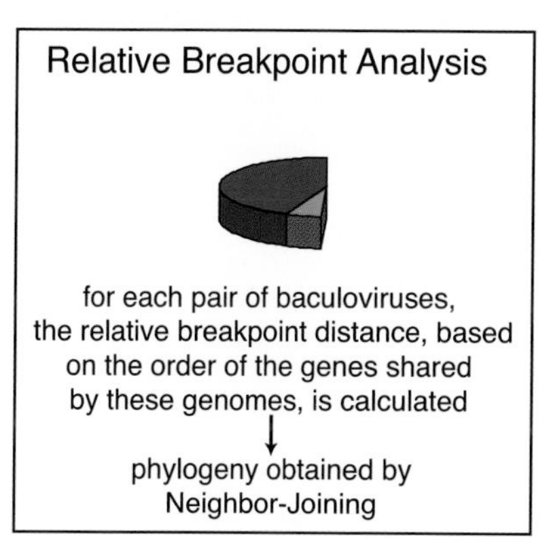

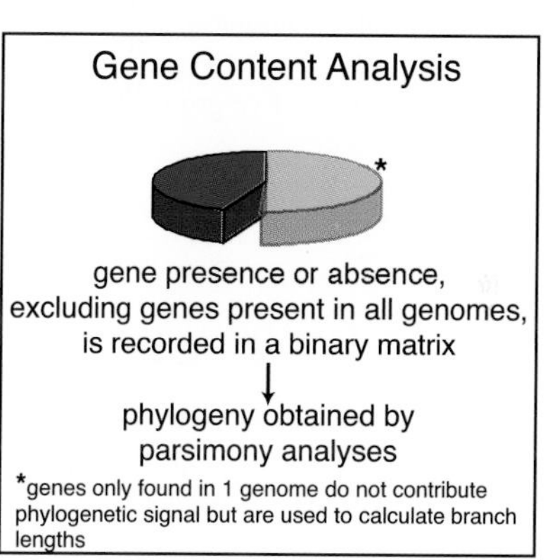

Figure 1 Gene data sets used in phylogenetic analyses. (*A*) Classes of baculovirus genes. (*B*) Homolog groups used for different phylogenetic methods.

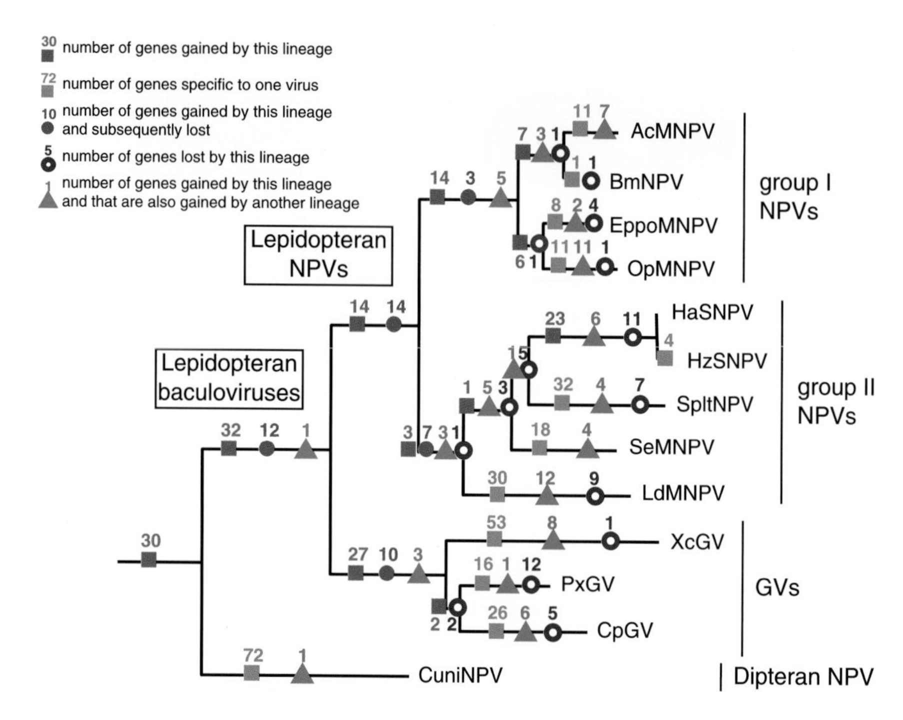

Figure 3 Gene content mapping. Most parsimonious hypothesis of changes in gene content during baculovirus evolution. The gene content data set is mapped onto the tree derived from the analysis of combined sequences of the 30 shared genes. Genes have been subdivided into five groups based on their patterns of acquisition and loss.

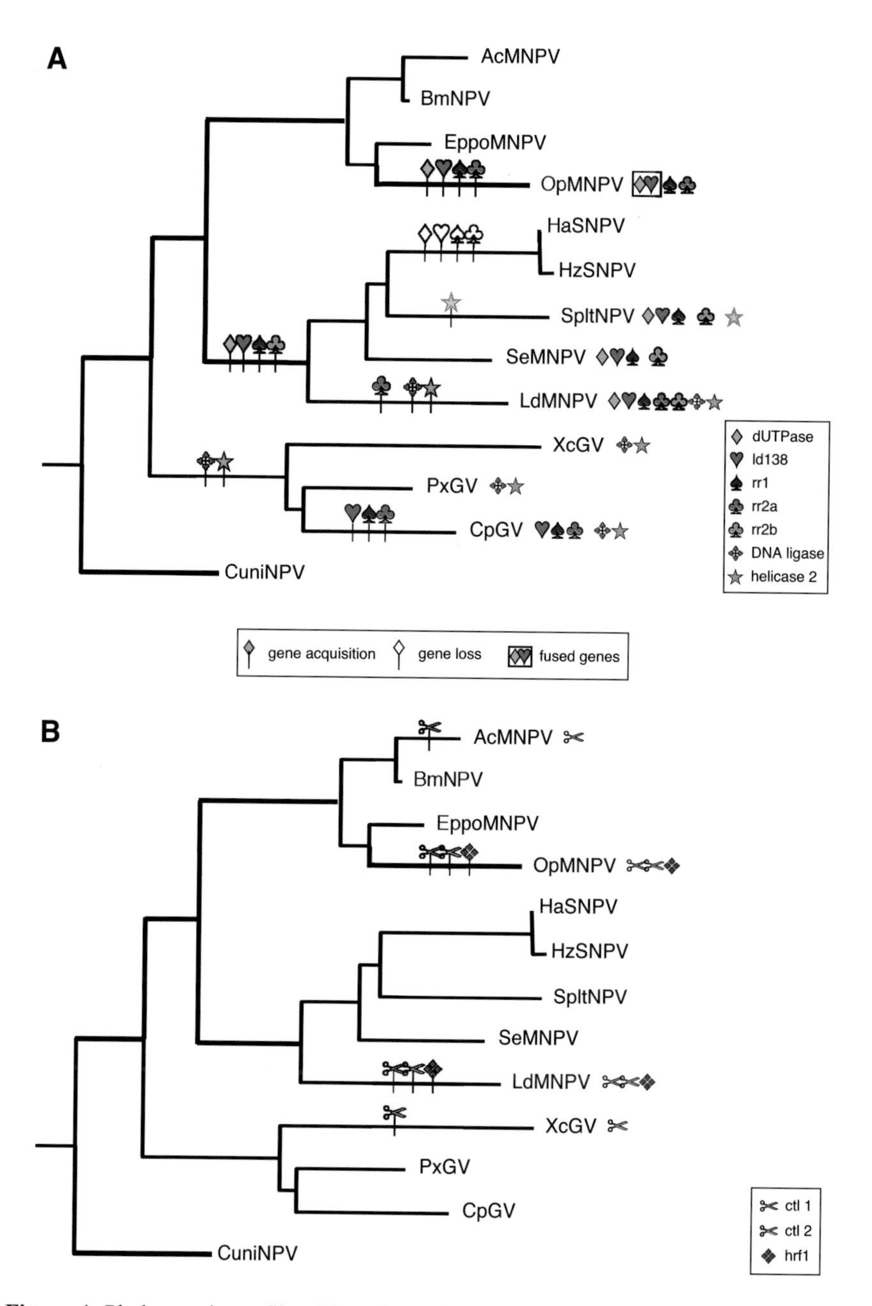

Figure 4 Phylogenetic profiles. The gains and losses of (*A*) *dUTPase*, *ld138*, *rr1*, *rr2a*, *rr2b*, *DNA ligase*, *helicase 2*, and (*B*) *hrf1*, *ctl1*, *ctl2* are mapped onto the tree derived from the analysis of combined sequences of the 30 shared genes. The presence of these genes in individual baculovirus genomes is summarized on the right of the figure.

link harder to envision. An alternative explanation is that the physical structure of this region of the genome is essential for virus replication and must be preserved. More detailed transcriptional mapping and functional analysis will be required to understand the significance of this striking conservation of gene order.

SUMMARY

The comparison of complete baculovirus genome sequences has provided extensive evidence for the fluid nature of these genomes. The evolutionary history of baculoviruses has been characterized by widespread genome rearrangements and frequent gains and losses of genes. However, despite this fluidity, there is ample information inherent in the complete genome sequence data to allow the construction of a robust, well-supported, and plausible tree describing baculovirus evolution. This tree indicates that the current subdivision of the family Baculoviridae into two genera, *Nucleopolyhedrovirus* and *Granulovirus*, should be revised. NPVs from lepidopteran and dipteran hosts are clearly more different from each other than lepidopterans NPVs are from GVs. They should not be grouped together in the same genus. The baculovirus phylogeny can now serve as the framework to place data from the comparison of baculovirus genomes into an evolutionary context. We have shown how comparative genomics can provide a fascinating view of the history of gene acquisitions and losses that have accompanied baculovirus evolution. The significance of most of these changes is still a matter of speculation owing to the lack of functional information for many of the genes. However, as our knowledge of baculovirus gene function grows, comparative genomics approaches will yield increasingly profound insights into baculovirus biology, baculovirus-host interactions, and the factors that have shaped and driven baculovirus evolution.

ACKNOWLEDGMENTS

We thank Renée Lapointe, Teresa Luque, Isabelle Nobiron, Yeon-Ho Je, and Carlos Lopez-Vaamonde for discussions about the manuscript. Natural Environment Research Council CASE studentship award GT04/99/TS/142 supported E.A. Herniou.

The *Annual Review of Entomology* is online at http://ento.annualreviews.org

LITERATURE CITED

1. Afonso CL, Tulman ER, Lu Z, Balinsky CA, Moser BA, et al. 2001. Genome sequence of a baculovirus pathogenic for *Culex nigripalpus*. *J. Virol.* 75:11157–65
2. Ahrens CH, Russell RLQ, Funk CJ, Evans JT, Harwood SH, Rohrmann GF. 1997. The sequence of the *Orgyia pseudotsugata* multicapsid nuclear polyhedrosis virus genome. *Virology* 229:381–99
3. Albà MM, Das R, Orengo CA, Kellam P. 2001. Genomewide function

conservation and phylogeny in *Herpesviridae*. *Genome Res.* 11:43–54
4. Alm RA, Bina J, Andrews BM, Doig P, Hancock REW, Trust TJ. 2000. Comparative genomics of *Helicobacter pylori*: analysis of the outer membrane protein families. *Infect. Immun.* 68:4155–68
5. Ayres MD, Howard SC, Kuzio J, Lopez-Ferber M, Possee RD. 1994. The complete DNA sequence of *Autographa californica* nuclear polyhedrosis virus. *Virology* 202:586–605
6. Barrett JW, Lauzon HAM, Mercuri PS, Krell P, Sohi SS, Arif B. 1996. The putative LEF-1 proteins from two distinct *Choristoneura fumiferana* multiple nucleopolyhedroviruses share domain homology to eukaryotic primases. *Virus Genes* 13:229–37
7. Belyavskyi M, Braunagel S, Summers M. 1998. The structural protein ODV-EC27 of *Autographa californica* nucleopolyhedrovirus is a multifunctional viral cyclin. *Proc. Natl. Acad. Sci. USA* 95:11205–10
8. Benz GA. 1986. Introduction: historical perspective. In *The Biology of Baculoviruses*, ed. RR Granados, BA Federici, pp. 1–35. Boca Raton, Florida: CRC
9. Bideshi DK, Bigot Y, Federici BA. 2000. Molecular characterization and phylogenetic analysis of the *Harrisina brillians* granulovirus granulin gene. *Arch. Virol.* 145:1933–45
10. Blanchette M, Kunisawa T, Sankoff D. 1999. Gene order breakpoint evidence in animal mitochondrial phylogeny. *J. Mol. Evol.* 49:193–203
11. Blissard GW, Black B, Crook N, Keddie BA, Possee R, et al. 2000. Family *Baculoviridae*. In *Virus Taxonomy: Seventh Report of the International Committee on Taxonomy of Viruses*, ed. MHV Van Regenmortel, CM Fauquet, DHL Bishop, EB Carstens, MK Estes, et al., pp. 195–202. San Diego: Academic
12. Blissard GW, Quant RRL, Rohrmann GF, Beaudreau GS. 1989. Nucleotide sequence, transcriptional mapping, and temporal expression of the gene encoding p39 a major structural protein of the multicapsid nuclear polyhedrosis virus of *Orgyia pseudotsugata*. *Virology* 168:354–62
13. Blissard GW, Rohrmann GF. 1989. Location sequence transcriptional mapping and temporal expression of the gp64 envelope glycoprotein gene of the *Orgyia pseudotsugata* multicapsid nuclear polyhedrosis virus. *Virology* 170:537–55
14. Blissard GW, Rohrmann GF. 1990. Baculovirus diversity and molecular biology. *Annu. Rev. Entomol.* 35:127–55
15. Blissard GW, Wenz JR. 1992. Baculovirus gp64 envelope glycoprotein is sufficient to mediate pH-dependent membrane fusion. *J. Virol.* 66:6829–35
16. Boore JL, Brown WM. 1998. Big trees from litle genomes: mitochondrial gene order as a phylogenetic tool. *Curr. Opin. Genet. Dev.* 8:668–74
17. Boore JL, Collins TM, Stanton D, Daehler LL, Brown WM. 1995. Deducing the pattern of arthropod phylogeny from mitochondrial DNA rearrangements. *Nature* 376:163–65
18. Bork P, Dandekar T, Diaz-Lazcoz Y, Eisenhaber F, Huynen M, Yuan YP. 1998. Predicting function: from genes to genomes and back. *J. Mol. Biol.* 283: 707–25
19. Bourque G, Pevzner PA. 2002. Genome-scale evolution: reconstructing gene orders in ancestral species. *Genome Res.* 12:26–36
20. Braunagel SC, Elton DM, Ma H, Summers MD. 1996. Identification and analysis of an *Autographa californica* nuclear polyhedrosis virus structural protein of the occluded-derived virus envelope: ODV-E56. *Virology* 217:97–110
21. Braunagel SC, He H, Ramamurthy P, Summers MD. 1996. Transcription, translation and cellular localization of three *Autographa californica* nuclear polyhedrosis virus structural proteins:

ODV-E18, ODV-E35 and ODV-EC27. *Virology* 222:100–14
22. Bulach DM, Kumar CA, Zaia A, Liang B, Tribe DE. 1999. Group II nucleopolyhedrovirus subgroups revealed by phylogenetic analysis of polyhedrin and DNA polymerases gene sequences. *J. Invertebr. Pathol.* 73:59–73
23. Chen C-J, Quentin ME, Brennan LA, Kukel C, Thiem SM. 1998. *Lymantria dispar* nucleopolyhedrovirus *hrf*-1 expands the larval host range of *Autographa californica* nucleopolyhedrovirus. *J. Virol.* 72:2526–31
24. Chen F, Suttle CA. 1996. Evolutionary relationships among large double-stranded DNA viruses that infect microalgae and other organisms as inferred from DNA polymerase genes. *Virology* 219:170–78
25. Chen X, Ijkel WFJ, Dominy C, Zanotto P, Hashimoto Y, et al. 1999. Identification, sequence analysis and phylogeny of the *lef-2* gene of *Helicoverpa armigera* single-nucleocapsid baculovirus. *Virus Res.* 65:21–32
26. Chen X, Ijkel WFJ, Tarchini R, Sun X, Sandbrink H, et al. 2001. The sequence of the *Helicoverpa armigera* single nucleocapsid nucleopolyhedrovirus genome. *J. Gen. Virol.* 82:241–57
27. Chen X, Zhang W-J, Wong J, Chun G, Lu A, et al. 2002. Comparative analysis of the complete genome sequences of *Helicoverpa zea* and *Helicoverpa armigera* single-nucleocapsid nucleopolyhedroviruses. *J. Gen. Virol.* 83:673–84
28. Chen XW, Hu ZH, Jehle JA, Zhang YQ, Vlak JM. 1997. Analysis of the ecdysteroid UDP-glucosyltransferase gene of *Heliothis armigera* single-nucleocapsid baculovirus. *Virus Genes* 15:219–25
29. Clarke EE, Tristem M, Cory JS, O'Reilly DR. 1996. Characterization of the ecdysteroid UDP-glucosyltransferase gene from *Mamestra brassicae* nucleopolyhedrosis virus. *J. Gen. Virol.* 77:2865–71
30. Clem RJ. 1997. Regulation of programmed cell death by baculoviruses. See Ref. 82, pp. 237–61
31. Clem RJ, Miller LK. 1994. Control of programmed cell death by the baculovirus genes *p35* and *iap*. *Mol. Cell. Biol.* 14:5212–22
32. Cowan P, Bulach D, Goodge K, Robertson A, Tribe DE. 1994. Nucleotide sequence of the polyhedrin gene region of *Helicoverpa zea* single nucleocapsid nuclear polyhedrosis virus: placement of the virus in lepidopteran nuclear polyhedrosis virus group II. *J. Gen. Virol.* 75:3211–18
33. Croizier G, Ribeiro HCT. 1992. Recombination as a possible major cause of genetic heterogeneity in *Anticarsia gemmatalis* nuclear polyhedrosis virus populations. *Virus Res.* 26:183–96
34. Dandekar T, Schuster S, Snel B, Huynen M, Bork P. 1999. Pathway alignment: application to the comparative analysis of glycolytic enzymes. *Biochem. J.* 343:115–24
35. Dandekar T, Snel B, Huynen M, Bork P. 1998. Conservation of gene order: a fingerprint of proteins that physically interact. *Trends Biochem. Sci.* 23:324–28
36. Eisen JA. 2002. Brouhaha over the other yeast. *Nature* 415:845–48
37. Eisenberg D, Marcotte EM, Xenarios I, Yeates TO. 2000. Protein function in the post-genomic era. *Nature* 405:823–26
38. Evans JT, Leisy DJ, Rohrmann GF. 1997. Characterization of the interaction between the baculovirus replication factors LEF-1 and LEF-2. *J. Virol.* 71:3114–19
39. Faulkner P, Kuzio J, Williams GV, Wilson JA. 1997. Analysis of p74, a PDV envelope protein of *Autographa californica* nucleopolyhedrovirus required for occlusion body infectivity *in vivo*. *J. Gen. Virol.* 78:3091–100
40. Fitz-Gibbon ST, House CH. 1999. Whole genome-based phylogenetic analysis of free living microorganisms. *Nucleic Acids Res.* 27:4218–22
41. Fraser MJ, Cary L, Boonvisudhi K, Wang

H-GH. 1995. Assay for movement of lepidopteran transposon IFP2 in insect cells using a baculovirus genome as a target DNA. *Virology* 211:397–407
42. Friesen PD. 1997. Regulation of baculovirus early gene expression. See Ref. 82, pp. 141–70
43. Friesen PD, Nissen MS. 1990. Gene organization and transcription of TED, a lepidopteran retrotransposon integrated within the baculovirus genome. *Mol. Cell. Biol.* 10:3067–77
44. Fuchs LY, Woods MS, Weaver RF. 1983. Viral transcription during *Autographa californica* nuclear polyhedrosis virus infection: a novel RNA polymerase induced in infected *Spodoptera frugiperda* cells. *J. Virol.* 43:641–46
45. Funk CJ, Braunagel SC, Rohrman GF. 1997. Baculovirus structure. See Ref. 82, pp. 7–27
46. Galperin MY, Koonin EV. 2000. Who's your neighbor? New computational approaches for functional genomics. *Nat. Biotechnol.* 18:609–13
47. Gomi S, Majima K, Maeda S. 1999. Sequence analysis of the genome of *Bombyx mori* nucleopolyhedrovirus. *J. Gen. Virol.* 80:1323–37
48. Guarino L, Jin J, Dong W. 1998. Guanylyltransferase activity of the LEF-4 subunit of baculovirus RNA polymerase. *J. Virol.* 72:10003–10
49. Hannenhalli S, Chappey C, Koonin EV, Pevzner PA. 1995. Genome sequence comparison and scenarios for gene rearrangements—a test-case. *Genomics* 30:299–311
50. Hashimoto Y, Hayakawa T, Ueno Y, Fujita T, Sano Y, Matsumoto T. 2000. Sequence analysis of the *Plutella xylostella* granulovirus genome. *Virology* 275:358–72
51. Hayakawa T, Ko R, Okano K, Seong SI, Goto C, Maeda S. 1999. Sequence analysis of the *Xestia c-nigrum* granulovirus genome. *Virology* 262:277–97
52. Hayakawa T, Rohrmann GF, Hashimoto Y. 2000. Patterns of genome organization and content in lepidopteran baculoviruses. *Virology* 278:1–12
53. Herniou EA, Luque T, Chen X, Vlak JM, Winstanley D, et al. 2001. Use of whole genome sequence data to infer baculovirus phylogeny. *J. Virol.* 75:8117–26
54. Hu ZH, Arif BM, Jin F, Martens JWM, Chen XW, et al. 1998. Distinct gene arrangement in the *Buzura suppressaria* single-nucleocapsid nucleopolyhedrovirus genome. *J. Gen. Virol.* 79: 2841–51
55. Hyink O, Dellow RA, Olsen MJ, Caradoc-Davies KMB, Drake K, et al. 2002. Whole genome analysis of the *Epiphyas postvittana* nucleopolyhedrovirus. *J. Gen. Virol.* 83:959–73
56. Ijkel W, van Strien EA, Heldens JG, Broer R, Zuidema D, et al. 1999. Sequence and organization of the *Spodoptera exigua* multicapsid nucleopolyhedrovirus genome. *J. Gen. Virol.* 80: 3289–304
57. Ijkel WFJ, Westenberg M, Goldbach RW, Blissard GW, Vlak JM, Zuidema D. 2000. A novel baculovirus envelope fusion protein with a proprotein convertase cleavage site. *Virology* 275:30–41
58. Iyer LM, Aravind L, Koonin E. 2001. Common origin of four diverse families of large eukaryotic DNA viruses. *J. Virol.* 75:11720–34
59. Jehle JA, Fritsch E, Nickel A, Huber J, Backhaus H. 1995. TC14.7: a novel lepidopteran transposon found in *Cydia pomonella* granulosis virus. *Virology* 207: 369–79
60. Jehle JA, Nickel A, Vlak JM, Backhaus H. 1998. Horizontal escape of the novel Tc10-like lepidopteran transposon TCp3.2 into *Cydia pomonella* granulovirus. *J. Mol. Evol.* 46:215–24
61. Jin J, Dong W, Guarino LA. 1998. The LEF-4 subunit of baculovirus RNA polymerase has RNA 5′-triphosphatase and ATPase activities. *J. Virol.* 72:10011–19

62. Jin T, Qi Y, Liu D, Su F. 1999. Nucleotide sequence of a 5892 base pairs fragment of the LsMNPV genome and phylogenetic analysis of LsMNPV. *Virus Genes* 18:265–76
63. Kang W, Tristem M, Maeda S, Crook NE, O'Reilly DR. 1998. Identification and characterization of the *Cydia pomonella* granulovirus cathepsin and chitinase genes. *J. Gen. Virol.* 79:2283–92
64. Ko R, Okano K, Maeda S. 2000. Structural and functional analysis of the *Xestia c-nigrum* granulovirus matrix metalloproteinase. *J. Virol.* 74:11240–46
65. Kondo A, Maeda S. 1991. Host range expansion by recombination of the baculoviruses *Bombyx mori* nuclear polyhedrosis virus and *Autographa californica* nuclear polyehdrosis virus. *J. Virol.* 65:3625–32
66. Koonin EV, Aravind L, Kondrashov AS. 2000. The impact of comparative genomics on our understanding of evolution. *Cell* 101:573–76
67. Kreitman M, Gaasterland T. 2001. Genomes and evolution—ships in the night? *Curr. Opin. Genet. Dev.* 11:609–11
68. Kuzio J, Jaques R, Faulkner P. 1989. Identification of p74, a gene essential for virulence of baculovirus occlusion bodies. *Virology* 173:759–63
69. Kuzio J, Pearson MN, Harwood SH, Funk CJ, Evans JT, et al. 1999. Sequence and analysis of the genome of a baculovirus pathogenic for *Lymantria dispar*. *Virology* 253:17–34
70. Le TH, Blair D, Agatsuma T, Humair PF, Campbell NJH, et al. 2000. Phylogenies inferred from mitochondrial gene orders—a cautionary tale from the parasitic flatworms. *Mol. Biol. Evol.* 17:1123–25
71. Li CB, Li ZF, Yan QS, Hu GD, Pang Y. 2001. Cloning and sequencing of *Spodoptera litura* multicapsid nucleopolyhedrovirus gp37 gene cluster. *Prog. Biochem. Biophys.* 28:677–82
72. Lin J, Gerstein M. 2000. Whole-genome trees based on the occurrence of folds and orthologs: implications for comparing genomes on different levels. *Genome Res.* 10:808–18
73. Liu J, Maruniak JE. 1999. Molecular characterization of genes in the GP41 region of baculoviruses and phylogenetic analysis based upon GP41 and polyhedrin genes. *Virus Res.* 64:187–96
74. Lu A, Miller LK. 1994. Identification of three late expression factor genes within the 33.8- to 43.4-map-unit region of *Autographa californica* nuclear polyhedrosis virus. *J. Virol.* 68:6710–18
75. Luque T, Finch R, Crook N, O'Reilly DR, Winstanley D. 2001. The complete sequence of the *Cydia pomonella* granulovirus genome. *J. Gen. Virol.* 82:2531–47
76. Malik HS, Henikoff S, Eickbush TH. 2000. Poised for contagion: evolutionary origins of the infectious abilities of invertebrate retroviruses. *Genome Res.* 10:1307–18
77. Marcotte EM, Pellegrini M, Ng HL, Rice DW, Yeates TO, Eisenberg D. 1999. Detecting protein function and protein-protein interactions from genome sequences. *Science* 285:751–53
78. Martignoni ME, Iwai PJ. 1986. A catalogue of viral diseases of insects, mites, and ticks. Gen. Tech. Rep. PNW-195. Portland, OR: US Dep. Agric., For. Serv., Pac. Northwest Res. Stn. 51 pp. 4th ed.
79. McGeoch DJ, Dolan A, Ralph AC. 2000. Toward a comprehensive phylogeny for mammalian and avian herpesviruses. *J. Virol.* 74:10401–6
80. McLachlin JR, Miller LK. 1994. Identification and characterization of *vlf-1*, a baculovirus gene involved in very late gene expression. *J. Virol.* 68:7746–56
81. Miller LK. 1996. Insect viruses. In *Fields Virology*, ed. BN Fields, DM Knipe,

PM Howley, pp. 533–56. Philadelphia: Lippincott-Raven
82. Miller LK, ed. 1997. *The Baculoviruses*. New York: Plenum. 447 pp.
83. Montague MG, Hutchison CA. 2000. Gene content phylogeny of herpesviruses. *Proc. Natl. Acad. Sci. USA* 97:5334–39
84. Moscardi F. 1999. Assessment of the application of baculoviruses for control of Lepidoptera. *Annu. Rev. Entomol.* 44:257–89
85. Moser BA, Becnel JJ, White SE, Afonso C, Kutish G, et al. 2001. Morphological and molecular evidence that *Culex nigripalpus* baculovirus is an unusual member of the family Baculoviridae. *J. Gen. Virol.* 82:283–97
86. Nikaido M, Kawai K, Cao Y, Harada M, Tomita S, et al. 2001. Maximum likelihood analysis of the complete mitochondrial genomes of eutherians and a reevaluation of the phylogeny of bats and insectivores. *J. Mol. Evol.* 53:508–16
87. Olszewski J, Miller LK. 1997. A role for baculovirus GP41 in budded virus production. *Virology* 233:292–301
88. Olszewski J, Miller LK. 1997. Identification and characterization of a baculovirus structural protein, VP1054, required for nucleocapsid formation. *J. Virol.* 71:5040–50
89. O'Reilly DR. 1997. Auxiliary genes of baculoviruses. See Ref. 82, pp. 267–300
90. Page RDM. 2000. Extracting species trees from complex gene trees: reconciled trees and vertebrate phylogeny. *Mol. Phyl. Evol.* 14:89–106
91. Pang Y, Yu JX, Wang LH, Hu XH, Bao WD, et al. 2001. Sequence analysis of the *Spodoptera litura* multicapsid nucleopolyhedrovirus genome. *Virology* 287:391–404
92. Passarelli AL, Todd JW, Miller LK. 1994. A baculovirus gene involved in late gene expression predicts a large polypeptide with a conserved motif of RNA polymerases. *J. Virol.* 68:4673–78
93. Pearson MN, Groten C, Rohrmann GF. 2000. Identification of the *Lymantria dispar* nucleopolyhedrovirus envelope fusion protein provides evidence for a phylogenetic division of the Baculoviridae. *J. Virol.* 74:6126–31
94. Pearson MN, Rohrmann GF. 1998. Characterisation of a baculovirus-encoded ATP-dependent DNA ligase. *J. Virol.* 72:9142–49
94a. Pearson MN, Rohrmann GF. 2002. Transfer, incorporation, and substitution of envelope fusion proteins among members of the *Baculoviridae*, *Orthomyxoviridae*, and *Metaviridae* (insect rectrovirus) families. *J. Virol.* 76:5301–4
95. Pearson MN, Russell RLQ, Rohrmann GF. 2001. Characterization of a baculovirus-encoded protein that is associated with infected-cell membranes and budded virions. *Virology* 291:22–31
96. Pearson MN, Russell RLQ, Rohrmann GF, Beaudreau GS. 1988. p39, a major baculovirus structural protein: immunocytochemical characterization and genetic location. *Virology* 167:407–13
97. Pellegrini M, Marcotte EM, Thompson MJ, Eisenberg D, Yeates TO. 1999. Assigning protein functions by comparative genome analysis: protein phylogenetic profiles. *Proc. Natl. Acad. Sci. USA* 96:4285–88
98. Pellock BJ, Lu A, Meagher RB, Weise MJ, Miller LK. 1996. Sequence, function and phylogenetic analysis of an ascovirus DNA polymerase gene. *Virology* 216:146–57
99. Reichard P. 2002. Ribonucleotide reductases: the evolution of allosteric regulation. *Arch. Biochem. Biophys.* 397:149–55
100. Rohrmann GF. 1999. Nuclear polyhedrosis viruses. See Ref. 114a, pp. 146–52
101. Rohrmann GF, Karplus PA. 2001. Relatedness of baculovirus and gypsy

retrotransposon envelope proteins. *BMC Evol. Biol.* 1:1–9

102. Roncarati R, Knebel-Mörsdorf D. 1997. Identification of the early actin-rearrangement-inducing factor gene, *arif-1*, from *Autographa californica* multicapsid nuclear polyhedrosis virus. *J. Virol.* 71:7933–41
103. Russell RLQ, Rohrmann GF. 1997. Characterization of P91, a protein associated with virions of an *Orgyia pseudotsugata* baculovirus. *Virology* 233:210–33
104. Sicheritz-Pontén T, Andersson SGE. 2001. A phylogenomic approach to microbial evolution. *Nucleic Acids Res.* 29:545–52
105. Snel B, Bork P, Huynen MA. 1999. Genome phylogeny based on gene content. *Nat. Genet.* 21:108–10
106. Snel B, Bork P, Huynen MA. 2002. Genomes in flux: the evolution of archeal and proteobacterial gene content. *Genome Res.* 12:17–25
107. Tamames J, Gonzalez-Moreno M, Mingorance J, Valencia A, Vicente M. 2001. Bringing gene order into bacterial shape. *Trends Genet.* 17:124–26
108. Tekaia F, Lazcano A, Dujon B. 1999. The genomic tree as revealed from whole proteome comparisons. *Genome Res* 9:550–57
109. Thiem SM, Du XL, Quentin ME, Berner MM. 1996. Identification of a baculovirus gene that promotes *Autographa californica* nuclear polyhedrosis virus replication in a nonpermissive insect cell line. *J. Virol.* 70:2221–29
110. Thiem SM, Miller LK. 1989. Identification, sequence, and transcriptional mapping of the major capsid protein gene of the baculovirus *Autographa californica* nuclear polyhedrosis virus. *J. Virol.* 63:2008–18
111. Thornton JW, DeSalle R. 2000. Gene family evolution and homology: genomics meets phylogenetics. *Annu. Rev. Genomics Hum. Genet.* 1:41–73
112. Tweeten KA, Bulla LA Jr, Consigli RA. 1980. Characterization of an extremely basic protein derived from granulosis virus nucleocapsids. *J. Virol.* 33:866–76
113. Van Regenmortel MHV, Fauquet CM, Bishop DHL, Carstens EB, Estes ME, et al. 2000. *Virus Taxonomy: Seventh Report of the International Committee on Taxonomy of Viruses*. San Diego: Academic. 1162 pp.
114. Vassylyev DG, Morikawa K. 1996. Precluding uracyl from DNA. *Curr. Biol.* 4:1381–85

114a. Webster RG, Granoff A, eds. 1994. *Encyclopedia of Virology*. London: Academic. 2000 pp.

114b. Westenberg M, Wang H, Ijkel WFJ, Goldbach RW, Vlack JM, Zuidema D. 2002. Furin is involved in baculovirus envelope protein activation. *J. Virol.* 76:178–84

115. Whitford M, Faulkner P. 1992. Nucleotide sequence and transcriptional analysis of a gene encoding gp41, a structural glycoprotein of the baculovirus *Autographa californica* nuclear polyhedrosis virus. *J. Virol.* 66:4763–68
116. Whitford M, Stewart S, Kuzio J, Faulkner P. 1989. Identification and sequence analysis of a gene encoding gp67, an abundant envelope glycoprotein of the baculovirus *Autographa californica* nuclear polyhedrosis virus. *J. Virol.* 63:1393–99
117. Williams GV, Faulkner P. 1997. Cytological changes and viral morphogenesis during baculovirus infection. See Ref. 82, pp. 61–107
118. Wilson ME, Mainprize TH, Friesen PD, Miller LK. 1987. Location, transcription and sequence of a baculovirus gene encoding a small arginine-rich polypeptide. *J. Virol.* 61:661–66
119. Winstanley D, O'Reilly D. 1999. Granuloviruses. See Ref. 114a, pp. 140–46
120. Woese CR, Kandler O, Wheelis ML.

1990. Towards a natural system of organisms—proposal for the domains Archaea, Bacteria, and Eucarya. *Proc. Natl. Acad. Sci. USA* 87:4576–79

121. Yang S, Miller LK. 1999. Activation of baculovirus very late promoters by interaction with very late factor 1. *J. Virol.* 73:3404–9

122. Zanotto PMD, Kessing BD, Maruniak JE. 1993. Phylogenetic interrelationships among baculoviruses: evolutionary rates and host associations. *J. Invertebr. Pathol.* 62:147–64

Annu. Rev. Entomol. 2003. 48:235–60
doi: 10.1146/annurev.ento.48.091801.112624

First published online as a Review in Advance on August 27, 2002

GENOMICS IN PURE AND APPLIED ENTOMOLOGY

David G. Heckel
Centre for Environmental Stress and Adaptation Research, Department of Genetics, The University of Melbourne, Parkville, Victoria 3010, Australia; e-mail: dheckel@unimelb.edu.au

Key Words linkage map, physical map, EST, genome sequence, bioinformatics

■ **Abstract** Genomics is the study of the structure and function of the genome: the set of genetic information encoded in the DNA of the nucleus and organelles of an organism. It is a dynamic field that combines traditional paths of inquiry with new approaches that would have been impossible without recent technological developments. Much of the recent focus has been on obtaining the sequence of entire genomes, determining the order and organization of the genes, and developing libraries that provide immediate physical access to any desired DNA fragment. This has enabled functional studies on a genome-wide level, including analysis of the genetic basis of complex traits, quantification of global patterns of gene expression, and systematic gene disruption projects. The successful contribution of genomics to problems in applied entomology requires the cooperation of the private and public sectors to build upon the knowledge derived from the *Drosophila* genome and effectively develop models for other insect Orders.

CONTENTS

INTRODUCTION

The genome is the set of all genetic information of an organism encoded in the DNA of the nucleus and organelles. Genomics is the study of the structure, function, and evolution of entire genomes. Related concepts include the transcriptome (all RNA transcripts), the proteome (all proteins), and the metabolome (all metabolic pathways). This bold emphasis on "all" marks a new scientific revolution in biology, a paradigm shift made possible by stunning technological advances. Two perspectives have emerged: (*a*) Regarding the genome as the sum of its parts, genomics advances from the partial descriptions of the past to a completeness possible only when all of its constituent genes are known. (*b*) Viewing the genome as a unitary whole, genomics seeks to reveal and understand new properties that emerge at this higher level, when different genomes are compared and the forces shaping their evolution are studied.

In this review, I summarize the current state of genome projects in arthropods (mostly insects) in the public domain as of mid-March 2002. Much related research has been conducted by private companies and is not available in the public literature or databases. It is a pity that this review must omit this vast and valuable information, much of which could be released to the public domain once it has served its purpose. Indeed, I argue that the key to successful application of genomics is to forge workable public-private partnerships to make as much sequence data available as soon as possible, to encourage a wider research community to proceed with insect functional genomics.

STRUCTURAL GENOMICS

Assembling the DNA sequence of the entire genome—the ultimate description of its structure—has been long regarded as the Holy Grail of structural genomics. However, the entire sequence would be meaningless without additional information (called sequence annotation) attached to the bits in the assembly. In a fully sequenced and annotated genome, the occurrence of genes, introns, exons, promoters, enhancers, repressors, scaffold sequences, insulator elements, transposable elements, telomeres, centromeres, genetic markers, chromosomal breakpoints, and other features is marked out. We now consider three complementary approaches to this goal.

Genetic Linkage Maps

A linkage map is a set of genetic markers arranged in known sequence and spacing along linear linkage groups that correspond to chromosomes. Markers include morphological mutants, allozymes, RFLPs (restriction fragment length polymorphisms), microsatellites, RAPDs (randomly amplified polymorphic DNA), AFLPs (amplified fragment length polymorphisms), and SNPs (single nucleotide polymorphisms) [terms defined in (13)]. Crosses are performed and marker polymorphisms scored in parents and offspring. Statistical methods are used to infer gene order and estimate genetic distances between markers, resulting in a set of linkage groups. This arrangement of genes is one of the emergent properties of genomes that may be compared across different species if enough "anchor loci" are scored in common. This is usually impossible for maps based on "anonymous" markers such as RAPDs and AFLPs but is achievable using RFLPs with conserved protein-coding genes as probes. Insect linkage mapping since 1993 (47) is reviewed here.

DIPTERA In *Drosophila melanogaster*, the classical linkage-mapping approach (invented by Sturtevant for analysis of mutations in that species) produced a detailed map early in the history of genetics. In comparing different *Drosophila* species, Muller was the first to recognize that gene content, but not order, is conserved within chromosomal arms (now called Muller elements) owing to extensive paracentric inversions. This pre-genomic discovery is being extended and modified in the genomic era as well. Another recent trend has been to use the chromosome sequence to design sets of evenly spaced DNA-based markers. The resulting panels of microsatellites (102) and SNPs (11, 54) lend themselves to rapid scoring by high-throughput methods.

A recent marked increase in activity in mosquito genetics (104) was initiated by the first RFLP map for a mosquito, *Aedes aegypti* (105). The same RFLP probes were then mapped in other species: *Armigeres subalbatus* showed a high degree of conservation of marker order (33); somewhat less was seen in *Culex pipiens* (78) and *Cx. tritaeniorhynchus* (79). Use of SSCP (single-strand conformational polymorphism) to score variation in *Ae. aegypti* cDNA clones has increased the density of the map and even identified some areas of conservation with *D. melanogaster* (37). An earlier application of SSCP to RAPD products enabled efficient map construction in *Ae. aegypti* (5) and *Ae. albopictus* (81); however, the anonymous nature of the markers did not permit genome comparisons to be made. Microsatellites were useful in establishing linkage maps for *Anopheles gambiae* (128).

A linkage map based on RFLPs with cDNA clones in the apple maggot fly *Rhagoletis pomonella* (97) has also enabled comparison with *D. melanogaster*, revealing conservation of gene content (but not gene order) within the Muller elements (98). This confirms an earlier suggestion (35) that this mode of chromosomal evolution is typical of Diptera as a whole.

LEPIDOPTERA By the mid-1960s, an extensive classical map based on more than 200 morphological mutations had been constructed for the domesticated silkworm *Bombyx mori*. More recently, integration of molecular maps based on RFLPs (110), RAPDs (92, 126), and microsatellites (95) has been facilitated by use of the same inbred strains C108 and p50 to construct the F_2 mapping populations. Not yet integrated into that framework is a separate low-density RAPD map (61) and a high-density AFLP map (117) constructed using different strains. In other Lepidoptera, no classical maps existed. An integrated map with allozymes, RFLPs, and RAPDs was constructed for tobacco budworm *Heliothis virescens* (48) and used to study genes conferring resistance to chemical insecticides and *Bacillus thuringiensis* (Bt) toxins. An AFLP-based map of cotton bollworm *Helicoverpa armigera* was developed for a similar purpose (48); comparison of these species using anchor loci is currently in progress. An AFLP-based map in diamondback moth *Plutella xylostella* enabled mapping of a Bt resistance gene present in several populations worldwide (49).

HYMENOPTERA Complex behaviors in the honey bee *Apis mellifera*, subject of many classical studies in the past (90), can now be studied in more detail thanks in part to the development of a high-density linkage map based on RAPDs (59). Information loss owing to dominance in diploids was avoided by the use of haploid drones, and dense marker coverage was aided by high recombination rates. A RAPD map of the bumble bee *Bombus terrestris* (39) showed much lower recombination. RAPD maps for parasitoid wasps include *Bracon* species "near *hebetor*" (52), *Trichogramma brassicae* (70), two species in the genus *Nasonia* (40), and an SSCP-RAPD map for *Bracon hebetor* (5). Many of the RAPDs used in mapping turnip sawfly *Athalia rosae* were also sequenced and converted into sequence-tagged site (STS) markers (83).

COLEOPTERA AND OTHERS The red flour beetle *Tribolium castaneum* has long been studied by geneticists, and a new RAPD map (9) now complements the classical one. An AFLP map of the Colorado potato beetle *Leptinotarsa decemlineata* (45) is the first for this agricultural pest, which has developed resistance to a number of insecticides. Other insect groups with recent maps include two species of crickets (22, 108) and an aphid (46). The first map for a crustacean has been developed with AFLPs for the black tiger shrimp *Penaeus monodon* (124).

Genomic Libraries and Physical Maps

A physical map is a set of contigs (overlapping DNA fragments or clones) arranged in chromosomal order that cover each chromosome and ultimately the genome. The key to physical map construction has been the ability to clone large fragments of DNA in cosmid, YAC (yeast artificial chromosome), BAC (bacterial artificial chromosome), PAC (P1 artificial chromosome), or P1 vectors, as well as efficiently determining the overlap patterns of tens of thousands of such clones. Physical maps may be used as the substrate for whole-genome sequencing or to rapidly isolate any desired sequence for further manipulations, such as positional cloning. DNA-based markers are used as anchors to integrate genetic and physical maps.

DIPTERA In situ hybridization of DNA clones directly to polytene chromosome preparations has enabled special methods for constructing and correlating physical and genetic maps in Diptera by localizing clones without the need for polymorphism, meiotic recombination, or mapping families. Extensive physical mapping using all available techniques has been conducted with *D. melanogaster*.

Contigs of specific regions of interest were the first to be developed, e.g., the 2.9-MB ADH region for sequencing and detailed annotation (6). Larger contigs of cosmids on the X chromosome (75), BACs on chromosomes 2 and 3 (53), and P1 clones overall (67) were constructed later. Interspecific comparisons include *D. melanogaster* cosmids and cDNA probes hybridized to polytene chromosomes of *D. repleta*, *D. buzzattii*, and *D. virilis* (94), and P1 clones of *D. virilis* onto *D. montana* (121). To facilitate genome comparisons within the genus *Drosophila*, four species have been chosen (*erecta*, *pseudoobscura*, *littoralis*, and *willistoni*), representing a series of increasing evolutionary divergence. Fosmid libraries have been constructed from all four, and clones have been spotted onto high-density filters for screening by hybridization.

In the mosquito *Ae. aegypti*, mapped RFLP markers identify cosmid or cDNA clones that were further localized by in situ hybridization (16). Progress toward a physical map of *An. gambiae* was accelerated by hybridizing cDNAs, cosmids, and genetically mapped RAPD products to polytene chromosomes of itself (28, 30) and of *An. albimanus* (25). BAC libraries to support the genome sequencing effort have been developed (23). In a detailed comparison of *D. melanogaster* and *An. gambiae*, comparative hybridization mapping of 113 pairs of putative orthologs identified by whole-genome sequencing has shown that although the Muller elements are broadly conserved, 25%–50% of the genes are not found within the homologous element. Within each element there is considerable reshuffling, with only a suggestion of conserved gene order at the smallest scale (14).

Other notable physical mapping efforts in Diptera include the use of in situ hybridization to overcome the difficult genetic system of Hessian fly *Mayetiola destructor* (111) in studies of genes controlling virulence on varieties of wheat. A set of cosmid clones hybridized to metaphase chromosomes of the biting midge *Culicoides variipennis* has initiated physical map construction in this vector of bluetongue virus of livestock (84).

LEPIDOPTERA AND HYMENOPTERA In *B. mori*, BAC libraries have been constructed utilizing the same two strains used in the unified linkage map (125), and contigs have been used to confirm previous observations of tight genetic linkage. A separate BAC library with 11-fold coverage has been constructed from the p50 strain (BACPAC Resource Center). A BAC library of *A. mellifera* has been constructed, comprising about 37,000 clones with 23-fold coverage of the 180-MB genome (Clemson University Genomics Institute).

EST Sequencing Projects

An EST (expressed sequence tag) database is a set of single-pass sequencing reads from ends of random clones in a cDNA library. This captures a partial snapshot of the transcriptome in the tissue or developmental stage from which the library was constructed. Libraries may be non-normalized, in which clone abundance reflects natural abundance of the corresponding mRNA or normalized to increase the representation of rare mRNAs. Public-domain EST sequences are deposited in a special database (dbEST) with information on species and tissue of origin but usually without further annotation. Table 1 shows all the arthropods represented in dbEST as of March 22, 2002. Most of these EST projects have not yet been described in published articles.

DIPTERA The Berkeley Drosophila Genome Project (BDGP) developed the majority of ESTs from *D. melanogaster*, mostly from entire organisms rather than specific tissues. This dataset is being used to assemble a nonredundant set of full-length cDNAs of all expressed genes (99). A separate effort on testis (4) has yielded many ESTs not represented in the whole-organism libraries. Several ESTs from the male accessory gland of *D. simulans*, compared with homologs in *D. melanogaster*, have provided evidence for rapid divergence by natural selection (115).

The ESTs from *An. gambiae* include approximately 6000 clones derived from hemocyte-like cell lines, used in a search for genes involved in the immune response to *Plasmodium* infection (29). Eighty thousand more are from whole adults. ESTs of sandfly *Lutzomyia longipalpis*, a vector of *Leishmania*, have been described but not yet deposited into dbEST (93).

LEPIDOPTERA The most diverse EST project among arthropods has been undertaken in *B. mori*, where the silkworm's large body size facilitates dissection into separate tissues. More than 30 cDNA libraries have yielded about 1000 ESTs each, including wing discs at six different timepoints in larval development, ovary-derived cell lines before and three timepoints after inoculation with nucleopolyhedrosis virus, male versus female larval fat body, and embryos destined to enter or bypass diapause. Some of these ESTs have been deposited into dbEST but many are still only available on a separate web site, SilkBase. The use of these data in studying the response of cell cultures to viral infection (87) and in the identification of cuticle protein genes (116) has been described. Of a number of other projects in different Lepidoptera, the first to be published describes ESTs from antennae of *Manduca sexta* (96).

HYMENOPTERA AND COLEOPTERA The largest collection of ESTs from *A. mellifera* comes from normalized or subtracted libraries from adult brain, producing more than 20,000 sequences. All have been deposited into dbEST and are documented in an independent database, Bee-ESTdb (123a). ESTs from seven beetle species specifically chosen for their taxonomic diversity have been reported (118). This approach shows that even small cDNA libraries can be valuable in phylogenomics, revealing genes conserved across the insect Orders, as well as those more specific to Coleoptera.

OTHER ARTHROPODS Additional efforts include the cattle tick *Boophilus microplus* (26), black tiger shrimp *Penaeus monodon* (71), and a project aimed at discovery of immunity genes from two species of *Litopenaeus* shrimp (42). EST projects from nondipteran insect orders or other arthropods typically yield

TABLE 1 Arthropod representation in dbEST as of March 22, 2002

Species	Common name	Number of ESTs	Libraries represented
Aedes aegypti	Yellow fever mosquito	1523	Adult, adult midgut, adult female malpighian tubules + gut, antennae
Amblyomma americanum	Lone star tick	1971	Larva, adult, fed male salivary gland, unfed male salivary gland
Anagasta (Ephestia) kuehniella	Flour moth	12	Last larval instar
Anopheles gambiae	African malaria mosquito	94,032	Immune competent cell line, adult head, adult, adult (blood-fed)
Anopheles stephensi	Mosquito	24	Adult female post-infection (Plasmodium)
Antheraea yamamai	Japanese oak silk moth	610	Posterior silkgland
Aphis gossypii	Cotton aphid	1	(not specified)
Apis mellifera	Honey bee	15,309	Larva, adult brain, adult brain (normalized), male antennae, ovary
Araneus ventricosus	Spider	234	(not specified)
Artemia parthenogenetica	Brine shrimp	3	Male, female
Bombyx mandarina	Wild silkworm	226	Whole body
Bombyx mori	Domestic silkworm	14,849	Ovary cell line (BmNPV infected), pupal brain, compound eye, embryo (diapause-destined or cancelled), larval midgut, middle silkgland, posterior silkgland, pheromone gland, testis, wing disk (various stages)
Boophilus microplus	Cattle tick	143	Larva

(*Continued*)

TABLE 1 (*Continued*)

Species	Common name	Number of ESTs	Libraries represented
Circulifer tenellus	Beet leafhopper	2	(not specified)
Culex pipiens pallens	Mosquito	40	Female (subtracted, insecticide resistant)
Drosophila melanogaster	Fruit fly	255,456	Embryo, larvae + pupae, adult head, adult brain, testes, ovary, Schneider cells
Drosophila simulans	Fruit fly	259	Male accessory gland
Galleria mellonella	Wax moth	26	Pupal epidermis, larval hemocyte + fat body, larva
Haematobia irritans	Horn fly	43	Diapausing pupae, adult (subtracted, insecticide resistant)
Homarus americanus	American lobster	7	Olfactory organ
Ixodes ricinus	Deer tick	3	Female salivary gland
Litopenaeus (Penaeus) setiferus	Atlantic white shrimp	1042	Hemocytes and hepatopancreas
Litopenaeus (Penaeus) vannamei	Pacific white shrimp	874	Hemocytes and hepatopancreas, Taura Syndrome Virus-challenged whole-body
Manduca sexta	Tobacco hornworm	1626	Male antennae, female antennae, larval fat body (subtr., bacterial challenge)
Marsupenaeus (Penaeus) japonicus	Kuruma prawn	1170	Uninfected adult hemocytes, viral-infected adult hemocytes
Musca domestica	House fly	2	Early embryo
Penaeus monodon	Black tiger shrimp	673	Hemocyte, cephalothorax, eyestalk, pleopod
Planococcus lilacinus	Lilac mealybug	1	(not specified)
Pyrocoelia rufa	Firefly	71	Whole body
Rhagoletis pomonella	Apple maggot	296	Adult
Sarcoptes scabiei	Parasitic mite	891	Mixed lifestage
Trichoplusia ni	Cabbage looper	34	Fifth instar larval gut
		391,453	

a high fraction (20%–40%) of sequences with no obvious *Drosophila* homolog. The *Drosophila* homolog may be misannotated, may have diverged in sequence much faster than other arthropods, or may have been lost from the genome (123a). Many of these ESTs, however, do have identifiable homologs in libraries from other species in the same order. This has implications for the role of *Drosophila* as a genomic model for insects or arthropods in general.

Mitochondrial Genome Sequencing

Mitochondrial DNA-encoded genes have played an important role in studies of molecular systematics, population genetics, and evolution for at least 25 years. Once a daunting task, sequencing entire 15–18 kb mitochondrial genomes is now easily accomplished. The order and orientation of genes within the circular mtDNA is now recognized to be more variable within arthropods than previously thought, as more crustaceans and chelicerates have been examined (Table S1, see the Supplemental Material link in the online version of this chapter or at http://www.annualreviews.org/).

Nuclear Genome Sequencing

Regarded as impossible or infeasible until only recently, sequencing of eukaryotic genomes is now merely extremely complicated and expensive. The basic technology is still limited to obtaining sequencing reads smaller than 1 kb and so these must be joined together somehow to cover entire chromosomes. The "clone-by-clone" method assembles a physical map first, determines the set of clones covering the entire map with minimal overlap, and sequences each clone. The "whole-genome shotgun" (WGS) method involves sequencing a large number of randomly distributed clones and merging them all together simultaneously. This succeeds for small microbial genomes with mostly single-copy DNA, but its effectiveness in large genomes with lots of repetitive DNA is still a matter of debate (122). Fortunately for entomology, the fruit fly was chosen as the guinea pig for this strategy.

DIPTERA The sequence of most of the 120-MB euchromatic portion of the 180-MB genome of *D. melanogaster* was publicly released in March 2000 as a collaborative effort involving the newly formed company Celera Genomics and the ongoing publicly funded projects (3). The sequence was determined by a combination of WGS and sequencing of mapped BAC, P1, and cosmid clones. Several gaps remain, and many genes located in heterochromatic regions have been missed. (The 60 MB of heterochromatin consists largely of centromeric or telomeric repeats that are difficult to clone and sequence.) To make sense of the sequence, a group of about 40 experts participated in an "annotation jamboree" prior to sequence release. The initially estimated number of protein coding genes was about 13,600, surprisingly smaller than that of the roundworm *Caenorhabditis elegans*. The availability of the genome sequence has had an enormous impact, starting with its use in a landmark paper on comparative genomics of eukaryotes (100).

In *An. gambiae*, sequence from the ends of about 30,000 BAC clones totaling more than 14 MB has been obtained. A project to sequence the entire 280-MB genome was announced in 1999 and funded in 2001. The sequencing has already been completed and 10-fold WGS coverage has been assembled (51). In March 2002, Celera deposited 69,714 unannotated scaffold sequences from this assembly (some as large as 10–20 MB) in GenBank. An unanticipated result is the presence of several regions where sequence alignment problems suggest occurrence of two separate haplotypes in the PEST strain that was sequenced. Two independent annotations have been promised by the end of 2002 (51).

LEPIDTOPERA In *B. mori*, about 320 kb of the Z chromosome from a region containing an ortholog of the structural protein kettin has been sequenced, revealing 14 predicted genes, including a cluster of 4 showing a pattern of synteny with *Drosophila* chromosome 3 (68). In April 2002, Genoptera (a joint venture between Bayer and Exelixis) announced by press release that they had sequenced the genome of *H. virescens* and identified 90% of the genes. The extent of the genome coverage and availability of the sequence had not been made public at the time this review was completed.

HYMENOPTERA Genome sequencing of *A. mellifera* was announced to be a high priority of the U.S. National Human Genome Research Institute in May 2002. The sequence and annotation will be in the public domain.

Sequencing technologies are improving and the cost per base is still falling, making it easier to sequence more arthropods in the future. However, the total world sequencing capacity will probably not increase greatly. Competition for access to this capacity for other vertebrates, crop plants, and fungal pathogens will be stiff. Well-coordinated international projects that serve a large scientific community while promoting economic benefits of some kind will be required if we are to see more insect genomes sequenced in the future.

FUNCTIONAL GENOMICS

As structural genomics determines the sequence, functional genomics attempts to determine what it means. Meaning is sought at many levels, from the biochemical properties of individual gene products, to their interaction in metabolic processes and signaling pathways, to the specification of differentiation and form, and to traits of the organism that interface with the environment and other organisms, all of which change in the process of evolution. All these endeavors were pursued before genomics and would continue successfully without it. Yet the emphasis on asking "all"-type questions and devising technological approaches to answering them has produced the irresistible paradigm shift of seeking answers on a global scale. Functional genomics views this from many different perspectives.

QTL Mapping

One perspective starts from complex traits of whole organisms and attempts to work toward the underlying genes—a genetical approach. The extension of linkage mapping from discrete markers or single-gene effects to quantitative traits that may be controlled by many genes is nowadays referred to as QTL (quantitative trait locus) mapping. A dense linkage map with many markers is used in combination with a cross between two strains that differ with respect to one or more traits. Genes (QTLs) underlying these trait differences are identified by discovering chromosomal regions between the markers that have a high association with components of the trait in offspring of the cross. A high-density linkage map enables the entire genome to be scanned for these QTLs, facilitating positional or map-assisted cloning of the genes responsible.

DIPTERA Complex phenotypes have been extensively analyzed in *D. melanogaster*, starting with the classical quantitative traits of bristle number (74). In this species in situ hybridization to transposable elements in polytene chromosomes is often used in place of conventional markers, e.g., the distribution of *roo* elements was used for high-resolution mapping of QTLs affecting sternopleural bristle number (44). QTLs affecting adult longevity (72) and those that do so in a sex-specific manner (85) have also been mapped. QTLs have been identified that exhibit genotype *x* environment interactions for fitness (36), sensory bristle number (43), and adult life span (120). Even the level of *copia* retrotransposon activity in inbred lines (86) and traits related to the function of monoamine receptors (7) have been analyzed with the QTL approach. Other species of *Drosophila* have been mapped as well, often with interspecific crosses. Larvae of *D. sechellia* are resistant to octanoic acid produced by their host plant. Using maps based on interspecific backcrosses to octanoic-acid-susceptible species, QTLs for this resistance were identified (63). QTLs underlying interspecific differences in male genital morphology between *D. simulans* and *D. mauritiana* have been identified in hybrid crosses (127), as well as genes affecting other male cuticular and bristle traits (119).

The genetic basis of vectorial competence in mosquitoes has been of great interest from a public health perspective. The greatest variety of information comes from studies of *Ae. aegypti*, including susceptibility to the filarial worm *Brugia malayi* (10), the malarial parasite *Plasmodium gallinaceum* (106), or to both (107), and susceptibility to dengue virus (15). Interestingly, some of the genes involved with different parasites seem to occur at the same location, although the low recombination rate in mosquitoes makes it difficult to distinguish adjacent QTLs from each other. Genes affecting the encapsulation reaction involved in resistance to *Plasmodium* have been mapped in *An. gambiae* (129). Attempts at positional cloning of three such genes (24) have been complicated by the extremely low rates of recombination. QTLs affecting refractoriness of *Anopheles stephensi* to the human malaria parasite *Plasmodium falciparum* (32) have also been identified.

Mapping of genes controlling virulence of Hessian fly toward susceptible varieties of wheat has been accomplished (103).

HYMENOPTERA Much recent progress has been made in mapping genes influencing behavior of the honey bee, by means of exacting experimental designs exploiting haplodiploid genetics and the evaluation of traits at the colony level. QTLs affecting pollen-hoarding (60, 89), stinging behavior and body size (58), alarm pheromone levels (57), and various measures of learning performance (21) have all been mapped, and positional cloning has begun. The interval containing the major gene underlying complementary sex-determination has been narrowed in honey bee (12), bumble bee (39), and parasitic wasp *Bracon* (52). A RAPD map was used to identify interactions among genes causing reproductive incompatibility in hybrids of the parasitic wasps *Nasonia vitripennis* and *N. giraulti* (40).

LEPIDOPTERA AND OTHER ORDERS Two genes conferring reduced susceptibility to densonucleosis virus have been mapped in *B. mori* (1, 2). QTL mapping in *H. virescens* was crucial in the map-assisted cloning of *BtR-4*, the first Bt resistance gene cloned from insects (41). There has been an increasing trend for evolutionary biologists to construct AFLP maps specifically to investigate the genetic architecture of traits associated with speciation. The genetic basis of reproductive incompatibility (i.e., whether females utilize sperm from heterospecific males) has been investigated by QTL mapping in a hybrid zone of *Allonemobius* crickets (55). In the pea aphid *Acyrthosiphon pisum*, QTLs for performance on different plant hosts and QTLs for habitat choice were linked (46). The genetic architecture of a large difference in male song pulse rate between two recently diverged endemic Hawaiian cricket species (genus *Laupala*) has been dissected by QTL analysis (108).

Despite the low resolution of many of these mapping efforts, they are most useful in characterizing genetic architecture and additionally represent a first step toward identifying and ultimately cloning specific genes responsible for an impressive variety of phenotypes. Positional cloning is becoming increasingly feasible with the development of more physical maps; but the most rapid progress initially will be made with the candidate gene approach (34). By using the results of QTL analysis to eliminate most of the genome from consideration, effort can be more efficiently focused on testing the few likely candidates in the near vicinity of mapped QTLs.

Microarrays

Another approach to functional genomics starts with the transcriptome. If an EST project can be likened to a snapshot, a global-expression study with microarrays is a movie or a sculpture. In such a study, a regular array of tiny spots is constructed on a microscope slide or silicon chip, each spot with DNA from a different gene. mRNA in a sample from a particular cell type, tissue, or developmental stage is labeled and incubated with the array under conditions that allow hybridization of a sequence only to its exact complement. After background correction and normalization, the signal intensity of each gene-spot is taken as a measure of how abundantly its mRNA was expressed in the sample. Of course, this does not directly account for expression on the protein level, for which the techniques of proteomics would be required. However, it does provide a global view, revealing interesting patterns that may be followed up by other experiments.

DIPTERA Microarrays displaying the majority of expressed genes of *D. melanogaster* have been employed to investigate

- genome-wide changes in gene expression over time, during the course of embryogenesis (38), metamorphosis (123), adult aging (130), or during the circadian cycle (76);
- patterns of gene expression in specific tissues, such as follicle cells (18) or testis (4); and
- the dynamics of response to environmental stress, such as the oxidative stress caused by paraquat, a generator of free radicals (130), or the immune response to microbial infection (27, 62).

HYMENOPTERA AND LEPIDOPTERA In *A. mellifera*, arrays displaying 300 cDNAs from queen-biased or worker-biased subtractive cDNA libraries have been used to examine the divergence in gene expression between larvae raised as workers or as queens (31). Microarrays with 2500 cDNAs from an adult brain library were probed with mRNA from newly emerged workers, experienced foragers, or bees treated with caffeine, which accelerates associative olfactory learning (69). Pilot studies have recently been done with >7000 cDNAs, many of them annotated (123a). Arrays of a nonredundant set of 6000 of the *B. mori* ESTs were probed with mRNA isolated from wing discs at seven stages of development, and wing discs cultured with 20-hydroxyecdysone, to examine changes in gene expression during metamorphosis (88).

Systematic Gene Disruption

A third approach to functional genomics starts with the genes, and attempts to knock them out one by one to see the effects. Random gene inactivation by mutagenesis has long been used in genetics, but adapting these methods to the systematic and targeted genomic approach is far from simple. Nevertheless, genome-wide mutant collections are being assembled for many model organisms (22a). We briefly describe two techniques most likely to be useful in the near future for arthropods.

INSERTIONAL MUTAGENESIS BY TRANSPOSABLE ELEMENTS When transposable elements are mobilized, they can jump into or near genes and inactivate them, often causing observable mutations. Isolating genomic DNA and screening fragments for the presence of the transposable element can then identify the inactivated gene. This process of "cloning by transposon tagging" can be extended to the genomic level by accumulating enough mutant strains to account for one inactivation event per gene. Obtaining the genomic sequence that flanks each newly mobilized transposable element immediately localizes it, at least in a fully sequenced genome. The P-element is the most widely used transposon in *D. melanogaster* for insertional mutagenesis. A large number of P-element insertion lines have been created in a systematic attempt at inactivating all the genes (112). These have been useful, even though P-elements have certain disadvantages, e.g., a tendency to insert in the 5′ flanking regions of genes rather than the coding sequence. Improved elements are now being engineered with properties that are more desirable. This general approach is feasible for only a large research community that can sustain the expense of creating, mapping, and maintaining thousands of lines.

TRANSCRIPTIONAL SILENCING When presented to cells that are actively transcribing a particular gene, double-stranded RNA (dsRNA) with the same sequence will trigger the destruction of that transcript. This silencing or RNA inhibition (RNAi) may spread to adjacent cells and persist even after cell growth and division would have diluted the original dsRNA out. Although long known in plants and fungi, RNAi in animals was discovered only five years ago, and elucidation of its mechanism and applications is one of the most exciting areas of molecular biology today (20). The RNAi phenomenon is observed in *D. melanogaster* and can be used to analyze gene function in vivo by mechanical injection of dsRNA (64) or by transformation with constructs that express it (65). The latter method would enable tissue-specific or inducible expression of the dsRNA construct that is far more precise and elegant than the injection needle. Undoubtedly for *Drosophila* there will be a systematic gene-by-gene knockout program using RNAi such as is under way for the nematode *C. elegans* (66).

The most exciting aspect of RNAi is how readily it can be used in any animal, genome project or not. This tool has rapidly been taken up by many groups and already applied to studies of the house fly *Musca domestica* (109), the cyclorrhaphan fly *Megaselia abdita* (113), the cockroach *Periplaneta americana* (75a), pupal hemocytes of the flesh fly *Sarcophaga peregrina* (82), and a cultured cell line from *An. gambiae* (73). RNA interference has also been demonstrated in *Tribolium* (17), the milkweed bug *Oncopeltus fasciatus* (56), and the spider *Cupiennius salei* (101, 114). Transgenerational effects (suppression of genes in offspring of injected parents) have been documented in *Tribolium* (19) and *Hyalophora cecropia* moths (11a). RNAi does not always work, and its limitations are not fully understood. But in these early days it appears to be a good candidate for genome-wide or more limited approaches to functional genomics of sequenced organisms by gene disruption, requiring only knowledge of the sequence of the gene to be silenced.

DATABASES AND BIOINFORMATICS

Storing, accessing, and analyzing genomic datasets that accumulate at faster-than-exponential rates is difficult and complicated, at best. The interested reader may consult (80) for a general treatment. We confine our description to databases most relevant to the material reviewed above that are accessible by anyone with a connection to the World Wide Web and suitable Internet browser software. URL addresses are given in Table 2.

GenBank, the database maintained by the U.S. National Center for Biotechnology Information (NCBI), is the pre-eminent public repository for sequence information. Nucleotide or protein sequences may be searched for by key word,

TABLE 2 World Wide Web resources for insect genomics

Description	URL address
General	
Insect Genome Databases	http://www.hgmp.mrc.ac.uk/GenomeWeb/insect-gen-db.html
National Center for Biotechnology Information	http://www.ncbi.nlm.nih.gov/
Drosophila	
FlyBase @ flybase.bio.indiana.edu	http://flybase.bio.indiana.edu/
GadFly: Genome Annotation Database of Drosophila	http://www.fruitfly.org/annot/index.html
BDGP: Home	http://www.fruitfly.org/
Drosophila melanogaster @ NCBI	http://www.ncbi.nlm.nih.gov/PMGifs/Genomes/7227.html
fly.txt (SwissProt Index + FlyBase Cross-Refs)	http://kr.expasy.org/cgi-bin/lists?fly.txt
TIGR Drosophila gene index	http://www.tigr.org/tdb/dgi/
Mosquitoes	
ENSEMBL Mosquito Genome Server	http://www.ensembl.org/Anopheles_gambiae/
The Mosquito Genomics WWW Server	http://mosquito.colostate.edu/
An Anopheles Database (AnoDB)	http://konops.anodb.gr/AnoDB/
BBMI Anopheles gambiae Genome Page	http://bioweb.pasteur.fr/BBMI/index.html
TIGR Anopheles gambiae gene index	http://www.tigr.org/tdb/aggi/
Others	
Honey Bee Brain EST Project	http://titan.biotec.uiuc.edu/bee/honeybee_project.htm
SilkBase (ESTs of Bombyx mori)	http://www.ab.a.u-tokyo.ac.jp/silkbase/
KSU Tribolium Genetics Program	http://www.ksu.edu/tribolium/
Drosophila Microarray Data	
Access the Metamorphosis Database	http://quantgen.med.yale.edu/
Drosophila Immune Response	http://www.fruitfly.org/expression/immunity/
Materials	
BACPAC Resources	http://www.chori.org/bacpac/
CUGI (Clemson University Genomics Institute)	http://www.genome.clemson.edu/
Malaria Research & Reference Reagent Resource Ctr.	http://www.malaria.atcc.org

including species, gene, or protein name or function. For searching by sequence similarity (BLAST searches), GenBank is divided into subdatabases that must be tried separately, including the largest section "nr"; the Celera dataset in "Drosophila genome"; the *An. gambiae* genomic sequence in "wgs_anopheles"; the nonhuman, nonmouse ESTs in "est_others"; the short genomic landmark sequences in "db-STS"; and the self-explanatory "patent."

Information on *D. melanogaster* is found on four major web sites. (*a*) FlyBase contains information on maps, genes, sequences, stocks, transgenes, transposons, aberrations, anatomy, images, literature references, and names and affiliations of researchers, and a useful direct link to The Interactive Fly. (*b*) GadFly presents the Celera annotation of the genome sequence and its revisions, which can be queried by gene name, map position, molecular function, InterPro motif name, or precomputed similarity search. (*c*) The BDGP site contains all known *Drosophila* sequences in searchable format; analysis tools enabling prediction of genes, promoters, and transcription factor binding sites; descriptions of ongoing projects; and access to materials. GadFly and BDGP both utilize the GeneSeen Map Viewer, which presents a graphical view of the chromosome and can zoom down to the individual gene and nucleotide level, showing intron/exon boundaries, matches to ESTs, and results of BLAST searches of GenBank, P-elements, and other repeated sequences. (*d*) The *Drosophila* home page at NCBI features the Entrez Map Viewer, links to similar proteins of known three-dimensional structures, precomputed BLAST results, LocusLink, a single query interface to curated sequence, and descriptive information about genetic loci.

A preliminary annotation of the *An. gambiae* genome can be found on the ENSEMBL web site. The Mosquito Genomics WWW Server and the Anopheles Database (AnoDB) provide general information on mosquitoes. *An. gambiae* BAC-end sequences can be searched from the BBMI page at the Pasteur Institute. The Institute for Genomic Research (TIGR) maintains gene indices for *D. melanogaster*, *An. gambiae*, and *A. mellifera*. To create these indices, ESTs in the public databases were grouped into clusters representing transcripts of individual genes. Each cluster is represented by a consensus sequence, the supporting EST assembly, links to the contributing sequences, expression information on the frequencies of ESTs in different libraries, precomputed BLAST results, Gene Ontology classification terms, and map location.

SilkBase has the most complete collection of *B. mori* ESTs. These can be searched by BLASTN or TBLASTN, as well as by clone name and key word. Results of precomputed BLAST searches for each EST provide links to external databases such as Swiss-Prot and PIR. A first attempt at clustering ESTs has been carried out, with representation of each EST sequence across all 32 libraries depicted in tabular form. BEE-dbEST (123a) enables BLAST searches of the individual 15,000 bee-brain ESTs or the 9000 consensus sequences, each corresponding to an EST cluster derived from a separate gene. Search by key word is also possible by navigating through the Gene Ontology classification system. Materials, including many of the BAC libraries, high-density filters, clones, and libraries mentioned in this section, are obtainable from the BACPAC, CUGI, and MR4 sites.

THE GENOMIC APPROACH AND APPLIED ENTOMOLOGY

There are a number of obvious applications of genomics to problems caused by arthropod pests of agriculture and vectors of disease. The large financial investment required for the "big-science" approach has generally limited participation to large multinational crop protection industries. Most of them have had well-developed research programs in all of the following areas for several years.

Applications of Sequencing Projects

Sequencing of cDNA libraries and genomes of pest organisms is regarded as the key to gene discovery, the identification of novel targets for new insecticides. Gene recognition is a more accurate term, for it relies largely on sequence similarity to previously discovered and characterized genes in other organisms. Most insecticides currently in use have targets in the nervous system that are highly conserved in all animals. But the approach can also identify genes or gene families that are excellent candidates, yet not so highly conserved—potentially offering vulnerable targets for pest-specific insecticides with novel modes of action. Sequencing the *H. virescens* genome was explicitly justified in this manner. Genes in arthropod vectors involved in the immune response to the disease organisms they carry are another prime target of gene discovery, one of the justifications for the *An. gambiae* project. Since the majority of eukaryotic genomes is noncoding DNA, EST projects followed by targeted full-length cDNA sequencing is the most efficient approach. Genome sequencing would be useful mostly for sorting out multigene families and ensuring that underexpressed genes are not missed.

Applications of Functional Genomics

The microarray format lends itself to analysis of the transcriptome's response to pesticides, climatic stress, host plant defensive compounds, infection by viruses and microbial pathogens, and any number of phenomena affecting the pest status of arthropods. The responses of gene families known to detoxify insecticides, e.g., cytochromes P-450 in *C. elegans* (77) and *D. melanogaster* (N. Tijet & R. Feyereisen, personal communication), have already been studied using microarrays. Sublethal effects of novel insecticides may shed light on their mode of action and likely mechanisms of resistance. The outlines of hierarchical control of gene expression and networks of interacting genes may start to emerge. This highly exploratory approach will yield an overabundance of results, and the challenge remains of efficiently following up on those with special promise. Narrowing the field to a chosen few genes to be subjected to other evaluation and validation tests in a high-throughput testing pipeline is the crucial next step.

Genetic transformation of insects is another technique that will greatly affect the future role of genomics in applied entomology. Now a routine tool in *Drosophila*, stable insertion of foreign genes into the germ line of other insects has been achieved only recently (8, 50). Engineering malaria-refractory mosquitoes (62a) is an oft-cited goal that will certainly bring some benefits to understanding regardless of the attainability of population replacement in the field. Transformation technology will make genome-wide gene inactivation and enhancer trapping more feasible in pest species, although RNAi will also play a major role.

The Role of Model Systems

What is the relevance of the *Drosophila* genome to the rest of entomology? The validity of the model system approach to practical problem solving is the issue here. For fundamental processes that are conserved within arthropods, or all animals, using the powerful tools of *Drosophila* genetics and genomics to get the answers may be the best way. But not all arthropods are the same; indeed, one of the most exciting areas of entomology today is determining why they are different—the genetic and developmental mechanisms of evolutionary diversification in the basic body plan. Genomics technology now permits a decentralization of the model system approach, and this should be vigorously pursued. The path forward is easiest to see for the structural side. Genome sequencing projects for two models within each of the Lepidoptera, Coleoptera, Hemiptera, Hymenoptera, and Orthoptera at the least would allow for the first time a global view of the similarities and differences at the Order level. Shallow but taxonomically diverse EST projects within each Order would greatly flesh out this picture.

Public/Private Partnerships

Despite the potential benefits, public interest for funding such a decentralized model system approach in insect genomics is low. A recent meeting that showcased past achievements and attempted to marshal support for new projects concluded with a lukewarm assessment of the likelihood of new governmental funding (91). There are some bright spots, e.g., a recent U.S. National Science Foundation program to fund the construction of BAC libraries in a diverse collection of organisms and the high priority accorded to honey bee genome sequencing by the U.S. National Human Genome Research Institute. But in agricultural genomics at least, projects on crop plants and their bacterial and fungal pathogens are much better supported than anything on insects. As more powerful genomic tools become freely available in these systems, they become more attractive systems for researchers to work on, and more money flows to them as scientific activity increases. Arthropods are in last place in the genomics race, and the gap is widening.

If research on arthropods is to keep up, there must be enough genomic information available to prime the pump so that new sequencing proposals are competitive in the international scientific arena. Much of this information has already been obtained but is not yet in the public databases. Industries in the private sector, especially crop protection companies, would be doing a great service to the general effort in insect genomics to release their vast data on EST projects and genome sequencing once it has served their purposes. The random nature of EST projects suggests that at least 95% of the information obtained has no real commercial value. The development of public/private partnerships to ensure that these data can be released for universal and fair access, and effectively used to marshal more public support for insect genomics, should be a high priority. There is ample precedent for this model in plant genomics. Everyone stands to benefit from the increased research activity of a rejuvenated entomology community, just as everyone, including private industry, has benefited from the free and open nature of *Drosophila* research.

ACKNOWLEDGMENTS

Thanks to Charles Robin and Gene Robinson for reading the manuscript and helpful discussion. Supported by CESAR (Centre for Environmental Stress and Adaptation Research), a Special Research Centre funded by the Australian Research Council.

The *Annual Review of Entomology* is online at http://ento.annualreviews.org

LITERATURE CITED

1. Abe H, Harada T, Kanehara M, Shimada T, Ohbayashi F, Oshiki T. 1998. Genetic mapping of RAPD markers linked to the densonucleosis refractoriness gene, *nsd-1*, in the silkworm, *Bombyx mori*. *Genes Genet. Syst.* 73:237–42
2. Abe H, Sugasaki T, Kanehara M, Shimada T, Gomi SJ, et al. 2000. Identification and genetic mapping of RAPD markers linked to the densonucleosis refractoriness gene, *nsd-2*, in the silkworm, *Bombyx mori*. *Genes Genet. Syst.* 75:93–96
3. Adams MD, Celniker SE, Holt RA, Evans CA, Gocayne JD, et al. 2000. The genome sequence of *Drosophila melanogaster*. *Science* 287:2185–95
4. Andrews J, Bouffard GG, Cheadle C, Lu JN, Becker KG, Oliver B. 2000. Gene discovery using computational and microarray analysis of transcription in the *Drosophila melanogaster* testis. *Genome Res.* 10:2030–43
5. Antolin MF, Bosio CF, Cotton J, Sweeney W, Strand MR, Black WC. 1996. Intensive linkage mapping in a wasp (*Bracon hebetor*) and a mosquito (*Aedes aegypti*) with single-strand conformation polymorphism analysis of random amplified polymorphic DNA markers. *Genetics* 143:1727–38
6. Ashburner M, Misra S, Roote J, Lewis SE, Blazej R, et al. 1999. An exploration of the sequence of a 2.9-Mb region of the genome of *Drosophila melanogaster*: the *Adh* region. *Genetics* 153:179–219

7. Ashton K, Wagoner AP, Carrillo R, Gibson G. 2001. Quantitative trait loci for the monoamine-related traits heart rate and headless behavior in *Drosophila melanogaster*. *Genetics* 157:283–94
8. Atkinson PW, Michel K. 2002. What's buzzing? Mosquito genomics and transgenic mosquitoes. *Genesis* 32:42–48
9. Beeman RW, Brown SJ. 1999. RAPD-based genetic linkage maps of *Tribolium castaneum*. *Genetics* 153:333–38
10. Beerntsen BT, Severson DW, Klinkhammer JA, Kassner VA, Christensen BM. 1995. *Aedes aegypti*—a quantitative trait locus (QTL) influencing filarial worm intensity is linked to QTL for susceptibility to other mosquito-borne pathogens. *Exp. Parasitol.* 81:355–62
11. Berger J, Suzuki T, Senti KA, Stubbs J, Schaffner G, Dickson BJ. 2001. Genetic mapping with SNP markers in *Drosophila*. *Nat. Genet.* 29:475–81

11a. Bettencourt R, Terenius O, Faye I. 2002. *Hemolin* gene silencing by ds-RNA injected into Cecropia pupae is lethal to next generation embryos. *Insect Mol. Biol.* 11:267–71

12. Beye M, Hunt GJ, Page RE, Fondrk MK, Grohmann L, Moritz RFA. 1999. Unusually high recombination rate detected in the sex locus region of the honey bee (*Apis mellifera*). *Genetics* 153:1701–8
13. Black WC, Baer CF, Antolin MF, DuTeau NM. 2001. Population genomics: genome-wide sampling of insect populations. *Annu. Rev. Entomol.* 46:441–69
14. Bolshakov VN, Topalis P, Blass C, Kokoza E, della Torre A, et al. 2002. A comparative genomic analysis of two distant diptera, the fruit fly, *Drosophila melanogaster*, and the malaria mosquito, *Anopheles gambiae*. *Genome Res.* 12: 57–66
15. Bosio CF, Fulton RE, Salasek ML, Beaty BJ, Black WC. 2000. Quantitative trait loci that control vector competence for dengue-2 virus in the mosquito *Aedes aegypti*. *Genetics* 156:687–98
16. Brown SE, Severson DW, Smith LA, Knudson DL. 2001. Integration of the *Aedes aegypti* mosquito genetic linkage and physical maps. *Genetics* 157:1299–305
17. Brown SJ, Mahaffey JP, Lorenzen MD, Denell RE, Mahaffey JW. 1999. Using RNAi to investigate orthologous homeotic gene function during development of distantly related insects. *Evol. Dev.* 1:11–15
18. Bryant Z, Subrahmanyan L, Tworoger M, LaTray L, Liu CR, et al. 1999. Characterization of differentially expressed genes in purified *Drosophila* follicle cells: toward a general strategy for cell type-specific developmental analysis. *Proc. Natl. Acad. Sci. USA* 96:5559–64
19. Bucher G, Scholten J, Klingler M. 2002. Parental RNAi in *Tribolium* (Coleoptera). *Curr. Biol.* 12:R85–86
20. Carthew RW. 2001. Gene silencing by double-stranded RNA. *Curr. Opin. Cell Biol.* 13:244–48
21. Chandra SBC, Hunt GJ, Cobey S, Smith BH. 2001. Quantitative trait loci associated with reversal learning and latent inhibition in honeybees (*Apis mellifera*). *Behav. Genet.* 31:275–85
22. Chu JM, Howard DJ. 1998. Genetic linkage maps of the ground crickets *Allonemobius fasciatus* and *Allonemobius socius* using RAPD and allozyme markers. *Genome* 41:841–47

22a. Coelho PSR, Kumar A, Snyder M. 2000. Genome-wide mutant collections: toolboxes for functional genomics. *Curr. Opin. Microbiol.* 3:309–15

23. Collins FH, Kamau L, Ranson HA, Vulule JM. 2000. Molecular entomology and prospects for malaria control. *Bull. WHO* 78:1412–23
24. Collins FH, Zheng L, Paskewitz SM, Kafatos FC. 1997. Progress in the map-based cloning of the *Anopheles gambiae*

genes responsible for the encapsulation of malarial parasites. *Ann. Trop. Med. Parasitol.* 91:517–21
25. Cornel AJ, Collins FH. 2000. Maintenance of chromosome arm integrity between two *Anopheles* mosquito subgenera. *J. Hered.* 91:364–70
26. Crampton AL, Miller C, Baxter GD, Barker SC. 1998. Expressed sequenced tags and new genes from the cattle tick, *Boophilus microplus*. *Exp. Appl. Acarol.* 22:177–86
27. De Gregorio E, Spellman PT, Rubin GM, Lemaitre B. 2001. Genome-wide analysis of the *Drosophila* immune response by using oligonucleotide microarrays. *Proc. Natl. Acad. Sci. USA* 98:12590–95
28. della Torre A, Favia G, Mariotti G, Coluzzi M, Mathiopoulos KD. 1996. Physical map of the malaria vector *Anopheles gambiae*. *Genetics* 143:1307–11
29. Dimopoulos G, Casavant TL, Chang SR, Scheetz T, Roberts C, et al. 2000. *Anopheles gambiae* pilot gene discovery project: identification of mosquito innate immunity genes from expressed sequence tags generated from immune-competent cell lines. *Proc. Natl. Acad. Sci. USA* 97:6619–24
30. Dimopoulos G, Zheng LB, Kumar V, della Torre A, Kafatos FC, Louis C. 1996. Integrated genetic map of *Anopheles gambiae*—use of RAPD polymorphisms for genetic, cytogenetic and STS landmarks. *Genetics* 143:953–60
31. Evans JD, Wheeler DE. 2000. Expression profiles during honeybee caste determination. *Genome Biol.* 2:research-0001.1-.6. http://genomebiology.com/2000/2/1/research/0001/
32. Feldmann AM, van Gemert GJ, van de Vegte-Bolmer MG, Jansen RC. 1998. Genetics of refractoriness to *Plasmodium falciparum* in the mosquito *Anopheles stephensi*. *Med. Vet. Entomol.* 12:302–12
33. Ferdig MT, Taft AS, Severson DW, Christensen BM. 1998. Development of a comparative genetic linkage map for *Armigeres subalbatus* using *Aedes aegypti* RFLP markers. *PCR Methods Appl.* 8:41–47
34. Flint J, Mott R. 2001. Finding the molecular basis of quantitative traits: successes and pitfalls. *Nat. Rev. Genet.* 2:437–45
35. Foster GG, Whitten MJ, Konovalov C, Arnold JTA, Maffi G. 1981. Autosomal genetic maps of the Australian sheep blowfly *Lucilia cuprina dorsalis* (Diptera: Calliphoridae) and possible correlations with linkage maps of *Musca domestica* and *Drosophila melanogaster*. *Genet. Res.* 37:55–70
36. Fry JD, Nuzhdin SV, Pasyukova EG, McKay TFC. 1998. QTL mapping of genotype-environment interaction for fitness in *Drosophila melanogaster*. *Genet. Res.* 71:133–41
37. Fulton RE, Salasek ML, DuTeau NM, Black WC. 2001. SSCP analysis of cDNA markers provides a dense linkage map of the *Aedes aegypti* genome. *Genetics* 158:715–26
38. Furlong EEM, Andersen EC, Null B, White KP, Scott MP. 2001. Patterns of gene expression during *Drosophila* mesoderm development. *Science* 293:1629–33
39. Gadau J, Gerloff CU, Kruger N, Chan H, Schmid-Hempel P, et al. 2001. A linkage analysis of sex determination in *Bombus terrestris* (L.) (Hymenoptera: Apidae). *Heredity* 87:234–42
40. Gadau J, Page RE, Werren JH. 1999. Mapping of hybrid incompatibility loci in *Nasonia*. *Genetics* 153:1731–41
41. Gahan LJ, Gould F, Heckel DG. 2001. Identification of a gene associated with Bt resistance in *Heliothis virescens*. *Science* 293:857–60
42. Gross PS, Bartlett TC, Browdy CL, Chapman RW, Warr GW. 2001. Immune gene discovery by expressed sequence tag analysis of hemocytes and hepatopancreas in the Pacific White Shrimp, *Litopenaeus vannamei*, and the Atlantic

White Shrimp, *L. setiferus*. *Dev. Comp. Immunol.* 25:565–77
43. Gurganus MC, Fry JD, Nuzhdin SV, Pasyukova EG, Lyman RF, Mackay TFC. 1998. Genotype-environment interaction at quantitative trait loci affecting sensory bristle number in *Drosophila melanogaster*. *Genetics* 149:1883–98
44. Gurganus MC, Nuzhdin SV, Leips JW, Mackay TFC. 1999. High-resolution mapping of quantitative trait loci for sternopleural bristle number in *Drosophila melanogaster*. *Genetics* 152: 1585–604
45. Hawthorne DJ. 2001. AFLP-based genetic linkage map of the Colorado potato beetle *Leptinotarsa decemlineata*: sex chromosomes and a pyrethroid-resistance candidate gene. *Genetics* 158:695–700
46. Hawthorne DJ, Via S. 2001. Genetic linkage of ecological specialization and reproductive isolation in pea aphids. *Nature* 412:904–7
47. Heckel DG. 1993. Comparative genetic linkage mapping in insects. *Annu. Rev. Entomol.* 38:381–408
48. Heckel DG, Gahan LJ, Daly JC, Trowell S. 1998. A genomic approach to understanding *Heliothis* and *Helicoverpa* resistance to chemical and biological insecticides. *Philos. Trans. R. Soc. London Ser. B* 353:1713–22
49. Heckel DG, Gahan LJ, Liu YB, Tabashnik BE. 1999. Genetic mapping of resistance to *Bacillus thuringiensis* toxins in diamondback moth using biphasic linkage analysis. *Proc. Natl. Acad. Sci. USA* 96:8373–77
50. Heinrich JC, Li X, Henry RA, Haack N, Stringfellow L, et al. 2002. Germ-line transformation of the Australian sheep blow fly *Lucilia cuprina*. *Insect Mol. Biol.* 11:1–10
51. Hoffman SL, Subramanian GM, Collins FH, Venter JC. 2002. *Plasmodium*, human and *Anopheles* genomics and malaria. *Nature* 415:702–9
52. Holloway AK, Strand MR, Black WC, Antolin MF. 2000. Linkage analysis of sex determination in *Bracon* sp. near *hebetor* (Hymenoptera: Braconidae). *Genetics* 154:205–12
53. Hoskins RA, Nelson CR, Berman BP, Laverty TR, George RA, et al. 2000. A BAC-based physical map of the major autosomes of *Drosophila melanogaster*. *Science* 287:2271–74
54. Hoskins RA, Phan AC, Naeemuddin M, Mapa FA, Ruddy DA, et al. 2001. Single nucleotide polymorphism markers for genetic mapping in *Drosophila melanogaster*. *Genome Res.* 11:1100–13
55. Howard DJ, Marshall JL, Hampton DD, Britch SC, Draney ML, et al. 2002. The genetics of reproductive isolation: a retrospective and prospective look with comments on ground crickets. *Am. Nat.* 159:S8–21
56. Hughes CL, Kaufman TC. 2000. RNAi analysis of *Deformed*, *proboscipedia* and *Sex combs reduced* in the milkweed bug *Oncopeltus fasciatus*: novel roles for *Hox* genes in the Hemipteran head. *Development* 127:3683–94
57. Hunt GJ, Collins AM, Rivera R, Page RE, Guzman-Novoa E. 1999. Quantitative trait loci influencing honeybee alarm pheromone levels. *J. Hered.* 90:585–89
58. Hunt GJ, Guzmán-Novoa E, Fondrk MK, Page RE. 1998. Quantitative trait loci for honey bee stinging behavior and body size. *Genetics* 148:1203–13
59. Hunt GJ, Page RE. 1995. Linkage map of the honey bee, *Apis mellifera*, based on RAPD markers. *Genetics* 139:1371–82
60. Hunt GJ, Page RE, Fondrk MK, Dullum CJ. 1995. Major quantitative trait loci affecting honey bee foraging behavior. *Genetics* 141:1537–45
61. Hwang JS, Lee JS, Goo TW, Lee SM, Kang HA, et al. 1998. Construction of linkage map of the silkworm, *Bombyx mori*, using RAPD markers. *Kor. J. Genet.* 20:147–53

62. Irving P, Troxler L, Heuer TS, Belvin M, Kopczynski C, et al. 2001. A genome-wide analysis of immune responses in *Drosophila*. *Proc. Natl. Acad. Sci. USA* 98:15119–24
62a. Ito J, Ghosh A, Moreira LA, Wimmer EA, Jacobs-Lorena M. 2002. Transgenic anopheline mosquitoes impaired in transmission of a malaria parasite. *Nature* 417:452–55
63. Jones CD. 2001. The genetic basis of larval resistance to a host plant toxin in *Drosophila sechellia*. *Genet. Res.* 78:225–33
64. Kennerdell JR, Carthew RW. 1998. Use of dsRNA-mediated genetic interference to demonstrate that *frizzled* and *frizzled 2* act in the *wingless* pathway. *Cell* 95:1017–26
65. Kennerdell JR, Carthew RW. 2000. Heritable gene silencing in *Drosophila* using double-stranded RNA. *Nat. Biotechnol.* 18:896–98
66. Kim SK. 2001. Functional genomics: The worm scores a knockout. *Curr. Biol.* 11:R85–87
67. Kimmerly W, Stultz K, Lewis S, Lewis K, Lustre V, et al. 1996. A P1-based physical map of the *Drosophila* euchromatic genome. *PCR Methods Appl.* 6:414–30
68. Koike Y, Mita K, Maeda S, Osoegawa K, DeJong P, Shimada T. 2003. Genomic sequence of 320 kb containing a *kettin* orthologue on the Z chromosome in *Bombyx mori*. *Mol. Gen. Genomics* 269:In press
69. Kucharski R, Maleszka R. 2002. Evaluation of differential gene expression during behavioral development in the honeybee using microarrays and northern blots. *Genome Biol.* 3:research0007.1-.9 http://genomebiology.com/2002/3/2/research/0007/
70. Laurent V, Wajnberg E, Mangin B, Schiex T, Gaspin C, Vanlerberghe-Masutti F. 1998. A composite genetic map of the parasitoid wasp *Trichogramma brassicae* based on RAPD markers. *Genetics* 150:275–82
71. Lehnert SA, Wilson KJ, Byrne K, Moore SS. 1999. Tissue-specific expressed sequence tags from the black tiger shrimp *Penaeus monodon*. *Mar. Biotechnol.* 1:465–76
72. Leips J, Mackay TFC. 2000. Quantitative trait loci for life span in *Drosophila melanogaster*: interactions with genetic background and larval density. *Genetics* 155:1773–88
73. Levashina EA, Moita LF, Blandin S, Vriend G, Lagueux M, Kafatos FC. 2001. Conserved role of a complement-like protein in phagocytosis revealed by dsRNA knockout in cultured cells of the mosquito, *Anopheles gambiae*. *Cell* 104:709–18
74. Mackay TFC. 2001. Quantitative trait loci in *Drosophila*. *Nat. Rev. Genet.* 2:11–20
75. Madueno E, Papagiannakis G, Rimmington G, Saunders RDC, Savakis C, et al. 1995. A physical map of the X chromosome of *Drosophila melanogaster*-cosmid contigs and sequence tagged sites. *Genetics* 139:1631–47
75a. Marie B, Bacon JP, Blagburn JM. 2000. Double-stranded RNA interference shows that *Engrailed* controls the synaptic specificity of identified sensory neurons. *Curr. Biol.* 10:289–92
76. McDonald MJ, Rosbach M. 2001. Microarray analysis and organization of circadian gene expression in *Drosophila*. *Cell* 107:567–78
77. Menzel R, Bogaert T, Achazi R. 2001. A systematic gene expression screen of *Caenorhabditis elegans* cytochrome P450 genes reveals *CYP35* as strongly xenobiotic inducible. *Arch. Biochem. Biophys.* 395:158–68
78. Mori A, Severson DW, Christenson BM. 1999. Comparative linkage maps for the mosquitoes (*Culex pipiens* and *Aedes aegypti*) based on common RFLP loci. *J. Hered.* 90:160–64

79. Mori A, Tomita T, Hidoh O, Kono Y, Severson DW. 2001. Comparative linkage map development and identification of an autosomal locus for insensitive acetylcholinesterase-mediated insecticide resistance in *Culex tritaeniorhynchus*. *Insect Mol. Biol.* 10:197–203
80. Mount DW. 2001. *Bioinformatics: Sequence and Genome Analysis*. Cold Spring Harbor, NY: Cold Spring Harbor Lab. Press. 564 pp.
81. Mutebi JP, Black WC, Bosio CF, Sweeney WP, Craig GB. 1997. Linkage map for the Asian tiger mosquito [*Aedes (Stegomyia) albopictus*] based on SSCP analysis of RAPD markers. *J. Hered.* 88:489–94
82. Nishikawa T, Natori S. 2001. Targeted disruption of a pupal hemocyte protein of *Sarcophaga* by RNA interference. *Eur. J. Biochem.* 268:5295–99
83. Nishimori Y, Lee JM, Sumitani M, Hatakeyama M, Oishi K. 2000. A linkage map of the turnip sawfly *Athalia rosae* (Hymenoptera: Symphyta) based on random amplified polymorphic DNAs. *Genes Genet. Syst.* 75:159–66
84. Nunamaker RA, Brown SE, McHolland LE, Tabachnick WJ, Knudson DL. 1999. First-generation physical map of the *Culicoides variipennis* (Diptera: Ceratopogonidae) genome. *J. Med. Entomol.* 36:771–75
85. Nuzhdin SV, Pasyukova EG, Dilda CL, Zeng ZB, Mackay TFC. 1997. Sex-specific quantitative trait loci affecting longevity in *Drosophila melanogaster*. *Proc. Natl. Acad. Sci. USA* 94:9734–39
86. Nuzhdin SV, Pasyukova EG, Morozova EA, Flavell AJ. 1998. Quantitative genetic analysis of *copia* retrotransposon activity in inbred *Drosophila melanogaster* lines. *Genetics* 150:755–66
87. Okano K, Shimada T, Mita K, Maeda S. 2001. Comparative expressed-sequence-tag analysis of differential gene expression profiles in BmNPV-infected BmN cells. *Virology* 282:348–56
88. Ote M, Mita K, Kawasaki H, Kinisho H, Seki M. et al. 2002. DNA microarray analysis of wing discs in *Bombyx mori* during development. *Abstracts, Fourth Int. Symp. Mol. Insect Sci.*, Tucson, p. 81 (Abstr.)
89. Page RE, Fondrk MK, Hunt GJ, Guzmán-Novoa E, Humphries MA, et al. 2000. Genetic dissection of honeybee (*Apis mellifera* L.) foraging behavior. *J. Hered.* 91:474–79
90. Page RE, Gadau J, Beye M. 2002. The emergence of hymenopteran genetics. *Genetics* 160:375–79
91. Pennisi E. 2001. Genome sequencing—insects rank low among genome priorities. *Science* 294:1261–62
92. Promboon A, Shimada T, Fujiwara H, Kobayashi M. 1995. Linkage map of random amplified polymorphic DNAs (RAPDs) in the silkworm, *Bombyx mori*. *Genet. Res.* 66:1–7
93. Ramalho-Ortigao JM, Temporal P, de Oliveira SMP, Barbosa AF, Vilela ML, et al. 2001. Characterization of constitutive and putative differentially expressed mRNAs by means of expressed sequence tags, differential display reverse transcriptase-PCR and randomly amplified polymorphic DNA-PCR from the sand fly vector *Lutzomyia longipalpis*. *Mem. Inst. Oswaldo Cruz* 96:105–11
94. Ranz JM, Caceres M, Ruiz A. 1999. Comparative mapping of cosmids and gene clones from a 1.6 Mb chromosomal region of *Drosophila melanogaster* in three species of the distantly related subgenus *Drosophila*. *Chromosoma* 108:32–43
95. Reddy KD, Abraham EG, Nagaraju J. 1999. Microsatellites in the silkworm, *Bombyx mori*: abundance, polymorphism, and strain characterization. *Genome* 42:1057–65
96. Robertson HM, Martos R, Sears CR, Todres EZ, Walden KKO, Nardi JB. 1999. Diversity of odourant binding proteins revealed by an expressed sequence tag

project on male *Manduca sexta* moth antennae. *Insect Mol. Biol.* 8:501–18
97. Roethele JB, Feder JL, Berlocher SH, Kreitman ME, Lashkari DA. 1997. Toward a molecular genetic linkage map for the apple maggot fly (Diptera: Tephritidae): comparison of alternative strategies. *Ann. Entomol. Soc. Am.* 90:470–79
98. Roethele JB, Romero-Severson J, Feder JL. 2001. Evidence for broad-scale conservation of linkage map relationships between *Rhagoletis pomonella* (Diptera: Tephritidae) and *Drosophila melanogaster* (Diptera: Drosophilidae). *Ann. Entomol. Soc. Am.* 94:936–47
99. Rubin GM, Hong L, Brokstein P, Evans-Holm M, Frise E, et al. 2000. A *Drosophila* complementary DNA resource. *Science* 287:2222–24
100. Rubin GM, Lewis EB. 2000. A brief history of *Drosophila*'s contributions to genome research. *Science* 287:2216–18
101. Schoppmeier M, Damen WGM. 2001. Double-stranded RNA interference in the spider *Cupiennius salei*: the role of *Distal-less* is evolutionarily conserved in arthropod appendage formation. *Dev. Genes Evol.* 211:76–82
102. Schug MD, Wetterstrand KA, Gaudette MS, Lim RH, Hutter CM, Aquadro CF. 1998. The distribution and frequency of microsatellite loci in *Drosophila melanogaster*. *Mol. Ecol.* 7:57–70
103. Schulte SJ, Rider SD, Hatchett JH, Stuart JJ. 1999. Molecular genetic mapping of three X-linked avirulence genes, *vH6*, *vH9* and *vH13*, in the Hessian fly. *Genome* 42:821–28
104. Severson DW, Brown SE, Knudson DL. 2001. Genetic and physical mapping in mosquitoes: molecular approaches. *Annu. Rev. Entomol.* 46:183–219
105. Severson DW, Mori A, Zhang Y, Christensen BM. 1993. Linkage map for *Aedes aegypti* using restriction fragment length polymorphisms. *J. Hered.* 84:241–47
106. Severson DW, Thathy V, Mori A, Zhang Y, Christensen BM. 1995. Restriction fragment length polymorphism mapping of quantitative trait loci for malaria parasite susceptibility in the mosquito *Aedes aegypti*. *Genetics* 139:1711–17
107. Severson DW, Zaitlin D, Kassner VA. 1999. Targeted identification of markers linked to malaria and filarioid nematode parasite resistance genes in the mosquito *Aedes aegypti*. *Genet. Res.* 73:217–24
108. Shaw KL, Parsons YM. 2002. Divergence of mate recognition behavior and its consequences for genetic architectures of speciation. *Am. Nat.* 159:S61–75
109. Shaw PJ, Salameh A, McGregor AP, Bala S, Dover GA. 2001. Divergent structure and function of the *bicoid* gene in Muscoidea fly species. *Evol. Dev.* 3:251–62
110. Shi JR, Heckel DG, Goldsmith MR. 1995. A genetic linkage map for the domesticated silkworm, *Bombyx mori*, based on restriction fragment length polymorphisms. *Genet. Res.* 66:109–26
111. Shukle RH, Stuart JJ. 1995. Physical mapping of DNA sequences in the Hessian fly, *Mayetiola destructor*. *J. Hered.* 86:1–5
112. Spradling AC, Stern D, Beaton A, Rhem EJ, Laverty T, et al. 1999. The Berkeley *Drosophila* Genome Project gene disruption project: single P-element insertions mutating 25% of vital *Drosophila* genes. *Genetics* 153:135–77
113. Stauber M, Taubert H, Schmidt-Ott U. 2000. Function of *bicoid* and *hunchback* homologs in the basal cyclorrhaphan fly *Megaselia* (Phoridae). *Proc. Natl. Acad. Sci. USA* 97:10844–49
114. Stollewerk A, Weller M, Tautz D. 2001. Neurogenesis in the spider *Cupiennius salei*. *Development* 128:2673–88
115. Swanson WJ, Clark AG, Waldrip-Dail HM, Wolfner MF, Aquadro CF. 2001. Evolutionary EST analysis identifies rapidly evolving male reproductive proteins in *Drosophila*. *Proc. Natl. Acad. Sci. USA* 98:7375–79
116. Takeda M, Mita K, Quan GX, Shimada

T, Okano K, et al. 2001. Mass isolation of cuticle protein cDNAs from wing discs of *Bombyx mori* and their characterizations. *Insect Biochem. Mol. Biol.* 31:1019–28
117. Tan YD, Wan CL, Zhu YF, Lu C, Xiang ZH, Deng HW. 2001. An amplified fragment length polymorphism map of the silkworm. *Genetics* 157:1277–84
118. Theodorides K, De Riva A, Gómez-Zurita J, Foster PG, Vogler AP. 2002. Comparison of EST libraries from seven beetle species: towards a framework for phylogenomics of the Coleoptera. *Insect Mol. Biol.* 11:In press
119. True JR, Liu JJ, Stam LF, Zeng ZB, Laurie CC. 1997. Quantitative genetic analysis of divergence in male secondary sexual traits between *Drosophila simulans* and *Drosophila mauritiana. Evolution* 51:816–32
120. Vieira C, Pasyukova EG, Zeng ZB, Hackett JB, Lyman RF, Mackay TFC. 2000. Genotype-environment interaction for quantitative trait loci affecting life span in *Drosophila melanogaster. Genetics* 154:213–27
121. Vieira J, Vieira CP, Hartl DL, Lozovskaya ER. 1997. A framework physical map of *Drosophila virilis* based on P1 clones: applications in genome evolution. *Chromosoma* 106:99–107
122. Waterston RH, Lander ES, Sulston JE. 2002. On the sequencing of the human genome. *Proc. Natl. Acad. Sci. USA* 99:3712–16
123. White KP, Rifkin SA, Hurban P, Hogness DS. 1999. Microarray analysis of *Drosophila* development during metamorphosis. *Science* 286:2179–84
123a. Whitfield CW, Band M, Bonaldo MF, Kumar CG, Liu L et al. 2002. Annotated expressed sequence tags and cDNA microarrays for studies of brain and behavior in the honey bee. *Genome Res.* 12: 555–66
124. Wilson K, Li YT, Whan V, Lehnert S, Byrne K, et al. 2002. Genetic mapping of the black tiger shrimp *Penaeus monodon* with amplified fragment length polymorphism. *Aquaculture* 204:297–309
125. Wu C, Asakawa S, Shimizu N, Kawasaki S, Yasukochi Y. 1999. Construction and characterization of bacterial artificial chromosome libraries from the silkworm, *Bombyx mori. Mol. Gen. Genet.* 261:698–706
126. Yasukochi Y. 1998. A dense genetic map of the silkworm, *Bombyx mori*, covering all chromosomes based on 1018 molecular markers. *Genetics* 150:1513–25
127. Zeng ZB, Liu JJ, Stam LF, Kao CH, Mercer JM, Laurie CC. 2000. Genetic architecture of a morphological shape difference between two *Drosophila* species. *Genetics* 154:299–310
128. Zheng LB, Benedict MO, Cornel AJ, Collins FH, Kafatos FC. 1996. An integrated genetic map of the African human malaria vector mosquito, *Anopheles gambiae. Genetics* 143:941–52
129. Zheng LB, Cornel AJ, Wang R, Erfle H, Voss H, et al. 1997. Quantitative trait loci for refractoriness of *Anopheles gambiae* to *Plasmodium cynomolgi* b. *Science* 276:425–28
130. Zou S, Meadows S, Sharp L, Jan LY, Jan YN. 2000. Genome-wide study of aging and oxidative stress response in *Drosophila melanogaster. Proc. Natl. Acad. Sci. USA* 97:13726–31

Annu. Rev. Entomol. 2003. 48:261–81
doi: 10.1146/annurev.ento.48.091801.112639
First published online as a Review in Advance on August 27, 2002

MANAGEMENT OF AGRICULTURAL INSECTS WITH PHYSICAL CONTROL METHODS*

Charles Vincent,[1] Guy Hallman,[2] Bernard Panneton,[1] and Francis Fleurat-Lessard[3]

[1]*Horticultural Research and Development Centre, Agriculture and Agri-Food Canada, 430 Gouin Blvd., Saint-Jean-sur-Richelieu, Quebec, Canada J3B 3E6; e-mail: vincentch@agr.gc.ca; pannetonb@agr.gc.ca*

[2]*USDA-ARS, 2413 East Highway 83, Weslaco, Texas 78596; e-mail: ghallman@weslaco.ars.usda.gov*

[3]*Laboratory for Post-Harvest Biology and Technology, INRA, 71 Edouard Bourleaux Avenue, P. O. Box 81, F-33883 Villenave d'Ornon, France; e-mail: francis.fleurat-lessard@bordeaux.inra.fr*

Key Words integrated pest management, mechanical control, pneumatic control, thermal control, radio frequency, impacting machine

■ **Abstract** Ideally, integrated pest management should rely on an array of tactics. In reality, the main technologies in use are synthetic pesticides. Because of well-documented problems with reliance on synthetic pesticides, viable alternatives are sorely needed. Physical controls can be classified as passive (e.g., trenches, fences, organic mulch, particle films, inert dusts, and oils), active (e.g., mechanical, polishing, pneumatic, impact, and thermal), and miscellaneous (e.g., cold storage, heated air, flaming, hot-water immersion). Some physical methods such as oils have been used successfully for preharvest treatments for decades. Another recently developed method for preharvest situations is particle films. As we move from production to the consumer, legal constraints restrict the number of options available. Consequently, several physical control methods are used in postharvest situations. Two noteworthy examples are the entoleter, an impacting machine used to crush all insect stages in flour, and hot-water immersion of mangoes, used to kill tephritid fruit fly immatures in fruit. The future of physical control methods will be influenced by sociolegal issues and by new developments in basic and applied research.

CONTENTS

INTRODUCTION

There is a need to reduce the negative impacts of pest control methods on the environment. Increased concerns about the potential effects of pesticides on health, the reduction in arable land per capita (79), and the evolution of pest complexes likely to be accelerated by climate changes also contribute to change in plant protection practices. Insecticides are still widely used; however, more than 540 insect species are resistant to synthetic insecticides (71). Other drawbacks of synthetic insecticides include resurgence and outbreaks of secondary pests and harmful effects on nontarget organisms (82). This situation creates a demand for alternative control methods, including physical controls.

Metcalf et al. (72) wrote that physical controls "are in general costly in time and labor, often do not destroy the pest until much damage has been done, and rarely give adequate or commercial control." However, recent advances in physical controls and the restrictions placed on many chemical controls have resulted in a notable increase in research and application of physical controls. This review complements the work of Banks (6), Hallman & Denlinger (48), Oseto (80), and Vincent et al. (101) by offering a critical assessment of physical controls with the objective of formulating recommendations for further research and applications.

CONSIDERATIONS ON PHYSICAL CONTROL METHODS IN RELATION TO AGRICULTURAL PLANT PROTECTION

In physical control methods, the physical environment of the pest is modified in such a way that the insects no longer pose a threat to the agricultural crop. This can be achieved by generating stress levels ranging from agitation to death or by using devices such as physical barriers that protect produce or plants from infestation.

Many physical control methods target an ensemble of physiological and behavioral processes, whereas chemical methods have well-defined and limited modes of action.

In this review, physical control methods are grouped under two main classes, passive and active; a miscellaneous category groups those that do not readily fit this classification (follow the Supplemental Material link in the online version of this chapter or at http://www.annualreviews.org/). The active class is further subdivided into mechanical, thermal, and electromagnetic techniques. Passive methods do not require additional input after establishment to be effective over a given period. The efficacy of active methods depends on continued input over the period of control. The level of control achieved is related to the amount and intensity of the input.

Effective physical control methods protect plants during the entire season from emergence to postharvest. However, postharvest conditions are better suited to physical control methods because the environment is rather confined, the material is of high economic value, and the use of insecticides is frequently inappropriate or even unlawful. Physical control methods, such as cold, heat, and ionizing radiation, are used extensively as postharvest quarantine treatments where disinfestation of a given pest at a predetermined level of control must be achieved (47).

PASSIVE METHODS

Physical barriers may be defined as any living or nonliving material used to restrict movements or to delineate a space. They encompass a number of methods compatible with several other control methods (12). The economics of deploying barriers is closely related to the spatial scale. In that respect, it is easier to protect a stored product than a crop grown over large field areas. In the field, one major challenge is to deploy either degradable or nondegradable barriers that can be dismantled, and possibly reused, at low cost.

Trenches

Trenches to intercept walking insects such as the chinch bug were implemented as early as 1895 (72). Recently, several papers relevant to the Colorado potato beetle/potato system were published (12). A V-shaped trench lined with plastic retained up to 95% of Colorado potato beetles (74). The efficiency of the method depends on the density of the overwintered beetles and on the proportion of flying versus walking individuals (103) as well as on physical characteristics of the trenches. The furrow should be at least 25 cm deep with sides sloping at angles >45°. Adults that fall in trenches covered with dust have little chance of escaping because they cannot walk up the dusty walls and because they rarely fly before walking up to the top of plants or structures (12). Rainfall has little effect on the efficacy of the trench to retain beetles. A machine has been designed to install the plastic trenches (74). In field conditions of New Brunswick, Canada, a reduction of 48% in immigrating (overwintered) adults was observed. For an 8-ha field, the cost is recovered if one insecticide treatment is saved. An aboveground version has comparable efficiency, can be reused up to 10 years, and is suited for small fields of high-value crops (11). The rate of adoption of this technology has steadily increased in the 1990s, but interest decreased as an effective insecticide and transgenic plants appeared on the market (12).

Fences

Fencing is particularly relevant to exclude low-flying insects (e.g., anthomyiids) from annual crops where few chemicals are available and the crop value is high (e.g., onion and cole crops) (12). Fences 1 m high exclude 80% of flying female cabbage flies, *Delia radicum*. Height of the fences is critical and is limited by cost and resistance to wind. Although cabbage maggot flies can be captured up to 180 cm above ground level, Vernon & Mackenzie (98) adopted fences 90 cm high as an optimal fencing method. Overhangs (25 cm) decreased cabbage maggot fly, *Delia radicum*, trap catches inside the fenced plots and reduced damage to the crop (14). If vegetable crops are strategically segregated by fences and properly rotated, the effectiveness of exclusion fences improves over time, partly because the fence congregates natural enemies of anthomyiid adults. One drawback of fencing is that excluded individuals that are good flyers can attack a nearby but unprotected crop. Also, individuals that overcome the barrier and are confined within the enclosed area may damage the fenced crop.

Organic Mulch

Straw mulch indirectly affects Colorado potato beetle populations and significantly reduces damage (108) by favoring several species of its egg and larval predators: *Coleomegilla maculata*, *Hippodamia convergens*, *Chrysopa carnea*, and *Perillus bioculatus* (16). Although the yield of potato fields is higher in mulched than in nonmulched plantings or when straw mulch is incorporated into an insecticide program (16), the cost of using straw mulch may be prohibitive to nonorganic growers (30). The interaction of straw mulch and *Bacillus thuringiensis* subsp. *tenebrionis* sprays is positive and gives results comparable to insecticidal treatments because the two technologies do not interfere with one another (17). Straw mulch is agronomically and environmentally sound and can be useful as part of an insecticide resistance management program.

Mulches from Artificial Materials

Various protective materials such as paper or plastic sheets or aluminized films can be used for mulching. The primary objective of mulching is to improve productivity and reach harvest at an early date. It is usually used for high-value crops. Mulches from artificial materials can be designed for pest control. For example, plastic materials can be of such color as to modify the spectrum of incident light to alter a given insect behavior. Thrips are attracted to blue, black, and white (24), and aphids to yellow and blue (9, 24). Aluminized materials can attract some insect species while repelling others (7). The repellent properties have been related to the reflection of ultraviolet light at wavelengths <390 nm (62). In strawberry fields, reflective mulch has shown some potential by increasing productivity of the plants and by reducing damage from the tarnished plant bug, *Lygus lineolaris* (92). The use of mulches should be studied and implemented using a system approach looking not only at the impact on insect pests but also at the impact on weeds, other insects, diseases, nematodes, and yield. If one component (e.g., early harvest) provides an economic justification, then there is an opportunity for designing mulch that will have a positive impact on other segments such as insect control. Machinery to extract and roll mulch films is becoming available, and photobiodegradable materials are being developed (7).

Particle Films

Road dust drifting on crops can have a negative effect on natural enemies (29). The recent development of sprayable formulations of kaolin under the generic name "particle film technology" (43) fueled interest in this method by showing broader insecticidal activity. Several mechanisms are at play. Pear psylla, *Cacopsylla pyricola*, adults confined on a treated [hydrophobic particle film (PF)] surface become coated with tiny particles that interfere with visual cues (89). Adult behavior is disrupted to the point where they are unable to feed. Spirea aphids, *Aphis spiraecola*, lost footing and fell off the treated plant, and damage by the potato leafhopper, *Empoasca fabae*, was significantly reduced (43). Hydrophobic PF also deterred feeding and oviposition of the citrus root weevil, *Diaprepes abbreviatus* (65). Kaolin sprays reduced neonate walking speed, which reduced the rate at which neonates infested fruit, and oviposition by female codling moth, *Cydia pomonella* (96), and oblique-banded leafroller, *Choristoneura rosaceana* (58).

Hydrophobic PF have been superseded by hydrophilic PF. They have the same effects as mentioned above on pear psylla (89). One commercial formulation, Surround®, represents approximately 30% of all insecticides used in pear in the United States (G. Puterka, personal communication). In pear field trials, hydrophilic PF significantly reduced pear psylla populations and oviposition, plum curculio, *Conotrachelus nenuphar*, oviposition scars, but not codling moth damage. The treatments also increased fruit quality and yield. A potential limit of PF is the adhesion of films under heavy rain. This has been addressed by the development of new formulations, notably by using a custom spreader-sticker. Suggested negative impact of PF on natural enemies (96, 58) should be researched. Standard spraying equipment is used, and growers do not have to invest in specially dedicated machines. Owing to the wide range of insect taxa affected by PF and plant diseases (89), and to their positive effects on fruit physiology and quality (42), there should be more scientific and commercial development in PF in the near future.

Inert Dusts

There has been substantial research and development on inert dusts for two decades, resulting in the registration and commercial use of several inert dust formulations

(32, 44, 60). There are many kinds of inert dusts: lime, common salt, sand, kaolin, paddy husk ash, wood ash, clays, diatomaceous earths (ca. 90% SiO_2), synthetic and precipitated silicates (ca. 98% SiO_2), and silica aerogels (44). Because of their low mammalian toxicity, they are used to protect stored grains against a number of coleopteran pests. Diatomaceous earth is classified as a "generally recognized as safe" (GRAS) food additive by the U.S. Environmental Protection Agency (60). Inert dusts exert their effects slowly through several mechanisms that result in dehydration, notably by adsorption of cuticular lipids and, less importantly, by abrasion (29, 60). As some insects move into the grain storage area or within the stored grain, behavior also is a factor (32). There are large behavioral and physiological differences in susceptibility among species. Because of their mode of action, high relative humidity ($>$70%) or $>$14% water content in stored grain reduces their insecticidal effect. Diatomaceous earth collected from various parts of the world shows differences in diatom species, physical properties, and insecticidal efficacy (60), thus complicating the standardization of commercial formulations. Effective diatomaceous earths have $>$80% SiO_2, a pH $<$8.5, and a tapped density (a technical term refering to the specific mass for finely ground powder such as diatomaceous earth, measured from standard "tap flow" withdrawal from a bulk into 1 L vessel) of below 300 $g.L^{-1}$ (59). Problems associated with the use of diatomaceous earth in large scale operations are (*a*) machine abrasion; (*b*) a reduction in the bulk density (hectoliter mass), which is a measure of grain quality; (*c*) grain fluidity; (*d*) decrease in qualities such as color and presence of foreign material; and (*e*) health hazards (respiratory diseases) (44). Slurry formulations mitigate the latter problem for the treatment of structures (32). Korunic & Ormesher (61) demonstrated that, when exposed to diatomaceous earth for 5–7 generations, populations of *Tribolium castaneum*, *Cryptolestes ferrugineus*, and *Rhizopertha dominica* became less susceptible. More experiments should be done to make a definitive statement concerning the development of resistance by stored-product insects.

Trapping

As a management tool, perimeter trapping has been successful in intercepting flying dipterans that invade orchards from neighboring host plants [*Rhagoletis pomonella* in apple orchards (88, 15) and *Ceratitis capitata* in plum, persimmon, and pear orchards (22)]. Factors contributing to success are low infestation of pests, high density of traps, availability of an attractant, absence of nearby host plants, and trap maintenance. The use of dry, nonsticky traps (22) facilitates operations and lowers costs. It is likely that perimeter trapping will be researched and implemented in other crop and geographical situations. Mass trapping of stored-product moths has been successful with low densities of pyralid moths in storage buildings (i.e., in a tobacco warehouse infested by *Ephestia elutella*) (34, 41).

Oils

Mineral oils have been used alone or in combination with synthetic insecticides for a century to control soft-bodied arthropod pests of fruit trees. To date, no resistance has been reported. Although oils act primarily at contact sites by obstruction of the respiratory system (hypoxia), they may also act as an oviposition repellent. Several arthropod taxa are affected, e.g., mites, scale insects, mealybugs, psyllids, aphids, leafhoppers, and some lepidopteran pests such as eggs of codling moths. Because they have low residual activity, they are relatively harmless to beneficials. The most important factors explaining the pesticidal performance of oil formulations are chemical composition, paraffin (optimally of molecular weight C_{20}–C_{25}) and unsaturated compounds, and the equivalent n-paraffin carbon number (56). To minimize damage when applying oil sprays, it is recommended to avoid spraying when trees are stressed or when temperatures are too high or too low (25). Mineral oil is a reliable physical control method that is still evolving today. For example, horticultural mineral oils are optimized to have both good adjuvant and insecticidal activities and off-target spray drift management (56). Research and development of vegetable oil formulations to control arthropods are ongoing (40) and promising, particularly in organic produce markets.

Surfactants and Soaps

Surfactants may have direct or indirect effects on soft-bodied arthropods. For example, Cowles et al. (23) showed that trisiloxane, generally considered as an inert ingredient, either suffocates or disrupts important physiological processes in the two-spotted spider mite, *Tetranychus urticae*. Owing to their surfactant properties, they work as soaps, presumably allowing interaction of water with arthropod cuticles and causing drowning by permitting water to infiltrate tracheae or peritremes. They also may impair nerve cell functions (54). When assayed against citrus leafminer, *Phyllocnistis citrella*, larvae, Silwet L-77 (an organosilicone molecule) had insecticidal effects alone, and because of its surfactant effect, it might increase the insecticidal effect of *Bacillus thuringiensis* var. *kurstaki* (93). Soaps have been used to kill soft-bodied insects, and their mode of action and low residual activity resemble those of surfactants. Not all soaps have insecticidal properties and, consequently, special formulations with optimal properties should be used. Several insecticidal soaps are registered and commercially available, mainly for urban markets, as found on the World Wide Web.

ACTIVE METHODS

Mechanical

CLEANING Cleaning is a common postharvest treatment. When prescribed as a quarantine treatment, it is usually followed by inspection by the importing regulatory agencies to ensure that the cleaning was successful in removing undesired pests and debris. Cleaning alone may be insufficient and often needs to be followed by another treatment.

One variation of cleaning is the soapy water and wax treatment accepted by the United States against the false spider mite, *Brevipalpus chilensis*, in Chile. It consists of immersion of cherimoyas and limes in soapy water for 20 sec, a rinse, and then immersion in a wax coating (3). The wax physically immobilizes any mites that remain after washing. This treatment shows promise against a number of small, surface-infesting pests of fruits.

PRESSURE Baling hay at a pressure of 10.3 MPa for one day killed 100% of cereal leaf beetle, *Oulema melanopus* (106). However, compression of hay must currently be accompanied by phosphine fumigation (2.12 g/m^3 for 3 days at >21°C) to be accepted by Canada. Baling hay at 7.85 MPa plus fumigation with phosphine (2.12 g/m^3 for 3 days at >21°C) for 7 days satisfies Japanese requirements of hay at risk for importation of Hessian fly, *Mayetiola destructor* (107).

POLISHING Polishing is an industrial process consisting of rubbing off the pericarp of rice grains. Polishing to a weight loss of 11% of rice grains causes 40% acute mortality in rice weevil, *Sitophilus oryzae*, eggs, followed by another 40% mortality owing to the poor suitability of polished rice grains (67). Experiments on the joint use of rice polishing and two pteromalid parasites concluded that *Lariophagus distinguendus* is less affected by polished rice than *Anisopteromalus calandrae* (68).

SOUND Sounds at frequencies <20 Hz can be defined as infrasound, and ultrasound at frequencies higher (>~16 kHz) than human audibility (105). Sound propagating in a medium is attenuated at a rate approximately proportional to frequency. Ultrasounds propagate well under water but not in air.

All insects contain microscopic stable gas bodies that can oscillate under the influence of ultrasound. Abnormal development of *Drosophila melanogaster* resulted from these oscillations (105). Studies by Belton (8) on the use of ultrasound to protect corn from corn borers and those by Payne & Shorey (83) on the effect of ultrasound on oviposition by cabbage loopers showed promise. Ultrasonic pest control devices are available on the market; however, claims supporting ultrasonic devices in the elimination of insect pests are baseless (http://www.ent.iastate.edu/ipm/hortnews/2001/8-24-2001/ultrasonic.html).

As ultrasound transmits well through water, its use in postharvest while produce is being washed by immersion in water could be easily implemented. However, this approach was not effective in treating asparagus spears for thrips (97).

Acoustical sensors can be used to automatically monitor insect populations in stored grain. Timely insect detection and control can improve food safety and reduce the use of insecticides (52). Ultrasonic waves successfully controlled *Sitophilus granarius* adults inside a wheat grain mass (87).

STALK DESTRUCTION In central Texas, stalk shredding of cotton plants before winter results in 85%–90% larval mortality of pink bollworm, *Pectinophora gossypiella* (2). Winter burial of stalk by plowing causes 75%–80% larval mortality. Physical methods, combined with early crop production practices, allow management of the pest at acceptable population levels (2). In Nigeria, burning the stalk after harvest can destroy 100% of larvae of the sorghum stem borer, *Busseola fusca* (1). However, the practice is not implemented because the peasants use the stalk for other purposes, e.g., roofing material. Partial burning immediately after harvest to cure the stalk kills approximately 95% of larvae and would allow the use of stalk as roofing material.

PNEUMATIC Insect pests can be dislodged from plants using blown or aspirated air (57). Blowing uses energy more efficiently to dislodge insects from plants, but blown insects must be collected. Work has recently been done on *Lygus* spp. on strawberry, Colorado potato beetle (99, 63), *Liriomyza trifolii*, and *L. huidobrensis* on celery, and *Bemisia tabaci* on melons (104). Mobile insects, such as *Lygus* or *Bemisia*, are more likely to be efficiently removed than insects that cling to plants, such as Colorado potato beetle adults and larvae (73). Aspirating inlets can be located closer to plants that are flexible, resulting in a more efficient aspiration (100). Several points need to be studied if the efficiency of pneumatic technology is to be improved. Engineering should focus on aerodynamics, i.e., airflow rate and speed, design of inlets and design of control units. Work rate (ha per h) must be optimized (100). Research on insect behavior is also the key to improving efficiency. For example, both nymphs and adults of the tarnished plant bug, *Lygus lineolaris*, are present on strawberry at any time of the day and can be vacuumed at any time (90). Pneumatic control is nonspecific, and its effect on beneficials needs to be evaluated. In potato fields, it negatively affects diurnal predators belonging to the taxa Arachnida, Chrysopidae, and Coccinellidae, while having no effect on generalist-nocturnal predators (e.g., Carabidae and harvestmen) (63). No significant effect was measured for the *B. tabaci* parasites *Eretmocerus mundus* and *Encarsia lutea* (104). Shortly before the passage of a vacuum machine, 19% of pollinators flew away and, of the individuals remaining on the plants, 61% were aspirated (21). In potato fields, several passes of a vacuum machine did not spread the PSTV (potato spindle tuber viroid) and the PVX (potato mosaic virus) viruses, even if large proportions of infected plants were treated (10).

MECHANICAL IMPACT Disinfestation of wheat grain or wheat flour in mills by impacting machines (entoleter) has been routinely used for more than 50 years in the flour industry to destroy all insect stages (86).

Thermal

Temperature control is widely used in postharvest to slow down degradation of produce caused by physiological processes, pathogens, and insects. For control of insects, both high (37) and low temperatures (30a) can be effective. Temperature, rate of temperature change, and duration of exposure are contributing factors. The biological effects of temperature can be summarized on a thermobiological scale (37). In the field, thermal control is more complicated to implement because heat transfer is difficult to control and the differences in thermosensitivity between crops and pests can be subtle (64). The use of kinetic models to describe intrinsic thermal mortality is a promising approach to implement high-temperature short-time thermal treatments in a quarantine context (95).

COLD STORAGE One of the oldest and most widely used quarantine treatments is storage of fresh commodities at −0.6° to 3.3°C for 7 to 90 days, depending on the pest and temperature. It is used on a wide variety of fruits and vegetables (3). Advantages of cold treatment are its tolerance by a wide range of commodities, including many tropical fruits, and the fact that some fruits, such as apples, are stored for long periods of time at low temperatures lethal to insects to increase the marketing season. Also, unlike most other treatments, cold storage can be applied after the commodities have been packed and in slow transit such as in ships. The chief disadvantage is the length of the treatment period required.

Freezing is used for commodities that will be processed, such as fruit pulp as well as fresh fruit and vegetables directed to the consumer market. Freezing for at least one day will generally kill most insects that are not in diapause. Quick freezing at temperatures ≤ -15°C usually will kill diapausing insects.

HEATED AIR Quarantine treatments using heated air were used during the first Mediterranean fruit fly infestation in Florida in 1929. These treatments expose commodities to air at temperatures in the range of 43°–52°C from a few to many hours. The treatment can be done in one third the time if air is forced through the fruit load. The speed of treatment is dependent on temperature, size of individual commodities, density and arrangement of the load, air speed through the load, and moisture content of the air. These factors, combined with the generation of heat shock proteins modifying the susceptibility of both the insect and the living commodity to heat (46), contribute to the variability in the results of heated-air treatments in terms of efficacy and commodity quality. Preconditioning of fruits by a mild heat treatment before the pesticidal treatment reduces damage to commodities. However, that same pretreatment may also make the pest more tolerant of the pesticidal treatment (51). Heated-air treatments are generally not well tolerated by temperate fruits.

Dry air treatments at temperatures higher than those used for fresh produce (up to 100°C) are used to treat agricultural products such as meal, grain, straw, and dried plants for ornamental purposes and construction (such as pallets).

HOT-WATER IMMERSION Immersion at 43°–55°C ranging from a few minutes to a few hours is used to kill a variety of arthropods and nematodes on plant-propagative materials. Hot-water immersion is a simple, economical, and rapid treatment. It has been used to disinfest tephritid fruit flies virtually from all mangoes entering the United States since 1987 (3). Many fruits, especially temperate ones, are damaged when immersed in water hot enough to kill insects. This damage has been alleviated in some cases by gradual heating (70). As with heated-air treatments, tolerance of hot-water immersion can be increased by exposing some fruit to 25°–46°C for 0.5–72 h prior to the heat quarantine treatment (55). A variation of hot-water immersion is a hot-water drench, which is proposed to cause less damage to fruits than total immersion (13).

FLAMING When treating with a propane flamer, thermosensitivity of the target insect and the potato plant must be such as to allow the destruction or impairment of the pests without harming the crop (28). Temperature of exposure is used as an indicator of thermosensitivity, e.g., 70°C for Colorado potato beetle adults. Eggs, larvae, and adults are either injured or killed. Leaf shielding lowers egg mortality rates by 5%–20% and, for a given temperature and exposure duration, mortality is higher in spring than later when the canopy is developed. Because flaming is associated with soil compaction due to the need for repeating treatments, has no residual activity in the field, and above all, has negative environmental effects (i.e., release of combustion by-products: CO, CO_2, nitrous and sulphur oxides) that can be as important as those associated with pesticide use (64), its potential as an environmentally friendly alternative is questionable.

STEAMING In the field, the effects of steaming on insects are similar to those of flaming. Based on the fact that legs can be damaged by exposing them to temperatures from 68° to 75°C and that muscles of the legs are inactivated by dipping the insect from 0.2 to 0.4 sec in hot water (84), steam has been researched in laboratory and field conditions to impair locomotion of Colorado potato beetle adults (85). The proportion of adults impaired is positively correlated with temperature and exposure time. Because only 35% of adults are critically injured at the maximum acceptable temperature (ca. 79°C) for the potato plant (85), it is unlikely that this method could be of commercial use. Furthermore, steaming requires a large water supply that increases soil compaction, equipment, and operational costs (64).

SOLAR HEATING A solar heater made of dark cloth and translucent plastic sheeting has been tested for bruchid control in stored grains (75). The solar heater reaches temperatures >60°C. As all stages of the cowpea weevil, *Callosobruchus maculatus*, inside grains are destroyed at temperature >60°C for approximately 100 min, this method should prove useful, particularly for developing countries where the cost of energy is prohibitive. All stages of *Callosobruchus* spp. were killed when pigeonpea, *Cajanus cajan*, was solarized in polyethylene bags in India (20). Grain germination was minimally affected by solarization in the these two studies.

Electromagnetic Radiation

Electromagnetic radiation transfers energy from a source to a target without a need for an energy transfer fluid. Electromagnetic energy can be absorbed by ionizing atoms at the target site or by inducing vibration of charged particles within the matter, thus increasing temperature because of internal friction (66).

IRRADIATION Ionizing radiation provided by cobalt 60, cesium 137, or linear accelerators is an effective quarantine treatment that has a different measure of efficacy than all other treatments that have been used commercially (47). Heat produced by ionizing radiation does not contribute to insect control. To provide acute mortality of insects (the measure of efficacy of most quarantine treatments) with radiation requires higher doses than those tolerated by fresh commodities. However, radiation is effective at stopping development or providing sterility at doses tolerated by fresh commodities, and prevention of the establishment of exotic organisms does not require acute mortality. In 1995, papayas and other fruits began to be shipped from Hawaii to the mainland United States for irradiation and marketing. In August 2000, an electron beam facility built in Hawaii began treating and shipping fruits, and shipment of fruits for irradiation on the mainland ceased. In 1999, guavas irradiated against infestation by the Caribbean fruit fly began to be shipped from Florida to Texas and California; in 2000, Florida sweet potatoes irradiated against sweetpotato weevil began to be shipped to California. On limited occasions, South Africa has irradiated imported grapes and other fresh commodities that were considered a phytosanitary risk. Use of the treatment is expected to rise strongly if APHIS approves irradiation protocols for imported fruits first proposed in May 2000 (4).

RADIO FREQUENCY HEATING The radio frequency (RF) part of the electromagnetic spectrum spans roughly from 3 kHz to 300 GHz. These are nonionizing waves. RF transfers energy faster and more efficiently than heated air or water treatments. Although RF energy has been known to kill insects for more than 70 years and much research has been conducted on its effects (49, 50, 77), it has rarely been used on a commercial scale as an insect-control technique. RF heating would most likely work as a quarantine treatment against pests in dried products, such as walnuts (102), where the insect would have a much higher moisture content than its host and thus be more susceptible to RF heating, especially at lower frequencies of 10–100 MHz (78). Furthermore, dried products are more tolerant than fresh produce to the high temperature spikes that may occur with RF heating. When RF heating is studied as a quarantine treatment for pests of fresh commodities, many complicating factors affecting efficacy arise, such as moisture on the fruit surface that may conflict with microwave energy coupling, whether the insects were on the surface of the fruit or right under the surface, or minor differences in fruit size (53). Major challenges in developing effective RF quarantine treatments are providing uniform heating throughout the commodity and developing means to monitor and control end product temperature (95).

Little has been published on the effects of RF heating on insect physiology and histology. A notable exception is the finding that, after sublethal exposures, the yellow mealworm, *Tenebrio molitor*, last instar larvae and pupae produced adults at the end of their developmental cycle with malformations caused by the overheating of heat-sensitive tissues and cells (such as those forming imaginal discs) before ecdysis (39).

INFRARED HEATING Infrared radiation can disinfest grain provided that the product is exposed in a thin (ca. 2 cm) layer (19).

Miscellaneous Treatments

FLOODING Flooding is used as a standard agronomic practice in cranberry production, and its insecticidal value against a number of insects was recognized more than 70 years ago. Two types of management are used in cranberry plantations (5). "Early water" is defined as a bed where the winter flooding, used to protect the plant from winter injury, is removed in March without further flooding. "Late water" is flooding for 30 days from mid-April to mid-May to manage cranberry fruitworm, *Acrobasis vaccinii*, southern red mite, *Oligonychus ilicis*, and early-season cutworms. Late water significantly reduced cranberry fruitworm egg populations compared to early water. One additional benefit is that late water controls cranberry fruit rot. Flooding can only be used where water is abundant and where the crop would tolerate it for a prolonged period.

OVERHEAD IRRIGATION Overhead irrigation of watercress at night reduced the number of eggs laid by the diamondback moth, *Plutella xylostella* (94). The experiments could not preclude other possible mechanisms such as interference with egg hatching, larval development, pupation, adult emergence, and mating. In apple orchards of Washington, overhead watering significantly decreased codling moth, *Cydia pomonella*, flight, oviposition, and egg and larval survival. Limitations of the methods include damage caused to fruit by poor-quality water, limited availability of water in some regions, and, depending on timing and quantities of water use, increase in apple scab, *Venturia inaequalis*.

Combination of Methods

Physical methods can be used simultaneously or in sequence, especially if there are synergistic effects. A few selected examples follow.

HEAT AND CONTROLLED ATMOSPHERES Low-oxygen and high–carbon dioxide atmospheres in airtight enclosures have a higher efficacy rate when temperature is elevated to levels that cause hyperactivity (26). The reduction of time needed for the disinfestation process is significant even for species that are concentration insensitive to carbon dioxide such as *Tribolium confusum* (18). The synergy between these two types of physical stresses has been demonstrated with other stored-product species (35, 38).

HIGH PRESSURE AND MODIFIED ATMOSPHERES Combining pressure ranging from 2 to 5 MPa in an autoclave with a carbon dioxide–enriched atmosphere allows a complete disinfestation of raw material (e.g., pet food, spices, aromatic plants) packaged in non-airtight enclosures in less than 4 h (91).

MODIFIED ATMOSPHERE AND PACKAGING Modified atmospheres with high (50%–60% v/v) carbon dioxide content or low (<1% v/v) oxygen content are effective methods when food products are stored in an airtight enclosure or in an insect-resistant packaging film (36). However, due to the difficulty of remaining airtight for days or even months to achieve complete insect kill (35), only such methods are used with high-value commodities such as dried fruits or cut flowers.

CONCLUSION

The chain of food production constitutes a continuum with increased legal restrictions as the produce reaches the consumer. The replacement of one technology by another (e.g., chemical control by physical control) can be based upon several considerations, among which economics plays a major role. In preharvest large-scale situations, commercial growers can choose from several options that have, from a technical point of view, their own relative merits (82). The implementation of physical control technologies in preharvest situations has been hampered by several factors, including cost relative to competing technologies, technical difficulties in implementing the strategy, availability of products, and dependence on chemicals (81). Some methods, such as steaming of Colorado potato beetle in the field, simply lack technical efficacy (64). Other methods, such as field vacuuming, can be improved further by fundamental engineering and entomological research (99). Other methods (e.g., trenches) are technically and environmentally acceptable but are impractical and costly relative to an efficacious chemical. Many physical methods rely on energy transfer by diffusion, convection, or radiation. In the field, this is a major obstacle because it is difficult to use the energy efficiently without excessive losses to nontarget substrates. For example, when using flaming against the Colorado potato beetle, only a minute amount of the total energy is used in killing the insect compared to what is being lost heating air, soil, and plants. Targeting is therefore a major challenge in developing physical control methods for field use. The same difficulty exists with insecticide sprays where it has been estimated that less than 0.03% of the foliar spray against aphids on field beans is effectively used for killing the pest (69).

As we move up the chain, increased legal regulations restrict the number of options. In postharvest situations, especially stored grains, the use of pesticides is highly restricted and, consequently, physical control methods are widely used in these situations (e.g., entoleters, inert dusts, modified atmospheric storage). It is noteworthy that most successes and implementation of physical control methods actually occur in postharvest situations.

In physical control methods as in any components related to integrated pest management, the tandem science and technology should work hand in hand. Basic science should not be overlooked, and several fundamental questions remain to be answered. Can insects become resistant to some physical control methods? Studies on heat shock proteins (45) suggest that their modulation could be explained by genetic factors. Likewise, it is not certain if insects can become resistant to inert dusts (61). If the control method allows the insect to adapt and reproduce, such as anthomyiid adults flying over fences (12), then genetics is at play and resistance can develop. Except for a few studies related to the effects of vacuum on insect pollinators (21) and predators (63), little is known about the effects on nontarget organisms, especially in preharvest situations. Little is known also about the effects of physical control methods on plant diseases. Physical stresses induce general mechanisms of physiological responses that, in most cases, have nonspecific targets or receptors, in contrast to the chemicals. This "global physiological response" is a complex phenomenon that has received little attention in insects and associated host plants, except for drought stress in infested plants or cold acclimation in stored-grain insects (31). Consequently, to improve the efficiency of physical control methods and to respect the integrity of host plants, these physiological mechanisms must be identified and quantified in both pest targets and host plants. Basic research on the combined or synergistic effects of two or more methods is also needed.

From a technological point of view, advances in computer technologies (e.g., expert systems) in the past decade created favorable conditions to develop applications of physical methods. For example, modeling the process of energy transfer to target insect pests and host plants may foster progress in simulating practical situations (95). Laboratory pilot experiments currently allow an accurate assessment of the efficiency-to-cost ratio for each novel technique. Thus, the potential of physical control techniques can or should be entirely reassessed in light of recent computer capabilities and high-speed-response sensor technologies.

The sociolegal context of plant protection is key to the adoption of new technologies, and it is likely to evolve in the near future. For example, countries that signed the Montreal protocol are to replace methyl bromide with alternative methods by 2005 (33), offering a good opportunity for physical control methods. The United States is expected to adopt new standards in plant protection, including irradiation of foodstuff for quarantine, and to meet the new requirements of the Food Quality and Protection Act (FQPA).

Of particular concern is the widening technological and economic gaps between developed and developing countries (27). Although agronomic and entomological problems of temperate and tropical countries differ, the adoption of standards to fulfill the demands of global markets should not be at the expense of peasant agriculture. Technical solutions to manage insects should fit the socioeconomic and ecological realities of the countries. For example, Navarro & Noyes (76) provide standardized solutions to manage insects by aeration, but they also discuss aeration methods based on geographically diverse climatic conditions. Because their success generally depends on context, many physical control methods should be as useful and desirable in developed and developing countries.

ACKNOWLEDGMENTS

We thank Gilles Boiteau, Alan L. Knight, Gary Puterka, Anne Averill, Paul G. Fields, Eric Lucas, Victoria Yokoyama, Bob Vernon, Phyllis Weintraub, Yvan Pelletier, and Donald F. Jacques for sharing information or commenting on an early version of the manuscript, and France Labrèche for reading the last version of the manuscript. Thanks to Benoit Rancourt and David Biron for technical help and Dr. Y. Martel (AAC-Ottawa) for facilitating an INRA-Agriculture and Agri-Food Canada exchange program and financing a trip to USDA-ARS-Weslaco, Texas. This is contribution 335/2002.08.01/R of the Horticultural Research and Development Center, Agriculture and Agri-Food Canada, Saint-Jean-sur-Richelieu.

The *Annual Review of Entomology* is online at http://ento.annualreviews.org

LITERATURE CITED

1. Adesiyun AA, Ajayi O. 1980. Control of the sorghum stem borer, *Busseola fusca*, by partial burning of the stalks. *Trop. Pest. Manage.* 26:113–17
2. Adkisson PL, Wilkes LH, Cochran BJ. 1960. Stalk shredding and plowing as methods for controlling the pink bollworm, *Pectinophora gossypiella*. *J. Econ. Entomol.* 53:436–39
3. APHIS. 1998. *Treatment Manual*. Frederick, MD: USDA
4. APHIS. 2000. Irradiation phytosanitary treatment of imported fruits and vegetables. *Fed. Regist.* 65:34113–25
5. Averill AL, Sylvia MM, Kusek CS, DeMoranville CJ. 1997. Flooding in cranberry to minimize insecticide and fungicide inputs. *Am. J. Altern. Agric.* 12:50–54
6. Banks HJ. 1976. Physical control of insects-recent developments. *J. Aust. Entomol. Soc.* 15:89–100
7. Bégin S, Dubé SL, Calandriello J. 2001. Mulching and plasticulture. See Ref. 101, pp. 215–23
8. Belton P. 1962. A field test on the use of sound to repel the European corn borer. *Entomol. Exp. Appl.* 5:281–88
9. Black LL. 1980. "Aluminium" mulch: less virus disease, higher vegetable yields. *LA Agric.* 23:16–18
10. Boiteau G, Misener GC, Singh RP, Bernard G. 1992. Evaluation of a vacuum collector for insect pest control in potato. *Am. Potato J.* 69:157–66
11. Boiteau G, Osborn WPL. 1999. Comparison of plastic-lined trenches and extruded plastic traps for controlling *Leptinotarsa decemlineata* (Coleoptera: Chrysomelidae). *Can. Entomol.* 131:567–72
12. Boiteau G, Vernon RS. 2001. Physical barriers for the control of insect pests. See Ref. 101, pp. 224–47
13. Bollen AF, De la Rue BT. 1999. Hydrodynamic heat transfer—a technique for disinfestation. *Postharvest Biol. Technol.* 17:133–41
14. Bomford MK, Vernon RS, Päts P. 2000. Importance of overhangs on the efficacy of exclusion fences for managing cabbage flies (Diptera: Anthomyiidae). *Environ. Entomol.* 29:795–99
15. Bostanian NJ, Vincent C, Chouinard G, Racette G. 1999. Managing apple maggot, *Rhagoletis pomonella* (Diptera: Tephritidae), by perimeter trapping. *Phytoprotection* 80:21–33
16. Brust GE. 1994. Natural enemies in straw-mulch reduce Colorado potato beetle populations and damage in potato. *Biol. Control* 4:163–69
17. Brust GE. 1996. Interaction of mulch and *Bacillus thuringiensis* subsp. *tenebrionis* on Colorado potato beetle (Coleoptera: Chrysomelidae) populations and damage in potato. *J. Econ. Entomol.* 89:467–74
18. Buscarlet LA. 1993. Study on the influence of temperature on the mortality of *Tribolium confusum* J. exposed to carbon dioxide or nitrogen. *Z. Naturforsch.* 48c: 590–94
19. Busnel RG. 1953. Application du chauffage par infra-rouge à la destruction des parasites dans divers produits agricoles.

C.R. 3th Int. Congr. Electrotherm. 427(4): 907–12
20. Chauhan YS, Ghaffar MA. 2002. Solar heating of seeds—a low cost method to control bruchid (*Callosobruchus* spp.) attack during storage of pigeonpea. *J. Stored Prod. Res.* 38:87–91
21. Chiasson H, Vincent C, de Oliveira D. 1997. Effect of an insect vacuum device on strawberry pollinators. *Acta. Hortic.* 437:373–77
22. Cohen H, Yuval B. 2000. Perimeter trapping strategy to reduce Mediterranean fruit fly (Diptera: Tephritidae) damage on different host species in Israel. *J. Econ. Entomol.* 93:721–25
23. Cowles RS, Cowles EA, McDermott AM, Ramoutar D. 2000. "Inert" formulation ingredients with activity: toxicity of trisiloxane surfactant solutions to twospotted spider mites (Acari: Tetranychidae). *J. Econ. Entomol.* 93:180–88
24. Csizinsky AA, Schuster DJ, Kring JB. 1990. Effect of mulch color on tomato yields and on insect vectors. *Hortscience* 25:1131
25. Davidson NA, Dibble JE, Flint ML, Marer PJ, Guye A. 1991. Managing insects and mites with spray oils. *IPM Educ. Publ., Univ. Calif. Public* 3347
26. Denlinger DL, Yocum GD. 1998. Physiology of heat sensitivity. See Ref. 48, pp. 7–53
27. Donahaye EJ. 2000. Current status of non-residual control methods against stored product pests. *Crop. Prot.* 19:571–76
28. Duchesne RM, Laguë C, Khelifi M, Gill J. 2001. Thermal control of the Colorado potato beetle. See Ref. 101, pp. 61–73
29. Ebeling W. 1971. Sorptive dusts for pest control. *Annu. Rev. Entomol.* 16:123–58
30. Ferro D. 1996. Mechanical and physical control of the Colorado potato beetle and aphids. In *Lutte aux Insectes Nuisibles de la Pomme de Terre*, ed. RM Duchesne, G Boiteau, pp. 53–67. Québec, Can.: Agric. Agri-Food Can. 204 pp.
30a. Fields PG. 2001. Control of insects in post-harvest: low temperature. See Ref. 101, pp. 95–107
31. Fields PG, Fleurat-Lessard F, Lavenseau L, Febvay G, Peypelut L, et al. 1998. The effect of cold acclimation and deacclimation on cold tolerance, trehalose and free amino acid levels in *Sitophilus granarius* and *Cryptolestes ferrugineus* (Coleoptera). *J. Insect Physiol.* 44:955–65
32. Fields PG, Korunic Z, Fleurat-Lessard F. 2001. Control of insects in post-harvest: inert dusts and mechanical means. See Ref. 101. pp. 248–57
33. Fields PG, White NDG. 2002. Alternatives to methyl bromide treatments for stored-product and quarantine insects. *Annu. Rev. Entomol.* 47:331–59
34. Fleurat-Lessard F. 1986. Utilisation d'un attractif de synthèse pour la surveillance et le piègeage des pyrales *Phycitinae* dans les locaux de stockage et de conditionnement de denrées alimentaires végétales. *Agronomie* 6:567–73
35. Fleurat-Lessard F. 1990. Effect of modified atmospheres on insect and mites infesting stored products. In *Food Preservation by Modified Atmospheres*, ed. M Calderon, R Barkai-Golan, pp. 21–38. Boca Raton, FL: CRC
36. Fleurat-Lessard F. 1990. Résistance des emballages de denrées alimentaires aux perforations par des insectes nuisibles. 1. Méthodes d'étude et protocoles expérimentaux. *Sci. Aliment.* 10:5–16
37. Fleurat-Lessard F, Le Torc'h JM. 2001. Control of insects in post-harvest: high temperature and inert atmospheres. See Ref. 101, pp. 74–94
38. Fleurat-Lessard F, Le Torc'h JM, Marchegay G. 1998. Effect of temperature on insecticidal efficiency of hypercarbic atmospheres against three insect species of packaged foodstuffs. In *Proc. 7th Int. Working Conf. Stored Prod. Prot., Beijing*, ed. Z Jin, Q Liang, Y Liang, X Tan, L Guan, 1:676–84. Chengdu, PR. China: Sichuan Publ. Sci. Tech.

39. Fleurat-Lessard F, Lesbats M, Lavenseau L, Cangardel H, Moreau R, et al. 1979. Biological effects of microwaves on two insects *Tenebrio molitor* L. (Col.: Tenebrionidae) and *Pieris brassicae* L. (Lep.: Pieridae). *Ann. Zool. Ecol. Anim.* 11:457–78
40. Gauvrit C, Cabanne F. 2002. Huiles végétales et monoterpènes en formulations phytosanitaires. In *Biopesticides d'Origine Végétale*, ed. C Regnault-Roger, BJR Philogène, C Vincent, pp. 285–300. Paris: Lavoisier Tech. Doc. 338 pp.
41. Genève R, Delon R, Fleurat-Lessard F, Hicaubé D, Le Torc'h JM. 1991. Intérêt des pièges à phéromones pour la surveillance des populations d' *Ephestia elutella* dans les entrepôts de tabac d'importation. In *Proc. 5th Int. Working Conf. Stored Prod. Prot.*, ed. F Fleurat-Lessard, P Ducom, 2:1331–40. Bordeaux, France: INRA
42. Glenn DM, Puterka GJ, Drake S, Unruh TR, Knight AL, et al. 2001. Particle film application influences apple leaf physiology, fruit yield and fruit quality. *J. Am. Soc. Hortic. Sci.* 126:175–81
43. Glenn DM, Puterka GM, VanderZwet T, Byers RE, Feldhake C. 1999. Hydrophobic particle films: a new paradigm for suppression of arthropod pests and plant diseases. *J. Econ. Entomol.* 92:759–71
44. Golob P. 1997. Current and future perspective for inert dusts for control of stored product insects. *J. Stored Prod. Res.* 33:69–79
45. Goto SG, Kimura MT. 1998. Heat- and cold-shock responses and temperature adaptations in subtropical and temperate species of *Drosophila*. *J. Insect Physiol.* 44:1233–39
46. Hallman GJ. 2000. Factors affecting quarantine heat treatment efficacy. *Postharvest Biol. Technol.* 21:95–101
47. Hallman GJ. 2001. Irradiation as a quarantine treatment. In *Food Irradiation: Principles and Applications*, ed. R Molins, pp. 113–30. New York: Wiley
48. Hallman GJ, Denlinger DL, eds. 1998. *Temperature Sensitivity in Insects and Application in Integrated Pest Management*. Boulder, CO: Westview. 320 pp.
49. Hallman GJ, Sharp JL. 1994. Radio frequency heat treatments. In *Quarantine Treatments of Pests of Food Plants*, ed. JL Sharp, GJ Hallman, pp. 165–70. Boulder, CO: Westview
50. Halverson SL, Burkholder WE, Bigelow TS, Plarre R, Booske JH, et al. 1997. Recent advances in the control of insects in stored products with microwaves. ASAE No. 976098. 16 pp.
51. Hara AH, Hata TY, Hu BKS, Tsang MMC. 1997. Hot-air induced thermotolerance of red ginger flowers and mealybugs to postharvest hot-water immersion. *Postharvest Biol. Technol.* 12:101–8
52. Hickling R, Wei W, Hagstrum DW. 1998. *Studies of sound transmission in various types of stored grain for acoustic detection of insects*. http://www.nal.usda.gov/ttic/tektran/data/000006/76/0000067610.html
53. Ikediala JN, Tang J, Neven LG, Drake SR. 1999. Quarantine treatment of cherries using 915 MHz microwaves: temperature mapping, codling moth mortality and fruit quality. *Postharvest Biol. Technol.* 16:127–37
54. Imai T, Tsuchiya S, Morita K, Fujimori T. 1994. Surface tension-dependant surfactant toxicity on the green peach aphid, *Myzus persicae* (Sulzer) (Hemiptera). *Appl. Entomol. Zool.* 29:389–93
55. Jacobi KK, MacRae EA, Hetherington SE. 2001. Loss of heat tolerance in 'Kensington' mango fruit following heat treatment. *Postharvest Biol. Technol.* 21:321–30
56. Jacques DF, Kuhlmann B. 2002. Exxon Esso experience with horticultural mineral oils. In *Spray Oils Beyond 2000*, ed. GAC Beattie, DM Watson, ML Steven, DJ Rae, RN Spooner-Hart, pp. 39–51. Hawkesbury, Aust.: Univ. West. Syd.
57. Khelifi M, Laguë C, Lacasse B. 2001.

Pneumatic control of insects in plant protection. See Ref. 101, pp. 261–69

58. Knight AL, Unruh TH, Christianson BA, Puterka GJ, Glenn DM. 2000. Effects of a kaolin-based particle film on obliquebanded leafroller (Lepidoptera: Tortricidae). *J. Econ. Entomol.* 93:744–49
59. Korunic Z. 1997. Rapid assessment of the insecticidal value of diatomaceous earths without conducting bioassays. *J. Stored Prod. Res.* 33:219–29
60. Korunic Z. 1998. Diatomaceous earths, a group of natural insecticides. *J. Stored Prod. Res.* 34:87–97
61. Korunic Z, Ormesher P. 1998. Evaluation of a standardised testing of diatomaceous earth. In *Proc. 7th Int. Working Conf. Stored Prod. Prot., Beijing*, ed. Z Jin, Q Liang, Y Liang, X Tan, D Guan, pp. 738–44. Chengdu, PR China: Sichuan Publ. Sci. Technol.
62. Kring JB, Schuster DJ. 1992. Management of insects on pepper and tomato with UV-reflective mulches. *Fla. Entomol.* 75:119–29
63. Lacasse B, Laguë C, Roy PM, Khelifi M, Bourassa S, et al. 2001. Pneumatic control of Colorado potato beetle. See Ref. 101, pp. 282–93
64. Lague C, Gill J, Péloquin G. 2001. Thermal control in plant protection. See Ref. 101, pp. 35–46
65. Lapointe SL. 2000. Particle film deters oviposition by *Diaprepes abbreviatus* (Coleoptera: Curculionidae). *J. Econ. Entomol.* 93:1459–63
66. Lewandowski J. 2001. Electromagnetic radiation for plant protection. See Ref. 101, pp. 111–24
67. Lucas E, Riudavets J. 2000. Lethal and sublethal effects of rice polishing process on *Sitophilus oryzae* (Coleoptera: Curculionidae). *J. Econ. Entomol.* 93:1837–41
68. Lucas E, Riudavets J. 2002. Biological and mechanical control of *Sitophilus oryzae* (Coleoptera: Curculionidae) in rice. *J. Stored Prod. Res.* 38:293–304
69. Matthews GA, ed. 1992. *Pesticide Application Methods*. New York: Longman Sci. Tech.
70. McGuire RG. 1991. Market quality of grapefruit after heat quarantine treatments. *Hortscience* 26:1393–95
71. Metcalf RL. 1994. Insecticides in pest management. See Ref. 72a, pp. 245–314
72. Metcalf CL, Flint WP, Metcalf RL. 1962. *Destructive and Useful Insects, Their Habits and Control*. New York: McGraw Hill. 1087 pp. 4th ed.

72a. Metcalf RL, Luckmann WH, eds. 1994. *Introduction to Insect Pest Management*. New York: Wiley. 650 pp. 3rd ed.

73. Misener GC, Boiteau G. 1993. Holding capacity of the Colorado potato beetle to potato leaves and plastic surfaces. *Can. Agric. Eng.* 35:27–31
74. Misener GC, Boiteau G, McMillan LP. 1993. A plastic-lining trenching device for the control of Colorado potato beetle: beetle excluder. *Am. Potato J.* 70:903–8
75. Murdock LL, Shade RE. 1991. Eradication of cowpea weevil (Coleoptera: Bruchidae) in cowpeas by solar heating. *Am. Entomol.* 37:228–31
76. Navarro S, Noyes RT, eds. 2000. *The Mechanics and Physics of Modern Grain Aeration Management*. Boca Raton, FL: CRC. 672 pp.
77. Nelson SO. 1995. *Assessment of RF and microwave electric energy for stored-grain insect control*. Presented at ASAE Annu. Int. Meet. Paper No. 956527. 16 pp.
78. Nelson SO, Bartley PG Jr, Lawrence KC. 1998. RF and microwave dielectric properties of stored-grain insects and their implications for potential insect control. *Trans. ASAE* 41:685–92
79. Novartis 1997. *Le Livre vert du Maïs Cb*. St-Sauveur, France: Novartis Seeds. 109 pp.
80. Oseto CY. 2000. Physical control of insects. In *Insect Pest Management, Techniques for Environmental Protection*, ed.

JE Rechcigl, NA Rechcigl, pp. 25–100. Boca Raton, FL: Lewis. 392 pp.

81. Panneton B, Vincent C, Fleurat-Lessard F. 2001. Current status and prospects for the use of physical control in crop protection. See Ref. 101, pp. 303–9
82. Panneton B, Vincent C, Fleurat-Lessard F. 2001. Plant protection and physical control methods: the need to protect crop plants. See Ref. 101, pp. 9–32
83. Payne TL, Shorey HH. 1968. Pulsed ultrasonic sound for control of oviposition by cabbage looper moths. *J. Econ. Entomol.* 61:3–7
84. Pelletier Y, McLeod CD, Bernard G. 1995. Description of sublethal injuries caused to the Colorado potato beetle by propane flamer treatment. *J. Econ. Entomol.* 88:1203–5
85. Pelletier Y, Misener GC, McMillan LP. 1998. Steam as an alternative control method for the management of Colorado potato beetles. *Can. Agric. Eng.* 40:17–21
86. Plarre R, Reichmuth C. 2000. Impact. In *Alternatives to Pesticides in Stored-Product IPM*, ed. BH Subramanyam, DW Hagstrum, pp. 401–17. Boston: Kluwer
87. Pradzynska A. 1982. The suitability of ultrasound for controlling stored pest. *Prace-Naukowe Inst. Roslin.* 24:77–90
88. Prokopy RJ, Croft BA. 1994. Apple insect pest management. See Ref. 72a, pp. 543–85
89. Puterka GJ, Glenn DM, Sekutowski DG, Unruh TR, Jones SK. 2000. Progress toward liquid formulations of particle films for insect and disease control in pear. *Environ. Entomol.* 29:329–39
90. Rancourt B, Vincent C, de Oliveira D. 2000. Circadian activity of *Lygus lineolaris* (Hemiptera: Miridae) and effectiveness of sampling procedures in strawberry fields. *J. Econ. Entomol.* 93:1160–66
91. Reichmuth C, Wohlgemuth R. 1994. Carbon dioxide under high pressure of 15 bar and 20 bar to control the eggs of the Indian meal moth *Plodia interpunctella* (Hübner) (Lepidoptera: Pyralidae) as the most tolerant stage at 25°C. In *Stored Product Protection*, ed. E Highley, J Wright, HJ Banks, BR Champ, 1:163–72. Wallingford, UK: CABI
92. Rhainds M, Kovach J, Dosa EL, English-Loeb G. 2001. Impact of reflective mulch on yield of strawberry plants and incidence of damage by tarnished plant bug (Heteroptera: Miridae). *J. Econ. Entomol.* 94:1477–84
93. Shapiro JP, Schroeder WJ, Stansly PA. 1998. Bioassay and efficacy of *Bacillus thuringiensis* and organosilicone surfactant against the citrus leafminer (Lepidoptera: Phyllocnistidae). *Fla. Entomol.* 81:201–10
94. Tabashnik BE, Mau RFL. 1986. Suppression of diamondback moth (Lepidoptera: Plutellidae) oviposition by overhead irrigation. *J. Econ. Entomol.* 79:189–91
95. Tang J, Ikediala JN, Wang S, Hansen JD, Cavalieri RP. 2000. High-temperature-short-time thermal quarantine methods. *Postharvest Biol. Technol.* 21:129–45
96. Unruh TR, Knight AL, Upton J, Glenn DM, Puterka GJ. 2000. Particle films for suppression of the codling moth (Lepidoptera: Tortiricidae) in apple and pear orchards. *J. Econ. Entomol.* 93:737–43
97. van Epenhuijsen CW, Koolaard JP, Potter JF. 1997. Energy, ultrasound and chemical treatments for the disinfestation of fresh asparagus spears. *Proc. 50th N. Z. Plant Prot. Conf.* 50:436–31
98. Vernon RS, Mackenzie JR. 1998. The effects of exclusion fences on the colonization of rutabagas by cabbage flies (Diptera: Anthomyiidae). *Can. Entomol.* 130:153–62
99. Vincent C, Boiteau G. 2001. Pneumatic control of agricultural pests. See Ref. 101, pp. 270–81
100. Vincent C, Chagnon R. 2000. Vacuuming tarnished plant bug on strawberry: a bench study of operational parameters versus insect behavior. *Entomol. Exp. Appl.* 96: 347–54
101. Vincent C, Panneton B, Fleurat-Lessard

F, eds. 2001. *Physical Control in Plant Protection*. Berlin/Paris: Springer/INRA. 329 pp.

102. Wang S, Ikediala JN, Tang J, Hansen JD, Mitcham E, et al. 2001. Radio frequency treatments to control codling moth in in-shell walnuts. *Postharvest Biol. Technol.* 22:29–38
103. Weber DC, Ferro DN, Buonaccorsi J, Hazzard RV. 1994. Disrupting spring colonization of Colorado potato beetle to nonrotated potato fields. *Entomol. Exp. Appl.* 73:39–50
104. Weintraub PG, Horowitz AR. 2001. Vacuuming insect pests: the Israeli experience. See Ref. 101, pp. 294–302
105. WHO 1982. *Ultrasound, Environmental Health Criteria 22*. Geneva: WHO. 199 pp.
106. Yokoyama VY, Miller GT. 2002. Bale compression and hydrogen phosphide fumigation to control cereal leaf beetle (Coleoptera: Chrysomelidae) in exported rye straw. *J. Econ. Entomol.* 95:513–19
107. Yokoyama VY, Miller GT, Hartsell PL, Eli T. 1999. On-site confirmatory test, film wrapped bales, and shipping conditions of a multiple quarantine treatment to control hessian fly (Diptera: Cecidomyiidae) in compressed hay. *J. Econ. Entomol.* 92:1206–11
108. Zehnder GW, Hough-Goldstein J. 1990. Colorado potato beetle (Coleoptera: Chrysomelidae) population development and effects on yield of potatoes with and without straw mulch. *J. Econ. Entomol.* 83:1982–87

Annu. Rev. Entomol. 2003. 48:283–306
doi: 10.1146/annurev.ento.48.091801.112611

First published online as a Review in Advance on August 28, 2002

COMPARATIVE SOCIAL BIOLOGY OF BASAL TAXA OF ANTS AND TERMITES

Barbara L. Thorne
Department of Entomology, University of Maryland, College Park, Maryland 20742; e-mail: bt24@umail.umd.edu

James F. A. Traniello
Department of Biology, Boston University, Boston, Massachusetts 02215; e-mail: jft@bu.edu

Key Words Formicidae, Isoptera, eusocial evolution, colony structure, division of labor

■ **Abstract** Lacking a comprehensive fossil record, solitary representatives of the taxa, and/or a definitive phylogeny of closely related insects, comparison of the life history and social biology of basal, living groups is one of the few available options for developing inferences regarding the early eusocial evolution of ants and termites. Comparisons of a select group of basal formicid and isopteran taxa suggest that the reproductive organization of colonies and their patterns of division of labor were particularly influenced, in both groups, by nesting and feeding ecology. Opportunities for serial inheritance of the nest structure and colony population by kin may have been significant in the evolution of multiple reproductive forms and options. Disease has been a significant factor in the evolution of social organization in ants and termites, but the adaptive mechanisms of infection control differ. Evaluations of the convergent and divergent social biology of the two taxa can generate novel domains of research and testable hypotheses.

CONTENTS

INTRODUCTION

The study of the evolution of complex traits is greatly facilitated by the existence of phylogenetic intermediates that express gradual transitions in character states, and the comparative analysis of these transitions often provides the most compelling data for revealing patterns and developing robust hypotheses regarding selective factors that influence evolutionary change. Such stepping stone intermediates, however, are rarely present as a relatively complete fossil or living series, reflecting incomplete preservation and discovery of fossils, species extinction, the evanescence of annectant forms, and the fact that evolution is not always a gradual process. Eusociality is a highly complex trait of profound evolutionary interest because of the existence of subfertile or sterile colony members. Comparative studies of sister groups and basal taxa have been insightful in examining the evolution of eusociality in clades of bees and wasps because modern species show a cline of life histories that range from solitary to eusocial. Similar cladistic analyses are absent in ants and termites because all the roughly 10,200 species of living ants and over 2600 species of extant termites are eusocial, and solitary ancestors are sufficiently distant to obscure the linkages among selective regimes.

The structural elements of social organization in the Hymenoptera and Isoptera are highly convergent. Unlike the haplodiploid Hymenoptera, however, both sexes of termites are diploid, rendering explanations for eusocial evolution based on asymmetries in genetic relatedness generated by meiosis and fertilization inapplicable to termites. Nevertheless, the similarities and differences in the preadaptive characteristics of each group and ecological forces that impelled the evolution of social organization may offer significant sociobiological insight. Lacking the opportunity for comparative study within taxa, here we explore commonalities and contrasts in the life history, colony structure, reproductive dynamics, and socioecology of the most primitive living lineages of ants and termites. Although these phylogenetically divergent insects differ in fundamental ways (such as holometaboly in Hymenoptera and hemimetaboly in Isoptera), eusociality is based on the elaboration of family units in both groups, and eusocial evolution may be constructively discerned through focused comparative assessments. Observations on the biology of extant taxa cannot be used to definitively reconstruct ancestral states prior to the evolution of worker subfertility or sterility and thus cannot appropriately be used to test hypotheses or predictions regarding the evolution of eusociality. Once protoants or prototermites crossed the threshold of eusociality, life history constraints, especially those related to reproductive division of labor, may have been essentially irreversible. Data that allow specific comparisons are not always available, so a collateral goal of this paper is to identify domains of research that would further advance such an approach. We begin with an overview of the phylogenetic origin of ants and termites, identify the basal taxa considered in the review, and justify the inclusions of those groups central to our comparative analysis.

ORIGIN OF ANTS AND BASAL ANT SYSTEMATICS

The origins of the formicid theme of social organization have been sought in the vespoid wasps. There is a void in social behavior between basal ants and their closest vespoid relatives, although the fossil record offers some evidence of how and when ants attained their morphological distinctiveness and suggests a basic timeline for the emergence of the socially advanced groups. The hypothetical ancestral vespoid wasps are thought to be linked to ants through the subfamily Sphecomyrminae, the pleisiomorphic sister group to all ants, with its extinct Cretaceous fossil genera *Sphecomyrma* and *Cretomyrma* (124). *Sphecomyrma freyi*, dating from New Jersey amber of the late middle Cretaceous, exhibits a constellation of nonsocial wasp and ant traits: short bidentate mandibles, a reduced and wingless thorax, a petiolar constriction, and significantly, what appears to be a metapleural gland (33). It is considered the "nearly perfect link between some of the modern ants and the nonsocial aculeate wasps" (33, p. 23). *Kyromyrma neffi*, the first specimen of an extant ant subfamily (the Formicinae), also collected from the New Jersey amber (c. 92 million years ago), has an acidopore and is 50 mya older than *Sphecomyrma* (23). This suggests a divergence of the basal lineages of ants from the Sphecomyrminae approximately 105–110 mya. Further details of the fossil record and adaptive radiation of ants are given in Hölldobler & Wilson (33) and Crozier et al. (14), which provide molecular data dating the origin of ants to the Jurassic.

The basal division of the 17 ant subfamilies (5) separates the Myrmicinae, Pseudomyrmecinae, Nothomyrmeciinae, Myrmeciinae, Formicinae, and the Dolichoderinae from the remaining subfamilies. The Nothomyrmeciinae, Myrmeciinae, and Ponerinae include genera considered pivotal in ant social evolution because of their comparatively primitive morphology and social organization (33). The final basal group of ants is the Aneuretinae, a formicoid complex subfamily once global in distribution but today represented by a single species, *Aneuretus simoni*, found in limited areas of Sri Lanka (41).

The subfamily Nothomyrmeciinae is monotypic, known only from the single extant and elusive *Nothomyrmecia macrops* Clark from Australia. Its rediscovery in 1978 (104) was somewhat akin to finding the "Holy Grail" of myrmecology, and the collection and observation of queenright colonies made possible detailed accounts of the social organization of this relict species (31, 104). The basal characteristics of this ant include a wasp-like morphology (104), an exceptionally high level of inactivity, and low levels of social exchanges among workers in their small colonies (40). Queens do not receive food or other preferential treatment; indeed, queens and workers rarely interact. Remarkably, queens living in intact laboratory colonies collect and feed on insect prey on their own. The subfamily Myrmiciinae is represented by the Australian bulldog ants of the genus *Myrmecia*, which forms colonies of 600–900 workers (25). Colonies can be founded independently by single queens or polygynously by groups of females (12).

The Ponerinae is a large and diverse subfamily whose representatives display a mixture of basal and derived morphologies and social characters. Primitive ponerines include *Amblyopone*, an ant that exhibits morphologically and behaviorally primitive traits, although other genera in the tribe Amblyoponini show highly derivative characteristics. Because of the great diversity of ponerine ants, including numerous species with clearly derived traits (81, 82), we concentrate on the more primitive forms, using *A. pallipes* as a model, while noting that other species of *Amblyopone* may vary widely in their biology and even include queenless forms (37). In *A. pallipes*, alate queens discard their wings to establish new colonies and forage during the colony foundation stage (33). Nests, which house small colonies averaging roughly a dozen workers, are composed of simple chambers and galleries in soil and decayed wood. Populations of *A. pallipes* are patchy but can be locally abundant (112).

ORIGIN OF ISOPTERA AND INTERFAMILIAL RELATIONSHIPS

The higher-level phylogeny of termites has received considerable interest in the past decade. Although monophyly of the Dictyoptera is accepted (28, 50), relationships among the dictyopteran orders Blattaria, Mantodea, and Isoptera are not fully resolved (6, 17, 42, 44–47, 51, 57, 59, 108). Despite the topological uncertainty regarding whether cockroaches or some lineage(s) of a paraphyletic cockroach clade are the sister group to termites, there is consensus that study of the life history and social organization of the relict wood roach genus *Cryptocercus* provides constructive comparison and potential insights into the biology of prototermites and potential selective forces favoring the evolution of eusociality (10, 67, 68, 107). To date, the fossil record exposes no missing links that indicate intermediate stages between the orders, so identifying the most immediate ancestors of Isoptera, and gleaning the hints that they might reveal regarding the transition from solitary to eusocial life histories, has been impossible.

The early evolution and intrafamilial relationships of Isoptera also are not fully understood, but several lines of evidence identify the most basal lineages and provide increasing definition of their phylogeny. The earliest known fossil termites are from the Cretaceous and are representatives of the Hodotermitidae, Termopsidae, and possibly Mastotermitidae. These Mesozoic termites are distinctly primitive but reasonably diversified, suggesting an origin of the order in the Upper Jurassic (109). Hodotermitidae, represented in modern fauna by three genera (19 species) of highly specialized "harvester" termites, has the oldest described fossil (~130 mya) (52) and a total of six genera (seven species) in the Cretaceous. Although the foraging behaviors and colony organization of extant hodotermitids are derived, they retain pleisiomorphic morphological characters (18).

Termopsidae, the sister group of Hodotermitidae [(16, 18, 73); but see (105)], is represented by at least four known genera (five species) in the Cretaceous and five modern genera (20 species, the "dampwood" termites). Termopsids, especially the relict Himalayan *Archotermopsis wroughtoni* Desneux, are considered by many to be the most primitive living termites with respect to colony size, social organization, nesting biology, and caste polyphenism (35, 71, 101, 105, 107, 108).

Mastotermitidae, apparently represented in the Cretaceous by two genera and radiating broadly by the Tertiary (4 genera with more than 20 species known from Australia, Europe, North and South America, and the Caribbean) (109), now exists as only a single species, *Mastotermes darwiniensis* Froggatt, with a natural distribution in moist, tropical regions of Northern Australia. Mastotermitidae is viewed uncontroversially as the most basal living lineage within Isoptera and as the sister taxon to all other living termites (16, 17, 43, 47, 48). *M. darwiniensis* has distinct pleisiomorphic characters, but it also features a number of highly derived characteristics. For example, *M. darwiniensis* has an early and apparently irreversible split in development of nondispersive forms, soldiers secrete a defensive chemical, male reproductives have a unique type of multiflagellate sperm, colony population sizes can be large (several million individuals), gallery construction occurs within nests, and extensive foraging tunnels connect food sources located away from the nest (21, 53, 105, 108). *M. darwiniensis* thus exemplifies a common evolutionary pattern: It retains some primitive features but also has apomorphic anatomical and life history elements.

Along with these three confirmed ancient families, some classic (48, 75) and one recent family-level phylogeny (43) place Kalotermitidae, including the "drywood" termites along with some dampwood species (53), as among the most basal clades. Kalotermitids do not appear in the fossil record until the Paleocene; there are 446 modern species in 21 genera (11). Current hypotheses of relationship among the four basal termite families Mastotermitidae, Hodotermitidae, Termopsidae, and Kalotermitidae differ only in the position of Kalotermitidae. Taxonomic sampling issues and lack of integration between morphological and molecular studies have impeded resolution of family-level phylogenies, but for the purposes of this paper we assume that Mastotermitidae, Hodotermitidae, and Termopsidae comprise the most basal living termite clades. This is in accordance with recent phylogenetic analyses that differ in topology, but include the same three families as most basal [((((((T, R), S), K), (Tp, H)), M), B)[1] (105); (((((((T, R), S), K), Tp), H), M), B)] [Donovan et al. (16) and Eggleton's (17) "majority consensus rule" phylogeny].

---

[1]B, Blattaria; M, Mastotermitidae; H, Hodotermitidae; Tp, Termopsidae; K, Kalotermitidae; S, Serritermitidae; R, Rhinotermitidae; T, Termitidae.

ANCESTRAL ECOLOGY OF TERMITES: FOCUS ON TERMOPSIDAE

Because social organization of living members of the Mastotermitidae and Hodotermitidae appears to be derived, we typically draw inferences regarding ancestral socioecology from life history patterns of modern Termopsidae. This interpretation has been broadly held among termitologists (18, 72, 76, 101, 107), although there is some controversy regarding whether the developmental flexibility typical of Termopsidae is an ancestral or derived characteristic of termites. The traditional view is that "true workers," i.e., individuals that diverge early and irreversibly from the imaginal line, are a derived feature in termites (1, 70, 72, 76). According to this view, the worker caste developed at least three times independently because true workers occur in Mastotermitidae, Hodotermitidae, Serritermitidae, Rhinotermitidae, and Termitidae (3, 29, 66, 76, 77, 85, 86, 107, 117). In Termopsidae, Kalotermitidae, and the most primitive Rhinotermitidae, helpers have marked developmental flexibility throughout their lives; all individuals except soldiers may differentiate into reproductives (70, 76, 87, 101, 107) or undergo regressive molts to revert from the nymphal line into "pseudergates" (76).

Based on hypotheses of interfamilial phylogenetic relationships, however, Thompson et al. (105) follow Watson & Sewell (118, 119) in supporting irreversible worker differentiation as an ancestral element of termite social evolution rather than a derived, phylogenetic state. It is difficult to evaluate this postulate, however, because there are so few living representatives of taxa key to this interpretation, i.e., the families Mastotermitidae and Hodotermitidae, and those species that exist are highly derived in other social attributes (4, 18, 48, 49, 101, 107). Nesting and feeding habits may drive the evolution of social behaviors, obscuring phylogenetic analyses based on presumed homologous traits. For example, true workers are invariably found in species that forage away from the nest exploiting multiple resources, and helpers with lifelong flexible developmental options occur in "one-piece nesting" groups that consume only the wood in which they live and therefore face eventual resource limitation and instability (1, 3, 29, 53). This correlation suggests biological significance between termite nesting biology and presence or absence of true workers in modern species. The ancestral worker hypothesis (105, 118, 119) thus carries linked implications, suggesting for example that organized foraging away from the nest is an ancestral trait and that the "one-piece nesters" (1) with minimal nest architecture and foraging restricted to the nest wood is secondarily derived. Eggleton (17) rationally advocates resolution of phylogenies before attempting to map social, behavioral, developmental, or biogeographic characters.

INFERENCES REGARDING ANCESTRAL LIFE HISTORIES

Considering extant basal ants and termites as "windows" into ancestral life histories, it is apparent that individual species in either taxon rarely provide an entirely credible model reflecting the biology of the group early after the evolution of eusociality. Modern species belonging to even the most basal lineages have blended assemblages of primitive and derived traits, thus confounding interpretations. DNA sequences help resolve this issue for phylogenetic analyses, but no methodological safety net exists for evaluating social evolution because homology of behavioral traits can be difficult to verify and may be influenced by ecology or other derived life history attributes. We are thus left to draw inferences based on suites of characters considered to be primitive, compiled from a number of living taxa to yield a composite of likely traits and ancestral ecology of extinct lineages relatively close to the cusp of eusocial evolution. Our comparisons of likely ancestral character complexes from ants and termites ideally will yield productive insights regarding both commonalities and differences, and therefore potentially significant influences, favoring the evolution of eusociality in these insects. We focus on four broad and interrelated areas: reproductive plasticity, division of labor, foraging biology, and evolutionary pathobiology. We then conclude with a discussion of potential commonalities influencing the evolution of eusociality in these groups.

REPRODUCTIVE PLASTICITY IN ANTS AND TERMITES

Colony Structure, Gynes, and Replacement Reproductives

Developing a conceptual framework for the comparative analysis of reproductive variability in ants and termites has historically been impeded by the number and complexity of fertile and sterile forms, the existence of anatomical and physiological intermediates, nomenclature differences, and semantic controversy. To facilitate comparison and clarify our discussion we catalog the types of reproductives found in each group using currently recognized terminology (33, 88, 106).

In ants, reproductive division of labor presents itself in the typical dimorphic queen and worker castes: The queen is derived from the dispersing alate form, establishes a new colony, and is distinguished from her daughters by size, the extent of ovarian development and behavior. Among basal ant species, the wings may be reduced [as in the case of *Nothomyrmecia* (31)] and size differences may be limited to a somewhat broader thorax bearing the scars of the wings that are discarded following the dispersal flight. Some basal ants have fertile forms (ergatogynes) that are morphologically intermediate between independent, dealate colony-founding queens and workers and inseminated workers (gamergates). Ergatoid queens, which are found in some species of *Myrmecia* (13, 26), have a greater number of ovarioles than workers do, a filled spermatheca, and may replace a typical queen. Some ponerine ants, including *Amblyopone* (37, 83), are queenless. Reproduction by gamergates, which possess a functional spermatheca and are inseminated, occurs in these species. Colonies having gamergates occupy stressful environments, have reduced dispersal, mate within or nearby the nest, and reproduce by fission, as may colonies with ergatoid queens (83).

Recent research (96, 97) has begun to uncover unexpected and exciting details of the reproductive and genetic organization of colonies of *Nothomyrmecia macrops*. This basal ant is facultatively polyandrous; sampled queens were singly or multiply mated to unrelated males, with an overall average of 1.37 matings per queen (96). Worker nestmates are related by $b = 0.61 +/- 0.03$. Workers appear to be incapable of laying eggs. The mechanism of queen replacement in colonies of *N. macrops* is rare among ants and bears some resemblance to the pattern of colony inheritance exhibited by some basal termites. Although newly inseminated queens found *N. macrops* colonies monogynously, comparisons of worker and queen genotypes in some sampled colonies contained resident queens that were the sisters rather than the mothers of workers. Furthermore, larvae were genetically identified as the queen's progeny and not the offspring of reproductive workers. The likely explanation for this genetic structure is that the original colony-founding queen had died and been replaced by one of her daughters.

The colony life cycle of *N. macrops* has been reconstructed as follows (96, 97): New queens, one of which may inherit the parental colony, are produced from overwintering larvae that can develop from eggs laid in the autumn into gynes during the following year, even in the event of death of the queen mother. A replacement queen can produce sexual offspring in her first year. Overall, *N. macrops* illustrates a low level of serial polygyny; primarily daughters, but at least occasionally unrelated queens, are adopted by orphaned colonies. Under the condition of colony inheritance by daughters, inclusive fitness benefits extend to the original colony-founding queen (through the rearing of grand-offspring following her death) as well as to workers (through the production of nieces and nephews). The brachyptery of new queens may reflect limited dispersal and a reproductive strategy designed to favor replacement of the mother queen by her daughters. Ecological constraints such as habitat patchiness, nest site limitation, and the risk-prone foraging behavior of the partially claustral founding queens may have favored colony inheritance in *N. macrops*.

Does the presence of such reproductive flexibility in one of the most primitive extant ants accurately reflect an ancestral condition? Although it has been argued that the brachypterous queens of *N. macrops* favored the evolution of daughter replacement (97), it is also possible that brachyptery evolved concomitantly with daughter adoption under the selective pressure of dispersal-related mortality. Again, we note the difficulties inherent in analyzing the evolution of social traits in basal species whose biology may be a constellation of primitive and derived characters.

Several types of reproductives exist in termites. The terms *king* and *queen* typically refer to the colony-founding male and female. These *primary* reproductives are imagoes (alates) that drop their wings after pairing. Founding pairs in basal groups are nearly always monogamous, although there are some records of associated groups of primary reproductives (27). *Neotenics* are termite reproductives that are not derived from alates, but differentiate within their natal colony, breeding with a parent, sibling, or other inbred relative. Neotenic differentiation typically occurs upon death or senescence of the founding reproductive of the same sex (55). Multiple neotenics of each sex develop, persisting as typically consanguineous reproductive groups in most basal termites (21) but surviving as only one pair in Kalotermitidae (72). Neoteny, literally meaning reproduction as an immature, is related to hemimetaboly, requiring one or two molts that modify morphology and produce functional sex organs (72, 106). In *Mastotermes*, neotenics may develop from workers (119); in termopsids neotenics may form from any individual in instar four or above (except soldiers or imagoes) (71), although no true neotenics are known in *Archotermopsis* (35, 89). Soldier neotenics occur in six species of termopsids (64, 107). In *Archotermopsis*, the gonads of all soldiers are as well developed as in alates (35).

All offspring helpers in termopsid families (except soldiers) retain the capacity to differentiate into fertile reproductives (in the case of termites, either alates or neotenics). They are thus poised to potentially inherit their parents' resources of a nest, food, and established family (65, 107). In such a system of serial reproductive inheritance by kin, as in the cases of ergatogyne and gamergate ants reproducing in their natal nest, all colony members gain inclusive fitness benefits and some individuals attain direct fitness advantages. These cumulative fitness components may well exceed average individual fitness prospects of dispersing, fertile offspring in a similar, solitary species, thus favoring helpers that remain in their natal colony. In *Zootermopsis*, numerous colonies may be initiated in the same log, eventually resulting in intercolony interactions, which can lead to death of reproductives and opportunities for replacement by neotenics (107).

In ancestral groups, the reproductive skew between reproductives and helpers may have been less discrete. Imms (35) reported that worker-like individuals of *Archotermopsis wroughtoni* have extensive gonad development and a fat body equivalent to alates. He observed a captive worker-like *A. wroughtoni* lay seven eggs. The eggs did not develop normally, but whether due to sterility, lack of fertilization, or laboratory conditions is unknown. Eusociality itself is viewed as a continuously varying categorization depending on the portion of progeny that reduces or foregoes reproduction (101, 103).

The possibility of merged or indistinct colonies functioning within single pieces of wood has been raised several times. Concerning *Archotermopsis*, Imms (35, p. 126) observed, "I have, on several occasions, come across three or four queens with a single large colony of ova and larvae, which probably represent several colonies which have become confluent." Fused colonies or colony complexes have also been suggested in *Stolotermes* (20, 63, 110) and *Zootermopsis* (B.L. Thorne, personal observation). These observations and their generality, context, and implications are difficult to evaluate; identification of discrete but adjacent colonies within a log is often impossible. Sufficient descriptive evidence exists, however, to encourage genetic examination of these circumstances, especially relationships among the reproductives found with the possibly merged groups. Recent work on *M. darwiniensis* (21) suggests that although neotenics within a colony are often inbred, they sometimes originate from more than two genetic lineages, as has been indicated in some more derived termites (8). Extensive study on colony genetic structure and the possibility of merging or introduction of foreign reproductives is required for both basal ants and termites.

Reproductive Conflict

Conflict among nestmates whose fitness interests are incongruent is common in social insects. The apparent rarity of nuptial flights in some basal ant genera (13), the presence of fertile helpers in both ants and termites, and multiple replacement reproductives raise the possibility of intracolony reproductive conflict. Reproductive conflict is manifest in oophagy, the existence of inhibitory pheromones, dominance structures, mutilation, and policing behaviors (37, 62). In basal ants, larval hemolymph feeding by queens has been described in *Prionopelta* and *Amblyopone silvestrii* (32, 38, 60) and has been interpreted as a form of queen nutrition, although the behavior could also represent a mechanism to regulate reproductive capability.

Because of the monogynous and monoandrous organization of basal termite societies, conflicts similar to those observed in ants would not be expected until colony members approach a state of reproductive competence; then policing or other related mechanisms of reproductive competition might be evident in species with flexible development because nearly all individuals have the potential to differentiate into reproductive forms. Roisin (85) cites reports of intracolony mutilation in termopsids, kalotermitids, and some rhinotermitids and proposes that competition among late instar helpers, including nymphs attempting to become alates might explain such behaviors. He suggests that siblings bite wing pads, which causes some individuals to deflect from alate development, creating "lower status" helpers with reduced chances of future dispersal. Subsequent wing bud regeneration and formation of a normal alate is possible, but with delay and additional molts (107, 118). Roisin (85) proposes that the mutilated "losers" in intracolony conflicts formed the original helpers in termites. The contexts under which primitive termites lose wing buds need to be better understood before this hypothesis, or the implications of mutilation behaviors in termites, can be rigorously evaluated (107). For example, wing bud scars in termopsids are often due to self-induced abscission rather than mutilation by colony members (35, 107). Research on complete colonies of *Zootermopsis* in the laboratory suggests that self-abscission occurs when there are opportunities to become a replacement reproductive, perhaps inducing pre-alates to shed wing pads and differentiate into a neotenic in the natal colony (107).

DIVISION OF LABOR

The primary axis of division of labor in basal ant and termite species is reproductive, but colonies theoretically may partition tasks according to the size and age of subfertile or sterile individuals. Pheromones, temperature, and nutrition direct caste expression in both ants and termites (33, 34, 71, 72). Historical factors, including development, cause ant and termite castes to form in fundamentally different ways and therefore potentially preadapt these two groups toward disparate mechanisms of task partitioning. In ants, morphological variation is generated through allometry within a single adult instar; size variation and polymorphism in termites is found across instars, from immature through imago. Although the caste systems of termites with true workers show a strong convergence with ants, similarities among basal species may be obscure because of the prevalence of monomorphism in ant workers. Due to hemimetabolous development, immature termites contribute to colony needs as juveniles, whereas ant larvae are seemingly unable to meet labor demands unless they are involved in food processing and nutrient distribution.

Basal ant species such as *Amblyopone pallipes* have small colonies, and activity is restricted to a limited number of nest chambers and associated tunnels where prey capture occurs (112). Workers, which hunt vermiform arthropod prey, initiate foraging soon after eclosion. Foraging and brood-care are codependent tasks because larvae are carried to freshly paralyzed prey where they feed directly. Brood-care is thus reminiscent of the direct provisioning habits of solitary wasps, and the same individual often performs both foraging and brood-care tasks. *A. pallipes* lacks age-based division of labor (111), but interspecific comparisons of polyethism in *Amblyopone* suggest that age-related division of labor might take on elements similar to that of higher ants (61), although the reasons for such differences are unclear. Colony demography, feeding specialization, and the retention of ancestral behavioral traits seem to be important determinants in division of labor in ants (40, 111, 113). The degree of sociality, which varies in ants, may also influence patterns of division of labor in *A. pallipes* and *N. macrops*, in which queen-worker and worker-worker interactions are rare and polyethism is lacking (40). In addition, some *Amblyopone* species are queenless but contain multiple inseminated gamergates that form dominance hierarchies (37), which could influence division of labor (84). Among basal ants, *Aneuretus simoni* exhibits an age-related polyethism that foreshadows the form of temporal task partitioning typical of ants of the higher subfamilies.

Like primitive ants, basal termopsid species have small colonies (35, 53), activity is limited to nest galleries, and there has been no indication of age-based division of labor (94). It has been hypothesized that termite caste systems should be fully discretized due to hemimetabolous development (79). Noirot (72, p. 9) notes that the combination of helpers of both sexes and hemimetaboly gives termites, in comparison to Hymenoptera, "many more possibilities for the diversification of polymorphism and consequently, for its adaptations." Termites may also advance, regress, or retard their metabolic development (71, 76, 87) to respond to colony needs or individual fitness initiatives. During its postembryonic development, an individual termite, especially in Termopsidae and Kalotermitidae, may "belong to different physical castes in succession" (72, p. 8), possibly terminating by becoming a soldier or reproductive. Noirot & Bordereau (74) termed this pattern *temporal polymorphism*, juxtaposed with *temporal polyethism*, or change in task functions of a worker during its lifetime, as is characteristic in Hymenoptera and derived termites (80, 123). Flexibility in metamorphosis in basal termite species might provide a mechanism of task switching similar to, but less rapid than, the patterns of behavioral acceleration and regression seen in more advanced species of the social Hymenoptera. The multiple age (instar) cohorts generated by hemimetaboly would divide tasks along a finely graded scale, resulting in the evolution of one caste per task (79). Yet if first and second instar larvae are inactive (94), gradual metamorphosis may in essence yield a caste distribution that resembles only moderate polymorphism, although the duration between molts would seemingly provide ample time for temporal specialization. In any case, the ergonomically adaptive nature of polyethism in basal termites is virtually unknown. There is some suggestion that demography serves a function in infection control (95).

Termite soldiers are without equivalent in Hymenoptera (72); basal ants have few allometric size variants such as majors and minors, with the exception of *Aneuretus simoni* (33, 113), although large workers, possessing a disproportionately large number of ovarioles, have been described in *Myrmecia* (39). Soldier termites appear to be monophyletic (71, 76). In Mastotermitidae and Hodotermitidae, soldiers have "continuous" polymorphism because they originate from successive and numerous worker instars (72). In Termopsidae and Kalotermitidae, all except the youngest larvae [termite terminology uses "larvae" to describe apterous immatures differentiating along a nonreproductive pathway (87, 106)], all nymphs (termites with wing pads), and all pseudergates can produce soldiers with a tendency toward a later origin, and therefore larger soldiers, in older colonies (71, 72). The first termite soldiers may have had functional gonads, as in extant *Archotermopsis* (35), but it is unknown whether soldiers appeared and were selected for as a defensive caste or as replacement or supplementary reproductives as in modern neotenic soldiers (64, 86, 107).

Single-piece nest species, such as the dampwood genus *Zootermopsis*, provide an opportunity to examine the significance of the spatial organization of tasks to the evolution of division of labor. Brood-care and foraging both occur within the same piece of wood; in ants and multiple-piece nesting termites the nursery and foraging are separated inside and outside of the nest. Maturing termite larvae are likely to eclose in the proximity of the primary reproductives and egg pile and could care for reproductives early in life and transition to nonbrood-care tasks such as nest maintenance and feeding at more distal sites before they develop into reproductive forms and leave the labor force. However, *Zootermopsis* seems to show no temporal polyethism (94); third through seventh instar larvae attend to tasks with no apparent bias.

Reproductive plasticity may also influence polyethism in Termopsidae (76). The ability of larvae to achieve reproductive status in the natal nest and potential conflicts with siblings could reduce selection for behavioral schedules that enhance colony-level fitness at the expense of individual reproductive success. The reproductive plasticity of lower termites could cause individuals to remain near the egg pile where they might deposit their own eggs or engage in oophagy, performing brood-care as they age, rather than providing labor at other work sites. West-Eberhard (121, 122) offered a similar argument concerning temporal polyethism in the social Hymenoptera, suggesting that worker reproduction should result in a brood-care bias toward newly eclosed adults that have functional ovaries. If reproductive competency in termites increases with age, older larvae could be predisposed to brood-care behavior or at least be spatially associated with brood in basal isopteran species. In any case, comparisons of polyethism among termite species indicate that worker sterility and temporal division of labor are correlated (94, 114). The loss of reproductive options among workers and foraging ecology of termites have been prerequisites for the evolution of termite polyethism. A comprehensive theory for the evolution of age-related division of labor in termites requires an understanding of how and when individuals can become reproductively competent and of a species' foraging ecology.

FORAGING BIOLOGY

Striking variation is seen in the foraging biology of the basal ants; a spectrum of ancestral and derived habits has been documented (33). In some species, solitary huntresses search for arthropod prey in subterranean soil galleries and tunnels in decayed wood. In a manner reminiscent of their wasp ancestors, the sting injects paralytic venom into prey, which are subsequently transported to the nest. Feeding specialization and recruitment communication are diverse within and between genera in the tribe Amblyoponini. *Amblyopone pallipes*, for example, solitarily hunts prey such as geophilid centipedes, whereas other species may cooperate in prey capture and transport (36). Amblyoponine species, as well as species in more advanced ponerine tribes, may specialize on certain prey. *Prionopelta amabilis* workers, for example, feed exclusively on campodeid diplurans (33). Other basal ants such as *N. macrops* and *N. myrmecia* forage epigaeically as solitary individuals and use the sting to paralyze prey (31, 33). *A. simoni* workers also use the sting to subdue prey, but supplement their diet with carbohydrate foods such as decaying fruit, and have well-developed chemical trail communication (113).

The foraging ecology of basal termites, like basal ants, reflects the feeding habits of their solitary and subsocial ancestors. Termopsids are one-piece nesters (2, 53, 69), living in and consuming their host log. They do not forage away from the nest wood, and colonies do not leave one stump or log to occupy another. The galleries resulting from the consumption of their host wood become nest chambers, partitioned only by fecal pellet walls. The entire life cycle of most colony members transpires within a single piece of wood. Mature colonies produce fertile offspring (alates) seasonally, and many individuals within the colony differentiate into alates and disperse when resources in the nest wood are depleted (78).

Although termites prefer nutritionally valuable food sources that are low in secondary plant compounds (114), Termopsidae appear to have a limited array of mechanisms that could be implemented to harvest energetically rich cellulose sources. A choice mechanism exhibited by *Zootermopsis nevadensis* involves the foraging discrimination of colony-founding alates that settle on and defend nitrogen-rich wood cambium (53, 100). Whether workers in established colonies direct feeding at nutritionally rich sites in their nest log is unknown. If food selection occurs, it is likely regulated by secretions of the sternal gland, the source of trail pheromones in termites (114).

EVOLUTIONARY PATHOBIOLOGY

Many social insects nest and feed in soil and decayed-wood environments where diverse, abundant, and potentially pathenogenic microbial communities flourish. Group living may compound mortality risks through the interindividual transmission of infection (92). Adapting to disease has long been considered a major event in the evolution of sociality and the diversification of the ants (123), but only recently has the evolutionary significance of social insect pathobiology been the focus of empirical and theoretical investigation, primarily in Hymenoptera (24, 98, 99).

Ants have adapted to the constraints of living in infectious environments through the powerful antibiotic secretions of the metapleural gland (30, 33, 123). The metapleural gland is phylogenetically ancient, appearing in the extinct *Sphecomyrma* and found today in all ant subfamilies. Its evolution is considered to have been critically important to the ecological dominance of the ants (33, 123). Research on disease defenses in *Myrmecia* (58) suggests that the metapleural gland is highly significant but perhaps not the sole mechanism of infection control in basal species. Metapleural gland secretions alone, nevertheless, provide extraordinarily efficacious control of microbes.

Termites, like ants, nest and feed in areas where microbes thrive, and it is likely that pathogens have influenced their social biology. Termite life history traits (monogamy and long life span) as well as several characteristics of their host/pathogen relationships (likelihood of vertical and horizontal transmission among genetically related individuals, probability of prolonged contact with infection agents, and disease transfer through trophallactic exchanges) suggest that key aspects of termite biology could reflect adaptations to reduce pathogen virulence (19, 91). In basal isopteran species, disease resistance represents a confluence of the behavioral, physiological, and biochemical adaptations that characterized the solitary and/or presocial dictyopteran ancestors of termites and the newly adaptive mechanisms of infection control that accompanied their transition to eusociality. Although susceptibility to disease transmission was likely a cost of termite sociality, *Zootermopsis angusticollis* shows a number of infection-control adaptations such as allogrooming (92), colony demography (95), inducible humoral defenses (90), and the "social transfer" of immunocompetence (115). *Z. angusticollis* also communicates information about the presence of pathogens (90) and has antimicrobial exudates (92). In contrast to ants, termites appear to lack a metapleural gland equivalent (22), perhaps because the evolution of potent antimicrobial defenses was compromised by the need to protect their antibiotic-sensitive cellulose-digesting symbionts (92).

The life histories of primitive termites, which feature outbreeding by alates and inbreeding by offspring, may have in part allowed these insects to escape from or adapt to pathogens and lower disease risk (91). In contrast to the hypothesis that genetic similarity fosters the spread of disease in a colony (24), Lewis (54) proposed that pathogen and parasite avoidance, operating through the preferential association of relatives, could be a driving force for sociality. In this model, the spatial association of relatives and kin-directed altruism lowers the probability of infection by an unfamiliar pathogen, favoring reproduction of group members and the maintenance of its kin structure. In basal termites, the cycle of inbreeding by offspring is punctuated by the introduction of new genes through outbreeding, which may enhance disease resistance. Genetic studies on *Z. nevadensis* support alate outbreeding (102). In light of Lewis' model, inbreeding could also be considered as having a function in the avoidance of new and unfamiliar pathogens because it would favor the continued association of relatives. Based on the observation that primary reproductives of *Z. angusticollis* have significantly lower mortality when paired with sibling rather than distantly collected, nonsibling mates, outbreeding depression could occur if infections are transferred through social contact between males and females (93). Infections can be transferred socially between mates (91).

An alternative disease-related explanation for both outbreeding and inbreeding basal termites concerns selection for genetic variation and the maintenance of adaptive genotypes. Cycles of outbreeding by founding reproductives could infuse colonies with genotypes that vary in disease resistance. As colonies mature, coevolutionary interactions could result in selection of the most resistant host genotypes and/or the least virulent pathogens. Some individuals bearing these adapted genotypes may differentiate into inbreeding reproductives, thus maintaining the resistant trait in their offspring.

DISCUSSION

Insect social systems are shaped from the inertial, phylogenetic properties of species and their interaction with environmental forces. Despite the fundamental differences between Formicidae and Isoptera, such as ploidy and holo- versus hemimetaboly, formulating comparisons between basal taxa of these two diverse and entirely eusocial clades reveal commonalities and potentially constructive insights into their early eusocial evolution. This process is not as satisfying as examining evolutionary grades in a group containing a spectrum of solitary through eusocial species such as wasps or bees (9, 123) or in a clade with recent and repeated evolution of eusociality such as halictid bees (15), but no alternative approach exists in extant ants and termites.

Here we attempt to understand the relative contribution and significance of phylogenetic history and ecology in the evolution of colony structure in ants and termites by examining their commonalties and divergences, restricting our analysis to those species that appear to most closely approximate nascent eusocial forms. In doing so we acknowledge the inference restrictions inherent in using extant taxa to reconstruct the social past and the limitations imposed upon comparisons of analogous systems, as well as confounding issues surrounding the existence of both primitive and derived traits in modern basal species. After careful consideration of these caveats, however, we remain confident that students of both isopteran and hymenopteran societies can mutually benefit from an understanding of the predispositions and ecologies that have guided social evolution in each group, and that our preliminary attempt at a synthesis of the two literatures will encourage productive discussion and collaboration.

Reproductive Structure and Division of Labor

For basal ant and termite species, we suggest that the reproductive organization of colonies and patterns of division of labor were affected by suites of ecological factors that operated at various stages of colony life histories. We identify nesting and feeding ecology as the environmental influences that impelled the adaptive modification of reproductive organization and division of labor in both groups. In basal ants and termites, the nest (including the colony it houses) represents a resource that provides nutrition, a structured environment to rear offspring, and a labor force. In termites with the most primitive social structure, the nest and the food source are the same piece of wood. In some basal ants the nest is a collection of gallery systems and chambers from which foraging excursions are conducted over short distances within restricted areas. The nutritional aspect of the ant nest lies in the quality of its foraging territory. Nest structure is the result of prior colony labor. The nest, in the broad sense including ant foraging territory, is thus a valuable resource. Opportunities to inherit the nest and colony may have been significant in the evolution of multiple reproductive forms and options in both taxa.

Ancestral ants likely foraged relatively close to the nest and had little if any polyethism because of the sequential unity of prey paralysis, transport, and direct provisioning of larvae (33, 111). Because of small colony size, relatively synchrous brood development cycles, and large prey size, foraging excursions may have been few in number, close to the nest, and within the confines of subterranean galleries or tunnels and crevices in decayed wood. Predation rates in species such as *Amblyopone* may be low in comparison to the higher ants, which have large colony size and forage epigaeically at greater distances from the nest. Our model assumes that *Amblyopone*, rather than *Nothomyrmecia*, represents a closer approximation to the biology of incipient eusocial ants. In basal termites the nest is the food source; in basal ants the distinction between nest and foraging territory is minor. The expansion of the diet, foraging away from the nest chamber, and increased colony size in termites is correlated with the evolution of a true worker caste and division of labor (3, 53). It is a reasonable hypothesis that spatial separation of nest and feeding territory and their unambiguous discrimination was also of significance in the evolution of social organization of ants. Together with nest structure, age-related changes in the reproductive physiology of helpers, predation, and decreased reproductive competition were likely interrelated and important determinants of polyethic task schedules and division of labor, although understanding the influence of these factors requires further detailed investigation. It is a challenge, for example, to explain why *Nothomyrmecia*, which forages away from the nest, shows no division of labor (40). Perhaps ancestral social states, including limited interaction and cooperation, were retained in this relict ant.

Serial Reproductive Inheritance by Kin

Similarities in reproductive structure found in some basal ants and termites are striking and potentially revealing. Recent genetic work on the primitive ant *N. macrops* suggests colony inheritance by daughters (97). Other basal ants such as *Myrmecia* and perhaps some *Amblyopone* have life histories in which family members, even helper family members in some cases, may become reproductives within the parental nest. These dynamics and the associated resource and inclusive fitness advantages of nest inheritance in ants have similarities with the developmental plasticity and colony inheritance characteristic of replacement (neotenic) reproductives in basal termites. The common feature of serial reproductive inheritance by kin means that all colony members gain enhanced inclusive fitness benefits, and some individuals acquire direct fitness advantages. These cumulative fitness components may well exceed average individual fitness prospects of dispersing, fertile offspring in a similar, solitary species, thus favoring helpers that remain in the natal colony.

The influence of potential reproductive opportunities on the evolution of helping behavior have long been recognized and are compounded by inheritance of resources and the fitness advantages of reproducing in the natal nest. West-Eberhard (120, p. 853) observed, "... Michener has long insisted that helping behavior without altruism can occur if male production by 'workers' ... is important enough (56). The significance of this argument has not generally been appreciated. Whether among relatives or not, as long as a female has "hope" of laying eggs—at least some small probability of future reproduction—her participation in the worker tasks can be viewed as possibly or partially an investment in her own reproductive future.... As long as a certain percentage of functional workers ultimately lays some eggs, then every worker—even those which never do lay eggs—can be considered 'hopeful' in the sense of having a certain probability of reproduction." The "hopeful reproductive" dynamic is influenced by colony size. As colony size increases, individual workers have a lower chance of becoming replacement reproductives (7). The most primitive ants and termites have small family sizes (33, 35), thus favoring opportunities for individual offspring to become reproductives.

Disease Risk, the Evolution of Resistance, and Social Organization

In ancestral ants, the problems that disease risk posed for coloniality may have been fully solved by one key innovation: the evolution of the metapleural gland in ancestral species. This gland is well developed in all basal ants studied to date (30, 33). In basal termites, the ability to control infection with powerful antibiotic secretions biochemically similar to those secreted by the ant metapleural gland was likely compromised by the need to maintain gut symbionts (92). In contrast to the chemical mode of infection control in ants, cycles of inbreeding and outbreeding in basal termites may have resulted in the selection and maintenance of disease-adapted genotypes. We note, however, that we do not identify disease as the sole factor influencing termite life cycles and acknowledge that other factors were also significant (101, 107). Nevertheless, the dispersal of reproductive forms from the parental nest and the colonization of new food source nest sites may have involved local adaptations to pathogens that were generated and preserved by outbreeding and inbreeding, respectively.

Conclusion and Prospects

In both ants and termites, eusociality was probably fostered by a suite of contributing factors and the interacting selective pressures that they generated. Haplodiploidy, maternal care, and female-biased sex ratios have favored the evolution of eusociality in Hymenoptera [reviewed in (33, 116)]. Kin-based explanations anchored by inclusive fitness pay-offs are also the premise of the nonmutually exclusive theories explaining the evolution of eusociality in the diploid-diploid Isoptera. These include cyclic inbreeding (66), shift in dependent care (67), intragroup conflict (85), disease resistance (92, 95, 115), and predispositions related to ecological and life history attributes that favored helping behavior and reproductive skew (101, 107).

In addition to expanded study of key primitive taxa, priority domains of future research on basal ants and termites center on further understanding their mating and reproductive biology. Specifically, topics should include: (*a*) determination of relatedness among mates and number of mates of individuals that inherit colonies (e.g., daughter queens of *Nothomyrmecia*, ergatogynes, gamergates, and termite neotenics); (*b*) study of fertility or subfertility of workers and soldiers in *Archotermopsis*; (*c*) investigation of the possibility of foreign reproductives joining an existing colony, either through immigration or fusion; (*d*) further research on the circumstances surrounding policing and mutilation, and their implications for reproductive conflict; and (*e*) detailed studies of the mechanisms and organization of division of labor. Expanded knowledge of these subjects/areas will provide a stronger foundation for resolving reproductive patterns and their fitness implications for reproductives and helpers, thus facilitating broader synthesis of patterns of evolution in ants, termites, and other eusocial animals.

ACKNOWLEDGMENTS

We thank anonymous reviewers for constructive suggestions on the manuscript.

The *Annual Review of Entomology* is online at http://ento.annualreviews.org

LITERATURE CITED

1. Abe T. 1987. Evolution of life types in termites. In *Evolution and Coadaptation in Biotic Communities*, ed. S Kawano, JH Connell, T Hidaka, pp. 125–48. Tokyo: Univ. Tokyo Press
2. Abe T. 1990. Evolution of worker caste in termites. In *Social Insects and the Environments,* ed. GK Veeresh, B Mallik, CA Viraktamath, pp. 29–30. New Delhi: Oxford & IBH
3. Abe T. 1991. Ecological factors associated with the evolution of worker and soldier castes in termites. *Annu. Rev. Entomol.* 9:101–7
4. Abe T, Bignell DE, Higashi M, eds. 2000. *Termites: Evolution, Sociality, Symbioses, Ecology.* Dordrecht: Kluwer. 466 pp.
5. Baroni-Urbani UC, Bolton B, Ward PS. 1992. The internal phylogeny of ants (Hymenoptera:Formicidae). *Syst. Entomol.* 17:301–29
6. Boudreaux HB. 1979. *Arthropod Phylogeny with Special Reference to Insects.* New York: Wiley. 320 pp.
7. Bourke AFG. 1999. Colony size, social complexity and reproductive conflict in social insects. *J. Evol. Biol.* 12:245–57
8. Bulmer MS, Adams ES, Traniello JFA. 2001. Variation in colony structure in the subterranean termite *Reticulitermes flavipes. Behav. Ecol. Sociobiol.* 49:236–43
9. Choe J, Crespi B, eds. 1997. *The Evolution of Social Behavior in Insects and Arachnids.* Cambridge, UK: Cambridge Univ. Press. 541 pp.
10. Cleveland LR, Hall SR, Sanders EP, Collier J. 1934. The wood-feeding roach *Cryptocercus*, its protozoa, and the symbiosis between protozoa and roach. *Mem. Am. Acad. Arts Sci.* 17:185–342
11. Constantino R. 2002. Online termite database. http://www.unb.br/ib/zoo/docente/constant/catal/catnew.html
12. Craig R, Crozier RH. 1979. Relatedness in the polygynous ant *Myrmecia pilosula. Evolution* 33:335–41
13. Crosland MWJ, Crozier RH, Jefferson E. 1988. Aspects of the biology of the primitive ant genus *Myrmecia* F. (Hymenoptera:Formicidae). *J. Aust. Entomol. Soc.* 27:305–9
14. Crozier RH, Jermin LS, Chiotis M. 1997. Molecular evidence for a Jurassic origin of ants. *Naturwissenschaften* 84:22–23
15. Danforth BN. 2002. Evolution of sociality in a primitively eusocial lineage of bees. *Proc. Natl. Acad. Sci. USA* 99:286–90
16. Donovan SE, Jones DT, Sands WA, Eggleton P. 2000. Morphological phylogenetics of termites (Isoptera). *Biol. J. Linn. Soc.* 70:467–513
17. Eggleton P. 2001. Termites and trees: a review of recent advances in termite phylogenetics. *Insectes Soc.* 48:187–93
18. Emerson AE, Krishna K. 1975. The termite family Serritermitidae (Isoptera). *Am. Mus. Nov.* 2570:1–31
19. Ewald PW. 1994. *The Evolution of Infectious Disease.* New York: Oxford Univ. Press. 298 pp.
20. Gay FJ, Calaby JH. 1970. Termites of the Australian region. See Ref. 49, pp. 393–448
21. Goodisman MAD, Crozier RH. 2002. Population and colony genetic structure of the primitive termite *Mastotermes darwiniensis. Evolution* 56:70–83
22. Grassé PP. 1982. *Termitologia, Vol. I.* Paris: Masson. 676 pp.
23. Grimaldi D, Agosti D. 1997. A formicine

in New Jersey Cretaceous amber (Hymenoptera:Formicidae) and early evolution of the ants. *Proc. Natl. Acad. Sci. USA* 97:13678–83

24. Hamilton WD. 1987. Kinship, recognition, disease, and intelligence: constraints on social evolution. In *Animal Societies: Theories and Facts*, ed. Y Ito, JL Brown, J Kikkawa, pp. 81–102. Tokyo: Jpn. Sci. Soc. Press
25. Haskins CP, Haskins EF. 1950. Notes on the biology and social behavior of the archaic primitive ants of the genera *Myrmecia* and *Promyrmecia*. *Ann. Entomol. Soc. Am.* 43:461–91
26. Haskins CP, Haskins EF. 1955. The pattern of colony foundation in the archaic ant *Myrmecia regularis*. *Insectes Soc.* 2:115–26
27. Heath H. 1902. The habits of California termites. *Biol. Bull.* 4:47–63
28. Hennig W. 1981. *Insect Phylogeny.* New York: Wiley. 514 pp.
29. Higashi M, Yamamura N, Abe T, Burns TP. 1991. Why don't all termite species have a sterile worker caste? *Proc. R. Soc. London* 246:25–29
30. Hölldobler B, Engel-Siegel H. 1984. On the metapleural gland of ants. *Psyche* 91:201–24
31. Hölldobler B, Taylor R. 1983. A behavioral study of the primitive ant *Nothomyrmecia macrops* Clark. *Insectes Soc.* 30:384–401
32. Hölldobler B, Wilson EO. 1986. Ecology and behavior of the primitive crypobiotic ant *Prionopelta amabilis*. *Insectes Soc.* 33:45–58
33. Hölldobler B, Wilson, EO. 1990. *The Ants*. Cambridge: Harvard Univ. Press. 732 pp.
34. Hunt JH, Nalepa CA, eds. 1994. *Nourishment and Evolution in Insect Societies*. Boulder, CO: Westview. 449 pp.
35. Imms AD. 1919. On the structure and biology of *Archotermopsis*, together with descriptions of new species of intestinal protozoa, and general observations on the Isoptera. *Philos. Trans. R. Soc. London Ser. B*. 209:75–180
36. Ito F. 1993. Observation of group recruitment to prey in a primitive ponerine ant, *Amblyopone* sp. (reclinata group) (Hymenoptera; Formicidae). *Insectes Soc.* 40:163–67
37. Ito F. 1993. Social organization in a primitive ponerine ant—queenless reproduction, dominance hierarchy and functional polygyny in *Amblyopone* sp. (*reclinata* group) (Hymenoptera: Formicidae: Ponerinae). *J. Nat. Hist.* 27:1315–24
38. Ito F, Billen J. 1998. Larval hemolymph feeding and oophagy: behavior of queen and workers in the primitive ponerine ant *Prionopelta kraepelini* (Hymenoptera: Formicidae). *Belg. J. Zool.* 128:201–9
39. Ito F, Sugiura N, Higashi S. 1994. Worker polymorphism in the red-head bulldog ant (Hymenoptera: Formicidae), with description of nest structure and colony composition. *Ann. Entomol. Soc. Am.* 87: 337–41
40. Jaisson P, Fresneau D, Taylor RW, Lenoir A. 1992. Social organization in some primitive Australian ants. I. *Nothomyrmecia macrops*. *Insectes Soc.* 39:425–38
41. Jayasuriya AK, Traniello JFA. 1985. The biology of the primitive ant *Aneuretus simoni* (Emery) (Formicidae:Aneuretinae). I. Distribution, abundance, colony structure and foraging ecology. *Insectes Soc.* 32:363–74
42. Kambhampati S. 1995. A phylogeny of cockroaches and related insects based on DNA-sequence of mitochondrial ribosomal-RNA genes. *Proc. Natl. Acad. Sci. USA* 92:2017–20
43. Kambhampati S, Eggleton P. 2000. Taxonomy and phylogeny of termites. See Ref. 4, pp. 1–23
44. Klass KD. 1997. The external male genitalia and the phylogeny of *Blattaria* and *Mantodea*. *Bonn. Zool. Monogr.* 42:1–341
45. Klass KD. 1998. The ovipositor of Dictyoptera (Insecta): homology and

ground-plan of the main elements. *Zool. Anz.* 236:69–101
46. Klass KD. 1998. The proventriculus of the Dicondylia, with comments on evolution and phylogeny in Dictyoptera and Odonata (Insecta). *Zool. Anz.* 237:15–42
47. Klass KD. 2000. The male abdomen of the relic termite *Mastotermes darwiniensis* (Insecta: Isoptera: Mastotermitidae). *Zool. Anz.* 239:231–62
48. Krishna K. 1970. Taxonomy, phylogeny, and distribution of termites. See Ref. 49, pp. 127–52
49. Krishna K, Weesner FM, eds. 1970. *Biology of Termites, Vol. 2.* New York: Academic. 643 pp.
50. Kristensen NP. 1975. The phylogeny of hexapod 'orders.' A critical review of recent accounts. *Z. Zool. Syst. Evol.* 13:1–44
51. Kristensen NP. 1995. Forty years' insect phylogenetic systematics. *Zool. Beitr.* 36:83–124
52. Lacasa-Ruiz A, Martinez-Delclòs X. 1986. *Meiatermes*: nuevo género fosil de insecto Isoptero (Hodotermitidae) de las calizas Neocomienses del Montsec (Provincia de Lérida, Espana). *Publ. Inst. d'Estidis Ilerdencs Diput. Prov. Lléida.* pp. 5–65
53. Lenz M. 1994. Food resources, colony growth and caste development in wood-feeding termites. See Ref. 34, pp. 159–209
54. Lewis K. 1998. Pathogen resistance as the origin of kin altruism. *J. Theor. Biol.* 193:359–63
55. Light SF. 1943. The determination of caste in social insects. *Q. Rev. Biol.* 18:46–63
56. Lin N, Michener CD. 1972. Evolution of sociality in insects. *Q. Rev. Biol.* 47:131–59
57. Lo N, Tokuda G, Watanabe H, Rose M, Slaytor M, et al. 2000. Evidence from multiple gene sequences indicates that termites evolved from wood-feeding cockroaches. *Curr. Biol.* 10:801–4
58. Mackintosh JA, Veal DA, Beattie AJ, Gooley AA. 1998. Isolation from an ant *Myrmecia gulosa* of two inducible O-glycosylated proline-rich antibacterial peptides. *J. Biol. Chem.* 273:6139–43
59. Maekawa K, Kitade O, Matsumoto T. 1999. Molecular phylogeny of orthopteroid insects based on the mitochondrial cytochrome oxidase II gene. *Zool. Sci.* 16:175–84
60. Masuko K. 1986. Larval hemolymph feeding: a nondestructive parental cannibalism in the primitive ant *Amblyopone silvestrii* Wheeler. *Behav. Ecol. Sociobiol.* 19:249–55
61. Masuko K. 1996. Temporal division of labor among wokers in the ponerine ant *Amblyopone silvestrii. Sociobiology* 28:131–51
62. Monnin T, Ratnieks FLW. 2001. Policing in queenless ponerine ants. *Behav. Ecol. Sociobiol.* 50:97–108
63. Morgan FD. 1959. The ecology and external morphology of *Stolotermes ruficeps* Brauer (Isoptera: Hodotermitidae). *Trans. R. Soc. N. Z.* 86:155–95
64. Myles TG. 1986. Reproductive soldiers in the Termopsidae (Isoptera). *Pan Pac. Entomol.* 62:293–99
65. Myles TG. 1988. Resource inheritance in social evolution from termites to man. In *The Ecology of Social Behavior*, ed. CN Slobodchikoff, pp. 379–423. New York: Academic
66. Myles TG, Nutting WL. 1988. Termite eusocial evolution: a re-examination of Bartz's hypothesis and assumptions. *Q. Rev. Biol.* 63:1–23
67. Nalepa CA. 1994. Nourishment and the origin of termite eusociality. See Ref. 34, pp. 57–104
68. Nalepa CA, Bandi C. 2000. Characterizing the ancestors: paedomorphosis and termite evolution. See Ref. 4, pp. 53–75
69. Noirot C. 1970. The nests of termites. See Ref. 49, pp. 73–125
70. Noirot C. 1982. La caste des ouvriers, élément majeur du succès évolutif des termites. *Riv. Biol.* 72:157–95

71. Noirot C. 1985. Pathways of caste development in the lower termites. See Ref. 117, pp. 41–57
72. Noirot C. 1989. Social structure in termite societies. *Ethol. Ecol. Evol.* 1:1–17
73. Noirot C. 1995. The sternal glands of termites—segmental pattern, phylogenetic implications. *Insectes Soc.* 42:321–23
74. Noirot C, Bordereau C. 1989. Termite polymorphism and morphogenetic hormones. In *Morphogenetic Hormones of Arthropods*, ed. AP Gupta, pp. 293–324. New Brunswick: Rutgers Univ. Press
75. Noirot C, Noirot-Timothée C. 1977. Fine structure of the rectum in termites (Isoptera): a comparative study. *Tissue Cell* 9:693–710
76. Noirot C, Pasteels JM. 1987. Ontogenetic development and evolution of the worker caste in termites. *Experientia* 43:851–60
77. Noirot C, Pasteels JM. 1988. The worker caste is polyphyletic in termites. *Sociobiology* 14:15–20
78. Nutting WL. 1969. Flight and colony foundation. In *Biology of Termites, Vol. 1*, ed. K Krishna, FM Weesner, pp. 233–82. New York: Academic
79. Oster G, Wilson EO. 1978. *Caste and Ecology in the Social Insects*. Princeton: Princeton Univ. Press. 352 pp.
80. Pasteels JM. 1965. Polyéthisme chez les ouvriers de *Nasutitermes lujae* (Termitidae Isoptères). *Biol. Gabonica* 1:191–205
81. Peeters C. 1993. Monogyny and polygyny in ponerine ants with and without queens. In *Queen Number and Sociality in Insects*, ed. L Keller, pp. 234–61. New York: Oxford Univ. Press
82. Peeters C. 1997. Morphologically "primitive" ants: comparative review of social characters, and the importance of queen-worker dimorphism. See Ref. 9, pp. 372–91
83. Peeters C, Ito F. 2001. Colony dispersal and the evolution of queen morphology in social Hymenoptera. *Annu. Rev. Entomol.* 46:601–30
84. Powell S, Tschinkel WR 1999. Ritualized conflict in *Odontomachus brunneus* and the generation of interaction-based task allocation: a new organizational mechanism in ants. *Anim. Behav.* 58:965–72
85. Roisin Y. 1994. Intragroup conflicts and the evolution of sterile castes in termites. *Am. Nat.* 143:751–65
86. Roisin Y. 1999. Philopatric reproduction, a prime mover in the evolution of termite sociality? *Insectes Soc.* 46:297–305
87. Roisin Y. 2000. Diversity and the evolution of caste patterns. See Ref. 4, pp. 95–119
88. Roisin Y. 2001. Caste sex ratios, sex linkage, and reproductive strategies in termites. *Insectes Soc.* 48:224–30
89. Roonwal ML, Bose G, Verma SC. 1984. The Himalayan termite, *Archotermopsis wroughtoni* (synonyms *radcliffei* and *deodarae*). Identity, distribution and biology. *Rec. Zool. Surv. India* 81:315–38
90. Rosengaus RB, Jordan C, Lefebvre ML, Traniello JFA. 1999. Pathogen alarm behavior in a termite: a new form of communication in social insects. *Naturwissenschaften* 86:544–48
91. Rosengaus RB, Lefebvre ML, Carlock DM, Traniello JFA. 2000. Socially transmitted disease in adult reproductive pairs of the dampwood termite *Zootermopsis angusticollis*. *Ethol. Ecol. Evol.* 12:419–33
92. Rosengaus RB, Maxmen AB, Coates LE, Traniello JFA. 1998. Disease resistance: a benefit of sociality in the dampwood termite *Zootermopsis angusticollis* (Isoptera:Termopsidae). *Behav. Ecol. Sociobiol.* 44:125–34
93. Rosengaus RB, Traniello JFA. 1993. Disease risk as a cost of outbreeding in the termite *Zootermopsis angusticollis*. *Proc. Natl. Acad. Sci. USA* 90:6641–45
94. Rosengaus RB, Traniello JFA. 1993. Temporal polyethism in incipient colonies of the primitive termite *Zootermopsis angusticollis*: a single multi-age caste. *J. Insect Behav.* 6:237–52

95. Rosengaus RB, Traniello JFA. 2001. Disease susceptibility and the adaptive nature of colony demography in the dampwood termite *Zootermopsis angusticollis*. *Behav. Ecol. Sociobiol.* 50:546–56
96. Sanetra M, Crozier RH. 2001. Polyandry and colony genetic structure in the primitive ant *Nothomyrmecia macrops*. *J. Evol. Biol.* 14:368–78
97. Sanetra M, Crozier RH. 2002. Daughters inherit colonies from mothers in the 'living-fossil' ant *Nothomyrmecia macrops*. *Naturwissenschaften* 89:71–74
98. Schmid-Hempel P. 1998. *Parasites in Social Insects*. Princeton: Princeton Univ. Press. 409 pp.
99. Schmid-Hempel P, Crozier RH. 1999. Polyandry versus polygyny versus parasites. *Philos. Trans. R. Soc. London B* 354:507–15
100. Shellman-Reeve JS. 1994. Limited nutrients in a dampwood termite—nest preference, competition and cooperative nest defense. *J. Anim. Ecol.* 63:921–32
101. Shellman-Reeve JS. 1997. The spectrum of eusociality in termites. See Ref. 9, pp. 52–93
102. Shellman-Reeve JS. 2001. Genetic relatedness and partner preference in a monogamous, wood-dwelling termite. *Anim. Behav.* 61:869–76
103. Sherman PW, Lacey EA, Reeve HK, Keller L. 1995. The eusociality continuum. *Behav. Ecol.* 6:102–8
104. Taylor R. 1978. *Nothomyrmecia macrops*: a living fossil ant rediscovered. *Science* 201:979–85
105. Thompson GJ, Kitade O, Lo N, Crozier RH. 2000. Phylogenetic evidence for a single, ancestral origin of a 'true' worker caste in termites. *J. Evol. Biol.* 13:869–81
106. Thorne BL. 1996. Termite terminology. *Sociobiology* 28:253–63
107. Thorne BL. 1997. Evolution of eusociality in termites. *Annu. Rev. Ecol. Syst.* 28:27–54
108. Thorne BL, Carpenter JM. 1992. Phylogeny of the Dictyoptera. *Syst. Entomol.* 17:253–68
109. Thorne BL, Grimaldi DA, Krishna K. 2000. Early fossil history of the termites. See Ref. 4, pp. 77–93
110. Thorne BL, Lenz M. 2001. Population and colony structure of *Stolotermes inopinus* and *S. ruficeps* (Isoptera: Stolotermitinae) in New Zealand. *N. Z. Entomol.* 24:63–70
111. Traniello JFA. 1978. Caste in a primitive ant: absence of age polyethism in *Amblyopone*. *Science* 202:770–72
112. Traniello JFA. 1982. Population structure and social organization in the primitive ant *Amblyopone pallipes* (Hymenoptera: Formicidae). *Psyche* 89:65–80
113. Traniello JFA, Jayasuriya AK. 1985. The biology of the primitive ant *Aneuretus simoni* (Emery). II. The social ethogram and division of labor. *Insectes Soc.* 32:375–88
114. Traniello JFA, Leuthold R. 2000. The behavioral ecology of foraging in termites. See Ref. 4, pp. 141–68
115. Traniello JFA, Rosengaus RB, Savoie K. 2002. The development of immunocompetence in a social insect: evidence for the group facilitation of disease resistance. *Proc. Natl. Acad. Sci. USA* 99:6838–42
116. Wade MJ. 2001. Maternal effect genes and the evolution of sociality in haplo-diploid organisms. *Evolution* 55:453–58
117. Watson JAL, Okot-Kotber BM, Noirot C, eds. 1985. *Caste Differentiation in Social Insects*. Oxford: Pergamon. 405 pp.
118. Watson JAL, Sewell JJ. 1981. The origin and evolution of caste systems in termites. *Sociobiology* 6:101–18
119. Watson JAL, Sewell JJ. 1985. Caste development in *Mastotermes* and *Kalotermes*: which is primitive? See Ref. 117, pp. 27–40
120. West-Eberhard MJ. 1978. Polygyny and the evolution of social behavior in wasps. *J. Kans. Entomol. Soc.* 51:832–56
121. West-Eberhard MJ. 1979. Sexual selection, social competition and evolution. *Proc. Am. Philos. Soc.* 123:222–34

122. West-Eberhard MJ. 1981. Intragroup selection and the evolution of insect societies. In *Natural Selection and Social Behavior*, ed. RD Alexander, DW Tinkle, pp. 3–17. New York: Chiron
123. Wilson EO. 1971. *The Insect Societies.* Cambridge, MA: Harvard Univ. Press. 548 pp.
124. Wilson EO. 1987. The earliest known ants: an analysis of the Cretaceous species and an inference concerning their social organization. *Paleobiology* 13:44–53

Annu. Rev. Entomol. 2003. 48:307–37
doi: 10.1146/annurev.ento.48.091801.112728
First published online as a Review in Advance on October 16, 2002

THE ASCENDANCY OF *AMBLYOMMA AMERICANUM* AS A VECTOR OF PATHOGENS AFFECTING HUMANS IN THE UNITED STATES*

James E. Childs and Christopher D. Paddock
Viral and Rickettsial Zoonoses Branch, Division of Viral and Rickettsial Diseases, National Center for Infectious Diseases, Centers for Disease Control and Prevention, Atlanta, Georgia 30333; e-mail: jchilds@cdc.gov; cpaddock@cdc.gov

Key Words Ehrlichioses, epidemiology, tick-borne diseases, emerging diseases, zoonotic diseases

■ **Abstract** Until the 1990s, *Amblyomma americanum* was regarded primarily as a nuisance species, but a tick of minor importance as a vector of zoonotic pathogens affecting humans. With the recent discoveries of *Ehrlichia chaffeensis*, *Ehrlichia ewingii*, and "*Borrelia lonestari*," the public health relevance of lone star ticks is no longer in question. During the next 25 years, the number of cases of human disease caused by *A. americanum*-associated pathogens will probably increase. Based on current trajectories and historic precedents, the increase will be primarily driven by biological and environmental factors that alter the geographic distribution and intensity of transmission of zoonotic pathogens. Sociologic and demographic changes that influence the likelihood of highly susceptible humans coming into contact with infected lone star ticks, in addition to advances in diagnostic capabilities and national surveillance efforts, will also contribute to the anticipated increase in the number of recognized cases of disease.

CONTENTS

INTRODUCTION

Until relatively recently, the lone star tick, *Amblyomma americanum* (L.), was regarded as the pathogen-poor relative of the other common species of human-biting ixodid ticks inhabiting North America (125). Despite its distinction as the first tick to be described in the United States in 1754 and its reputation as a major pest to humans and livestock (62), the lone star tick's position as principal vector for any human disease was not convincingly demonstrated until the early 1990s.

Dermacentor variabilis in the eastern United States and *Dermacentor andersoni* in the western United States held principal claim for the transmission of *Rickettsia rickettsii* (20), the etiologic agent of Rocky Mountain spotted fever (RMSF), the most commonly fatal tick-borne disease in the Western Hemisphere. Similarly, *Ixodes scapularis* in the eastern and north-central United States and *Ixodes pacificus* in the western United States were firmly established as the principal vectors of *Borrelia burgdorferi* (78), the etiologic agent of Lyme disease, the most frequently reported vector-borne disease in the United States. In contrast, the primary human and veterinary health concerns regarding *A. americanum* were founded upon its aggressive and nondiscriminatory biting habits at all life stages, resulting in its notorious reputation as a nuisance species (13, 62).

There have been several occasions when *A. americanum* appeared to be the natural suspect in situations involving outbreaks or sporadic occurrences of human disease in which other tick vectors could be effectively eliminated. The most famous of these outbreaks was the mysterious "Bullis fever" that swept through a company of soldiers, stationed at Fort Sam Houston, who had participated in maneuvers at Camp Bullis, Texas, during the spring of 1942. Over 1000 cases of an acute febrile illness, accompanied by severe headache, marked lymphadenopathy, weakness, nausea, and vomiting, developed among the soldiers, all of who had received multiple tick bites during the days preceding their illness (148). At the time of hospitalization, several of the ill soldiers still had ticks attached to them that were identified as lone star ticks. James M. Brennan, a medical entomologist who investigated the site, commented about the innumerable abundance of lone star ticks: "The writer could find no records in literature, through correspondence, or from verbal information, of a greater concentration of this species elsewhere in the United States" (17). Brennan noted that on July 24, 1943, four men collected 4086 adult *Amblyomma* from a single location without moving (17). Serologic and animal inoculation tests ruled out *Coxiella burnetii* (the agent of Q fever), *R. rickettsii*, and *Rickettsia typhi* (the agent of murine typhus) as causes of Bullis fever (148). Rickettsiae were reported to have been isolated from the blood and lymph nodes of patients with Bullis fever and from emulsions of *A. americanum* (4). The putative agent of Bullis fever was named "*Rickettsia texiana*" (4), although no isolate exists today. Bullis fever apparently vanished after 1943; however, speculation about the nature of the causative agent continues (58). The lone star tick would have to wait another 50 years to unequivocally obtain principal vector status for an infectious agent of humans.

In this review we focus on the accumulating data that incriminate *A. americanum* as an important vector of zoonotic pathogens of humans, in particular, concentrating on the *Amblyomma*-associated ehrlichioses. We summarize information on the population dynamics of this tick and how its geographic distribution and population density have been influenced by corresponding changes occurring among its principal vertebrate hosts. Last, we describe a variety of additional factors that have contributed to the increasing recognition of the public health significance of human diseases associated with this tick and speculate on future trends in the incidence of disease.

NATURAL HISTORY OF *AMBLYOMMA AMERICANUM*

A. americanum is a three-host, non-nidicolous tick distributed from west-central Texas, north to Iowa, and eastward in a broad belt spanning the southeastern United States. Along the Atlantic Coast, the range of this species extends through coastal areas of New England as far north as Maine (71). Sustainable lone star tick populations may also occur or exist transiently in foci well outside their well-established range. Historical records (64) and isolated reports of lone star ticks from western and upper-midwestern states (95) could reflect established regional populations or ticks unintentionally transported on humans with a recent history of travel.

A. americanum is found predominantly in woodland habitats, particularly young second-growth forests with dense underbrush (62). The abundance of lone star tick populations is influenced largely by the availability of suitable animal hosts for the life stages of the tick and by the availability of habitats with physiographic features that offer protection for hosts and guard against desiccation of the tick. In this context, white-tailed deer represent a preeminent host for *A. americanum* because they provide blood meal sources for all three stages of *A. americanum* and generally deposit engorged ticks in wooded habitats that maximize tick survival (99). Lone star ticks are aggressive nonspecific feeders and bite humans at all three stages. Similarly, few mammals or birds are exempt as potential hosts for one or more stages of this tick. Adult *A. americanum* feed on medium- and large-sized mammals, and larvae and nymphs infest various ground-feeding birds, medium- and large-sized mammals, and, on occasion, small mammals (76). Although the host range for lone star ticks is vast, *A. americanum* exhibits considerable dependence on larger wildlife species as hosts. A parameter to quantify the relative qualities of a host for *A. americanum* ticks seeking a blood meal was developed by Mount et al. (99) as the base- or intrinsic host-finding rate. These rates were derived from published values and varied with the size, habits, attractiveness, and suitability of a particular host for tick feeding and as rates specific to tick stage. Estimates of the intrinsic host-finding rates for larval ticks are estimated to be >20-fold higher for white-tailed deer than for small mammals and birds and >5-fold higher for white-tailed deer than for medium-sized mammals (99). Intrinsic host-finding rates for adult lone star ticks are even more disparate and are estimated to be >400-fold higher for white-tailed deer than for medium-sized mammals (99). These data suggest that in the absence of large mammalian hosts, *A. americanum* populations will decline and densities of ticks on medium-sized mammals and birds will also diminish (99).

Within their geographic range, lone star ticks are often the most common tick submitted for identification or reported by humans parasitized by a tick. When newspaper advertisements and public awareness posters in Georgia and South Carolina from 1990 through 1995 solicited tick submissions, 83% ($N = 913$) of the submitted ticks were *A. americanum*, including 231 adults, 262 nymphs, and 265 larvae (52). In an investigation that provided an epidemiologic link between *A. americanum* and an erythema migrans-like rash illness in North Carolina, 97% ($N = 588$) of the ticks collected from vegetation and 95% ($N = 197$) of the ticks attached to humans were lone star ticks (72).

Adult and nymphal lone star ticks are generally most active during April through June and decline markedly in abundance and activity as summer progresses (36, 62). The early-season activity of adult and nymphal ticks, which precedes that of larvae, increases the probability of acquisition of a pathogen by larval ticks at the first blood meal. *A. americanum* overwinters as replete larvae, unfed or replete nymphs, or unfed adults (62).

BACTERIA (OTHER THAN EHRLICHIAE) ASSOCIATED WITH *AMBLYOMMA AMERICANUM*

Various bacteria have been isolated or detected from *A. americanum* (Table 1). At least five are agents of disease in humans. Some of the bacteria listed in Table 1 have been isolated only from ticks (e.g., WB-8-2 and MOAa agents) and are of unknown pathogenicity in humans. Others are believed to cause human infection on the basis of serologic reactivity to their specific antigens [e.g., 85-1034, "*R.*

TABLE 1 Bacteria isolated or identified from *Amblyomma americanum*

Bacterial agent	Disease in humans	Comments	Reference
Ehrlichia chaffeensis	Human monocytic ehrlichiosis (HME)	The most severe of the three ehrlichioses of humans in the United States. Underreported and probably as common as Rocky Mountain spotted fever.	(2, 53)
Ehrlichia ewingii	*E. ewingii* ehrlichiosis	Most commonly diagnosed in immunosuppressed persons. Less than 20 cases documented.	(19, 108)
Rickettsia rickettsii	Rocky Mountain spotted fever	Role of lone star ticks in transmission is uncertain, as recent surveys have not identified *R. rickettsii* in ticks.	(59, 113)
Coxiella burnetii	Q fever	Tick transmission is not thought to play a significant role in human disease, although many species of ticks are naturally infected.	(32, 113)
Francisella tularensis	Tularemia	Tick transmission remains important in endemic occurrence. Other routes of transmission, such as direct contact with wild rabbits, are also significant.	(66, 136)
"*Borrelia lonestari*"	Probable cause of southern tick-associated rash illness	Likely to become recognized as a common disease where lone star ticks exist in high numbers. Agent as yet uncultivable.	(9, 70)
85-1034 ("*Rickettsia amblyommii*")	Possible mild spotted fever rickettsiosis	Association with human disease based on serologic reactivity only.	(33)
Rickettsia parkeri	None described	Originally isolated from *Amblyomma maculatum* in Texas.	(59, 112)
WB-8-2	None described	Nonpathogenic or mildly pathogenic in guinea pigs and meadow voles. Most closely related to MOAa and *Rickettsia montana*.	(22, 146)
MOAa	None described	Most closely related to WB-8-2 and *Rickettsia montana*.	(146)

amblyommii" (33)] or on the basis of identification of presumed pathogen DNA in samples from clinically ill persons [e.g., *Borrelia lonestari* (70)].

Francisella tularensis

The potential link between a tick vector and the transmission of *Francisella tularensis* was first recognized in the late 1940s in Arkansas where it became apparent that most tularemia cases were occurring from April through September, when rabbit hunting and direct contact with rabbits was rare but tick bites were common (145). In the early 1950s, *F. tularensis* was isolated from lone star ticks collected from Arkansas (66). Although the prevalence of *F. tularensis* among field-collected *A. americanum* ticks was low, estimated at 0.04% among ticks from Arkansas (1.9% of 576 pools of lone star ticks composed of 28,661 individuals were positive), investigations suggested that ticks were involved in the transmission of tularemia to dogs and potentially to cattle (23). Transstadial transmission of *F. tularensis* was subsequently demonstrated by experimental infection of *A. americanum* (65). Tick bite continues to be strongly associated with the occurrence of tularemia in the United States. In a series of 1026 cases of tularemia reported from 1981 to 1987 from Arkansas, Kansas, Louisiana, Missouri, Oklahoma, and Texas, 63% of cases reported an attached tick, while only 23% had exposure to rabbits (136).

Rickettsia rickettsii

Data linking *A. americanum* to the transmission of *Rickettsia rickettsii* and this tick's involvement in the epidemiology of RMSF in humans are largely circumstantial (57). The first guinea pig isolations of a spotted fever group rickettsiae (SFGR) believed to be *R. rickettsii* were made in Texas in 1942 from samples collected from two fatal cases of spotted fever occurring at a location heavily infested by lone star ticks, specimens of which were submitted by the family of the decedents (5). The esteemed rickettsiologist R.R. Parker recovered rickettsiae he identified as *R. rickettsii* from unfed *A. americanum* nymphs collected from Oklahoma in 1942 (113). Other investigators reported that emulsions produced from lone star ticks collected in Texas were highly virulent when inoculated into guinea pigs and presumptively identified *R. rickettsii* (6). However, more recent attempts to associate *A. americanum* as a potential vector of *R. rickettsii* have been unsuccessful. Burgdorfer et al. (22) failed to identify *R. rickettsii* among 1700 lone star ticks collected in Arkansas, South Carolina, and Tennessee, including ticks collected at sites where RMSF was endemic, although they did identify a high prevalence of a rickettsiae they designated as the WB-8-2 agent (Table 1). Similarly, Goddard & Norment (59) tested 3067 adult *A. americanum* collected from Mississippi, Kentucky, Oklahoma, and Texas between 1983 and 1984. Although a variety of tests yielded evidence of infection by different SFGR, no ticks were found infected with *R. rickettsii* (Table 1). Definitive contemporary evidence incriminating *A. americanum* in the epidemiology of RMSF is lacking, and if natural infection with *R. rickettsii* occurs in this species, it is likely at a low prevalence.

Other Spotted Fever Group Rickettsiae

Several SFGR have been isolated from *A. americanum* collected in various regions of the United States (Table 1), although whether these various agents cause human disease requires more investigation. Some of these SFGR, such as *Rickettsia parkeri*, are known to cause mild illness when inoculated into guinea pigs (112), and others [isolate 85–1034 ("*R. amblyommii*")] possess specific antigens that are recognized by convalescent-phase serum obtained from humans recovering from illnesses temporally associated with tick bite (33). SFGR transmitted by *A. americanum* will eventually be isolated from sick humans, as has been reported in the past (5). Attack rates of illnesses of presumed tick origin yet of unproven etiology can be substantial in settings in which lone star ticks are the principal or only tick vector present (92).

Coxiella burnetii

Coxiella burnetii has been isolated or identified from many species of ticks around the world (118). *C. burnetii* has been isolated from nymphal and adult *A. americanum* collected in eastern Texas (111) and Mississippi (115). However, the role of ticks in transmission of *C. burnetii* to humans is believed to be minimal and largely confined to maintenance of natural transmission cycles among wildlife.

"*Borrelia lonestari*"

The occurrence of Lyme borreliosis in the southern United States has been a controversial topic. Although *I. scapularis* ticks and small rodents infected with *B. burgdorferi* can be found in southern states (104), naturally occurring human infection has never been demonstrated through isolation of the spirochete. However, beginning in the 1980s, an illness accompanied by a rash resembling erythema migrans (sometimes referred to as southern tick-associated rash illness or Masters' disease) was reported with increasing frequency among patients from Missouri (90), North Carolina (73, 79), and Maryland (9). The tick incriminated in these disease occurrences was *A. americanum*; although in experimental settings, it has not been demonstrated to act as a competent vector for *B. burgdorferi* (105). It is now known that *A. americanum* harbors a spirochete distinct from *B. burgdorferi*, provisionally named "*Borrelia lonestari*," that has not yet been cultivated (12). The DNA of *B. lonestari* has been identified from a skin biopsy obtained from an erythematous lesion where an attached *A. americanum* was present on a patient (70). Although the public health significance of *B. lonestari* is currently under investigation, it appears likely that this species is a cause of Lyme-like disease in the southern United States (9).

AMBLYOMMA AMERICANUM-ASSOCIATED EHRLICHIOSES

Historical Perspectives

Until the mid-1980s, bacteria in the genus *Ehrlichia* were not considered to cause human disease in the United States (92a), and studies of ehrlichioses were relegated predominantly to investigators and clinicians in veterinary sciences. Considerable literature existed on *Ehrlichia canis* and *Anaplasma* (formerly *Ehrlichia*) *phagocytophila* (45) tick-borne bacteria with cosmopolitan distributions causing moderate to severe febrile disease in dogs and ruminants, respectively. However, in 1986 a clinician viewing the peripheral blood smear of a critically ill man with an unexplained febrile illness noted unusual inclusions in several of the patient's white blood cells. These inclusions were subsequently identified as membrane-bound, tightly packed clusters of bacteria called "morulae," which are a characteristic feature of ehrlichiae. The patient had received multiple tick-bites while visiting northern Arkansas two weeks earlier and had been diagnosed presumptively with RMSF (87). Serologic studies later implicated an *Ehrlichia* species as the cause of this patient's severe disease. Within several years, additional patients were diagnosed with ehrlichiosis in the southeastern and south central United States; although these illnesses were initially ascribed to *E. canis* infection (87), the causative organism was subsequently identified as a new species and named *Ehrlichia chaffeensis* (2, 37). Disease caused by *E. chaffeensis* is most commonly referred to as human monocytic ehrlichiosis (HME).

Within the next 13 years, two additional *Ehrlichia* species were reported as agents of human disease in the United States, namely *Anaplasma phagocytophila* in 1994, the cause of human granulocytic ehrlichiosis (HGE) (11), and *Ehrlichia ewingii* in 1999, a second cause of granulocytic ehrlichiosis in humans (19). Through 2001, approximately 1150 cases of HME and 1220 cases of HGE were reported through national surveillance (26, 93) [Centers for Disease Control and Prevention (CDC), unpublished data]. About 20 cases of ehrlichiosis caused by *E. ewingii* have been identified to date [(19, 107, 108); CDC, unpublished data].

Evidence for Transmission of *E. chaffeensis* and *E. ewingii* by *A. americanum*

Within a few years of the initial description of human ehrlichiosis in the United States, a geographic pattern of cases emerged that approximated the recognized distribution of *A. americanum*, implicating this tick as a potential vector for *E. chaffeensis* (47). This hypothesis was strengthened when *E. chaffeensis* DNA was amplified from pools of *A. americanum* adults collected from several states where cases of disease had originated (3). Subsequent studies demonstrated experimental transmission of *E. chaffeensis* among white-tailed deer by adult and nymphal lone star ticks (49), and retrospective ecologic and serologic surveys identified temporal and spatial associations between lone star tick infestations and the presence and prevalence of antibodies reactive to *E. chaffeensis* in white-tailed deer populations (82).

DNA of *E. chaffeensis* has been detected in lone star ticks collected in at least 15 states in the southeastern, midwestern, and northeastern United States (3, 67, 85, 131, 147). The prevalence of infection in adult ticks tested individually by use of polymerase chain reaction (PCR) generally varies from about 5% to 15% among specimens collected from areas where the agent is endemic (67, 85, 147); however, these prevalence estimates are subject to variation due to different assays employed by different researchers and the intrinsic variability associated with cross-sectional sampling. Crude minimum infection rates (MIRs) determined from pools of adult ticks have generally ranged from 1% to 5%; however, this method often underestimates the true level of infection at a particular location (85, 131).

Infections have been reported in both adult male and female ticks. As expected, the prevalence of infection appears to be lower in immature stages of ticks than in adults. A sample composed of 81 pools of nymphal *A. americanum* (representing 2723 individual ticks) collected from Harford County, Maryland, showed an overall MIR of 0.8%: The MIR of adult ticks collected at the same location was 3.5% (132). Failure to detect *E. chaffeensis* in nymphs collected at sites with confirmed infections in adult *A. americanum* has also been described (3).

Little is known about the dynamics of infection of *E. chaffeensis* in *A. americanum* populations; however, the prevalence of infection appears to be spatially and temporally discontinuous. Surveys of ticks collected from nearby sampling sites or among ticks collected at the same site during different years revealed marked variability in infection prevalence. Similarly, infection may not always be evident among ticks at a specific location at a particular time of sampling (130, 147).

More than one species of *Ehrlichia* may be present in the same tick or circulate within the same population of ticks, and this may have consequences to host and vector that remain unexplored. An as-yet-unnamed *Ehrlichia* sp. infecting white-tailed deer has been detected in lone star ticks (85), and the DNA of *E. ewingii* has been amplified from questing adult and nymphal lone star ticks collected in North Carolina and Florida (133, 148a). Transstadial passage of *E. ewingii* within *A. americanum* with subsequent transmission to dogs provides further support for the contention that *A. americanum* is a key tick in the maintenance and transmission of several ehrlichiae pathogenic for humans (8).

The replication, growth, and development cycles of ehrlichiae in *A. americanum* and the exact mechanism(s) by which these bacteria are transmitted to the vertebrate host during feeding are unknown. Detection of ehrlichiae in questing nymphal and adult ticks and successful transmission of the pathogen between deer by nymphal and adult ticks infected during the previous life stage confirm that *E. chaffeensis* is passaged transstadially (49). Detection of *E. chaffeensis* in larval *A. americanum* has been described (131); however, there are no other data to suggest that transovarial transmission occurs.

Reservoir Hosts for *Amblyomma*-Transmitted Ehrlichiae

E. chaffeensis and presumably *E. ewingii* are maintained in nature as complex zoonoses, potentially involving a wide variety of vertebrates that can serve as competent reservoirs for the bacteria, as sources of blood for tick vectors, or as both. The ability to infect a broad range of hosts is generally regarded as an important factor in promoting the emergence of a zoonotic pathogen (43), and a parallel argument can be applied to the feeding habits of a vector. The catholic feeding proclivity of *A. americanum* for the blood of a wide range of mammalian and avian species is well documented (13, 62). Considerably less is known about which vertebrates can serve as competent reservoirs for ehrlichiae, although the available data suggest that *E. chaffeensis* may infect a wide host range.

White-Tailed Deer

The white-tailed deer (*Odocoileus virginianus*) is the sole vertebrate species currently recognized as a complete and sufficient host for maintaining the transmission cycle of *E. chaffeensis*. White-tailed deer are an important source of blood for adult and immature stages of *A. americanum* (13, 62). Field surveys of white-tailed deer from areas where lone star ticks occur have reported that 80%–100% of sampled animals were infested with all three stages of *A. americanum* ticks; average tick burdens (adults and nymphs) frequently exceed 300 per deer (16, 17). The number of larvae on white-tailed deer has been described as "... so numerous that counting was impracticable" in certain circumstances (17). One detailed monthly survey of white-tailed deer in Kentucky and Tennessee reported maximum monthly half-body densities of 1493 larval lone star ticks on deer with corresponding values of 480 nymphal and >200 adult ticks during peak months (15) (Figure 1*a*). Simulations modeling the density of adult *A. americanum* ticks as a function of white-tailed deer density indicate that this keystone species exerts a profound effect on tick populations (99) (Figure 1*b*).

White-tailed deer are naturally infected with *E. chaffeensis* in the southeastern United States, as determined on the basis of PCR detection (85) and isolation of the organism (84). In addition, deer experimentally infected with *E. chaffeensis* remain bacteremic for at least 24 days (40) and can infect laboratory-reared larval and nymphal *A. americanum*, which maintain infections transstadially (49).

The prevalence of *E. chaffeensis* infections among populations of white-tailed deer in nature is difficult to determine. Antibody surveys have demonstrated a high prevalence of antibody reactive to *E. chaffeensis* antigens among white-tailed deer populations (frequently >50% of deer at sites where any antibody-positive animals were present) (39, 68, 85, 100), and field data have confirmed a site-specific correlation between antibody prevalence and presence of *A. americanum* (81). However, deer can be infected singly or in combination with several *Ehrlichia* species that are related antigenically to various degrees (41, 80, 148b). Studies using only serologic testing cannot routinely distinguish between antibodies resulting from *E. chaffeensis* infection from those resulting from infection with the antigenically related white-tailed deer agent or *E. ewingii*.

Other Wildlife

Coyotes serve as hosts for all stages of *A. americanum* (14, 31). In one of the infrequent surveys to quantify tick loads on this canid, coyotes were identified as the relatively most important host of adult *A. americanum* among 13 species of mammals infested with lone star ticks in Oklahoma: 4 of 6 coyotes were infested with 182 adult, 115 nymphal, and 108 larval lone star ticks (75). Coyotes naturally infected with *E. chaffeensis* in Oklahoma have been identified at a high prevalence (15/21; 71%) by use of PCR, suggesting that these animals could be a significant reservoir for *E. chaffeensis* (74a).

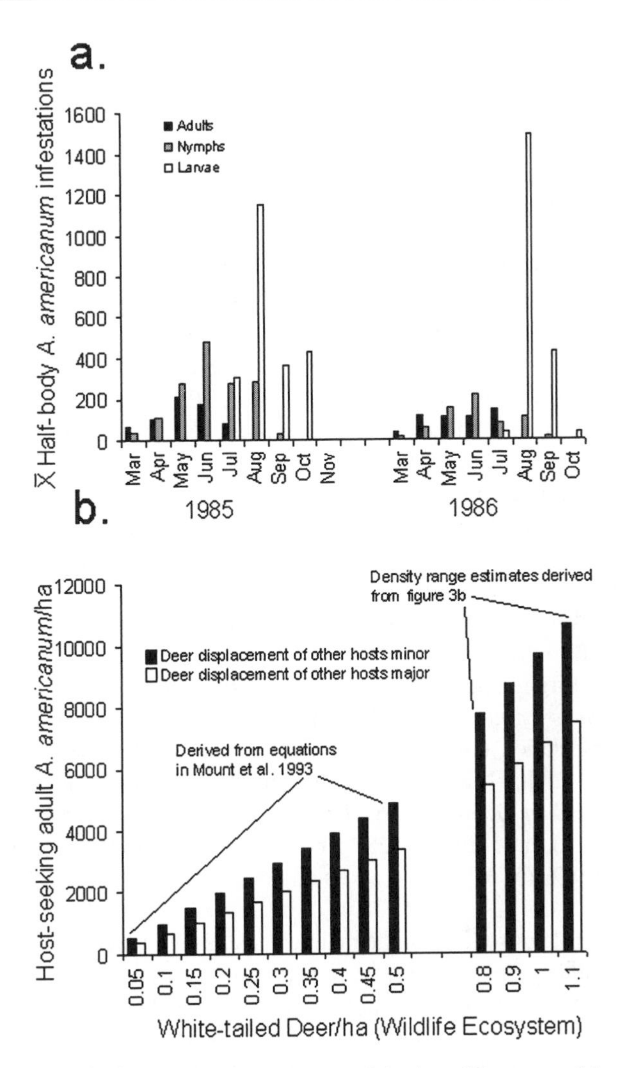

Figure 1 (*a*) Seasonal activity and abundance of the three life stages of *A. americanum* ticks infesting white-tailed deer in Kentucky and Tennessee, 1985–1986. Figure drawn from data published in (15). (*b*) Simulated effect of *A. americanum* population density as a function of white-tailed deer density. Figure drawn from equations published in (99). The higher-density estimates were derived from data shown in Figure 3*b* (see color insert).

Red foxes (*Vulpes vulpes*) can serve as hosts for all stages of *A. americanum* (139), but counts of *A. americanum* from red foxes are low relative to deer (128). Red foxes were susceptible to infection with a white-tailed deer isolate of *E. chaffeensis* (15B-WTD-GA strain), and ehrlichiae could be reisolated from the blood of experimentally infected animals for 14 days after infection (35). Antibody reactive to *E. chaffeensis* has been detected in field surveys among both red and gray foxes (35), suggesting that these canids may play a role in the maintenance of *E. chaffeensis*.

Other mammals have been implicated as potential reservoirs for *E. chaffeensis* or antigenically related *Ehrlichia* spp. only through antibody surveys. Raccoons (*Procyon lotor*) are frequently parasitized by all life stages of *A. americanum* (13) and occur throughout much of North America. In addition, raccoons frequently reach their highest population densities in suburban and urban-park locations, so contact rates with humans and domestic pets are high (119). Average lone star tick infestation counts for raccoons ranged between 0 and 1.5 adults, 0 and 80 nymphs, and 0 and 383 larvae per raccoon for surveys conducted in Georgia (116), Kansas (18), Oklahoma (75), North Carolina (107), Tennessee (76, 149), Texas (17), and Virginia (128). However, the results from tick surveys should be interpreted cautiously, considering that the lowest and one of the highest *A. americanum* counts for raccoons came from two sites in different locations of Tennessee (76, 149). Antibody reactive to *E. chaffeensis* antigens was found in 20% ($N = 411$) of raccoons sampled from eight states (30), although PCR failed to amplify the causative species of *Ehrlichia*. A high prevalence of antibodies (20%) was also found among raccoons sampled from an *E. chaffeensis*-enzootic site in Georgia (85).

The Virginia opossum (*Didelphis virginianus*) can serve as a host for all stages of *A. americanum* but does not appear to be a preferred host. Average lone star tick infestation counts for opossums are relatively low and ranged between 0 and 0.2 adults, 0 and 1.7 nymphs, and 0 and 9 larvae per opossum for surveys conducted in Georgia (116), Kansas (18), Oklahoma (75), Tennessee (76, 149), Texas (17), and Virginia (128). Antibodies reactive to *E. chaffeensis* were identified among 8% ($N = 38$) of opossums sampled at an *E. chaffeensis* enzootic site (83).

The role of rodents in the maintenance of *E. chaffeensis* is unknown, although larval and nymphal *A. americanum* parasitize a number of rodent species (17, 62). Antibodies reactive to *E. chaffeensis* at reciprocal titers >80 were identified in 31 of 294 white-footed mice sampled from Connecticut (88); however, the causative agent was not identified. In contrast, no antibodies were detected among 281 rodents of eight species sampled from an *E. chaffeensis* enzootic site in the southern United States (83).

The role of birds as a natural reservoir for *E. chaffeensis* or *E. ewingii* has yet to be defined, although many ground-feeding species serve as important sources of blood for immature stages of *A. americanum* (62). An important example is the wild turkey (*Meleagris gallopavo*); *A. americanum* is called the "turkey tick" in some regions of the midwestern United States because of this close association (42). Average lone star tick infestation counts for turkeys ranged between 15 and 39 nymphs, and 0 and 46 larvae among surveys conducted in Oklahoma (75) and Texas (17), and studies from Kansas also indicate that the wild turkey is an important host for nymphal and larval *A. americanum* (96).

In Russia, ehrlichiae, including an organism identified as *E. chaffeensis*, have been identified by PCR from *I. ricinus* ticks recovered from several species of migratory passerine birds (1). Should it be established that birds do not serve as competent reservoir host for ehrlichiae, their importance as hosts and dispersal agents for ticks potentially infected with these pathogens will not be diminished. As an example, 10 of 46 bird species sampled from a Georgia barrier island were parasitized by nymphal or larval *A. americanum* at prevalences that frequently exceeded 50% from June to August (46).

Domestic Animals

When serving as a core component in the maintenance cycle of a zoonotic pathogen, perhaps no other animal demands as much public health attention as does the domestic dog. Their popularity as companion animals (>61,500,000 owned dogs in the United States as of March 1999; American Pet Association, http://www.apapets.com/) and their access to both tick-infested habitats and to human habitations make them a priority in investigations into the natural history of the human ehrlichioses. Dogs serve as hosts for all stages of *A. americanum* (13, 62) and can serve as a competent reservoir host for *E. chaffeensis* and *E. ewingii* (50).

Ehrlichial infections among dogs are common in many regions of the United States. In southeastern Virginia, 38% of sampled dogs ($N = 74$) had antibody reactive to *E. chaffeensis* antigens, and 8 of 19 had ehrlichial DNA in whole-blood samples tested by a nested PCR (38). A survey conducted in Oklahoma found similar results: 7 of 65 dogs had antibody reactive to *E. chaffeensis* antigens and 4 animals had *E. chaffeensis* DNA in whole-blood samples tested by a nested PCR (101). Caution is warranted when interpreting the results of surveys based solely on serologic testing because dogs may be concurrently infected with multiple species of *Ehrlichia* that are antigenically related (77).

Domestic goats (*Capra hircus*) serve as hosts for all life stages of *A. americanum* (13, 62). Although knowledge of goats as a reservoir species for *E. chaffeensis* in the United States is limited to a single report, a high prevalence of reactive antibody (28/38 animals; 74%) and presence of ehrlichial DNA in whole blood (6/38; 16%), as determined by PCR, suggest a potential reservoir role for this common animal (44). Of special interest was the isolation of *E. chaffeensis* from a single goat sampled at time points 40 days apart, suggesting a persistent bacteremia.

In summary, it appears certain that *E. chaffeensis* has a broad range of vertebrate hosts that can act as competent reservoir hosts for transmission of the bacterium to various stages of the lone star tick. At least three species of mammals in the order Carnivora (all in the family Canidae) and two in the order Artiodactyla (families Bovidae and Cervidae) have been infected by *E. chaffeensis* in natural or experimental settings. The white-tailed deer is a competent reservoir host for *E. chaffeensis* and a critical or even keystone host for *A. americanum*. Little is known about potential reservoirs for *E. ewingii* other than domestic dogs, although it is likely that one or more species of wildlife are involved in the maintenance of this agent.

EPIDEMIOLOGY OF *AMBLYOMMA*-ASSOCIATED HUMAN EHRLICHIOSES

Although *E. chaffeensis* was the first *Ehrlichia* identified as a human pathogen in the United States, our knowledge of the biology of *A. phagocytophila*, which was isolated several years later, is already more substantial. Epidemiologists and biologists have utilized the public health infrastructure and research findings accumulated over nearly two decades of study of Lyme borreliosis and *I. scapularis* in the northeastern and northcentral United States toward understanding the epidemiology of HGE.

In contrast to cases of HGE, most cases of HME and all cases of *E. ewingii* ehrlichiosis (EWE) have been reported from the southcentral and southeastern United States (53, 93, 108), where *A. americanum* reaches its greatest population densities (Figure 2*a*, see color insert). Mandated national surveillance and reporting of the ehrlichioses has been in effect only since 1999. Although reporting by individual states has been incomplete, these data indicate a region of highest risk ranging from central Texas through Oklahoma and Missouri east to Virginia and all states to the south (Figure 2*b*). Cases of HME are reported sporadically along the East Coast, most notably on the Atlantic coastal plain. DNA from *E. chaffeensis* has been recovered from *A. americanum* from as far north as Connecticut and Rhode Island (Figure 2*a*). Many human cases of ehrlichiosis are diagnosed by serologic testing, and antibodies resulting from infection with *E. ewingii* or *A. phagocytophila* can cross-react with *E. chaffeensis* antigen (19). In addition, the travel histories of persons suspected of having ehrlichiosis are usually not provided when serum samples are submitted for diagnostic evaluation to reference laboratories, such as CDC, so that some cases of ehrlichiosis relegated to a specific state in summary reports may have been imported (93). Finally, lone star ticks can be accidentally transported to or may exist in foci within states not considered within the range of *A. americanum* (Figure 2*a*). An analysis based on tick specimens parasitizing humans accessioned into the U.S. National Tick Collection identified lone star ticks attached to persons from Michigan, Nebraska, New Mexico, Wisconsin, and Wyoming (95).

Although *E. chaffeensis* has been isolated only from the United States and EWE is only documented from this country, there are data indicating that human infections with antigenically related ehrlichiae occur in Europe (102), Asia (63), South America (120), and Africa (142). Because the only proven tick vector of *E. chaffeensis* (i.e., *A. americanum*) is restricted to the New World, these findings suggest involvement of other tick species in the transmission of HME, the cosmopolitan distribution of ehrlichiae antigenically related to *E. chaffeensis*, or both. Ehrlichiae that are closely related or identical to *E. chaffeensis* have been identified from a variety of ticks collected in Asia (1, 127), although the significance of these findings for human disease is unclear.

Reliable incidence data on the *Amblyomma*-transmitted ehrlichioses derived from active, population-based surveillance are restricted to a few localities and do not exist for EWE. The estimated incidence for hospitalized cases of HME was 5.5 per 100,000 persons in southeastern Georgia, which was higher than that for RMSF during the same period (54). Estimates of 8 and 14 cases of HME per 100,000 persons during 1997 and 1998, respectively, were obtained by active surveillance for HME in southeast Missouri (103). The frequencies of occurrence of HME in cohorts of patients presenting with fever and a history of tick bite in Tennessee (7/38 patients) (129a) and central North Carolina (10/35) were nearly identical (24) and similar to the number of RMSF diagnoses. From these observations we conclude that where endemic, HME occurs at an incidence similar to that of RMSF, a well-known tick-borne disease considered uncommon but not rare.

HME is a highly seasonal disease. Although cases have been reported during March through November, about 70% of cases occur during May through July (53, 129). This seasonality corresponds to the peak feeding-activity periods of nymphal and adult *A. americanum* throughout much of their range (Figure 1*a*). Reports of HME into the late fall and winter are unusual but may be more common in the South (117).

HME is predominantly a disease of adults; most patients are >40 years of age, and in all age groups men are diagnosed with the disease more frequently than women (53). Of particular note is that HME in children is relatively rare: Among the first 250 reported cases of ehrlichiosis, <10% were in individuals 2 to 13 years of age (53). The reasons for this age distribution remain unclear, although the severity of HME correlates with immune function, which becomes increasingly impaired with age.

Most HME cases occur as sporadic infections. Recreational or occupational activities that place individuals in tick-infested habitats are well-documented risk factors for infection. A recollection of recent tick bite was reported by 68% of ehrlichiosis cases in a national survey (53), but that figure can exceed 80% in specific investigations (103, 129). Rare outbreaks of HME have occurred among golfers living in a retirement community in Tennessee (129) and among military personnel participating in field training exercises (92).

Because lone star ticks transmit several bacterial pathogens, coinfections may occur. In North Carolina, a concurrent infection with *E. chaffeensis* and SFGR has been diagnosed (126), and several seroepidemiologic studies have demonstrated simultaneous seroconversions to *E. chaffeensis* and SFGR among military personnel exposed to *A. americanum*-infested habitats (92).

FACTORS IN THE EMERGENCE OF *AMBLYOMMA*-ASSOCIATED EHRLICHIOSES

Recent Clinical Recognition or New Diseases?

It is likely that the observation of morulae in white blood cells of a patient with an unexplained febrile illness was noted long before the formal discovery of ehrlichiosis in 1986. Peripheral blood smear evaluation was the standard of hematologic evaluation before widespread availability of automated cytometry, and morulae would have probably been noted, but their connection to an ehrlichia was either missed or left uninvestigated. In 2002, even experienced medical technologists have trouble differentiating true morulae from intraleukocytic inclusions associated with unrelated infectious or noninfectious conditions.

Is there something unique about the clinical presentation of the ehrlichioses that would have made these diseases stand out and led to their recognition as distinct and novel human diseases? In fact, the early disease manifestations of HME and EWE are relatively nonspecific and present a diagnostic challenge even to physicians knowledgeable about tick-borne diseases (143). As disease progresses, involvement of multiple organ systems may complicate the clinical course and result in various life-threatening scenarios. However, the clinical course is often nondescript when complicating underlying factors, such as immunosuppression, are lacking (108), suggesting that the ehrlichioses, had they been occurring at appreciably lower incidence than at the present time, would have been difficult to identify as unique disease entities. The protean clinical presentations range from generalized and relatively vague initial symptoms to more targeted complaints, and initial diagnoses have included "viral syndrome," upper respiratory infection, pneumonia, meningoencephalitis, cholecystitis, pharyngitis, urinary tract infection, epididymitis, or prostatitis (48, 108, 109). Even the hematologic abnormalities associated with HME (e.g., leukopenia and thrombocytopenia) are consistent with alternative diagnoses, such as sepsis, thrombotic thrombocytopenic purpura, or hematologic neoplasia (69, 89). The broad differential, coupled to a low incidence of disease, would have made the ehrlichioses difficult to identify until a certain threshold frequency of human disease was crossed. Although the *Amblyomma*-associated ehrlichioses undoubtedly occurred in the past, we believe that factors affecting the incidence and distribution of HME and EWE have made it increasingly likely that these diseases would be documented.

The Concept of Emerging Disease

The term "emerging infection" has been overused; however, the designation as originally intended (i.e., to signify new diseases or preexisting diseases that are rapidly increasing in incidence) still has relevance. Although changes in public health surveillance practices and the availability of diagnostic assays play a major role in determining disease incidence, there are also a number of factors independent of human activities that can radically influence the emergence of a zoonotic disease (Table 2). Zoonoses are diseases of animals that are transmissible to humans, and the epidemiology of vector-borne zoonoses must be understood within the context of natural maintenance cycles of pathogens that involve wildlife and arthropods. Understanding the natural history of zoonotic agents is no academic luxury because effective control of the human diseases caused by these pathogens frequently hinges on targeting the vector or reservoir populations. In addition, human disease is an insensitive indicator of the magnitude of the zoonotic-pathogen reservoir and

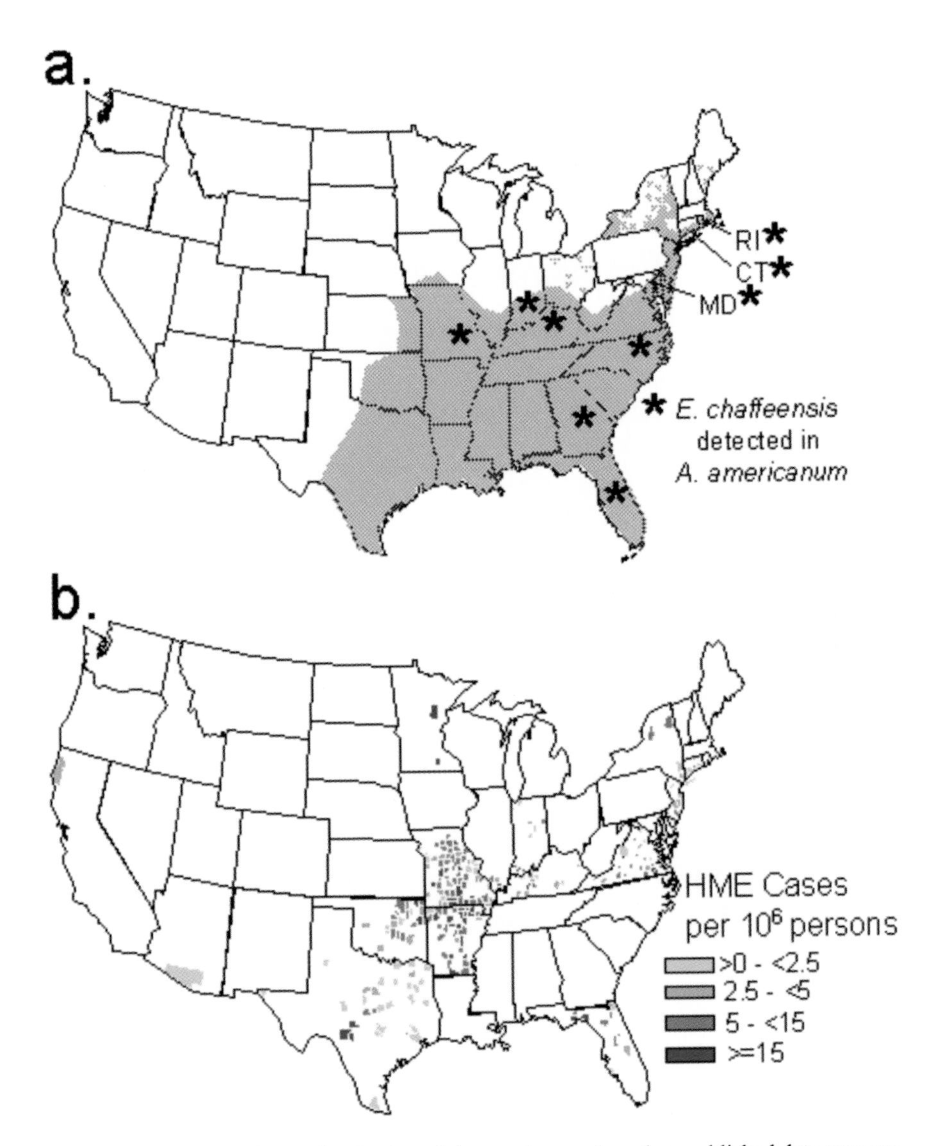

Figure 2 (*a*) Approximate distribution of *A. americanum* based on published data or maps (13, 71, 94). (*b*) Average annual incidence of HME, 1998–2000, based on states reporting data to the level of county (ehrlichiosis is not notifiable in all states where *E. chaffeensis* and *E. ewingii* are endemic) (26, 93).

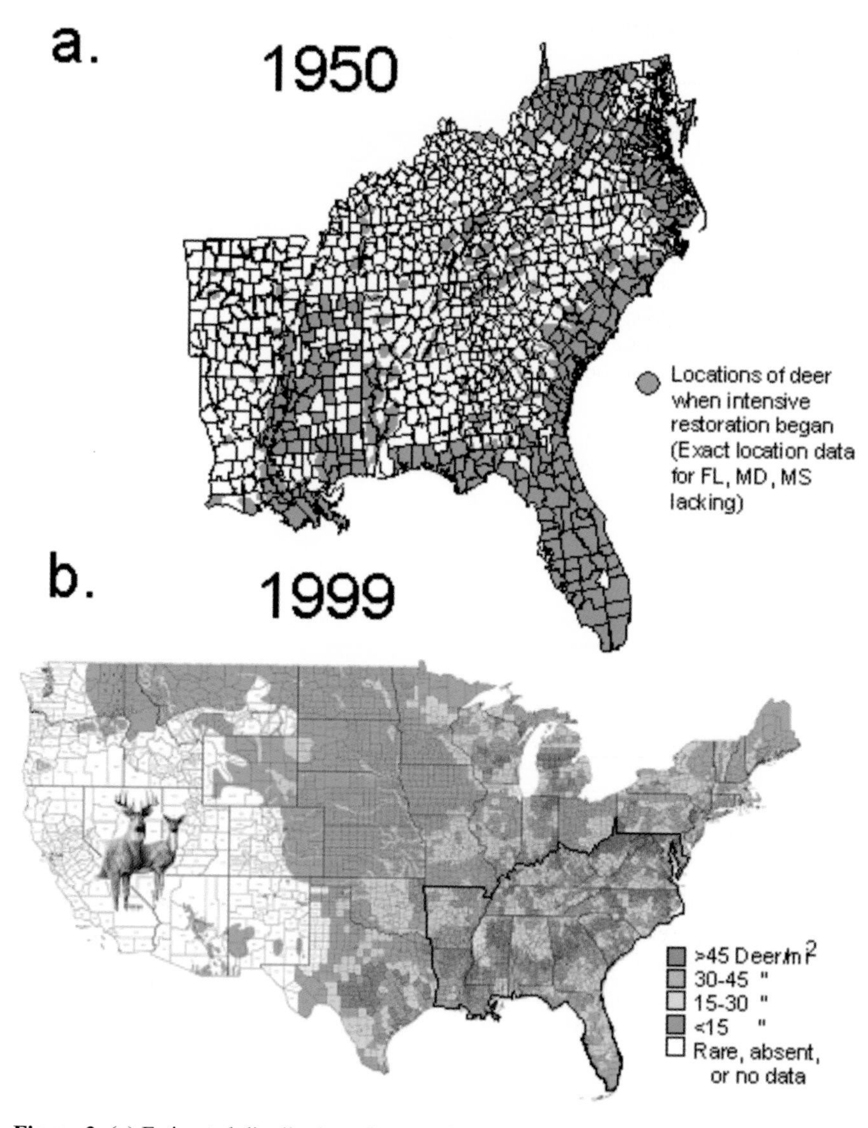

Figure 3 (*a*) Estimated distribution of extant deer populations in the southeastern United States circa 1950. Map was redrawn from maps produced by the Southeastern Cooperative Wildlife Disease Study (SCWDS), Athens, Georgia, from data compiled by State Game and Fish Biologists of the Southeastern Region. Maps may differ from previously published maps (74) due to problems in resolution. (*b*) Estimated white-tailed deer density, 1999. Map provided and published with permission of Quality Deer Management Association, Watkinsville, Georgia, 30677 (http://www.qdma.com/).

TABLE 2 Factors in the emergence of *Amblyomma*-associated zoonoses with emphasis on the ehrlichioses

Factor in emergence	Example	Reference
Vector dynamics	Increase in *A. americanum* population density	(56)
	Increase in geographic distribution of *A. americanum*	(71, 94)
Reservoir host dynamics	Increase in population density and geographic distribution of vertebrate host populations (especially white-tailed deer, turkeys) that serve as hosts for *A. americanum*	(91, 94)
	Increase in competent reservoir host (e.g., white-tailed deer) populations for *E. chaffeensis* and *E. ewingii*	(91, 137, 148b)
Human behavior	Increased human contact with natural foci of infection through recreation or occupation	(109, 129)
	Improved surveillance and reporting	(93)
	Habitat modification and climate change	(60, 106)
Human demographics	Increasing size of human population >40 years of age	(140)
	Increasing size and longevity of immunocompromised populations	(25, 110, 122)
	Population shift to rural environments	(86, 122)

the dynamics of transmission in nature cycles. These aspects of the maintenance cycle are usually "silent" and cannot be appreciated without special study.

Several recent reviews have examined factors relevant to the emergence of infectious diseases among wildlife (34, 43) and vector-borne diseases of humans (144). We focus our discussion on the various factors influencing the maintenance of ehrlichial pathogens in nature and consider how dynamic changes in these factors could drive emergence by affecting the frequency and severity of the corresponding human diseases and the geographic region affected by endemic disease. We first consider the complex interactions of tick vectors and vertebrate hosts that are sensitive to environmental influences that can drive epizootics. The emergence and spread of Lyme disease in the eastern United States is a classic example of how changes in environmental conditions have influenced patterns in vertebrate-host and vector-tick distribution and population densities to affect the incidence of a human disease (137). In addition, changes in the patterns of susceptibility within a population can be a critical factor in disease emergence, both in increasing the opportunity for sporadic transmission of pathogens to humans (98) and in the dynamics of epizootics occurring among wildlife populations (43).

Dynamics of Reservoir Host and *A. americanum* Populations

The greatest influence on the emergence of *Amblyomma*-associated ehrlichioses has been the explosive growth of white-tailed deer populations in the United States. Lone star ticks were identified as the most common species of tick (73.3% of 367 attached ticks) parasitizing humans in a survey of tick attachment sites on humans conducted in Georgia and South Carolina from 1995 to 1998 (51). However, a similar study conducted in South Carolina during 1973–1974 found that *A. americanum* contributed only 7.7% ($N = 220$) of the sample of human-biting ticks (21). Increases in white-tailed deer populations have been suggested as a primary factor in driving this increase in lone star tick abundance (52).

The dramatic rise in white-tailed deer numbers was preceded by the reforestation of extensive tracks of land originally abandoned by westward-bound emigrants in the early 1800s (91). Recovery of deer populations was not an inevitable or monotonic process, and a period of intense overharvesting kept populations at historic lows until the early 1900s, when most deer herds in the southeastern United States reached their nadir (Figure 3*a*, see color insert). From these historic lows, the number of white-tailed deer increased about 50-fold during the twentieth century, from an estimated 350,000 animals in 1900 to at least 17 million animals by the mid-1990s (91). This remarkable increase in population has been matched by an equally impressive range expansion throughout most suitable habitats in the eastern half of the United States (Figure 3*b*) (Quality Deer Management Association, Watkinsville, Georgia, http://www.qdma.com/). Similar links between white-tailed deer and increases in the number and range of *I. scapularis* and the emergence of Lyme disease, babesiosis, and HGE have been described (137).

Deer were nearly extirpated from northeastern states until the reintroduction efforts of the 1930s (121). In the southeastern United States, where this mammal is closely linked to increases in the abundance and expanded geographic range of the lone star tick (52, 99), similar decline and resurgence occurred. As an example, by 1920 native deer were considered extirpated from nearly all of western Virginia, and in the Tidewater area remnant deer herds remained only in remote areas inaccessible to humans (74). An intensive campaign of white-tailed deer restoration resulted in repopulation of most of the state by 1970 (74). The deer population of the southern United States achieved its current level and distribution only within the past few decades, and population numbers have continued to increase as assessed by white-tailed deer harvest numbers (123) and counts of deer-vehicle collisions (91). The impact of white-tailed deer on the dynamics of lone star tick populations has been discussed (see also Figure 1b).

Other important vertebrate hosts for *A. americanum* and potential reservoirs for *E. chaffeensis* and possibly *E. ewingii* have undergone similarly impressive increases in population growth and geographic distribution. Coyotes have expanded their range throughout North America since the 1800s, a time when they were restricted to the Great Plains and the western United States (97). In the southeastern United States, the number of coyotes has increased dramatically, as evidenced by an increase in harvest of these animals in Mississippi from 500 in 1975 to 40,000 in 1988 (97). These carnivores exist in most habitats and have become established in suburban and urban locations where contacts and attacks on humans and domestic animals are increasingly reported (10). These developments increase the potential for infected lone star ticks to be seeded into the peri-domestic environment.

Other nonmammalian vertebrates that are considered important sources of blood for lone star ticks but whose competence as reservoirs for ehrlichiae is unknown have also experienced increases in geographic range and population numbers. For example, the dynamical change in geographic distribution and abundance of the wild turkey paralleled that described for white-tailed deer. By the early to mid-1800s, populations were extirpated from New England, and by 1920 only 21 of the original 39 states with turkeys had remaining populations (72). Between 1959 and 1990, estimates of the turkey population of the United States increased from 500,000 to 3.5 million (72), and as determined by the basis of hunter kills, population numbers continue to increase at a high rate (94).

As wild turkey populations have increased throughout their historic geographic range, they are credited with reintroducing and increasing population densities of the lone star tick. Concurrent expansion and increases in turkey and lone star tick populations have been reported at the extremes of their known range in New York to the north (94) and Kansas to the west (Figure 2*a*) (96). Restoration programs have introduced turkeys into every state except Alaska, including 10 states considered outside of the ancestral range of this bird (72). It appears certain that the geographic range of *A. americanum* will continue to expand with the success of this host. It is also likely that in some of these newly colonized locations, infections in humans with tick-transmitted ehrlichiae will occur.

Improved Diagnostics and Surveillance

The development and increasing availability of diagnostic reagents, changes in surveillance activities, and requirements for national notification have had a major impact on our understanding of the epidemiology and emergence of HME (27). The number of cases of ehrlichiosis reported to state health departments and to CDC increased from 69 in 1994 to 363 in 2000 [(93); CDC, unpublished data]. However, diagnostic tests based on serology alone are not sensitive indicators of disease early in the clinical course (28), and reporting remains inconsistent or nonexistent in several southern states where HME and EWE are of special concern. Enhanced surveillance and education programs are required to raise the level of diagnostic suspicion for HME and EWE in order to provide details as to the full spectrum of disease, and this will undoubtedly lead to a better appreciation for the public health impact of HME and elucidation of the epidemiology of EWE.

Expansion of Highly Susceptible Human Subpopulations

One fundamental factor contributing to the emergence of new pathogens and diseases has been change in host susceptibility, operating at the population level through immunosuppression. The various means by which large segments of the human population may become immunosuppressed include aging, malignancy, and infectious causes (98). Demographics indicate that the U.S. population is becoming increasingly weighted toward the older age groups, and the ehrlichioses

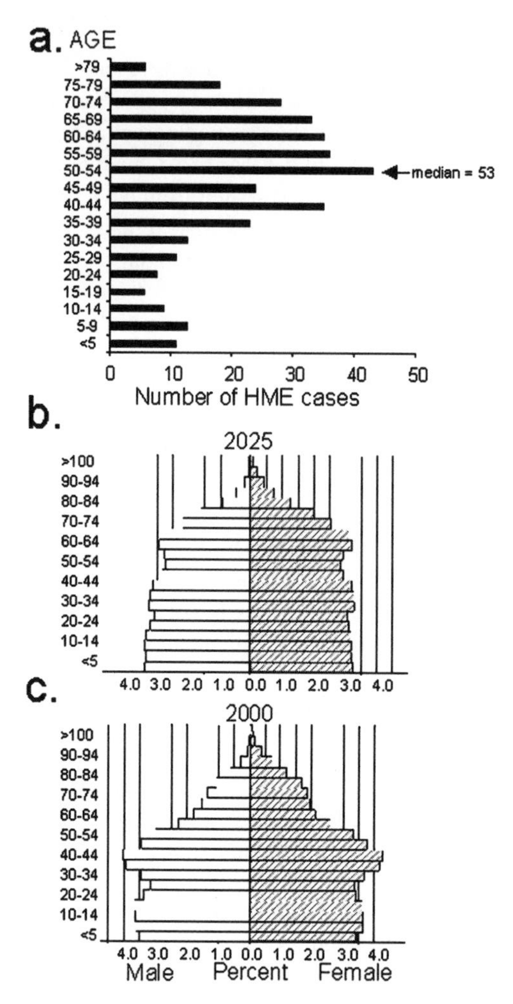

Figure 4 (*a*) Human monocytic ehrlichiosis (HME) is predominantly a disease of adults and the elderly (CDC, previously unpublished data). (*b*) Estimated changes in the population structure of the United States between 2025 and (*c*) 2000 Data from (141).

appear to be predominantly diseases of adults and especially severe in the elderly (Figure 4*a*). In one of the first epidemiologic investigations of 149 patients with HME, increasing age (>60 years) and delay (>8 days) in effective antibiotic treatment were the only independent risk factors for severe or fatal illness (53). Aging and the associated "immunosenescence" of the immune system are a well-described but imperfectly understood phenomenon (114). The United States Census estimates that the percentage of the population >45 years of age will increase from 34.9% in 2000 to 41.3% in 2025 (Figure 4*b*,*c*) (141). Coupled with longer life expectancy is better general heath, which encourages many older persons to enjoy outdoor activities that bring them into contact with ticks (129). This combination of factors indicates that the number and severity, if not the overall incidence, of HME and EWE cases will increase as the size of the most susceptible sector of the population expands to unprecedented levels.

Because ehrlichiae are obligately intracellular pathogens, it is likely that intact humoral and cell-mediated immunity are essential for successful clearance of *E. chaffeensis*. Severe and fatal HME have been described repeatedly in persons with compromised immunity from human immunodeficiency virus (HIV) disease (108) and immunosuppressive therapies (7, 124). Although the absolute prevalence of HIV among persons in nonmetropolitan areas remains significantly lower than in urban centers, the number of HIV-infected persons residing in nonmetropolitan areas has increased most rapidly since the 1980s (55). The expansion of HIV into rural populations is particularly evident in the southeastern United States (29), where the risk of HME and EWE is greatest (93). New therapies for HIV have braked the progression of HIV infection in many patients, permitting a level of health that allows occupational and recreational activities not previously possible. Many of these activities (e.g., hunting, hiking, camping, or working outdoors) place these patients at increased risk for tick bites and have been directly linked with acquisition of *E. chaffeensis* in HIV-infected persons (108).

Other noninfectious causes of immunosuppression have also been identified as increasing the risk of ehrlichiosis and the potential for severe disease. The number of persons living with transplants and receiving potent immunosuppressive drugs will approach 200,000 by the end of 2002 (United Network for Organ Sharing, http://www.unos.org/framede??fault.asp). Given the relatively small number of transplant recipients, it is even more remarkable that HME and EWE have been diagnosed in transplant patients (19, 124).

Technology, Land Use, and Human Activities

Many factors contributing to the increasing size of the segment of our population most susceptible to infectious diseases, such as HME and EWE, are the direct result of technological advances in medicine, pharmacology, and public health practice. Important examples already discussed include improvements to general health (e.g., through better nutrition and childhood immunizations) that have led to greater longevity and the aging of the U.S. population and increases in the number of persons on potent immunosuppressive drugs as part of medical therapy (98).

The reforestation of the eastern United States and the emergence of zoonotic pathogens associated with white-tailed deer and the ticks that feed on deer have been discussed. However, the effect of these factors was magnified by a dramatic reversal of a demographic trend in the United States that first became noticeable in the twentieth decennial census conducted in 1980. In contrast to historical and global trends, the U.S. population between 1970 and 1980 grew faster in nonmetropolitan regions (17.1%) than in metropolitan regions (10.0%), with much of the growth occurring in true rural environments rather than suburbia (86). From 1990 to 1999, the growth rate of nonmetropolitan areas, although still substantial, had slowed to 7%, whereas metropolitan areas grew by 10% (141). The convergence of population growth in rural environments, the increase in the numbers and geographic distribution of tick vectors and reservoir hosts, and the increase in the proportion of the population considered to be highly prone to more severe disease helped drive the emergence of *Amblyomma*-associated zoonoses.

The local species diversity available to a pathogen can play a role in either promoting or hindering its persistence and rate of geographic spread (135). The impact of humans in altering the components of ecological communities has been dramatic. For example, humans have reduced the numbers of large predators, such as wolves (*Canis lupus*) and mountain lions (*Felis concolor*), throughout North America. Removing predators from predator-prey cycles can have effects beyond the direct effects on competitors and prey populations. The geographic spread and increase in population densities of white-tailed deer and the white-footed mouse (*Peromyscus leucopus*) have been facilitated by removal of predators, and this diminished biodiversity may have contributed to the rapid emergence of a human disease in the case of Lyme borreliosis (106).

Recreational opportunities associated with rural or suburban living frequently have been implicated as risk factors for acquiring tick-borne diseases. In 1947, Topping (138) identified golf as a potential risk factor for RMSF and postulated that the incidence of the disease would be greatest among less skilled golfers, who would spend more time in the rough. Nearly 50 years later, Standaert et al. (129) demonstrated just such an association when an outbreak of HME occurred in a retirement golf community in Tennessee. Golfers with average scores >100 had 2.4-fold-greater odds of having antibody to *E. chaffeensis* than those with scores <100 (129). In addition, individuals with a habit of retrieving a lost ball had 3.7-fold-greater odds of having antibody to *E. chaffeensis* than those golfers who simply used a new ball. Between 1970 and 2000, the number of persons playing golf in the United States more than doubled from about 11 million to almost 27 million, and the number of golf courses increased from 10,848 to 17,108; approximately 27% of golfers are age 50 or over (National Golf Association, http://www.ngf.org/faq/#1/).

CONCLUSIONS AND FUTURE PROSPECTS

During the next 25 years, the number of cases of disease caused by *A. americanum*-associated pathogens will likely continue to increase, partly because of biologic factors that influence the likelihood of susceptible humans coming into contact with infected lone star ticks and partly because of increasing physician awareness and our ability to diagnose these diseases. However, the many biological, sociological, and environmental factors that drive the distribution and intensity of transmission of zoonotic pathogens are not static, and certainly other wildcards exist that may change projections of the frequency of illness in the human population. One such factor is the contentious issue of climate change and its potential effect on vector-borne diseases. The ecology of arthropod vectors and their hosts and the resulting transmission dynamics of the pathogens they transmit are strongly influenced by climatic factors (60). Although the impact of climate change on the ehrlichioses is unknown, models suggest that another tick-borne disease, RMSF, which has an incidence similar to HME, may decline if average summertime temperatures increase in the southeastern United States, reducing survival of *D. variabilis* ticks (61). From a medical entomologic perspective, the impact of *A. americanum* on human health has changed dramatically in a relatively short span of time since 1986. Revisiting this topic in future years should be informative and instructive.

The *Annual Review of Entomology* is online at http://ento.annualreviews.org

LITERATURE CITED

1. Alekseev AN, Dubinina HV, Semenov AV, Bolshakov CV. 2001. Evidence of ehrlichiosis agents found in ticks (Acari: Ixodidae) collected from migratory birds. *J. Med. Entomol.* 38:471–74
2. Anderson BE, Dawson JE, Jones DC, Wilson KH. 1991. *Ehrlichia chaffeensis*, a new species associated with human ehrlichoisis. *J. Clin. Microbiol.* 29:2838–42
3. Anderson BE, Sims KG, Olson JG, Childs JE, Piesman JF, et al. 1993. *Amblyomma americanum*: a potential vector of human ehrlichiosis. *Am. J. Trop. Med. Hyg.* 49:239–44
4. Anigstein L, Anigstein D. 1975. A review of the evidence in retrospect for a rickettsial etiology in Bullis fever. *Tex. Rep. Biol. Med.* 33:201–11
5. Anigstein L, Bader MN. 1942. New epidemiological aspect of spotted fever in the gulf coast of Texas. *Science* 96:357–58
6. Anigstein L, Bader MN. 1943. Investigations on rickettsial diseases in Texas, part 4. Experimental study on Bullis fever. *Tex. Rep. Biol. Med.* 1:380–409
7. Antony SJ, Dummer JS, Hunter E. 1995. Human ehrlichiosis in a liver transplant recipient. *Transplantation* 60:879–81
8. Anziani OS, Ewing SA, Barker RW. 1990. Experimental transmission of a granulocytic form of the tribe Ehrlichieae by *Dermacentor variabilis* and *Amblyomma americanum* to dogs. *Am. J. Vet. Res.* 51:929–31
9. Armstrong PM, Brunet LR, Spielman A, Telford SR III. 2001. Risk of Lyme disease: perceptions of residents of a Lone Star tick-infested community. *Bull. WHO* 79:916–25
10. Baker RO, Timm RM. 1998. Management of conflicts between urban coyotes and humans in southern California. *Proc. Vert. Pest Conf.* 18:288–312
11. Bakken JS, Dumler JS, Chen SM, Eckman MR, Van Etta LL, Walker DH. 1994. Human granulocytic ehrlichiosis in the upper midwest United States. A new species emerging? *JAMA* 272:212–18
12. Barbour AG, Maupin GO, Teltow GJ, Carter CJ, Piesman J. 1996. Identification of an uncultivable *Borrelia* species in the hard tick *Amblyomma americanum*: possible agent of a Lyme disease-like illness. *J. Infect Dis.* 173:403–9
13. Bishopp FC, Trembley HL. 1945. Distribution and hosts of certain North American ticks. *J. Parasitol.* 31:1–54
14. Bloemer SR, Zimmerman RH. 1988. Ixodid ticks on the coyote and gray fox at Land Between the Lakes, Kentucky-Tennessee, and implications for tick dispersal. *J. Med. Entomol.* 25:5–8

15. Bloemer SR, Zimmerman RH, Fairbanks K. 1988. Abundance, attachment sites, and density estimators of lone star ticks (Acari: Ixodidae) infesting white-tailed deer. *J. Med. Entomol.* 25:295–300
16. Bolte JR, Hair JA, Fletcher J. 1970. White-tailed deer mortality following tissue destruction induced by lone star ticks. *J. Wildl. Manag.* 34:546–52
17. Brennan JM. 1945. Field investigations pertinent to Bullis fever. The lone star tick, *Amblyomma americanum* (Linnaeus 1758). Notes and observations from Camp Bullis, Texas. *Tex. Rep. Biol. Med.* 3:204–26
18. Brillhart DB, Fox LB, Upton SJ. 1994. Ticks (Acari: Ixodidae) collected from small and medium-sized Kansas mammals. *J. Med. Entomol.* 31:500–4
19. Buller RS, Arens M, Hmiel SP, Paddock CD, Sumner JW, et al. 1999. *Ehrlichia ewingii*, a newly recognized agent of human ehrlichiosis. *N. Engl. J. Med.* 341:148–55
20. Burgdorfer W. 1975. A review of Rocky Mountain spotted fever (tick-borne typhus), its agent, and its tick vectors in the United States. *J. Med. Entomol.* 12:269–78
21. Burgdorfer W, Adkins TR Jr, Priester LE. 1975. Rocky Mountain spotted fever (tick-borne typhus) in South Carolina: an educational program and tick/rickettsial survey in 1973 and 1974. *Am. J. Trop. Med. Hyg.* 24:866–72
22. Burgdorfer W, Hayes SF, Thomas LA, Lancaster JL Jr. 1981. A new spotted fever group rickettsia from the lone star tick, *Amblyomma americanum*. In *Rickettsiae and Rickettsial Diseases*, ed. W Burgdorfer, RL Anacker, pp. 595–602. New York: Academic
23. Calhoun EL, Alford HI. 1955. Incidence of tularemia and Rocky Mountain spotted fever among common ticks of Arkansas. *Am. J. Trop. Med. Hyg.* 4:310–17
24. Carpenter CF, Gandhi TK, Kong LK, Corey GR, Chen SM, et al. 1999. The incidence of ehrlichial and rickettsial infection in patients with unexplained fever and recent history of tick bite in central North Carolina. *J. Infect. Dis.* 180:900–3
25. Centers for Disease Control and Prevention. 2000. *HIV/AIDS Surveillance Rep. 12(2)1–44*, CDC, Atlanta, GA
26. Centers for Disease Control and Prevention. 2001. Summary of notifiable diseases, United States 1999. *MMWR Morb. Mortal. Wkly. Rep.* 48:1–104
27. Childs JE, McQuiston JH, Sumner JW, Nicholson WL, Comer JA, et al. 1999. Human monocytic ehrlichiosis due to *Ehrlichia chaffeensis*: how do we count the cases? In *Rickettsiae and Rickettsial Diseases at the Turn of the Third Millenium*, ed. D Raoult, P Brouqui, pp. 287–93. Paris: Elsevier
28. Childs JE, Sumner JW, Nicholson WL, Massung RF, Standaert SM, Paddock CD. 1999. Outcome of diagnostic tests using samples from patients with culture-proven human monocytic ehrlichiosis: implications for surveillance. *J. Clin. Microbiol.* 37:2997–3000
29. Cohn SE, Klein JD, Mohr JE, van der Horst CM, Weber DJ. 1994. The geography of AIDS: patterns of urban and rural migration. *South. Med. J.* 87:599–606
30. Comer JA, Nicholson WL, Paddock CD, Sumner JW, Childs JE. 2000. Detection of antibodies reactive with *Ehrlichia chaffeensis* in the raccoon. *J. Wildl. Dis.* 36:705–12
31. Cooley RA, Kohls GM. 1944. The genus *Amblyomma* (Ixodidae) in the U.S. *J. Parasitol.* 30:77–111
32. Cox HR. 1940. *Rickettsia diaporica* and American Q fever. *Am. J. Trop. Med. Hyg.* 20:463–69
33. Dasch GA, Kelly DJ, Richards AL, Sanchez JL, Rives CC. 1993. Western blotting analysis of sera from military personnel exhibiting serological reactivity to spotted fever group rickettsiae. *Am. Soc. Trop. Med. Hyg.* 49(Suppl. 3):220

34. Daszak P, Cunningham AA, Hyatt AD. 2000. Emerging infectious diseases of wildlife—threats to biodiversity and human health. *Science* 287:443–49
35. Davidson WR, Lockhart JM, Stallknecht DE, Howerth EW. 1999. Susceptibility of red and gray foxes to infection by *Ehrlichia chaffeensis*. *J. Wildl. Dis.* 35: 696–702
36. Davidson WR, Siefken DA, Creekmore LH. 1994. Seasonal and annual abundance of *Amblyomma americanum* (Acari: Ixodidae) in central Georgia. *J. Med. Entomol.* 31:67–71
37. Dawson JE, Anderson BE, Fishbein DB, Sanchez JL, Goldsmith CS, et al. 1991. Isolation and characterization of an *Ehrlichia* sp. from a patient diagnosed with human ehrlichiosis. *J. Clin. Microbiol.* 29:2741–45
38. Dawson JE, Biggie KL, Warner CK, Cookson K, Jenkins S, et al. 1996. Polymerase chain reaction evidence of *Ehrlichia chaffeensis*, etiologic agent of human ehrlichiosis, in dogs from southeast Virginia. *Am. J. Vet. Res.* 57:1175–79
39. Dawson JE, Childs JE, Biggie KL, Moore C, Stallknecht D, et al. 1994. White-tailed deer as a potential reservoir of *Ehrlichia* spp. *J. Wildl. Dis.* 30:162–68
40. Dawson JE, Stallknecht DE, Howerth EW, Warner C, Biggie K, et al. 1994. Susceptibility of white-tailed deer (*Odocoileus virginianus*) to infection with *Ehrlichia chaffeensis*, the etiologic agent of human ehrlichiosis. *J. Clin. Microbiol.* 32:2725–28
41. Dawson JE, Warner CK, Baker V, Ewing SA, Stallknecht DE, et al. 1996. *Ehrlichia*-like 16S rDNA sequence from wild white-tailed deer (*Odocoileus virginianus*). *J. Parasitol.* 82:52–58
42. Demaree HA Jr. 1986. Ticks of Indiana. *Pittman-Robertson Bull.* 16:1–178
43. Dobson A, Foufopoulos J. 2001. Emerging infectious pathogens of wildlife. *Philos. Trans. R. Soc. London Ser. B.* 356: 1001–12
44. Dugan VG, Little SE, Stallknecht DE, Beall AD. 2000. Natural infection of domestic goats with *Ehrlichia chaffeensis*. *J. Clin. Microbiol.* 38:448–49
45. Dumler JS, Barbet AF, Bekker CP, Dasch GA, Palmer GH, et al. 2001. Reorganization of genera in the families Rickettsiaceae and Anaplasmataceae in the order Rickettsiales: unification of some species of *Ehrlichia* with *Anaplasma*, *Cowdria* with *Ehrlichia* and *Ehrlichia* with *Neorickettsia*, descriptions of six new species combinations and designation of *Ehrlichia equi* and 'HGE agent' as subjective synonyms of *Ehrlichia phagocytophila*. *Int. J. Syst. Evol. Microbiol.* 51:2145–65
46. Durden LA, Oliver JH Jr, Kinsey AA. 2001. Ticks (Acari: Ixodidae) and spirochetes (Spirochaetaceae: Spirochaetales) recovered from birds on a Georgia Barrier Island. *J. Med. Entomol.* 38:231–36
47. Eng TR, Harkess JR, Fishbein DB, Dawson JE, et al. 1990. Epidemiologic, clinical, and laboratory findings of human ehrlichiosis in the United States, 1988. *JAMA* 264:2251–58
48. Everett ED, Evans KA, Henry RB, McDonald G. 1994. Human ehrlichiosis in adults after tick exposure; diagnosis using polymerase chain reaction. *Ann. Intern. Med.* 120:730–35
49. Ewing SA, Dawson JE, Kocan AA, Barker RW, Warner CK, et al. 1995. Experimental transmission of *Ehrlichia chaffeensis* (Rickettsiales: Ehrlichieae) among white-tailed deer by *Amblyomma americanum* (Acari: Ixodidae). *J. Med. Entomol.* 32:368–74
50. Ewing SA, Roberson WR, Buckner RG, Hayat CS. 1971. A new strain of *Ehrlichia canis*. *J. Am. Vet. Med. Assoc.* 159:1771–74
51. Felz MW, Durden LA. 1999. Attachment

sites of four tick species (Acari: Ixodidae) parasitizing humans in Georgia and South Carolina. *J. Med. Entomol.* 36:361–64

52. Felz MW, Durden LA, Oliver JH. 1996. Ticks parasitizing humans in Georgia and South Carolina. *J. Parasitol.* 82:505–8
53. Fishbein DB, Dawson JE, Robinson LE. 1994. Human ehrlichiosis in the United States, 1985 to 1990. *Ann. Intern. Med.* 120:736–43
54. Fishbein DB, Kemp A, Dawson JE, Greene NR, Redus MA, Fields DH. 1989. Human ehrlichiosis: prospective active surveillance in febrile hospitalized patients. *J. Infect. Dis.* 160:803–9
55. Gardner LI, Brundage JF, Burke DS, McNeil JG, Visintine R, Miller RN. 1989. Evidence for spread of the human immunodeficiency virus epidemic into low prevalence areas of the United States. *J. Acquir. Immune Defic. Syndr.* 2:521–32
56. Ginsberg HS, Ewing CP, O'Connell AF, Bosler EM, Daley JG, Sayre MW. 1991. Increased population densities of *Amblyomma americanum* (Acari: Ixodidae) on Long Island, New York. *J. Parasitol.* 77:493–95
57. Goddard J. 1987. A review of the diseases harbored and transmitted by the lone star tick, *Amblyomma americanum* (L.). *South. Entomol.* 12:158–71
58. Goddard J. 1988. Was Bullis fever actually ehrlichiosis? JAMA 260:3006–7
59. Goddard J, Norment BR. 1986. Spotted fever group rickettsiae in the lone star tick, *Amblyomma americanum* (Acari: Ixodidae). *J. Med. Entomol.* 23:465–72
60. Gubler DJ, Reiter P, Ebi KL, Yap W, Nasci R, Patz JA. 2001. Climate variability and change in the United States: potential impacts on vector- and rodent-borne diseases. *Environ. Health Perspect.* 109(Suppl. 2):223–33
61. Haile DG. 1989. *Computer simulation of the effects of changes in weather patterns on vector-borne disease transmission. Rep. U.S. Environ. Prot. Agency 230-05-89-057, Appendix G*, U.S. Environ. Prot. Agency, Washington, DC
62. Hair JA, Howell DE. 1970. *Lone star ticks. Their biology and control in Ozark recreation areas. Rep. Bull. B-679*, Oklahoma State Univ., Agric. Exp. Sta., Okla
63. Heppner DG, Wongsrichanalai C, Walsh DS, McDaniel P, Eamsila C, et al. 1997. Human ehrlichiosis in Thailand. *Lancet* 350:785–86
64. Hooker WA, Bishopp FC, Wood HP. 1912. The life history and bionomics of some North American ticks. *Rep. Bureau Entomol.-Bull.* No. 106. U.S. Dep. Agric., Washington, DC
65. Hopla CE. 1953. Experimental studies on tick transmission of tularemia organisms. *Am. J. Hyg.* 58:101–8
66. Hopla CE, Downs CM. 1953. The isolation of *Bacterium tularense* from the tick *Amblyomma americanum. J. Kans. Entomol. Soc.* 26:71–72
67. Ijdo JW, Wu C, Magnarelli LA, Stafford KC, Anderson JF, Fikrig E. 2000. Detection of *Ehrlichia chaffeensis* DNA in *Amblyomma americanum* ticks in Connecticut and Rhode Island. *J. Clin. Microbiol.* 38:4655–56
68. Irving RP, Pinger RR, Vann CN, Olesen JB, Steiner FE. 2000. Distribution of *Ehrlichia chaffeensis* (Rickettsiales: Rickettsiaeceae) in *Amblyomma americanum* in southern Indiana and prevalence of *E. chaffeensis*–reactive antibodies in white-tailed deer in Indiana and Ohio in 1998. *J. Med. Entomol.* 37:595–600
69. Jackson RT, Jackson JW. 1997. Ehrlichiosis with systemic sepsis syndrome. *Tenn. Med.* 90:185–86
70. James AM, Liveris D, Wormser GP, Schwartz I, Montecalvo MA, Johnson BJ. 2001. *Borrelia lonestari* infection after a bite by an *Amblyomma americanum* tick. *J. Infect. Dis.* 183:1810–14
71. Keirans JE, Lacombe EH. 1998. First records of *Amblyomma americanum*,

Ixodes (Ixodes) *dentatus*, and *Ixodes* (Ceratixodes) *uriae* (Acari : Ixodidae) from Maine. *J. Parasitol.* 84:629–31

72. Kennamer JE, Kennamer M, Brenneman R. 1992. History. In *The Wild Turkey: Biology and Management*, ed. JG Dickson, pp. 6–17. Mechanicsburg, PA: Stackpole Books
73. Kirkland KB, Klimko TB, Meriwether RA, Schriefer M, Levin M, et al. 1997. *Erythema migrans*-like rash illness at a camp in North Carolina: a new tick-borne disease? *Arch. Intern. Med.* 157:2635–41
74. Knox WM. 1997. Historical changes in the abundance and distribution of deer in Virginia. See Ref. 93a, pp. 27–36

74a. Kocan A, Levesque GC, Whitworth LC, Murphy GL, Ewing SA, Barker RW. 2000. Naturally occurring *Ehrlichia chaffeensis* infection in coyotes from Oklahoma. *Energy Infect. Dis.* 6:477–80

75. Koch HG, Dunn JE. 1980. Ticks collected from small and medium-sized wildlife hosts in Leflore County, Oklahoma. *South. Nat.* 5:214–21
76. Kollars TM. 1993. Ticks (Acari: Ixodidae) infesting medium-sized wild mammals in southwestern Tennessee. *J. Med. Entomol.* 30:896–900
77. Kordick SK, Breitschwerdt EB, Hegarty BC, Southwick KL, Colitz CM, et al. 1999. Coinfection with multiple tick-borne pathogens in a Walker Hound kennel in North Carolina. *J. Clin. Microbiol.* 37:2631–38
78. Lane RS, Piesman J, Burgdorfer W. 1991. Lyme borreliosis: relation of its causative agent to its vectors and hosts in North America and Europe. *Annu. Rev. Entomol.* 36:587–609
79. Levine JF, Sonenshine DE, Nicholson WL, Turner RT. 1991. *Borrelia burgdorferi* in ticks (Acari: Ixodidae) from coastal Virginia. *J. Med. Entomol.* 28:668–74
80. Little SE, Stallknecht DE, Lockhart JM, Dawson JE, Davidson WR. 1998. Natural coinfection of a white-tailed deer (*Odocoileus virginianus*) population with three *Ehrlichia* spp. *J. Parasitol.* 84:897–901
81. Lockhart JM, Davidson WR, Dawson JE, Stallknecht DE. 1995. Temporal association of *Amblyomma americanum* with the presence of *Ehrlichia chaffeensis* reactive antibodies in white-tailed deer. *J. Wildl. Dis.* 31:119–24
82. Lockhart JM, Davidson WR, Stallknecht DE, Dawson JE. 1996. Site-specific geographic association between *Amblyomma americanum* (Acari: Ixodidae) infestations and *Ehrlichia chaffeensis*-reactive (Rickettsiales: Ehrlichieae) antibodies in white-tailed deer. *J. Med. Entomol.* 33:153–58
83. Lockhart JM, Davidson WR, Stallknecht DE, Dawson JE. 1998. Lack of seroreactivity to *Ehrlichia chaffeensis* among rodent populations. *J. Wildl. Dis.* 34:392–96
84. Lockhart JM, Davidson WR, Stallknecht DE, Dawson JE, Howerth EW. 1997. Isolation of *Ehrlichia chaffeensis* from wild white-tailed deer (*Odocoileus virginianus*) confirms their role as natural reservoir hosts. *J. Clin. Microbiol.* 35:1681–86
85. Lockhart JM, Davidson WR, Stallknecht DE, Dawson JE, Little SE. 1997. Natural history of *Ehrlichia chaffeensis* (Rickettsiales: Ehrlichieae) in the piedmont physiographic province of Georgia. *J. Parasitol.* 83:887–94
86. Long L, DeAre D. 1982. Repopulating the countryside: a 1980 census trend. *Science* 217:1111–16
87. Maeda K, Markowitz N, Hawley RC, Ristic M, Cox D, McDade JE. 1987. Human infection with *Ehrlichia canis*, a leukocytic rickettsia. *N. Engl. J. Med.* 316:853–56
88. Magnarelli LA, Anderson JF, Stafford KC, Dumler JS. 1997. Antibodies to multiple tick-borne pathogens of babesiosis, ehrlichiosis, and Lyme borreliosis

in white-footed mice. *J. Wildl. Dis.* 33:466–73

89. Marty AM, Dumler JS, Imes G, Brusman HP, Smrkovski LL, Frisman DM. 1995. Ehrlichiosis mimicking thrombotic thrombocytopenic purpura. Case report and pathological correlation. *Hum. Pathol.* 26:920–25
90. Masters EJ, Donnell HD, Fobbs M. 1994. Missouri Lyme disease: 1989 through 1992. *J. Spir. Tick-Borne Dis.* 1:12–17
91. McCabe TR, McCabe RE. 1997. Recounting whitetails past. See Ref. 93a, pp. 11–26
92. McCall CL, Curns AT, Singleton JS, Comer JA, Olson JG, et al. 2001. Fort Chaffee revisited; the epidemiology of tickborne diseases at a persistent focus. *Vect. Borne Zoonot. Dis.* 2:119–27

92a. McDade JE. 1990. Ehrlichiosis—a disease of animals and humans. *J. Infect. Dis.* 161:609–17

93. McQuiston JH, Paddock CD, Holman RC, Childs JE. 1999. The human ehrlichioses in the United States. *Emerg. Infect. Dis.* 5:635–42

93a. McShea WJ, Underwood HB, Rappole JH, eds. 1997. *The Science of Over Abundance: Deer Ecology and Population Management*. Washington, DC: Smithson. Inst. Press. 432 pp.

94. Means RG, White DJ. 1997. New distribution records of *Amblyomma americanum* (L.) (Acari: Ixodidae) in New York State. *J. Vector Ecol.* 22:133–45
95. Merten HA, Durden LA. 2000. A state-by-state survey of ticks recorded from humans in the United States. *J. Vector Ecol.* 25:102–13
96. Mock DE, Applegate RD, Fox LB. 2001. Preliminary survey of ticks (Acari: Ixodidae) parasitizing wild turkeys (Aves: Phasianidae) in eastern Kansas. *J. Med. Entomol.* 38:118–21
97. Moore GC, Parker GR. 1992. Colonization by the eastern coyote (Canis latrans). In *Ecology and Management of the Eastern Coyote*, ed. AH Boer, pp. 21–37. Fredericton, N.B.: Wildl. Res. Unit, Univ. New Brunswick
98. Morris JG Jr, Potter M. 1997. Emergence of new pathogens as a function of changes in host susceptibility. *Emerg. Infect. Dis.* 3:435–41
99. Mount GA, Haile DG, Barnard DR, Daniels E. 1993. New version of LSTSIM for computer simulation of *Amblyomma americanum* (Acari: Ixodidae) population dynamics. *J. Med. Entomol.* 30:843–57
100. Mueller-Anneling L, Gilchrist MJ, Thorne PS. 2000. *Ehrlichia chaffeensis* antibodies in white-tailed deer, Iowa, 1994 and 1996. *Emerg. Infect. Dis.* 6:397–400
101. Murphy GL, Ewing SA, Whitworth LC, Fox JC, Kocan AA. 1998. A molecular and serologic survey of *Ehrlichia canis*, *E. chaffeensis*, and *E. ewingii* in dogs and ticks from Oklahoma. *Vet. Parasitol.* 79:325–39
102. Nuti M, Serafini DA, Bassetti D, Ghionni A, Russino F, et al. 1998. *Ehrlichia* infection in Italy. *Emerg. Infect. Dis.* 4:663–65
103. Olano JP, Masters E, Cullman L, Hogrefe W, Yu XJ, Walker DH. 1999. Human monocytotropic ehrlichiosis (HME): epidemiological, clinical and laboratory diagnosis of a newly emergent infection in the United States. In *Rickettsiae and Rickettsial Diseases at the Turn of the Third Millenium*, ed. D Raoult, P Brouqui, pp. 262–68. Paris: Elsevier
104. Oliver JH Jr, Chandler FW Jr, James AM, Sanders FH Jr, Hutcheson HJ, et al. 1995. Natural occurrence and characterization of the Lyme disease spirochete, *Borrelia burgdorferi*, in cotton rats (*Sigmodon hispidus*) from Georgia and Florida. *J. Parasitol.* 81:30–36
105. Oliver JH Jr, Chandler FW Jr, Luttrell MP, James AM, Stallknecht DE, et al. 1993. Isolation and transmission of the Lyme disease spirochete from the

southeastern United States. *Proc. Natl. Acad. Sci. USA* 90:7371–75

106. Ostfeld RS, Keesing F. 2000. Biodiversity and disease risk: the case of Lyme disease. *Conserv. Biol.* 14:722–28
107. Ouellette J, Apperson CS, Howard P, Evans TL, Levine JF. 1997. Tick-raccoon associations and the potential for Lyme disease spirochete transmission in the coastal plain of North Carolina. *J. Wildl. Dis.* 33:28–39
108. Paddock CD, Folk SM, Shore GM, Machado LJ, Huycke MM, et al. 2001. Infections with *Ehrlichia chaffeensis* and *Ehrlichia ewingii* in persons coinfected with HIV. *Clin. Infect. Dis.* 33:1586–94
109. Paddock CD, Sumner JW, Shore GM, Bartley DC, Elie RC, et al. 1997. Isolation and characterization of *Ehrlichia chaffeensis* strains from patients with fatal ehrlichiosis. *J. Clin. Microbiol.* 35:2496–502
110. Palella FJ, Delaney KM, Moorman AC, Loveless MO, Fuhrer J, et al. 1998. Declining morbidity and mortality among patients with advanced human immunodeficiency virus infection. HIV outpatient study investigators. *N. Engl. J. Med.* 338:853–60
111. Parker RR, Kohls GM. 1943. American Q fever: the occurrence of *Rickettsia diaporica* in *Amblyomma americanum* from eastern Texas. *Pub. Health Rep.* 58:1510–11
112. Parker RR, Kohls GM, Cox GW, Davis GE. 1939. Observations on an infectious agent from *Amblyomma maculatum*. *Pub. Health Rep.* 54:1482–84
113. Parker RR, Kohls GM, Steinhaus EA. 1943. Rocky Mountain spotted fever: spontaneous infection in *Amblyomma americanum*. *Pub. Health Rep.* 58:721–29
114. Pawelec G, Solana R, Remarque E, Mariani E. 1998. Impact of aging on innate immunity. *J. Leukoc. Biol.* 64:703–12
115. Philip CB, White JS. 1955. Disease agents recovered incidental to a tick survey of the Mississippi Gulf Coast. *J. Econ. Entomol.* 48:396–99
116. Pung OJ, Durden LA, Banks CW, Jones DN. 1994. Ectoparasites of opossums and raccoons in southeastern Georgia. *J. Med. Entomol.* 31:915–19
117. Rawlings J. 1996. Human ehrlichiosis in Texas. *J. Spir. Tick-Borne Dis.* 3:94–97
118. Rehacek J. 1989. Ecological relationships between ticks and rickettsiae. *Eur. J. Epidemiol.* 5:407–13
119. Riley SPD, Hadidian J, Manski DA. 1998. Population density, survival, and rabies in raccoons in an urban national park. *Can. J. Zool.* 76:1153–64
120. Ripoll CM, Remondegui CE, Ordonez G, Arazamendi R, Fusaro H, et al. 1999. Evidence of rickettsial spotted fever and ehrlichial infections in a subtropical territory of Jujuy, Argentina. *Am. J. Trop. Med. Hyg.* 61:350–54
121. Rue LL. 1978. *The Deer of North America*. New York: Crown Publ.
122. Rumley RL, Shappley NC, Waivers LE, Esinhart JD. 1991. AIDS in rural eastern North Carolina—patient migration: a rural AIDS burden. *AIDS* 5:1373–78
123. Rutberg AT. 1997. The science of deer management; an animal welfare perspective. See Ref. 93a, pp. 37–54
124. Sadikot R, Shaver MJ, Reeves WB. 1999. *Ehrlichia chaffeensis* in a renal transplant recipient. *Am. J. Nephrol.* 19:674–76
125. Schultze TL, Bosler EM. 1996. Another look at the potential role of *Amblyomma americanum* in the transmission of tick-borne disease. *J. Spir. Tick-Borne Dis.* 3:113–15
126. Sexton DJ, Corey GR, Carpenter C, Kong LQ, Gandhi T, et al. 1998. Dual infection with *Ehrlichia chaffeensis* and a spotted fever group rickettsia: a case report. *Emerg. Infect. Dis.* 4:311–16
127. Shibata S, Kawahara M, Rikihisa Y, Fujita H, Watanabe Y, et al. 2000. New *Ehrlichia* species closely related

to *Ehrlichia chaffeensis* isolated from *Ixodes ovatus* ticks in Japan. *J.Clin. Microbiol.* 38:1331–38

128. Sonenshine DE, Stout IJ. 1971. Ticks infesting medium-sized wild mammals in two forest localities in Virginia (Acarina: Ixodidae). *J. Med. Entomol.* 8:217–27

129. Standaert SM, Dawson JE, Schaffner W, Childs JE, Biggie KL, et al. 1995. Ehrlichiosis in a golf-oriented retirement community. *N. Engl. J. Med.* 333:420–25

129a. Standaert SM, Yu T, Scott MA, Childs JE, Paddock CD, et al. 2000. Primary isolation of *Ehrlichia chaffeensis* from patients with febrile illnesses: clinical and molecular characteristics. *J. Infect. Dis.* 181:1082–88

130. Steiner FE, Pinger RR, Vann CN. 1999. Infection rates of *Amblyomma americanum* (Acari: Ixodidae) by *Ehrlichia chaffeensis* (Rickettsiales: Ehrlichieae) and prevalence of *E. chaffeensis*-reactive antibodies in white-tailed deer in southern Indiana, 1997. *J. Med. Entomol.* 36:715–19

131. Stromdahl EY, Evans SR, O'Brien JJ, Gutierrez AG. 2001. Prevalence of infection in ticks submitted to the human tick test kit program of the U.S. Army Center for Health Promotion and Preventive Medicine. *J. Med. Entomol.* 38:67–74

132. Stromdahl EY, Randolph MP, O'Brien JJ, Gutierrez AG. 2000. *Ehrlichia chaffeensis* (Rickettsiales: Ehrlichieae) infection in *Amblyomma americanum* (Acari: Ixodidae) at Aberdeen Proving Ground, Maryland. *J. Med. Entomol.* 37:349–56

133. Sumner JW, McKechnie D, Janowski D, Paddock CD. 2000. American Society for Rickettsiology. Presented at 15th Meet., Captiva Flor. 51 (Abstr.)

134. Sumner JW, Storch GA, Buller RS, Liddell AM, Stockham SL, et al. 2000. PCR amplification and phylogenetic analysis of groESL operon sequences from *Ehrlichia ewingii* and *Ehrlichia muris*. *J. Clin. Microbiol.* 38:2746–49

135. Tabor GM, Ostfeld RS, Poss M, Dobson AP, Aguirre AA. 2001. Conservation biology and the health sciences: defining the research priorities of conservation medicine. In *Conservation Biology*, ed. ME Soulé, GH Orians, pp. 155–73. Washington, DC: Island

136. Taylor JP, Istre GR, McChesney TC, Satalowich FT, Parker RL, McFarland LM. 1991. Epidemiologic characteristics of human tularemia in the southwest-central states, 1981–1987. *Am. J. Epidemiol.* 133:1032–38

137. Thompson C, Spielman A, Krause PJ. 2001. Coinfecting deer-associated zoonoses: lyme disease, babesiosis, and ehrlichiosis. *Clin. Infect. Dis.* 33:676–85

138. Topping NH. 1947. The epidemiology of Rocky Mountain spotted fever. *N. Y. State J. Med.* 1585–87

139. Tugwell P, Lancaster JL Jr. 1962. Results of a tick-host study in northwest Arkansas. *J. Kans. Entomol. Soc.* 35:202–11

140. U.S. Census Bureau. 1999. Statistical Abstract of the United States. Washington, DC: U.S. Census Bureau. pp. 1–1005

141. U.S. Census Bureau. 2001. Current population reports, series P23–205, population profile of the United States, 1999. Washington, DC: U.S. Gov Print. Off.

142. Uhaa IJ, MacLean JD, Green CR, Fishbein DB. 1992. A case of human ehrlichiosis acquired in Mali: clinical and laboratory findings. *Am. J. Trop. Med. Hyg.* 46:161–64

143. Walker DH. 2000. Diagnosing human ehrlichioses: current status and recommendations. *ASM News* 66:287–89

144. Walker DH, Barbour AG, Oliver JH, Lane RS, Dumler JS, et al. 1996. Emerging bacterial zoonotic and vector-borne diseases. Ecological and epidemiological factors. *JAMA* 275:463–69

145. Washburn AM, Tuohy JH. 1949. The

changing picture of tularemia transmission in Arkansas. *South. Med. J.* 42:60–62

146. Weller SJ, Baldridge GD, Munderloh UG, Noda H, Simser J, Kurtti TJ. 1998. Phylogenetic placement of rickettsiae from the ticks *Amblyomma americanum* and *Ixodes scapularis*. *J.Clin. Microbiol.* 36:1305–17
147. Whitlock JE, Fang QQ, Durden LA, Oliver JH Jr. 2000. Prevalence of *Ehrlichia chaffeensis* (Rickettsiales: Rickettsiaceae) in *Amblyomma americanum* (Acari: Ixodidae) from the Georgia coast and Barrier Islands. *J. Med. Entomol.* 37:276–80
148. Woodland JC, McDowell MM, Richards JT. 1943. Bullis fever (Lone Star tick fever-tick fever). *JAMA* 122:1156–60

148a. Wolf L, McPherson T, Harrison B, Engber B, Anderson A, Whitt P. 2000. Prevalence of *Ehrlichia ewingii* in *Amblyomma americanum* in North Carolina. *J. Clin. Microbiol.* 38:2795

148b. Yabsley MJ, Varela AS, Tate CM, Dugan VG, Stallknecht DE, et al. 2002. *Ehrlichia ewingii* infection in white-tailed deer (*Odocoileus virginianus*). *Emerg. Infect. Dis.* 8:668–71

149. Zimmerman RH, McWherter GR, Bloemer SR. 1988. Medium-sized mammal hosts of *Amblyomma americanum* and *Dermacentor variabilis* (Acari: Ixodidae) at Land Between the Lakes, Tennessee, and effects of integrated tick management on host infestations. *J. Med. Entomol.* 25:461–66

Annu. Rev. Entomol. 2003. 48:339–64
doi: 10.1146/annurev.ento.48.091801.112731

First published online as a Review in Advance on August 28, 2002

SELECTIVE TOXICITY OF NEONICOTINOIDS ATTRIBUTABLE TO SPECIFICITY OF INSECT AND MAMMALIAN NICOTINIC RECEPTORS

Motohiro Tomizawa and John E. Casida
Environmental Chemistry and Toxicology Laboratory, Department of Environmental Science, Policy and Management, University of California, Berkeley, California 94720-3112; e-mail: tomizawa@nature.berkeley.edu; ectl@nature.berkeley.edu

Key Words imidacloprid, neonicotinoid insecticides, nicotine, nicotinic acetylcholine receptor, molecular features conferring selectivity

■ **Abstract** Neonicotinoids, the most important new class of synthetic insecticides of the past three decades, are used to control sucking insects both on plants and on companion animals. Imidacloprid (the principal example), nitenpyram, acetamiprid, thiacloprid, thiamethoxam, and others act as agonists at the insect nicotinic acetylcholine receptor (nAChR). The botanical insecticide nicotine acts at the same target without the neonicotinoid level of effectiveness or safety. Fundamental differences between the nAChRs of insects and mammals confer remarkable selectivity for the neonicotinoids. Whereas ionized nicotine binds at an anionic subsite in the mammalian nAChR, the negatively tipped ("magic" nitro or cyano) neonicotinoids interact with a proposed unique subsite consisting of cationic amino acid residue(s) in the insect nAChR. Knowledge reviewed here of the functional architecture and molecular aspects of the insect and mammalian nAChRs and their neonicotinoid-binding site lays the foundation for continued development and use of this new class of safe and effective insecticides.

CONTENTS

INTRODUCTION

The neonicotinoids are the most important new class of synthetic insecticides of the past three decades. Although related to nicotine in action, and partially in structure, the neonicotinoids originated instead from screening novel synthetic chemicals to discover a lead compound (44). Once optimized to imidacloprid (IMI) and analogs (34), the neonicotinoids joined the earlier chlorinated hydrocarbons, organophosphorus compounds, methylcarbamates, and pyrethroids to constitute the five principal types of active ingredients, all of which are neuroactive insecticides (11). Neonicotinoids are increasingly used for systemic control of plant-sucking insects, replacing the organophosphorus compounds and methylcarbamates, which have decreased effectiveness because of resistance or increased restrictions due to toxicological considerations. Neonicotinoids are also important in animal health care. These developments were possible because of the selective toxicity of the neonicotinoids (125), which is attributable to the specificity of insect and mammalian nicotinic receptors as reviewed here.

NEONICOTINOIDS AND NICOTINOIDS

Neonicotinoids

Nithiazine (Figure 1), a nitromethylene heterocycle discovered by Soloway and colleagues (44, 99, 100) of the former Shell Development Company in California, was the first prototype of the highly insecticidal neonicotinoids. The lead compound was 2-(dibromonitromethyl)-3-methylpyridine (25), recognized by Shell scientists in 1970 to have unexpected although modest insecticidal activity. Structural optimization led to nithiazine with higher insecticidal activity than parathion against house fly adults (*Musca domestica*) and much higher activity against corn earworm larvae (*Helicoverpa zea*), combined with good systemic action in plants and low mammalian toxicity (44, 99, 100). Unfortunately, nithiazine showed high

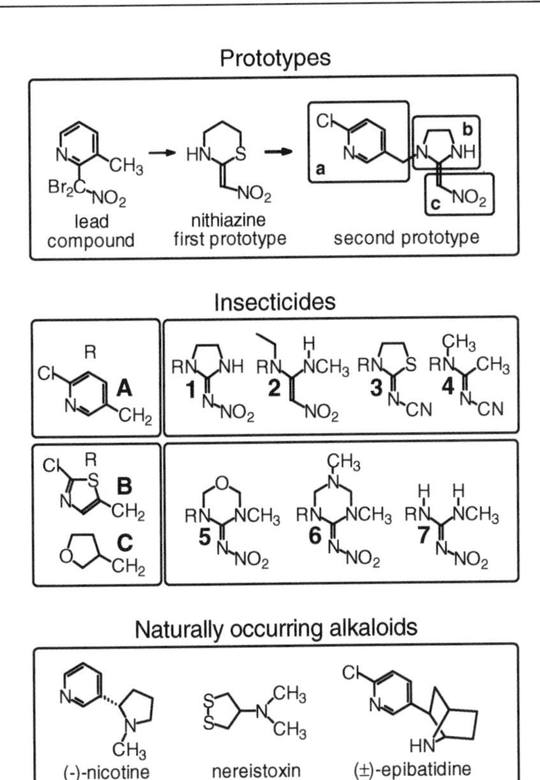

Figure 1 Phylogeny of neonicotinoid insecticides. Three delineated portions (*a–c*) of the second prototype indicate the key segments for neonicotinoid design. Four other commercial neonicotinoids are nitenpyram (A2), thiacloprid (A3), acetamiprid (A4), and thiamethoxam (B5). Three developmental compounds are AKD-1022 (B6), chlothianidin (B7), and dinotefuran (C7). Structures are also given for naturally occuring alkaloids affecting nicotinic receptor function. Although the second prototype is shown originating from nithiazine, and might be considered as a theoretical hybrid with nicotine, this compound and ultimately IMI were not derived in this way as discussed in the text.

photolability (41, 99, 100), and photostabilization as the *N*-formyl derivative was not adequate for practical application (44). Nithiazine ultimately was commercialized as the active ingredient of a house fly trap device for poultry and animal husbandry (44).

IMI (A1 in Figure 1), discovered by Kagabu and coworkers of Nihon Tokushu Noyaku Seizo in Japan (presently Nihon Bayer Agrochem), was optimized from studies on nithiazine starting in 1979 (34). The green rice leafhopper (*Nephotettix cincticeps*) was selected instead of a lepidopterous larva for studying structure-activity relationships because it is a major hemipteran pest on rice and vegetables in Japan. The nithiazine hydrothiazine ring was replaced with other *N*-heterocycles and substituents were introduced on the nitrogen atom (34). Activity was enhanced first with the *N*-(4-chlorobenzyl) and then the *N*-(3-pyridinylmethyl) moieties (39). Interestingly, even though IMI was not modeled on nicotine, the *N*-(3-pyridinyl)methyl group was known earlier to be an essential structural moiety of the insecticidal nicotinoids (123). In 1984, these developments led to 1-(6-chloropyridin-3-ylmethyl)-2-nitromethylene-imidazolidine (referred to here as the second prototype) with over 100-fold-higher activity than nithiazine against *Nephotettix* (using a strain resistant to organophosphorus compounds, methylcarbamates, and pyrethroids) (34, 39). Finally, replacement of the nitromethylene by nitroimine yielded IMI in February 1985 (34, 65) with 62- to >3000-fold-higher insecticidal activity, depending on species, than that of (−)-nicotine (Table 1). IMI was introduced into the market in 1991 and quickly became one of the most important insect control chemicals (34, 68).

The discovery of IMI prompted intensive research and development programs by many companies in which the three portions of the molecule shown in the second prototype were varied (Figure 1, *a–c*). Four other neonicotinoids are currently on the market as follows: nitenpyram (A2) (62), thiacloprid (A3) (22, 98), acetamiprid (A4) (102), and thiamethoxam (B5) (58). There are also three developmental compounds, i.e., AKD-1022 (B6) (58), chlothianidin (B7) (68, 119), and dinotefuran (C7) (43, 63, 64) (Figure 1).

The physicochemical properties of the neonicotinoids played an important role in their development. The principal target pests are aphids, leafhoppers, whiteflies, and other sucking insects due to the excellent plant-mobile (systemic) property conferred by the moderate water solubility. Thiacloprid, acetamiprid, and the nitromethylene-thiazolidine analog of IMI with relatively high hydrophobicity show lepidoptericidal activities against codling moth (*Cydia pomonella*), tobacco cutworm (*Spodoptera litura*), diamondback moth (*Plutella xylostella*), rice leafroller (*Cnaphalocrocis medinalis*), peach fruit moth (*Carposina niponensis*), cabbage moth (*Mamestra brassicae*), or Oriental fruit moth (*Grapholitha molesta*) (22, 97, 122). Photostability is an important factor in field performance of the

TABLE 1 Potency increase for IMI over (−)-nicotine against three insect pests

	Insecticidal activity		
Insect	**(−)-Nicotine**	**IMI**	**Potency increase**
Myzus persicae[a]	4.5	0.073	62
Nephotettix cincticeps[b]	>1000	0.32	>3100
Musca domestica[c]	>50	0.045	>1100

[a] LC_{50} (ppm) with modified aphid dip test (73).

[b] LC_{90} (ppm) with seedling treatment (from Reference 65 and personal communication with S. Kagabu).

[c] LD_{50} (μg/g) with intrathoracic injection following synergist pretreatment (111).

neonicotinoids. The nitromethylene chromophore of nithiazine and IMI analogs ($\lambda_{max} > 320$ nm) absorbs strongly in the sunlight range of 290 to 400 nm and rapidly decomposes to noninsecticidal compounds. Replacing nitromethylene with a nitroimine or cyanoimine substituent ($\lambda_{max} < 270$ nm) avoids the absorption of sunlight and consequently confers prolonged residual activity. In this series, primary excitation by ultraviolet absorption is the initiator of intramolecular donor-acceptor electron transfer. The energy gap from the ground state to the single excited state is in the order of cyanoimine > nitroimine > nitromethylene (34, 35, 38).

Nicotinoids and Other Alkaloids

Among the large number and variety of alkaloids (117, 118), three are of special interest because of their similarity to the neonicotinoids in structure or action (Figure 1). (−)-Nicotine in tobacco extracts has been used as an aphicide since at least 1690 and is still used today (particularly in China), despite relatively low potency and a narrow spectrum of insecticidal activity (Table 1). The insecticidal activity of nicotine is improved in synthetic nicotinoids such as dihydronicotyrine and *N,N*-disubstituted 3-pyridinylmethylamines (123, 124) but not to the degree required for commercialization. Nereistoxin from the marine worm *Lumbriconereis heteropoda* served as the lead to develop the synthetic insecticides cartap, bensultap, and thiocyclam, which also modulate nicotinic receptors but by a mechanism different from that of IMI (23, 83, 96, 112, 113). The use of nicotine and the nereistoxin analogs rapidly declined after introduction of the neonicotinoids. Epibatidine isolated from the skin of an Ecuadorean frog (*Epipedobates tricolor*) contains the 6-chloropyridin-3-yl moiety (101) in common with IMI (Figure 1) and displays ultrahigh agonist potency at the mammalian nicotinic receptor (3).

Terms and Designations

The term neonicotinoid was originally proposed for IMI and related insecticidal compounds with a structural similarity to nicotine and a common mode of action (116). Neonicotinoids and nicotinoids (123, 125) are defined here by their common structural features and action as agonists at the nicotinic acetylcholine receptor (nAChR) with further differentiation by their ionization at physiological pH and target site specificity between insects and mammals, i.e., the neonicotinoids are not ionized and selective for the insect nAChR, and the nicotinoids are ionized and selective for the mammalian nAChR. The classical neonicotinoid is IMI and nicotinoid is nicotine. The nicotinoids include anabasine (known as neonicotine in 1931), nornicotine, and synthetic analogs of similar properties. On this basis, epibatidine and desnitro-IMI (discussed later), despite their 6-chloropyridin-3-yl moiety, have nicotinoid-type action.

A variety of terms have been used to subdivide the neonicotinoids based on unique structural types (Figure 1): chloronicotinyls (A), chlorothiazolyls (B), and tetrahydrofuranyl (C); nitroimines or nitroguanidines (1 and 5–7), nitromethylenes (the second prototype and 2), cyanoimines (3 and 4), and nitromethylene heterocycle (nithiazine); first generation (with moiety A) and second generation (with moieties B and C).

MODE OF ACTION

Target

Acetylcholine (ACh) is the endogenous agonist and excitatory neurotransmitter of the cholinergic nervous system. Neurotransmission through the nicotinic cholinergic synapse is mediated in two steps (Figure 2). First, ACh is released from the presynaptic membrane by exocytosis and interacts with the binding site located at the extracellular domain of the nAChR/ion channel complex. Second, a conformational change of the receptor molecule leads to opening the ion channel, promoting the influx of extracellular Na^+ and efflux of intracellular K^+ to disrupt the equilibrium status of the membrane potential. In insects, the nAChR is widely and predominantly distributed in the neuropil regions of the central nervous system (CNS). It is not only responsible for rapid neurotransmission but it is also an important target for insecticide action.

Penetration into the Central Nervous System

Nicotinoids with insecticidal activity such as nicotine, nornicotine, and anabasine have a basic nitrogen atom that is mostly protonated (89%, 99%, and 98% ionization, respectively) under physiological conditions (123). Ionized nicotine is poor in penetrating the "ion-impermeable" barrier surrounding the insect CNS (76). On the other hand, the cationic nature of nicotine (as with the ammonium head of ACh) is an essential requirement to interact with the nAChR. The inferior insecticidal

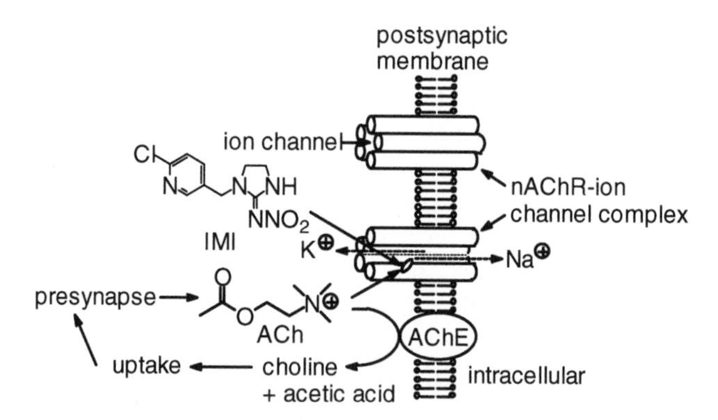

Figure 2 Cholinergic neurotransmission mediated by the nAChR on the postsynaptic membrane. The neurotransmitter ACh released presynaptically binds to the nAChR, leading to activation of the ion channel. ACh is then hydrolyzed by acetylcholinesterase (AChE).

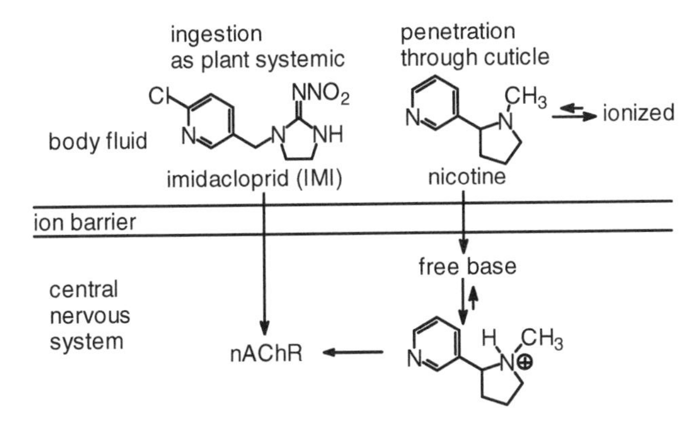

Figure 3 Comparison between imidacloprid and nicotine in their insecticidal action.

activity of nicotine is therefore attributed to these two contradictory features (115, 123, 124, 126). The penetration of neonicotinoids into the insect CNS is related to their hydrophobicity and is greater than that of the ionizable nicotine (126). Thus, the neonicotinoids overcome the ionization dilemma of nicotine in their translocation and target site interaction (111) (Figure 3) (Table 1).

Neonicotinoid-Binding Site

In insects, neonicotinoid insecticides act as agonists at the postsynaptic nAChR with much higher affinity (nanomolar level) than that of nicotine (micromolar level). The neonicotinoid-binding site in insects is the same as or closely coupled to that of ACh, nicotine, and α-bungarotoxin (α-BGT, a toxin from the elapid snake *Bungarus multicinctus*) (4, 75, 84, 87, 115) and is most commonly determined with [^{3}H]IMI as the radioligand (48), which displays saturable and reversible binding with fast kinetics (54). The high-affinity [^{3}H]IMI-binding site is conserved in neonicotinoid sensitivity and specificity (structure-activity relationships) across a broad range of insects including the green peach aphid (*Myzus persicae*), cowpea aphid (*Aphis craccivora*), *Nephotettix*, glassy-winged sharpshooter (*Homalodisca coagulata*), whitefly (*Bemisia argentifolii*), American cockroach (*Periplaneta americana*), migratory locust (*Locusta migratoria*), tobacco hornworm (*Manduca sexta*), *Musca*, fruit fly (*Drosophila melanogaster*), and honey bee (*Apis mellifera*) (15, 51, 69, 109, 114, 121, 128).

Comparing binding affinities of neonicotinoids at the insect nAChR with their intrinsic insecticidal activity validates the toxicological relevance of the target site. This was shown with *Musca* using a series of neonicotinoids and correlating displacement of [^{3}H]IMI binding with injected knockdown activity for insects pretreated with a synergist to block CYP450-mediated metabolic detoxification ($r = 0.90$, $n = 22$) (55, 110). It was verified with *P. americana* correlating frequency of spontaneous discharges in nerve cords with minimum lethal doses ($r = 0.84$, which was increased to $r = 0.94$ with a hydrophobicity parameter; $n = 16$) (75).

VERTEBRATE NICOTINIC ACETYLCHOLINE RECEPTORS

Functional Architecture and Diversity

The functional architecture and diversity of nAChRs are much better understood for vertebrates than for insects. In vertebrates, the nAChR is a pentameric transmembrane complex in the superfamily of neurotransmitter-gated ion channels including the γ-aminobutyric acid, glycine, and serotonin receptors. Sequences for each nAChR subunit predict hydrophilic extracellular domains containing a binding site for cholinergic ligands and four transmembrane hydrophobic segments. The second domain of the five subunits is considered to form the lumen of the ion channel and the binding sites for ion channel blockers (1, 2, 17, 103).

The vertebrate nAChR consists of diverse subtypes assembled as five subunits in combinations from ten α (α1–10), four β (β1–4), δ, γ, and ε subunits (Figure 4). The skeletal muscle or electric ray (*Torpedo*) subtype is made up of two α1 subunits and one each of β1, δ, and γ (or ε in adult muscle) subunits and is best defined relative to the ligand-binding site environment (1, 2). Neuronal nAChR subtypes expressed in vertebrate brain and ganglia are assembled in combinations of α2–10 and β2–4 and are pharmacologically classified into two branches based on sensitivity to α-BGT. The α-BGT-insensitive branch is made up of subtypes with combinations of α2–6 and β2–4 subunits. Of these, the multiple $\alpha3\beta2$[and/or $\beta4$]$\alpha5$ subtype is mainly found in ganglia, and $\alpha4\beta2$ is the most abundant subtype in brain. The $\alpha4\beta2$ receptor consists of two α4 and three β2 subunits and represents >90% of the tritiated ligand- (e.g., [^{3}H]nicotine) binding sites in brain. The α7–10 subunits are involved in the α-BGT-sensitive subtypes in brain, ganglia, and the limbic system, and the amount of α7 receptor is comparable to that of the $\alpha4\beta2$ subtype in brain and is far predominant over the $\alpha3\beta2$[and/or $\beta4$]$\alpha5$ subtype in ganglia. The native α7 receptor in mammalian central and peripheral nervous

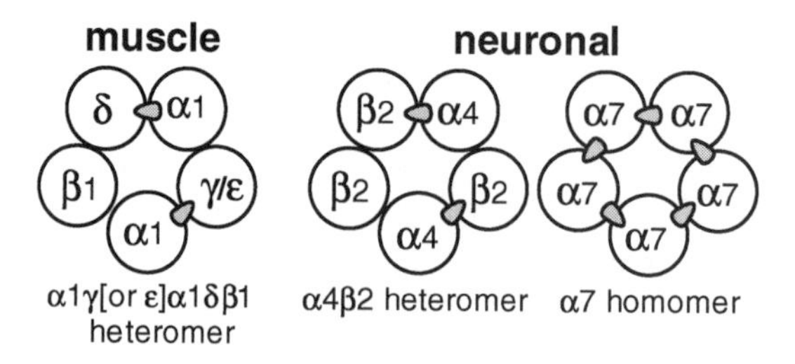

Figure 4 Functional assembly of vertebrate nicotinic receptor subtypes consisting of different subunit combinations viewed in cross-section from the top. *Wedges* designate binding sites for cholinergic ligands at the interface between subunits.

systems is considered to be homomeric status, although the $\alpha7\alpha8$ heteromeric receptor with unidentified subunit(s) exists in chick brain and retina. The $\alpha8$ subunit is found only in avian brain, retina, or ganglia, and the α9–10 in rat vestibular and cochlea mechanosensory hair cells or olfactory bulb. The $\alpha9$ subunit forms a functional homomeric receptor. The $\alpha10$-containing receptor is functional only on concomitant assembly with an $\alpha9$ subunit (16, 24, 31, 79, 103).

Nicotinic Receptors as Targets for Pharmaceuticals

The mammalian nAChR is the target for potential therapeutic agents for diverse neurological disorders as follows: nonopioid and non-anti-inflammatory analgesia, neurodegenerative diseases (Alzheimer's and Parkinson's diseases), cognitive dysfunction, schizophrenia, depression, epilepsy, attention-deficit hyperactivity, Tourette's syndrome, and anxiety (16, 52, 56). Because the agonist- or drug-binding site is localized at the interface region between subunits, the specific subunit combinations confer differences in sensitivity to ACh and/or in pharmacological profiles among the receptor subtypes (16, 31, 56) (Figure 4). One aspect of this research is the discovery of highly subtype-selective agonists and antagonists (56). Nicotine has modest analgesic activity of short duration while epibatidine displays an analgesic efficacy in rodents comparable to that of morphine at a 200-fold-lower dose and an ultrahigh affinity to the $\alpha4\beta2$ nAChR (3), which is probably the primary target for nicotine-induced analgesia (16, 56).

INSECT NICOTINIC ACETYLCHOLINE RECEPTORS

Molecular Biology

Several genes encoding ligand-binding α-type and structural β- (or non-α−) type subunits have been cloned in *Drosophila* and other insects, but the functional architecture and diversity of insect nAChRs are poorly defined (26, 107). In *Drosophila*, the identified components are encoded by four genes for the α-type subunits (7, 32, 45, 53, 85, 89) and two genes for the β-type subunits (27, 47, 86) (Table 2). Four additional candidate genes for nAChR subunits are predicted from *Drosophila* genome data (53). On expression in *Xenopus* oocytes, human embryonic kidney 293 cells or *Drosophila* S2 cells, the four α subunit genes (for ALS, Dα2, Dα3, and Dα4) and two β subunit genes (for ARD and SBD), singly or in various combinations, never produce any electrophysiological response or [^{3}H]epibatidine binding. However, functional ion channel property or [^{3}H]epibatidine-binding activity is clearly observed when any of the four α subunits is coexpressed with chick or rat $\beta2$ or with rat $\beta4$ subunits (5, 45, 47, 89).

In *Myzus*, six genes are cloned [five encoding α-type subunits (Mpα1–5) and the sixth a β-type subunit (Mpβ1)] (Table 2) and expressed in *Xenopus* oocytes (93) or the *Drosophila* S2 cell line (29, 30). Mpα1 and Mpα2 form functional homomeric receptors (93), but no specific [^{3}H]epibatidine binding is detected with any of the

TABLE 2 Identified nicotinic receptor subunits from *Drosophila* and *Myzus*[a]

	Molecular mass, kDa, or amino acid (aa) length	
Species and subunit[b]	**Core**[c]	**Mature or native**[d]
Drosophila		
ALS (Dα1)	62	80, 69
Dα2 (SAD)	61	66, 65
Dα3	85.4	105
Dα4	568aa	53
ARD (Dβ1)	57.3	50, 45
SBD (Dβ2)	57.3	60, 61
Myzus		
Mpα1	532aa	?
Mpα2	571aa	45?
Mpα3	60	45?
Mpα4	59	?
Mpα5	482aa	?
Mpβ1	56	48

[a]See text for references.

[b]Based on conserved sequence and northern analysis of transcripts.

[c]Deduced from gene sequence (excluding glycosylated region).

[d]Based on immunology and protein biochemistry approaches.

four α-type subunits (Mpα1–4) (29). [^{3}H]Epibatidine-binding activity is observed when any one of Mpα1, Mpα2, or Mpα3 is coexpressed with the rat β2 subunit, but the Mpα4/rat β2 receptor does not confer this activity (29). A polyclonal antiserum pAb24 immunoprecipitates the Mpβ1 gene product (48 kDa) (Table 2) (30). Although the Mpβ1 subunit seems to coassemble with any one of Mpα1, Mpα2, or rat α4 subunits, these three hybrid receptors are not functional because no radioligand binding is detected (30). This strongly suggests that additional *Myzus* nAChR subunits remain to be identified.

Other genes encoding insect nAChR subunits are known from *Locusta* (28), desert locust (*Schistocerca gregaria*) (59), and *M. sexta* (21), although their diversity and native structure are not clarified (26, 107).

Immunology and Protein Biochemistry

Immunology and protein biochemistry approaches identify several mature or native *Drosophila* nAChR subunits (Table 2). Anti-ALS, anti-Dα2, and anti-SBD recognize protein bands of 80, 65, and 60 kDa, respectively, on immunodetection by each corresponding antibody. In addition, anti-ALS and anti-Dα2 coprecipitate

SBD protein, and vice versa, while anti-SBD coprecipitates ALS or Dα2 protein (13, 88). The molecular masses of the immunodetected 65- and 60-kDa proteins match those of the Dα2 and SBD core proteins excluding the glycosylated region (61 or 57.3 kDa, respectively, calculated from the sequence) (13, 88). Moreover, the subunits ALS, Dα2, and SBD can be copurified by α-BGT affinity chromatography. Interestingly, the genes encoding these three subunits appear to be directly linked in the *Drosophila* genome at region 96A of the third chromosome (13). However, the 80-kDa protein is not a good match in molecular mass to ALS core protein (62 kDa), implying that this apparent discrepancy may be due to extensive *N*-glycosylation (13, 88). Immunohistochemical localization of ALS, Dα2, and SBD subunits shows similar distribution patterns, particularly in the adult optic lobe (13, 33, 92). Both the ALS and Dα2 subunits are functionally coassembled with chick β2 subunit in *Xenopus* oocytes within a single receptor complex (88). Anti-Dα3 and anti-ARD corecognize distinct structures in the distal lamina of *Drosophila* CNS, but antibodies against ALS and Dα2 do not. Anti-Dα3 coprecipitates ARD protein and anti-ARD also coprecipitates Dα3 protein (12), although their functional coexpression does not lead to binding activity for four nicotinic radioligands (46). Further, SBD and ARD appear to form an additional receptor complex without participation of the other three α-type subunits (13). The ARD protein immunodetected by anti-ARD appears as double bands at 50/45 kDa (12), a molecular mass distinctly lower than that calculated from the sequence (57.3 kDa excluding glycosylated region). Transfection of Dα4 gene into *Drosophila* S2 cells gives a gene product with a mass of 53 kDa (45).

Native *Drosophila* and *Musca* nAChRs have been isolated by affinity chromatography on both neonicotinoid- and α-BGT-agarose matrices using a neonicotinoid insecticide for specific displacement (109). Three putative subunits are obtained, in each case with molecular masses of 69, 66, and 61 kDa comparable to those deduced from gene sequences. These three isolated polypeptides might correspond to those of the *Drosophila* ALS, Dα2, and SBD subunits (80, 65, and 60 kDa, respectively, based on immunodetection experiments), although there are small differences in their masses due probably to limitations in the precision of SDS-PAGE separation, particularly for the lower migration area (Table 2) (13, 88).

The neonicotinoid-binding site environment is partially defined using two neonicotinoid photoaffinity probes (36, 50, 114) that recognize a ~66, ~66, ~45, or ~56 kDa polypeptide in *Drosophila*, *Musca*, *Myzus*, or *Homalodisca*, respectively. Photoaffinity labeling is inhibited by cholinergic ligands including nicotine, carbachol, α-BGT, and *d*-tubocurarine and the insecticides IMI and acetamiprid, i.e., these labeled polypeptides are pharmacologically consistent with the ligand- and insecticide-binding subunits of the insect nAChRs (104, 114). The labeled 66-kDa subunit of *Drosophila* coincides in its molecular mass with the 65-kDa band raised by immunoblotting (88), suggesting that the photoaffinity-labeled 66-kDa polypeptide is Dα2 as a main neonicotinoid-binding subunit in *Drosophila*. Interestingly, a [^{125}I]neonicotinoid probe labels both the primary subunit at 66 kDa and the complementary subunit at 61 kDa of the native *Drosophila* nAChR isolated by neonicotinoid-agarose affinity chromatography (104), suggesting that the insecticide-binding site may be localized at the interface between the two subunits by analogy with other ligands in vertebrate nAChRs (2).

Subtype Diversity

Insect nAChR subtypes with different subunit combinations may have distinctive pharmacological profiles. In hybrid nAChRs consisting of *Drosophila* ALS and chick $\beta 2$ or *Drosophila* Dα3 and chick $\beta 2$ subunits expressed in *Xenopus* oocytes, ACh-evoked current is blocked by α-BGT, whereas the electrophysiological response in a hybrid receptor from Dα2 and chick $\beta 2$ subunits is not (5, 89). Similarly, [^{125}I]α-BGT recognizes either ALS/rat $\beta 2$ or Dα3/rat $\beta 2$ but not Dα2/rat $\beta 2$ hybrid receptor expressed in *Drosophila* S2 clonal cell line. On the other hand, [^{3}H]IMI and [^{3}H]epibatidine bind to all of the above three hybrid receptors with high affinities (46). A Dα4/rat $\beta 2$ hybrid receptor shows [^{3}H]epibatidine but not [^{125}I]α-BGT-binding activity (45). Immunology approaches suggest two native *Drosophila* subtypes consisting of ALS/Dα2/SBD and Dα3/ARD (12, 13). Also in *Myzus* homomeric receptor expressed in *Xenopus* oocytes, a nicotine-induced current is antagonized by α-BGT in Mpα1 but not Mpα2. In addition, α-BGT does not block a nicotine-evoked current in the Mpα2/chick $\beta 2$ receptor (93). Furthermore, hybrid receptors assembled from Mpα1, 2, or 3 with rat $\beta 2$ subunit in *Drosophila* S2 cells display different properties in their sensitivity to [^{3}H]IMI, i.e., [^{3}H]IMI binds to both the Mpα2/rat $\beta 2$ and Mpα3/rat $\beta 2$ receptors but not to the Mpα1/rat $\beta 2$ receptor. In contrast, [^{3}H]epibatidine binds to all three hybrid receptors (29). Although coexpression of the insect α-type subunit with the vertebrate β-type subunit constitutes the best available model at present, these hybrid receptors may not faithfully reflect the insect nAChRs. In *Periplaneta*, two types of nAChRs, based on sensitivity to α-BGT, exist in the dorsal unpaired median neuron, and both types are sensitive to IMI (8). Further, the α-BGT-insensitive receptor is subdivided into two types with different IMI sensitivities (18). Interestingly, the neonicotinoids and related compounds act on multiple mammalian nAChR subtypes with differential selectivity conferred by only minor structural modifications (105). Therefore, subtype or subunit selectivity is also expected in the insect nAChR subunits.

MOLECULAR BASIS FOR ACTIVITY AND SELECTIVITY

Target Site Selectivity

Neonicotinoids are more toxic to aphids, leafhoppers, and other sensitive insects than to mammals (34, 94, 95). This insect/mammalian selectivity is also evident at the target site level (Table 3) and is probably a general feature of the neonicotinoids (111). Thus, the insect nAChR is of similar neonicotinoid sensitivity and structure-activity relationships in all the insects examined by [^{3}H]IMI binding (15, 51, 55, 69, 109, 114, 121, 128). In contrast, neonicotinoids have little or

TABLE 3 Molecular features conferring selective toxicity and selectivity for insect versus mammalian nicotinic receptors

	Insects[b]		Mammals[c]	
Compound[a]	*Drosophila* nAChR IC_{50}, nM[d]	Toxicity to house fly LD_{50}, $\mu g/g$[e]	$\alpha4\beta2$ nAChR IC_{50}, nM[f]	Toxicity to mouse LD_{50}, mg/kg[g]
Selective for insect				
IMI	4.6	0.05	2600	45
thiacloprid	2.7	0.03	860	28
Selective for mammal				
desnitro-IMI	1500	>5	8.2	8.0
descyano-thiacloprid	200	>5	4.4	1.1
(−)-nicotine	4000	>50	7.0	7.0
(±)-epibatidine	430	>25	0.037	0.08

[a]Other mammalian nAChR subtypes give IC_{50} values for the 6 tabulated compounds in sequence as follows: on $\alpha1\gamma\alpha1\delta\beta1$ subtype assayed with [^{125}I]α-BGT binding (in μM) >300, 120, 11, 6.0, 21, and 0.16, respectively; on $\alpha7$ subtype assayed with [^{125}I]α-BGT binding (in μM) 270, 100, 9.9, 6.0, 21 and 0.031, respectively; on $\alpha3\beta2$[and/or $\beta4$]$\alpha5$ subtypes assayed with [^{3}H]nicotine binding (in μM) 14 for IMI, 0.014 for desnitro-IMI, and 0.045 for nicotine (111).

[b]From (111).

[c]From (108).

[d]Assayed with [^{3}H]IMI binding.

[e]Intrathoracic injection with synergist (CYP450 inhibitor) pretreatment.

[f]Assayed with [^{3}H]nicotine binding.

[g]Intraperitoneal route.

no effect on binding activity on any of the following vertebrate nAChRs: *Torpedo* electric organ (113); rodent brain (14, 105, 127); mammalian neuroblastoma (N1E-115 and SH-SY5Y) (105, 131), muscle (BC3H1) (131), medulloblastoma (TE671) (111), and pheochromocytoma (PC12) cells (66); and recombinant subtypes expressed in *Xenopus oocytes*, mouse fibroblast, or insect SF9 cell (20, 60, 106, 126). On the other hand, the nicotinoids are more toxic to mammals than insects and more potent at $\alpha4\beta2$ than *Drosophila* nAChRs (Table 3). This comparison includes IMI and thiacloprid versus their *N*-unsubstituted imines and is therefore clearly due to the $=NNO_2$ or $=NCN$ versus $=NH$ moiety.

Structural Requirements: "Magic Nitro or Cyano"

The neonicotinoids bind to the nAChR at the same site as (or a coupled site to) that for nicotine and ACh, leading to structural comparisons with these agonists. X-ray crystal structure analysis of neonicotinoids indicates that the distances between the van der Waals surface of the nitrogen atom of the 3-pyridinylmethyl moiety and the atomic center of the 1-nitrogen (N-1) of the imidazolidine are 5.45–6.06 Å. This range coincides with the distance between the ammonium nitrogen and carboxyl oxygen of ACh and between the nitrogen atoms of nicotine (5.9 Å) (37), although the affinity of nicotine to the insect receptor is much lower than that of IMI. N-1 of the imidazolidine ring is conferred a partial positive charge (δ^+) by an electron-withdrawing nitro substituent and is suggested to interact with the anionic subsite of the insect but not the mammalian nAChR (116, 127). This binding model is supported by computational chemistry (61, 77). Alternatively, the distance between N-1 of the imidazolidine and an oxygen atom of the nitro group is also 5.9 Å, suggesting these two points are important for binding interaction without the involvement of the pyridine nitrogen atom (37). However, the role of N-1 on binding interaction and/or selectivity remains unclear because the calculated charge status of this nitrogen is rather marginally positive and does not correlate with the binding affinity (67, 111). Furthermore, N-1 can be replaced by a carbon atom with retention of significant insecticidal activity (6).

With a different approach, the single substituent change of replacing $=NNO_2$ by $=NH$ in the IMI series or $=NCN$ by $=NH$ in the thiacloprid series gives dramatically opposite properties in selectivity for insect versus mammalian nAChRs (Table 3) (Figure 5). The critical feature for insect nAChR binding appears to be the negatively charged tip or region of the nitro or cyano group. The nitrogen atom of the 3-pyridine ring is also important for the insect nAChR because nithiazine and 2-nitromethylene-imidazolidine show considerably lower binding affinity than analogs with the 3-pyridinylmethyl moiety (54, 112, 115). In contrast, with the mammalian nAChR, the nitrogen atom of the *N*-unsubstituted imines (i.e., desnitro or descyano) serves as a proton acceptor as with the ammonium nitrogen in nicotine and epibatidine, generating a positive charge on the nitrogen and/or carbon atom of the imine moiety ($^+C\text{-}NH_2 \leftrightarrow C{=}^+NH_2$) under physiological conditions. Consequently, as a simple principle, the "magic nitro or cyano" substituent confers the differential selectivity of neonicotinoids for insect versus mammalian nAChRs (111).

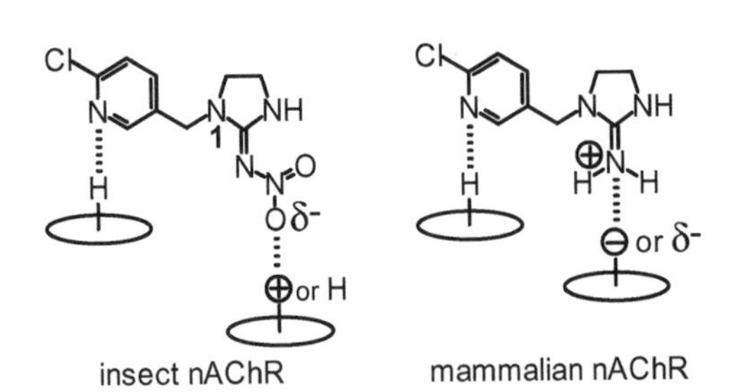

Figure 5 Molecular features of neonicotinoids represented by IMI and nicotinoids represented by desnitro-IMI conferring selectivity for insect versus mammalian nicotinic receptors. In neonicotinoids, the negatively charged tip or region can also be a cyano substituent. According to this model, the optimal intramolecular distance between the two pharmacophores is different for neonicotinoids and the insect nAChR than for nicotinoids and the mammalian nAChR.

Binding Subsites

The molecular features of neonicotinoids conferring selectivity for insect versus mammalian receptors can be rationalized on the basis of fundamental structural differences in their subsite(s) as shown in Figure 5 (111). We propose that in the insect nAChR the negatively charged tip interacts with cationic amino acid residue(s) such as lysine, arginine, or histidine in the neonicotinoid-binding region. These residues form a putative cationic subsite or provide a hydrogen-bonding point. Lysine and arginine residues are prominent in the relevant 149–220 (extracellular) region from the N termini of *Drosophila* α and β subunits and one or two histidine residues are also found. It is further proposed that with mammalian nAChRs the iminium cation of the desnitro- or descyano-neonicotinoid or ammonium nitrogen atom of nicotine or epibatidine interacts with the anionic subsite accompanied by association of the 3-pyridinyl nitrogen atom with the hydrogen-bonding subsite. There are two hypotheses relative to amino acid residue(s) in the anionic subsite. The first suggests cation interaction with the carboxyl anion from glutamate and aspartate residues on the γ or δ subunit (the agonist-binding site is localized at the interface between $\alpha 1$-γ or $\alpha 1$-δ subunits) (40). The second proposes cation π-electron interaction with the aromatic amino acid residues of tryptophan and tyrosine on the $\alpha 1$ subunit as the possible anionic subsite [the quaternary ammonium head makes van der Waals contact with the π-electron (δ^-) of the aromatic ring] (130). Photoaffinity labeling by nicotinic ligands of several aromatic amino acid residues may support this hypothesis (1, 2).

TOXICOLOGY MECHANISMS

Neonicotinoid Action at Vertebrate Nicotinic Receptors

Sheets (94, 95) reviews the mammalian toxicology of neonicotinoids on a general basis, and so only specific aspects are dealt with here. Their toxicity to mammals is considered to be primarily due to action at the $\alpha 4\beta 2$ nAChR in brain based on the centrally mediated toxicity signs or behaviors as well as the agonist action (108). Interestingly, minor structural modifications in neonicotinoids confer differential subtype selectivity for mammalian nAChRs; for example, insecticidal neonicotinoids with a nitromethylene substituent display higher affinity than that of nicotine to the $\alpha 7$ subtype (105). Toxicological evaluations of insecticide safety should consider the subtype level as well as the organismal effect.

Understanding the toxicology of neonicotinoids versus nicotinoids is facilitated by comparing the insecticides IMI and thiacloprid with their *N*-unsubstituted imine derivatives desnitro-IMI and descyano-thiacloprid, respectively. The imine derivatives are more toxic to mice than the parent insecticides and show high affinity to and/or agonist potency at the mammalian $\alpha 4\beta 2$ nAChR subtype (and $\alpha 1\gamma\alpha 1\delta\beta 1$, $\alpha 3\beta 2$[and/or $\beta 4$]$\alpha 5$, and $\alpha 7$ subtypes) comparable to or greater than that of nicotine (20, 49, 105, 106, 108, 111) (Table 3). The binding affinities or agonist potencies with the above four subtypes correlate well with the intraperitoneal (ip) toxicity to mice (108). Chronic exposure to the neonicotinoid insecticides or *N*-unsubstituted imines, just as with nicotine, upregulates the $\alpha 4\beta 2$ nAChR numbers in M10 cells without altering the sensitivity of the binding site (106). The analgesic activity of nicotine and epibatidine assayed with mouse hotplate and abdominal constriction tests is not evident with the neonicotinoid insecticides and their *N*-unsubstituted imines with one interesting exception. An insecticidal neonicotinoid analog (the "second prototype" with tetrahydropyrimidine instead of imidazolidine) (Figure 1) has weak agonist potency at the $\alpha 4\beta 2$ nAChR, yet is as potent as nicotine in inducing analgesic activity, and the effect persists longer than that of nicotine or epibatidine. Furthermore, mecamylamine prevents antinociception induced by nicotine but not by the neonicotinoid, suggesting a different mechanism of action or primary target from that of nicotine and epibatidine (108).

Metabolism

The neonicotinoids are moderately soluble in water, nonionized, not readily hydrolyzed at physiological pH values, and biodegradable, so they do not accumulate in mammals or through food chains (82). Their metabolic fate is most fully reported for IMI, which has a variety of functional groups that undergo degradation (Figure 6) to a large number of metabolites formed by multiple pathways, both alternative and sequential (82, 94, 95). The same or similar metabolites (except for some of the conjugates) are found in rats, goats, hens, plants, and soils (82, 94, 95), with many of these products also produced on photolysis. The initial reactions of imidazolidine oxidation and nitro reduction are carried out by human CYP3A4 and other less active CYP450 isoforms (90) while a microsomal "neonicotinoid nitro reductase" with NADPH forms the hydrazone (moiety c becomes $=N\text{–}NH_2$) (91), which undergoes further reactions. Metabolism of the nitroimine thiamethoxam

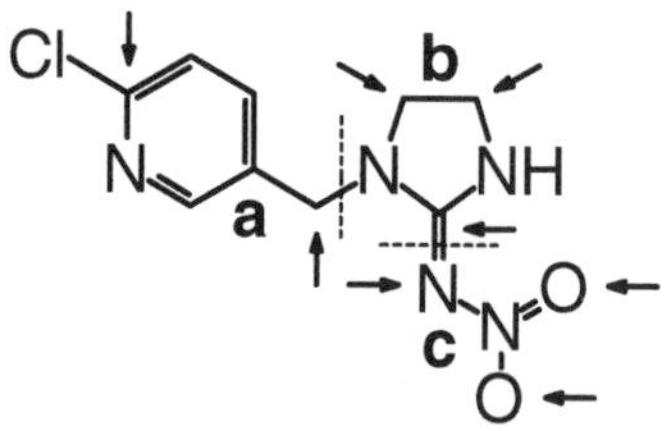

Figure 6 Metabolism of IMI in mammals, plants, and soils showing regions (*a–c*) and positions (*arrows*) of metabolic attack. (*a*) Chloropyridinylmethyl moiety undergoes oxidative cleavage and conjugation, with glutathione displacing chlorine. (*b*) Imidazolidine moiety undergoes ethylene hydroxylation, dehydrogenation, and cleavage. (*c*) Nitroimine moiety undergoes nitro reduction and cleavage of $N\text{–}NO_2$ and $C{=}N$ bonds. Combinations of these steps plus conjugation form excreted or terminal metabolites.

is similar in rats, goats, and hens (82). The nitromethylene analogs nithiazine in *Musca* (81) and nitenpyram in plants (82) undergo oxidation at the nitromethylene carbon followed by decarboxylation. In one of many pathways for thiacloprid metabolism in animals, the cyanoimine moiety is hydrolyzed to the corresponding amide derivative, which is then *N*-hydroxylated (42).

Neonicotinoid metabolism involves mostly detoxification reactions evident in three ways: the brief period for poisoning signs at sublethal doses in both insects and mammals, the synergistic effects of detoxification inhibitors on insecticidal action, and the potency of the metabolites relative to the parent compounds at nicotinic receptors. Thus, CYP450-inhibiting piperonyl butoxide, sesamex, and *O*-propyl *O*-(2-propynyl) phenylphosphonate strongly synergize the toxicity of IMI and several neonicotinoids in *Musca*, *Periplaneta*, and/or the German cockroach (*Blattella germanica*) (55, 75, 81, 120, 126). IMI undergoes several bioactivation reactions relative to potency at the insect nAChRs and insecticidal activity to *Myzus*, *Aphis*, *Bemisia*, *Musca*, and *Apis*, i.e., the active metabolites are monohydroxy and dehydro in moiety b and nitroso (=NNO) in moiety c (69, 72, 74) (Figure 6). The *N*-methyl (N-3) derivative of IMI undergoes bioactivation on desmethylation (126). Thiamethoxam is rather modest in potency at the [^{3}H]IMI-binding site and is proposed to act in a different manner (57, 121) but its *N*-desmethyl derivative has high affinity consistent with that of IMI (128). Analogously, the *N*-desmethyl derivative of nitenpyram is a potent insecticide (62) and displays high affinity to the insect nAChRs from *Musca* and *Drosophila*(109). The receptor potency of dinotefuran does not fully reflect the insecticidal activity, again suggesting but not in itself establishing possible metabolic activation (64, 128). Interestingly, the same metabolite [*N*-unsubstituted imine (=NH) in moiety c] may be a detoxification product in insects and an activation product in mammals (14, 55, 74). Thus, desnitro-IMI and descyano-thiacloprid are much more toxic to mammals and active on mammalian nAChRs than the corresponding insect systems (111).

Resistance in Pests

Three types of resistance to IMI have been observed under field conditions. The first is for some insects feeding on tobacco, presumably as a cross-resistance associated with evolutionary selection for nicotine tolerance. A Japanese strain of a tobacco-feeding form of *M. persicae*, a species closely related to *M. nicotianae*, also displays a low level of sensitivity to IMI (73). A French strain of *M. nicotianae* showed 192-fold and >22-fold resistance to IMI and nicotine, respectively, compared to susceptible *M. persicae* (70). A positive relationship between IMI and nicotine resistance has also been demonstrated for *Bemisia* (9). This difference between tolerant or resistant and susceptible *Myzus* is not due to modified nAChR sensitivity to nicotine and IMI (71, 73), and the nAChR of nicotine-insensitive *Manduca* has the same sensitivity to cholinergic ligands and IMI as that of nAChRs from other susceptible insects (21). The second type of resistance is from preexisting metabolic and excretion patterns selected by previous exposure to other insecticides, illustrated here by Colorado potato beetle (*Leptinotarsa decemlineata*) adults and larvae with 100- and 13-fold resistance to IMI, respectively (78, 129). Several neonicotinoid-resistant strains of *M. persicae* are more strongly synergized by piperonyl butoxide than the susceptible strain (I. Denholm, personal communication). DDT resistance in *Drosophila* correlates with overexpression of *Cyp6g1* (an isozyme of CYP450), which appears to confer cross-resistance to IMI (19). The third and greatest concern is for the type of resistance from long-term selection with IMI and other neonicotinoids. IMI resistance of at least 15-fold was observed in field-collected *B. tabaci* from the Almeria region of Spain (10) and *B. argentifolii* from the Imperial Valley in California (80). Continuous laboratory selection with IMI elevated the resistance of *B. argentifolii* to >80-fold after 24 generations (80). This indicates that IMI resistance genes exist in some whitefly populations at high frequency. Increased levels of resistance to neonicotinoids in some pests with concomitant control failures will inevitably result from continued selection pressure. Although a modified nAChR due to neonicotinoid selection has not yet been reported, it remains to be seen if metabolic or target site resistance will ultimately prove to be of greatest importance.

FUTURE PROSPECTS

About 90% of the synthetic organic insecticides and acaricides, by market share, are nerve poisons acting on only four targets: acetylcholinesterase (AChE) for organophosphorus compounds and methylcarbamates, the voltage-dependent sodium ion channel for DDT and pyrethroids, nAChR for the botanical nicotine and most recently synthetic neonicotinoids, and the γ-aminobutyric acid (GABA)-gated chloride channel for polychlorocycloalkanes and fipronil (Table 4). Compounds acting as developmental inhibitors may have outstanding potency and selectivity but are often not preferred because of their slow action. In the market value comparison, chitin synthesis inhibition (e.g., diflubenzuron) is currently the most important non-nerve target followed by NADH dehydrogenase (also known as NADH: ubiquinone oxidoreductase) (e.g., rotenone) and ATPase (e.g., cyhexatin), with minor use of uncouplers of oxidative phosphorylation (e.g., dinitrophenol derivatives) and ecdysone receptor agonists (e.g., tebufenozide). From 1987 to 1997, the use of compounds acting at the cholinergic nAChR shifted from sixth to third [(68); R. Nauen, personal communication] in overall ranking, in the most part replacing AChE inhibitors, and this trend is expected to continue with possibly 15% world market share and third ranking for the neonicotinoids by 2005.

The long-term future of neonicotinoids will depend on continued evidence for the human and environmental safety of current compounds, including low toxicity to predators, parasites, and pollinators, no adverse environmental distribution, and fate. It will be enhanced by the discovery of new compounds with a broader spectrum of useful properties including control of lepidopterous larvae and pest strains resistant to earlier analogs. These biological features must be combined with suitable hydrophilicity for transport in plants, hydrophobicity for contact

TABLE 4 Market share of insecticides and acaricides by mode of action [updated from (68)]

Mode of action	Percent[a] 1987	1997
Nerve		
acetylcholinesterase	67.1	53.4
voltage-dependent Na^+ channel	15.6	20.0
nAChR	1.5	8.6
GABA-gated Cl^- channel	4.7	7.3
octopamine receptor	0.5	0.4
Other		
chitin biosynthesis	2.0	2.6
NADH dehydrogenase	0	1.3
ATPase	3.0	1.3
uncouplers	0	0.4
ecdysone receptor	0	0.3

The world insecticide market is about 7 billion dollars annually, with little change in the past decade.

[a]The remaining compounds are fumigants.

activity, and photostability for residual efficacy. Much has been learned about neonicotinoids in the first decade of their use and about the nicotinic receptor as a target for selective toxicity between insects and mammals. The benefits of neonicotinoids in crop protection and animal health can be enjoyed for many decades ahead with attention to their proper use in pest management systems that delay or circumvent the development of resistance in pest insects.

ACKNOWLEDGMENTS

Neonicotinoid research in the Berkeley laboratory was supported in part by grant R01 ES08424 from the National Institute of Environmental Health Sciences, the National Institutes of Health. The authors received valuable advice and assistance from our former or current colleagues S. Kagabu, N. Zhang, and G.B. Quistad.

The *Annual Review of Entomology* is online at http://ento.annualreviews.org

LITERATURE CITED

1. Arias HR. 1997. Topology of ligand binding sites on the nicotinic acetylcholine receptor. *Brain Res. Rev.* 25:133–91
2. Arias HR. 2000. Localization of agonist and competitive antagonist binding sites on nicotinic acetylcholine receptors. *Neurochem. Int.* 36:595–645
3. Badio B, Daly JW. 1994. Epibatidine, a potent analgesic and nicotinic agonist. *Mol. Pharmacol.* 45:563–69

4. Bai D, Lummis SCR, Leicht W, Breer H, Sattelle DB. 1991. Actions of imidacloprid and a related nitromethylene on cholinergic receptors of an identified insect motor neurone. *Pestic. Sci.* 33:197–204
5. Bertrand D, Ballivet M, Gomez M, Bertrand S, Phannavong B, Gundelfinger ED. 1994. Physiological properties of neuronal nicotinic receptors reconstituted from the vertebrate $\beta 2$ subunit and *Drosophila* α subunits. *Eur. J. Neurosci.* 6:869–75
6. Boëlle J, Schneider R, Gérardin P, Loubinoux B, Maienfisch P, Rindlisbacher A. 1998. Synthesis and insecticidal evaluation of imidacloprid analogs. *Pestic. Sci.* 54:304–7
7. Bossy B, Ballivet M, Spierer P. 1988. Conservation of neural nicotinic acetylcholine receptors from *Drosophila* to vertebrate central nervous systems. *EMBO J.* 7:611–18
8. Buckingham SD, Lapied B, Le Corronc H, Grolleau F, Sattelle DB. 1997. Imidacloprid actions on insect neuronal acetylcholine receptors. *J. Exp. Biol.* 200:2685–92
9. Cahill M, Denholm I. 1999. Managing resistance to the chloronicotinyl insecticides—rhetoric or reality? See Ref. 125, pp. 253–70
10. Cahill M, Gorman K, Day S, Denholm I, Elbert A, Nauen R. 1996. Baseline determination and detection of resistance to imidacloprid in *Bemisia tabaci* (Homoptera: Aleyrodidae). *Bull. Entomol. Res.* 86:343–49
11. Casida JE, Quistad GB. 1998. Golden age of insecticide research: past, present, or future? *Annu. Rev. Entomol.* 43:1–16
12. Chamaon K, Schulz R, Smalla K-H, Seidel B, Gundelfinger ED. 2000. Neuronal nicotinic acetylcholine receptors of *Drosophila melanogaster*: the α-subunit Dα3 and the β-subunit ARD co-assemble within the same receptor complex. *FEBS Lett.* 482:189–92
13. Chamaon K, Smalla K-H, Thomas U, Gundelfinger ED. 2002. Nicotinic acetylcholine receptors of *Drosophila*: three subunits encoded by genomically linked genes can co-assemble into the same receptor complex. *J. Neurochem.* 80:149–57
14. Chao S-L, Casida JE. 1997. Interaction of imidacloprid metabolites and analogs with the nicotinic acetylcholine receptor of mouse brain in relation to toxicity. *Pestic. Biochem. Physiol.* 58:77–88
15. Chao S-L, Dennehy TJ, Casida JE. 1997. Whitefly (Hemiptera: Aleyrodidae) binding site for imidacloprid and related insecticides: a putative nicotinic acetylcholine receptor. *J. Econ. Entomol.* 90:879–82
16. Cordero-Erausquin M, Marubio LM, Klink R, Changeux J-P. 2000. Nicotinic receptor function: new perspectives from knockout mice. *Trends Pharmacol. Sci.* 21:211–17
17. Corringer J-P, Le Novère N, Changeux J-P. 2000. Nicotinic receptors at the amino acid level. *Annu. Rev. Pharmacol. Toxicol.* 40:431–58
18. Courjaret R, Lapied B. 2001. Complex intracellular messenger pathways regulate one type of neuronal α-bungarotoxin-resistant nicotinic acetylcholine receptors expressed in insect neurosecretory cells (dorsal unpaired median neurons). *Mol. Pharmacol.* 60:80–91
19. Daborn P, Boundy S, Yen J, Pittendrigh B, ffrench-Constant R. 2001. DDT resistance in *Drosophila* correlates with *Cyp6g1* over-expression and confers cross-resistance to the neonicotinoid imidacloprid. *Mol. Genet. Genomics* 266: 556–63
20. D'Amour KA, Casida JE. 1999. Desnitroimidacloprid and nicotine binding site in rat recombinant $\alpha 4\beta 2$ neuronal nicotinic acetylcholine receptor. *Pestic. Biochem. Physiol.* 64:55–61
21. Eastham HM, Lind RJ, Eastlake JL, Clarke BS, Towner P, et al. 1998. Characterization of a nicotinic acetylcholine

receptor from the insect *Manduca sexta*. *Eur. J. Neurosci.* 10:879–89
22. Elbert A, Buchholz A, Ebbinghaus-Kintscher U, Erdelen C, Nauen R, Schnorbach H-J. 2001. The biological profile of thiacloprid—a new chloronicotinyl insecticide. *Pflanzenschutz-Nachr. Bayer* 54:185–208
23. Eldefrawi AT, Bakry NM, Eldefrawi ME, Tsai M-C, Albuquerque AX. 1980. Nereistoxin interaction with the acetylcholine receptor-ionic channel complex. *Mol. Pharmacol.* 17:172–79
24. Elgoyhen AB, Vetter DE, Katz E, Rothlin CV, Heinemann SF, Boulter J. 2001. α10: a determinant of nicotinic cholinergic receptor function in mammalian vestibular and cochlear mechanosensory hair cells. *Proc. Natl. Acad. Sci. USA* 98:3501–6
25. Feuer H, Lawrence JP. 1969. The alkyl nitrate nitration of active methylene compounds. VI. A new synthesis of α-nitroalkyl heterocyclics. *J. Am. Chem. Soc.* 91:1856–57
26. Gundelfinger ED, Schulz R. 2000. Insect nicotinic acetylcholine receptors: genes, structure, physiological and pharmacological properties. In *Handbook of Experimental Pharmacology*. Vol. 144: *Neuronal Nicotinic Receptors*, ed. F Clementi, D Fornasari, C Gotti, pp. 497–521. Berlin: Springer
27. Hermans-Borgmeyer I, Zopf D, Ryseck R-P, Hovemann B, Betz H, Gundelfinger ED. 1986. Primary structure of a developmentally regulated nicotinic acetylcholine receptor protein from *Drosophila*. *EMBO J.* 5:1503–8
28. Hermsen B, Stetzer E, Thees R, Heiermann R, Schrattenholz A, et al. 1998. Neuronal nicotinic receptors in the locust *Locusta migratoria*. *J. Biol. Chem.* 273:18394–404
29. Huang Y, Williamson MS, Devonshire AL, Windass JD, Lansdell SJ, Millar NS. 1999. Molecular characterization and imidacloprid sensitivity of nicotinic acetylcholine receptor subunits from the peach-potato aphid *Myzus persicae*. *J. Neurochem.* 73:380–89
30. Huang Y, Williamson MS, Devonshire AL, Windass JD, Lansdell SJ, Millar NS. 2000. Cloning, heterologous expression and co-assembly of Mpβ1, a nicotinic acetylcholine receptor subunit from the aphid *Myzus persicae*. *Neurosci. Lett.* 284:116–20
31. Itier V, Bertrand D. 2001. Neuronal nicotinic receptors: from protein structure to function. *FEBS Lett.* 504:118–25
32. Jonas P, Baumann A, Merz B, Gundelfinger ED. 1990. Structure and developmental expression of the Dα2 gene encoding a novel nicotinic acetylcholine receptor protein of *Drosophila melanogaster*. *FEBS Lett.* 269:264–68
33. Jonas PE, Phannavong B, Schuster R, Schröder C, Gundelfinger ED. 1994. Expression of the ligand-binding nicotinic acetylcholine receptor subunit Dα2 in the *Drosophila* central nervous system. *J. Neurobiol.* 25:1494–508
34. Kagabu S. 1997. Chloronicotinyl insecticides—discovery, application and future perspective. *Rev. Toxicol.* 1:75–129
35. Kagabu S, Akagi T. 1997. Quantum chemical consideration of photostability of imidacloprid and related compounds. *J. Pestic. Sci.* 22:84–89
36. Kagabu S, Maienfisch P, Zhang A, Granda-Minones J, Haettenschwiler J, et al. 2000. 5-Azidoimidacloprid and an acyclic analogue as candidate photoaffinity probes for mammalian and insect nicotinic acetylcholine receptors. *J. Med. Chem.* 43:5003–9
37. Kagabu S, Matsuno H. 1997. Chloronicotinyl insecticides. 8. Crystal and molecular structures of imidacloprid and analogous compounds. *J. Agric. Food Chem.* 45:276–81
38. Kagabu S, Medej S. 1995. Stability comparison of imidacloprid and related compounds under simulated sunlight, hydrolysis conditions, and to oxygen. *Biosci. Biotechnol. Biochem.* 59:980–85

39. Kagabu S, Moriya K, Shibuya K, Hattori Y, Tsuboi S, Shiokawa K. 1992. 1-(6-Halonicotinyl)-2-nitromethylene-imidazolidines as potential new insecticides. *Biosci. Biotechnol. Biochem.* 56:362–63
40. Karlin A, Akabas MH. 1995. Toward a structural basis for the function of nicotinic acetylcholine receptors and their cousins. *Neuron* 15:1231–44
41. Kleier D, Holden I, Casida JE, Ruzo LO. 1985. Novel photoreactions of an insecticidal nitromethylene heterocycle. *J. Agric. Food Chem.* 33:998–1000
42. Klein O. 2001. Behaviour of thiacloprid (YRC 2894) in plants and animals. *Pflanzenschutz-Nachr. Bayer* 54:209–40
43. Kodaka K, Kinoshita K, Wakita T, Yamada E, Kawahara N, Yasui N. 1998. MTI-446: a novel systemic insect control compound. *Proc. Brighton Crop Prot. Conf. Pests Dis.* 1:21–26
44. Kollmeyer WD, Flattum RF, Foster JP, Powell JE, Schroeder ME, Soloway SB. 1999. Discovery of the nitromethylene heterocycle insecticides. See Ref. 125, pp. 71–89
45. Lansdell SJ, Millar NS. 2000. Cloning and heterologous expression of Dα4, a *Drosophila* neuronal nicotinic acetylcholine receptor subunit: identification of an alternative exon influencing the efficiency of subunit assembly. *Neuropharmacology* 39:2604–14
46. Lansdell SJ, Millar NS. 2000. The influence of nicotinic receptor subunit composition upon agonist, α-bungarotoxin and insecticide (imidacloprid) binding affinity. *Neuropharmacology* 39:671–79
47. Lansdell SJ, Schmitt B, Betz H, Sattelle DB, Millar NS. 1997. Temperature-sensitive expression of *Drosophila* neuronal nicotinic acetylcholine receptors. *J. Neurochem.* 68:1812–19
48. Latli B, Casida JE. 1992. [^{3}H]Imidacloprid: synthesis of a candidate radioligand for the nicotinic acetylcholine receptor. *J. Label. Compd. Radiopharm.* 31:609–13
49. Latli B, D'Amour KA, Casida JE. 1999. Novel and potent 6-chloro-3-pyridinyl ligands for the $\alpha4\beta2$ neuronal nicotinic acetylcholine receptor. *J. Med. Chem.* 42:2227–34
50. Latli B, Tomizawa M, Casida JE. 1997. Synthesis of a novel [^{125}I]neonicotinoid photoaffinity probe for the *Drosophila* nicotinic acetylcholine receptor. *Bioconjug. Chem.* 8:7–14
51. Lind RJ, Clough MS, Reynolds SE, Earley FGP. 1998. [^{3}H]Imidacloprid labels high- and low-affinity nicotinic acetylcholine receptor-like binding sites in the aphid *Myzus persicae* (Hemiptera: Aphididae). *Pestic. Biochem. Physiol.* 62:3–14
52. Lindstrom J. 1997. Nicotinic acetylcholine receptors in health and disease. *Mol. Neurobiol.* 15:193–222
53. Littleton JT, Ganetzky B. 2000. Ion channels and synaptic organization: analysis of the *Drosophila* genome. *Neuron* 26:35–43
54. Liu M-Y, Casida JE. 1993. High affinity binding of [^{3}H]imidacloprid in the insect acetylcholine receptor. *Pestic. Biochem. Physiol.* 46:40–46
55. Liu M-Y, Lanford J, Casida JE. 1993. Relevance of [^{3}H]imidacloprid binding site in house fly head acetylcholine receptor to insecticidal activity of 2-nitromethylene- and 2-nitroimino-imidazolidines. *Pestic. Biochem. Physiol.* 46:200–6
56. Lloyd GK, Williams M. 2000. Neuronal nicotinic acetylcholine receptors as novel drug targets. *J. Pharmacol. Exp. Ther.* 292:461–67
57. Maienfisch P, Angst M, Brandl F, Fischer W, Hofer D, et al. 2001. Chemistry and biology of thiamethoxam: a second generation neonicotinoid. *Pest Manage. Sci.* 57:906–13
58. Maienfisch P, Brandl F, Kobel W, Rindlisbacher A, Senn R. 1999. CGA 293'343: a novel, broad-spectrum neonicotinoid insecticide. See Ref. 125, pp. 177–209
59. Marshall J, Buckingham SD, Shingai R,

Lunt GG, Goosey MW, et al. 1990. Sequence and functional expression of a single α subunit of an insect nicotinic acetylcholine receptor. *EMBO J.* 9:4391–98
60. Matsuda K, Buckingham SD, Freeman JC, Squire MD, Baylis HA, Sattelle DB. 1998. Effects of the α subunit on imidacloprid sensitivity of recombinant nicotinic acetylcholine receptors. *Br. J. Pharmacol.* 123:518–24
61. Matsuda K, Buckingham SD, Kleier D, Rauh JJ, Grauso M, Sattelle DB. 2001. Neonicotinoids: insecticides acting on insect nicotinic acetylcholine receptors. *Trends Pharmacol. Sci.* 22:573–80
62. Minamida I, Iwanaga K, Tabuchi T, Aoki I, Fusaka T, et al. 1993. Synthesis and insecticidal activity of acyclic nitroethene compounds containing a heteroarylmethylamino group. *J. Pestic. Sci.* 18:41–48
63. Mitsui Toatsu Kagaku. 1996. *Jpn. Kokai Tokkyo Koho* JP8-48683
64. Mori K, Okumoto T, Kawahara N, Ozoe Y. 2002. Interaction of dinotefuran and its analogues with nicotinic acetylcholine receptors of cockroach nerve cords. *Pest Manage. Sci.* 58:190–96
65. Moriya K, Shibuya K, Hattori Y, Tsuboi S, Shiokawa K, Kagabu S. 1992. 1-(6-Chloronicotinyl)-2-nitroimino-imidazolidines and related compounds as potential new insecticides. *Biosci. Biotechnol. Biochem.* 56:364–65
66. Nagata K, Aistrup GL, Song JH, Narahashi T. 1996. Subconductance-state currents generated by imidacloprid at the nicotinic acetylcholine receptor of PC 12 cells. *NeuroReport* 7:1025–28
67. Nakayama A, Sukekawa M. 1998. Quantitative correlation between molecular similarity and receptor-binding activity of neonicotinoid insecticides. *Pestic. Sci.* 52:104–10
68. Nauen R, Ebbinghaus-Kintscher U, Elbert A, Jeschke P, Tietjen K. 2001. Acetylcholine receptors as sites for developing neonicotinoid insecticides. In *Biochemical Sites of Insecticide Action and Resistance*, ed. I Ishaaya, pp. 77–105. Berlin: Springer-Verlag
69. Nauen R, Ebbinghaus-Kintscher U, Schmuck R. 2001. Toxicity and nicotinic acetylcholine receptor interaction of imidacloprid and its metabolites in *Apis mellifera* (Hymenoptera: Apidae). *Pest Manage. Sci.* 57:577–86
70. Nauen R, Elbert A. 1997. Apparent tolerance of a field-collected strain of *Myzus nicotianae* to imidacloprid due to strong antifeeding responses. *Pestic. Sci.* 49:252–58
71. Nauen R, Hungenberg H, Tollo B, Tietjen K, Elbert A. 1998. Antifeedant effect, biological efficacy and high affinity binding of imidacloprid to acetylcholine receptors in *Myzus persicae* and *Myzus nicotianae*. *Pestic. Sci.* 53:133–40
72. Nauen R, Reckmann U, Armborst S, Stupp H-P, Elbert A. 1999. Whitefly-active metabolites of imidacloprid: biological efficacy and translocation in cotton plants. *Pestic. Sci.* 55:265–71
73. Nauen R, Strobel J, Tietjen K, Otsu Y, Erdelen C, Elbert A. 1996. Aphicidal activity of imidacloprid against a tobacco feeding strain of *Myzus persicae* (Homoptera: Aphididae) from Japan closely related to *Myzus nicotianae* and highly resistant to carbamates and organophosphates. *Bull. Entomol. Res.* 86:165–71
74. Nauen R, Tietjen K, Wagner K, Elbert A. 1998. Efficacy of plant metabolites of imidacloprid against *Myzus persicae* and *Aphis gossypii* (Homoptera: Aphididae). *Pestic. Sci.* 52:53–57
75. Nishimura K, Kanda Y, Okazawa A, Ueno T. 1994. Relationship between insecticidal and neurophysiological activities of imidacloprid and related compounds. *Pestic. Biochem. Physiol.* 50:51–59
76. O'Brien RD. 1967. *Insecticides: Action and Metabolism*. New York: Academic. 332 pp.
77. Okazawa A, Akamatsu M, Ohoka A,

Nishiwaki H, Cho W-J, et al. 1998. Prediction of the binding mode of imidacloprid and related compounds to house-fly head acetylcholine receptors using three-dimensional QSAR analysis. *Pestic. Sci.* 54:134–44
78. Olson ER, Dively GP, Nelson JO. 2000. Baseline susceptibility to imidacloprid and cross resistance patterns in Colorado potato beetle (Coleptera: Chrysomelidae) populations. *J. Econ. Entomol.* 93:447–58
79. Paterson D, Nordberg A. 2000. Neuronal nicotinic receptors in the human brain. *Prog. Neurobiol.* 61:75–111
80. Prabhaker N, Toscano NC, Castle SJ, Henneberry TJ. 1997. Selection of imidacloprid resistance in silverleaf whiteflies from the Imperial Valley and development of a hydroponic bioassay for resistance monitoring. *Pestic. Sci.* 51:419–28
81. Reed WT, Erlam GJ. 1978. The house fly metabolism of nitromethylene insecticides. See Ref. 93a, pp. 159–69
82. Roberts TR, Hutson DH, eds. 1999. *Metabolic Pathways of Agrochemicals*. Pt. 2: *Insecticides and Fungicides*. Cambridge, UK: R. Soc. Chem. 1475 pp.
83. Sakai M. 1967. Studies on the insecticidal action of nereistoxin, 4-*N,N*,-dimethylamino-1,2,-dithiolane. V. Blocking action on the cockroach ganglion. *Botyu-Kagaku* 32:21–33
84. Sattelle DB, Buckingham SD, Wafford KA, Sherby SM, Bakry NM, et al. 1989. Actions of the insecticide 2(nitromethylene)tetrahydro-1,3-thiazine on insect and vertebrate nicotinic acetylcholine receptors. *Proc. R. Soc. London Ser. B* 237:501–14
85. Sawruk E, Schloss P, Betz H, Schmitt B. 1990. Heterogeneity of *Drosophila* nicotinic acetylcholine receptors: SAD, a novel developmentally regulated α-subunit. *EMBO J.* 9:2671–77
86. Sawruk E, Udri C, Betz H, Schmitt B. 1990. SBD, a novel structural subunit of the *Drosophila* nicotinic acetylcholine receptor, shares its genomic localization with two α-subunits. *FEBS Lett.* 273:177–81
87. Schroeder ME, Flattum RF. 1984. The mode of action and neurotoxic properties of the nitromethylene heterocycle insecticides. *Pestic. Biochem. Physiol.* 22:148–60
88. Schulz R, Bertrand S, Chamaon K, Smalla K-H, Gundelfinger ED, Bertrand D. 2000. Neuronal nicotinic acetylcholine receptors from *Drosophila*: two different types of α subunits coassemble within the same receptor complex. *J. Neurochem.* 74:2537–46
89. Schulz R, Sawruk E, Mülhardt C, Bertrand S, Baumann A, et al. 1998. Dα3, a new functional α subunit of nicotinic acetylcholine receptors from *Drosophila*. *J. Neurochem.* 71:853–62
90. Schulz-Jander D, Casida JE. 2002. Imidacloprid insecticide metabolism: human cytochrome P450 isozymes differ in selectivity for imidazolidine oxidation versus nitroimine reduction. *Toxicol. Lett.* 132:65–70
91. Schulz-Jander DA, Leimkuehler WM, Casida JE. 2002. Neonicotinoid insecticides: reduction and cleavage of imidacloprid nitroimine substituent by liver microsomal and cytosolic enzymes. *Chem. Res. Toxicol.* 15:1158–65
92. Schuster R, Phannavong B, Schröder C, Gundelfinger ED. 1993. Immunohistochemical localization of a ligand-binding and a structural subunit of nicotinic acetylcholine receptors in the central nervous system of *Drosophila melanogaster*. *J. Comp. Neurol.* 335:149–62
93. Sgard F, Fraser SP, Katkowska MJ, Djamgoz MBA, Dunbar SJ, Windass JD. 1998. Cloning and functional characterization of two novel nicotinic acetylcholine receptor α subunits from the insect pest *Myzus persicae*. *J. Neurochem.* 71:903–12
93a. Shankland DL, Hollingworth RM, Smyth T Jr, eds. 1978. *Pesticide and Venom Neurotoxicity*. New York: Plenum

94. Sheets LP. 2001. Imidacloprid: a neonicotinoid insecticide. In *Handbook of Pesticide Toxicology*, Vol. 2, ed. R Krieger, pp. 1123–30. San Diego: Academic
95. Sheets LP. 2002. The neonicotinoid insecticides. In *Handbook of Neurotoxicology*, Vol. 1, ed. EJ Massaro, pp. 79–87. Totowa, NJ: Humana
96. Sherby SM, Eldefrawi AT, David JA, Sattelle DB, Eldefrawi ME. 1986 Interactions of charatoxins and nereistoxin with the nicotinic acetylcholine-receptors of insect CNS and *Torpedo* electric organ. *Arch. Insect Biochem. Physiol.* 3:431–45
97. Shiokawa K, Moriya K, Shibuya K, Hattori Y, Tsuboi S, Kagabu S. 1992. 3-(-6-Chloronicotinyl)-2-nitromethylene-thiazolidine as a new class of insecticide acting against Lepidoptera species. *Biosci. Biotechnol. Biochem.* 56:1364–65
98. Shiokawa K, Tsuboi S, Kagabu S, Sasaki S, Moriya K, Hattori Y. 1987. *Jpn. Kokai Tokkyo Koho* JP62-207266
99. Soloway SB, Henry AC, Kollmeyer WD, Padgett WM, Powell JE, et al. 1978. Nitromethylene heterocycles as insecticides. See Ref. 93a, pp. 153–58
100. Soloway SB, Henry AC, Kollmeyer WD, Padgett WM, Powell JE, et al. 1979. Nitromethylene insecticides. In *Advances in Pesticide Science*, Vol. 2, ed. H Geisbüehler, pp. 206–17. Oxford: Pergamon
101. Spande TF, Garraffo HM, Edwards MW, Yeh HJC, Pannell L, Daly JW. 1992. Epibatidine: a novel (chloropyridyl)azabicycloheptane with potent analgesic activity from Ecuadorean poison frog. *J. Am. Chem. Soc.* 114:3475–78
102. Takahashi H, Mitsui J, Takakusa N, Matsuda M, Yoneda H, et al. 1992. NI-25, a new type of synthetic and broad spectrum insecticide. *Proc. Brighton Crop Prot. Conf. Pest Dis.* 1:89–96
103. Tassonyi E, Charpantier E, Muller D, Dumont L, Bertrand D. 2002. The role of nicotinic acetylcholine receptors in the mechanisms of anesthesia. *Brain Res. Bull.* 57:133–50
104. Tomizawa M, Casida JE. 1997. [^{125}I]Azidonicotinoid photoaffinity labeling of insecticide-binding subunit of *Drosophila* nicotinic acetylcholine receptor. *Neurosci. Lett.* 237:61–64
105. Tomizawa M, Casida JE. 1999. Minor structural changes in nicotinoid insecticides confer differential subtype selectivity for mammalian nicotinic acetylcholine receptors. *Br. J. Pharmacol.* 127:115–22
106. Tomizawa M, Casida JE. 2000. Imidacloprid, thiacloprid, and their imine derivatives up-regulate the $\alpha 4\beta 2$ nicotinic acetylcholine receptor in M10 cells. *Toxicol. Appl. Pharmacol.* 169:114–20
107. Tomizawa M, Casida JE. 2001. Structure and diversity of insect nicotinic acetylcholine receptors. *Pest Manage. Sci.* 57:914–22
108. Tomizawa M, Cowan A, Casida JE. 2001. Analgesic and toxic effects of neonicotinoid insecticides in mice. *Toxicol. Appl. Pharmacol.* 177:77–83
109. Tomizawa M, Latli B, Casida JE. 1996. Novel neonicotinoid-agarose affinity column for *Drosophila* and *Musca* nicotinic acetylcholine receptors. *J. Neurochem.* 67:1669–76
110. Tomizawa M, Latli B, Casida JE. 1999. Structure and function of insect nicotinic acetylcholine receptors studied with nicotinoid insecticide affinity probes. See Ref. 125, pp. 271–92
111. Tomizawa M, Lee DL, Casida JE. 2000. Neonicotinoid insecticides: molecular features conferring selectivity for insect versus mammalian nicotinic receptors. *J. Agric. Food. Chem.* 48:6016–24
112. Tomizawa M, Otsuka H, Miyamoto T, Eldefrawi ME, Yamamoto I. 1995. Pharmacological characteristics of insect nicotinic acetylcholine receptor with its ion channel and comparison of the effect of nicotinoids and neonicotinoids. *J. Pestic. Sci.* 20:57–64
113. Tomizawa M, Otsuka H, Miyamoto T,

Yamamoto I. 1995. Pharmacological effects of imidacloprid and its related compounds on the nicotinic acetylcholine receptor with its ion channel from the *Torpedo* electric organ. *J. Pestic. Sci.* 20: 49–56
114. Tomizawa M, Wen Z, Chin H-L, Morimoto H, Kayser H, Casida JE. 2001. Photoaffinity labeling of insect nicotinic acetylcholine receptors with a novel [^{3}H]azidoneonicotinoid. *J. Neurochem.* 78:1359–66
115. Tomizawa M, Yamamoto I. 1992. Binding of nicotinoids and the related compounds to the insect nicotinic acetylcholine receptor. *J. Pestic. Sci.* 17:231–36
116. Tomizawa M, Yamamoto I. 1993. Structure-activity relationships of nicotinoids and imidacloprid analogs. *J. Pestic. Sci.* 18:91–98
117. Ujiváry I. 1999. Nicotine and other insecticidal alkaloids. See Ref. 125, pp. 29–69
118. Ujiváry I. 2001. Pest control agents from natural products. In *Handbook of Pesticide Toxicology*. Vol. 1: *Principles*, ed. RI Krieger, pp. 109–79. San Diego: Academic
119. Uneme H, Iwanaga K, Higuchi N, Kando Y, Okauchi T, et al. 1998. Synthesis and insecticidal activity of nitroguanidine derivatives. *9th IUPAC Congr. Pestic. Chem.* 1:1D-009 (Abstr.)
120. Wen Z, Scott JG. 1997. Cross-resistance to imidacloprid in strains of German cockroach (*Blattella germanica*) and house fly (*Musca domestica*). *Pestic. Sci.* 49:367–71
121. Wiesner P, Kayser H. 2000. Characterization of nicotinic acetylcholine receptors from the insects *Aphis craccivora*, *Myzus persicae* and *Locusta migratoria* by radioligand binding assays: relation to thiamethoxam action. *J. Biochem. Mol. Toxicol.* 14:221–30
122. Yamada T, Takahashi H, Hatano R. 1999. A novel insecticide, acetamiprid. See Ref. 125, pp. 149–76
123. Yamamoto I. 1965. Nicotinoids as insecticides. In *Advances in Pest Control Research*, Vol. 6, ed. RL Metcalf, pp. 231–60. New York: Wiley
124. Yamamoto I. 1999. Nicotine to nicotinoids: 1962 to 1997. See Ref. 125, pp. 3–27
125. Yamamoto I, Casida JE, eds. 1999. *Nicotinoid Insecticides and the Nicotinic Acetylcholine Receptor*. Tokyo: Springer-Verlag. 300 pp.
126. Yamamoto I, Tomizawa M, Saito T, Miyamoto T, Walcott EC, Sumikawa K. 1998. Structural factors contributing to insecticidal and selective actions of neonicotinoids. *Arch. Insect Biochem. Physiol.* 37:24–32
127. Yamamoto I, Yabuta G, Tomizawa M, Saito T, Miyamoto T, Kagabu S. 1995. Molecular mechanism for selective toxicity of nicotinoids and neonicotinoids. *J. Pestic. Sci.* 20:33–40
128. Zhang A, Kayser H, Maienfisch P, Casida JE. 2000. Insect nicotinic acetylcholine receptor: conserved neonicotinoid specificity of [^{3}H]imidacloprid binding site. *J. Neurochem.* 75:1294–303
129. Zhao J-Z, Bishop BA, Grafius EJ. 2000. Inheritance and synergism of resistance to imidacloprid in the Colorado potato beetle (Coleoptera: Chrysomelidae). *J. Econ. Entomol.* 93:1508–14
130. Zhong W, Gallivan JP, Zhang Y, Li L, Lester HA, Dougherty DA. 1998. From *ab initio* quantum mechanics to molecular neurobiology: a cation-π binding site in the nicotinic receptor. *Proc. Natl. Acad. Sci. USA* 95:12088–93
131. Zwart R, Oortgiesen M, Vijverberg HPM. 1994. Nitromethylene heterocycles: selective agonists of nicotinic receptors in locust neurons compared to mouse N1E-115 and BC3H1 cells. *Pestic. Biochem. Physiol.* 48:202–13

Annu. Rev. Entomol. 2003. 48:365–96
doi: 10.1146/annurev.ento.48.060402.102800
First published online as a Review in Advance on August 28, 2002

NONTARGET EFFECTS—THE ACHILLES' HEEL OF BIOLOGICAL CONTROL? Retrospective Analyses to Reduce Risk Associated with Biocontrol Introductions*

S.M. Louda,[1] R.W. Pemberton,[2] M.T. Johnson,[3] and P.A. Follett[4]

[1]*School of Biological Sciences, University of Nebraska, Lincoln, Nebraska 68588-0118; e-mail: slouda@unl.edu*

[2]*Invasive Plant Research Laboratory, USDA Agricultural Research Service, 3205 College Ave. Ft. Lauderdale, Florida 33314; e-mail: bobpem@eemail.com*

[3]*Institute of Pacific Islands Forestry, USDA Forest Service, P.O. Box 236, Volcano, Hawaii 96785; e-mail: tracyjohnson@fs.fed.us*

[4]*U.S. Pacific Basin Agricultural Research Center, USDA, Agricultural Research Service, P.O. Box 4459, Hilo, Hawaii 96720; e-mail: pfollett@pbarc.ars.usda.gov*

Key Words classical biological control, ecological risk, weed control, insect control

■ **Abstract** Controversy exists over ecological risks in classical biological control. We reviewed 10 projects with quantitative data on nontarget effects. Ten patterns emerged: (*a*) Relatives of the pest are most likely to be attacked; (*b*) host-specificity testing defines physiological host range, but not ecological range; (*c*) prediction of ecological consequences requires population data; (*d*) level of impact varied, often in relation to environmental conditions; (*e*) information on magnitude of nontarget impact is sparse; (*f*) attack on rare native species can accelerate their decline; (*g*) nontarget effects can be indirect; (*h*) agents disperse from agroecosystems; (*i*) whole assemblages of species can be perturbed; and (*j*) no evidence on adaptation is available in these cases. The review leads to six recommendations: Avoid using generalists or adventive species; expand host-specificity testing; incorporate more ecological information; consider ecological risk in target selection; prioritize agents; and pursue genetic data on adaptation. We conclude that retrospective analyses suggest clear ways to further increase future safety of biocontrol.

CONTENTS

INTRODUCTION

The risks to native species associated with classical biological control, the introduction of exotic natural enemies to control alien pest species, have been debated (36, 38, 70, 71, 73, 114, 115, 122–124, 129). The issue is not new (1, 60, 61, 96, 97, 111, 113, 125). However, the debate has intensified in light of increasing evidence of nontarget host use by biocontrol agents. In this review we examined 10 cases in detail. These cases were reviewed because nontarget impacts were quantified. They do not constitute a random sample, and so do not estimate the frequency of nontarget effects. Our aims are to summarize the data available, evaluate the patterns, and define further research needed. Our premise is that such retrospective analysis can provide important information on traits associated with environmental risk and suggest protocols or research that would continue to increase the safety of biological control.

Debate over nontarget effects of biocontrol agents has polarized biologists. On the one hand, many practitioners view biological control as a progressive, environmentally benign alternative to chemicals (62, 80, 124). When successful, biological control leads to long-term reduction in pest numbers, potential for wide-ranging control, elimination of chemical residues, and a low economic cost/benefit ratio. On the other hand, many ecologists view the intentional introductions of alien species into complex biological communities as a threat to their structure and dynamics. Major concerns include the irreversibility of introductions, potential for host switching, dispersal into nonagricultural habitats, lack of research on both efficacy and ecological impacts, possibility of evolutionary adaptation to new hosts, and difficulty of predicting interaction outcomes in complex systems (60, 61, 71–74, 77, 97, 108, 111, 112, 114, 122, 123). In fact, traits viewed as advantageous in biocontrol, such as capacity for self-replication, rapid increase, and high dispersal, are also traits that enhance the probability of unexpected ecological effects when native species are within the potential host range.

As invasive species proliferate and pressure to control them grows, the need to understand the ecological effects of biological control increases. Environmental protection laws in many countries restrict the introduction of exotic species into natural ecosystems unless there is evidence that such introduction will not have adverse effects on those systems (6, 7, 86). Adverse effects on native species, representing a continuum from small effects to extinctions (31, 37), must be quantified to be evaluated. Thus, there is a real need for increased scientific attention toward the measurement and prediction of impacts on target and nontarget species.

CASE HISTORIES IN BIOLOGICAL CONTROL OF WEEDS

Biological control of weeds has a long recorded history, starting with redistribution in 1832 of *Dactylopius* scales to control weedy prickly pear cacti in South Africa (66). Since then, 153 insects have been released in the continental United States and Hawai'i against 53 target weeds (66). Estimates of success for these weed control projects vary from 41% with evidence of some control (93) to 20% with significant control (136). Nontarget feeding, oviposition, and development are reported on 41 native plants, by 15 of the 112 biocontrol insects (13.4%) established against weeds in Hawai'i, the continental United States, and the Caribbean (99). Plants closely related to the targeted weed were more vulnerable than distantly related species (99). However, neither the quantity of feeding nor the ecological ramifications have been studied in most cases. We review the three cases for which the most quantitative data exist.

Rhinocyllus conicus Against Exotic Thistles

INTRODUCTION The flower head weevil, *R. conicus* (Curculionidae), indigenous in Eurasia, has been released in Argentina, Australia, New Zealand, and North America against weedy thistles (Asteraceae, Carduinae), especially *Carduus nutans* (41, 141). In the United States, it was redistributed freely until August 2000, when the USDA APHIS Plant Protection and Quarantine removed *R. conicus* from the list of insects preapproved for interstate movement into new areas (T. Horner, personal communication).

EVALUATION In the field in Europe, hosts included 7 out of 9 *Carduus* spp. (77.8%), 4 out of 17 *Cirsium* spp. (23.5%), plus *Silybum marianum* and *Onopordum acanthium* (141). In the laboratory, no-choice feeding tests found that adults did not feed on cultivated non-Astereae but accepted 12 of 16 European thistles. Leaves of the one North American species tested (*Cirsium undulatum*) were accepted inconsistently. In choice tests, *R. conicus* adults preferred leaves of *C. nutans* over 9 of 11 other thistles (141). Adult oviposition and larval development of *R. conicus* were evaluated in a no-choice cage test using four European species (*Carduus nutans*, *C. personata*; *Cirsium arvense*, *C. palustre*) but no North American species. All four species received eggs and supported complete development; however, adults emerging from the *Cirsium* spp. were 10% smaller than their parents from *C. nutans*. Although these data suggested nontarget effects could occur, no major impacts were expected because of (*a*) generally lower preference for *Cirsium* spp. than for *Carduus nutans*, (*b*) higher larval mortality and smaller adult size on *Cirsium* spp. than on *Carduus* spp., and (*c*) low population densities of the North American *Cirsium* spp. (141). Side effects, if any, on native thistles were even considered a bonus by some ranchers (D. Schröder, personal communication).

NONTARGET EFFECTS Use of North American *Cirsium* spp. by *R. conicus* was reported soon after its introduction in 1969. Rees (102) found *R. conicus* developing on *C. undulatum*. Earlier, Maw [reported in (141)] found it developing on *C. flodmanii*. Goeden & Ricker [(42) and references therein] and Turner et al. (126) reported *R. conicus* developed in the flower heads of 17 California *Cirsium* spp. (57% of species sampled), including 3 rare ones. Unexpectedly, weevils from some nontarget *Cirsium* spp. were larger than those from *Carduus* spp. (126). By 2001, *R. conicus* was reported using 22 of the 90+ North American *Cirsium* spp. (99). By 1993, *R. conicus* invaded sand prairie sites in Nebraska where it adopted two well-studied native thistles as hosts in the absence of exotic thistles (68, 70, 74, 77). Prior experiments had shown that native floral insects limited seed, seedlings, population density, and lifetime fitness of one of these, Platte thistle (*Cirsium canescens*) (76). By 1996, the addition of *R. conicus* to the inflorescence guild further reduced seed production by 85.9% (74, 77). The density of Platte thistle in demography plots then declined dramatically (72). In addition, the simultaneous decrease in a native floral-feeding tephritid, *Paracantha culta*, suggested *R. conicus* also may have major indirect ecological effects (73, 74, 77). These outcomes are consistent with the recent theoretical prediction that "shared predation modules" present an indirect risk to nontarget species and their fauna, even at sites away from the area of control (57). The availability of long-term data on interacting populations prior to host range expansion of *R. conicus* onto North American thistles provides unique documentation for ecological aspects of nontarget effects.

IMPLICATIONS This case illustrates the early lack of concern over potential nontarget effects on noneconomic native species. Additionally, although the host range expansion onto North American thistles is consistent with the host-specificity tests, those tests did not pinpoint the risk to secondary hosts in the absence of the preferred host. Clearly, low native plant densities and lack of habitat overlap with targeted hosts were not sufficient to prevent harm. The influence of variation in availability of the preferred host and potential indirect effects were not evaluated. Finally, as expected, the introduced insect was not restricted to the release sites, but in this case *R. conicus* dispersed into a remote native habitat without the targeted weed.

Larinus planus Against Canada Thistle

INTRODUCTION This European thistle weevil was first found in the northeast United States in 1971 in fields where *Altica carduorum* (Chrysomelidae) was released against Canada thistle, *Cirsium arvense* (135). In spite of contemporary tests in 1990, which showed that native thistles could be acceptable though less preferred (= "secondary") hosts (82), *L. planus* was distributed by state and federal agencies into six western states and British Columbia in 1990–2000 (75).

EVALUATION In the 1960s, six species of *Larinus*, including *L. planus* (= *L. carolinae*), were screened as potential biocontrol agents (140). All six European *Cirsium* spp. offered to *L. planus* were acceptable food plants. No North American species were tested. Because *Larinus* spp. had greater diet breadth than *R. conicus*, no official introductions were made into North America. After *L. planus* was found in the northeast United States, it was re-evaluated using contemporary protocols as a prerequisite to deliberate release in Canada (82). Adult preference and larval performance were tested on agronomic plants and native thistles (*Cirsium andrewsii*, *C. flodmanii*, *C. foliosum*, *C. hookerianum*, *C. undulatum*). The average amount of adult feeding in no-choice and choice tests appeared lower on native thistles than on Canada thistle (82), but was not significantly different (75). In no-choice oviposition tests, *Cirsium drummondii* received no eggs in a small test ($n = 4$ female-days), but both *C. flodmanii* and *C. undulatum* were ovipositional and developmental hosts in larger tests (82). However, no emergence of new adults occurred from these native species under the test conditions. McClay (82) concluded in 1990 that *L. planus* was "unlikely to form significant populations on them [native North American thistles]" and that "the redistribution of *L. planus* to Alberta, and other areas of North America where *C. arvense* is a problem, should be considered."

NONTARGET EFFECTS This weevil was released in the 1990s on state and federal lands in Colorado, a state with many native *Cirsium* spp. (131), at least two of which are considered rare. In 1999, *L. planus* was found feeding on a uncommon native species, Tracy's thistle (*C. undulatum* var. *tracyi*), in western Colorado. A subsequent study found that 75% of the main seed-producing flower heads were destroyed by *L. planus* (75). At the same time, this agent had no effect on Canada thistle nearby (5.2% total seed damaged, with no evidence of *L. planus*). In 2001, *L. planus* also was reared from the flower heads of three native thistles collected in Oregon: *Cirsium brevistylum*, *C. remotifolium*, and *C. undulatum* (128). Apparently, *L. planus* was discovered in Oregon in 1993 and, subsequently, limited redistribution and introduction occurred [(75); E.M. Coombs, personal communication]. Oregon releases have stopped (E.M. Coombs, personal communication).

IMPLICATIONS This case demonstrates that redistribution of an adventive exotic species for biological control can have undesirable effects. Release into new, geographically disjunct ecosystems occurred in spite of evidence that native species were acceptable feeding, ovipositional, and possibly developmental hosts. Contemporary host-specificity tests were not sufficient to predict relative use among acceptable hosts in the field. Because permits for interstate transport and release of *L. planus* apparently were issued without formal evaluation by the main scientific advisory group for biological control introductions (USDA, APHIS, Technical Advisory Group), the case suggests that more oversight of movement of adventive exotic species for biological control is warranted.

Cactoblastis cactorum Against *Opuntia* spp. (Prickly Pear)

INTRODUCTION This Argentine pyralid moth, now used worldwide (66), was introduced first into Australia in 1926 and quickly controlled exotic *Opuntia* spp. (23). *Cactoblastis cactorum* was introduced to Nevis Island in the Caribbean in 1957 against weedy native *Opuntia* spp. (116). By 1960, densities of one native shrub (*O. triancantha*) and several "tall" *Opuntia* spp. were lower (116). This moth is now widespread in the Caribbean, including the Bahamas and Cuba (8); the magnitude of its ecological impact there still needs to be evaluated. Introduction of *C. cactorum* into the continental United States was considered in the 1960s; no release was made out of concern both for economic prickly pears in Mexico (F.J. Bennett, personal communication) and for potential effects on *Opuntia* spp. forage and wildlife support functions (62). The issue arose again in the 1980s, by inclusion of *Opuntia* spp. on a list of rangeland weeds for control (97). Such targeting of native plants for biological control was challenged because both their ecological and economic functions could be irreversibly impaired by the spread of agents beyond areas in which the plants were pests (97).

EVALUATION Field studies of cactus-feeding insect herbivores in the Western Hemisphere identified host associations (23). *C. cactorum* completed development on all prickly pears, except one (*O. sulphurea*). One exception to the pattern of narrow host specificity defined by genus was a species of *Cleistocactus*, in a different subfamily (Cactoideae) than *Opuntia* (Opuntiioideae), on which feeding but no development was observed (23). No-choice starvation tests indicated that *C. cactorum* could feed but did not develop on plants other than cacti. Potential conservation effects were not considered at the time of introduction into the Caribbean nor studied afterward (F.J. Bennett, personal communication). Retrospective tests of host specificity with six native Florida *Opuntia* spp. indicated that all of these species support complete development (64).

NONTARGET EFFECTS In 1989, *C. cactorum* was discovered in Florida (8), where it now develops on five of six *Opuntia* spp. (64). Up to 90% of *O. stricta* studied were damaged by *C. cactorum*, and 15% of the monitored plants died (65). The adoption of Florida's *Opuntia* spp. is not surprising because the moth uses almost all *Opuntia* spp. within its native range (23). As with many accidental introductions, how

C. cactorum reached Florida is unclear (98, 139). The quantitative effect of *C. cactorum* on nontarget cacti is unknown. Given its dramatic impact on native *Opuntia* spp. on Nevis Island (116), it is likely that some Caribbean species have been damaged seriously, some possibly driven close to extinction on Nevis and Grand Cayman Islands (8). In Florida, restoration of *Opuntia corallicola* (= *O. spinosissima*), an endemic species already reduced by habitat loss, is hindered by infestation by *C. cactorum* (64). *O. stricta*, one of the cacti under *C. cactorum* control in Australia, is native to Florida and is a common host. Additionally, *O. stricta* occurs all around the Gulf of Mexico, so this plant could serve as a bridge for *C. cactorum* to expand its range, potentially threatening the large complex of native *Opuntia* spp. in the southwestern United States and Mexico (100, 139). By 1999, the moth had spread to southern Georgia (121). Indirect effects of *C. cactorum* on other species also are unknown. Some Caribbean iguanid lizards use *Opuntia* spp. for food (121), and a variety of native insects are associated with *Opuntia* spp. A native moth, *Melitara dentata* (Pyralidae), limited the growth and density of *O. fragilis*, a native of sand prairie in the upper Great Plains (14), suggesting the spread of *C. cactorum* could disrupt such limiting interactions between native *Opuntia* spp. and their associated insects.

IMPLICATIONS Relatively specialized feeders, such as *C. cactorum*, can still pose an ecological threat. Use of this moth against prickly pear cacti in parts of the world with no native *Opuntia* spp. has caused no known nontarget effects. Yet, use of this moth led to damage of native *Opuntia* spp. in the Caribbean, in close proximity to continental areas with many more native relatives. Thus, the same agent can be either relatively safe or relatively dangerous, depending on geographic region, ecosystem, and number of native relatives in the flora.

CASE HISTORIES IN BIOLOGICAL CONTROL OF INSECTS

Since the 1888 introduction of the vedalia beetle against cottony-cushion scale in California, more than 5000 releases of insects for classical biological control have been documented (49, 79). Establishment frequency is estimated to be 34%–50% (49, 50). Successful control of targeted insects resulted in 3% of introductions, and an additional 11% resulted in partial control (78). Use of native nontarget species has been recorded for 7% of 59 predators and 10% of 115 parasitoids introduced for insect control in Hawai'i (39) and for 16% of 313 parasitoids of holometabolous pests in North America (54). Worldwide, a recent review (79) identified 92 cases (1.7%) of introductions for which nontarget effects were reported. Effects in most cases were minor or minimally documented, but 18 introductions had evidence of significant negative impacts on nontarget populations. Because nontarget feeding by agents for insect biocontrol is greatly under-reported, Lynch & Thomas (79) further estimated that as many as 11% of past introductions for insect biocontrol may have had serious nontarget effects. We review the seven cases that have received the most detailed quantitative study.

Microctonus spp. Against Forage Weevils in New Zealand

INTRODUCTION Two braconid parasitoids have been introduced into New Zealand for control of pest weevils on forage. *Microctonus aethiopoides* was released in 1982 against *Sitona discoides* in alfalfa (lucerne), and *M. hyperodae* was released in 1991 against *Listronotus bonariensis* in pasture (6, 7, 43, 44).

EVALUATION *M. aethiopoides* was released after minimal host testing (7). At the time, it was thought to attack only two genera, *Sitona* and *Hypera* (67). Conversely, extensive host specificity and suitability tests were conducted with *M. hyperodae* before release (43). Of the 24 weevil species tested, 4 native species were physiologically suitable hosts, but the potential nontarget hosts produced only 18.7% as many parasitoids compared to the target. A native minor pest, *Irenimus aequalis*, was the best nontarget host. Earlier, under field conditions in Western Patagonia, *M. hyperodae* was found attacking only *L. bonariensis* and not its three coexisting congeners (67).

NONTARGET EFFECTS Retrospective analysis was initiated to determine the ability of laboratory host range testing to predict field host range (6). In laboratory tests, *M. aethiopoides* had a broader host range, attacking 12 of 13 species compared to 7 of 30 species attacked by *M. hyperodae*. Among the species attacked by *M. aethiopoides*, mean parasitism was 58% for nontarget species and 62% for targeted species compared to 13% in nontarget species and 61% in targeted species attacked by *M. hyperodae* (5). Cresswell (18) showed that in the laboratory *M. aethiopoides* can complete successive generations on the native weevil *Nicaeana cervina*, and female wasps that first oviposited on *N. cervina* selected this species over the targeted host, *S. discoides*, for subsequent ovipositions. In the field, *M. aethiopoides* parasitized 16 of 48 species of weevils, whereas *M. hyperodae* parasitized only 3. Species attacked by *M. aethiopoides* included four indigenous and four exotic weevil genera, including the thistle biocontrol agent *R. conicus*. Average parasitism among attacked nontarget species was higher by *M. aethiopoides* than by *M. hyperodae* (23% versus 2%), and *M. aethiopoides* attacked nontarget species at more sites (17 versus 2 of 33 sites) [(6); B.I.P. Barratt, personal communication]. In a survey of *M. aethiopoides* parasitism along an elevational gradient, average parasitism of native weevils (Brachycerinae) by *M. aethiopoides* was 64% in alfalfa (450 m), 23% in surrounding pastoral habitats, 4%–10% in overgrown native grasslands (620–780 m), and 2% in a subalpine habitat (850 m) (B.I.P. Barratt, personal communication). Thus, *M. aethiopoides* showed relatively broad host range, and it parasitized multiple nontarget species outside the target host environment, although at lower rates. *M. hyperodae* demonstrated a higher level of specificity, as predicted, and it parasitized multiple nontarget species only in the target's environment.

IMPLICATIONS The case shows that prerelease laboratory host-specificity tests can be useful in predicting postrelease specificity in the field ("realized host range"). Also, the introduced biocontrol agents are not restricted to the habitat of the target species where they were released; rather they invaded remote native habitats. Finally, environmental gradients can provide spatial refuges that dampen the impact of the biocontrol agent on some nontarget species, such as in natural habitats at higher elevation.

Introduced Parasitoids Against Tephritid Fruit Flies in Hawai'i

INTRODUCTION Between 1913 and 1950, over 30 parasitoids were introduced into Hawai'i for control of three exotic pest fruit flies (*Ceratitis capitata*, *Bactrocera dorsalis*, *B. curcurbitae*) (10, 133). Six opiine braconids and a eulophid (*Tetrastichus giffardianus*) became widely established and contributed to pest suppression. For example, the braconid *Fopius arisanus* killed over 90% of oriental fruit fly (*B. dorsalis*) eggs in guava, *Psidium guajava* (88).

EVALUATION At the time of introduction, host-acceptance tests were conducted with the targeted species, but not with the 33 other tephritids in Hawai'i (26). The other tephritids are gall makers and flower head feeders and include 26 endemic species as well as 5 intentionally and 2 inadvertently introduced weed control agents.

NONTARGET EFFECTS Nontarget risk to endemic tephritids in Hawai'i was assessed recently. When several introduced fruit fly parasitoids were exposed in cages to flower heads of *Dubautia menziesii* containing the endemic fly *Trupanea dubautiae*, little ovipositional activity occurred (24, 27, 28). In another test, *Diachasmimorpha tryoni* wasps probed galls containing lantana gall fly (*Eutreta xanthochaeta*) less than they probed coffee berries containing *C. capitata*, although learning modified this response (25). In the field, the parasitoid *D. longicaudata*, attracted to decaying fruit infested with fruit fly larvae, visited more nontarget stem galls when decaying fruit was absent (26). In small cages *D. longicaudata* parasitized 54% of the *E. xanthochaeta* larvae in stem galls, whereas in large field cages and in the field <1% of fly larvae in galls were attacked (26). In field surveys, no deliberately introduced parasitoids were recovered from native tephritids (29). However, the adventive opiine braconid *Habrocytus elevatus* as well as the eulophid *Euderus metallicus* were found attacking weed agents (lantana gall fly; Eupatorium gall fly, *Procecidochares utilis*) and native *Trupanea* spp.; *H. elevatus* also was found attacking an adventive tephritid used in weed biocontrol, *Ensina sonchii* (24, 29). A detailed budget of population losses of lantana gall fly

(*E. xanthochaeta*) larvae was conducted to assess the role of fruit fly parasitoids in its population dynamics by using the recruitment method to quantify mortality (30). *D. tryoni* accounted for >86% of total parasitism but caused only a 10% increase in total mortality. Percentage parasitism by *D. tryoni* varied significantly among habitats.

IMPLICATIONS These studies highlight the importance of microhabitat selection by parasitoids and, therefore, the importance of ecologically relevant host-specificity tests. The difference in vulnerability between gall-forming and flower head-feeding fruit fly species suggests that host testing should include ecologically distinct groups of potential nontarget species. It is also clear that the outcome of host-specificity tests can be influenced strongly by the test conditions (cage size, choice versus no-choice options, starvation, physiological age). Because learning can play a role in host range adaptation in insect parasitoids, simple evaluations of host and habitat preference may underestimate the magnitude of risk to the less preferred native hosts from attack by introduced natural enemies. Finally, detailed life table analysis can play an important part in impact assessment.

Compsilura concinnata Against Gypsy Moth

INTRODUCTION Beginning in 1906, this tachinid fly, a polyphagous parasitoid of Lepidoptera, was introduced into North America against the gypsy moth (*Lymantria dispar*) and the browntail moth (*Euproctis chrysorrhoea*). *L. dispar* is univoltine, whereas *Compsilura concinnata* is multivoltine. Despite knowledge that this parasitoid depended upon nontarget species to complete multiple generations per year and despite early documentation of nontarget attack on native insects, *C. concinnata* was released repeatedly until 1986 (11, 19, 104, 130).

EVALUATION *C. concinnata* was poorly studied in Europe before its importation, but it was known to be polyphagous, gregarious, multivoltine, and highly vagile (19).

NONTARGET EFFECTS The ability of *C. concinnata* to utilize a wide range of species, including over 200 spp. of Lepidoptera and Hymenoptera, became clear soon after initial release (3, 19, 130). However, this polyphagy was considered a bonus, allowing the parasitoid to spread ahead of the gypsy moth and perhaps slow its invasion (130). In 1915, the USDA initiated a 15-year program to determine the native hosts of the introduced parasitoids (104). Over 300,000 field-collected lepidopteran larvae were reared to survey parasitoids. For the macrolepidopterans that were well sampled (n = 164 spp.), 66.9% of the species were attacked by *C. concinnata*. Also, *C. concinnata* was present in 79.6% of the 93 macrolepidopteran species collected >9 years, and in 91.5% of the 59 species that were collected >9 years and had total collections of >200 larvae. Giant silkworm moths (Saturniidae) were among those attacked (104). In fact, the range caterpillar

Hemileuca oliviae, a native saturniid pest of grazing lands in the southwestern United States, was a target of some releases. In the 1950s, some saturniid populations in the northeastern United States appeared to undergo a massive, rapid decline (106). Although populations of most species occupy their precrash range again, reported densities are thought to be lower (106). Three species (*H. maia maia*, *Eacles imperialis*, *Anisota stigma*) have been placed on several state endangered species lists (11), and two *Citheronia* spp. apparently have been extirpated in the northeastern United States (106).

Boettner et al. (11) argued that parasitism by *C. concinnata* likely contributed significantly to the decline in native silk moths. Earlier, Stamp & Bowers (119) suggested that *C. concinnata* could be an important mortality factor for *H. lucina*. Using laboratory-reared cohorts placed as sentinel larvae, Boettner et al. (11) found that *C. concinnata* caused 81% mortality in *Hyalophora cecropia* and 67.5% in *Callosamia promethea* in Massachusetts. In addition, *C. concinnata* was reared from 36% of wild-collected *H. maia maia* (11). Levels of parasitism actually may be higher because these studies did not include late instars. However, other factors also could be involved in the declines, including habitat loss (127), successional change (120), and aerial spraying of *Bacillus thuringiensis* for gypsy moth control (51). In sum, the evidence is clear that *C. concinnata* is a persistent and substantial source of mortality for native Lepidoptera in the northeastern United States, and the recent evidence suggests it could be seriously harming some species. However, further research is needed to quantify nontarget population impacts and to explain mechanisms underlying the peristence of *H. cecropia*, *C. promethea*, and *H. maia maia* populations even after 95 years of nontarget attack by *C. concinnata*. Compensatory and persistence mechanisms for the populations sustaining high levels of parasitism are not known but could include low rates of added mortality from parasitism, life history strategies (127), and the presence of refuges (132).

IMPLICATIONS This case history shows that polyphagy, combined with multivoltinism and high vagility, represents a risky set of traits for a prospective biocontrol agent in terms of potential for nontarget effects. It also suggests that nontarget effects can continue over time, in this case almost 100 years. With generalists such as *C. concinnata*, there is no opportunity for density-dependent mechanisms to reduce its population level when a specific prey species becomes rare because it can maintain its population on other prey species. This trait, considered advantageous initially, increases the potential for detrimental effects on nontarget species, particularly less common species whose populations can be swamped by large numbers of the biocontrol agent.

Parasitoids Against *Nezara viridula* in Hawai'i

INTRODUCTION The polyphagous crop pest *Nezara viridula* (Pentatomidae) invaded Hawai'i in 1961 (20). Following unsuccessful attempts at eradication, biocontrol agents used previously in Australia were introduced. Three agents were established by 1963: the egg parasitoid *Trissolcus basalis* (Scelionidae) and two adult parasitoids, *Trichopoda pilipes* and *T. pennipes* (Tachinidae) (20). Declines in native Hawaiian stink bugs, including the koa bug (*Coleotichus blackburniae*: Scutelleridae), have been attributed to the introduction of these parasitoids (61).

EVALUATION Laboratory tests and field observations around the time of introduction showed that both the tachinids and the egg parasitoid could locate and develop on *C. blackburniae* (20). Recent studies showed that *T. basalis*, maintained on *N. viridula* in the laboratory, accepted eggs of *N. viridula* and *C. blackburniae* equally in choice and no-choice tests, independent of arena size (M.T. Johnson, unpublished data). Behavioral observations in petri dishes suggested that *T. basalis* used cues for host acceptance similarly in both hosts. However, *C. blackburniae* appeared to be less suitable because many parasitized eggs died without developing parasitoids (M.T. Johnson, unpublished data).

NONTARGET EFFECTS Examination of museum specimens collected between 1965 and 1995 for attached *Trichopoda* egg shells revealed attacks on *N. viridula* (17%, $n = 302$ specimens), *C. blackburniae* (8%, $n = 107$), and three alien pentatomids, but not on native pentatomids in the genus *Oechalia* ($n = 96$) (37). The low numbers and the haphazard nature of such collections limit these data as a precise historical record of population impacts (37). A recent two-year field study confirmed that *T. basalis* and *T. pilipes* attacked *C. blackburniae* on all four islands surveyed (M.T. Johnson, unpublished data). Parasitism of *C. blackburniae* eggs and sentinel *N. viridula* eggs occurred at levels up to 20% at low elevations on an alien host plant (*Acacia confusa*), but parasitism was low (0%–2%) at higher elevations on the koa bug's two native host plants (*A. koa*, *Dodonaea viscosa*). Contemporary life table analyses indicate that accidentally introduced natural enemies now have a greater impact on *C. blackburniae* populations than do biocontrol agents (M.T. Johnson, unpublished data). The most common egg parasitoid found was an accidentally introduced eupelmid, *Anastatus* sp., not *T. basalis*. Also, egg predation by other accidentally introduced species, primarily ants and the spider *Cheiracanthium mordax*, was a more important source of mortality (10%–80%) than parasitism (M.T. Johnson, unpublished data). Parasitism of adult *C. blackburniae* by *T. pilipes* averaged 7% across sites on Hawai'i, with the highest levels (25%–42%) observed on native host plants $\geq$10 km away from any agricultural areas, levels comparable to those for *N. viridula* adults in agricultural areas. It is not known whether *T. pilipes* persists on *C. blackburniae* at these remote sites or migrates from areas where *N. viridula* is abundant. Although *T. pilipes* uses the male aggregation pheromone in host finding (52), large proportions of *C. blackburniae* females (up to 37%) and fifth instars (up to 26%) were sometimes parasitized. The impact of tachinid parasitism on population growth is potentially more severe for *C. blackburniae* than for *N. viridula* because *C. blackburniae* females produce smaller egg masses (32 versus 70 eggs per mass) and have slower ovarial development (M.T. Johnson & A. Taylor, unpublished data).

IMPLICATIONS This case illustrates the value of life table analysis in impact assessment. Evaluation of nontarget impacts 30 years after release, however, is difficult because the relative contribution of various mortality sources may have changed. Generalist predators now cause the highest mortality to *C. blackburniae* in all habitats on Hawai'i, but at least one introduced biocontrol agent appears to have a significant impact on populations in one native habitat. Most of the evidence for population decline in *C. blackburniae* to date comes from Oahu, where displacement of native host plants and invasions by alien generalist predators also could be involved in the decline of native insects (37). Because biocontrol agents were shown to attack *C. blackburniae* in the laboratory at the time of introduction, this case study confirms the validity of prerelease testing of native species to predict potential nontarget hosts.

Parasitoids Against *Pieris rapae* in New England

INTRODUCTION *Pieris rapae*, a pest of crucifers, invaded North America (Canada) from Europe in 1860 (16). The braconid parasitoid, *Cotesia glomerata*, was released against *P. rapae* in the 1880s, although it may have arrived earlier in parasitized hosts (107). *P. napi oleracea*, a native species in northeastern United States and eastern Canada, underwent a range reduction in the late 1800s (107, 109). A recent field study attempted to determine if parasitoids played a role in this decline (R.G. Van Driesche, personal communication).

EVALUATION Potential nontarget effects of *C. glomerata*, and other parasitoids introduced against *P. rapae* in the United States, on native pierids such as *P. napi* were not evaluated prior to release.

NONTARGET EFFECTS Artificially placed (sentinel) larvae of *P. napi* and *P. rapae* were used to examine parasitism by *C. glomerata* and another introduced wasp (*C. rubecula*) in Vermont, where *P. napi* persists, and in western Massachusetts, where *P. napi* is now rare or extinct (J. Benson et al., unpublished data). *P. napi* overwinters as a pupa and spends its first generation on *Cardamine* (=*Dentaria*) *diphylla* in wooded habitats and subsequent generations in open fields, whereas *P. rapae* inhabits open fields exclusively. *Cotesia* parasitoids attacked sentinel larvae of both *Pieris* spp. more frequently in fields than in forests; attack by *C. glomerata* was more frequent on *P. napi* than on *P. rapae* when both species were presented in fields (J. Benson et al., unpublished data). Because parasitism rates were similar in Vermont and western Massachusetts, they hypothesized that differences in host diapause and life history accounted for differences in the impact of added parasitism. If the *P. napi* population in Vermont commits a greater percentage of its first generation woodland population to overwintering diapause than does the population in western Massachusetts, then this allocation may buffer the effects in Vermont of heavy second-generation mortality imposed by the acquired parasitoid in open meadows in both locales. Sentinel larvae (*P. rapae*, *P. napi*) also were used to evaluate the role of *C. glomerata* in the range contraction of another native pierid, *P. virginiensis*, in New York and Ontario (9). Even though both *C. glomerata* and *C. rubecula* parasitize and successfully develop in *P. virginiensis* in the laboratory, no parasitism was detected on sentinel larvae on two host plants, *Brassica oleracea* or *C. diphylla*, within the woodland habitat of *P. virginiensis* (9), suggesting parasitism is not a likely explanation of the observed range contraction.

IMPLICATIONS Parasitoid foraging behavior and host life history variation in this case appear to have influenced the effect of the introduced parasitoids on native, nontarget species. *Cotesia* parasitoids do not forage in closed woodland, so this habitat provides the strictly woodland species, *P. virginiensis*, with an absolute refuge even though it is an acceptable host in the laboratory. Similarly, woodland provides the first generation of *P. napi* a refuge from parasitism. However, the second generation of *P. napi* suffers significant mortality when it attempts to reproduce in open areas where *Cotesia* parasitoids are active. If life history variation occurs in the portion of the host population so exposed, then the outcome of the added mortality may vary over the range of the nontarget host species, as appears likely for *P. napi*. This case underscores the need for information about agent foraging behavior and host life history variation to supplement physiological host range data.

Coccinella septempunctata (C7) Against Aphids in North America

INTRODUCTION Roughly 150,000 ladybird beetles of the Palearctic species *Coccinella septempunctata* (C7) were released on aphids in crops in 12 states and Nova Scotia from 1957 to 1971 (103). Despite this effort, establishment of C7 probably occurred accidentally in 1973 in New Jersey and Quebec (2, 21). Over 500,000 beetles were redistributed in 1974–1978 from New Jersey to 20 states (2, 103). Distribution efforts expanded to western states, especially after the 1986 invasion by the Russian wheat aphid, *Diuraphis noxia* (45, 101). Meanwhile, C7 populations increased dramatically in the eastern and central United States (89, 103), fueling concern over possible competitive displacement of native coccinellids (33). Federal programs have stopped dispersing alien coccinellids, but private efforts likely continue (63).

EVALUATION Early programs paid little attention to potential impacts on nontarget species (123). In the Russian wheat aphid program, prerelease studies were skipped in the interest of responding rapidly to the economic threat (101). Yet, coccinellids were perhaps the least promising agents because their foraging effectiveness is limited by the Russian wheat aphid's habit of feeding within tightly curled leaves (137). Recent studies of nontarget effects of alien coccinellids have focused almost entirely on possible impacts on native coccinellids in agricultural systems (90). Although impacts on other coccinellids were not predicted explicitly, they are not surprising in light of C7's dominant role among aphidophagous coccinellids in Europe (2, 56). Numerical superiority of C7 in North America was expected to improve pest control (21). Although some coccinellids are broadly polyphagous, C7 population dynamics correlate closely with aphid prey (56). However, Horn (59) argued that C7 could threaten an endangered butterfly in Ohio because it was found in the same habitat and fed on eggs of a congener in the laboratory.

NONTARGET EFFECTS Field evidence for displacement of native coccinellids by C7 is correlative. By the 1990s, C7 had become the dominant coccinellid in a variety of habitats in the United States (2, 13, 45, 87, 134). One of the most severe declines coinciding with a rise of C7 appears to have occurred for *C. novemnotata* in the Northeast. A common species historically, *C. novemnotata* was collected infrequently from 1973–1985 as C7 spread, and only five times after 1985 in spite of extensive searches (134). In eastern South Dakota, populations of two native species (*Adalia bipunctata*, *C. transversoguttata*) declined from historical levels as C7 invaded, but other species appeared unaffected (32). Population growth of C7 did not increase overall coccinellid densities, which suggests that aphid control did not increase (32, 90). Because *A. bipunctata* is better adapted to woodland, negative impacts of C7 on this species are likely to be limited to areas with little forest cover; impacts of C7 on *C. transversoguttata* are more likely to be regional because of broad overlap in habitat and ecological traits (90). Arrival of C7 did not appear to increase competition for prey among coccinellids in Utah because adult body sizes of five native species did not decline (34). Interaction between larvae of C7 and the native *Coleomegilla maculata* in laboratory arenas led to lower survival of *C. maculata*, but only at low prey densities, reflecting either competition for prey or intraguild predation (92). In contrast, no negative interactions were detected between these species in field cages (91). The possibility of indirect impacts by C7 on other biological control agents was investigated in alfalfa (35). Adding C7 to caged field plots reduced numbers of alfalfa weevil larvae slightly, but it also greatly decreased parasitism of the weevil by an introduced ichneumonid parasitoid.

IMPLICATIONS This case study illustrates potential consequences of using predators with relatively broad diets. In addition to C7, five other alien coccinellids established in the United States raise similar concerns over nontarget effects (21). For example, *Harmonia axyridis* can displace competitors, including C7, from some habitats (13, 17). Gaps in knowledge of coccinellid ecology greatly limit our ability to predict the outcome of such complex interactions (90). Ongoing coccinellid invasions present opportunities to measure effects before and after agents arrive in new areas (17). The lack of published studies of nontarget impacts of introduced coccinellids on noneconomic insect species suggests another important avenue for future research.

Parasitoid Infiltration of a Native Food Web in Hawai'i

INTRODUCTION At least 84 parasitoids of lepidopteran pests have been released in Hawai'i, and 32 became established (39, 53). These parasitoids have been suspected of having severe impacts on native moth populations (40, 138). Although targeted pests were concentrated in agricultural areas at lower elevations (<1000 m), naturalists have worried that biocontrol agents may invade native habitats at higher elevation (60). Henneman & Memmott (55) recently advocated and used a novel food web approach to quantify the penetration of exotic parasitoids, including biological control agents, into the lepidopteran assemblage in a remote native Hawaiian forest.

EVALUATION Most biocontrol parasitoids against lepidopteran pests in Hawai'i (61 out of 84 species) were released prior to 1960 (53), and they were not screened for host specificity (39). At the time, broad host range was considered advantageous in a parasitoid because it allowed attack on multiple pests and persistence on alternative hosts when targeted pests were rare. Since 1960, the trend in Hawai'i has been to release more specialized parasitoids (39).

NONTARGET EFFECTS In the high, remote Alaka'i Swamp (1200 m elevation) on the island of Kaua'i, Henneman & Memmott (55) collected leaf-feeding caterpillars from all plant species in two replicate 0.5 ha plots in the native forest over two summer seasons. Moths or parasitoids were reared to adulthood and identified. The collections ($n = 2112$ larvae) contained 58 moth species (93% native) from 60 plant species (85% native). Out of 216 individual parasitoids reared, 83% belonged to one of three species of biocontrol agents, 14% were accidental immigrants (five spp.), and only 3% were native (five spp.). Two braconid parasitoids (*Meteorus laphygmae*, *Cotesia marginiventris*), introduced for biological control in 1942, dominated the guild. They were reared from several native moth species in six and three families. A third biocontrol agent (*Eriborus sinicus*: Ichneumonidae) was reared from three native tortricoid species. Although several of these collections represent new host records, all three agents were known to attack native Lepidoptera (39). Overall, parasitism based on the emergence of adult parasitoids was approximately 10% each year, but parasitism by biocontrol agents reached 28% in some native species. Attack rate on geometrids, determined by dissecting larvae, was 22.6%, nearly twice the 11.6% alien parasitoid emergence rate (M.L. Henneman, personal communication). The potential also exists for indirect effects, via competition with native parasitoids. The native parasitoids were rare. Four species were reared only once from native hosts (M.L. Henneman, personal communication). However, two of five species shared their lepidopteran host species with a biocontrol agent, and three of five shared their host with an accidentally introduced parasitoid (55).

IMPLICATIONS This study provides some support for the hypothesis that introduced biocontrol agents contributed to the reduction of native Lepidoptera in

Hawai'i, while documenting agent spread from targeted agricultural habitats into native habitats at higher elevations. The measured rates of parasitism (10% average, up to 28%) may substantially underestimate nontarget impacts because some parasitoids killed their host but died before emerging (55). The population effect of these parasitism rates is unknown. The data are insufficient to determine impact on a species-by-species basis. Furthermore, interaction intensities may have changed since the time of introduction and dispersal. Finally, this case demonstrates the insights on community-level impacts of biological control agents available from a food web approach.

EMERGENT PATTERNS

Although the available data are insufficient for a quantitative meta-analysis of specific hypotheses, some clear consistencies emerged from our comparative review of the most intensively quantified cases of nontarget effects reported in weed and insect biocontrol.

1. **Native species most closely related to the targeted species are most likely to be attacked.** Nontarget feeding among our case histories in weed biocontrol was confined to plants in the same tribe or genus as the targeted weed, supporting previous findings (99). Among our case histories in insect biocontrol, nontarget attack occurred over a broader taxonomic range, from subfamily to order (Table 1). A contributing factor is that host specificity and restricted host range have not been the norm until recently in insect biocontrol. Relatedness to a target nevertheless appeared to be a good indicator of potential risk to nontarget insects in most cases. For example, host-specificity tests of weevil parasitoids (*Microctonus* spp.) in New Zealand were consistent at the subfamily level with host utilization realized in the field.
2. **Host-specificity testing determines physiological host range but not ecological range.** Ecologists have been concerned about the power of prerelease evaluations of host specificity to predict nontarget effects (4, 61, 114, 121). Our review supports the current paradigm that field surveys of hosts in the native range plus host-specificity tests provide crucial information on physiological host range, particularly when related native species are included in the testing protocol (Table 1). However, lack of complete larval development on natives (82), considered definitive evidence that the probability of ecological impact is low (81, 83, 108), fell short of predicting host use and impact of *L. planus* in the field (75). Evaluation of larval development necessarily reflects the conditions under which it is done; and, in the case of *L. planus*, test conditions apparently did not simulate field environmental conditions that subsequently allowed complete development of *L. planus* on a native host plant. However, in the case of *P. virginiensis*, while host-specificity tests indicated inclusion in the physiological range of *Cotesia* spp., nonoverlapping habitat use by the native butterfly and the introduced parasitoids

TABLE 1 Synthesis of patterns and information across case histories

Characteristic	Weed control projects			Insect control projects						
	Rhinocyllus	*Larinus*	*Cactoblastis*	*Microctonus*	*Tephritids*	*Compsilura*	*Nezara*	*Pieris*	*Coccinella* (C7)	HI Foodweb
Control of target: s = substantial, p = partial, n = negligible	p	n	s	p	p	p	p	p	n	p
Evidence of population-level nontarget effects	x	x	x			x			x	x
Host-specificity testing										
Prerelease	x	x	x	x						
Concurrent with releases	x					x	x			
In retrospect	x	x	x	x	x	x	x	x		
Not done									x	x
Native species included in prerelease tests										
Tested intensively		x (late)								
Some/partial testing	x	x (early)		x						
Indirect effects potential also evaluated										
Results of host-specificity tests with native species										
No evidence of acceptance of native species					x					
Small amount of acceptance (no-choice only)				x						
Significant acceptance (choice test use too)	x	x					x	x		
Evidence of larval development on native	x	x	x	x			x	x		

Evidence of nontarget feeding/attack:										
Consistent with natural history in native range	x	x	x	x	x	x	x	x	x	x
Consistent with host-specificity data	x	x	x	x	x		x			
Relative amount/impact greater than expected	x	x							x	
No-choice test informative on nontarget use	x	x	x	x	x		x			
Ecological factors in nontarget host use										
Habitat not restricted to agronomic release habitat	x	x	x	x	x	x	x		x	x
Attack or impact varies with habitat	x			x			x	x	x	
Elevational refuge from attack for some species	x	x		x		x	x			
Phenological synchrony of agent and nontarget	x	x		x				x		
Degree of relatedness of nontarget to target	tribe	tribe	genus	subfam.	family	2 orders	superfam.	family	class	suborder
Low host density did not prevent attack	x	x	x	x		x				x
Diet breadth:										
Stenophagous (one genus)			x					x (*C.r.*)		
Oligophagous (multiple genera in one family)	x	x		x	x					
Polyphagous (multiple families, or more)						x	x	x (*C.g.*)	x	x

reduced ecological risk. The case histories illustrate that a variety of factors can influence field host use (Table 1): phenological synchrony (*R. conicus*, *L. planus*), host and agent dispersal (*R. conicus*, stink bug parasitoids, pierid parasitoids), habitat type (*Microctonus* spp., pierid parasitoids), life history variation (pierids), and learning (tephritid parasitoids). Thus, although host-specificity tests in general accurately identified potential host range, insect preference based on those tests failed to predict the magnitude of nontarget risk to native host species in the field.

3. **Prediction of ecological consequences of nontarget attack requires more types of studies.** The case histories suggest that quantification of potential nontarget risk will require better information for agents on the determinants of (*a*) host finding and choice, (*b*) dispersal and dispersal limitation, (*c*) use of alternate host resources in the field, (*d*) insect population growth and stability, (*e*) insect impact on plant population growth and density, and (*f*) environmental effects on the intensity of the interactions. Host finding and host choice often were influenced by environmental conditions that determined the strength of actual versus potential interactions. Information on the conditions under which alternative hosts are used would provide important information relevant for prediction of potential nontarget use in the field. For example for *L. planus*, the earlier flowering phenology of Tracy's thistle, compared to its targeted Canada thistle, at high elevation in Colorado apparently overrode any preference for Canada thistle because the availability of Tracy's thistle flower heads was synchronized better with the weevil's oviposition period. A lack of synchrony in flowering with *R. conicus*'s oviposition periods also may help explain why many native North American *Cirsium* thistles are not significant hosts of the weevil (100). Degree of overlap of host and parasitoid populations helps explain differences in *Cotesia* spp. effects on native pierids (9). More data on the ecological parameters influencing host use and insect population growth, for example from the indigenous region, should provide a stronger basis for evaluation of the probability of major nontarget impacts on native species accepted as secondary hosts in testing. Currently, estimation of this probability largely depends on expert opinion.
4. **Documented impacts on nontarget species vary from negligible to devastating.** The cases we reviewed, those with the most quantitative data, illustrate the range of effects discovered to date. In some cases, the impacts appear minimal, occurring at levels that suggest no effect on host densities in the long run. For example, fruit fly parasitoids intentionally introduced into Hawai'i appear to be insignificant sources of mortality for the native tephritid *T. dubautiae*. On the other hand, impacts on nontargets can be severe, depressing populations of sparse regional plant species (*R. conicus* on Platte thistle) or enhancing extinction risk of a rare species (*C. cactorum* on *O. corallicola* in Florida). Major impacts were associated with redistribution of inadvertently established insects (e.g., *L. planus* for Canada thistle control). Some studies (*R. conicus* on thistles, *Compsilura* on silk moths)

measured attacks on nontarget species reaching levels that could cause long-term population declines and even extirpation from local communities. The only case with quantitative preinvasion data (*R. conicus*) demonstrates that such interactions can cause a severe decline in a native population within its characteristic habitat. In general, however, nontarget effects have been evaluated only after the fact or measured over too short a period to confidently predict the outcomes over time. Long-term monitoring and life table analyses may improve our assessment of the consequences of nontarget use. However, studies conducted decades after introduction, e.g., Hawaiian parasitoids in native forest, will not provide evidence of severe population dynamic effects if those effects occurred near the onset of the invasion into native systems.

5. **Data on magnitude of nontarget impact is sparse.** Use of nontarget species by introduced biocontrol organisms is not by itself evidence of an adverse impact upon a host population. We found that nontarget feeding clearly added to mortality for nontarget insects and decreased growth or reproductive success of nontarget plants. In most cases, however, few of the studies included data with which to estimate the magnitude of the effect of the documented nontarget attack on the population density and growth rate of the native species. A clear exception was the *R. conicus* case, where "before" and "after" data on Platte thistle's vital rates allow estimation of the impact on population parameters. Other cases suggested that life table analyses provide important information with which to initiate assessment of population consequences, e.g., parasitoid impact on the koa bug or on native pierid butterflies. The cases also suggest that population response data are needed to anticipate the quantitative effects of feeding in the field when there is laboratory evidence that native species are within the potential host range (41). Population response data include population growth rates, substitutability among mortality factors, spatial scale of population interactions, variation of life history strategies, compensatory responses, and facultative increases in reproductive rates. Methods by which such information can be acquired include carefully designed experiments in large field cages in which host plant phenology and other environmentally driven traits are varied experimentally, quantitative studies of variable host plant use and impact in the indigenous region (110), and demographic modeling, such as that used in retrospective studies of the tephritid parasitoids in Hawai'i. These data could be used to prioritize candidate agents by maximal likely effectiveness as well as minimal nontarget risk.
6. **Nontarget attack on rare native species can accelerate their decline and enhance their risk of extinction.** High population levels developed by biological control agents may lead to unexpected pressure on uncommon nontarget hosts. Theory indicates that a natural enemy that maintains a high population on its targeted host can extirpate its less common nontarget host species (12, 58). The cases of *Cactoblastis cactorum* and tachinid parasitoids of *Nezara viridula* suggest the potential for this kind of interaction. This risk

is expected to increase if a biocontrol agent causes only a minor reduction in the fitness of its target because then both target and agent can persist at relatively high levels while putting pressure on a rarer nontarget species (94). As an example, *Larinus planus*, which is not a narrow specialist among thistles, is having a minor impact on its targeted host but a major impact on seed production of Tracy's thistle (75). These results support the theoretical arguments for choosing narrow specialists with high virulence against their target as biocontrol agents and rejecting those expected to have relatively minor impacts (84, 85).

7. **Nontarget effects of biocontrol species can be indirect, as well as direct, via food webs and cross-linkages.** Indirect effects are those mediated through effects on another species, such as by competitive resource depletion and by shared hosts or natural enemies. A recent example is the indirect effect of *Urophora* spp., released for knapweed control; it appears to alter plant competitive hierarchies (15), while also augmenting deer mouse populations that subsequently affect the native vegetation (95). Host-specificity testing is designed to detect direct interactions, leaving other less obvious effects to go unevaluated. Some of the case histories demonstrate or suggest ripple effects via indirect interactions. For example, behavior and numbers of the native tephritid fly (*Paracantha culta*), which depends on Platte thistle flower heads in its first generation in sand prairie, have been affected by *R. conicus* [(72); S.M. Louda et al., unpublished data]. Among the insect cases, the effect of exotic introduced coccinellids on native species provides an example of potential indirect effects via competition for shared aphid resources as well as potential direct effects via intraguild predation.
8. **Biocontrol agents can infiltrate natural areas away from targeted agroecosystems.** The case studies legitimize the concern of conservation biologists over the ability of biocontrol agents to disperse into native habitats. The study of Henneman & Memmott (55) provides a clear example of the extent of such penetration in a remote native habitat by introduced parasitoids. Similarly, the cases of *R. conicus* on Platte thistle, *C. cactorum* on the rare *O. corallicola*, *M. aethiopoides* on weevils, and *T. pilipes* on koa bug illustrate that biocontrol agents are able to disperse, locate, and reproduce on nontarget hosts at considerable distances from concentrations of their targeted host. Estimation of likely natural dispersal, for example based on habitat range within the indigenous environment, would improve assessment of ecological risk of biological control.
9. **Biocontrol agents have the potential to perturb whole guilds and assemblages of nontarget organisms.** Among our case studies, the most obvious mechanism for community impact appears to be extreme polyphagy. For example, *C. concinnata* in the northeastern United States and generalist parasitoids in Hawai'i are implicated in declines in nontarget lepidopteran populations. Also, the *R. conicus* case demonstrates effects on the feeding

guild into which it penetrated. In most cases, however, without better data on food webs and population sizes prior to nontarget use, the magnitude of the perturbation for interconnected species remains unknown. The *C. cactorum* case suggests that a relatively specialized (genus-level) feeder also can have far-reaching ecological impacts, in this case by its threat to the diverse assemblage of related native North American *Opuntia* spp. and their interacting dependent species. Another possible mechanism for community-level effects involves nontarget impact on a keystone species (105). In theory, this could lead to a cascade of detrimental effects on associated native species (113, 114). Our case studies provided no evidence of keystone effects, reflecting the fact that little information exists on the ecological roles of most native insect species.

10. **Evidence on adaptive change is missing in these case histories.** Although exotic organisms have the potential of evolving to utilize new hosts where introduced (46–48, 113, 117), in none of the cases we reviewed were such adaptations studied directly. The evidence available in these case histories suggests that most host shifts involved preadaptation, rather than newly evolved ability, to utilize nontarget hosts. However, because the population genetic evidence necessary to determine if adaptive change is occurring was absent, this issue presents an opportunity for important future research.

RECOMMENDATIONS BASED ON THE CASE HISTORIES REVIEWED

1. **Avoid the use of exotic generalist predators and parasitoids.** Generalists, such as *Coccinella septempunctata* and *Compsilura concinnata*, utilize a greater number of nontarget species and therefore carry a greater chance of direct and indirect nontarget effects. If the goal is to maximize the predictive power of prerelease studies of candidate agents, then assessment of potential nontarget impacts is greatly simplified by selecting agents with narrow host range.
2. **Expand host-specificity testing.** The case histories provide strong support for the suggestion that host-specificity testing is highly informative on physiological host range, if the tests include potential hosts. This was true for insect cases in which such data were gathered, as well as for the weed cases, supporting the increasing use of such tests in insect control programs. Measures of host specificity across an array of potential hosts that were indicative that nontarget use was possible or likely include feeding preference with no-choice (starvation) tests, in addition to choice tests; oviposition under choice and no-choice conditions; and subsequent larval development and eclosion on an ecologically based set of potential hosts. Because host-specificity preference tests, especially choice tests, did not predict actual host impact in many

cases, these tests need to be supplemented with information on ecological range.

3. **Utilize more ecological information to increase precision of risk assessment for potential host species.** Ecological parameters were significant in many cases in determining the outcome of interactions of introduced agents with potential nontarget species in the field. Thus, the influence of ecological factors on potential nontarget effects needs to be better evaluated for environmental risk assessment. Prerelease field research on candidate biological control agents within their native ranges could provide additional insight into realized host breadth under a range of environmental conditions. Climate matching of the recipient system and the native range may help predict dispersal and potential geographic spread, as was the case for *C. cactorum.* Furthermore, because the cases illustrate that natural dispersal and spread of natural enemies into climatically suitable habitats is likely, more information on habitat preference, host-finding behavior, and limiting conditions for growth would enable better assessment of potential nontarget host attack and spread.
4. **Incorporate population-level measurements of ecological risk.** A consistent theme in the case histories was that environmental conditions influence host utilization and population impact, sometimes contradicting predictions based on host-preference testing. Habitat, phenology, and elevation were critically important predictors of ecological host range and use of native hosts in several cases. Thus, our review suggests that experimental and field research on parameters underlying population growth rates and factors influencing interaction strengths under field conditions would improve prediction of population response when host-specificity tests indicate that native species are potential hosts.
5. **Add ecological risk criteria to target selection.** As pointed out elsewhere (69, 71, 77, 84, 85, 99, 100, 123), not all pest species are appropriate targets for biological control. Some may be associated with high agricultural or ecological risks. For example, targeting weeds with close native relatives, or targeting native weeds with dependent food webs of native species, sharply increases ecological risk. Redistribution of accidentally established insects for biocontrol also appears risky. Furthermore, not all pest populations are likely to be controlled by natural enemies. The possibility that specialist biological control agents that are both effective and safe may not exist needs to be incorporated into target selection and program planning.
6. **Prioritize host-specific agents according to their predicted effectiveness.** It is striking that the pressure assumed to be exerted by the introduced insect often appeared ineffective or only partly effective in providing control (Table 1). Given the accumulating evidence that unanticipated ecological side effects are possible, the benefits to be gained need better quantification. The case histories provide support for further development of protocols, such as

impact studies in candidate agents' native ranges and in large field cages, to evaluate potential effectiveness of agents as well as their potential ecological effects. Such studies would allow for the choice of the most effective agents, allowing intensive screening for those least likely to impose measurable nontarget effects.

ACKNOWLEDGMENTS

The authors would like to thank all their collaborators over the years, whose energy and camaraderie made the science fun, and all of the funding agencies that made the original studies possible. We are also grateful to G.H. Boettner, T.D. Center, E.M. Coombs, M.L. Henneman, J.C. Herr, L.G. Higley, W.W. Hoback, R.A. Huffbauer, A. Joern, A.S. McClay, T.A. Rand, D. Schröder, U. Schaffner, and R.G. Van Driesche for helpful comments on the original manuscript.

The *Annual Review of Entomology* is online at http://ento.annualreviews.org

LITERATURE CITED

1. Andres LA. 1985. Interactions of *Chrysolina quadrigemina* and *Hypericum* spp. in California. See Ref. 22, pp. 235–39
2. Angalet GW, Tropp JM, Eggert AN. 1979. *Coccinella septempunctata* in the United States: recolonizations and notes on its ecology. *Environ. Entomol.* 8:896–901
3. Arnaud PH. 1978. *A Host-Parasite Catalog of North American Tachinidae (Diptera)*. Washington, DC: U.S. Dep. Agric. Misc. Publ. No. 1319. 860 pp.
4. Arnett AE, Louda SM. 2002. Re-test of *Rhinocyllus conicus* host specificity, and the prediction of ecological risk in biological control. *Biol. Conserv.* 106:151–57
5. Barratt BIP, Evans AA, Ferguson CM, Barker GM, McNeill MR, Phillips CB. 1997. Laboratory nontarget host range of the introduced parasitoids *Microctonus aethiopoides* and *M. hyperodae* (Hymenoptera: Braconidae) compared with field parasitism in New Zealand. *Environ. Entomol.* 26(3):694–702
6. Barratt BIP, Ferguson CM, Goldson SL, Phillips CM, Hannah DJ. 1999. Predicting the risk from biological control agent introductions: a New Zealand approach. See Ref. 36, pp. 59–75
7. Barratt BIP, Ferguson CM, McNeill MR, Goldson SL. 1999. Parasitoid host specificity testing to predict host range. In *Recommendations for Host Specificity Testing Procedures in Australasia—Towards Improved Assays for Biological Control Agents*, ed. TM Withers, L Barton-Browne, JN Stanley, pp. 70–83. Brisbane, Austr.: CRC Trop. Pest Manag.
8. Bennett FD, Habeck DH. 1995. *Cactoblastis cactorum*: a successful weed control agent in the Caribbean, now a pest in Florida. *Proc. 8th Int. Symp. Biol. Control Weeds, 2–7 February 1992*, ed. ES Delfosse, RR Scott, pp. 21–6. Canterbury, N. Z.: Lincoln Univ.
9. Benson J, Pasqualte A, Van Driesche RG, Elkinton J. 2002. Assessment of risk posed by introduced braconid wasps to *Pieris virginiensis*, a native woodland butterfly in New England. *Biol. Control.* In press
10. Bess HA, van den Bosch R, Haramoto FH. 1961. Fruit fly parasites and their

activities in Hawaii. *Proc. Hawaii. Entomol. Soc.* 17:367–78

11. Boettner GH, Elkinton JS, Boettner CJ. 2000. Effects of a biological control introduction on three nontarget native species of Saturniid moths. *Conserv. Biol.* 14:1798–806
12. Bowers RG, Begon M. 1991. A host-host pathogen model with free-living infective stages, applicable to microbial pest control. *J. Theor. Biol.* 148:303–29
13. Brown MW, Miller SS. 1998. Coccinellidae (Coleoptera) in apple orchards of eastern West Virginia and the impact of invasion by *Harmonia axyridis*. *Entomol. News* 109:136–42
14. Burger JC, Louda SM. 1994. Indirect vs. direct effects of grasses on growth of a cactus (*Opuntia fragilis*): insect herbivory vs. competition. *Oecologia* 99:79–87
15. Callaway RM, De Luca TH, Belliveau WM. 1999. Biological-control herbivores increase competitive ability of the noxious weed *Centaurea maculosa*. *Ecology* 80:1196–201
16. Clausen CP. 1978. *Introduced Parasites and Predators of Arthropod Pests and Weeds: A World Review*. Washington, DC: Supt. Doc., U.S. Gov. Print. Off.
17. Colunga-Garcia M, Gage SH. 1998. Arrival, establishment, and habitat use of the multicolored Asian lady beetle (Coleoptera: Coccinellidae) in a Michigan landscape. *Environ. Entomol.* 27:1574–80
18. Cresswell AS. 1999. *Aspects of target and non-target host selection by the parasitoid* Microctonus aethiopoides *Loan (Hymenoptera: Braconidae)*. M.S. thesis, Univ. Otago, Dunedin, New Zealand. 126 pp.
19. Culver JJ. 1919. A study of *Compsilura concinnata*, an imported tachinid parasite of the gipsy moth and brown-tail moth. *Bull. 766*. Washington, DC: U.S. Dep. Agric.
20. Davis CJ. 1964. The introduction, propagation, liberation, and establishment of parasites to control *Nezara viridula* variety *smaragdula* (Fabricius) in Hawaii (Heteroptera: Pentatomidae). *Proc. Hawaii. Entomol. Soc.* 18:369–75
21. Day WH, Prokrym DR, Ellis DR, Chianese RJ. 1994. The known distribution of the predator *Propylea quatuordecimpunctata* (Coleoptera: Coccinellidae) in the United States, and thoughts on the origin of this species and five other exotic lady beetles in eastern North America. *Entomol. News* 105:244–56
22. Delfosse ES, ed. 1985. *Proceedings of the 6th International Symposium on the Biological Control of Weeds, 19–25 August 1984*. Ottawa: Can.: Agric. Can.
23. Dodd AP. 1940. *The Biological Campaign Against Prickly Pear*. Brisbane, Aust.: Commonw. Prickly Pear Board. 177 pp.
24. Duan JJ, Messing RH. 1998. Effect of *Tetrastichus giffardianus* (Hymenoptera: Eulophidae) on non-target flowerhead-feeding tephritids (Diptera: Tephritidae). *Environ. Entomol.* 27:1022–28
25. Duan JJ, Messing RH. 1999. Effects of origin and experience on patterns of host acceptance by the Opiine parasitoid *Diachasmimorpha tryoni*. *Ecol. Entomol.* 24:284–91
26. Duan JJ, Messing RH. 1999. Evaluating nontarget effects of classical biological control: fruit fly parasitoids in Hawaii as a case study. See Ref. 36, pp. 95–109
27. Duan JJ, Messing RH. 2000. Effect of *Diachasmimorpha tryoni* on two non-target flowerhead-feeding tephritids. *BioControl* 45:113–25
28. Duan JJ, Messing RH. 2000. Host specificity tests of *Diachasmimorpha kraussii* (Hymenoptera: Braconidae), a newly introduced opiine fly parasitoid with four non-target tephritids in Hawaii. *Biol. Control* 19:28–34
29. Duan JJ, Purcell MF, Messing RH. 1996. Parasitoids of non-target tephritid flies in Hawaii: implications for biological control of fruit fly pests. *Entomophaga* 41:245–56
30. Duan JJ, Purcell MF, Messing RH. 1998.

Association of the Opiine parasitoid *Diachasmimorpha tryoni* (Hymenoptera: Braconidae) with the lantana gall fly (Diptera: Tephritidae) on Kauai. *Environ. Entomol.* 27:419–26
31. Ehler LE. 1999. Critical issues related to nontarget effects in classical biological control of insects. See Ref. 36, pp. 3–13
32. Elliott NC, Kieckhefer RW, Kauffman WC. 1996. Effects of an invading coccinellid on native coccinellids in an agricultural landscape. *Oecologia* 105:537–44
33. Evans EW. 1991. Intra versus interspecific interactions among lady beetles (Coleoptera: Coccinellidae) attacking aphids. *Oecologia* 87:401–8
34. Evans EW. 2000. Morphology of invasion: body size patterns associated with the establishment of *Coccinella septempunctata* in western North America. *Eur. J. Entomol.* 97:469–74
35. Evans EW, England S. 1996. Indirect interactions in biological control of insects: pests and natural enemies in alfalfa. *Ecol. Appl.* 920–30
36. Follett P, Duan J. 1999. *Nontarget Effects of Biological Control.* Dortrecht/Boston/London: Kluwer. 316 pp.
37. Follett PA, Duan JJ, Messing RH, Jones VP. 2000. Parasitoid drift after biological control introductions: re-examining Pandora's box. *Am. Entomol.* 46:82–94
38. Frank JH. 1998. How risky is biological control? Comment. *Ecology* 79:1829–34
39. Funasaki GY, Lai P-Y, Nakahara LM, Beardsley JW, Ota AK. 1988. A review of biological control introductions in Hawaii: 1890–1985. *Proc. Hawaii. Entomol. Soc.* 28:105–60
40. Gagne WC, Howarth FG. 1985. Conservation status of endemic Hawaiian Lepidoptera. *Proc. 3rd Congr. Eur. Lepidopterol., Cambridge UK*, ed. J Heath, pp. 74–84. Karlsruhe, Germany: Soc. Eur. Lepidopterogica
41. Gassmann A, Louda SM. 2001. *Rhinocyllus conicus*: initial evaluation and subsequent ecological impacts in North America. See Ref. 129, pp. 147–83
42. Goeden RD, Ricker DW. 1987. Phytophagous insect faunas of native *Cirsium* thistles, *C. mohavense*, *C. neomexicanum*, and *C. nidulum*, in the Mojave Desert of southern California. *Ann. Entomol. Soc. Am.* 80:161–75
43. Goldson SL, McNeill MR, Phillips CB, Barratt BIP. 1992. Host specificity testing and suitability of the parasitoid *Microctonus aethiopoides* (Hym: Braconidae) as a biological control agent of *Listronotus bonariensis* (Col.: Curculionidae) in New Zealand. *Entomophaga* 37:483–98
44. Goldson SL, McNeill MR, Proffitt GM, Barker GM, Addison PJ, et al. 1993. Systematic mass rearing and release of *Microctonus hyperodae* (Hym.: Braconidae, Euphorinae), a parasitoid of the Argentine stem weevil *Listronotus bonariensis* (Col.: Curculionidae) and records of its establishment in New Zealand. *Entomophaga* 38:1–10
45. Gordon RD, Vandenberg N. 1991. Field guide to recently introduced species of Coccinellidae (Coleoptera) in North America, with a revised key to North American genera of Coccellini. *Proc. Entomol. Soc. Wash.* 93:845–64
46. Gould F. 1979. Rapid host range evolution in a population of the phytophagous mite *Tetranychus urticae* Koch. *Evolution* 33:791–802
47. Gould F. 1988. Genetics of pairwise and multispecies plant-herbivore coevolution. In *Chemical Mediation of Coevolution*, ed. KC Spencer, pp. 13–55. New York: Academic
48. Gould F. 1991. The evolutionary potential of crop pests. *Am. Sci.* 79:496–507
49. Greathead DJ, Greathead AH. 1992. Biological control of insect pests by insect parasitoids and predators: the BIOCAT database. *Biocontrol News Inform.* 13:61–68
50. Hall RW, Ehler LE. 1979. Rate of establishment of natural enemies in classical

biological control. *Bull. Entomol. Soc. Am.* 25:280–82

51. Hall S, Sullivan P, Schweitzer DF. 2000. *Assessment of Risk to Non-Target Macro-Moths after Btk Application to Asian Gypsy Moth in the Cape Fear Region of North Carolina.* Washington, DC: U.S. Dep. Agric.
52. Harris VE, Todd JW. 1980. Male-mediated aggregation of male, female, and 5th instar southern green stink bugs and concomitant attraction of a tachinid parasite, *Trichopoda pennipes. Entomol. Exp. Appl.* 27:117–26
53. Hawaii Department of Agriculture. 1983. *List of biological control introductions in Hawaii.* Hawaii Dep. Agric., Plant Quar. Off., Honolulu, HI. Unpublished report
54. Hawkins BA, Marino PC. 1997. The colonization of native phytophagous insects in North America by exotic parasitoids. *Oecologia* 112:566–71
55. Henneman ML, Memmott J. 2001. Infiltration of a Hawaiian community by introduced biological control agents. *Science* 293:1314–16
56. Hodek I, Honek A. 1996. *Ecology of Coccinellidae.* Dordrecht, The Netherlands: Kluwer
57. Holt RD, Hochberg ME. 2001. Indirect interactions, community modules and biological control: a theoretical perspective. See Ref. 129, pp. 13–37
58. Holt RD, Pickering J. 1985. Infectious disease and species coexistence: a model of Lotka-Volterra form. *Am. Nat.* 126:196–211
59. Horn D. 1991. Potential impact of *Coccinella septempunctata* on endangered Lycaenidae (Lepidoptera) in northwestern Ohio, USA. In *Behavior and Impact of Aphidophaga*, ed. L Polgar, RJ Chambers, AFG Dixon, I Hodek, pp. 159–62. The Hague: SPB Acad. Publ.
60. Howarth FG. 1983. Classical biocontrol: panacea or Pandora's box? *Proc. Hawaii. Entomol. Soc.* 24:239–44
61. Howarth FG. 1991. Environmental impacts of classical biological control. *Annu. Rev. Entomol.* 36:485–509
62. Huffaker CB. 1964. Fundamentals of biological weed control. In *Control of Insect Pests and Weeds*, ed. P DeBach, E Schlinger, pp. 74–117. London, UK: Chapman & Hall
63. Hunter CD. 1997. *Suppliers of Beneficial Organisms in North America.* Sacramento, CA: Calif. Environ. Prot. Agency
64. Johnson DM, Stiling PD. 1996. Host specificity of *Cactoblastis cactorum* (Lepidoptera: Pyralidae), an exotic *Opuntia*-feeding moth, in Florida. *Environ. Entomol.* 25:743–48
65. Johnson DM, Stiling PD. 1998. Distribution and dispersal of *Cactoblastis cactorum* (Lepidoptera: Pyralidae), an exotic *Opuntia*-feeding moth, in Florida. *Fla. Entomol.* 81:12–22
66. Julien MH, Griffiths MW. 1998. *Biological Control of Weeds: A World Catalogue of Agents and Their Target Weeds.* Wallingford, UK: CAB Int.
67. Loan CC, Lloyd DC. 1974. Description and field biology of *Microctonus hyperodae* Loan n. sp. (Hymenoptera: Braconidae, Eurphorinae) a parasite of *Hyperodes bonariensis* in South America (Coleoptera: Curculionidae). *Entomophaga* 19:7–12
68. Louda SM. 1998. Population growth of *Rhinocyllus conicus* (Coleoptera: Curculionidae) on two species of native thistles in prairie. *Environ. Entomol.* 27:834–41
69. Louda SM. 1999. Ecology of interactions needed in biological control practice and policy. *Bull. Br. Ecol. Soc.* 29:8–11
70. Louda SM. 1999. Negative ecological effects of the musk thistle biocontrol agent, *Rhinocyllus conicus* Fröl. See Ref. 36, pp. 215–43
71. Louda SM. 2000. *Rhinocyllus conicus*—insights to improve predictability and minimize risk of biological control of weeds. See Ref. 118, pp. 187–93
72. Louda SM, Arnett AE. 2000. Predicting

non-target ecological effects of biological control agents: evidence from *Rhinocyllus conicus*. See Ref. 118, pp. 551–67
73. Louda SM, Arnett AE, Rand TA, Russell FL. 2003. Fighting fire with fire: invasiveness of some biological control insects challenges adequacy of ecological risk assessment and regulation. *Conserv. Biol.* 17: In press
74. Louda SM, Kendall D, Connor J, Simberloff D. 1997. Ecological effects of an insect introduced for the biological control of weeds. *Science* 277:1088–90
75. Louda SM, O'Brien CW. 2002. Unexpected ecological effects of distributing the exotic weevil, *Larinus planus* (F.), for the biological control of Canada thistle. *Conserv. Biol.* 16:717–27
76. Louda SM, Potvin MA. 1995. Effect of inflorescence-feeding insects in the demography and lifetime fitness of a native plant. *Ecology* 76:229–45
77. Louda SM, Simberloff D, Boettner G, Connor J, Kendall D, Arnett AE. 1998. Insights from data on the nontarget effects of the flowerhead weevil. *Biocontrol News Inform.* 26:70N–71N
78. Lynch LD, Hokkanen HMT, Babendreier D, Bigler F, Burgio G, et al. 2001. Insect biological control and non-target effects: a European perspective. See Ref. 129, pp. 99–125
79. Lynch LD, Thomas MB. 2000. Nontarget effects in the biological control of insects with insects, nematodes and microbial agents: the evidence. *Biocontrol News Inform.* 21:117–30
80. MacFadyen RE. 1998. Biological control of weeds. *Annu. Rev. Entomol.* 43:369–93
81. Marohasy J. 1998. The design and interpretation of host-specificity tests for weed biological control with particular reference to insect behavior. *Biocontrol News Inform.* 19:13N–20N
82. McClay AS. 1990. The potential of *Larinus planus* (Coleoptera: Curculionidae), an accidentally-introduced insect in North America, for biological control of *Cirsium arvense*. *Proc. 7th Int. Symp. Biol. Control Weeds, 6–11 March 1988*, ed. ES Delfosse, pp. 173–79. Rome, Italy: Inst. Sper. Patol. Veg.
83. McEvoy PB. 1996. Host specificity and biological pest control. *BioScience* 46:401–5
84. McEvoy PB, Coombs EM. 1999. A parsimonious approach to biological control of plant invaders. *Ecol. Appl.* 9:387–401
85. McEvoy PB, Coombs EM. 1999. Why things bite back: unintended consequences of biological weed control. See Ref. 36, pp. 167–94
86. Miller M, Aplet G. 1993. Biological control: a little knowledge is a dangerous thing. *Rutgers Law Rev.* 45:285–334
87. Nalepa CA, Ahlstrom KR, Nault BA, Williams JL. 1998. Mass appearance of lady beetles (Coleoptera: Coccinellidae) on North Carolina beaches. *Entomol. News* 109:277–81
88. Newell IM, Haramoto FH. 1968. Biotic factors influencing populations of *Dacus dorsalis* in Hawaii. *Proc. Hawaii. Entomol. Soc.* 20:81–139
89. Obrycki JJ, Bailey WC, Stoltenow CR, Puttler B, Carlson CE. 1987. Recovery of the sevenspotted lady beetle *Coccinella septempunctata* (Coleoptera: Coccinellidae) in Iowa and Missouri. *J. Kans. Entomol. Soc.* 60:584–88
90. Obrycki JJ, Elliott NC, Giles LL. 1999. Coccinellid introductions: potential for evaluation of nontarget effects. See Ref. 36, pp. 127–45
91. Obrycki JJ, Giles KL, Ormond AM. 1998. Experimental assessment of interactions between larval *Coleomegilla maculata* and *Coccinella septempunctata* (Coleoptera: Coccinellidae) in field cages. *Environ. Entomol.* 27:1280–88
92. Obrycki JJ, Giles KL, Ormord AM. 1998. Interactions between an introduced and indigenous coccinellid species at different prey densities. *Oecologia* 117:279–85
93. Office of Technology Assessment,

Congress of the US. 1995. *Biologically Based Technologies for Pest Control.* Washington, DC: U.S. Gov. Printing Off.

94. Onstad DW, McManus ML. 1996. Risks of host range expansion by parasites of insects. *BioScience* 46:430–35
95. Pearson DE, McKelvey KS, Ruggiero LF. 2000. Non-target effects of an introduced biological control agent on deer mouse ecology. *Oecologia* 122:121–28
96. Pemberton RW. 1985. Native plant considerations in the biological control of leafy spurge. See Ref. 22, pp. 365–90
97. Pemberton RW. 1985. Native weeds as candidates for biological control research. See Ref. 22, pp. 869–77
98. Pemberton RW. 1995. *Cactoblastis cactorum* (Lepidoptera: Pyralidae) in the United States: an immigrant biological control agent or an introduction of the nursery industry? *Am. Entomol.* 41:230–32
99. Pemberton RW. 2000. Predictable risk to native plants in weed biological control. *Oecologia* 125:489–94
100. Pemberton RW. 2000. Safety data crucial for biological control insect agents. *Science* 290:1896–97
101. Prokrym DR, Pike KS, Nelson DJ. 1998. Biological control of *Diuraphis noxia* (Homoptera: Aphididae): implementation and evaluation of natural enemies. In *Response Model for an Introduced Pest: The Russian Wheat Aphid*, ed. SS Quisenberry, FB Peairs, pp. 183–208. Lanham, MD: ESA
102. Rees NE. 1977. Impact of *Rhinocyllus conicus* on thistles in southwestern Montana. *Environ. Entomol.* 6:839–42
103. Schaefer PW, Dysart RJ, Specht HB. 1987. North American distribution of *Coccinella septempunctata* (Coleoptera: Coccinellidae) and its mass appearance in coastal Delaware. *Environ. Entomol.* 16:368–73
104. Schaffner JV, Griswold CL. 1934. *Macrolepidoptera and Their Parasites Reared from Field Collections in the Northeastern Part of the United States.* Publ. 188. Washington, DC: U.S. Dep. Agric.
105. Schreiner IH, Nafus DM. 1992. Changes in a moth community mediated by biological control of the dominant species. *Environ. Entomol.* 21:664–68
106. Schweitzer DF. 1988. Status of Saturniidae in the northeastern USA: a quick review. *News Lepidopt. Soc.* 1:4–5
107. Scudder SH. 1889. *The Butterflies of the Eastern United States and Canada with Special Reference to New England.* Cambridge, MA: Privately published. 1775 pp.
108. Secord D, Kareiva P. 1996. Perils and pitfalls in the host specificity paradigm. *BioScience* 46:448–53
109. Shapiro AM. 1974. Butterflies and skippers of New York State. *Search Agric. (Entomol.)* 4:1, 20–1, 49
110. Sheppard AW, Woodburn T. 1996. Population regulation in insects used to control thistles: can this predict effectiveness? In *Frontiers of Population Ecology*, ed. RB Floyd, AW Sheppard, PJ De Barro, pp. 277–90. Melbourne, Aust.: CSIRO
111. Simberloff D. 1981. Community effects of introduced species. In *Biotic Crises in Ecological and Evolutionary Time*, ed. TH Nitecki, pp. 53–81. New York, NY: Academic
112. Simberloff D. 1991. Keystone species and community effects of biological introductions. In *Assessing Ecological Risks of Biotechnology*, ed. L Ginzburg, pp. 1–19. Boston: Butterworth-Heinemann
113. Simberloff D. 1992. Conservation of pristine habitats and unintended effects of biological control. In *Selection Criteria and Ecological Consequences of Importing Natural Enemies*, ed. WC Kauffman, JE Nechols, pp. 103–17. Lanham, MD: ESA
114. Simberloff D, Stiling P. 1996. How risky is biological control? *Ecology* 77:1965–74
115. Simberloff D, Stiling P. 1998. How risky is biological control? Reply. *Ecology* 79:1834–36

116. Simmonds FJ, Bennett FD. 1966. Biological control of *Opuntia* spp. by *Cactoblastis cactorum* in the Leeward Islands (West Indies). *Entomophaga* 11:183–89
117. Singer MC, Ng D, Vasco D, Thomas CD. 1992. Rapidly evolving associations among oviposition preferences fail to constrain evolution of insect diet. *Am. Nat.* 139:9–20
118. Spencer NR, ed. 2000. *Proceedings of the 10th International Symposium on the Biological Control of Weeds, 4–14 July 1999*. Bozeman, MT: Montana State Univ.
119. Stamp NE, Bowers MD. 1990. Parasitism of New England buckmoth caterpillars (*Hemileuca lucina*: Saturniidae) by tachinid flies. *J. Lepidopt. Soc.* 35:199–200
120. Sternberg JG, Waldbauer GP, Scarborough AG. 1981. Distribution of *Cecropia* moth (Saturniidae) in central Illinois: a study in urban ecology. *J. Lepidopt. Soc.* 35:304–20
121. Stiling P, Simberloff D. 1999. The frequency and strength of non-target effects of invertebrate biological control agents. See Ref. 36, pp. 31–43
122. Strong DR, Pemberton RW. 2000. Biological control of invading species—risk and reform. *Science* 288:169–79
123. Strong DR, Pemberton RW. 2001. Food webs, risks of alien enemies, and reform of biological control. See Ref. 129, pp. 57–79
124. Thomas MB, Willis AJ. 1998. Biocontrol—risky but necessary? *Trends Ecol. Evol.* 13:325–29
125. Turner CE. 1985. Conflicting interests and biological control of weeds. See Ref. 22, pp. 203–25
126. Turner CE, Pemberton RW, Rosenthal SS. 1987. Host utilization of native *Cirsium* thistles (Asteraceae) by the introduced weevil *Rhinocyllus conicus* (Coleoptera: Curculionidae) in California. *Environ. Entomol.* 16:111–15
127. Tuskes PM, Tuttle JP, Collins MM. 1996. *The Wild Silkmoths of North America: A Natural History of the Saturniidae of the United States and Canada*. Ithaca, NY: Cornell Univ. Press. 250 pp.
128. Villegas B. 2001. Thistle control projects. In *Biological Control Program Annual Summary, 2000*, ed. DM Woods, pp. 76–77. Sacramento, CA: Calif. Dep. Food Agric., Plant Health Pest Prev. Serv.
129. Wajnberg E, Scott JK, Quimby PC. 2001. *Evaluating Indirect Ecological Effects of Biological Control*. Wallingford, Oxon, UK: CABI. 261 pp.
130. Webber RT, Schaffner JV. 1926. *Host Relations of* Compsilura concinnata *Meigen, An Important Tachinid Parasite of the Gipsy Moth and Brown-Tail Moth*. Bull. 1363. Washington, DC: U.S. Dep. Agric.
131. Weber WA. 1987. *Colorado Flora: Western Slope*. Boulder, CO: Colorado Assoc. Univ. Press
132. Weseloh RW. 1982. Implications of tree microhabitat preferences of *Compsilura concinnata* (Diptera: Tachinidae) for its effectiveness as a gypsy moth parasitoid. *Can. Entomol.* 114:617–22
133. Wharton RA. 1989. Classical biological control of fruit infesting Tephritidae. In *World Crop Pests: Fruit Flies—Their Biology, Natural Enemies, and Control*, ed. AS Robinson, G Hooper, pp. 303–13. Amsterdam: Elsevier
134. Wheeler AG Jr, Hoebeke ER. 1995. *Coccinella novemnotata* in northeastern North America: historical occurrence and current status (Coleoptera: Coccinellidae). *Proc. Entomol. Soc. Wash.* 97:701–16
135. White JC. 1972. A European weevil, *Larinus carolinae* Oliver, collected in Maryland. *Coop. Econ. Insect Rep.* 22:418
136. Williamson M, Fitter A. 1996. The varying success of invaders. *Ecology* 77:1661–66
137. Wraight SP, Poprawski TJ, Meyer WL, Peairs FB. 1993. Natural enemies of Russian wheat aphid (Homoptera: Aphididae) and associated cereal aphid species in spring-planted wheat and barley in Colorado. *Environ. Entomol.* 22:1383–91

138. Zimmerman EC. 1978. *Insects of Hawaii.* Honolulu, HI: Univ. Hawaii Press
139. Zimmermann HG, Moran VC, Hoffmann J. 2001. The renowned cactus moth, *Cactoblastis cactorum* (Lepidoptera: Pyralidae): its natural history and threat to native *Opuntia* floras in Mexico and the United States of America. *Fla. Entomol.* 84:543–51
140. Zwölfer H, Frick KE, Andres LA. 1971. A study of the host plant relationships of European members of the genus *Larinus* (Col.: Curculionidae). *Commonw. Inst. Biol. Control Tech. Bull.* 14:97–143
141. Zwölfer H, Harris P. 1984. Biology and host specificity of *Rhinocyllus conicus* (Froel.) (Col.: Curculionidae), a successful agent for biocontrol of the thistle, *Carduus nutans* L. *Z. Ang. Entomol.* 97:36–62

Annu. Rev. Entomol. 2003. 48:397–423
doi: 10.1146/annurev.ento.48.091801.112703

First published online as a Review in Advance on September 3, 2002

THE EVOLUTION OF ALTERNATIVE GENETIC SYSTEMS IN INSECTS

Benjamin B. Normark
Department of Entomology, University of Massachusetts, Amherst, Massachusetts 01003; e-mail: bnormark@ent.umass.edu

Key Words evolution of sex, haplodiploidy, parthenogenesis, paternal genome elimination, thelytoky

■ **Abstract** There are three major classes of insect genetic systems: those with diploid males (diplodiploidy), those with effectively haploid males (haplodiploidy), and those without males (thelytoky). Mixed systems, involving cyclic or facultative switching between thelytoky and either of the other systems, also occur. I present a classification of the genetic systems of insects and estimate the number of evolutionary transitions between them that have occurred. Obligate thelytoky has arisen from each of the other systems, and there is evidence that over 900 such origins have occurred. The number of origins of facultative thelytoky and the number of reversions from obligate thelytoky to facultative and cyclic thelytoky are difficult to estimate. The other transitions are few in number: five origins of cyclic thelytoky, eight origins of obligate haplodiploidy (including paternal genome elimination), the strange case of *Micromalthus*, and the two reversions from haplodiploidy to diplodiploidy in scale insects. Available evidence tends to support W.D. Hamilton's hypothesis that maternally transmitted endosymbionts have been involved in the origins of haplodiploidy. Bizarre systems of extrazygotic inheritance in Sternorrhyncha are not easily accommodated into any existing classification of genetic systems.

CONTENTS

INTRODUCTION

The laws of genetics are not universal: They vary from species to species. The adaptive significance of this variation is poorly understood and constitutes a central problem in evolutionary biology (3, 64, 69, 82, 113). A staggering range of different genetic systems is found within insects, and the purpose of this review is to survey this diversity and to characterize the broad pattern of its evolution. I do not review the ecological patterns of the distribution of parthenogenesis or the hypotheses for the adaptive significance of sexuality, both of which have received thorough attention elsewhere (7, 53, 80). I briefly review the major classes of genetic systems and then discuss the patterns of evolutionary transition between them, giving case histories for the rarest transitions. Finally, I discuss strange patterns of extrazygotic inheritance that do not fit into existing classifications. Throughout, I highlight recent progress as well as outstanding problems that remain.

THE MAJOR GENETIC SYSTEMS OF INSECTS

A classification of the major genetic systems of insects is presented in the Appendix. It is useful to recognize three basic classes of systems: those with diploid males, those with effectively haploid males, and those without males. I refer to these as diplodiploidy, haplodiploidy, and thelytoky (see Appendix, below). There are also mixed systems that can be seen as straddling the boundaries between these three basic classes (Figure 1).

Hermaphroditism

Before discussing the diversity of genetic systems that occur in insects, I must acknowledge one class of systems that is conspicuous by its absence: hermaphroditism. Although extremely important in the multicellular biota in general (35), hermaphroditism (monoecy) is almost unknown in insects. It has been reported from one genus of scale insects (see Additional Paternal Genomes, below). Otherwise, apart from occasional developmental anomalies, insects appear to be obligately gonochoristic (dioecious), with individuals differentiating either as males or as females. This apparent constraint on the evolution of insect genetic systems, in spite of the high frequency of hermaphroditism in close insect relatives such as Crustacea (77), is one of many unexplained phenomena in this field of inquiry.

Diplodiploidy

Diplodiploidy, or amphimixis with diploid males (Appendix, I.A), needs little introduction. It characterizes all mammals, all birds, most other vertebrates, and a large majority of insect species. Diplodiploidy is the most thoroughly sexual class of genetic systems, in that the complete cycle of meiosis and syngamy occurs in each generation. Every individual has two parents and carries a diploid genome consisting of one haploid genome from each parent. Each parent contributes to each offspring a recombined haploid genome in which maternally derived and paternally derived chromosome regions have equal probabilities of being represented. This review of "alternative genetic systems" focuses on those systems that lack diploid males, so little more will be said about diplodiploidy. However, many of the evolutionary dynamics of alternative genetic systems depend on features of the diplodiploid systems from which they arise (27, 149), and the interplay between the details of diplodiploid systems (especially sex-determining mechanisms and extent of intersexual conflict) and the evolutionary dynamics of alternative systems arising from them is likely to be a fruitful area of future research.

Thelytoky

In this review I use the term thelytoky in the broad sense to refer to genetic systems in which females transmit only maternal genes and produce only daughters. Thus, it encompasses systems that completely lack males as well as the rarer systems in which mating with males of a related species is necessary to initiate development. This category of genetic systems is ill served by existing terminology. The most popular term for these genetic systems is parthenogenesis, but the etymology of this term (virgin birth) has encouraged scholars to apply it as well to arrhenotoky (virgin birth of males) and deny its application to sperm-dependent systems. Clonality is another candidate term, but it does not really apply to automictic systems

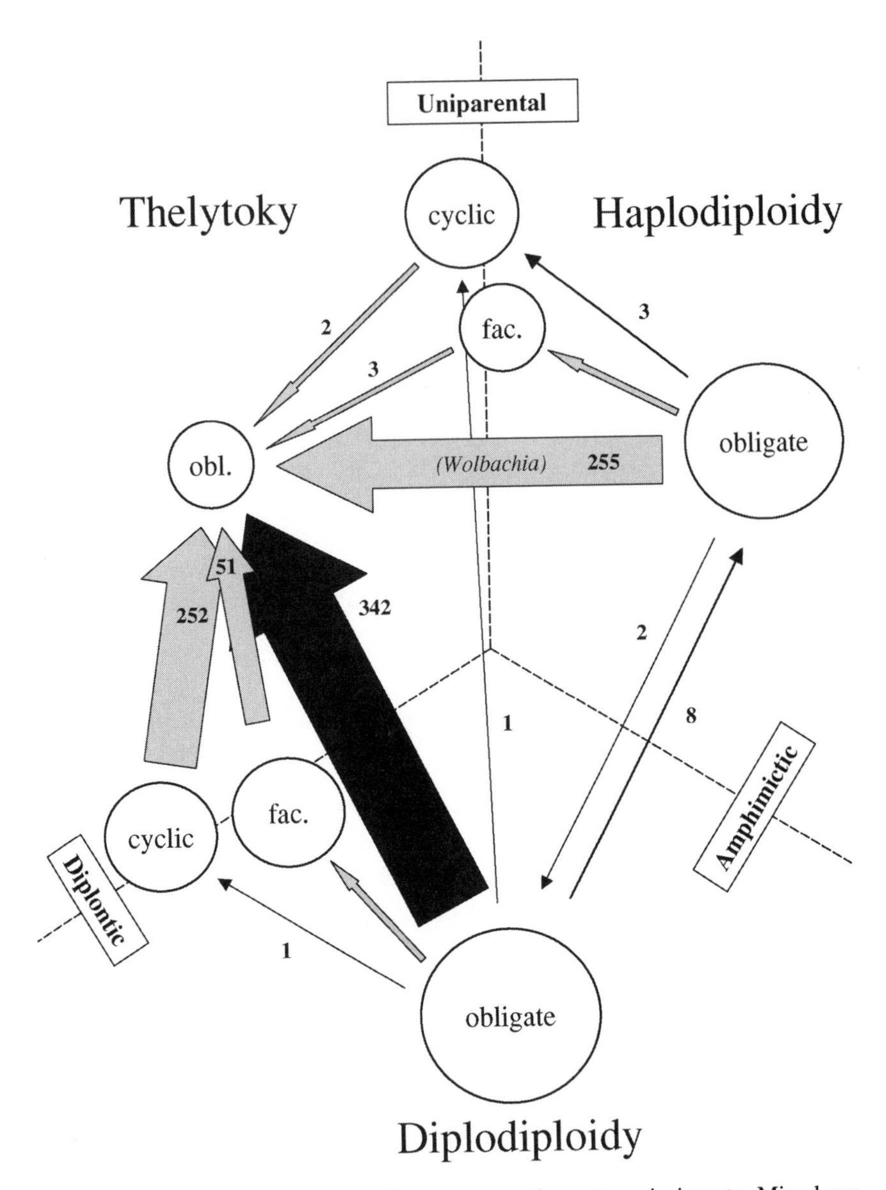

Figure 1 Evolutionary transitions between genetic systems in insects. Mixed systems, cyclic and facultative (fac.), are drawn at the boundaries between the systems they alternate between. For definitions, see the Appendix. The *gray block arrows* represent transitions that are relatively likely to be reversible. The other *arrows* represent transitions for which reversal is relatively rare or unlikely. The *circles* are drawn with diameters roughly proportional to the logarithm of the estimated total species diversity of lineages having that genetic system. To estimate the number of transitions to obligate (obl.) thelytoky, I compiled a database of over 900 insect species in which the existence of at least one obligately thelytokous lineage has been reported or can be inferred. A list of these species and bibliographic references is available as Supplemental Material: Follow the Supplemental Material link on the Annual Reviews homepage at http://www.annualreviews.org/. The other transitions are discussed in the text.

(Appendix, II.A.2) in which recombination occurs. The term thelytoky has the disadvantage of being unfamiliar to most biologists, but it may be gaining currency because of its widespread use in the burgeoning literature on *Wolbachia*, intracellular bacteria that often convert one form of parthenogenesis into another when they convert arrhenotokous haplodiploidy (Appendix I.B.1) to thelytoky.

The major features of thelytoky are (*a*) reproductive efficiency, with little or no energy wasted on "all this silly rigamarole of sex" (61), and (*b*) a lack of recombination between the genomes of different individuals. Thelytoky occurs frequently in many groups of organisms but is especially frequent in small invertebrates inhabiting freshwater and terrestrial environments (7, 33, 78, 134), and the large number of thelytokous insect lineages reflects this pattern.

Haplodiploidy

The best-known form of haplodiploidy is arrhenotokous haplodiploidy, in which males develop from unfertilized eggs (Appendix, I.B.1). This system is often called simply arrhenotoky, meaning virgin birth of males. But virgin females give birth to diploid males in some mixed genetic systems such as that of aphids, and this is also called arrhenotoky (14). In addition to arrhenotokous haplodiploidy, there is another class of genetic systems in which males begin life as diploid zygotes but ultimately produce sperm that carry only their mother's genome (Appendix, I.B.2). Here I refer to these as systems of paternal genome elimination (PGE) (67). PGE is identical to arrhenotokous haplodiploidy in terms of transmission genetics. Other authors have referred collectively to all systems with haplodiploid transmission genetics as "male haploidy" (26) or "uniparental male systems" (27), but here I use the term haplodiploidy in this broad sense to cover all such systems. Previously the term haplodiploidy has been used in the entomological literature as a simple synonym for arrhenotokous haplodiploidy and in the acarological literature to refer collectively to those systems in which males have a haploid soma (41, 110), arrhenotokous haplodiploidy, and embryonic PGE (Appendix, I.B.2.a).

Mixed Systems

This is an extremely heterogeneous category of genetic systems, covering all those complex systems in which there is an alternation (either facultatively or cyclically) between two of the different genetic systems described above. No system of alternation between diplodiploidy and haplodiploidy has ever been described, even though diploid males can occur at high frequencies in otherwise haplodiploid populations (119) and can be fertile (83). Rather, it is always thelytoky that alternates with one of the other systems.

A simplified classification of mixed genetic systems is presented in the Appendix, focusing on whether thelytoky alternates with diplodiploidy or haplodiploidy, and whether the alternation is cyclic or facultative. Because these mixed systems embody transitions between genetic systems, they are potentially a rich source of insights into the evolutionary history of such transitions. There have been relatively few origins of cyclic thelytoky. An excellent review of these has been published (66), to which I add some updated information below. Facultative thelytoky presents a more difficult problem, in part because it is more difficult to detect. A classification of modes of facultative thelytoky and an attempt to identify the evolutionary origins of facultative thelytoky in insects would be most valuable and is beyond the scope of this review.

MAJOR PATTERNS IN THE EVOLUTION OF INSECT GENETIC SYSTEMS

The apparent patterns of transition between genetic systems in insects are summarized in Figure 1. This figure relies on estimates of quantities that are in many cases poorly known. The overwhelming trend apparent from the figure is that the most frequent genetic-system transitions seen in insects are transitions to obligate all-female systems from each of the other (partially or wholly amphimictic) systems. Although virtually every student of genetic systems has been convinced of the reality of this trend (7, 71, 100, 134), many details are difficult to establish.

The gray arrows in Figure 1 indicate transitions that are likely to be relatively easily reversible. The transitions from mixed systems (the cyclically and facultatively amphimictic systems arrayed along the dotted border lines) to obligate parthenogenesis are expected to be reversible (at least early in their evolutionary history) because mixed systems are all capable of some kind of reversion to amphimixis following one or more parthenogenetic generations. For instance, a temperate aphid lineage transported to the tropics may become obligately apomictic (Appendix, II.A.1) in the absence of a photoperiodic cue to switch to amphimixis, but descendents reinvading the temperate zone may readily recover a sexual phase. Transitions from arrhenotokous haplodiploidy to obligate thelytoky are also expected to be subject to reversal because in the ancestral arrhenotokous system a virgin female is capable of producing males and this capacity may in principle be switched back on. In the case of *Wolbachia*-induced thelytoky, this reversal has been effected experimentally in many cases (22, 148). Of course, not all the transitions encompassed by the gray arrows are reversible. There are several cases of *Wolbachia*-associated thelytoky in which 'curing' the *Wolbachia* infection leads to sterility but not to a reversal to arrhenotokous haplodiploidy (22). And there are also several aphid species in which extensive experimental efforts have failed to elicit a switch to a sexual phase (59, 126).

The large black arrow in Figure 1 represents the largest class of genetic-system transitions, from obligate diplodiploidy to obligate thelytoky, and indicates that these are relatively unlikely to revert to amphimixis. For instance, the majority of these transitions occur in weevils, in which the obligately all-female lineage is typically triploid or has an even higher ploidy level, which makes meiosis extremely problematic. Also, these weevils are derived from lineages in which the males had Y chromosomes, chromosomes that do not occur in females and are therefore extinct. Of course it might be possible to recapture a Y chromosome from a related species or otherwise reinvent the male, but the obstacles to reversion are generally much higher in these cases than in those indicated by the gray arrows (28, 107).

ORIGINS OF OBLIGATE THELYTOKY

The standard model of the evolution of obligately thelytokous lineages holds that they arise more or less frequently from amphimictic populations and may be briefly favored by selection, but that, on an evolutionary timescale, they quickly go extinct. This pattern is reflected in Figure 1. In spite of the large number of thelytokous insect lineages and their importance in the historical discovery of parthenogenesis (7, 101, 141), in recent years other groups such as vertebrates and aquatic invertebrates have been more prominent in studies of obligate thelytoky (2, 34, 46, 50, 78, 140). But there are a few areas in which a good deal of progress has been made lately, and some of these are discussed below.

Conversion of Arrhenotokous Haplodiploidy to Thelytoky by *Wolbachia*

Fields of scientific inquiry undergo periodic revolutions, precipitated by new theoretical frameworks or new technologies. It is much rarer for a field to be revolutionized by the discovery of a simple, long-overlooked fact. But this is what has happened to the study of insect genetic systems following the 1990 discovery that experimental applications of antibiotics can convert thelytokous wasp lineages back to arrhenotokous haplodiploidy (133). By 2000, nearly one half of all new journal articles on alternative insect genetic systems were devoted to the effects of *Wolbachia* or other microbial agents, and this proportion shows no signs of declining. I do not dwell on this copious new literature because it has been amply reviewed elsewhere (22, 36, 148).

The fervor for *Wolbachia* studies is justified, as their advent represents an unusual forward leap, both theoretically and empirically, for the study of insect genetic systems. Note, however, that microbial agents have been found to induce only 1 of the 20 types of genetic-system transitions diagrammed in Figure 1: the transition from obligate haplodiploidy (specifically arrhenotoky) to obligate thelytoky. And, indeed, the expression parthenogenesis-inducing or parthenogenesis induction in reference to *Wolbachia* is misleading because there are no established cases of a nonparthenogenetic lineage rendered parthenogenetic by *Wolbachia*. What *Wolbachia* does in these cases is convert a lineage in which males are produced parthenogenetically to one in which females are produced parthenogenetically, by feminizing (through genome doubling) parthenogenetically produced sons.

It is possible that *Wolbachia* or other maternally inherited elements are involved in other genetic-system transitions as well, but such involvement has yet to be clearly demonstrated. In the case of transitions from mixed haplodiploid-thelytokous systems to obligate thelytoky, studies have found no evidence for *Wolbachia* involvement (132, 147). Possibly, mixed genetic systems are more likely to be robust to manipulation by *Wolbachia* because (*a*) selection on maternal elements to induce obligate thelytoky is weaker in a mixed system than in an obligately sexual system (a mixed system realizes some of the twofold fitness advantage of thelytoky, so selection to induce further thelytoky is less than twofold), and (*b*) mixed systems must have specific adaptations for the recovery of the sexual phase of the life cycle.

Probably the best candidate for another genetic-system transition induced by microbes is the most frequent transition observed: the origin of obligate thelytoky from obligate diplodiploidy (Figure 1). There is evidence that *Wolbachia* plays a role in the origins of parthenogenesis in weevils (104, 131, 148). However, in this case, and perhaps more generally for ancestrally diplodiploid lineages, there is little scope for a role of *Wolbachia* in maintaining thelytoky. *Wolbachia*-infected parthenogenetic weevils are polyploid, and male weevils typically have a Y chromosome (128). Thus, there is little chance that administration of antibiotics (to XXX females) will result in viable male (XY) offspring. Understanding the historical role of *Wolbachia* in the origin of such diplodiploid-derived parthenogenetic lineages is an area for future research.

Contagious Thelytoky

Classically, origins of thelytoky have been modeled as independent spontaneous mutations arising within sexual populations. Recently, two cases have been described in which sexual populations acquire thelytoky from preexisting conspecific thelytokous lineages. One of these involves *Wolbachia*. The phylogenetic incongruence between *Wolbachia* and their hosts has widely been interpreted as indicating frequent horizontal transfer of *Wolbachia* between widely disparate groups of insects (150). Experimental transfers have typically resulted in unstable, symptom-free infections (22). But in *Trichogramma kaykai*, an egg parasite, uninfected larvae sharing an egg with *Wolbachia*-infected conspecifics readily become infected, and within a few generations their progeny can be entirely thelytokous (70).

A different mechanism for the transfer of obligate thelytoky has recently been described in aphids as a second form of contagious parthenogenesis (126). Some aphid lineages that never produce sexual females do produce some males, which can apparently introduce genes for obligate thelytoky into otherwise sexual (cyclic diplodiploid-thelytokous) populations. Similar mechanisms for the spread of thelytoky have long been thought to occur in other cyclically parthenogenetic and hermaphroditic taxa, but only recently has there been evidence that it occurs in an insect, the aphid *Rhopalosiphum padi* (47).

Hybridization and Thelytoky

Our understanding of the role of hybridization in the origin of obligately thelytokous insect lineages has lagged far behind our understanding of the same question in vertebrates. Virtually all thelytokous vertebrates are of hybrid origin (2). In insects,

Wolbachia-induced thelytoky has nothing to do with hybridization (22, 148), and neither do most losses of the sexual phase in aphids (14, 59, 126). Nonetheless, it is possible that most or all origins of obligate thelytoky from obligate diplodiploidy (the only type of transition seen in vertebrates) do involve hybridization, although we know frustratingly little about this topic, with information available for only a few species. In a pioneering 1978 study, Drosopoulos (48) synthesized a hybrid triploid pseudogamous leafhopper lineage in the laboratory. The majority of transitions from obligate diplodiploidy to obligate thelytoky occur in broad-nosed weevils (Curculionidae: Entimini) (97), and there is good evidence of hybrid origin of one weevil species (131, 137).

Hybridization is also clearly involved in at least some cases of transition from facultative diplodiploidy-thelytoky to obligate thelytoky, as has been demonstrated for two species in the stick insect genus *Bacillus* (91). Several lines of evidence have demonstrated the hybrid ancestry of the apomictic species *Bacillus whitei* and *B. lynceorum*. Studies of *Bacillus* have also resulted in the first discovery of hybridogenesis (previously known only in vertebrates) in insects, as well as the discovery of androgenesis, which is the formation of new individuals by the fusion of two sperm (92, 136).

Hybridization with sexual lineages may also be involved in the subsequent evolution of obligately thelytokous lineages (89, 123). Genes from sexual lineages may introgress into obligately thelytokous lineages (8), or occasionally genes from obligately thelytokous lineages may introgress into sexual populations, as in the bizarre case of androgenesis (91, 136).

Genome Evolution

We still do not understand the major consequences of the loss of sex for the evolution of eukaryote genomes. For instance, it has not yet been possible to determine whether deleterious mutations are continually accumulating in the genomes of thelytokous lineages, as implied by some theories of sex (80, 108), or how transposable elements typically respond to the loss of sex (1).

Diploidy, the presence of exactly two copies of each autosome and each of its constituent loci, is a necessary feature of some stage of every sexual life cycle. In the mechanics of the cell cycle, diploidy is both an artifact of syngamy and an adaptation for meiosis. The origin of obligate apomixis coincides with a loss of the cytogenetic constraint to remain diploid. A nondiploid structure has long been known for the karyotypes of some obligately apomictic aphid lineages (11, 12), and the karyotypes of apomictic aphids can evolve rapidly (105, 129, 151). The diploid structure of the ribosomal DNA array is typically lost early in the history of apomictic lineages of aphids (15) and may also be lost in stick insects (94), although in both thelytokous aphids and stick insects the evolution of rDNA arrays can be complex (19, 94). Comparative studies of other repetitive sequences in both groups are ongoing (16–18, 90, 93, 95, 124, 130).

Persistence of Obligately Thelytokous Lineages

Although most obligately thelytokous lineages are apparently of recent origin (64, 100), in the past few years there has been considerable interest in the possibility of identifying "ancient asexuals" (78, 87, 96). Until recently, it had been thought that the highest-ranking (and thus potentially most ancient) strictly thelytokous taxon of insects was the aphid tribe Tramini (59, 126, 145). However, the recent discovery of sexual Tramini has shown this to be incorrect (13).

None of the most ancient asexuals (tens of millions of years old) are insects (78). However, if "ancient" is defined as more than one million years (My) old (87), then there are a number of candidates. The nonhybrid automictic stick insect species *Bacillus atticus* may have diverged from its closest sexual relative about 15 My ago (91), and thelytokous lineages within the stick insect genus *Timema* may be more than 1 My old (87, 122). For one thelytokous weevil lineage, an age of 2 My has been estimated (104), although possibly with some history of introgression from sexual lineages (107). In aphids, the thelytokous species *Tuberolachnus salignus* shows hints of an accumulation of deleterious mutations, possibly as a consequence of long-term thelytoky (108). Even within the aphid species *Rhopalosiphum padi*, some clones appear to be hundreds of thousands of years old (47, 125).

ORIGINS OF CYCLIC THELYTOKY

In contrast to the origins of obligate thelytoky (which are many) and the origins of facultative thelytoky (which are hard to assess), the origins of cyclic thelytoky are relatively few and well defined. Of the eight origins of cyclic thelytoky in animals, five occurred in insects. Here I briefly discuss each of these, focusing on improvements in phylogenetic resolution that have occurred since the excellent review by Hebert (66). Cyclic thelytoky requires unmated females to produce males, which places severe constraints on the sex determination system (66) and helps explain the apparent association between haplodiploidy and cyclic thelytoky evident in Figure 1.

Aphids

The single origin of cyclic alternation between thelytoky and diplodiploidy occurred in the branch leading to all extant Aphidoidea (Phylloxeridae, Adelgidae, and Aphididae), sister group of the Coccoidea (scale insects). Aphidoids thus account for the entire pathway in Figure 1 connecting this system to others, and hence for 3 of the 20 types of genetic-system transitions seen in insects. Apomictically produced aphid females with XX sex chromosome karyotypes produce XO males by eliminating one X chromosome. In contrast to the PGE displayed by some other cyclically thelytokous insects (sciarids and possibly cecidomyiids) and by close relatives of aphids (most scale insects), the elimination of the X chromosome has recently been shown to be random with respect to parent of origin (152).

Micromalthus debilis

Micromalthus debilis is typically paedogenetic, consisting entirely of thelytokous larviform females (112). Under poor culture conditions, haploid male and diploid female imagoes are produced. The origin of *M. debilis* is depicted with the longest arrow in Figure 1, crossing the center of the diagram. *M. debilis* is a single species that is not closely related to anything else; recent evidence suggests that it is a basal lineage within Archostemata, the basal-most suborder of beetles (9). Every other extant lineage in that region of insect phylogeny is obligately diplodiploid; however, it seems likely that the lineage leading to *M. debilis* must have passed through a period of obligate haplodiploidy before arriving at the complex system seen today (in which case Figure 1 would show nine origins of obligate haplodiploidy and four transitions from obligate haplodiploidy to cyclic haplodiploidy-thelytoky).

Oak and Maple Gall Wasps

Cyclic thelytoky characterizes a clade within the gall wasps (Cynipidae) consisting of the two tribes Cynipini (oak gall wasps) and Pediaspidini (maple gall wasps) (115, 132). This is the only group of metazoans in which there is obligate alternation between thelytokous and amphimictic generations (66). Stone et al. (132) provide an excellent recent review of their biology.

Heteropezini

Heteropezini is a tribe of cecidomyiid midges within the subfamily Porricondylinae (51). Its most likely sister group is the tribe Winnertziini (45, 155). Like *Micromalthus*, they are paedogenetic, with thelytokous generations becoming reproductively mature at a subadult stage. This stage varies between species, some reproducing as larvae and others as pupae (155). The developmental mechanisms of paedogenesis in *Heteropeza* have recently been studied by Hodin & Riddiford (68).

Mycophila + *Tekomyia*

Cyclic, paedogenetic thelytoky has arisen in a second group of cecidomyiid midges (51), in the tribe Aprionini (72, 74) within the subfamily Lestremiini (73). The developmental mechanism of paedogenesis in *Mycophila* is substantially different from that in *Heteropeza* (68).

THE ORIGINS OF HAPLODIPLOIDY

A great deal of importance has been placed on the ecological context in which obligate thelytoky tends to arise (7, 53, 141); indeed, the ecological constraints on the distribution of thelytokous lineages has received considerably more theoretical and empirical attention than the cytogenetic constraints on their origins (7, 29). In the case of haplodiploid lineages, the opposite is true. A number of authors have considered the genetic and cytogenetic constraints on origins of haplodiploidy

(21, 27, 56, 127), but their ecological distribution has received less attention for several good reasons. Thelytokous lineages are many and recent, providing abundant evidence of the ecological situations in which they tend to arise. Their relative cytogenetic simplicity and wide phylogenetic scatter imply relatively few constraints on their origins. Haplodiploid lineages, in contrast, are few and ancient. The ecological circumstances of their origins are often obscure, and the unlikelihood of evolving viable haploid males would seem to be severe.

And yet, if one does investigate the ecological circumstances in which haplodiploidy tends to arise, a fairly clear pattern emerges. In insects, haplodiploidy arises in lineages that use woody plant stems as a food source, either wood or sap. W.D. Hamilton (62) noted the tendency for haplodiploidy to arise in insects that live under the bark of dead trees: Of the nine origins in insects, discussed below, six probably occurred in association with dead wood. The other three origins occurred in insects that feed on phloem sap. Hamilton (63) also noted another common feature of these lineages that might shed light on the origins of haplodiploidy: Woody plant stems are a nutritionally poor resource, and insects that rely on them usually also rely on maternally inherited bacteria.

The nature of the connection between maternally transmitted bacteria and haplodiploidy is unclear. Hamilton speculated that haplodiploidy might be the final outcome of a history of conflict over sex determination between intracellular bacteria and their hosts. In Hamilton's scenario, endosymbionts seek to feminize their hosts by attacking and eliminating male-determining sex chromosomes; hosts respond by moving and multiplying the sex-determining elements across the genome. This in turn gives the bacteria more targets for elimination, until finally all the surviving autosomes behave like X chromosomes and sex determination rests on chromosome dosage alone.

Micromalthus debilis

Micromalthus debilis, whose phylogenetic affinities and life cycle are discussed above, feeds on rotting wood, as do most of its relatives in Archostemata. In all stages of the life cycle, gram-positive bacteria are abundant in the fat body and hemolymph. Their maternal transmission is indicated by their presence in ovarioles and developing eggs (85).

Xyleborini and Relatives

Arrhenotokous lineages in the scolytine beetle tribes Xyleborini and Dryocoetini have recently been shown to represent one large haplodiploid clade, with over 1000 species (75, 76, 106). The transition to haplodiploidy occurred in a phloem-feeding lineage. Many phloem-feeding bark beetles are associated with a complex microbiota of bacteria and fungi for which they act as vectors. Some species harbor transovarially transmitted endosymbiotic bacteria (25, 109). There is some experimental evidence that at least one of the arrhenotokous species is unable to reproduce when treated with antibiotics (109).

Hypothenemus and Relatives

Outside of the arrhenotokous clade discussed above, there is just one species of bark beetles (Coleoptera: Curculionidae: Scolytinae) in which a male-haploid system has been demonstrated. This is the species *Hypothenemus hampei*, the coffee berry borer, in which a system of germline PGE (Appendix) was recently reported (24). However, this species is nested within what appears to be a much larger clade of inbreeding species, and it is likely that male haploidy is common to all of them. This putative clade contains about 190 species and consists of the genera *Hypothenemus*, *Cryptocarenus*, *Trischidias*, and *Periocryphalus* (79). The sister group of this clade is unknown. It is traditionally placed in the tribe Cryphalini, but recent molecular studies have cast doubt on the monophyly of Cryphalini without producing strong support for any alternative grouping (49). Like many bark beetles (including *Coccotrypes*), *H. hampei* has transovarially transmitted bacteria concentrated in the Malpighian tubules (25).

Sciaridae

Sciarid fungus gnats have a system of germline PGE (67). The term genomic imprinting was originally coined to describe chromosome behavior in *Sciara* (88). The PGE system of sciarids is apparently unique in that the paternal autosomes are expressed (or at least not heterochromatinized) in males, in contrast to the inactivation or early elimination of paternal autosomes in scale insects, *Hypothenemus*, and mites. Among the many other unusual features of sciarid genetic systems is a tendency to produce unisexual broods, either all male or all female (114). Some sciarids are commercially important as pests in mushroom houses (65), and there are ongoing empirical and theoretical (57, 121) studies of their genetics (4, 10, 102) and cytology (20). The ancestral ecology of Sciaridae is probably feeding on fungi in association with rotting wood (99). Intracellular bacteria are not known from Sciaridae (25), and it is unclear whether the fungi on which they depend are vectored by adult sciarids and thus potentially "inherited."

Hymenoptera

The order Hymenoptera is the most species-rich haplodiploid clade. The vast majority of hymenopteran species are strictly arrhenotokous. Although the Hymenoptera have diversified into a vast array of ecological niches and are best known as parasitoids, predators, and nectar-feeders, the larvae of most of the basal lineages of Hymenoptera (sawflies) feed on dead wood (116, 142) and harbor intracellular maternally inherited bacteria (25). Hymenoptera arose early in the history of the Holometabola (84). The larval habit of feeding in rotting wood may be primitive in Holometabola, and arrhenotokous haplodiploidy in both Hymenoptera and *Micromalthus* may be a legacy of the ancestral holometabolan habitat (62).

Thysanoptera

The thrips, like the Hymenoptera, comprise a primitively arrhenotokous order (66a). Feeding on fungi associated with dead wood appears to be the primitive habit in Thysanoptera, exemplified by the extant lineage Merothripidae (103a). Some fungus-feeding thrips have been thought to harbor maternally transmitted bacterial endosymbionts (25), but this may be an error (139a).

Aleyrodidae

Whiteflies comprise an arrhenotokous family within the hemipteran suborder Sternorrhyncha. The most likely sister group of the Aleyrodoidea is Coccoidea (scale insects) + Aphidoidea (aphids) (32, 144). Virtually all Sternorrhyncha feed on phloem sap and rely on maternally inherited bacterial endosymbionts in specialized organs termed bacteriomes or mycetomes (5), and there have been numerous recent studies of the phylogeny and physiology of whitefly endosymbionts (31, 38, 39, 103, 135).

Iceryini

Iceryini is an arrhenotokous tribe of scale insects (111) (Sternorrhyncha: Coccoidea). The phylogenetic position of Iceryini within the otherwise diplodiploid family Margarodidae is not well resolved (55). Some of iceryines are androdioecious, consisting of hermaphroditic individuals and males (54, 111, 120), a system otherwise unknown in insects and rare in arthropods (146). Like most Sternorrhyncha, iceryines harbor maternally transmitted intracellular bacteria. Among the tribes of the family Margarodidae, there is a great diversity of bacteriome configurations and modes of transmission (25). Little is known about these, except what Buchner discovered in a series of beautiful histological studies in the 1960s (138).

Most Neococcoidea

Most of the extant species of scale insects (Sternorrhyncha: Coccoidea) have a system of PGE. The basal families of scale insects (Margarodidae, Ortheziidae, Phenacoleachiidae, Putoidae) are primitively diplodiploid, but the majority of scale insect families (and a large majority of species) comprise a monophyletic clade characterized by PGE (37). Strikingly similar habits and morphology are conserved across the origin of the PGE clade, making possible a fairly clear picture of the ancestral situation in which PGE arose. The sister group of the PGE clade is the family Putoidae, and the basal lineage of the PGE clade is the family Pseudococcidae. Both putoids and pseudococcids are "mealybugs" and have only recently been recognized as separate families. Both harbor bacteriome-associated, maternally transmitted, intracellular bacteria. The bacterial endosymbionts of pseudococcids have recently been found to include the only known example of a bacterium endosymbiotic within another bacterium (143). Little is known of the endosymbionts of putoids.

There appears to be some evolutionary instability in the PGE system of scale insects. Embryonic PGE has evolved from germline PGE repeatedly within the family Diaspididae (67). More strikingly, the only two known cases of reversion from haplodiploidy back to diplodiploidy have been found in scale insects, in the genera *Stictococcus* and *Lachnodius* (37, 111). *Stictococcus* has an unusual system of sex determination in which embryos that receive bacteria develop as females and those that do not develop as males (25), supporting Hamilton's conjecture of a role for endosymbionts in the evolution of haplodiploidy. *Stictococcus* is also unusual in that embryos are nourished via a placenta, tissue derived from the embryo that digests the maternal nurse cells (25). This coincidence between the loss of PGE and the gain of a placenta in the evolutionary lineage leading to *Stictococcus* tends to cast doubt on Haig's (56, 58) contention that there is no adaptive connection between the genomic imprinting seen in paternal genome–eliminating insects and the genomic imprinting seen in placental mammals and plants.

Other Probable Cases

CECIDOMYIIDAE A diversity of complex chromosome cycles is found in cecidomyiid midges (Diptera). One common feature of these is that one set of chromosomes is eliminated during spermatogenesis (23), as in Sciaridae. It seems likely that it is the paternal genome that is eliminated, as in Sciaridae and other chromosome-elimination systems, but this has never been clearly demonstrated in Cecidomyiidae (23, 27). Chromosome elimination had been suggested as a synapomorphy uniting a monophyletic Sciaridae + Cecidomyiidae (153), but a more-thorough analysis has indicated that these are not sister taxa but instead represent independent origins of unusual genetic systems (98) in a similar ecological situation (99). A similar case of possibly paternal chromosome elimination has recently been found, slightly outside the Insecta proper, in the collembolan suborder Symphypleona (43, 44).

FOUR SMALL CLADES OF SCOLYTINE BEETLES Extreme inbreeding (regular brother-sister mating) has frequently evolved in association with haplodiploidy (60, 154). Of the clades discussed above, four (Hymenoptera, Thysanoptera, and the two bark beetle clades) include extreme inbreeders. Hamilton (60) described a set of characteristics correlated with extreme inbreeding and haplodiploidy, including extremely female-biased sex ratios and small flightless males. He predicted that taxa with this suite of characters would turn out to be haplodiploid, and where this prediction has been tested it has turned out to be correct (24, 40). According to Hamilton's criteria we can predict that haplodiploidy will be found in four additional clades within the beetle subfamily Scolytinae (Curculionidae): throughout the genera *Premnobius* (106) and *Sueus* and within the genera *Araptus* and *Bothrosternus* (79).

EXTRAZYGOTIC INHERITANCE

It is a general feature of metazoan life cycles that they contain a single-celled zygote stage. Even in vegetative reproduction, mediated by many-celled propagules, there is still an occasional passage through a single-celled zygote (30). This fact is implicit in classifications of metazoan genetic systems, such as that in the Appendix, which essentially are classifications of modes of zygote production (81).

However, there are some insect genetic systems that, remarkably, lack a one-celled zygote stage and in which new individuals are founded by two genetically different lineages of cells or of nuclei. Such developmental modes do not fit neatly into any classification of modes of zygote production. Here I classify these according to whether they have additional paternal genomes (in the form of sperm pronuclei, even when these are inherited from the mother) or an additional maternal genome (in the form of maternal polar bodies or cell lineages derived from the mother). These unusual developmental modes are found only in the hemipteran suborder Sternorrhyncha.

Additional Paternal Genomes

Iceryine scale insects have been classified as hermaphrodites because unmated, anatomically female individuals contain active sperm that fertilize their eggs to produce diploid female progeny. Careful study of *Icerya purchasi* by Royer (120) revealed that multiple sperm pronuclei are transovarially transmitted from mothers to daughters and that they autonomously proliferate in the bodies of the daughters and migrate into the next generation of eggs. Hermaphroditism may be an inadequate term to describe this situation. It could alternatively be characterized as involving "totipotent sperm" or "a permanent cancer of the germline." *I. purchasi* is primarily selfing. However, if there is occasional outcrossing, then there will be competition between the indigenous lineages of sperm pronuclei (those inherited from an individual's parent) and the newly introduced lineages (those received from its mate) for representation in the offspring and in future generations. The fact that this remarkable system characterizes a widespread economic pest (*I. purchasi* is the cottony-cushion scale) raises hope for the practicability of its further study.

Additional Maternal Genomes

Sometimes the transmission of endosymbionts from mother to offspring in the Sternorrhyncha is accompanied by the transmission of entire cells from the maternal bacteriome. This process has been studied recently in whiteflies (39, 135), in which maternal bacteriocytes eventually break down and release their endosymbiotic bacteria (25, 135). The putoid mealybugs, sister group to the large PGE clade of scale insects, also transmit intact bacteriocytes from mother to offspring. However, in the case of the putoids, these maternal bacteriocytes never break down and release their endosymbionts. They proliferate to form the bacteriome of the offspring and persist from generation to generation (25).

Intriguingly, the shift from diplodiploidy to haplodiploidy in scale insects seems to have coincided with a revolution in the mode of formation of the bacteriome. The formation of the bacteriome in the haplodiploid (PGE) pseudococcid mealybugs is different from that in the diplodiploid putoids, but it still involves the inheritance of additional maternal genomes. In typical metazoan oogenesis, meiosis occurs and results in one haploid egg nucleus and three additional haploid genomes in the form of polar bodies. Typically, the polar bodies degenerate. But in Pseudococcidae, the polar bodies do not degenerate. They migrate into the developing embryo and (sometimes following fusion with a cleavage nucleus from the embryo) proliferate to form the embryo's bacteriome (139). This is the only case in nature of maternal polar bodies forming a permanent part of the soma of their offspring. The closest parallel may be the role of polar bodies in the angiosperm seed, which is also (like the bacteriome) involved in nutrition, but which (in contrast to the bacteriome) is active only for a short time early in the ontogeny of the offspring and is never fully integrated with it physiologically.

If a placenta is an outgrowth of an embryo that invades the body of its mother, then the bacteriome of Pseudococcidae might be termed an antiplacenta, an outgrowth of a mother that invades the body of its offspring. Thus, the origin of haplodiploidy in scale insects is coincident with the origin of an antiplacenta. Recall that in the scale insect genus *Stictococcus*, there is a reversal in the direction of flow of tissues and gene products, with the origin of a normal placenta. And recall that this coincides with the reversion of haplodiploidy back to diplodiploidy. Possibly, mother-offspring interactions can help explain the differential fate of paternal versus maternal genes in paternal genome–eliminating insects, as they have in the case of genomic imprinting in mammals and angiosperms (58).

PROSPECTS

Insects have long played a critical role in furthering our understanding of the diversity of genetic systems and their evolution, from the first demonstration of parthenogenesis in aphids by Bonnet (134) to the recent elucidation of intragenomic conflict by Hamilton (60) and his followers (149). The adaptive significance of the laws of genetics remains a huge unsolved problem, which insects will no doubt be instrumental in helping us solve. In addition to the usual advantages of rearability and manipulability that have made model systems of the likes of *Drosophila*, insects have additional advantages that to date have been underexploited. One of these is the enormous diversity of insect genetic systems, whose faint outlines have been traced in this review. Another is the affinity between some of these truly strange animals and our agricultural ecosystems.

The low diversity and frequent disturbance of agricultural systems seems conducive to a number of clonal pests (6, 7, 86). There is a remarkable congruence between the theoretical question of what makes clones vulnerable to extinction and the practical questions of how to precipitate extinctions of clonal pests while preserving the vitality of increasingly clonal crops. A number of agricultural systems are characterized by genetic system variation on three trophic levels. Consider an armored scale insect species such as *Aspidiotus nerii*, the oleander scale. It consists of sexual and clonal lineages (52), attacks a mosaic of sexual and clonal *Citrus* and other tree species, and is attacked by sexual and clonal *Aphytis* and other (often *Wolbachia*-infected) parasitic wasps (22, 117, 118). It will likely be from systems such as this, offering the possibility of switching Mendel's laws on and off in the various players in an ecosystem, that we will ultimately learn what Mendel's laws are for.

APPENDIX: CLASSIFICATION OF THE MAJOR GENETIC SYSTEMS OF INSECTS

A somewhat more detailed classification, with additional taxonomic information and references, is available via the Supplemental Material link on the Annual Reviews homepage at http://www.annualreviews.org/. The most important references are (7, 27, 66, 67, 134).

I. **Obligate amphimixis** (sex, sexuality). Every female inherits one haploid genome from her mother and one haploid genome from her father.

- (A) **Diplodiploidy** (diploid-male systems). Every male inherits one haploid genome from his mother and one haploid genome from his father, and these two haploid genomes have equal probability of transmission through his sperm. Found in large majority of all insect species; ancestral system in all orders except Thysanoptera and Hemiptera.
- (B) **Haplodiploidy** (haploid-male systems, male haploidy, uniparental-male systems). A male transmits only his mother's genome.
  - (1) **Arrhenotokous haplodiploidy** (Arrhenotoky). Every male develops from an unfertilized egg and has only a haploid genome inherited from his mother. Found in Thysanoptera, Hymenoptera, Hemiptera, and Coleoptera.
  - (2) **Paternal genome elimination** [(**PGE**), paternal genome loss]. Every male develops from a zygote containing one haploid genome from his mother and one haploid genome from his father, but only the maternal genome is transmitted through his sperm. Found in Hemiptera, Coleoptera, and Diptera.

II. **Thelytoky** (all-female systems, uniparental reproduction, parthenogenesis). A female transmits only her mother's genome; no sons are produced, only daughters.

- (A) **Thelytokous parthenogenesis.** No mating occurs. There are no males.

(1) **Apomixis** (ameiotic parthenogenesis, strict clonality). Eggs are produced mitotically. Found in several orders.

(2) **Automixis** (meiotic parthenogenesis). Meiosis occurs, but ploidy of eggs is restored through any of various mechanisms. Found in several orders.

(B) **Sperm-dependent thelytoky** (obligate mating). Mating (with males of a related amphimictic population) is necessary to initiate development.

(1) **Pseudogamy** (gynogenesis). Sperm activate development but the sperm nucleus does not fuse with the egg nucleus. Only maternal genes are transmitted to offspring (all daughters).

(a) **Apomictic pseudogamy.** Known in Hemiptera.

(b) **Automictic pseudogamy.** Known in Coleoptera and Lepidoptera.

(2) **Hybridogenesis** (hemiclonal inheritance). Syngamy occurs and the paternal genome is present and active in soma of offspring (all daughters) but is eliminated during oogenesis. Known in Phasmatodea.

III. **Mixed systems.** Regular or irregular alternation between different genetic systems, typically between amphimixis and thelytoky.

(A) **Thelytoky (or polyembryony) alternating with haplodiploidy.**

(1) **Cyclic alternation.**

(a) **Polyembryony.** Clonal proliferation of embryos (male or female) by fission. Found in Hymenoptera.

(b) **Cyclic haplodiploidy-thelytoky** (cyclic parthenogenesis). Found in Hymenoptera, Coleoptera, and Diptera.

(2) **Facultative haplodiploidy-thelytoky** (facultative parthenogenesis). Any system in which reproduction may be either amphimictic or thelytokous. In most of these systems, reproduction is typically amphimictic, but unmated females may produce some viable offspring by thelytoky. Found in Hymenoptera and Hemiptera.

(B) **Thelytoky alternating with diplodiploidy**

(1) **Cyclic diplodiploidy-thelytoky** (cyclic parthenogenesis). Found in Hemiptera.

(2) **Facultative diplodiploidy-thelytoky** (facultative parthenogenesis). Found in several orders.

ACKNOWLEDGMENTS

I am grateful to the authors of the great treatises on the diversity of animal genetic systems, M.J.D. White, G. Bell, J.J. Bull, E. Suomalainen, A. Saura, and J. Lokki. I am also grateful to everyone who has helped me find out about weird bugs, including N.A. Moran, R.G. Harrison, N.E. Pierce, B.J. Crespi, B.D. Farrell, A.A. Lanteri, B.H. Jordal, P.J. Gullan, J. Seger, D. Haig, D.R. Miller, L. Cook, M. Sorin, L. Magnano, T. Kondo, T.J.C. Anderson, O.P. Judson, W.D. Hamilton, and

R.D.K. Normark. Thanks to D.N. Ferro and my other UMass colleagues for the supportive working environment, and to J.A. Rosenheim for encouraging me to write this review. N.A. Johnson, R.L. Blackman, B.S. Heming, L.M. Provencher, B. Sello, M.E. Gruwell, and two anonymous reviewers provided helpful criticism of the manuscript.

The *Annual Review of Entomology* is online at http://ento.annualreviews.org

LITERATURE CITED

1. Arkhipova I, Meselson M. 2000. Transposable elements in sexual and ancient asexual taxa. *Proc. Natl. Acad. Sci. USA* 97:14473–77
2. Avise JC. 1994. *Molecular Markers, Natural History, and Evolution*. New York: Chapman & Hall
3. Barton NH, Charlesworth B. 1998. Why sex and recombination? *Science* 281:1980–82
4. Basso LR, Vasconcelos C, Fontes AM, Hartfelder K, Silva JA, et al. 2002. The induction of DNA puff BhC4-1 gene is a late response to the increase in 20-hydroxyecdysone titers in last instar dipteran larvae. *Mech. Dev.* 110:15–26
5. Baumann P, Moran NA, Baumann L. 2000. Bacteriocyte-associated endosymbionts of insects. In *The Prokaryotes*, ed. M Dworkin. New York: Springer. http://link.springer.de/link/service/books/10125/
6. Beardsley JW Jr, Gonzalez RH. 1975. The biology and ecology of armored scales. *Annu. Rev. Entomol.* 20:47–73
7. Bell G. 1982. *The Masterpiece of Nature*. Berkeley, CA: Univ. Calif. Press
8. Belshaw R, Quicke DL, Volkl W, Godfray HCJ. 1999. Molecular markers indicate rare sex in a predominantly asexual parasitoid wasp. *Evolution* 53:1189–99
9. Beutel RG, Haas F. 2000. Phylogenetic relationships of the suborders of Coleoptera (Insecta). *Cladistics* 16:103–41
10. Bielinsky AK, Blitzblau H, Beall EL, Ezrokhi M, Smith HS, et al. 2001. Origin recognition complex binding to a metazoan replication origin. *Curr. Biol.* 11:1427–31
11. Blackman RL. 1980. Chromosome numbers in the Aphididae and their taxonomic significance. *Syst. Entomol.* 5:7–25
12. Blackman RL. 1990. The chromosomes of Lachnidae. *Acta Phytopathol. Entomol. Hung.* 25:273–82
13. Blackman RL, De Boise E, Czylok A. 2001. Occurrence of sexual morphs in *Trama troglodytes* von Heyden, 1837 (Hemiptera, Aphididae). *J. Nat. Hist.* 35:779–85
14. Blackman RL, Eastop VF. 2000. *Aphids on the World's Crops: An Identification and Information Guide*. Chichester: Wiley. 466 pp. 2nd ed.
15. Blackman RL, Spence JM. 1996. Ribosomal DNA is frequently concentrated on only one X chromosome in permanently apomictic aphids, but this does not inhibit male determination. *Chromosome Res.* 4:314–20
16. Blackman RL, Spence JM, Field LM, Devonshire AL. 1995. Chromosomal location of the amplified esterase genes conferring resistance to insecticides in *Myzus persicae* (Homoptera: Aphididae). *Heredity* 75:297–302
17. Blackman RL, Spence JM, Field LM, Devonshire AL. 1999. Variation in the chromosomal distribution of amplified esterase (FE4) genes in Greek field populations of *Myzus persicae* (Sulzer). *Heredity* 82:180–86

18. Blackman RL, Spence JM, Field LM, Javed N, Devine GJ, Devonshire AL. 1996. Inheritance of the amplified esterase genes responsible for insecticide resistance in *Myzus persicae* (Homoptera: Aphididae). *Heredity* 77:154–67
19. Blackman RL, Spence JM, Normark BB. 2000. High diversity of structurally heterozygous karyotypes and rDNA arrays in parthenogenetic aphids of the genus *Trama* (Aphididae: Lachninae). *Heredity* 84:254–60
20. Borges AR, Gaspar VP, Fernandez MA. 2000. Unequal X chromosomes in *Bradysia hygida* (Diptera: Sciaridae) females: karyotype assembly and morphometric analysis. *Genetica* 108:101–5
21. Borgia G. 1980. Evolution of haplodiploidy: models for inbred and outbred systems. *Theor. Popul. Biol.* 17:103–28
22. Braig HR, Turner B, Normark BB, Stouthamer R. 2002. Microorganism-induced parthenogenesis. In *Progress in Asexual Propagation and Reproductive Strategies*, ed. RN Hughes. Chichester: Wiley. In press
23. Brown SW, Chandra HS. 1977. Chromosome imprinting and the differential regulation of homologous chromosomes. In *Cell Biology: A Comprehensive Treatise*, ed. L Goldstein, DM Prescott, pp. 109–89. New York: Academic
24. Brun LO, Borsa P, Gaudichon V, Stuart JJ, Aronstein K, et al. 1995. Functional haplodiploidy. *Nature* 374:506
25. Buchner P. 1965. *Endosymbiosis of Animals with Plant-Like Micro-Organisms*. New York: Wiley
26. Bull JJ. 1979. An advantage for the evolution of male haploidy and systems with similar genetic transmission. *Heredity* 43:361–81
27. Bull JJ. 1983. *Evolution of Sex Determining Mechanisms*. Menlo Park, CA: Benjamin/Cummings
28. Bull JJ, Charnov EL. 1985. On irreversible evolution. *Evolution* 39:1149–55
29. Burt A. 2000. Perspective: sex, recombination, and the efficacy of selection—was Weismann right? *Evolution* 54:337–51
30. Buss LW. 1987. *The Evolution of Individuality*. Princeton: Princeton Univ. Press
31. Campbell BC. 1993. Congruent evolution between whiteflies (Homoptera: Aleyrodidae) and their bacterial endosymbionts based on respective 18S and 16S rDNAs. *Curr. Microbiol.* 26:129–32
32. Campbell BC, Steffen-Campbell JD, Gill RJ. 1996. Origin and radiation of whiteflies: an initial molecular phylogenetic assessment. In *Bemisia 1995: Taxonomy, Biology, Damage, Control and Management*, ed. D Gerling, RT Mayer, pp. 29–51. Andover, UK: Intercept
33. Chaplin JA, Havel JE, Hebert PDN. 1994. Sex and ostracods. *Trends Ecol. Evol.* 9:435–39
34. Chaplin JA, Hebert PDN. 1997. *Cyprinotus incongruens* (Ostracoda): an ancient asexual? *Mol. Ecol.* 6:155–68
35. Charnov EL. 1982. *The Theory of Sex Allocation*. Princeton: Princeton Univ. Press
36. Cook JM, Butcher RDJ. 1999. The transmission and effects of *Wolbachia* bacteria in parasitoids. *Res. Popul. Ecol.* 41:15–28
37. Cook LG, Gullan PJ, Trueman HE. 2002. A preliminary phylogeny of the scale insects (Hemiptera: Sternorrhyncha: Coccoidea) based on nuclear small-subunit ribosomal DNA. *Mol. Phylogenet. Evol.* In press
38. Costa HS, Henneberry TJ, Toscano NC. 1997. Effects of antibacterial materials on *Bemisia argentifolii* (Homoptera: Aleyrodidae) oviposition, growth, survival, and sex ratio. *J. Econ. Entomol.* 90:333–39
39. Costa HS, Toscano NC, Henneberry TJ. 1996. Mycetocyte inclusion in the

oocytes of *Bemisia argentifolii* (Homoptera, Aleyrodidae). *Ann. Entomol. Soc. Am.* 89:694–99
40. Crespi B. 1993. Sex allocation ratio selection in Thysanoptera. See Ref. 154, pp. 214–34
41. Cruickshank RH, Thomas RH. 1999. Evolution of haplodiploidy in dermanyssine mites (Acari: Mesostigmata). *Evolution* 53:1796–803
42. Deleted in proof
43. Dallai R, Fanciulli PP, Carapelli A, Frati F. 2001. Aberrant spermatogenesis and sex determination in Bourletiellidae (Hexapoda, Collembola), and their evolutionary significance. *Zoomorphology* 120:237–45
44. Dallai R, Fanciulli PP, Frati F. 2000. Aberrant spermatogenesis and the peculiar mechanism of sex determination in symphypleonan Collembola (Insecta). *J. Hered.* 91:351–58
45. Dallai R, Lupetti P, Frati F, Mamaev BM, Afzelius BA. 1996. Characteristics of sperm ultrastructure in the gall midges Porricondylinae (Insecta, Diptera, Cecidomyiidae), with phylogenetic considerations on the subfamily. *Zoomorphology* 116:85–94
46. Dawley RM, Bogart JP, eds. 1989. *Evolution and Ecology of Unisexual Vertebrates*. Albany: NY State Mus. Press. 302 pp.
47. Delmotte F, Leterme N, Bonhomme J, Rispe C, Simon J-C. 2001. Multiple routes to asexuality in an aphid species. *Proc. R. Soc. London Ser. B* 268:2291–99
48. Drosopoulos S. 1978. Loboratory synthesis of a pseudogamous triploid "species" of the genus *Muellerianella* (Homoptera: Delphacidae). *Evolution* 32:916–20
49. Farrell BD, Sequeira AS, O'Meara B, Normark BB, Chung JH, Jordal BH. 2001. The evolution of agriculture in beetles (Curculionidae: Scolytinae and Platypodinae). *Evolution* 55:2011–27
50. Fu J, MacCulloch RD, Murphy RW, Darevsky IS, Tuniyev BS. 2000. Allozyme variation patterns and multiple hybridization origins: clonal variation among four sibling parthenogenetic Caucasian rock lizards. *Genetica* 108:107–12
51. Gagné RJ. 1994. *The Gall Midges of the Neotropical Region*. Ithaca: Cornell Univ. Press. 352 pp.
52. Gerson U. 1990. Biosystematics. See Ref. 116a, pp. 129–34
53. Glesener RR, Tilman D. 1978. Sexuality and the components of environmental uncertainty: clues from geographic parthenogenesis in terrestrial animals. *Am. Nat.* 112:659–73
54. Gullan PJ. 1986. The biology of *Auloicerya acaciae* Morrison and Morrison—an Australian iceryine margarodid. *Boll. Lab. Entomol. Agraria Filippo Silvestri* 43:155–60
55. Gullan PJ, Sjaarda AW. 2001. Trans-Tasman *Platycoelostoma* Morrison (Hemiptera: Coccoidea: Margarodidae) on endemic Cupressaceae, and the phylogenetic history of margarodids. *Syst. Entomol.* 26:257–78
56. Haig D. 1993. The evolution of unusual chromosomal systems in coccoids: extraordinary sex ratios revisited. *J. Evol. Biol.* 6:69–77
57. Haig D. 1993. The evolution of unusual chromosomal systems in sciarid flies: intragenomic conflict and the sex raio. *J. Evol. Biol.* 6:249–61
58. Haig D. 2000. The kinship theory of genomic imprinting. *Annu. Rev. Ecol. Syst.* 31:9–32
59. Hales DF, Tomiuk J, Woehrmann K, Sunnucks P. 1997. Evolutionary and genetic aspects of aphid biology: a review. *Eur. J. Entomol.* 94:1–55
60. Hamilton WD. 1967. Extraordinary sex ratios. *Science* 156:477–88
61. Hamilton WD. 1975. Gamblers since life began: barnacles, aphids, elms. *Q. Rev. Biol.* 50:175–80

62. Hamilton WD. 1978. Evolution and diversity under bark. In *Diversity of Insect Faunas*, ed. LA Mound, N Waloff, pp. 154–75. Oxford: Blackwell Sci.
63. Hamilton WD. 1993. Inbreeding in Egypt and in this book: a childish perspective. In *The Natural History of Inbreeding and Outbreeding: Theoretical and Empirical Perspectives*, ed. NW Thornhill, pp. 429–50. Chicago: Univ. Chicago Press
64. Hamilton WD. 2001. *Narrow Roads of Gene Land*, Vol. 2: *The Evolution of Sex*. Oxford, UK: Oxford Univ. Press
65. Harris MA, Gardner WA, Oetting RD. 1996. A review of the scientific literature on fungus gnats (Diptera: Sciaridae) in the genus *Bradysia*. *J. Entomol. Sci.* 31:252–76
66. Hebert PDN. 1987. Genotypic characteristics of cyclic parthenogens and their obligately asexual derivatives. In *The Evolution of Sex and Its Consequences*, ed. SJ Stearns, pp. 175–95. Basel: Birhäuser
66a. Heming B. 1995. History of the germ line in male and female thrips. In *Thrips Biology and Management*, ed. BL Parker, M Skinner, T Lewis, pp. 505–35. New York: Plenum
67. Herrick G, Seger J. 1999. Imprinting and paternal genome elimination in insects. In *Genomic Imprinting: An Interdisciplinary Approach*, ed. R Ohlsson, pp. 41–71. Berlin: Springer
68. Hodin J, Riddiford LM. 2000. Parallel alterations in the timing of ovarian ecdysone receptor and ultraspiracle expression characterize the independent evolution of larval reproduction in two species of gall midges (Diptera: Cecidomyiidae). *Dev. Genes Evol.* 210:358–72
69. Howard RS, Lively CM. 1994. Parasitism, mutation accumulation and the maintenance of sex. *Nature* 367:554–57
70. Huigens ME, Luck RF, Klaassen RHG, Maas MFPM, Timmermans MJTN, Stouthamer R. 2000. Infectious parthenogenesis. *Nature* 405:178–79
71. Hurst LD, Hamilton WD, Ladle RJ. 1992. Covert sex. *Trends Ecol. Evol.* 7:144–45
72. Jaschhof M. 1998. Revision der "Lestremiinae" (Diptera, Cecidomyiidae) der Holarktis. *Studia Dipterol. Suppl.* 4:1–552
73. Jaschhof M. 2000. Catotrichinae *subfam. n.*: a re-examination of higher classification in gall midges (Diptera: Cecidomyiidae). *Entomol. Sci.* 3:639–52
74. Jaschhof M, Hippa H. 1999. *Pseudoperomyia gen. n.* from Malaysia and the phylogeny of the Micromyidi (Diptera: Cecidomyiidae, Lestremiinae). *Beitr. Entomol.* 49:147–71
75. Jordal BH, Beaver RA, Normark BB, Farrell BD. 2002. Extraordinary sex ratios, and the evolution of male neoteny in sib-mating *Ozopemon* beetles. *Biol. J. Linn. Soc.* 75:353–60
76. Jordal BH, Normark BB, Farrell BD. 2000. Evolutionary radiation of an inbreeding haplodiploid beetle lineage (Curculionidae, Scolytinae). *Biol. J. Linn. Soc.* 71:483–99
77. Juchault P. 1999. Hermaphroditism and gonochorism. A new hypothesis on the evolution of sexuality in Crustacea. *C. R. Acad. Sci. Ser. III* 322:423–27
78. Judson OP, Normark BB. 1996. Ancient asexual scandals. *Trends Ecol. Evol.* 11:41–46
79. Kirkendall LR. 1993. Ecology and evolution of biased sex ratios in bark and ambrosia beetles. See Ref. 154, pp. 235–345
80. Kondrashov AS. 1993. Classification of hypotheses on the advantage of amphimixis. *J. Hered.* 84:372–87
81. Kondrashov AS. 1997. Evolutionary genetics of life cycles. *Annu. Rev. Ecol. Syst.* 28:391–435
82. Kondrashov AS. 2001. Sex and U. *Trends Genet.* 17:75–77
83. Krieger MJB, Ross KG, Chang CWY,

Keller L. 1999. Frequency and origin of triploidy in the fire ant *Solenopsis invicta*. *Heredity* 82:142–50
84. Kristensen NP. 1999. Phylogeny of the endopterygote insects, the most successful lineage of living organisms. *Eur. J. Entomol.* 96:237–53
85. Kühne H. 1972. Entwicklungsablauf und -stadien von *Micromalthus debilis* LeConte (Col., Micromalthidae) aus einer Laboratoriums-Population. *Z. Angew. Entomol.* 72:157–68
86. Lanteri AA, Normark BB. 1995. Parthenogenesis in the tribe Naupactini (Coleoptera: Curculionidae). *Ann. Entomol. Soc. Am.* 88:722–31
87. Law JH, Crespi B. 2002. Recent and ancient asexuality in *Timema* walkingsticks. *Evolution.* In press
88. Lloyd VK, Sinclair DA, Grigliatti TA. 1999. Genomic imprinting and position-effect variegation in *Drosophila melanogaster*. *Genetics* 151:1503–16
89. Lynch M. 1984. Destabilizing hybridization, general-purpose genotypes and geographical parthenogenesis. *Q. Rev. Biol.* 59:257–90
90. Mantovani B. 1998. Satellite sequence turnover in parthenogenetic systems: the apomictic triploid hybrid *Bacillus lynceorum* (Insecta, Phasmatodea). *Mol. Biol. Evol.* 15:1288–97
91. Mantovani B, Passamonti M, Scali V. 2001. The mitochondrial cytochrome oxidase II gene in *Bacillus* stick insects: ancestry of hybrids, androgenesis, and phylogenetic relationships. *Mol. Phylogenet. Evol.* 19:157–63
92. Mantovani B, Scali V. 1992. Hybridogenesis and androgenesis in the stick-insect *Bacillus rossius-grandii benazzii* (Insecta: Phasmatodea). *Evolution* 46:783–96
93. Mantovani B, Tinti F, Bachmann L, Scali V. 1997. The Bag320 satellite DNA family in *Bacillus* stick insects (Phasmatodea): different rates of molecular evolution of highly repetitive DNA in bisexual and parthenogenetic taxa. *Mol. Biol. Evol.* 14:1197–205
94. Marescalchi O, Scali V. 1997. Chromosomal and NOR patterns in the polyclonal stick insect *Bacillus atticus atticus* (Insecta: Phasmatodea). *Genome* 40:261–70
95. Marescalchi O, Scali V, Zuccotti M. 1998. Flow-cytometric analyses of intraspecific genome size variations in *Bacillus atticus* (Insecta, Phasmatodea). *Genome* 41:629–35
96. Mark Welch D, Meselson M. 2000. Evidence for the evolution of bdelloid rotifers without sexual reproduction or genetic exchange. *Science* 288:1211–15
97. Marvaldi AE. 1997. Higher level phylogeny of Curculionidae (Coleoptera: Curculionidae) based mainly on larval characters, with special reference to broad-nosed weevils. *Cladistics* 13:285–312
98. Matile L. 1990. Recherches sur la sytématique et l'évolution des Keroplatidae (Diptera, Mycetophiloidea). *Mém. Mus. Natl. Hist. Nat. Sér. A Zool.* 148:1–682
99. Matile L. 1997. Phylogeny and evolution of the larval diet in the Sciaroidea (Diptera, Bibionomorpha) since the Mesozoic. *Mem. Mus. Natl. Hist. Nat.* 173:273–303
100. Maynard Smith J. 1978. *The Evolution of Sex*. Cambridge, UK: Cambridge Univ. Press
101. Mayr E. 1963. *Animal Species and Evolution*. Cambridge, MA: Harvard Univ. Press
102. Mok EH, Smith HS, DiBartolomeis SM, Kerrebrock AW, Rothschild LJ, et al. 2001. Maintenance of the DNA puff expanded state is independent of active replication and transcription. *Chromosoma* 110:186–96
103. Morin S, Ghanim M, Sobol I, Czosnek H. 2000. The GroEL protein of the whitefly *Bemisia tabaci* interacts with the coat

protein of transmissible and nontransmissible begomoviruses in the yeast two-hybrid system. *Virology* 276:404–16
103a. Mound LA, Heming BS. 1991. Thysanoptera. In *The Insects of Australia: A Textbook for Students and Research Workers*, ed. CSIRO, 1:510–15. Ithaca, NY: Cornell Univ. Press
104. Normark BB. 1996. Phylogeny and evolution of parthenogenetic weevils of the *Aramigus tessellatus* species complex (Coleoptera: Curculionidae: Naupactini): evidence from mitochondrial DNA sequences. *Evolution* 50:734–45
105. Normark BB. 1999. Evolution in a putatively ancient asexual aphid lineage: recombination and rapid karyotype change. *Evolution* 53:1458–69
106. Normark BB, Jordal BH, Farrell BD. 1999. Origin of a haplodiploid beetle lineage. *Proc. R. Soc. London Ser. B* 266:2253–59
107. Normark BB, Lanteri AA. 1998. Incongruence between morphological and mitochondrial-DNA characters suggests hybrid origins of parthenogenetic weevil lineages (genus *Aramigus*). *Syst. Biol.* 47:459–78
108. Normark BB, Moran NA. 2000. Testing for the accumulation of deleterious mutations in asexual eukaryote genomes using molecular sequences. *J. Nat. Hist.* 34:1719–29
109. Norris DM, Chu H. 1980. Symbiote-dependent arrhenotokous parthenogenesis in the eukaryotic *Xyleborus*. In *Endocytobiology: Endosymbiosis and Cell Biology*, ed. W Schwemmler, HEA Schenk, pp. 453–60. Berlin: de Gruyter
110. Norton RA, Kethley JB, Johnston DE, O'Connor BM. 1993. Phylogenetic perspectives on genetic systems and reproductive modes of mites. See Ref. 154, pp. 8–99
111. Nur U. 1980. Evolution of unusual chromosome systems in scale insects (Coccoidea: Homoptera). In *Insect Cytogenetics*, ed. RL Blackman, GM Hewitt, M Ashburner, pp. 97–117. London: Blackwell Sci.
112. Pollock DA, Normark BB. 2002. The life cycle of *Micromalthus debilis* LeConte (Coleoptera: Archostemmata: Micromalthidae): historical review and evolutionary perspective. *J. Syst. Evol. Res.* 40:105–12
113. Rice WR, Chippindale AK. 2001. Sexual recombination and the power of natural selection. *Science* 294:555–59
114. Rocha LS, Perondini ALP. 2000. Analysis of the sex ratio in *Bradysia matogrossensis* (Diptera, Sciaridae). *Genet. Mol. Biol.* 23:97–103
115. Ronquist F. 1999. Phylogeny, classification and evolution of the Cynipoidea. *Zool. Scripta* 28:139–64
116. Ronquist F, Rasnitsyn AP, Roy A, Eriksson K, Lindgren M. 1999. Phylogeny of the Hymenoptera: a cladistic reanalysis of Rasnitsyn's (1988) data. *Zool. Scripta* 28:13–50
116a. Rosen D, ed. 1990. *Armored Scale Insects: Their Biology, Natural Enemies, and Control, Volume B*. Amsterdam: Elsevier
117. Rosen D, DeBach P. 1979. *Species of Aphytis of the World*. The Hague: Junk
118. Rosen D, DeBach P. 1990. Ectoparasites. See Ref. 116a, pp. 99–120
119. Roubik DW. 2001. Searching for genetic pattern in orchid bees: a reply to Takahashi et al. *Evolution* 55:1900–1
120. Royer M. 1975. Hermaphroditism in insects: studies on *Icerya purchasi*. In *Intersexuality in the Animal Kingdom*, ed. R Reinboth, pp. 135–45. Berlin: Springer
121. Sanchez L, Perondini ALP. 1999. Sex determination in sciarid flies: a model for the control of differential X-chromosome elimination. *J. Theor. Biol.* 197:247–59
122. Sandoval C, Carmean DA, Crespi BJ. 1998. Molecular phylogenetics of sexual and parthenogenetic *Timema* walking-sticks. *Proc. R. Soc. London Ser. B* 265: 589–95

123. Saura A, Lokki J, Suomalainen E. 1993. Origin of polyploidy in parthenogenetic weevils. *J. Theor. Biol.* 163:449–56
124. Scali V, Tinti F. 1999. Satellite DNA variation in parental and derived unisexual hybrids of *Bacillus* stick insects (Phasmatodea). *Insect Mol. Biol.* 8:557–64
125. Simon J-C, Martinez-Torres D, Latorre A, Moya A, Hebert PDN. 1996. Molecular characterization of cyclic and obligate parthenogens in the aphid *Rhopalosiphum padi* (L.). *Proc. R. Soc. London Ser. B* 263:481–86
126. Simon J-C, Rispe C, Sunnucks P. 2002. Ecology and evolution of sex in aphids. *Trends Ecol. Evol.* 17:34–39
127. Smith NGC. 2000. The evolution of haplodiploidy under inbreeding. *Heredity* 84:186–92
128. Smith SG, Virkki N. 1978. *Animal Cytogenetics, Volume 3 Insecta 5 Coleoptera.* Berlin: Gebrüder Borntraeger
129. Spence JM, Blackman RL. 2000. Inheritance and meiotic behaviour of a de novo chromosome fusion in the aphid *Myzus persicae* (Sulzer). *Chromosoma* 109:490–97
130. Spence JM, Blackman RL, Testa JM, Ready PD. 1998. A 169-base pair tandem repeat DNA marker for subtelomeric heterochromatin and chromosomal rearrangements in aphids of the *Myzus persicae* group. *Chromosome Res.* 6:167–75
131. Stenberg P, Terhivuo J, Lokki J, Saura A. 2000. Clone diversity in the polyploid weevil *Otiorhynchus scaber*. *Hereditas* 132:137–42
132. Stone GN, Schönrogge K, Atkinson RJ, Bellido D, Pujade-Villar J. 2002. The population biology of oak gall wasps (Hymenoptera: Cynipidae). *Annu. Rev. Entomol.* 47:633–68
133. Stouthamer R, Luck RF, Hamilton WD. 1990. Antibiotics cause parthenogenetic *Trichogramma* (Hymenoptera/Trichogrammatidae) to revert to sex. *Proc. Natl. Acad. Sci. USA* 87:2424–27
134. Suomalainen E, Saura A, Lokki J. 1987. *Cytology and Evolution in Parthenogenesis*. Boca Raton: CRC
135. Szklarzewicz T, Moskal A. 2001. Ultrastructure, distribution, and transmission of endosymbionts in the whitefly *Aleurochiton aceris* Modeer (Insecta, Hemiptera, Aleyrodinea). *Protoplasma* 218:45–53
136. Tinti F, Scali V. 1996. Androgenetics and triploids from an interacting parthenogenetic hybrid and its ancestors in stick insects. *Evolution* 50:1251–58
137. Tomiuk J, Loeschcke V. 1992. Evolution of parthenogenesis in the *Otiorhynchus scaber* complex. *Heredity* 68:391–98
138. Tremblay E. 1977. Advances in endosymbiont studies in Coccoidea. *Va. Polytech. Inst. State Univ. Res. Div. Bull.* 127:23–33
139. Tremblay E, Caltagirone LE. 1973. Fate of polar bodies in insects. *Annu. Rev. Entomol.* 18:421–44
139a. Tsutsumi T, Matsuzaki M, Haga K. 1994. New aspect of the "mycetome" of a thrips, *Bactrothrips brevitubus* Takahashi (Insecta: Thysanoptera). *J. Morphol.* 221:235–42
140. Van Raay TJ, Crease TJ. 1995. Mitochondrial DNA diversity in an apomictic *Daphnia* complex from the Canadian High Arctic. *Mol. Ecol.* 4:149–61
141. Vandel A. 1928. La parthénogenèse géographique: contribution à l'étude biologique et cytologique de la parthénogenèse naturelle. *Bull. Biol. France Belgique* 62:164–281
142. Vilhelmsen L. 1997. The phylogeny of lower Hymenoptera (Insecta), with a summary of the early evolutionary history of the order. *J. Zool. Syst. Evol. Res.* 35:49–70
143. von Dohlen CD, Kohler S, Alsop ST, McManus WR. 2001. Mealybug β-proteobacterial endosymbionts contain γ-proteobacterial symbionts. *Nature* 412:433–36

144. von Dohlen CD, Moran NA. 1995. Molecular phylogeny of the Homoptera: a paraphyletic taxon. *J. Mol. Evol.* 41: 211–23
145. Weeks AR, Hoffmann AA. 1998. Intense selection of mite clones in a heterogeneous environment. *Evolution* 52:1325–33
146. Weeks SC. 1999. Inbreeding depression in a self-compatible, androdioecious crustacean, *Eulimnadia texana*. *Evolution* 53:472–83
147. Wenseleers T, Billen J. 2000. No evidence for *Wolbachia*-induced parthenogenesis in the social Hymenoptera. *J. Evol. Biol.* 13:277–80
148. Werren JH. 1997. Biology of *Wolbachia*. *Annu. Rev. Entomol.* 42:587–609
149. Werren JH, Beukeboom LW. 1998. Sex determination, sex ratios, and genetic conflict. *Annu. Rev. Ecol. Syst.* 29:233–61
150. Werren JH, Zhang W, Guo LR. 1995. Evolution and phylogeny of *Wolbachia*: reproductive parasites of arthropods. *Proc. R. Soc. London Sci. Ser. B* 261:55–63
151. Wilson AC, Sunnucks P, Hales DF. 1999. Microevolution, low clonal diversity and genetic affinities of parthenogenetic *Sitobion* aphids in New Zealand. *Mol. Ecol.* 8:1655–66
152. Wilson ACC, Sunnucks P, Hales DF. 1997. Random loss of X chromosome at male determination in an aphid, *Sitobion* near *fragariae*, detected using an X-linked polymorphic microsatellite marker. *Genet. Res.* 69:233–36
153. Wood DM, Borkent A. 1989. Phylogeny and classification of the Nematocera. In *Manual of Nearctic Diptera, Vol. 3*, ed. JF McAlpine, DM Wood, pp. 1333–70. Ottawa: Agric. Can.
154. Wrensch DL, Ebbert MA. 1993. *Evolution and Diversity of Sex Ratio in Insects and Mites*. New York: Chapman & Hall
155. Wyatt IJ. 1967. Pupal paedogenesis in the Cecidomyiidae (Diptera). 3. A reclassification of the Heteropezini. *Trans. R. Entomol. Soc. London* 119:71–98

Annu. Rev. Entomol. 2003. 48:425–53
doi: 10.1146/annurev.ento.48.091801.112645

BIOCHEMISTRY AND MOLECULAR BIOLOGY OF DE NOVO ISOPRENOID PHEROMONE PRODUCTION IN THE SCOLYTIDAE

Steven J. Seybold
Departments of Entomology and Forest Resources, 219 Hodson Hall, 1980 Folwell Avenue, University of Minnesota, St. Paul, Minnesota 55108-6125; e-mail: sseybold@tc.umn.edu

Claus Tittiger
Department of Biochemistry/330, University of Nevada, Reno, Nevada 89557-0014; e-mail: crt@unr.edu

Key Words bark beetles, *Dendroctonus*, gene expression, HMG-CoA reductase, *Ips*, monoterpenoids

■ **Abstract** Recent application of biochemical and molecular techniques to study the genesis of scolytid aggregation pheromones has revealed that bark beetles are primarily responsible for the endogenous synthesis of widely occurring pheromone components such as ipsenol, ipsdienol, and frontalin. Because many of the chemical signals are isoprenoids, the roles of the mevalonate biosynthetic pathway and the enzyme HMG-CoA reductase (HMG-R) have been investigated. This has led to the identification of endothelial cells in the anterior midgut as the site of synthesis and to the concept that de novo pheromone biosynthesis is regulated in part by the positive effect of juvenile hormone III (JHIII) on gene expression for HMG-R. Both the pronounced regulation by JHIII and the expression pattern of eukaryotic *HMG-R* argue against synthesis of these pheromones by prokaryotes. As the mevalonate pathway and its regulation have been studied in few other insects, broader issues addressed through the study of scolytid pheromone biosynthesis include major step versus coordinate regulation of the pathway and a genomics approach to elucidating the entire pathway and the mode of action of JHIII.

CONTENTS

INTRODUCTION AND OVERVIEW

Isoprenoids comprise an abundant and richly diverse family of biologically active natural products with over 22,000 described compounds (37). Members of this family are united through their construction from the repeating isoprene (C_5) motif yielding end products ranging from the hemiterpenoids to linear and multicyclic polymeric isoprenoids such as dolichol, ubiquinone, and the steroids (29) (Figure 1). In most organisms, mevalonate is the defining intermediate of the isoprenoid biosynthetic pathway (29). During the past decade, the biochemical significance of isoprenoids has approached that of other lipids, carbohydrates, proteins, and nucleic acids. Whereas isoprenoid synthesis and function were once a curiosity known primarily to natural products chemists, the role of isoprenoids in the prenylation of membrane-associated proteins (e.g., cell signaling and human carcinogenesis) and the startling discovery of a new pathway to isopentenyl diphosphate in archaebacteria and plants (112) have positioned isoprenoid biochemistry and molecular biology centrally in the awareness of many chemically oriented biologists.

Although vertebrates direct large quantities of the carbon in the isoprenoid biosynthetic pathway into steroid (C27) end products, insects do not possess squalene synthase and thus cannot synthesize sterols de novo (35, 127). Therefore, the carbon flow through this pathway in the Insecta generally ends up in lower-molecular-weight products such as extra-organismal (51) or intra-organismal (114) molecules of chemical communication, or as key structural or functional components of metabolism (5). Examples of insectan isoprenoids range from hemiterpenoid (C_5) bark beetle pheromone components (the methyl butenols) (3, 126) to monoterpenoid (C_{10}) and sesquiterpenoid (C_{15}) aphid alarm pheromone components (100) or termite defense secretions (42, 43) (the pinenes and farnesenes) or the sesquiterpenoid juvenile hormones (JHs) (114) (Figure 1), to the polyprenolic ubiquinones and dolichols, which are key players in respiration and glycoprotein synthesis, respectively (5).

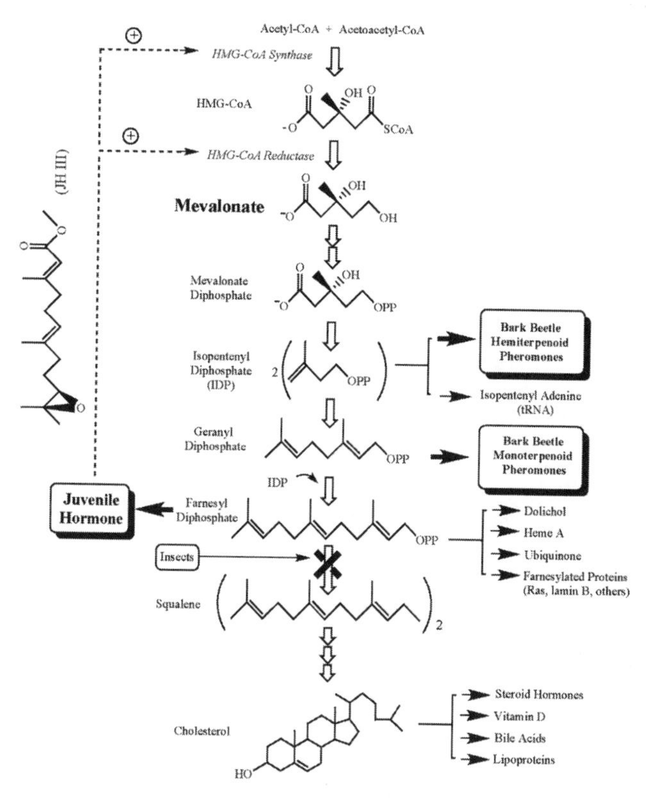

Figure 1 Isoprenoid biosynthetic pathway leading through mevalonate to hemiterpenoid and monoterpenoid pheromone products of the Scolytidae. The biosynthesis is regulated by juvenile hormone III, which is a sesquiterpenoid product of the same pathway [adapted from (118) and used with permission of the Entomological Society of Canada].

Perhaps the best-studied insect isoprenoids are the monoterpenoid aggregation pheromone components of bark beetles (Scolytidae, primarily *Ips* spp. and *Dendroctonus* spp.) and weevils (Curculionidae, primarily *Pissodes* spp. and *Anthonomus grandis* Boheman) (Figure 2). Examples include ipsenol and ipsdienol (119, 123, 147, 149), amitinol (50, 124), *E*-myrcenol (25, 95, 109), lanierone (130), *cis*- and *trans*-verbenol (104, 106, 123), verbenone (106), verbenene (58a, 59), pityol (49), sulcatol (27a), frontalin (4a, 83), lineatin (90a), and grandisol/grandisal (49, 67, 137) (Figure 2). The origin and activity of these compounds have attracted scientific attention for many years (1, 12, 66, 99, 120, 145), and these behavioral chemicals are starting to have extensive utility in pest management programs (125). The association of monoterpenoids with the Scolytidae is a consequence of both the endogenous metabolic activities of the beetles and their coevolutionary relationships with host conifers (118), nonhost angiosperms (68), and microorganisms (15, 16, 26, 75, 87, 88). Nonetheless, it has become increasingly clear that the

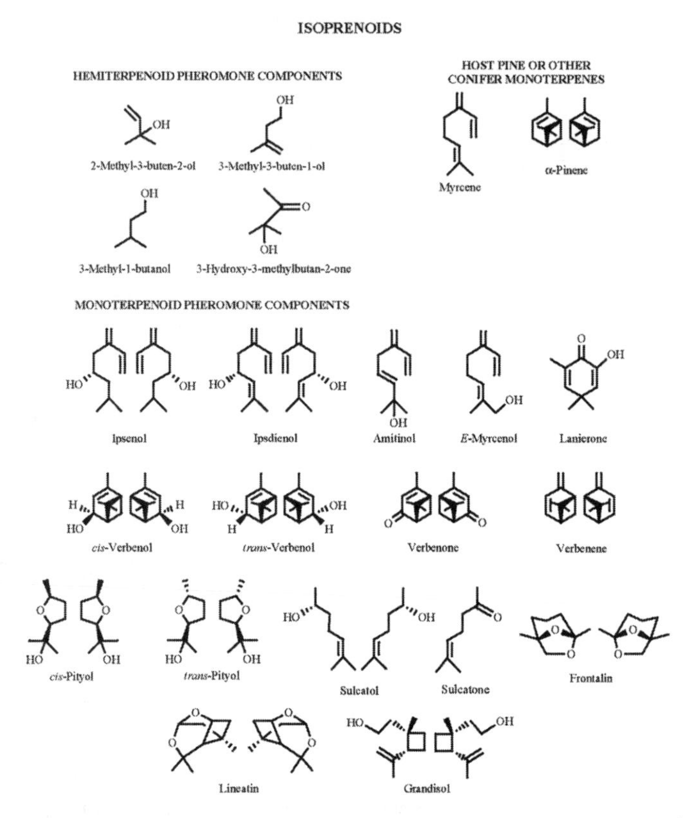

Figure 2 Isoprenoid pheromone components and host pine (or other confier) monoterpenes in the Scolytidae [adapted from (118) and used with permission of the Entomological Society of Canada].

synthesis of certain monoterpenoid bark beetle pheromones is a process whose control and outcome are intrinsically associated with insect tissues, cells, and genes. Aspects of the genesis of aggregation pheromones in the Scolytidae have been reviewed previously (10, 11, 23, 105, 118, 133, 139, 141, 146), but in this overview, we focus on the recent application of modern biochemical and molecular methods to the study of the origin of monoterpenoid pheromone components in this economically important group of forest insects.

BIOCHEMICAL STUDIES OF ISOPRENOID PHEROMONE PRODUCTION IN THE SCOLYTIDAE

Experimental approaches to the synthesis of isoprenoid pheromone components in living scolytids or their isolated tissues have involved the application of various unlabeled and isotopically labeled intermediates or precursors. These substrates have been administered in vivo through feeding, topical application, exposure in the volatile headspace, or injection. In a few cases, the precursors have been administered in vitro by incorporation into the tissue incubation medium. In the vast majority of published accounts, adult beetles were exposed to volatiles of precursors, although in more recent experiments, precursors have been injected through the cuticle. Ultimately, these studies have led to the realization that de novo synthesis of pheromone components may occur frequently in the Scolytidae. In hindsight, the biosynthesis of these isoprenoids from short-chain metabolic building blocks rather than from host precursors was foreshadowed in *Ips* spp. by numerous studies of the endocrine regulation of pheromone production. Pheromone components readily accumulate in beetle hindgut tissue following treatment with juvenile hormone III (JHIII) or its analogs (JHA) in the absence of host precursors (12, 32, 74, 107). Further work that estimated a deficit in the amount of available host precursors (19, 24) or that blocked a key enzyme in the de novo pathway (80) also adumbrated an inherently scolytid synthesis for these compounds.

In the earliest studies, a de novo synthesis was not presumed, and unlabeled host monoterpenes of pines (*Pinus* spp.) whose carbon skeletons matched those of the oxidatively derived behavioral chemicals were used as precursors (Figure 2). Hughes and colleagues (71, 74) first demonstrated that exposure to myrcene in the volatile headspace increased the amounts of ipsenol and ipsdienol in hindgut tissues of the male California fivespined ips, *Ips paraconfusus* Lanier, and related species. Subsequent studies with *I. paraconfusus* established a positive quantitative relationship between the mass of myrcene in the volatile headspace and the mass of ipsenol and ipsdienol in the posterior alimentary canal (27). The route of precursor entry from these in vivo myrcene exposure studies is unclear, but it may have involved passage through the tracheae and tracheoles into the cells and hemolymph of the hemocoel, absorption through the cuticle into the cells and hemolymph, or swallowing into the alimentary canal. Given that much more ipsenol and ipsdienol accumulated in the mid- and hindguts of male *I. paraconfusus* after they had fed on *Pinus ponderosa* Laws. phloem than after exposure to myrcene, Byers (19) suggested that under "natural conditions" (i.e., during feeding on the host) myrcene entered through the alimentary canal with the food bolus.

In later studies, unlabeled myrcene was delivered in the headspace to male *Dendroctonus* spp. (4, 20, 76, 109, 121), the male Eurasian spruce engraver, *Ips duplicatus* Sahlberg (78), the male larch engraver, *Ips cembrae* Heer (107), and the male pine engraver, *Ips pini* (Say) (90). These experiments resulted in the accumulation of ipsdienol and *E*-myrcenol (*Dendroctonus* spp. and *I. duplicatus*) or of ipsdienol and ipsenol (*I. cembrae* and *I. pini*) in hindgut tissues. In a creative approach, the quantity of myrcene in host phloem from five species of pines was used as a screening procedure to suggest that de novo synthesis of ipsenol and ipsdienol occurs in male *I. paraconfusus* (24). The enantiomeric composition of the compounds was also largely invariant when *I. paraconfusus* fed on a wider range of hosts, including nonpines (116). Besides myrcene, unlabeled ipsdienol has been applied topically or as a vapor to *Ips* spp. to confirm its conversion to ipsenol [*I. paraconfusus* and the eastern fivespined ips, *I. grandicollis* (Eichhoff) (71), *I. cembrae* (107), *I. paraconfusus*, and *I. pini* (90)].

The conversion of host α-pinene to the diastereomers of verbenol (and the enantiomers of verbenone) has also been studied using unlabeled materials (4, 17, 19, 21, 22, 59, 69, 70, 72). Hughes (69, 70) exposed adult *Dendroctonus* spp. to α-pinene vapors and found that primarily *trans*- and some *cis*-verbenol and verbenone accumulated in hindgut tissue. Topically applied α-pinene was also converted to *trans*-verbenol and other α-pinene derivatives by female mountain pine beetle, *D. ponderosae* Hopkins, and red turpentine beetle, *D. valens* LeConte (69). To further establish the precursor-product relationship between α-pinene and the verbenols, Byers (19) used unlabeled α-pinene to show that the mass of *cis*-verbenol (and to a lesser extent *trans*-verbenol and myrtenol) in hindgut tissue of male *I. paraconfusus* increased as the amount of α-pinene increased in the volatile headspace, and Hughes (69) showed a positive quantitative relationship between the duration of exposure to vapors of α-pinene and the amount of *trans*-verbenol present in the hindguts of female western pine beetle, *D. brevicomis* LeConte. In a study that used the enantiomeric composition of α-pinene as a rudimentary labeling technique, Renwick et al. (108) found that male *I. paraconfusus* produced *cis*-verbenol from the (−)-enantiomer of α-pinene and *trans*-verbenol from the (+)-enantiomer. Similar production of verbenol from α-pinene originating from the host, Norway spruce, *Picea abies*, occurs in the Eurasian spruce engraver, *I. typographus* (L.) (84, 89). In contrast, both sexes of *D. brevicomis*, the Jeffrey pine beetle, *D. jeffreyi* Hopkins, and *D. ponderosae* convert each enantiomer of α-pinene to the corresponding enantiomer of *trans*-verbenol (4, 21). In *D. jeffreyi* and *D. ponderosae*, the stereospecific conversion of α-pinene to *cis*-verbenol has less fidelity (4). The bioorganic reactions whereby monoterpenes are metabolized to oxygenated monoterpenoids (allylic alcohols, aldehydes, or ketones) in the Scolytidae have been outlined (52, 101) and are likely to involve P450 enzymes (45, 77, 143, 144).

To provide more convincing evidence of pheromone biosynthetic relationships in the Scolytidae, nonradioactive isotopes have been used to label precursors for mass spectrometric end product detection. ^{2}H-labeled myrcene, ipsdienol, and ipsdienone have been used to demonstrate that these monoterpenoids are involved in the late stages of the pathway leading to ipsdienol and ipsenol in *I. paraconfusus* (46, 47, 66) and *I. pini* (138). Gries et al. (59) exposed female *D. ponderosae* to volatiles of ^{2}H-labeled α-pinene [(−)-α-pinene-d_6] to prove the syntheses of *trans*-verbenol, verbenene, *p*-mentha-1,5,8-triene, and *p*-cymene from the deuterated precursor. Studies of bicyclic acetal synthesis in bark beetles have also utilized isotopic labeling with mass spectrometric detection. (*Z*)-6-nonen-2-one, the immediate metabolic precursor to the nonisoprenoid, *exo*-brevicomin, was identified in volatiles trapped from unfed male *D. ponderosae* (140), and the mechanism of in vivo cyclization of this acyclic precursor has also been determined using labeling studies with [O^{18}]oxygen gas (142). The mechanism involves an epoxidation of the C-6 double bond followed by cyclization to *exo*-brevicomin. In an analogous series of reactions, Perez et al. (98) demonstrated that ^{2}H-labeled 6-methyl-6-hepten-2-one was converted to ^{2}H-frontalin in *D. ponderosae* and in the spruce beetle, *D. rufipennis* (Kirby). There are no obvious host sources for the two ketone precursors noted above, but a related methyl ketone (6-methyl-5-hepten-2-one = sulcatone) appears to be synthesized by a fungal symbiont of the southern pine beetle, *D. frontalis* Zimmerman (14, 15). The early stages of the biosynthesis of *exo*-brevicomin have not been studied, but frontalin is clearly synthesized de novo via the isoprenoid pathway [see (4a) and below].

Incontrovertible evidence for the de novo synthesis of isoprenoids by scolytids comes from experiments using injected radiolabeled precursors such as ^{14}C-acetate or ^{14}C-mevalonolactone (a cyclic form of mevalonate), whose incorporation into end products is detected by liquid scintillation counting. If the end product is isolated from a tissue, whole-body, or headspace extract by chromatography [gas chromatography (GC) or high performance liquid chromatography (HPLC)] and then derivatized, the comparative GC or HPLC analyses of the labeled end product and its derivative with unlabeled standards provide compelling evidence for the precursor-product relationship. In the first application of radiotracers with scolytids, ^{14}C-mevalonolactone (but not ^{14}C-acetate) injected into unsexed *I. typographus* was incorporated into hindgut and headspace volatile extracts, and the radioactivity was associated with preparative GC fractions that coeluted with the hemiterpenoid pheromone component, 2-methyl-3-buten-2-ol (86). ^{14}C-glucose fed to these beetles, either on filter paper or in *Picea abies* logs, did not lead to incorporation of radioactivity into the hemiterpenoid. In other in vivo studies, ipsenol, ipsdienol, and amitinol in *I. paraconfusus* and ipsdienol and amitinol in *I. pini* have been labeled from ^{14}C-acetate (120, 131) or ^{14}C-mevalonolactone (90, 131). Following derivatization of the enantiomers of the alcohols to diastereomeric esters, HPLC analysis revealed that the enantiomeric compositions of the labeled alcohols matched the natural products (119, 120). ^{14}C-acetate and ^{14}C-mevalonolactone have also been incorporated into ipsenone (male *I. paraconfusus*) and ipsdienone (male *I. pini*) during an in vitro assay of homogenized tissue (79). This assay was later used to localize the synthesis of radiolabeled ipsenone to the metathorax of male *I. paraconfusus* (81), and isolated midgut tissue from this body region of male *I. pini* converted ^{14}C-acetate to ^{14}C-ipsdienol (61). Copious production of frontalin by male *D. jeffreyi* has made it an excellent experimental animal in which to study the biosynthesis of this bicyclic acetal, and ^{14}C-acetate, mevalonolactone, and isopentenol have each been injected into male *D. jeffreyi* with radiolabel incorporated into HPLC fractions associated with frontalin (4a). Incorporation of ^{13}C-acetate into frontalin by this insect was demonstrated unequivocally using isotope ratio monitoring mass spectrometry, a technique that holds great promise of direct assessment of pheromone biosynthesis. ^{3}H-leucine can also be used by male *D. jeffreyi* as a substrate for ^{3}H-frontalin synthesis (4a), leading to biosynthetic hypotheses involving fatty acid–like elongation of leucine or catabolism of leucine to acetyl-CoA followed by an isoprenoid synthesis via hydroxy-methylglutaryl-CoA (HMG-CoA) as alternative routes to frontalin in addition to the mevalonate pathway. Finally, isolated midgut tissue from male *D. jeffreyi* also converted ^{14}C-acetate to ^{14}C-frontalin (62).

The late-stage reactions of the isoprenoid pathway (those between isopentenyl diphosphate and the pheromone end products) (Figure 1) are beginning to be examined in *Ips* spp. (92, 118, 152). Activated geraniol (geranyl diphosphate) is expected to be central to this portion of the pathway. Wood et al. (148) found geranyl acetate in extracts of male boring dust (frass) from *I. paraconfusus*, whereas Ivarsson & Birgersson (78) reported small amounts of geraniol from hindgut extracts of male *I. duplicatus*. In the latter study, trends in accumulation of geraniol corresponded to treatments expected to stimulate and inhibit pheromone production. Homogenized tissue from male *I. pini* contained geranyl diphosphate in large quantities, indicative of activity of geranyl diphosphate synthase (152). Surprisingly, when whole-body extracts of *I. pini* from two populations were incubated with ^{3}H-geranyl diphosphate, the extracts of males sex specifically converted the substrate to ^{3}H-myrcene (92). This represents the first evidence for a monoterpene synthase and the biosynthesis of a monoterpene (sensu stricto) by a metazoan. In other late-stage reactions, the conversion of myrcene and ketone analogs of ipsenol and ipsdienol to the alcohols have been investigated in *I. paraconfusus* (46, 47, 78, 79, 81, 90, 138) and in *I. pini* (90, 79, 138). Much work remains to elucidate the "uniquely scolytid" biochemical transformations that promise to be found in the late-stage reactions of the mevalonate pathway.

ISOLATION AND CHARACTERIZATION OF GENES FOR ISOPRENOID PATHWAY ENZYMES IN THE SCOLYTIDAE

Background

Biochemical studies have demonstrated that the mevalonate pathway (Figure 1) is essential for de novo pheromone biosynthesis in several species of Scolytidae. In vertebrate systems, two enzymes, HMG-CoA reductase (HMG-R, EC 1.1.1.34) and HMG-CoA synthase (HMG-S, EC 4.1.3.5), are pivotal to the regulation of carbon flow through this pathway to sterol end products (58). Limited studies have also highlighted the significance of these enzymes to isoprenoid biosynthesis in other insects (18, 28, 30, 31, 55, 93, 93a). In addition to the classical precursor-product studies outlined in the previous section, another approach to elucidating the isoprenoid pathway in relationship to scolytid pheromone production is to investigate the molecular biology of HMG-R and HMG-S.

Isolation of HMG-R and HMG-S cDNAs

Until quite recently, the Scolytidae had no track record as research organisms for molecular studies (117). However, the development of polymerase chain reaction (PCR) methods have made nearly any group of organisms genetically tractable if a suitably homologous gene exists. The first cDNA fragment for a scolytid HMG-R was isolated from *I. paraconfusus* by 3′ RACE (rapid amplification of cDNA ends) PCR (134). Alignments of metazoan HMG-Rs in GenBank revealed numerous highly conserved regions, particularly in the catalytic (carboxy terminal) portion of the enzyme. Within the catalytic domain, a highly conserved octapeptide region containing three methionine codons and two 2-fold degenerate codons proved a superb choice for the design of the 32-fold degenerate PCR primer, "PARAHMD." Located within a kilobase of the 3′ utr (untranslated region), the site allowed convenient amplification of first-strand cDNA templates by 3′ RACE. Indeed, this primer was used to isolate HMG-R cDNA fragments from *I. paraconfusus* (134), *I. pini* (61), and *D. jeffreyi* (133a).

The cDNA fragments were then used to isolate their corresponding complete cDNAs by a combination of PCR and/or library screening. cDNA libraries were constructed for *I. pini* and *D. jeffreyi* and screened either by PCR or by classical plaque hybridizations. For *I. paraconfusus*, a cDNA library was not constructed; however, the 5′ end of the HMG-R cDNA was recovered using a combination of reverse transcriptase (RT)-PCR (with a degenerate primer near the 5′ end of the open reading frame) and ligation-anchored (LA)-PCR (136). The scolytid HMG-R cDNAs are available in GenBank (accession numbers AF071750 for *I. paraconfusus*, AF304440 for *I. pini*, and AF159136 for *D. jeffreyi*). A similar strategy was used to isolate HMG-S cDNA (AF166002) from *D. jeffreyi* (135). Degenerate primers based on conserved portions of the German cockroach, *Blattella germanica* (L.), HMG-S protein were used for RT-PCR of first-strand cDNA that had been primed with oligo-dT. As with HMG-Ss from *B. germanica* (18, 93a) and *D. melanogaster* Meigen (48), the absence of an amino-terminal targeting sequence in the cDNA for the scolytid enzyme predicts that the protein will be cytosolic.

The primary structures of bark beetle HMG-Rs inferred from the cDNAs are highly conserved when compared with other metazoan HMG-Rs (133a). The proteins have an amino-terminal membrane anchor consisting of eight transmembrane segments, followed by a short hydrophilic linker and a carboxy-terminal catalytic domain (64, 97). In mammalian HMG-R, membrane anchor structure is important because the anchor plays a critical role in regulating enzyme stability by sensing cholesterol concentrations in membranes (34, 85, 115). Because insects are cholesterol auxotrophs and sterol levels do not modulate *D. melanogaster* HMG-R activity (17a, 122), the membrane anchor structure may have been conserved between insects for another functional purpose such as subcellular targeting (55).

Gene Structure of *HMG-R*

Bark beetle *HMG-R* gene structures are of interest because of the roles that introns play in regulating mammalian HMG-R activity. The only bark beetle genomic *HMG-R* sequences reported so far are from *D. jeffreyi* (AF159137, AF159138, AH009515), and these correspond to the transcribed portion of the gene and do not include upstream control elements (133a).

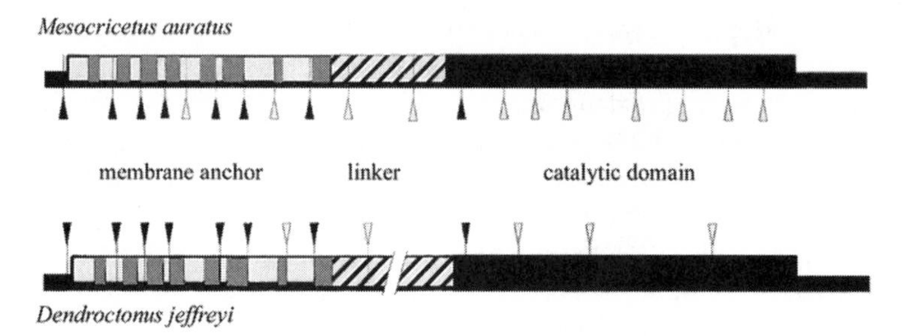

Figure 3 Comparison of intron splice sites in HMG-R genes from the Syrian hamster, *Mesocricetus auratus* (Waterhouse) (110) (above), and the Jeffrey pine beetle, *Dendroctonus jeffreyi* Hopkins (133a) (below). Untranslated regions (utrs) are indicated by *thin lines* and coding regions are indicated by *boxes*. *Boxes* representing membrane-spanning domains are shaded *gray*, hydrophilic loops are *white*, linker domains are *striped*, and catalytic domains are *black*. The *diagonal slash* in the *D. jeffreyi* gene indicates a gap inserted to maintain the alignment of the catalytic domains. Intron sites are indicated by *vertical lines capped with triangles*. *Black triangles* indicate conserved (identical or similar) intron sites between the two species, whereas *gray triangles* indicate unconserved intron sites.

D. jeffreyi HMG-R spans nearly 10 kb and is interrupted by 13 introns (Figure 3). When compared with HMG-Rs from the Syrian hamster, *Mesocricetus auratus* (Waterhouse)(110), and from humans, *Homo sapiens* L. (95a), there is an asymmetry in the conservation of the intron sites between the scolytid and mammalian genes. In the 5′ utr and the membrane anchor-encoding domain, seven of eight *D. jeffreyi* intron positions are conserved with their mammalian homologs, whereas in the linker and catalytic domains only one of five positions is conserved. This suggests an important functional role for most introns associated with the anchor domain. Furthermore, introns at the borders of transmembrane domains or linker regions may indicate the evolutionary assembly of the membrane anchor by exon shuffling (95a, 110).

In contrast to the suggested evolutionary role played by introns in the membrane anchor domain, the first intron, located in the 5′ utr, may have a current functional role. Alternative splicing of this intron leads to differential transcript stability in mammals (54, 110), and it is tempting to speculate that similar regulation occurs in bark beetles. Our own Northern blot experiments infrequently reveal longer-than-normal HMG-R mRNA signals in *I. pini*, although we have not confirmed if these are due to alternative splicing (C. Tittiger, unpublished data). Further, analyses of the first intron in *D. jeffreyi HMG-R* using three algorithms (FGENESH+, HSPL, and Gene Mark) predict various splice donor and acceptor sites (133a). Similarly, *HMG-S* in *D. melanogaster* is predicted to undergo alternative splicing at the first intron, located in the 5′ utr (48), and alternative splicing of introns in the 5′ utr of *M. auratus* cytosolic HMG-S hnRNA is common (56). Intron sites

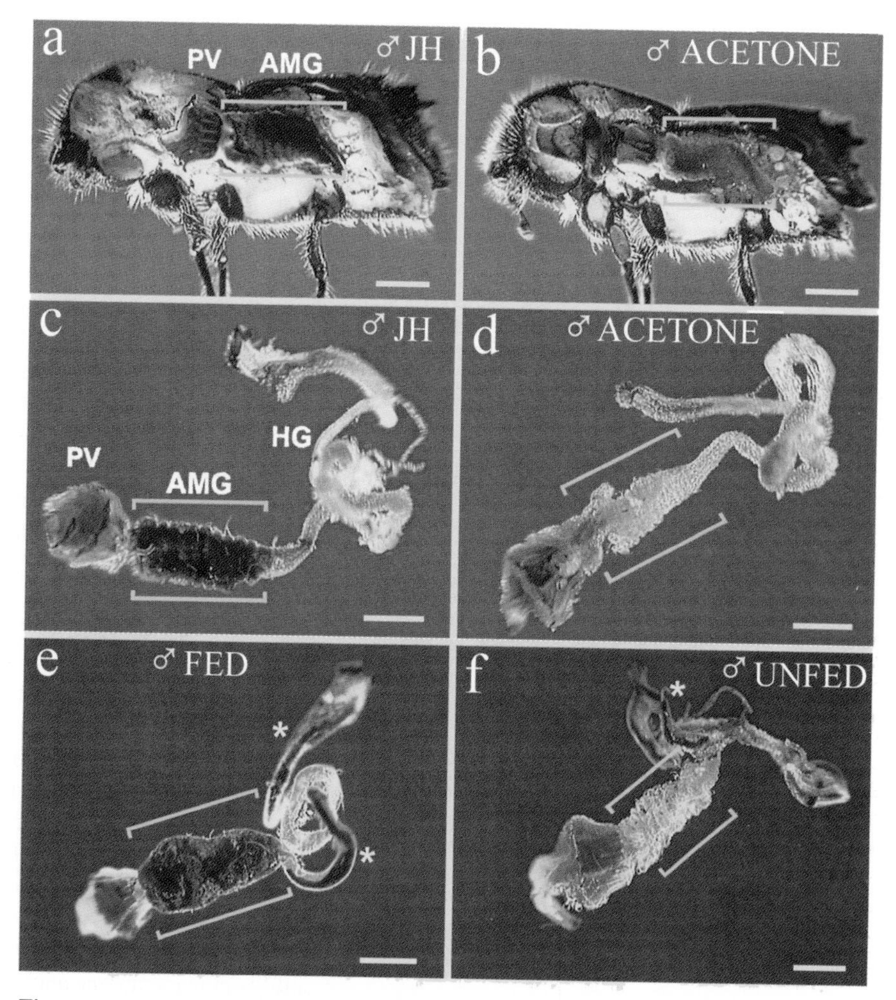

Figure 4 In situ hybridization of digoxigenin-labeled anti-sense *HMG-R* riboprobe with *Ips pini* (Say) whole-body, sagittal sections of (*a*) juvenile hormone III (JHIII)-treated males, (*b*) acetone-treated males, as well as of isolated alimentary canals of (*c*) JHIII-treated males, (*d*) acetone-treated males, (*e*) phloem-fed males, and (*f*) unfed males. Female and sense *HMG-R* male controls (both JHIII-treated) showed no hybridization (data not shown). Yellow brackets enclose the anterior midgut region in all panels. AMG, anterior midgut; HG, hindgut; PV, proventriculus; and *, nonspecific staining of fecal material in the posterior alimentary canal. Scale bars = 0.5 mm. Figure adapted from Figure 2 in Reference 61 with permission of the author and Springer-Verlag.

have yet to be mapped for any scolytid *HMG-S*. Given the precedents in other animals, posttranscriptional regulation of HMG-R, particularly given regulatory differences in pheromone biosynthesis between *I. paraconfusus* and *I. pini*, may depend on splicing patterns in the 5′ utr.

In other Metazoa studied so far, *HMG-R* is a single-copy gene, whereas *HMG-S* occurs as separate genes for mitochondrial and cytosolic enzymes (2). *D. jeffreyi* and *I. pini* have single-copy *HMG-R*s (133a), but we found evidence for only one copy of *HMG-S* in *D. jeffreyi* (135). There are two copies of *HMG-S* in *Blattella germanica*, but both encode cytosolic enzymes (18). Because one of the *B. germanica* genes evidently evolved recently by retrotransposition (30), and there is only one *HMG-S* in *D. jeffreyi* and *Drosophila* (48), most insect genomes may only contain single copies of *HMG-S*.

STUDIES OF THE SITE OF MONOTERPENOID PHEROMONE PRODUCTION

In most insects, pheromones are synthesized in specialized cells or tissues associated with the epidermis (8, 133). Localization of the site of monoterpenoid pheromone production in scolytids has proceeded using increasingly sophisticated methodology. Prior to isolation of the aggregation pheromone of *I. paraconfusus*, the response of female beetles in a laboratory walking bioassay was used to localize the attractant to the male hindgut (including the Malpighian tubules) and the fecal pellet that emanated from that body region (12, 102, 103, 148). Similar integrated anatomical/behavioral studies were conducted with the Douglas-fir beetle, *Dendroctonus pseudotsugae* Hopkins (153), and the striped ambrosia beetle, *Trypodendron lineatum* (Olivier) (13). Chemical detection of pheromone components by GC and GC-MS led to more definitive studies of the accumulation of monoterpenoids in the hemolymph of *D. ponderosae* and *D. valens* (69) and in various regions of the alimentary canal of *I. paraconfusus* and *D. brevicomis* (22) or throughout the body of *D. pseudotsugae* (91).

Knowledge of the involvement of the mevalonate pathway in pheromone biosynthesis in *I. paraconfusus*, *I. pini*, and *D. jeffreyi* has been exploited to further elucidate the site of biosynthesis. An in vitro bioassay of the synthesis of ipsenone and ipsdienone from ^{14}C-acetate in male *I. paraconfusus* demonstrated that of six body regions and organs radiolabeled product was most abundant in the metathorax (81). Surprisingly, when fat body, flight muscle, and abdomen were compared in the assay, flight muscle had the highest amount of incorporation.

Measurement of the expression of *HMG-R* has also been used effectively to delineate the site of isoprenoid (i.e., pheromone) biosynthesis in scolytids. Following treatment with JHIII, HMG-R mRNA accumulated primarily in the male thorax (not head or abdomen) of both *I. paraconfusus* (81) and *I. pini* (61). In male *D. jeffreyi*, minor variation in the sectioning method affected the distribution of the signal, but expression of both *HMG-S* (135) and *HMG-R* (62) was localized to the junction of the metathorax and abdomen.

The highest resolution mapping of *HMG-R* expression to date has been achieved in males of *I. pini* and *D. jeffreyi* by a modified in situ hybridization technique using antisense RNA for HMG-R (61, 62). Although the technique is not quantitative, the striking "beetle-on-the-half-shell" images (Figure 4*a*,*b*, see color insert) allow clear differentiation between high- and low-expressing tissues. In isolated alimentary canals of JHIII-treated and *Pinus jeffreyi* Grev. & Balf. phloem-fed male *I. pini*, the anterior midgut of these pheromone-producing insects specifically and strongly expressed *HMG-R* (Figure 4*c*,*e*). Anterior midgut tissues from acetone-treated and unfed male *I. pini* did not have elevated *HMG-R* mRNA levels (Figure 4*d*,*f*). Control alimentary canals from both JHIII-treated females incubated with antisense HMG-R and JHIII-treated males incubated with sense HMG-R did not show expression (61). A similar analysis of JHIII-treated male and female *D. jeffreyi* also confirmed that the male anterior midgut is the site of elevated *HMG-R* expression, and radiotracer incorporation studies with males of both *I. pini* and *D. jeffreyi* verified that the pheromone components ipsdienol and frontalin, respectively, were produced in this tissue (61, 62).

In males of both *I. pini* and *D. jeffreyi*, the JHIII-mediated increase in *HMG-R* expression is accompanied by a dramatic change in the subcellular structure of midgut cells. These putative pheromone-synthesizing cells have large crystalline arrays of smooth endoplasmic reticulum (SER) (96). Overproduction of SER is common in cholesterol-deprived mammalian cells that produce high levels of HMG-R (6, 33). By analogy, pheromone-synthesizing midgut cells likely have large amounts of crystalline SER because they are strongly expressing *HMG-R*, and they likely have correspondingly elevated HMG-R protein levels. Two conclusions can be drawn from recent efforts to localize pheromone biosynthesis in bark beetles: (*a*) There is not a pheromone gland per se. Rather, all cells in the midgut appear to be involved in pheromone production because the signals from in situ hybridizations are evenly distributed throughout the tissue (61, 62), and the SER arrays are observed in all midgut cells (96). Thus, cells in the midgut are actively involved in both digestion and pheromone synthesis and release. (*b*) De novo synthesis of pheromone components in the anterior midgut involves insect tissues, not symbiotic bacteria. Prokaryotic HMG-Rs are both structurally and mechanistically different from their eukaryotic homologs (53, 128), so activity, gene expression, and subcellular changes accompanying pheromone production can be attributed solely to scolytid tissues.

STUDIES OF THE REGULATION OF MONOTERPENOID PHEROMONE PRODUCTION

Background

In the earliest studies of isoprenoid pheromone production by male *I. paraconfusus*, both adult male maturity (9, 22, 150) and feeding in host phloem (145, 148) were noted as regulatory factors. Later research established that JHIII or fenoxycarb (a JHA) (12, 32, 74) regulates monoterpenoid pheromone production in *I. paraconfusus*. Further experiments with other *Ips* spp. (60, 78, 107), *Dendroctonus* spp. (4, 17, 36, 73), *Pityokteines* spp. (65), and *Scolytus* spp. (7) defined a central role for JHIII as a regulator for pheromone production throughout the Scolytidae.

Major Step Versus Coordinate Regulation of Isoprenoid Synthesis

In vertebrates, isoprenoid biosynthesis is negatively regulated by the titer of the primary product, cholesterol (58), and potentially by the titer of farnesol (38, 94). The enzymes HMG-S and HMG-R are the primary points of regulation in the pathway. Regulation of HMG-R is complex, involving coordinate changes in gene expression, processing and stability of RNA and protein via feedback inhibition of mevalonate pathway end products (58), as well as cross-regulation, which occurs when biochemical processes such as cellular energy charge (i.e., ATP level) and cellular oxygen availability modulate levels of HMG-R activity (64). In contrast, in scolytid pheromone biosynthesis, the primary product of the mevalonate pathway is likely a monoterpenoid (C_{10}), and the pathway is positively regulated by a sesquiterpenoid (C_{18}, JHIII) derived from farnesyl diphosphate (Figure 1). Ignoring the more complex levels of regulation known from vertebrates and yeast, one hypothetical regulatory paradigm in the scolytids is that JHIII regulates the pathway only at the level of HMG-S and HMG-R. Alternatively, all enzymes in the pathway may be regulated in a coordinate fashion by JHIII, with major regulation exerted through HMG-S and HMG-R. In the Scolytidae, HMG-S would likely be regulated to a lesser extent, reflecting the less prominent role that it has in controlling the mevalonate pathway in vertebrates (58).

Biochemical Studies of Regulation Through JHIII and Feeding

Ivarsson & Birgersson (78) used the HMG-R inhibitor compactin and the JHA methoprene to offer indirect evidence that JH regulates de novo pheromone biosynthesis in male *I. duplicatus*. Using [1-^{14}C]acetate and [2-^{14}C]mevalonolactone in vivo and L-[*methyl*-^{3}H]-methionine in vitro, Tillman et al. (131) evaluated the relationships among feeding on host (*P. jeffreyi*) phloem, JH biosynthesis, and de novo ipsdienol biosynthesis by male *I. pini*. HPLC analysis of corpora allata (CA) extracts demonstrated that the type of JH released by the CA in male *I. pini* is JHIII. Also, increasing incorporation of radiolabeled acetate into ipsdienol by male *I. pini* with increasing topical JHIII dose demonstrated unequivocally that JHIII regulates de novo pheromone production (131). However, incorporation of radiolabeled mevalonolactone into ipsdienol by male *I. pini* was not affected by increasing JHIII dose, suggesting that JH primarily influences enzymes prior to mevalonate in this pathway (i.e., HMG-S and HMG-R). In contrast, preliminary biochemical studies of some of the late-stage reactions in monoterpenoid biosynthesis suggest that the conversion of ^{3}H-geranyl diphosphate to ^{3}H-myrcene is upregulated in male *I. pini* by both JHIII treatment and feeding on *P. jeffreyi* phloem (92).

De novo ipsdienol biosynthesis by male *I. pini* was also stimulated by feeding on host phloem (131); thus, feeding may be the initial environmental cue that stimulates the intermediary biosynthesis and release of JH from the CA. This hypothesis was addressed using an in vitro assay that compared levels of JH released (likely biosynthesis) (44) from CA in unfed and previously fed male and female *I. pini*. The rate of JHIII release from the CA was significantly higher in male *I. pini* that had fed for 24 h relative to unfed (24-h incubated) males, whereas females displayed overall lower rates of JH release. No significant differences between fed and unfed females were found at the time points assayed. These radiolabeling studies collectively provide evidence for a behavioral and physiological sequence of events leading to feeding-induced de novo pheromone biosynthesis in male *I. pini*: (*a*) feeding on host phloem, (*b*) feeding-induced JHIII release (i.e., biosynthesis) by the CA, and (*c*) JHIII-stimulated de novo ipsdienol biosynthesis.

It is not clear whether JHIII alone is sufficient to upregulate de novo pheromone biosynthesis in all *Ips* spp. Following a series of decapitation and gland (CA and corpora cardiaca) implantation experiments, Hughes & Renwick (74) proposed that JH may act indirectly through a brain hormone to stimulate pheromone biosynthesis in male *I. paraconfusus*. To date, no one has conducted enzyme assays of purified HMG-S or HMG-R; however, Tillman et al. (132) monitored the conversion of ^{14}C-HMG-CoA to ^{14}C-mevalonate by crude microsomal preparations from JHIII-treated or phloem-fed male *I. paraconfusus* and *I. pini* to establish a measure of HMG-R activity. In this study, the effects of topical application of JHIII and feeding were compared with regard to pheromone production, HMG-R activity, and HMG-R transcript levels in both *I. paraconfusus* and *I. pini*. The comparative results of these studies indicate that (*a*) JHIII induces 150-fold more of the principal pheromone component of male *I. pini* (ipsdienol) than it does of the principal pheromone component of male *I. paraconfusus* (ipsenol), whereas feeding induces similar amounts of the principal pheromone components in each species; (*b*) JHIII stimulates HMG-R enzyme activity in male *I. pini*, but not in male *I. paraconfusus*, whereas feeding stimulates similar amounts of enzyme activity in each species; and (*c*) JHIII induces similar increases in the transcript for HMG-R in both species (Figure 5). Due to the difficulty of isolating mRNA from phloem-fed insects, the effects of feeding on transcription remain to be measured. Lu (90) repeated this study with decapitated male *I. paraconfusus* and male *I. pini* and largely validated the results of Tillman et al. (132). Additionally, Lu (90) showed that decapitation reduced the amount of HMG-R transcript, HMG-R activity, and de novo ipsdienol produced by *I. pini*, but only reduced the amount of HMG-R transcript in *I. paraconfusus*. In *I. paraconfusus*, although JHIII alone can upregulate the transcript for HMG-R, JHIII may act in concert with a second feeding-associated factor (hormone) to stimulate HMG-R translation, transcript or protein stability, enzyme activation, catalysis, and eventually, ipsdienol and ipsenol biosynthesis through full activity of HMG-R. This factor does not appear to be necessary in *I. pini*. It is surprising the degree to which the regulation of de novo pheromone biosynthesis differs in these two closely related species of Scolytidae, and one must be cautious

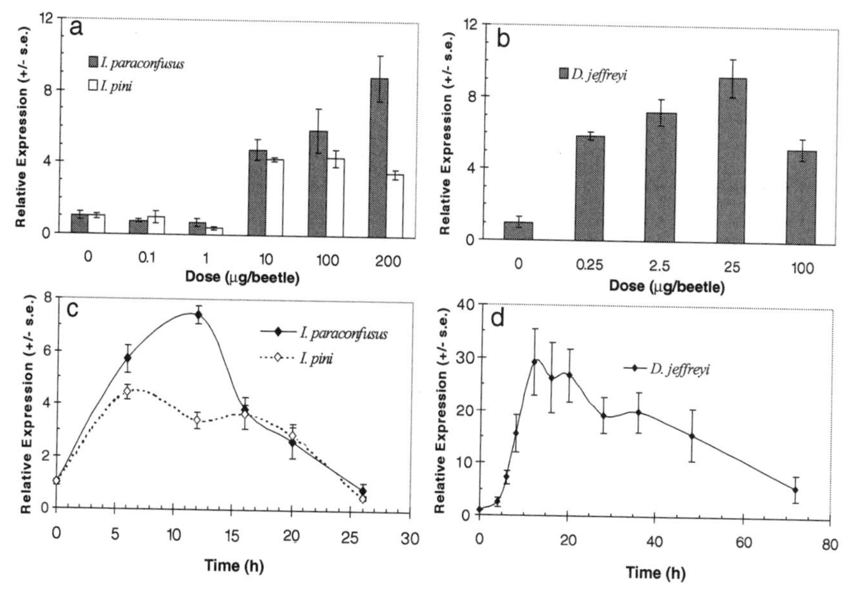

Figure 5 Summary of Northern experiments investigating juvenile hormone III (JHIII)-regulated HMG-R gene expression in male pine bark beetles. Dose response curves are presented for *(a) I. paraconfusus* and *I. pini* (samples taken 20 h following topical JHIII application) and for *(b) D. jeffreyi* (samples taken 20 h following JHIII application). Time response curves are presented for *(c) I. paraconfusus* and *I. pini* (following topical application of 10 mg JHIII) and for *(d) D. jeffreyi* (following application of 13.3 μg JHIII). All measurements were replicated three times [*Ips* spp.: 10 intact males/sample (132); *D. jeffreyi*: 5 isolated male thoraces/sample (4, 133a)]. Expression levels were normalized to mouse actin and control (0 μg JHIII or 0 h incubation) signals.

when extending information about pheromone regulation from one bark beetle species to another (132).

Molecular Studies of Regulation Through JHIII and Feeding

The first demonstration of the endocrine regulation of a pheromone biosynthetic gene in insects was a study of JHIII-regulated *HMG-R* mRNA levels in male *I. paraconfusus* (134). Both dose- and time-effects of JHIII treatment on *HMG-R* expression have been explored in *I. paraconfusus*, *I. pini* (132), and *D. jeffreyi* (4, 133a). Males of all three species elevate HMG-R mRNA levels in response to JHIII dose, but there are interspecific differences in the details of the responses (Figure 5). For example, although both species of *Ips* and *D. jeffreyi* show a general trend toward increased *HMG-R* expression with increasing JHIII dose, low doses ($\leq$1 μg) do not stimulate *HMG-R* expression in the *Ips* species, but they do in

D. jeffreyi. The highest JHIII doses tested (100–200 μg) depress *HMG-R* mRNA levels compared to maximal levels in *I. pini* and *D. jeffreyi*, but not in *I. paraconfusus*. Similarly, all three species show a pulse of elevated *HMG-R* expression over time in response to topical JHIII application, but peak expression levels are reached within 6 to 12 h and decline to basal levels within 26 h in *Ips* spp., whereas peak levels occur within 12 to 20 h, but remain above basal levels after 72 h in *D. jeffreyi*. The maximal induction level for *HMG-R* mRNA in *I. paraconfusus* and *I. pini* is ~8-fold above uninduced levels, whereas in *D. jeffreyi* levels can rise ~30-fold (Figure 5).

The expression of *HMG-S* has been investigated in male *D. jeffreyi* (135), and as is the case for *HMG-R*, *HMG-S* responds to JHIII in a dose- and time-dependent manner. Consistent with HMG-S playing a minor role in regulating the mevalonate pathway (58), maximal expression levels for *HMG-S* are only ~4-fold above uninduced levels. The coordinate regulation of transcription of *HMG-S* and *HMG-R* in *D. jeffreyi* implies that genes for other enzymes in the pheromone biosynthetic pathway may also be induced by JHIII. Indeed, preliminary studies of a putative geranyl diphosphate synthase gene in male *I. pini* suggest modest induction following JHIII treatment (152).

Because *HMG-R* (and in *D. jeffreyi*, *HMG-S*) mRNA levels depend on JHIII dose, their decline over time probably reflects decreasing JHIII titers within the insect. Active JHIII titers following topical application are unknown in the Scolytidae, but some fraction of the topically applied JHIII will likely be trapped on the cuticle or catabolized in the hemolymph before reaching the midgut (63). Also, although it is generally safe to assume that higher levels of HMG-S and HMG-R mRNA are due to increased transcription, the possibility of altered mRNA stability cannot be discounted and is yet to be tested.

Developmental expression profiles for *HMG-R* throughout the entire bark beetle life cycle have not been investigated, but the ability to respond to JHIII by elevating mevalonate pathway genes is presumed to be characteristic of mature, adult (i.e., pheromone-producing) males. The interaction of adult maturation with the ability of the mevalonate pathway to respond to JHIII has been studied in male *D. jeffreyi* that have eclosed, but have not yet emerged through the bark of their brood tree (defined as pre-emerged adults). These immature males do not have frontalin in their hindgut tissue (4); however, following treatment with JHIII, they show weak *HMG-R* expression and enzyme activity and produce a small quantity of frontalin. Groups of emerged, JHIII-treated male *D. jeffreyi* strongly express *HMG-R*, have high HMG-R activity levels, and produce comparatively large amounts of frontalin. These findings suggest that pre-emerged, JHIII-treated males may have an attenuated ability to respond to JHIII. However, there is wide variability in the system as individual pre-emerged, JHIII-treated males can have *HMG-R* mRNA levels ranging from undetectable to highly elevated.

The JHIII titers involved in these investigations, although artificial, tell us something about the ability of the insects to respond to the hormone. However, to learn about naturally occurring events, it is more physiologically relevant to study expression levels in response to feeding. Although feeding stimulates the activity of HMG-R in *I. paraconfusus* and *I. pini*, technical difficulties associated with the extraction of mRNA from phloem-fed beetles must be overcome to study the impact of feeding on *HMG-R* expression in *Ips* spp. or *D. jeffreyi* (132). Future progress in this area may involve work with artificial diets.

COMPLETE ELUCIDATION OF THE STRUCTURAL AND REGULATORY COMPONENTS OF THE ISOPRENOID PATHWAY IN THE SCOLYTIDAE

Biochemical Approaches

The synthesis of scolytid isoprenoid pheromones diverges from the mevalonate pathway at isopentenyl diphosphate and geranyl diphosphate (Figure 1). Thus, a complete description of the de novo biosynthetic pathway involves the substrates, activities, and regulatory factors from acetyl-CoA to end products such as 2-methyl-3-buten-2-ol, ipsdienol, or frontalin. In the case of the monoterpenoids, the late-stage reactions such as the synthesis of geranyl diphosphate, the conversions of geranyl diphosphate to myrcene or 6-methyl-6-hepten-2-one, and the stereospecific hydroxylation of myrcene or a substrate of similar oxidation state to yield *R*- or *S*-ipsdienol, and eventually, *S*-ipsenol are all expected to be highly novel and perhaps unique to the Scolytidae (118). Furthermore, there is likely to be considerable variability in the P450s involved in the hydroxylation reactions. For example, Lu (90) found that a California population of male *I. pini* hydroxylates myrcene in the volatile headspace to form racemic ipsdienol, whereas JHIII-treated or phloem-fed male *I. pini* produce the behaviorally active (−)-enantiomer. Thus, if myrcene is synthesized de novo as an intermediate in the pheromone biosynthetic pathway, enantio-specific hydroxylation to ipsdienol probably involves a dedicated enzyme distinct from that used for the detoxification of environmental myrcene from the coniferous host. In addition to discovering novel enzymatic activities, a complete understanding of the biosynthesis of scolytid pheromones may reveal the details of how JHIII acts on pheromone biosynthetic cells.

The classical approach to describing the pathway completely would be to test each substrate in vivo for conversion to the pheromone end product and then test the conversion of each substrate to succeeding substrates under in vivo and in vitro assay conditions. This would be followed by functional and structural characterization of each enzyme activity and its regulation. To date, the outlines of the complete pathway have been sketched through in vivo studies of the conversion of acetate, mevalonate, leucine, and isopentenol to ketone and alcohol end products and through in vitro analyses of the conversion of HMG-CoA to mevalonate and geranyl diphosphate to myrcene. These functional assays have been carried out in the presence or absence of regulatory factors such as JHIII or feeding.

Our current understanding of the effects of JHIII or feeding on HMG-R and HMG-S is limited to mRNA and enzyme activity levels (for HMG-R). We still know little about the proteins, which may be regulated through translational activity, protein stability, and/or enzyme activation (38, 113, 129). In *I. paraconfusus*, enzyme activity is not well correlated with HMG-R gene expression, suggesting that research at the protein level will be crucial. The similarity of scolytid HMG-R membrane anchor structure to vertebrate structure suggests a conserved functional role beyond targeting, perhaps also in regulated degradation. Because insects are sterol auxotrophs, farnesol may affect insect HMG-R stability, as has been suggested in mammals (38, 94). Furthermore, although cytological study (96) suggests that HMG-R protein is targeted to the SER, this has not yet been conclusively demonstrated. Given evidence from mammals (82), HMG-R may (also) be distributed in peroxisomes. Similarly, HMG-S is predicted to be a cytosolic enzyme, although insects appear to have the ability to synthesize ketone bodies and thus may also require a mitochondrial HMG-S (135). Immunochemical studies will be needed to quantify the levels of HMG-R and HMG-S from biochemical extracts of scolytids and to determine all subcellular locations for HMG-R, HMG-S, and other enzymes in the mevalonate pathway.

Genomics Approaches

Classical techniques could be exhaustively applied to reach a complete description of the mevalonate pathway in the Scolytidae, but limited amounts of available tissue, poorly defined genetics, and the absence in most cases of a clearly identified gene to study make this a daunting challenge. Fortunately, recently developed high-throughput sequencing and hybridization technologies, i.e., functional genomics, can be applied to any organism as a means to isolate relevant clones. The approach begins with isolating hundreds or thousands of cDNAs indiscriminately and then, through a series of steps including sequence analyses, microarray hybridization studies, and computer modeling, identifying a handful of cDNAs as potential candidates (111, 129a). Interesting clones from pheromone-biosynthesizing bark beetles may include, but are not limited to, pheromone biosynthetic enzymes, JHIII signaling components, and proteins involved in pheromone transport and secretion. The candidate genes can then be confirmed through classical functional assays and/or molecular biology experiments.

Given limited resources and the expense of the approach, *I. pini* was chosen as a model for all bark beetles because of its ready availability and vigorous pheromone production rate. An expressed sequence tag (EST) database is being produced from a cDNA library of midguts from JHIII-treated male *I. pini* (A. Eigenheer, C. Keeling & C. Tittiger, unpublished data). Given the strong induction of the pheromone biosynthetic pathway in midgut cells, there is a good probability that most of the relevant genes will be represented in a modestly sized (<1000 cDNAs) project. Indeed, interesting ESTs corresponding to mevalonate pathway enzymes and P450s were recovered during preliminary sequencing (40), and a large fraction corresponds to various digestive enzymes (e.g., cellulases and glucosidases), as would be expected from midgut cells (57). The convergence of the classical biochemical and genomics approaches should lead rapidly to a more complete understanding of how scolytids produce their monoterpenoid pheromone components.

CONCLUSIONS AND FUTURE IMPLICATIONS

There have been significant recent advances in our knowledge of pheromone production in the Scolytidae. However, most of this progress is based on the synthesis of only a few monoterpenoids (ipsenol, ipsdienol, and frontalin) from only a few species (*I. paraconfusus*, *I. pini*, and *D. jeffreyi*). There are both more monoterpenoids of interest (51) and nearly 6000 more species of Scolytidae (151) that might bear further investigation. Already, comparative biochemical studies have revealed major differences between the mechanism of regulation of pheromone biosynthesis in *I. paraconfusus* and *I. pini*. Particularly significant in these two species of *Ips* will be studies to determine the mechanism by which JHIII increases HMG-R transcript abundance (i.e., induces the rate of *HMG-R* transcription or increases mRNA stability) and the influence of JHIII on HMG-R protein localization, abundance, stability, and activity. Within the *Dendroctonus*, the biosynthesis of frontalin has been allocated through evolution to one sex or the other across various species in the genus, offering the potential to study mechanisms for sex-specific regulation of the isoprenoid pathway. The theme established for the three model isoprenoids and three scolytid species noted above will no doubt be subject to subtle but interesting variations as the biochemical and taxonomic diversity are explored.

Isoprenoid pheromone synthesis in the Scolytidae provides a plethora of unique target points for the control of these economically important taxa, and most of these unique reactions occur in the late stages of the pathway. The genomics approach will accelerate the gathering of the details of this portion of the pathway and its regulation through rapid identification of candidate sequences associated with feeding and JHIII treatment. The comprehensive, annotated, and publicly available EST database from *I. pini* midgut tissue will be the first of its kind in the Coleoptera and should establish *I. pini* as a model organism for the study of the synthesis of all short-chain isoprenoids. A particularly significant outcome of the concerted study of the isoprenoid pathway in bark beetles may be a better understanding of the mode of action of JHIII (39). In male *I. pini*, the consistently strong JHIII-induced responses of *HMG-R* transcription and pheromone production in a clearly defined tissue hold great promise for unraveling the mechanism of JHIII activation in a nondevelopmental system. Microarray experiments should identify various primary responder genes, and their 5′ flanking sequences could be isolated from a genomic library and compared with known JH response elements (154) for conserved motifs. These sequences should serve a starting point to identify and isolate control elements, specific transcription factors, and other components of the JH signaling apparatus responsible for regulating isoprenoid pheromone biosynthesis in the Scolytidae.

ACKNOWLEDGMENTS

We thank the USDA National Research Initiative Competitive Grants Program (#9302089, #9502551, #9702991, #97-35302-4223, #9802897, #98-35302-6997) for continuous funding of work on the biochemistry and molecular biology of aggregation pheromone production in *Ips* spp. from 1993 to 2000, the National Science Foundation (#IBN-972855/#IBN-9906530) for support of work on *Dendroctonus* pheromone biosynthesis, and the Human Frontier Science Program (#RGY0382). S.J. Seybold further acknowledges the roles of D.L. Wood, W. Francke, I. Kubo, and G.J. Blomquist as mentors in scolytid chemical ecology, bioorganic chemistry, and biochemistry and molecular biology, which have been applied to the area of scolytid pheromone biosynthesis. We are grateful to G.M. Hall and J.A. Tillman for assistance with graphics. This paper is a contribution of the Minnesota (Project MN-17-070) and Nevada (Manuscript # 0302331) Agricultural Experiment Stations and presents the results of work coordinated through Multi-state Project W-189, Biorational Methods for Insect Pest Management: Bioorganic and Molecular Approaches.

LITERATURE CITED

1. Anderson RF. 1948. Host selection by the pine engraver. *J. Econ. Entomol.* 41: 596–602
2. Ayté J, Gil-Gomez G, Haro D, Marrero PF, Hegardt FG. 1990. Rat mitochondrial and cytosolic 3-hydroxy-3-methylglutaryl-CoA synthases are encoded by two different genes. *Proc. Natl. Acad. Sci. USA* 87:3874–78
3. Bakke A, Frøyen P, Skattebøl L. 1977. Field response of a new pheromonal compound isolated from *Ips typographus*. *Naturwissenschaften* 64:98
4. Barkawi LS. 2002. *Biochemical and molecular studies of aggregation pheromones of bark beetles in the genus* Dendroctonus *(Coleoptera: Scolytidae), with special reference to the Jeffrey pine beetle*, Dendroctonus jeffreyi *Hopkins*. PhD thesis. Univ. Nevada, Reno. 193 pp.

4a. Barkawi LS, Francke W, Blomquist GJ, Seybold SJ. 2002. Frontalin: de novo biosynthesis of an aggregation pheromone component by *Dendroctonus* spp. bark beetles (Coleoptera: Scolytidae). *Insect Biochem. Mol. Biol.* In press

5. Beedle AS, Walton MJ, Goodwin TW. 1975. Isoprenoid biosynthesis in aseptic larvae of *Calliphora erythrocephala*. *Insect Biochem.* 5:465–72
6. Berciano MT, Fernandez R, Pena E, Calle E, Villagra NT, et al. 2000 Formation of intranuclear crystalloids and proliferation of the smooth endoplasmic reticulum in Schwann cells induced by tellurium treatment: association with overexpression of HMG CoA reductase and HMG CoA synthase mRNA. *Glia* 29:246–59
7. Blight MM, Wadhams LJ, Wenham MJ. 1979. Chemically mediated behavior in the large elm bark beetle, *Scolytus scolytus*. *Bull. Entomol. Soc. Am.* 25:122–24
8. Blum MS. 1985. Exocrine systems. In *Fundamentals of Insect Physiology*, ed. MS Blum, pp. 535–79. New York: Wiley
9. Borden JH. 1967. Factors influencing the response of *Ips confusus* (Coleoptera:

Scolytidae) to male attractant. *Can. Entomol.* 99:1164–93
10. Borden JH. 1982. Aggregation pheromones. In *Bark Beetles in North American Conifers*, ed. JB Mitton, KB Sturgeon, pp. 74–139. Austin: Univ. Texas Press
11. Borden JH. 1985. Aggregation pheromones. See Ref. 82a, 9:257–85
12. Borden JH, Nair KK, Slater CE. 1969. Synthetic juvenile hormone: induction of sex pheromone production in *Ips confusus*. *Science* 166:1626–27
13. Borden JH, Slater CE. 1969. Sex pheromone of *Trypodendron lineatum*: production in the female hindgut-Malpighian tubule region. *Ann. Entomol. Soc. Am.* 62:454–55
14. Brand JM, Barras SJ. 1977. The major volatile constituents of a basidiomycete associated with the southern pine beetle. *Lloydia* 40:398–400
15. Brand JM, Bracke JW, Britton LN, Markovetz AJ, Barras SJ. 1976. Bark beetle pheromones: production of verbenone by a mycangial fungus of *Dendroctonus frontalis*. *J. Chem. Ecol.* 2: 195–99
16. Brand JM, Bracke JW, Markovetz AJ, Wood DL, Browne LE. 1975. Production of verbenol pheromone by a bacterium isolated from bark beetles. *Nature* 254:136–37
17. Bridges JR. 1982. Effects of juvenile hormone on pheromone synthesis in *Dendroctonus frontalis*. *Environ. Entomol.* 11:417–20
17a. Brown K, Havel CM, Watson JA. 1983. Isoprene synthesis in isolated embryonic *Drosophila* cells: regulation of 3-hydroxy-3-methylglutaryl coenzyme A reductase activity. *J. Biol. Chem.* 258:8512–18
18. Buesa C, Martinez-Gonzales J, Casals N, Haro D, Piulachs M-D, et al. 1994. *Blattella germanica* has two HMG-CoA synthase genes. *J. Biol. Chem.* 269:11707–13
19. Byers JA. 1981. Pheromone biosynthesis in the bark beetle, *Ips paraconfusus*, during feeding or exposure to vapors of host plant precursors. *Insect Biochem.* 11:563–69
20. Byers JA. 1982. Male specific conversion of the host plant compound, myrcene, to the pheromone, (+)-ipsdienol, in the bark beetle, *Dendroctonus brevicomis*. *J. Chem. Ecol.* 8:363–71
21. Byers JA. 1983. Bark beetle conversion of a plant compound to a sex-specific inhibitor of pheromone attraction. *Science* 220:624–26
22. Byers JA. 1983. Influence of sex, maturity and host substances on pheromones in the guts of the bark beetles, *Ips paraconfusus* and *Dendroctonus brevicomis*. *J. Insect Physiol.* 29:5–13
23. Byers JA. 1989. Chemical ecology of bark beetles. *Experientia* 45:271–83
24. Byers JA, Birgersson G. 1990. Pheromone production in a bark beetle independent of myrcene precursor in host pine species. *Naturwissenschaften* 77: 385–87
25. Byers JA, Schlyter F, Birgersson G, Francke W. 1990. *E*-Myrcenol in *Ips duplicatus*: an aggregation pheromone component new for bark beetles. *Experientia* 46:1209–11
26. Byers JA, Wood DL. 1981. Antibiotic-induced inhibition of pheromone synthesis in a bark beetle. *Science* 213:763–64
27. Byers JA, Wood DL, Browne LE, Fish RH, Piatek B, et al. 1979. Relationship between a host plant compound, myrcene, and pheromone production in the bark beetle, *Ips paraconfusus*. *J. Insect Physiol.* 25:477–82
27a. Byrne KJ, Swigar AA, Silverstein RM, Borden JH, Stokkink E. 1974. Sulcatol: population aggregation pheromone in the scolytid beetle, *Gnathotrichus sulcatus*. *J. Insect Physiol.* 20:1895–900
28. Cabano J, Buesa C, Hegardt FG, Marrero PF. 1997. Catalytic properties of recombinant 3-hydroxy-3-methylglutaryl

coenzyme A synthase-1 from *Blattella germanica*. *Insect Biochem. Mol. Biol.* 27:499–505

29. Cane DE. 1999. Isoprenoid biosynthesis: overview. In *Comprehensive Natural Products Chemistry*. Vol 2: *Isoprenoids Including Carotenoids and Steroids*, ed. D Barton, K Nakanishi, O Meth-Cohn, pp. 1–13. Amsterdam: Elsevier
30. Casals N, Buesa C, Marrero PF, Belles X, Hegardt FG. 2001. 3-Hydroxy-3-methylglutaryl coenzyme A synthase-1 of *Blattella germanica* has structural and functional features of an active retrogene. *Insect Biochem. Mol. Biol.* 31:425–33
31. Casals N, Buesa C, Piulachs M-D, Cabano J, Marrero PF, et al. 1996. Coordinated expression and activity of 3-hydroxy-3-methylglutaryl coenzyme A synthase and reductase in the fat body of *Blattella germanica* (L.) during vitellogenesis. *Insect Biochem. Mol. Biol.* 26: 837–43
32. Chen NM, Borden JH, Pierce HD Jr. 1988. Effect of juvenile hormone analog, fenoxycarb, on pheromone production by *Ips paraconfusus* (Coleoptera: Scolytidae). *J. Chem. Ecol.* 14:1087–98
33. Chin DJ, Luskey KL, Anderson RGW, Faust JR, Goldstein JL, et al. 1982. Appearance of crystalloid endoplasmic reticulum in compactin-resistant Chinese hamster cells with a 500-fold increase in 3-hydroxy-3-methylglutaryl coenzyme A reductase. *Proc. Natl. Acad. Sci. USA* 79:1185–89
34. Chun KT, Simoni RD. 1992. The role of the membrane domain in the regulated degradation of 3-hydroxy-3-methylglutaryl coenzyme A reductase. *J. Biol. Chem.* 267:4236–46
35. Clark AJ, Bloch K. 1959. The absence of sterol synthesis in insects. *J. Biol. Chem.* 234:2578–82
36. Conn JE, Borden JH, Hunt DWA, Holman J, Whitney HS, et al. 1984. Pheromone production by axenically reared *Dendroctonus ponderosae* and *Ips paraconfusus* (Coleoptera: Scolytidae). *J. Chem. Ecol.* 10:281–90
37. Connolly JD, Hill RA. 1991. *Dictionary of Terpenoids*. London: Chapman & Hall
38. Correll CC, Ng L, Edwards PA. 1994. Identification of farnesol as the non-sterol derivative of mevalonic acid required for the accelerated degradation of 3-hydroxy-3-methylglutaryl coenzyme A reductase. *J. Biol. Chem.* 269:17390–93
39. Davey KG. 2000. The modes of action of juvenile hormones: some questions we ought to ask. *Insect Biochem. Mol. Biol.* 30:663–69
40. Eigenheer A, Blomquist GJ, Tittiger C. 2001. *Genomics of pheromone biosynthesis in* Ips pini. Presented at Annu. Meet. Int. Soc. Chem. Ecol., 18th, Lake Tahoe
41. Eisen MB, Spellman PT, Brown PO, Botstein D. 1998. Cluster analysis and display of genome-wide expression patterns. *Proc. Natl. Acad. Sci. USA* 95: 14863–68
42. Everaerts C, Bonnard O, Pasteels JM, Rosin Y, König WA. 1990. (+)-α-Pinene in the defensive secretion of *Nasutitermes princes* (Isoptera: Termitidae). *Experientia* 46:227–30
43. Everaerts C, Rosin Y, Le Quere J-L, Bonnard O, Pasteels JM. 1993. Sesquiterpenes in the frontal gland secretions of nasute soldier termites from New Guinea. *J. Chem. Ecol.* 19:2865–79
44. Feyereisen R. 1985. Regulation of juvenile hormone titer: synthesis. See Ref. 82a, 7:391–429
45. Feyereisen R. 1999. Insect P450 enzymes. *Annu. Rev. Entomol.* 44:507–33
46. Fish RH, Browne LE, Bergot BJ. 1984. Pheromone biosynthetic pathways: conversion of ipsdienone to (−)-ipsdienol, a mechanism for enantioselective reduction in the male bark beetle, *Ips paraconfusus*. *J. Chem. Ecol.* 10:1057–64
47. Fish RH, Browne LE, Wood DL, Hendry

LB. 1979. Pheromone biosynthetic pathways: conversions of deuterium labelled ipsdienol with sexual and enantioselectivity in *Ips paraconfusus* Lanier. *Tetrahedron Lett.* 17:1465–68

48. FlyBase. 1999. The Flybase database of the *Drosophila* genome projects and community literature. http://flybase.bio.indiana.edu/. *Nucleic Acids Res.* 27:85–88
49. Francke W, Pan ML, König WA, Mori K, Puapoomchareon P, et al. 1987. Identification of "pityol" and "grandisol" as pheromone components of the bark beetle, *Pityophthorus pityographus. Naturwissenschaften* 74:343–45
50. Francke W, Sauerwein P, Vité JP, Klimetzek D. 1980. The pheromone bouquet of *Ips amitinus. Naturwissenschaften* 67: 147–48
51. Francke W, Schulz S. 1999. Pheromones. In *Comprehensive Natural Products Chemistry*. Vol. 8: *Miscellaneous Natural Products Including Marine Natural Products, Pheromones, Plant Hormones and Aspects of Ecology*. ed. D Barton, K Nakanishi, O Meth-Cohn, pp. 197–261. Amsterdam: Elsevier
52. Francke W, Vité JP. 1983. Oxygenated terpenes in pheromone systems of bark beetles. *Z. Angew. Entomol.* 96:146–56
53. Friesen JA, Rodwell VW. 1997. Protein engineering of the HMG-CoA reductase of *Pseudomonas mevalonii*. Construction of mutant enzymes whose activity is regulated by phosphorylation and dephosphorylation. *Biochemistry* 36:2173–77
54. Gayen AK, Peffley DM. 1995. The length of 5′-untranslated leader sequences influences distribution of 3-hydroxy-3-methylglutaryl-coenzyme A reductase mRNA in polysomes: effects of lovastatin, oxysterols, and mevalonate. *Arch. Biochem. Biophys.* 322: 475–85
55. Gertler FB, Chiu C-Y, Richter-Mann L, Chin DJ. 1988. Developmental and metabolic regulation of the *Drosophila melanogaster* 3-hydroxy-3-methylglutaryl coenzyme A reductase. *Mol. Cell. Biol.* 8:2713–21
56. Gil G, Brown MS, Goldstein JL. 1986. Cytoplasmic 3-hydroxy-3-methylglutaryl coenzyme A synthase from the hamster. *J. Biol. Chem.* 261:3717–24
57. Girard C, Jouanin L. 1999. Molecular cloning of cDNAs encoding a range of digestive enzymes from a phytophagous beetle, *Phaedon cochleariae. Insect Biochem. Mol. Biol.* 29:1129–42
58. Goldstein JL, Brown MS. 1990. Regulation of the mevalonate pathway. *Nature* 343:425–30

58a. Gries G, Borden JH, Gries R, LaFontaine JP, Dixon EA, et al. 1992. 4-Methylene-6,6-dimethylbicyclo[3.1.1]hept-2-ene (verbenene): new aggregation pheromone of the scolytid beetle, *Dendroctonus rufipennis. Naturwissenschaften* 79:367–68

59. Gries G, Leufvén A, LaFontaine JP, Pierce HD Jr, Borden JH, et al. 1990. New metabolites of α-pinene produced by the mountain pine beetle, *Dendroctonus ponderosae* (Coleoptera: Scolytidae). *Insect Biochem.* 20:365–71
60. Hackstein E, Vité JP. 1978. Pheromone Biosynthese und Reizkette in der Besiedlung von Fichten durch den Buchdrucker *Ips typographus. Mitt. Dtsch. Ges. All. Ang. Entomol.* 1:185–88
61. Hall GM, Tittiger C, Andrews GL, Mastick GS, Kuenzli M, et al. 2002. Midgut tissue of male pine engraver, *Ips pini*, synthesizes monoterpenoid pheromone component ipsdienol de novo. *Naturwissenschaften* 89:79–83
62. Hall GM, Tittiger C, Blomquist GJ, Andrews G, Mastick G, et al. 2002. Male Jeffrey pine beetles, *Dendroctonus jeffreyi*, synthesize the pheromone component frontalin de novo in anterior midgut tissue. *Insect Biochem. Mol. Biol.* In press

63. Hammock BD. 1985. Regulation of juvenile hormone titer: degradation. See Ref. 82a, 7:431–72
64. Hampton R, Dimster-Denk D, Rine J. 1996. The biology of HMG-CoA reductase: the pros of contra-regulation. *Trends Biol. Sci.* 21:140–45
65. Harring CM. 1978. Aggregation pheromones of the European fir engraver beetles *Pityokteines curvidens*, *P. spinidens*, and *P. vorontzovi* and the role of juvenile hormone in pheromone biosynthesis. *Z. Angew. Entomol.* 85:281–317
66. Hendry LB, Piatek B, Browne LE, Wood DL, Byers JA, et al. 1980. In vivo conversion of a labelled host plant chemical to pheromones of the bark beetle, *Ips paraconfusus*. *Nature* 284:485
67. Hibbard BE, Webster FX. 1993. Enantiomeric composition of grandisol and grandisal produced by *Pissodes strobi* and *P. nemorensis* and their electroantennogram response to pure enantiomers. *J. Chem. Ecol.* 19:2129–41
68. Huber DPW, Gries R, Borden JH, Pierce HD Jr. 1999. Two pheromones of coniferophagous bark beetles found in the bark of nonhost angiosperms. *J. Chem. Ecol.* 25:805–16
69. Hughes PR. 1973. *Dendroctonus*: production of pheromones and related compounds in response to host monoterpenes. *Z. Angew. Entomol.* 73:294–312
70. Hughes PR. 1973. Effect of α-pinene exposure on *trans*-verbenol synthesis in *Dendroctonus ponderosae* Hopk. *Naturwissenschaften* 60:261–62
71. Hughes PR. 1974. Myrcene: a precursor of pheromones in *Ips* beetles. *J. Insect Physiol.* 20:1271–75
72. Hughes PR. 1975. Pheromones of *Dendroctonus*: origin of α-pinene oxidation products present in emergent adults. *J. Insect Physiol.* 21:687–91
73. Hughes PR, Renwick JAA. 1977. Hormonal and host factors stimulating pheromone synthesis in female western pine beetles, *Dendroctonus brevicomis*. *Physiol. Entomol.* 2:289–92
74. Hughes PR, Renwick JAA. 1977. Neural and hormonal control of pheromone biosynthesis in the bark beetle, *Ips paraconfusus*. *Physiol. Entomol.* 2:117–23
75. Hunt DWA, Borden JH. 1989. Conversion of verbenols to verbenone by yeasts isolated from *Dendroctonus ponderosae* (Coleoptera: Scolytidae). *J. Chem. Ecol.* 16:1385–97
76. Hunt DWA, Borden JH, Pierce HD Jr, Slessor KN, King GGS, et al. 1986. Sex-specific production of ipsdienol and myrcenol by *Dendroctonus ponderosae* (Coleoptera: Scolytidae) exposed to myrcene vapors. *J. Chem. Ecol.* 12:1579–86
77. Hunt DWA, Smirle MJ. 1988. Partial inhibition of pheromone production in *Dendroctonus ponderosae* (Coleoptera: Scolytidae) by polysubstrate monooxygenase inhibitors. *J. Chem. Ecol.* 14: 529–36
78. Ivarsson P, Birgersson G. 1995. Regulation and biosynthesis of pheromone components in the double spined bark beetle *Ips duplicatus* (Coleoptera: Scolytidae). *J. Insect Physiol.* 41:843–49
79. Ivarsson P, Blomquist GJ, Seybold SJ. 1997. In vitro production of the pheromone intermediates ipsdienone and ipsenone by the bark beetles *Ips pini* (Say) and *I. paraconfusus* Lanier (Coleoptera: Scolytidae). *Naturwissenschaften* 84:454–57
80. Ivarsson P, Schlyter F, Birgersson G. 1993. Demonstration of de novo pheromone biosynthesis in *Ips duplicatus* (Coleoptera: Scolytidae): inhibition of ipsdienol and *E*-myrcenol production by compactin. *Insect Biochem. Mol. Biol.* 23:655–62
81. Ivarsson P, Tittiger C, Blomquist C, Borgeson CE, Seybold SJ, et al. 1998. Pheromone precursor synthesis is localized in the metathorax of *Ips*

paraconfusus Lanier (Coleoptera: Scolytidae). *Naturwissenschaften* 85:507–11

82. Keller GA, Pazirandeh M, Krisans S. 1986. 3-hydroxy-3-methylglutaryl coenzyme A reductase localization in rat liver peroxisomes and microsomes of control and cholestyramine-treated animals: quantitative biochemical and immunoelectron microscopical analyses. *J. Cell Biol.* 103:875–86

82a. Kerkut GA, Gilbert LI, eds. 1985. *Comprehensive Insect Physiology, Biochemistry, and Pharmacology*, Vols. 7, 9, 10. Oxford: Pergamon. 564 pp. 735 pp. 715 pp.

83. Kinzer GW, Fentiman AF, Page TF Jr, Foltz RL, Vité JP, et al. 1969. Bark beetle attractants: identification, synthesis and field bioassay of a new compound isolated from *Dendroctonus*. *Nature* 221:477–78

84. Klimetzek D, Francke W. 1980. Relationship between enantiomeric composition of α-pinene in host trees and the production of verbenols in *Ips* species. *Experientia* 36:1343–44

85. Kumagai H, Chun KT, Simoni RD. 1995. Molecular dissection of the role of the membrane domain in the regulated degradation of 3-hydroxy-3-methylglutaryl coenzyme A reductase. *J. Biol. Chem.* 270:19107–13

86. Lanne BS, Ivarsson P, Johnson P, Bergström G, Wassgren AB. 1989. Biosynthesis of 2-methyl-3-buten-2-ol, a pheromone component of *Ips typographus* (Coleoptera: Scolytidae). *Insect Biochem.* 19:163–68

87. Leufvén A, Bergström G, Falsen E. 1984. Interconversion of verbenols and verbenone by identified yeasts isolated from the spruce bark beetle *Ips typographus*. *J. Chem. Ecol.* 10:1349–61

88. Leufvén A, Bergström G, Falsen E. 1988. Oxygenated monoterpenes produced by yeasts, isolated from *Ips typographus* (Coleoptera: Scolytidae) and grown in phloem medium. *J. Chem. Ecol.* 14:353–61

89. Lindström M, Norin T, Birgersson G, Schlyter F. 1989. Variation of enantiomeric composition of α-pinene in Norway spruce, *Picea abies*, and its influence on production of verbenol isomers by *Ips typographus* in the field. *J. Chem. Ecol.* 15:541–48

90. Lu F. 1999. *Origin and endocrine regulation of pheromone biosynthesis in the pine bark beetles,* Ips pini *(Say) and* Ips paraconfusus *Lanier (Coleoptera: Scolytidae)*. PhD thesis. Univ. Reno, Nevada. 152 pp.

90a. Macconnell JG, Borden JH, Silverstein RM, Stokkink E. 1977. Isolation and tentative identification of lineatin, a pheromone from the frass of *Trypodendron lineatum* (Coleoptera: Scolytidae). *J. Chem. Ecol.* 3:549–61

91. Madden JL, Pierce HD Jr, Borden JH, Butterfield A. 1988. Sites of production and occurrence of volatiles in Douglas-fir beetle, *Dendroctonus pseudotsugae* Hopkins. *J. Chem. Ecol.* 14:1305–17

92. Martin D, Bohlmann J, Gershenzon J, Francke W, Seybold SJ. 2001. *A novel sex-specific and inducible monoterpene synthase activity associated with the bark beetle,* Ips pini *(Say)*. Presented at Annu. Meet. Int. Soc. Chem. Ecol., 18th, Lake Tahoe

93. Martinez-Gonzalez J, Buesa C, Piulachs M-D, Belles X, Hegardt FG. 1993. Molecular cloning, developmental pattern and tissue expression of 3-hydroxy-3-methylglutaryl coenzyme A reductase of the cockroach, *Blattella germanica*. *Eur. J. Biochem.* 213:233–41

93a. Martinez-Gonzalez J, Buesa C, Piulachs M-D, Belles X, Hegardt F. 1993. 3-Hydroxy-3-methylglutaryl-coenzyme-A synthase from *Blattella germanica*. Cloning, expression, developmental pattern and tissue expression. *Eur. J. Biochem.* 217:691–99

94. Meigs TE, Simoni RD. 1997. Farnesol

as a regulator of HMG-CoA reductase degradation: characterization and role of farnesyl pyrophosphatase. *Arch. Biochem. Biophys.* 345:1–9

95. Miller DR, Gries G, Borden JH. 1990. *E*-myrcenol: a new pheromone for the pine engraver, *Ips pini* (Say) (Coleoptera: Scolytidae). *Can. Entomol.* 122:401–6

95a. Nakajima T, Iwaki K, Hamakubo T, Kodama T, Emi M. 2000. Genomic structure of the gene encoding human 3-hydroxy-3-methylglutaryl coenzyme A reductase: comparison of exon/intron organization of strol-sensing domains among four related genes. *J. Hum. Genet.* 45:284–89

96. Nardi JB, Young AG, Ujhelyi E, Tittiger C, Lehane MJ, Blomquist GJ. 2002. Specialization of midgut cells for synthesis of male isoprenoid pheromone components in two scolytid beetles, *Dendroctonus jeffreyi* and *Ips pini*. *Tissue Cell.* 34:221–31

97. Olender EH, Simoni RD. 1992. The intracellular targeting and membrane topology of 3-hydroxy-3-methylglutaryl coenzyme A reductase. *J. Biol. Chem.* 267:4223–35

98. Perez AL, Gries R, Gries G, Oelschlager AC. 1996. Transformation of presumptive precursors to frontalin and *exo*-brevicomin by bark beetles and the West Indian sugarcane weevil (Coleoptera). *Bioorg. Med. Chem.* 4:445–50

99. Person HL. 1931. Theory in explanation of the selection of certain trees by the western pine beetle. *J. For.* 29:696–99

100. Pickett JA, Griffiths DC. 1980. Composition of aphid alarm pheromones. *J. Chem. Ecol.* 6:349–60

101. Pierce HD Jr, Conn JE, Oehlschlager AC, Borden JH. 1987. Monoterpene metabolism in female mountain pine beetles, *Dendroctonus ponderosae* Hopkins, attacking ponderosa pine. *J. Chem. Ecol.* 13:1455–80

102. Pitman GB, Kliefoth RA, Vité JP. 1965. Studies on the pheromone of *Ips confusus* (LeConte). II. Further observations on the site of production. *Contrib. Boyce Thompson Inst. Plant Res.* 23:13–17

103. Pitman GB, Vité JP. 1963. Studies on the pheromone of *Ips confusus* (LeC.). I. Secondary sexual dimorphism in the hindgut epithelium. *Contrib. Boyce Thompson Inst. Plant Res.* 22:21–25

104. Pitman GB, Vité JP, Kinzer GW, Fentiman AF Jr. 1968. Bark beetle attractants: *trans*-verbenol isolated from *Dendroctonus*. *Nature* 218:168–69

105. Raffa KF, Phillips TW, Salom SM. 1993. Strategies and mechanisms of host colonization by bark beetles. In *Beetle-Pathogen Interactions in Conifer Forests*, ed. TD Schowalter, GM Filip, pp. 103–28. London: Academic

106. Renwick JAA. 1967. Identification of two oxygenated terpenes from the bark beetles *Dendroctonus frontalis* and *Dendroctonus brevicomis*. *Contrib. Boyce Thompson Inst. Plant Res.* 23:355–60

107. Renwick JAA, Dickens JC. 1979. Control of pheromone production in the bark beetle, *Ips cembrae*. *Physiol. Entomol.* 4:377–81

108. Renwick JAA, Hughes PR, Krull IS. 1976. Selective production of *cis*- and *trans*-verbenol from (−)- and (+)-α-pinene by a bark beetle. *Science* 191:99–201

109. Renwick JAA, Hughes PR, Pitman GB, Vité JP. 1976. Oxidation products of terpenes identified from *Dendroctonus* and *Ips* bark beetles. *J. Insect Physiol.* 22:725–27

110. Reynolds GA, Basu SK, Osborne TF, Chin DJ, Gil G, et al. 1984. HMG-CoA reductase: a negatively regulated gene with unusual promoter and 5′ untranslated regions. *Cell* 38:275–85

111. Robinson GE. 2002. *Honey bee brain EST project*. http://titan.biotec.uiuc.edu/bee/honeybee_project.htm

112. Rohmer M. 1999. A mevalonic-independent route to isopentenyl diphosphate. In *Comprehensive Natural Products Chemistry*. Vol 2: *Isoprenoids*

Including Carotenoids and Steroids, ed. D Barton, K Nakanishi, O Meth-Cohn, pp. 45–67. Amsterdam: Elsevier
113. Sato R, Goldstein JL, Brown MS. 1993. Replacement of serine-871 of hamster 3-hydroxy-3-methylglutaryl coenzyme A reductase prevents phosphorylation by AMP-activated kinase and blocks inhibition of sterol synthesis induced by ATP depletion. *Proc. Natl. Acad. Sci. USA* 90:9261–65
114. Schooley DA, Baker FC. 1985. Juvenile hormone biosynthesis. See Ref. 82a, 7:363–89
115. Sekler MS, Simoni RD. 1995. Mutation in the lumenal part of the membrane domain of HMG-CoA reductase alters its regulated degradation. *Biochem. Biophys. Res. Commun.* 206:186–93
116. Seybold SJ. 1992. *The role of chirality in the olfactory-directed aggregation behavior of pine engraver beetles in the genus* Ips *(Coleoptera: Scolytidae)*. PhD thesis. Univ. Calif., Berkeley. 355 pp.
117. Seybold SJ. 1992. The status of scolytid genetics: a reductionist's view. *Proc. Workshop on Bark Beetle Genet., May 17–18, 1992, Berkeley, CA*, pp. 12–13. USDA For. Serv. Gen. Tech. Rep. PSW-GTR-138.
118. Seybold SJ, Bohlmann J, Raffa KF. 2000. The biosynthesis of coniferophagous bark beetle pheromones and conifer isoprenoids: evolutionary perspective and synthesis. *Can. Entomol.* 132:697–753
119. Seybold SJ, Ohtsuka T, Wood DL, Kubo I. 1995. Enantiomeric composition of ipsdienol: a chemotaxonomic character for North American populations of *Ips* spp. in the *pini* subgeneric group (Coleoptera: Scolytidae). *J. Chem. Ecol.* 21:995–1016
120. Seybold SJ, Quilici DR, Tillman JA, Vanderwel D, Wood DL, et al. 1995. De novo biosynthesis of the aggregation pheromone components ipsenol and ipsdienol by the pine bark beetle, *Ips paraconfusus* Lanier and *Ips pini* (Say) (Coleoptera: Scolytidae). *Proc. Natl. Acad. Sci. USA* 92:8393–97
121. Seybold SJ, Teale SA, Wood DL, Zhang A, Webster FX, et al. 1992. The role of lanierone in the chemical ecology of *Ips pini* (Coleoptera: Scolytidae) in California. *J. Chem. Ecol.* 18:2305–29
122. Silberkang M, Havel CM, Friend DS, McCarthy BJ, Watson JA. 1983. Isoprene synthesis in isolated embryonic *Drosophila* cells: sterol independent eukaryotic cells. *J. Biol. Chem.* 258:8503–11
123. Silverstein RM, Rodin JO, Wood DL. 1966. Sex attractants in frass produced by male *Ips confusus* in ponderosa pine. *Science* 154:509–10
124. Silverstein RM, Rodin JO, Wood DL, Browne LE. 1966. Identification of the two new terpene alcohols from frass produced by *Ips confusus* in ponderosa pine. *Tetrahedron* 22:1929–36
125. Skillen EL, Berisford CW, Camann MA, Reardon RC. 1997. *Semiochemicals of forest and shade tree insects in North America and management applications. FHTET-96–15*, USDA For. Ser., For. Health Technol. Enterp. Team Publ. 182 pp.
126. Stoakley JT, Bakke A, Renwick JAA, Vité JP. 1978. The aggregation pheromone system of the larch bark beetle, *Ips cembrae* Heer. *Z. Angew. Entomol.* 86:174–77
127. Svoboda JA, Thompson MJ. 1985. Steroids. See Ref. 82a, 10:137–75
128. Tabernero L, Bochar DA, Rodwell VW, Stauffacher CV. 1999. Substrate-induced closure of the flap domain in the ternary complex structures provides insights into the mechanism of catalysis by 3-hydroxy-3-methylglutaryl-CoA reductase. *Proc. Natl. Acad. Sci. USA* 96:7167–71
129. Tanaka RD, Edwards PA, Lan SF, Fogelman AM. 1983. Regulation of 3-hydroxy-3-methylglutaryl coenzyme A reductase activity in avian myeloblasts.

Mode of action of 25-hydroxycholesterol. *J. Biol. Chem.* 258:13331–39

129a. Tautz D. 2002. Insects on the rise. *Trends Genet.* 18:179–80

130. Teale SA, Webster FX, Zhang A, Lanier GN. 1991. Lanierone: a new pheromone component from *Ips pini* (Coleoptera: Scolytidae) in New York. *J. Chem. Ecol.* 17:1159–76

131. Tillman JA, Holbrook GL, Dallara PL, Schal C, Wood DL, et al. 1998. Endocrine regulation of de novo aggregation pheromone biosynthesis in the pine engraver, *Ips pini* (Say) (Coleoptera: Scolytidae). *Insect Biochem. Mol. Biol.* 28:705–15

132. Tillman JA, Lu F, Donaldson Z, Dwinell SC, Tittiger C, et al. 2001. *Biochemical and molecular aspects of the regulation of pheromone biosynthesis in pine bark beetles.* Presented at Annu. Meet. Int. Soc. Chem. Ecol., 18th, Lake Tahoe

133. Tillman JA, Seybold SJ, Jurenka RA, Blomquist GJ. 1999. Insect pheromones—an overview of biosynthesis and endocrine regulation. *Insect Biochem. Mol. Biol.* 29:481–514

133a. Tittiger C, Barkawi LS, Bengoa CS, Blomquist GJ, Seybold SJ. 2002. Structure and juvenile hormone-mediated regulation of the HMG-CoA reductase gene from the Jeffrey pine beetle, *Dendroctonus jeffreyi. Mol. Cell. Endocrinol.* In press

134. Tittiger C, Blomquist GJ, Ivarsson P, Borgeson CE, Seybold SJ. 1999. Juvenile hormone regulation of HMG-R gene expression in the bark beetle, *Ips paraconfusus* (Coleoptera: Scolytidae): implications for male aggregation pheromone biosynthesis. *Cell. Mol. Life Sci.* 54:121–27

135. Tittiger C, O'Keeffe C, Bengoa CS, Barkawi LS, Seybold SJ, et al. 2000. Isolation and endocrine regulation of an HMG-CoA synthase cDNA from the male Jeffrey pine beetle, *Dendroctonus jeffreyi. Insect Biochem. Mol. Biol.* 30: 2103–11

136. Troutt AB, McHeyzer-Williams MG, Pulendran B, Nossal GJ. 1992. Ligation-anchored PCR: a simple amplification technique with single-sided specificity. *Proc. Natl. Acad. Sci. USA* 89:9823–25

137. Tumlinson JH, Hardee DD, Gueldner RC, Thompson AC, Hedin PA, et al. 1969. Sex pheromones produced by male boll weevil: isolation, identification, and synthesis. *Science* 166:1010–12

138. Vanderwel D. 1991. *Pheromone biosynthesis by selected species of grain and bark beetles.* PhD thesis. Simon Fraser Univ. 172 pp.

139. Vanderwel D. 1994. Factors affecting pheromone production in beetles. *Arch. Insect Biochem. Physiol.* 25:347–62

140. Vanderwel D, Gries G, Singh SM, Borden JH, Oehlschlager AC. 1992. (*E*)- and (*Z*)-6-Nonen-2-one: biosynthetic precursors of *endo*- and *exo*-brevicomin in two bark beetles (Coleoptera: Scolytidae). *J. Chem. Ecol.* 18:1389–404

141. Vanderwel D, Oehlschlager AC. 1987. Biosynthesis of pheromones and endocrine regulation of pheromone production in Coleoptera. In *Pheromone Biochemistry*, ed. GD Prestwich, GJ Blomquist, pp. 175–215. Orlando: Academic

142. Vanderwel D, Oehlschlager, AC. 1992. Mechanism of brevicomin biosynthesis from (Z)-6-nonen-2-one in a bark beetle. *J. Am. Chem. Soc.* 114:5081–86

143. White RA Jr, Agosin M, Franklin RT, Webb JW. 1980. Bark beetle pheromones: evidence for physiological synthesis mechanisms and their ecological implications. *Z. Angew. Entomol.* 90: 255–74

144. White RA Jr, Franklin RT, Agosin M. 1979. Conversion of α-pinene oxide by rat liver and the bark beetle *Dendroctonus terebrans* microsomal fractions. *Pest Biochem. Physiol.* 10:233–42

145. Wood DL. 1962. The attraction created by males of a bark beetle *Ips confusus* (LeConte) attacking ponderosa pine. *Pan-Pac. Entomol.* 38:141–45
146. Wood DL. 1982. The role of pheromones, kairomones, and allomones in the host selection and colonization behavior of bark beetles. *Annu. Rev. Entomol.* 27:411–16
147. Wood DL, Browne LE, Bedard WD, Tilden PE, Silverstein RM, et al. 1968. Response of *Ips confusus* to synthetic sex pheromones in nature. *Science* 159: 1373–74
148. Wood DL, Browne LE, Silverstein RM, Rodin JO. 1966. Sex pheromones of bark beetles. I. Mass production, bio-assay, source, and isolation of sex pheromone of *Ips confusus* (LeC.). *J. Insect Physiol.* 12:523–36
149. Wood DL, Stark RW, Silverstein RM, Rodin JO. 1967. Unique synergistic effects produced by the principal sex attractant compounds of *Ips confusus* (LeConte) (Coleoptera: Scolytidae). *Nature* 215:206
150. Wood DL, Vité JP. 1961. Studies on the host selection behavior of *Ips confusus* (LeConte) (Coleoptera: Scolytidae) attacking *Pinus ponderosa*. *Contrib. Boyce Thompson Inst. Plant Res.* 21:79–96
151. Wood SL, Bright DE. 1992. A catalog of Scolytidae and Platypodidae (Coleoptera), Part 2, Taxonomic index, Volume A. *Great Basin Naturalist* No. 13, 833 pp.
152. Young A, Tittiger C, Welch W, Blomquist GJ. 2001. *Monoterpenoid pheromone biosynthesis: fishing for the elusive geranyl diphosphate synthase in bark beetles*. Presented at Annu. Meet. Int. Soc. Chem. Ecol., 18th, Lake Tahoe
153. Zethner-Møller O, Rudinsky JA. 1967. Studies on the site of sex pheromone production in *Dendroctonus pseudotsugae* (Coleoptera: Scolytidae). *Ann. Entomol. Soc. Am.* 60:575–82
154. Zhang J, Saleh DS, Wyatt GR. 1996. Juvenile hormone regulation of an insect gene: a specific transcription factor and a DNA response element. *Mol. Cell. Endocrinol.* 122:15–20

Annu. Rev. Entomol. 2003. 48:455–84
doi: 10.1146/annurev.ento.48.091801.112629

First published online as a Review in Advance on October 17, 2002

CONTACT CHEMORECEPTION IN FEEDING BY PHYTOPHAGOUS INSECTS

R. F. Chapman
ARL Division of Neurobiology, University of Arizona, Tucson, Arizona 85721; e-mail: chapman@neurobio.arizona.edu

Key Words nutritional balance, secondary compound, evolution, phagostimulant, deterrent

■ **Abstract** Gustatory receptors associated with feeding in phytophagous insects are broadly categorized as phagostimulatory or deterrent. No phytophagous insect is known that tastes all its essential nutrients, and the ability to discriminate between nutrients is limited. The insects acquire a nutritional balance largely "adventitiously" because leaves have an appropriate chemical composition. Sugars are the most important phagostimulants. Plant secondary compounds are most often deterrent but stimulate phagostimulatory cells if they serve as host-indicating sign stimuli, or if they are sequestered for defense or used as pheromone precursors. The stimulating effects of chemicals are greatly affected by other chemicals in mixtures like those to which the sensilla are normally exposed. Host plant selection depends on the balance of phagostimulatory and deterrent inputs with, in some oligophagous and monophagous species, a dominating role of a host-related chemical. Evolution of phytophagy has probably involved a change in emphasis in the gustatory system, not fundamentally new developments. The precise role of the gustatory systems remains unclear. In grasshoppers, it probably governs food selection and the amounts eaten, but in caterpillars there is some evidence that central feedbacks are also involved in regulating the amount eaten.

CONTENTS

0066-4170/03/0107-0455$14.00

INTRODUCTION

Insect taste receptors are characterized by the possession of a small number (3 to 10) of sensory neurons within a hair or cone of cuticle with a single relatively large pore at the tip. Because in most cases the gustatory function is inferred from the structure rather than known from experiments, it is common to use the term uniporous sensilla, to distinguish them from olfactory sensilla that have many small pores (multiporous) (5). I define phytophagous insects as those that feed on living plant tissues. The terms "gustation" and "contact chemoreception" are used in preference to "taste" because, in insects, receptors with similar anatomy and physiological properties are often present on the antennae and tarsi, as well as on the mouthparts, and are involved in the regulation of oviposition (140) and sometimes mate selection, as well as in feeding.

Stimulating chemicals reach the dendrites through the terminal pore. When an insect is feeding, the sensilla on the mouthparts will often be bathed in fluid released from the food. The neurons respond to chemicals in solution, and their function is directly comparable with taste receptors in the mouths of vertebrates. In many situations, however, chemicals from dry surfaces stimulate insect contact chemoreceptors, and this is probably the normal way in which these receptors operate when an insect first encounters a leaf surface (33, 140). Presumably in order to detect chemicals on a dry surface, the compound is taken up by the material surrounding the tips of the dendrites that reaches and sometimes exudes from the terminal pore in the sensillum (159). If compounds are water soluble, they may simply dissolve in the material, but lipophilic compounds are probably transported by carrier proteins. Although there is still no conclusive evidence of this process, putative transport proteins have been described (6, 98).

Terminal pore sensilla can also respond to odors. In the tobacco hornworm, *Manduca sexta*, neurons in the lateral galeal sensilla respond to leaf odor (141), and terminal pore sensilla on the hindlegs of the desert locust, *Schistocerca gregaria*, respond to the odors of formic and acetic acids [(102, 103) see also (47)]. Terminal pore sensilla have numerous small pores more laterally on the hair in addition to the relatively large terminal pore (98, 125), and in the adult of the arctiid moth, *Rhodogastra bubo*, these pores are similar in appearance to the wall pores of multiporous (olfactory) sensilla (4). They probably permit the entry of chemicals into the receptor lymph cavity, and the sheath surrounding the dendrites is permeable at least to some molecules (125). Thus, there is a diffusion pathway from the outside environment via the lateral pores to the dendrites, giving rise to the suggestion that odor molecules might travel along this path to reach the dendrites (98). However, the terminal pore provides much more direct access to the dendrites, and molecules can almost certainly enter the terminal material from an aqueous solution, a solid surface, or from the air. Under most normal situations, molecules in the air (odors) are not at a high-enough concentration to produce a response. Whatever the route by which molecules reach a neuron, their subsequent neural processing will be the same.

Terminology

It is common practice to designate gustatory receptor cells according to the type of compound that stimulates them. Thus, there may be a sugar cell, an amino acid cell, a salt cell, and so on. This approach was rejected by Rees (106) and Simmonds & Blaney (129) because of its teleological implications, but their alternatives, type 1, 2, 3 or neuron A, B, C, have not found favor probably because of their complete lack of any indication of function. Such an approach makes comparisons between species, or even between sensilla on the same insect, impossible without detailed prior knowledge. With time, however, it has become increasingly apparent that individual gustatory cells may respond to more than one class of compound. For example, the water cell in the blow fly, *Protophormia terraenovae*, also responds to fructose (158); the sugar cell in the red turnip beetle, *Entomoscelis americana*, also responds to five amino acids (89); and a single cell in the caterpillar of the arctiid moth, *Grammia geneura*, responds to sucrose, glucose, several amino acids, and the iridoid glycoside, catalpol (14). This led Bernays & Chapman (13) to suggest that gustatory neurons should be designated according to their function rather than by the type of compound that stimulates them. Thus, in relation to feeding there would be phagostimulatory cells and deterrent cells. Activity of these in response to appropriate stimuli would enhance or reduce feeding, respectively. The functional description could then be extended, where necessary and appropriate, by reference to the stimulating compounds. This terminology is adopted in this review.

Phagostimulatory and deterrent neurons could be considered the basic labeled lines of the insect taste receptor system. They may respond to a broad range of compounds or only to specific sugars or sign stimuli, for example, and would then convey information concerning these particular substances to the central nervous system (CNS) separately from other kinds of information. They would then be labeled lines in the more commonly accepted sense, but the difference is only one of degree. Whether the insect uses this information in a different way from all the rest is not generally known, but that must be the case where secondary compounds act as sign stimuli.

MONITORING PLANT CHEMICALS

Nutrients

The primary function of the taste receptor system of any animal is to enable the organism to recognize food. It must, therefore, respond to nutrient compounds. The major nutrients are proteins, or free amino acids, from which the organism builds its own proteins, and carbohydrates, as a source of energy, and from which lipids can be derived. Phagostimulatory cells dedicated to the reception of some of these compounds are present in all phytophagous insects that have been studied and are almost certainly universally present.

PROTEINS There are virtually no data on the ability of phytophagous insects to taste proteins, but evidence from species with other feeding habits suggests that insects do not have taste receptor neurons that respond specifically to proteins. Nevertheless, proteins can be detected if they have terminals for which receptor sites exist on the neurons. Experimentally glycosylated bovine serum albumen stimulated the sugar-sensitive cell in sensilla of the American cockroach, *Periplaneta americana*, and bovine serum albumen also stimulates cells in the labellar sensilla of *P. terraenovae* (8, 80). In the latter species the response was produced by L-alanine, the C-terminal amino acid of bovine serum albumen. It is highly likely that phytophagous insects show similar responses if they have receptors that respond to the terminal acids of plant proteins. However, their importance as phagostimulants would depend on their concentrations relative to stimulatory free amino acids and sugars, and their effect is probably never more than marginal. The caterpillar of the large white butterfly, *Pieris brassicae*, does not respond behaviorally or physiologically to the sweet-tasting (for humans) protein thaumatin (117).

AMINO ACIDS Based on the small number of species examined, phytophagous insects have the same dietary requirements for 10 essential amino acids as most other insects (39). Aphids are exceptional because their symbionts manufacture essential amino acids from nonessential ones (51), and this is probably true of other species that harbor symbionts. It might be expected, then, that some taste receptors of all phytophagous insects would have the capacity to taste amino acids, and for a majority, the ability to recognize the essential acids would be particularly important.

Among the phytophagous insects that have been investigated neurophysiologically, none taste all the protein amino acids, and some apparently cannot taste any, although there is no insect in which a complete functional inventory of all sensilla is yet available. Neurons in the palp-tip sensilla of the migratory locust, *Locusta migratoria*, respond to alanine and/or serine, lysine, and leucine, but not to various other amino acids (17, 132). None of the work on acridids makes it possible to determine if these responses are from amino acid-specific neurons, but serine is a phagostimulant. Of the 17 other amino acids tested behaviorally, only proline is a phagostimulant at relevant concentrations (37).

In phytophagous Coleoptera, sensory responses to amino acids have been examined in the larvae and/or adults of three species of Chrysomelidae (36, 95, 89). In *E. americana*, five amino acids stimulate the same cell as that responding to sucrose in one galeal sensillum, and a somewhat similar cell occurs in the Colorado potato beetle, *Leptinotarsa decemlineata*. Cells in other sensilla of these two species also respond to some amino acids. Their specificity has not been determined but they could be amino acid specific. In *L. decemlineata*, only 3 of 23 different protein amino acids induced active biting and feeding in a majority of individuals when tested as 0.01 M solutions in agar/cellulose discs, and only 6 were phagostimulatory at 0.1 M (75).

No caterpillar so far investigated responds to all the amino acids, and no single amino acid is a major phagostimulant for all species. Indeed, the differences between species are striking. Three of the four amino acids that stimulate the neuron in the lateral sensillum most strongly in *P. brassicae* are the least stimulatory in *P. rapae* (150).

It is usually assumed that amino acids are phagostimulants (3), but they may also stimulate deterrent cells. This is true of methionine in *S. litura* (72), and both methionine and phenylalanine in *G. geneura* (13) and the cabbage moth, *Mamestra brassicae* (157). This is not to imply that these two aromatic amino acids are always deterrent. In the *Pieris* species, both stimulate a phagostimulatory cell in the lateral sensillum, as is also the case with methionine in *G. geneura*. In the spruce budworm, *Choristoneura fumiferana*, valine stimulates a lateral phagostimulatory cell but is a behavioral deterrent, implying that it stimulates a deterrent cell elsewhere (104).

Insects almost certainly cannot distinguish between amino acids that stimulate a single neuron. The firing patterns (change in action potential frequency with time) of all amino acids stimulating a phagostimulatory cell in the medial galeal sensillum of *G. geneura* are similar (13). Consequently, even if firing patterns could provide a means for distinguishing compounds, there is no basis for such differentiation. It may be possible for an insect to distinguish different amino acids if they stimulate different sensory neurons, as is sometimes the case (124). Caterpillars do have the neural capacity to distinguish some amino acids from other stimulating compounds because they have at least one phagostimulatory cell in the mouthpart sensilla dedicated to the detection of amino acids. This may also be true for the beetles, although the data currently available do not make this certain. In the Orthoptera, some amino acids are phagostimulatory, and it is possible that they can be distinguished from sugars, for example, on the basis of a system of across-fiber patterning, although there is currently no evidence for this.

The fact that caterpillars, and probably other phytophagous insects, have little or no capacity to distinguish between individual amino acids does not imply that differences in the response spectra of different species are unimportant. Clearly, an insect's nutritional needs for specific amino acids are not reflected in the sensitivity of their gustatory receptors. If this were the case, we should expect that they would exhibit sensitivity to all the essential amino acids, and this is not so. But the fact that different neurons, even in the same insect, respond to different amino acids implies that the insect uses this information. The occurrence of a neuron in *C. fumiferana* that responds only to proline gives rise to the suggestion that this enables the caterpillar to use proline as an indicator of water stress and to select appropriate foliage (104). Given the effects of mixtures on the responses of neurons, differences in neuronal sensitivity may in some way provide information about mixture composition, and this will affect amounts eaten. Such subtle effects are, however, beyond our current level of understanding.

It is difficult to assess the importance of the deterrent responses produced by methionine and phenylalanine in some caterpillars. Under experimental conditions, these may result in behavioral deterrence (13, 72), but their concentrations in plants are generally so low that they are unlikely to have any significant effect, especially in the complex mixtures present in plants. Wieczorek (157) suggests that they stimulate a glycoside (presumed deterrent) receptor in *M. brassicae* because their side chains have similarities with some glycosides. In this case, the fact that they stimulate the deterrent cell may be an accident of receptor structure rather than a functional adaptation.

CARBOHYDRATES Sugars are phagostimulants for all the phytophagous insects studied, and cells in one or more mouthpart sensilla of phytophagous larval and adult Coleoptera, larval Diptera, and larval Lepidoptera respond to sugars, usually including sucrose (74, 89, 91, 124). Caterpillars often have sugar-sensitive phagostimulatory cells in both medial and lateral galeal sensilla, and in these cases the cells differ in the ranges of compounds to which they are sensitive (124). These commonly include pentose sugars, although these are commonly believed to be without nutritional value to most insects (40). The threshold of these cells for sucrose varies from about 10^{-4} to 10^{-2} M, and the few published values for glucose and fructose are within this range. Sucrose levels in plants are usually greater than 10^{-3} M. There might be little adaptive value if the insects were sensitive to low sugar concentrations because these might be of little nutritional value or would require large amounts to be eaten to be useful to the insect.

Grasshoppers are unusual. Although sugars are behavioral phagostimulants (37), only Varanka (152) obtained results suggestive of a sugar-sensitive phagostimulatory cell on the maxillary palps of *L. migratoria*. Other neurophysiological studies on six species, including *L. migratoria*, have employed sucrose or fructose, but have failed to find a comparable cell, and some have found a reduction in action potential frequency as sugar concentration increased. In these cases, two cells were firing, and the response of one of these declined as concentration increased, creating the appearance of an inverse dose effect when the action potentials of the two cells were not distinguished. The response of one cell, presumed to be phagostimulatory, does not change with concentration, while the activity of the other, a putative deterrent cell, is decreased, resulting in a net increase in phagostimulatory input as concentration increases (31).

Inositol is a sugar alcohol present in cell membranes and acts as a second messenger probably in all insects. It is ubiquitous in plants and sometimes, at least, may be present as the free polyol in equivalent amounts to the three principal leaf sugars (101). Some insects synthesize it, but for *B. mori* it is an essential dietary component. *M. sexta* can use inositol as an energy source (100). Some caterpillars have a neuron in one or both galeal sensilla that responds specifically to this compound (59, 116).

INORGANIC SALTS The interpretation of data on the sensory responses of phytophagous insects to inorganic salts is still uncertain. A variety of inorganic ions are essential nutrients, and every phytophagous insect that has been examined electrophysiologically possesses sensory neurons responding to the chlorides of monovalent cations [for example, Orthoptera (155), Coleoptera (90), Lepidoptera (83)]. Often, more than one cell responds to inorganic salts. In caterpillars, two or three cells commonly fire in response to dilute solutions of NaCl, and in some species the number of cells firing increases with concentration (50, 121). In *E. americana* three or four of the four neurons in the galeal sensillum respond when the concentration of NaCl exceeds 200 mM (90). Two or three cells also respond to salt solutions in grasshoppers without any obvious tendency for different cells to respond at different concentrations (17, 155).

In a few caterpillars the response has been shown to be dose dependent (13). The total number of action potentials also increases with concentration in the American grasshopper, *Schistocerca americana*, but only in about half the salt-sensitive cells in *L. migratoria*. In the former, the same cells respond to NaCl and KCl, and also to sucrose despite the fact that KCl is a behavioral deterrent and the latter a phagostimulant [(32); R.F. Chapman, unpublished data]. One of the cells must be phagostimulatory and the other deterrent.

In the lateral galeal sensillum of *G. geneura*, two cells respond to KCl, two to a range of deterrent compounds, and two are phagostimulatory cells, although the sensillum probably contains only four gustatory neurons (13). This led to the suggestion that the cells normally responding to inorganic salts in caterpillars are deterrent cells that also often respond to other deterrent compounds. This does not imply that there may not be deterrent cells dedicated to salt perception, although in most studies the numbers of potential deterrent compounds studied has been too limited to make any such statement possible.

There are no good data that relate sensory input from inorganic salts to feeding behavior in phytophagous insects, but such data as there are show that, at least at the relatively high concentrations required to produce an electrophysiological response, these salts have a deterrent effect in long-term experiments (49, 83, 137, 145). Sometimes, however, inorganic salts do have a phagostimulatory effect, as in the case of puddling butterflies (138). It may be that so-called water cells are important in this respect. These cells have been most thoroughly studied in flies. They respond to water alone, and dilute salt solutions enhance the rate of firing (108), but above 5 mM the firing rate declines and the cell is completely inhibited by 50 mM NaCl and above. A direct link between water cells and drinking has been established in various flies, indicating that they have a phagostimulatory function (46, 63, 158). Water cells have been described in a number of caterpillar species (124) and are probably of widespread occurrence.

MINOR NUTRIENTS There is no electrophysiological evidence that any phytophagous insects possess phagostimulatory cells dedicated to the perception of minor nutrients such as vitamins and fatty acids, although few studies have been made. If these compounds have any sensory effect at all, it may be through their effects on the responses to other compounds. Surprisingly often, they seem to reduce phagostimulatory activity.

All phytophagous insects require a dietary source of sterols, although microorganisms may synthesize them in some species. None of the utilizable phytosterols, cholesterol, sitosterol, or stigmasterol, has been tested electrophysiologically. Behavioral evidence suggests that grasshoppers are unable to taste nutrient sterols (9, 67) that do not affect their food intake over long periods (37, 67). Hamamura et al. (62) indicate that β-sitosterol is a biting factor for *B. mori* larvae, but the data of Nayar & Fraenkel (99) contradict this and, indeed, suggest that both cholesterol and β-sitosterol tend to reduce the likelihood of feeding. Hsiao & Fraenkel (75) suggest that these two compounds induce biting by *L. decemlineata* but a chi^2 test on their data shows that the results do not differ from control values. Other examples in which sterols are stated to be phagostimulants are based on experiments lasting hours or days in which internal feedbacks coupled with learning could have produced the results (69, 86).

Among other phytosterols that are not, in general, utilizable by phytophagous insects, the widely occurring 20-hydroxyecdysone (also the active molting hormone of most insects) stimulates deterrent cells and is a behavioral deterrent for caterpillars of several species, while ecdysone (the precursor of 20-hydroxyecdysone) weakly stimulates a deterrent cell of *B. mori* (45, 83, 143). From the data available in insects from several taxa, it appears that insects do not taste nutrient sterols, while other phytosterols are often deterrent.

Plant acids are potential nutrients, but there is no evidence that they stimulate phagostimulatory cells in phytophagous insects. In *M. sexta*, they alter the activity of sensory cells through their effects on pH, and at high concentrations they cause a deterrent cell to fire (15). In mixtures with glucose or inositol, ascorbic acid causes a reduction in activity of the phagostimulatory cell.

ACHIEVING A BALANCE OF NUTRIENTS The sensory array of caterpillars, even though its properties are incompletely known, clearly does not permit the insects to select specific nutrients on the basis of their individual tastes. In addition, mixtures of compounds greatly modify the responses of individual neurons. How then, do these insects acquire the range of nutrients necessary for their growth and reproduction? Clearly, the answer must lie in the fact that plants, in general, contain all the nutrients necessary for an insect because the plant's own metabolism involves the same range of chemicals. Hence, by eating a leaf, a phytophagous insect acquires all the necessary dietary components, as was pointed out by Fraenkel (52a) long ago. All that is required is the necessary signal inducing the insect to eat. Because all phytophagous insects have gustatory cells responding to sugars and all green plants produce sugars in relatively large amounts as the end product of photosynthesis, it seems logical to conclude that sugars are commonly, and perhaps usually, the key phagostimulatory components. This is consistent with the observation that the sensory responses of *L. decemlineata* to the saps of several acceptable plants are dominated by a single cell in each of the galeal sensilla that is probably the same cell as that responding to sucrose and amino acids (94).

This does not mean that phytophagous insects are incapable of adjusting their food intake to achieve a balanced diet. This is clearly demonstrated in experiments with grasshoppers and caterpillars with respect to the balance of carbohydrate and protein (135, 144, 153) and also with respect to minor components (27, 146, 114). Changes in taste receptor sensitivity accompany deficiencies in protein and carbohydrate in a manner that might permit the insects to modify their food intake appropriately (1, 24, 131, 134). These changes almost certainly play a significant role in the locust because the insects started to feed when they first encountered the food appropriate for compensation (whereas by implication they did not do so when they encountered the inappropriate food) (131). These immediate differences strongly suggest that sensory responses govern the behavior.

However, the caterpillars of the Egyptian cotton leafworm, *Spodoptera littoralis*, started to feed whether or not the food contained the appropriate nutrients, and only later did they compensate for any nutritional imbalance (131). This suggests that sensory differences may not have had the dominant role in correcting a dietary imbalance that is seen in the locust.

Grasshoppers can select food containing an appropriate sterol despite the fact that they lack appropriate gustatory receptors. In this case, changes in the insect's selection are determined by internal feedbacks and selection of an appropriate diet results from a learned association of nutritional inadequacy with some other feature of the food (9).

The implication of these findings is that short-term post-absorptive feedbacks concerning the nutrient status of the insect coupled with learning are as important, and perhaps more important, than changes in the peripheral nervous system in enabling insects to regulate their nutrient intake. Whether the gustatory system is involved in such learning is unclear.

Plant Secondary Compounds

SECONDARY COMPOUNDS AS DETERRENTS Secondary compounds are probably produced by all plants, at least at some stage during their development (68). All phytophagous insects that have been examined respond behaviorally to some of these compounds, most of which produce a negative reaction in the insects. Reduction or complete inhibition of feeding has been demonstrated in Acrididae, Hemiptera, Coleoptera, larval Lepidoptera, and larval Hymenoptera (96, 105, 109, 124). In a few Orthoptera, a coleopteran, and a number of larval Lepidoptera, secondary compounds that inhibit feeding have been shown to stimulate gustatory receptor cells in the mouthpart sensilla, and in aphids similar cells are present on the tips of the antennae (32, 36, 105, 124). They are called phagodeterrent, or simply deterrent, cells. Where the responses of these cells have been examined more extensively, each cell is generally found to respond to secondary compounds from several different chemical classes, and they are regarded as generalist deterrent cells. Only in *P. brassicae*, however, have as many as 50 compounds been tested on a single sensillum (149). This general response presumably reflects the occurrence of different types of receptor molecules in the membrane of the neuronal dendrite, as appear to occur in mammals (2). The insect is probably unable to differentiate between the chemicals stimulating such a cell, although different firing patterns could provide a basis for distinction (60). Compounds may also be distinguishable if different deterrent receptor cells differ in the ranges of compounds to which they are sensitive. Examples are known in a number of caterpillars (12, 61, 149). In contrast to the presumed wide range of sensitivities of most deterrent cells, a cell in the lateral galeal sensillum of *P. brassicae* is considered to be a specialist deterrent cell (149). It is sensitive to some cardenolides, and although it does respond to some other compounds, the thresholds for these are at least two orders of magnitude higher.

Species with limited food ranges appear to be more sensitive behaviorally to deterrent secondary compounds than are polyphagous species. That is, feeding is reduced or inhibited by lower concentrations of these compounds in specialists than in generalists (10, 16, 52, 97). In *B. mori*, the deterrent cells of some polyphagous mutants are less sensitive to some compounds than those of wild-type larvae but are equally, or even more, sensitive to some others (7). In a comparison of the receptor sensitivity of two sister species of *Heliothis*, *subflexa* and *virescens*, one of which is monophagous and the other polyphagous, no differences were found in sensory responses to sucrose, inositol, or several different deterrent compounds, or in the interactions between the compounds, even though *H. subflexa* was behaviorally much more sensitive than *H. virescens* [(11); E.A. Bernays & R.F. Chapman, unpublished data]. Behavioral differences are, at least in this case, almost certainly the result of differences in the processing of information in the CNS rather than in the peripheral receptors.

The fact that a secondary compound is deterrent to an insect does not imply that the insect does not eat plants containing that compound. Whether or not it does so depends on its sensitivity to the compound, the background of other deterrent and phagostimulatory information in which it is perceived, and the degree of food deprivation incurred by the insect. Polyphagous insects habitually eat food containing compounds that, by themselves, are deterrent. This is clear from experiments on *Mamestra configurata* and *Trichoplusia ni*. Both are polyphagous but commonly feed on Brassicaceae despite the fact that sinigrin is deterrent to them and they have deterrent cells responding to the compounds (126–128).

The deterrent cells of species with more restricted diet breadths, however, usually do not respond to the secondary compounds of their hosts. This is most obvious in the brassica-feeding larvae of *Pieris* species (124), but is also seen in some species of *Yponomeuta* that are monophagous (147).

Mitchell and his coworkers (92, 139) consider that deterrent cells do not occur in the chrysomelid beetles they examined. However, cells in the galeal sensilla of the western corn rootworm, *Diabrotica virgifera*, which is also a chrysomelid, do respond to some deterrents (36), and it seems possible that Mitchell and his colleagues were misled by their focus on the glycoalkaloids of Solanaceae.

SECONDARY COMPOUNDS AS PHAGOSTIMULANTS Secondary compounds may also have positive effects on feeding, acting as phagostimulants. In contrast to deterrent cells, phagostimulatory cells that respond to deterrent compounds are usually sensitive to only one type of compound. The best-known examples are those in which particular secondary compounds act as sign stimuli for appropriate host plants. Gustatory neurons specific to glucosinolates are present in larval *Pieris* species (124). Eight caterpillar species that feed primarily or only on Rosaceae have a neuron that responds to sorbitol, a compound characteristic of Rosaceae, while species of *Yponomeuta* that feed only on spindle tree have a dulcitol-sensitive cell. Dulcitol occurs in the host, although it is not restricted to it (124, 147). In the case of *Yponomeuta*, the specificity of the cell has not been determined. Outside the Lepidoptera, a neuron in a tarsal sensillum of the beetle *Chrysolina brunsvicensis* responds to hypericin, a compound characteristic of its host plant, *Hypericum hirsutum* (107), and galeal sensilla of *E. americana* contain a glucosinolate-sensitive phagostimulatory cell (90). Given the great number of phytophagous insects and the widespread occurrence of host specificity, the number of examples of insect species with gustatory cells tuned to host chemicals is still small.

Some insects sequester secondary compounds for defense against predators or as pheromone precursors. Iridoid glycosides, for example, are sequestered by some polyphagous arctiid caterpillars (26). One of these, *G. geneura*, has a generalist phagostimulatory cell that responds to some sugars and amino acids and also to the iridoid, catalpol (14). Thus, the caterpillars perceive catalpol simply as a phagostimulant, and if the insect feeds on a catalpol-rich plant the phagostimulatory effect is presumably enhanced. Species of *Plantago* are favored seasonal hosts for this polyphagous insect and sometimes contain high concentrations of catalpol (26). *Diabrotica virgifera* has a cell responding to cucurbitacins that it sequesters, but whether this cell is dedicated to detection of cucurbitacins has not yet been determined (36).

Pyrrolizidine alkaloids are used by many arctiid species as pheromones, and the insects obtain the chemical precursors from host plants. Two species, *Utetheisa ornata*, monophagous on *Crotalaria*, and *Estigmene acrea*, which is polyphagous, are known to have gustatory neurons dedicated to detection of these chemicals on the mouthparts of the caterpillars. In *E. acrea*, two cells, one in each galeal sensillum, respond almost exclusively to pyrrolizidine alkaloids, and this is probably true also in *U. ornata* [(13a) E.A. Bernays, unpublished data]. *E. acrea* feeds on many plant species that do not contain pyrrolizidine alkaloids so that, unlike the situation in *Pieris*, possession of a highly specific taste receptor neuron does not limit the insect's host range. When the insect encounters a plant containing pyrrolizidine alkaloids, it feeds actively, but the circuitry in the CNS is clearly not such that input from these cells is necessary for feeding to occur.

THRESHOLDS OF DETECTION The sensitivity of the cells that respond to plant secondary compounds might be expected to vary according to their importance to the insect and the concentrations in which the compounds occur naturally. Figure 1 illustrates threshold values for cells responding to secondary compounds in caterpillars. The data are probably biased by the species studied and the variable extent of the studies; nevertheless, some pattern emerges. Thresholds for compounds that

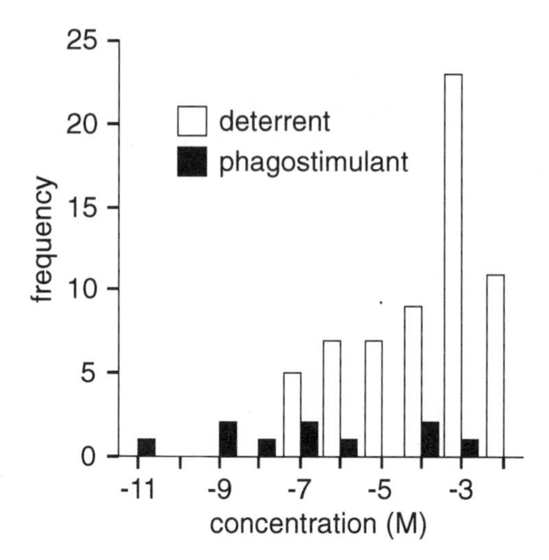

Figure 1 Threshold sensitivity to secondary compounds in the galeal receptors of caterpillars (7, 11, 14, 43, 45, 58, 72, 81–83, 112, 120, 126, 143, 148, 149, 151, 157).

stimulate a deterrent cell vary from 10^{-7} to 10^{-2} M, but a majority is in the range of 10^{-4} to 10^{-2} M. A single neuron may exhibit almost the whole of this range to different compounds. For example, the deterrent neuron in the medial galeal sensillum of *P. brassicae* has a threshold of about 10^{-3} M for chlorogenic and protocatechuic acids but of 10^{-7} M for azadirachtin. This low threshold may reflect the potential toxicity of the compound.

The thresholds for secondary compounds that stimulate phagostimulatory cells of caterpillars are usually much lower than for those that stimulate deterrent cells. They vary from 10^{-11} to 10^{-9} M for pyrrolizidine alkaloids in arctiids using them for pheromones (13a). However, the threshold of detection of catalpol by *G. geneura* is about 10^{-4} M. Catalpol is one of several compounds that this species may sequester, so its detection in small amounts is less critical and feeding on plants with low concentrations may not be advantageous. This pattern is mirrored by the response to sinigrin in different species. In six species where it stimulates deterrent cells, its threshold varies from 2×10^{-5} to 5×10^{-3} M, but in *P. brassicae*, and in the diamondback moth, *Plutella xylostella*, where it is a phagostimulant, the threshold is 10^{-7} M and 10^{-6} M, respectively (151). The threshold of detection for cucurbitacin by adult *D. virgifera* is less than 10^{-7} M (36). Changes in sensitivity of deterrent and phagostimulatory receptors occur in relation to age, nutrient status, and experience (21, 123, 130, 131).

DESENSITIZATION OF DETERRENT RECEPTORS There are several examples of insects habituating to plant secondary compounds, enabling them to eat food that initially was distasteful (142). In a few cases, habituation is known to be associated

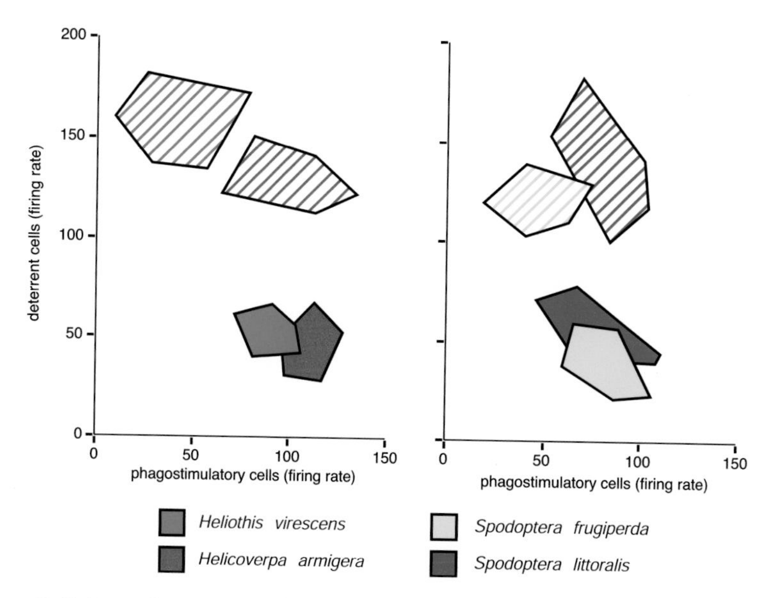

Figure 2 Pattern of response produced by the saps of host and nonhost plants in the galeal sensilla of four polyphagous noctuid caterpillars. Each caterpillar was stimulated with the saps of four plants that were readily eaten (*solid color*), and four that were rejected (*hatched color*). Each polygon encloses the sums of responses in the four phagostimulatory cells and the four deterrent cells in the two sensilla [after (129)].

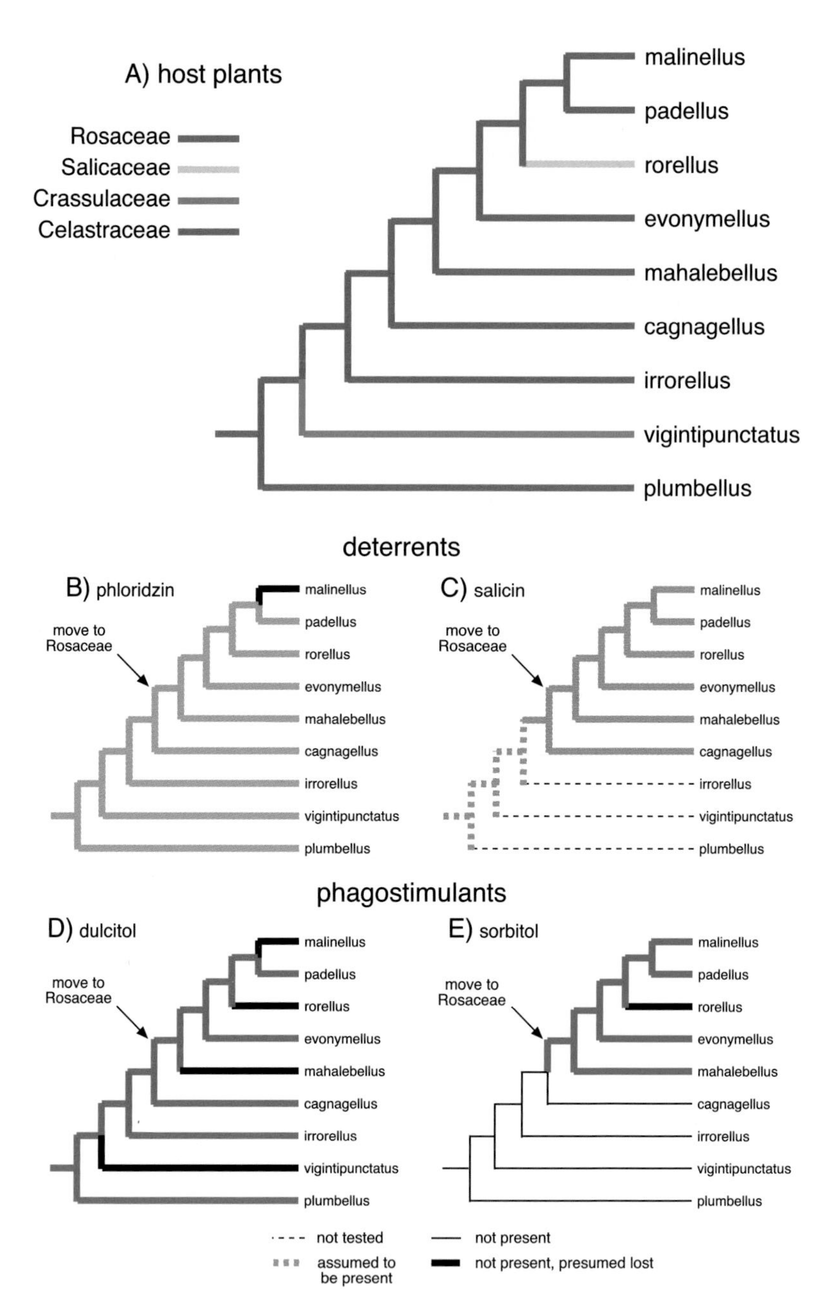

Figure 3 Phylogeny of gustatory receptor sensitivity in caterpillars of the genus *Yponomeuta*. (*A*) Phylogeny of some European species with their hosts. (*B,C*) Responses to two secondary compounds that are deterrent. (*D,E*) Responses to two secondary compounds that are phagostimulants. Thick colored lines in *B–E* show species with a sensory response to the specific chemical (53, 85, 147).

with the desensitization of deterrent receptors, and it has generally been assumed that this permits the insect to make the behavioral change from rejection to acceptance (115). After being reared on an artificial diet containing caffeine or salicin, caterpillars of *M. sexta* readily accept food containing them, even though both are deterrent (55, 56, 60, 115). Schoonhoven (115) found that the increase in acceptability of salicin-containing food was correlated with a reduction in the activity (desensitization) of the deterrent cells responding to this compound and concluded that the sensory change was responsible for the increase in amount eaten. Glendinning et al. (56), however, obtained only a marginal decrease in the sensory response to salicin despite a major change in acceptance and concluded that the behavioral change was mediated centrally rather than by peripheral changes in the taste receptors. Their data for caffeine, on the other hand, largely mirror the earlier results of Schoonhoven with salicin (57). Desensitization of the taste receptors is clearly not an essential component of habituation, and its importance where it does occur is open to question. Because the initiation of feeding is dependent on the sensory response, it may be that desensitization facilitates a more rapid response and an increase in the likelihood of accepting a previously distasteful food, but it is not essential for a change in food intake to occur.

The sensory response to salicin was reduced when *M. sexta* was habituated to caffeine (not salicin), indicating that the receptors for these chemicals share a common second messenger. The response to aristolochic acid, however, did not change, suggesting that although it stimulates the same cell, it acts via a separate second messenger cascade. Glendinning et al. (56) suggest that the importance of this relates to the toxicity of the compounds. Aristolochic acid is more toxic than salicin so that habituation to the former might be harmful to the insect, whereas habituation to salicin (and other comparable compounds) would enable the insect to eat plants that were distasteful but not toxic (55, 57).

EFFECTS OF MIXTURES

Gustatory sensilla always contain more than one chemosensory neuron, and they usually, perhaps always, contain both phagostimulatory and deterrent neurons. The galeal styloconic sensilla of many caterpillars, for example, contain two phagostimulatory cells and two deterrent cells (assuming that the salt response is deterrent) (124). When a sensillum encounters a mixture of compounds, as will normally be the case, the responses of the neurons within the sensillum do not necessarily directly reflect their responses to the individual compounds alone. Interactions between stimulating molecules may occur before any cell is stimulated, and there may also be some interaction between the neurons after electrical events have been initiated.

Binary mixtures of compounds that stimulate the same cell usually produce additive effects, although limited by the maximum firing rate of the cell. This is true for both phagostimulants and deterrents (13, 57, 58). However, more complex mixtures of amino acids, resembling the free amino acid composition of the host plants, are markedly less stimulating than expected from the effects of the individual components. For example, a mixture based on the free amino acid profile of *Plantago* contained 20 amino acids at concentrations ranging from 0.3 to 4.2 mM. Nine of these stimulate a phagostimulatory cell in the medial galeal sensillum of *G. geneura*, and the sum of the estimated effects of these 9 at the appropriate concentrations was in excess of 200 spikes/s^{-1}. The measured firing rate in response to the complete plant mixture was only about 50 spikes/s^{-1}. Similarly, the expected response from an amino acid–sensitive phagostimulatory cell in the lateral sensillum, stimulated by 11 of the amino acids in the mixture, was about 150 spikes/s^{-1}, and the measured response to the mixture was less than 20. The maximum firing rates of both cells were in excess of 100 spikes/s^{-1}, so the low rates of firing are the product of interactions between the chemicals at the dendrite (12). Similarly, in *Leptinotarsa decemlineata*, a mixture of five amino acids, all known to stimulate a cell in the lateral galeal sensillum, at a total concentration equivalent to that of the free amino acids in potato leaf produced a response of only 27 spikes/s^{-1}. Based on the firing rates in response to the individual amino acids, the theoretical expected response to the mixture was over 100 spikes/s^{-1} if their individual responses were summed, although this cell has a maximum firing rate close to 60 spikes/s^{-1} (87, 88). Even compounds that do not stimulate may reduce responses to stimulatory chemicals in the same class. Glutamic acid, which has no effect alone, reduces the response to serine by phagostimulatory cell 1 in the medial sensillum of *G. geneura*, although it has no effect on the response to sucrose that stimulates the same cell (13).

Mixtures of nutrient compounds that stimulate separate cells in a sensillum usually cause the cells to fire independently of each other, but sometimes produce synergy (50). This is also sometimes true of mixtures of nutrient and deterrent compounds. Sinigrin caused "a large increase" in the response to 0.1 M sucrose in the lateral galeal sensillum of the polyphagous caterpillar of *Isia isabella* (50). In *B. mori*, morin, which is not known to stimulate cells on the mouthparts, nevertheless causes a large increase in the amount of sugar eaten (77), presumably by synergizing the sensory response to sucrose. Typically, however, the interaction between a phagostimulant and a deterrent results in some concentration-dependent inhibition of the response (119). Toosendanin is a triterpenoid from *Melia toosendan* that inhibits feeding and stimulates a deterrent cell in the caterpillar of *P. brassicae* (81). Increasing the toosendanin concentration while keeping the concentration of sucrose constant causes a dose-dependent decrease in action potential production by the phagostimulatory cell. Conversely, the response to a deterrent may be inhibited. Sinigrin is a deterrent for many caterpillars and stimulates one or more deterrent cells (11, 126, 127). The activity of these cells is reduced in *M. configurata* and *T. ni* by high concentrations of sucrose (60 mM) or inositol (100 mM) at least when the sinigrin concentration is relatively low (2 mM) (128). Less complete data for other insects show that the effects of compounds in mixtures are dependent on concentration but the effects are not necessarily reciprocal (66).

Although these effects are widespread, the concentrations at which they occur often seem too high to have relevance in a natural situation. However, there are examples of interactions at relevant concentrations. Toosendanin, for example, causes some reduction in the response to sucrose at 10^{-9} M (122). Depression of the activity of presumed phagostimulatory cells by secondary compounds at concentrations below 1 mM have also been recorded in *S. americana* (32).

Where interactions occur, not all responses are affected; there may be differences between cells or even in the responses of one cell. Toosendanin at 10^{-4} M, for example, has no effect on the firing rate of the cell responding to proline, while the cell responding to sucrose in the same sensillum is inhibited (122). Glutamic acid inhibits the response of a phagostimulatory cell of *G. geneura* to serine, but the response to sucrose stimulating the same cell is unaffected (14).

The mechanisms involved in these interactions between chemicals at the periphery are not understood, but there are almost certainly several different ones. Mitchell & Harrison (92) consider the inhibition of nutrient responses in *L. decemlineata* to be damage effects. This may also be true in other cases where sensilla have been exposed to a chemical for extended periods (1 min or more). Warburganal, for example, suppresses the subsequent response of galeal sensilla of the African armyworm, *Spodoptera exempta*, to sucrose and inositol, and recovery of the response takes many minutes (84).

In most cases, however, recovery of response occurs without any delay, suggesting that the effects are not damaging to the cells. Nonstimulating nutrients may affect the activity of phagostimulatory cells by altering the cell environment. This appears to be the case with organic acids. Their effect in *Manduca sexta* is to alter the activity of other cells through their effects on pH (15), and in mixtures with glucose or inositol, ascorbic acid causes a reduction in activity of the phagostimulatory cells responding to these compounds. Other interactions probably involve the receptor sites on the neuronal cell membrane. The differential effects of glutamic acid on the responses to serine and sucrose by phagostimulatory cell 1 in the medial sensillum of *G. geneura* are an example of this (13). In lobsters, in which comparable interactions occur in the olfactory receptor system, some neurons have receptor sites that cause hyperpolarizations in addition to others that depolarize, and competitive interactions may also occur (38). A possibly analogous situation has been suggested for gustatory receptors of *D. virgifera* (97).

Such interaction is likely to be critically important in the normally feeding insect. Consequently, studies on the sensory effects of saps expressed from plants are critical additions to the work with single chemicals that is necessary for the basic understanding of the system. This was clearly understood by Mitchell and his associates in their studies on beetles (91, 93, 139). It is possible that the sensitivities of all the sensory neurons in the mouthparts are molded by the effects they produce in mixtures. A high sensitivity to a particular amino acid, for example, may not reflect a particular need for that acid but may affect the way the cell will respond to mixtures.

A different type of interaction was recorded from tibial sensilla of *S. americana* (156). Here, the train of action potentials produced by a deterrent cell was periodically interrupted by the activity of another cell in the same sensillum, and occasionally the firing pattern of the deterrent cell was completely disrupted.

Similar effects have been noted in other grasshoppers and on palp tip sensilla, although there is no evidence that the feeding behavior of the insects is affected (R.F. Chapman, unpublished). In this case the interaction between the cells occurs after stimulation and is one of an electrical/ionic nature.

HOST PLANT SELECTION

Feeding by phytophagous insects is governed by the balance of phagostimulatory and deterrent inputs, as was clearly recognized by Dethier (48) and Schoonhoven (118) and demonstrated in simplified models (25, 72, 83). Among polyphagous insects, this is probably the sole determinant of acceptance or rejection, as shown most clearly by the work of Simmonds & Blaney (129). Saps from acceptable and unacceptable plants produced broadly similar levels of stimulation in the phagostimulatory cells of the galeal sensilla of *Helicoverpa armigera* and *Spodoptera littoralis* (Figure 2, see color insert), but the response of the deterrent cells was markedly higher with unacceptable plants. There is a similar pattern in *Heliothis virescens* and the fall armyworm, *Spodoptera frugiperda*, although in these species phagostimulatory input also tends to be lower with the unacceptable plants. Although Simmonds & Blaney (129) did not consider the roles of other sensilla on the mouthparts, these would probably contribute to the pattern in an essentially similar manner. The contrast between net phagostimulatory inputs and net deterrent inputs governs the insect's responses. The model tacitly assumes that the peripheral inputs are treated equally within the CNS. There is no direct evidence for this, but indirect evidence from studies on *P. brassicae* suggests that inputs from deterrent cells are weighted more heavily in the CNS than are the inputs from phagostimulatory cells (118). This presumably means that synapses in the deterrent pathway give greater gain to the signal than do those in the phagostimulatory pathway. If this was also true in the insects examined by Simmonds & Blaney (129), the magnitude of the differences they observed would be increased. Any interactions between chemicals and neurons that may occur at the periphery will have the effect of altering the phagostimulatory or deterrent inputs and could significantly change the balance.

The much less extensive work on the polyphagous grasshopper, *S. americana*, is consistent with this pattern and provides some evidence concerning interaction at the periphery (32). Canavanine is a feeding deterrent but does not stimulate a deterrent cell. It does, however, suppress the response to sucrose so that, at least in a simple experimental situation, the insect receives no phagostimulatory input.

Food selection in oligophagous species also involves the balance of phagostimulatory and deterrent inputs (118) but, at least commonly, with the addition of a specific chemical sign stimulus that may be overriding. Both *P. rapae* and *M. sexta* are polyphagous at hatching and become oligophagous only after they have experienced their normal hosts (42, 110). Presumably, at first, the balance of phagostimulatory and deterrent inputs determines food intake. After the experience of a glucosinolate, in the case of *P. rapae*, or indiocide D, in the case of

M. sexta, the insects will only eat plants containing these compounds, species of Brassicaceae and Solanaceae, respectively (42, 110). Changes in the activity of the peripheral system occur in *M. sexta* (*P. rapae* has not been examined) and presumably contribute to the change in behavior when the insect becomes oligophagous, but the precise manner in which these effects are brought about is not yet known. Changes in the CNS, perhaps facilitation at synapses on the phagostimulatory input pathway, are possibly of major importance in the behavioral changes, perhaps in addition to changes at the periphery.

Chemical sign stimuli are also important in monophagous species, but they may act in different ways. *Yponomeuta cagnagellus* is monophagous on spindle (*Euonymus*) that contains the sugar alcohol, dulcitol. Dulcitol is a phagostimulant for this species that has a cell in each of the lateral and medial galeal sensilla responding to the chemical (70, 147). Although the specificity of these cells for dulcitol has not been rigorously examined, it is almost certainly high, and these cells probably have an overriding role in host plant selection. In *B. mori*, the picture is different. Sucrose alone is a phagostimulant stimulating cells in the lateral and epipharyngeal sensilla. Saps from most non-host plants stimulate a deterrent cell in the medial galeal sensillum, whereas that from mulberry, the host plant, and a few other plant species, do not. Input from these cells may prevent the insect from feeding on most plant species, and the absence of a deterrent response from mulberry is permissive, allowing the insect to feed (71). Mulberry leaves are characterized by the presence of morin, yet there is no evidence that a morin-specific phagostimulatory cell is present in the galeal sensilla (76) or in the epipharyngeal sensilla (7). Such a receptor may exist on the maxillary palp, but this has not been determined. Morin alone is not a behavioral phagostimulant, but it synergizes the behavioral effect of sucrose (77); perhaps it increases the responses of the sucrose-sensitive cells. The nonspecific odor of *n*-butyl propionate also synergizes the behavioral effects of sucrose and morin. Whether these effects alone are sufficient to provide the insect's specificity is still unknown. It remains possible that some minor component of mulberry leaves, comparable with those involved in host selection by *M. sexta* larvae or adult cabbage root fly, *Delia radicum* (42, 140), provides a key signal.

If the gustatory abilities of phytophagous beetles are similar to those of caterpillars, as studies on *D. virgifera* suggest (36), host plant selection probably occurs in a similar way. However, Mitchell and his associates suggest that potential host plants of three species of *Leptinotarsa* are indicated by the activity of a single phagostimulatory cell in each of the galeal sensilla. Plant secondary compounds may inhibit the activity of this cell, and the authors suggest that the variability of input from other taste receptor cells on all 15 galeal sensilla may itself provide a signal. These cells do not usually respond strongly to the saps of plants that are readily eaten, but if they do the variability across sensilla is low. (The study is based on differences between insects, but it assumes that the results are representative of the population of receptors on the galea of a single insect because these respond in a similar way to various stimuli). On plants that are not eaten, these secondary cells tend to respond more strongly, and with higher variability, and it is argued that the variability provides a signal of unsuitability (90, 93, 139).

ASSOCIATIVE LEARNING

There is some evidence that gustatory receptors are involved in associative learning in grasshoppers and caterpillars (9, 22, 23, 35). There is, however, some doubt about the conditioned stimulus. Because in grasshoppers the response is clearly associated with palpation while the antennae are not in contact with the surface, it has been assumed that gustatory receptors are involved. But small numbers of multiporous, putative olfactory sensilla are distributed among the gustatory receptors of the palp tip (19). Consequently, the possibility that the response involves olfactory learning has not been excluded. In caterpillars, the position of the antennae beneath the head, close to the leaf surface, also makes olfactory learning difficult to separate from gustatory learning.

One reason for suspecting that the gustatory system is not involved in learning is the circuitry within the CNS. Whereas input from olfactory receptor cells is integrated in the antennal lobes, which are closely associated with the mushroom bodies (73, 78), the axon terminals of the gustatory system are dispersed and make synaptic connections in the segmental ganglia (103, 113, 136). This does not mean that there are no links to the higher learning centers in the brain, but they are not known to exist. If they do not, it is not likely that gustatory learning can occur. However, the axons of at least some sensory cells on the galea and maxillary palp of caterpillars have a branch that extends to the brain, ending in the larval antennal center (79). The subsequent connections of these axons in the brain are not known, but they do provide a possible link to the mushroom bodies, although it is possible that they are the axons of mechanoreceptors, not chemoreceptors (93).

Whether the gustatory system of phytophagous insects may be involved in associative learning remains an open question.

EVOLUTION

Marked changes have occurred in insect gustatory systems in the course of evolution. The orthopteroid orders and Thysanura have large numbers of gustatory sensilla, and the sensilla contain relatively large numbers of sensory neurons (28). These features are thus basal insectan characters unrelated to feeding habit. Hemipteroid and holometabolous insects have far fewer sensilla on the mouthparts, and each sensillum commonly contains four chemosensory neurons. The reduction in number occurs across all orders, and again there is no reason to expect that it was in any way a direct consequence of phytophagy. It is possible, however, that the reduction in numbers had an impact on host plant selection. It is noteworthy that, whereas a majority of grasshoppers are polyphagous, the vast majority of phytophagous insects in all the other orders are specialists. In the grasshoppers, there is evidence that specialization is associated with a reduction in numbers of sensilla (28, 34).

Because of the complexity of the grasshopper gustatory system, there is virtually no work on cell specificity. The most extensive studies are those of Blaney (17, 18) who deliberately avoided cell typing, believing that it was the combined inputs from all cells in a sensillum that conveyed the information. While this now seems unlikely, most evidence suggests that grasshoppers lack a neuron that responds in a dose-dependent manner to sugars, and distinguishing between inorganic salts and sugars depends on the differential activity of two cells (31). Increased precision in decision making might require comparisons to be made between the inputs from numerous cells not just two, and detection of low concentrations may be achieved through summation. These may be contributing factors to the possession of a large number of sensilla, as has been suggested for Crustacea (30, 44). Making decisions rapidly on the basis of the few and variable numbers of action potentials that occur in the few milliseconds immediately following stimulation also requires numerous sensilla, especially in the case of the palp tip sensilla where the periods of contact with the leaf surface during palpation are less than 20 ms (20). Chapman (30) suggests that 20 sensilla may be close to the minimum number required to make a decision.

Gustatory neurons sensitive to sugars and amino acids are known from the firebrat, *Thermobia domestica* (Thysanura), and *Periplaneta brunnea* (Blattodea) representatives of more basal nonphytophagous insect taxa (64, 65). There are also records of behavioral feeding deterrence in cockroaches (154), but whether they also have deterrent gustatory cells that respond to these compounds is an open and critical question. Bacteria, protozoans, and nematodes have receptors for plant secondary compounds, and all animals require the ability to detect potential toxins in their food. There is no reason to regard phytophagous insects as unusual in this respect.

Consequently, possession of deterrent cells may be a basal condition in the insects that has perhaps become emphasized in the phytophagous species. It is the evolution of the deterrent response that is central to the evolution of insect/host plant interactions. The thesis that a change from polyphagy to oligophagy (or the reverse) involves a general increase (or reduction) in sensitivity to deterrent compounds is supported at the behavioral level (10, 16, 52). Physiologically, however, most evidence points to a change within the CNS rather than in the peripheral receptors as the first step in such a change. Such evidence as there is suggests that peripheral changes follow host plant switches, rather than being causative (10, 14, 36). Where insects have come to feed on a particular plant they tend to lose their deterrent response to the key secondary compounds in that plant. This is most clearly seen in the European small ermine moths (*Yponomeuta*) where changes in the sensory system can be followed through a group of closely related species with a known phylogeny. These species have evolved from an ancestor feeding on plant(s) in the Celastraceae. There has been one major host shift to Rosaceae and two minor ones to *Sedum* (Crassulaceae) and *Salix* (Salicaceae) (Figure 3*A*, see color insert)

(85). Phloridzin is probably restricted to *Malus* (apple) (41), which is eaten by *Y. malinellus*. Except for *Y. malinellus*, caterpillars of all the *Yponomeuta* species that have been studied have a neuron in the medial galeal sensillum that responds to phloridzin (Figure 3*B*) (147). Phloridzin is deterrent to these caterpillars, and clearly *Y. malinellus* has lost the capacity to detect the compound associated with its host. Whether this occurred before or after it moved on to *Malus* is not clear. Salicin is found only in Salicaceae, and only *Y. rorellus* among this group of *Yponomeuta* eats *Salix*. Caterpillars in all other species that have been tested have deterrent cells in the lateral and medial galeal sensilla that respond to salicin but *Y. rorellus* is an exception. It exhibits no response in the lateral sensillum where the deterrent cell must have lost its sensitivity; the medial sensillum retains its sensitivity but the response is weak compared with other species (Figure 3*C*) (147). *Y. rorellus* appears to be in the process of losing its sensitivity to salicin. Whether this process started before the host switch is not known, but the colonization of a previously unpalatable plant is likely to involve a prior reduction in sensitivity to its deterrent properties by at least some members of the insect population.

Some insects use host plant–specific compounds as sign stimuli. This could be achieved by expressing the appropriate receptor on a phagostimulatory cell instead of a deterrent cell. Perhaps *G. geneura* represents such a step where catalpol, a compound that is deterrent to some other arctiids and does stimulate a deterrent cell even in *G. geneura*, strongly stimulates the same phagostimulatory cell as sucrose and serine and is a behavioral phagostimulant (14). Emphasis of this pathway within the CNS, coupled with loss of receptors for the nutrients, could result in a cell dedicated to catalpol as a phagostimulant and with a dominant role in host plant selection, as glucosinolates are for the *Pieris* species.

Studies on *Yponomeuta* illustrate how additional changes could occur. Dulcitol and sorbitol are phagostimulants for some *Yponomeuta* species. Dulcitol occurs in a number of plant families and is abundant in the Celastraceae that are hosts of some *Yponomeuta*. Caterpillars of most *Yponomeuta* species have one or two phagostimulatory neurons in the lateral and medial galeal sensilla that respond to dulcitol irrespective of whether they feed on Celastraceae or Rosaceae (Figure 3*D*) (147). *Y. vigintipunctatus* clearly lost sensitivity in association with the move to *Sedum*, and so did *Y. mahalabellus*, *Y. rorellus*, and *Y. malinellus* when they moved on to their new hosts. None of these plants contains dulcitol, but the species of Rosaceae fed on by *Y. padellus* and *Y. evonymellus*, which retain their sensitivity to the chemical, do (53). Even *Y. padellus* and *Y. evonymellus*, however, retain sensitivity only in the lateral sensillum; two of the species feeding on Celastraceae have a dulcitol-sensitive phagostimulatory cell in each of the lateral and medial sensilla. The picture that emerges here is of a sensitivity that is retained as long as it has functional value but is otherwise lost.

The responses to sorbitol illustrate a different situation. Whereas the species in this study had sensitivity to phloridzin and salicin as deterrents, and to dulcitol as a phagostimulant, from the outset, and this sensitivity has subsequently been lost in some species, the sensitivity to sorbitol first appears with *Y. mahalabellus* (Figure 3*E*). The more basal species have no cells in the galeal sensilla that are sensitive to sorbitol. Sorbitol is only found in Rosaceae and is a phagostimulant for the *Yponomeuta* species feeding on plants in the family. It is a stereoisomer of dulcitol. *Y. evonymellus* is stimulated by dulcitol and sorbitol, and a phagostimulatory cell in the lateral galeal sensillum responds to both compounds indicating that it contains the receptors for the two chemicals (112). Dulcitol is present in small amounts in some Rosaceae (53) and could provide *Yponomeuta* feeding on Celastraceae with a bridge leading to the evolution of species feeding only on Rosaceae. Thus, it is conceivable and perhaps likely that the phagostimulatory response to sorbitol arose after an initial move to Rosaceae.

Host plant switches in *Yponomeuta* were probably initiated by adults ovipositing on the "wrong" plant rather than by the larvae. Such evidence as is available suggests that the adults do not use the same chemical cues as larvae in selecting their hosts (111), reinforcing the likelihood that the changes in larval sensory response were a consequence of moving rather than the cause.

The studies on *Yponomeuta* emphasize the value of a comparative approach based on phylogeny. Only from such studies will it be possible to make firm statements about evolutionary changes in the sensory system.

CONCLUSION: THE ROLE OF THE GUSTATORY SYSTEM IN FEEDING

Gustation has a key role in food acceptance or rejection by phytophagous insects, but its importance in subsequent feeding is not always clear. A model of feeding regulation in *L. migratoria* (133) shows that meal size is probably determined by the level of excitation in the CNS resulting from the net phagostimulatory input occurring at the start of a feed. Sustained feeding by these insects also requires continuous sensory input (29).

Among caterpillars, however, the position is less clear. There are sometimes strong correlations between the activities of phagostimulatory cells and amounts eaten, but sometimes there is a lack of correlation. Although the sensory responses of caterpillars do change after eating nutritionally unbalanced food in a manner that appears to favor selection of food that compensates for any deficiency, the immediate behavior does not necessarily reflect this. Although habituation to deterrent compounds is sometimes accompanied by sensory desensitization, this is not always the case. These apparent exceptions to the importance of gustatory input in governing amounts eaten by caterpillars perhaps indicate that central nervous phenomena, perhaps associated with rapid post-ingestive feedback mechanisms (54), are at least of equal importance to changes in the sensory system in these insects. This could be a reflection of the reduced numbers of gustatory neurons present on the mouthparts.

ACKNOWLEDGMENTS

Erich Städler kindly allowed me to see preprints of unpublished manuscripts, and, as always, I have been greatly aided by the continuous inputs of Elizabeth Bernays.

The *Annual Review of Entomology* is online at http://ento.annualreviews.org

LITERATURE CITED

1. Abisgold JD, Simpson SJ. 1988. The effect of dietary protein levels and haemolymph composition on the sensitivity of the maxillary palp chemoreceptors of locusts. *J. Exp. Biol.* 135:215–29
2. Adler E, Hoon MA, Mueller KL, Chandrashekar J, Ryba NJ, Zuker CS. 2000. A novel family of mammalian taste receptors. *Cell* 100:693–702
3. Albert PJ, Parisella S. 1988. Feeding preferences of eastern spruce budworm larvae in two-choice tests with extracts of mature foliage and with pure amino acids. *J. Chem. Ecol.* 14:1649–56
4. Altner H, Altner I. 1986. Sensilla with both terminal pore and wall pores on the proboscis of the moth, *Rhodogastria bubo* Walker (Lepidoptera: Arctiidae). *Zool. Anz.* 216:129–50
5. Altner H, Prillinger L. 1980. Ultrastructure of invertebrate chemo-, thermo-, and hygroreceptors and its functional significance. *Int. Rev. Cytol.* 67:69–139
6. Angeli S, Ceron F, Scaloni A, Monti M, Monteforti G, Minnocci A, et al. 1999. Purification, structural characterization, cloning and immunocytochemical localization of chemoreception proteins from *Schistocerca gregaria*. *Eur. J. Biochem.* 262:745–54
7. Asaoka K. 2000. Deficiency of gustatory sensitivity to some deterrent compounds in "polyphagous" mutant strains of the silkworm, *Bombyx mori*. *J. Comp. Physiol. A* 186:1011–18
8. Becker A, Peters W. 1989. Localization of sugar-binding sites in contact chemosensilla of *Periplaneta americana*. *J. Insect Physiol.* 35:239–50
9. Behmer ST, Elias DO, Bernays EA. 1999. Post-ingestive feedbacks and associative learning regulate the intake of unsuitable sterols in a generalist grasshopper. *J. Exp. Biol.* 202:739–48
10. Bernays EA, Chapman RF. 1987. The evolution of deterrent responses in plant-feeding insects. In *Perspectives in Chemoreception and Behavior*, ed. RF Chapman, EA Bernays, JG Stoffolano, pp. 159–73. New York: Springer
11. Bernays EA, Chapman RF. 2000. A neurophysiological study of sensitivity to a feeding deterrent in two sister species of *Heliothis* with different diet breadths. *J. Insect Physiol.* 46:905–12
12. Bernays EA, Chapman RF. 2001. Electrophysiological responses of taste cells to nutrient mixtures in the polyphagous caterpillar of *Grammia geneura*. *J. Comp. Physiol. A* 187:205–13
13. Bernays EA, Chapman RF. 2001. Taste cell responses in the polyphagous arctiid, *Grammia geneura*: towards a general pattern for caterpillars. *J. Insect Physiol.* 47:1029–43

13a. Bernays EA, Chapman RF, Hartmann T. 2002. A highly sensitive taste receptor cell for pyrrolizidine alkaloids in the lateral galeal sensillum of a polyphagous caterpillar, *Estigmene acraea*. *J. Comp. Physiol. A* In press

14. Bernays EA, Chapman RF, Singer MS. 2000. Sensitivity to chemically diverse phagostimulants in a single gustatory neuron of a polyphagous caterpillar. *J. Comp. Physiol. A* 186:13–19
15. Bernays EA, Glendinning JI, Chapman RF. 1998. Plant acids modulate

chemosensory responses in *Manduca sexta* larvae. *Physiol. Entomol.* 23:193–201
16. Bernays EA, Oppenheim S, Chapman RF, Kwon H, Gould F. 2000. Taste sensitivity of insect herbivores to deterrents is greater in specialists than in generalists: a behavioral test of the hypothesis with two closely related caterpillars. *J. Chem. Ecol.* 26:547–63
17. Blaney WM. 1974. Electrophysiological responses of the terminal sensilla on the maxillary palps of *Locusta migratoria* (L.) to some electrolytes and non-electrolytes. *J. Exp. Biol.* 60:275–93
18. Blaney WM. 1975. Behavioural and electrophysiological studies of taste discrimination by the maxillary palps of larvae of *Locusta migratoria* (L.). *J. Exp. Biol.* 62:555–69
19. Blaney WM. 1977. The ultrastructure of an olfactory sensillum on the maxillary palps of *Locusta migratoria* (L.). *Cell Tissue Res.* 184:397–409
20. Blaney WM, Chapman RF. 1970. The function of the maxillary palps of Acrididae (Orthoptera). *Entomol. Exp. Appl.* 13:363–76
21. Blaney WM, Schoonhoven LM, Simmonds MSJ. 1986. Sensitivity variations in chemoreceptors: a review. *Experientia* 42:13–19
22. Blaney WM, Simmonds MSJ. 1985. Food selection by locusts: the role of learning in rejection behaviour. *Entomol. Exp. Appl.* 39:273–78
23. Blaney WM, Simmonds MSJ. 1986. Learning in larval food selection: the role of plant surfaces. In *Insects and the Plant Surface*, ed B Juniper, R Southwood, pp. 342–44. London: Arnold
24. Blaney WM, Simmonds MSJ, Simpson SJ. 1990. Dietary selection behaviour: comparisons between locusts and caterpillars. *Symp. Biol. Hung.* 39:47–52
25. Blom F. 1978. Sensory activity and food intake: a study of input-output relationships in two phytophagous insects. *Neth. J. Zool.* 28:277–340
26. Bowers MD. 1991. Iridoid glycosides. In *Herbivores: Their Interactions with Secondary Plant Metabolites*, ed. GA Rosenthal, MR Berenbaum, 1:297–325. San Diego: Academic
27. Champagne DE, Bernays EA. 1991. Phytosterol unsuitability as a factor mediating food aversion learning in the grasshopper *Schistocerca americana*. *Physiol. Entomol.* 16:391–400
28. Chapman RF. 1982. Chemoreception: the significance of receptor numbers. *Adv. Insect Physiol.* 16:247–356
29. Chapman RF. 1982. Regulation of food intake by phytophagous insects. In *Exogenous and Endogenous Influences on Metabolic and Neural Control*, ed. ADF Addink, N Spronk, 1:19–30. Oxford: Pergamon
30. Chapman RF. 1988. Sensory aspects of host-plant recognition by Acridoidea: questions associated with the multiplicity of receptors and variability of response. *J. Insect Physiol.* 34:167–74
31. Chapman RF, Ascoli-Christensen A. 1999. Sensory coding in the grasshopper (Orthoptera: Acrididae) gustatory system. *Ann. Entomol. Soc. Am.* 92:873–79
32. Chapman RF, Ascoli-Christensen A, White PR. 1991. Sensory coding for feeding deterrence in the grasshopper *Schistocerca americana*. *J. Exp. Biol.* 158:241–59
33. Chapman RF, Bernays EA. 1989. Insect behavior at the leaf surface and learning as aspects of host plant selection. *Experientia* 45:215–22
34. Chapman RF, Fraser J. 1989. The chemosensory system of the monophagus grasshopper, *Bootettix argentatus* Bruner (Orthoptera: Acrididae). *Int. J. Insect Morphol. Embryol.* 18:111–18
35. Chapman RF, Sword G. 1993. The importance of palpation in food selection by a polyphagous grasshopper (Orthoptera: Acrididae). *J. Insect Behav.* 6:79–91

36. Chyb S, Eichenseer H, Hollister B, Mullin CA, Frazier JL. 1995. Identification of sensilla involved in taste mediation in adult western corn rootworm (*Diabrotica virgifera virgifera* LeConte). *J. Chem. Ecol.* 21:313–29
37. Cook AG. 1977. Nutrient chemicals as phagostimulants for *Locusta migratoria* (L.). *Ecol. Entomol.* 2:113–21
38. Cromarty SI, Derby CD. 1998. Inhibitory receptor binding events among the components of complex mixtures contribute to mixture suppression in responses of olfactory receptor neurons of spiny lobsters. *J. Comp. Physiol. A* 183:699–707
39. Dadd RH. 1977. Qualitative requirements and utilization of nutrients: insects. In *Handbook Series in Nutrition and Food.* Section D, Vol. 1: *Nutritional Requirements*, ed. M RechCigl, pp. 305–46. Cleveland: CRC
40. Dadd RH. 1985. Nutrition: organisms. See Ref. 79a, 4:311–90
41. Darnley Gibbs R. 1974. *Chemotaxonomy of Flowering Plants*, Vol. 1. Montreal: McGill-Queen's Univ. Press. 680 pp.
42. del Campo ML, Miles CI, Schroeder FC, Mueller C, Booker R, Renwick JA. 2001. Host recognition by the tobacco hornworm is mediated by a host plant compound. *Nature* 411:186–89
43. Den Otter CJ. 1992. Responses of the African armyworm and three species of borers to carbohydrates and phenolic substances: an electro- and behavioral physiological study. *Entomol. Exp. Appl.* 63:27–37
44. Derby CD, Steullet P. 2001. Why do animals have so many receptors? The role of multiple chemosensors in animal perception. *Biol. Bull.* 200:211–15
45. Descoins C, Marion-Poll F. 1999. Electrophysiological responses of gustatory sensilla of *Mamestra brassicae* (Lepidoptera, Noctuidae) larvae to three ecdysteroids: ecdysone, 20-hydroxyecdysone and ponasterone A. *J. Insect Physiol.* 45:871–76
46. Dethier VG. 1968. Chemosensory input and taste discrimination in the blowfly. *Science* 161:389–91
47. Dethier VG. 1972. Sensitivity of the contact chemoreceptors of the blowfly to vapors. *Proc. Natl. Acad. Sci. USA* 69:2189–92
48. Dethier VG. 1973. Electrophysiological studies of gustation in lepidopterous larvae II. Taste spectra in relation to foodplant discrimination. *Z. Vergl. Physiol.* 82:103–34
49. Dethier VG. 1977. The taste of salt. *Am. Sci.* 65:744–51
50. Dethier VG, Kuch JH. 1971. Electrophysiological studies of gustation in lepidopterous larvae. I. Comparative sensitivity to sugars, amino acids and glucosides. *Z. Vergl. Physiol.* 72:343–63
51. Douglas AE, Minto LB, Wilkinson TL. 2001. Quantifying nutrient production by the microbial symbionts in an aphid. *J. Exp. Biol.* 204:349–58
52. Eichenseer H, Mullin CA. 1997. Antifeedant comparisons of GABA/glycinergic antagonists for diabroticite leaf beetles (Coleoptera: Chrysomelidae) *J. Chem. Ecol.* 23:71–82
52a. Fraenkel G. 1959. The raison d'être of secondary plant substances. *Science* 129:1466–70
53. Fung SY. 1988. Sorbitol and dulcitol in some Celastraceous and Rosaceous plants, hosts of *Yponomeuta* species. *Biochem. Syst. Ecol.* 16:191–94
54. Glendinning JI. 1996. Is chemosensory input essential for the rapid rejection of toxic foods? *J. Exp. Biol.* 199:1523–34
55. Glendinning JI, Brown H, Capoor M, Davis A, Gbedemah A, Long E. 2001. A peripheral mechanism for behavioral adaptation to specific "bitter" taste stimuli in an insect. *J. Neurosci.* 21:3688–96
56. Glendinning JI, Domdon S, Long E. 2001. Selective adaptation to noxious foods by a herbivorous insect. *J. Exp. Biol.* 204:3355–67
57. Glendinning JI, Ensslen S, Eisenberg ME,

Weiskopf P. 1999. Diet-induced plasticity in the taste system of an insect: localization to a single transduction pathway in an identified taste cell. *J. Exp. Biol.* 202:2091–102

58. Glendinning JI, Hills TT. 1997. Electrophysiological evidence for two transduction pathways within a bitter-sensitive taste receptor. *J. Neurophysiol.* 78:734–45
59. Glendinning JI, Nelson NM, Bernays EA. 2000. How do inositol and glucose modulate feeding in *Manduca sexta* caterpillars? *J. Exp. Biol.* 203:1299–315
60. Glendinning JI, Tarre M, Asaoka K. 1999. Contributions of different bitter-sensitive taste cells to feeding inhibition in a caterpillar (*Manduca sexta*). *Behav. Neurosci.* 113:840–54
61. Glendinning JI, Valcic S, Timmermann BN. 1998. Maxillary palps can mediate taste rejection of plant allelochemicals by caterpillars. *J. Comp. Physiol. A* 183:35–43
62. Hamamura Y, Hayashiya K, Naito K-I, Matsuura K, Nishida J. 1962. Food selection by silkworm larvae. *Nature* 194:754–55
63. Hansen K, Wacht S, Seebauer H, Schnuch M. 1998. New aspects of chemoreception in flies. *A. NY Acad. Sci.* 855:143–47
64. Hansen-Delkeskamp E. 2001. Responsiveness of antennal taste hairs of the apterygotan insect, *Thermobia domestica* (Zygentoma); an electrophysiological investigation. *J. Insect Physiol.* 47:689–97
65. Hansen-Delkeskamp E, Hansen K. 1995. Responses and spike generation in the largest antennal taste hairs of *Periplaneta brunnea* Burm. *J. Insect Physiol.* 41:773–81
66. Hanson FE, Stitt J, Frazier J. 2002. Electrophysiological and behavioral responses of tobacco hornworm to gustatory stimulus mixtures. *Entomol. Exp. Appl.* In press
67. Harley KLS, Thorsteinson AJ. 1967. The influence of plant chemicals on the feeding behavior, development, and survival of the two-striped grasshopper, *Melanoplus bivittatus* (Say), Acrididae: Orthoptera. *Can. J. Zool.* 45:305–19
68. Hartmann T. 1996. Diversity and variability of plant secondary metabolism: a mechanistic view. *Entomol. Exp. Appl.* 80:177–88
69. Hedin PA, Miles LR, Thompson AC, Minyard JP. 1968. Constituent of a cotton bud. Formulation of a boll weevil feeding stimulant mixture. *J. Agric. Food Chem.* 16:505–13
70. Herrebout WM, Fung SY, Kooi RE. 1987. Sugar alcohols and host-plant selection in *Yponomeuta*. *Proc. 6th Int. Symp. Insect-Plant Relations*, pp. 257–60. Dordrecht: Junk
71. Hirao T, Arai N. 1991. On the role of gustatory recognition in host-plant selection by the silkworm, *Bombyx mori* L. *Jpn. J. Appl. Entomol. Zool.* 35:197–206
72. Hirao T, Arai N. 1993. Electrophysiological studies on gustatory responses in common cutworm larvae, *Spodoptera litura* Fabricius. *Jpn. J. Appl. Entomol. Zool.* 37:129–36
73. Homberg U, Christensen TA, Hildebrand JG. 1989. Structure and function of the deutocerebrum in insects. *Annu. Rev. Entomol.* 34:477–501
74. Honda I, Ishikawa Y. 1987. Electrophysiological studies on the dorsal and anterior organs of the onion fly larva, *Hylemya antiqua* Meigen (Diptera: Anthomyiidae). *Appl. Entomol. Zool.* 22:410–16
75. Hsiao TH, Fraenkel G. 1968. The influence of nutrient chemicals on the feeding behavior of the Colorado potato beetle, *Leptinotarsa decemlineata* (Coleoptera: Chrysomelidae). *Ann. Entomol. Soc. Am.* 61:44–54
76. Ishikawa S. 1966. Electrical response and function of a bitter substance receptor associated with the maxillary sensilla of the larva of the silkworm, *Bombyx mori* L. *J. Cell. Physiol.* 67:1–11

77. Ishikawa S, Hirao T, Arai N. 1969. Chemosensory basis of hostplant selection in the silkworm. *Entomol. Exp. Appl.* 12:544–54
78. Itagaki H, Hildebrand JG. 1990. Olfactory interneurons in the brain of the larval sphinx moth *Manduca sexta*. *J. Comp. Physiol. A* 167:309–20
79. Kent KS, Hildebrand JG. 1987. Cephalic sensory pathways in the central nervous system of larval *Manduca sexta* (Lepidoptera: Sphingidae). *Philos. Trans. R. Soc. London Ser. A* 315:1–36
79a. Kerkut G, Gilbert LI, eds. 1985. *Comprehensive Insect Physiology Biochemistry and Pharmacology*, Vols. 4, 6. Oxford: Pergamon. 639 pp., 710 pp.
80. Liscia A, Stoffolano JG, Tomassini Barbarossa I, Muroni P, Crnjar R. 1995. Sensitivity to albumen and its terminal amino acids of labellar taste chemoreceptors in *Protophormia terraenovae* (Diptera: Calliphoridae). *J. Insect Physiol.* 41:597–602
81. Luo Lin-er, van Loon JJA, Schoonhoven LM. 1995. Behavioural and sensory responses to some neem compounds by *Pieris brassicae* larvae. *Physiol. Entomol.* 20:134–40
82. Ma WC. 1969. Some properties of gustation in the larva of *Pieris brassicae*. *Entomol. Exp. Appl.* 12:584–90
83. Ma WC. 1972. Dynamics of feeding responses in *Pieris brassicae* Linn. as a function of chemosensory input: a behavioural, ultrastructural and electrophysiological study. *Meded. Landbouwhogesch. Wagening.* 72(11):1–162
84. Ma WC. 1977. Alterations of chemoreceptor function in armyworm larvae (*Spodoptera exempta*) by a plant-derived sesquiterpenoid and by sulfhydryl reagents. *Physiol. Entomol.* 2:199–207
85. Menken SBJ. 1996. Pattern and process in the evolution of insect-plant associations: *Yponomeuta* as an example. *Entomol. Exp. Appl.* 80:297–305
86. Meisner J, Ascher KRS, Lavie D. 1974. Phagostimulants for the larva of the potato tuber moth, *Gnorimoschema operculella* Zell. *Z. Angew. Entomol.* 77:77–106
87. Mitchell BK. 1974. Behavioural and electrophysiological investigations on the responses of larvae of the Colorado potato beetle (*Leptinotarsa decemlineata*) to amino acids. *Entomol. Exp. Appl.* 17:255–64
88. Mitchell BK. 1985. Specificity of an amino acid-sensitive cell in the adult Colorado beetle, *Leptinotarsa decemlineata*. *Physiol. Entomol.* 10:421–29
89. Mitchell BK, Gregory P. 1979. Physiology of the maxillary sugar sensitive cell in the red turnip beetle, *Entomoscelis americana*. *J. Comp. Physiol. A* 132:167–78
90. Mitchell BK, Gregory P. 1981. Physiology of the lateral galeal sensillum in red turnip beetle larvae (*Entomoscelis americana* Brown): responses to NaCl, glucosinolates and other glucosides. *J. Comp. Physiol. A* 144:495–501
91. Mitchell BK, Harrison GD. 1984. Characterization of galeal chemosensilla in the adult Colorado beetle, *Leptinotarsa decemlineata*. *Physiol. Entomol.* 9:49–56
92. Mitchell BK, Harrison GD. 1985. Effects of *Solanum* glycoalkaloids on chemosensilla in Colorado beetle. A mechanism of feeding deterrence. *J. Chem. Ecol.* 11:73–83
93. Mitchell BK, Itagaki H, Rivet M-P. 1999. Peripheral and central structures involved in insect gustation. *Microsc. Res. Tech.* 47:401–15
94. Mitchell BK, Rolseth BM, McCashin BG. 1990. Differential responses of galeal gustatory sensilla of the adult Colorado potato beetle, *Leptinotarsa decemlineata* (Say), to leaf saps from host and non-host plants. *Physiol. Entomol.* 15:61–72
95. Mitchell BK, Schoonhoven LM. 1974. Taste receptors in Colorado beetle larvae. *J. Insect Physiol.* 20:1787–93
95a. Morgan ED, Bhushan Mandava N, eds. 1990. *Handbook of Natural Pesticides.*

Vol. 6: *Insect Attractants and Repellents.* Boca Raton: CRC

96. Morgan ED, Warthen JD. 1990. Insect feeding deterrents. Part B: insect feeding deterrents (1980–1987). See Ref. 95a, 6:83–134
97. Mullin CA, Chyb S, Eichenseer H, Hollister B, Frazier JL. 1994. Neuroreceptor mechanisms in insect gustation: a pharmacological approach. *J. Insect Physiol.* 40:913–31
98. Nagnan-Le Meillour P, Cain AH, Jacquin-Joly E, Francois MC, Ramachandran S, et al. 2000. Chemosensory proteins from the proboscis of *Mamestra brassicae. Chem. Senses* 25:541–53
99. Nayar JK, Fraenkel G. 1962. The chemical basis of hostplant selection in the silkworm, *Bombyx mori* (L.). *J. Insect Physiol.* 8:505–25
100. Nelson NM. 1996. *Feeding and oviposition behavior of tobacco hornworms,* Manduca sexta, *in relation to myo-inositol.* MS thesis. Univ. Arizona, Tucson. 178 pp.
101. Nelson N, Bernays EA. 1998. Inositol in two host plants of *Manduca sexta. Entomol. Exp. Appl.* 88:189–91
102. Newland PL. 1998. Avoidance reflexes mediated by contact chemoreceptors on the legs of locust. *J. Exp. Biol.* 155:313–24
103. Newland PL. 1999. Processing gustatory information by spiking local interneurons in the locust. *J. Neurophysiol.* 82:3149–59
104. Panzuto M, Albert PJ. 1998. Chemoreception of amino acids by female fourth- and sixth-instar larvae of the spruce budworm. *Entomol. Exp. Appl.* 86:89–96
105. Powell G, Hardie J, Pickett JA. 1995. Behavioural evidence for detection of the repellent polygodial by aphid antennal tip sensilla. *Physiol. Entomol.* 20:141–46
106. Rees CJC. 1968. The effect of aqueous solutions of some 1:1 electrolytes on the electrical response of the type 1 ('salt') chemoreceptor cell in the labella of *Phormia. J. Insect Physiol.* 14:1331–64
107. Rees CJC. 1969. Chemoreceptor specificity associated with choice of feeding site by the beetle *Chrysolina brunsvicensis* on its foodplant, *Hypericum hirsutum. Entomol. Exp. Appl.* 12:565–83
108. Rees CJC. 1970. The primary process of reception in the type 3 ("water") receptor cell of the fly, *Phormia terraenovae. Proc. R. Soc. London Ser. B* 174:469–90
109. Renwick JAA. 2001. Variable diets and changing taste in plant-insect relationships. *J. Chem. Ecol.* 27:1063–76
110. Renwick JAA, Lopez K. 1999. Experienced-based food consumption by larvae of *Pieris rapae* : addiction to glucosinolates. *Entomol. Exp. Appl.* 91:51–58
111. Roessingh P, Hora KH, Fung SY, Peltenburg A. 2000. Host acceptance behaviour of the small ermine moth *Yponomeuta cagnagellus*: larvae and adults use different stimuli. *Chemoecology* 10:41–47
112. Roessingh P, Hora KH, van Loon JJA, Menken SBJ. 1999. Evolution of gustatory sensitivity in *Yponomeuta* caterpillars: sensitivity to the stereo-isomers dulcitol and sorbitol is localised in a single sensory cell. *J. Comp. Physiol. A* 184: 119–26
113. Rogers SM, Simpson SJ. 1999. Chemodiscriminatory neurones in the suboesophageal ganglion of *Locusta migratoria. Entomol. Exp. Appl.* 91:19–28
114. Schiff NM, Waldbauer GP, Friedman S. 1988. Dietary self-selection for vitamins and lipids by larvae of the corn earworm, *Heliothis zea. Entomol. Exp. Appl.* 46: 249–56
115. Schoonhoven LM. 1969. Sensitivity changes in some insect chemoreceptors and their effect on food selection behaviour. *Proc. Konink. Nederl. Akad. Wet. C* 72:491–98

116. Schoonhoven LM. 1973. Plant recognition by lepidopterous larvae. *Symp. R. Entomol. Soc. London* 6:87–99
117. Schoonhoven LM. 1974. Comparative aspects of taste receptor specificity. In *Transduction Mechanisms in Chemoreception*, ed. TM Poynder, pp. 189–201. London: Information Retr.
118. Schoonhoven LM. 1987. What makes a caterpillar eat? The sensory codes underlying feeding behaviour. In *Advances in Chemoreception and Behavior*, ed. RF Chapman, EA Bernays, JG Stoffolano, pp. 69–97. New York: Springer
119. Schoonhoven LM, Blaney WM, Simmonds MSJ. 1992. Sensory coding of feeding deterrents in phytophagous insects. In *Insect-Plant Interactions*, ed. E Bernays, 4:59–79. Boca Raton: CRC
120. Schoonhoven LM, Blom F. 1988. Chemoreception and feeding behaviour in a caterpillar: towards a model of brain functioning in insects. *Entomol. Exp. Appl.* 49:123–29
121. Schoonhoven LM, Dethier VG. 1966. Sensory aspects of host-plant discrimination by lepidopterous larvae. *Arch. Néerland. Zool.* 16:497–530
122. Schoonhoven LM, Luo Lin-er. 1994. Multiple mode of action of the feeding deterrent, toosendanin, on the sense of taste in *Pieris brassicae* larvae. *J. Comp. Physiol. A* 175:519–24
123. Schoonhoven LM, Simmonds MSJ, Blaney WM. 1991. Changes in responsiveness of the maxillary styloconic sensilla of *Spodoptera littoralis* to inositol and sinigrin correlate with feeding behavior during the final larval stadium. *J. Insect Physiol.* 37:261–68
124. Schoonhoven LM, van Loon JJA. 2002. An inventory of taste in caterpillars: each species its own key. *Acta Zool. Acad. Sci. Hung.* 48(Suppl. 1):215–63
125. Shields VDC. 1996. Comparative ultrastructure and diffusion pathways in styloconic sensilla on the maxillary galea of larval *Mamestra configurata* (Walker) (Lepidoptera: Noctuidae) and five other species. *J. Morphol.* 228:89–105
126. Shields VDC, Mitchell BK. 1995. Sinigrin as a feeding deterrent in two-crucifer-feeding, polyphagous lepidopterous species and the effects of feeding stimulant mixtures on deterrency. *Philos. Trans. R. Soc. London Ser. A* 347:439–46
127. Shields VDC, Mitchell BK. 1995. Responses of maxillary styloconic receptors to stimulation by sinigrin, sucrose and inositol in two crucifer-feeding, polyphagous lepidopterous species. *Philos. Trans. R. Soc. London Ser. A* 347:447–57
128. Shields VDC, Mitchell BK. 1995. The effect of phagostimulant mixtures on deterrent receptor(s) in two-crucifer-feeding, polyphagous lepidopterous species. *Philos. Trans. R. Soc. London Ser. A* 347:459–64
129. Simmonds MSJ, Blaney WM. 1990. Gustatory codes in lepidopterous larvae. *Symp. Biol. Hung.* 39:17–27
130. Simmonds MSJ, Schoonhoven LM, Blaney WM. 1991. Daily changes in the responsiveness of taste receptors correlate with feeding behaviour in larvae of *Spodoptera littoralis*. *Entomol. Exp. Appl.* 61:73–81
131. Simmonds MSJ, Simpson SJ, Blaney WM. 1992. Dietary selection behaviour in *Spodoptera littoralis*: the effects of conditioning diet and conditioning period on neural responsiveness and selection behaviour. *J. Exp. Biol.* 162:73–90
132. Simpson CL, Simpson SJ, Abisgold JD. 1990. The role of various amino acids in the protein compensatory response of *Locusta migratoria*. *Symp. Biol. Hung.* 39:39–52
133. Simpson SJ. 1995. Regulation of a meal: chewing insects. In *Regulatory Mechanisms in Insect Feeding*, ed. RF Chapman, G de Boer, pp. 137–56. New York: Chapman & Hall

134. Simpson SJ, James S, Simmonds MSJ, Blaney WM. 1991. Variation in chemosensitivity and the control of dietary selection behaviour in the locust. *Appetite* 17:141–54
135. Simpson SJ, Raubenheimer D. 1993. A multilevel analysis of feeding behavior—the geometry of nutritional decisions. *Philos. Trans. R. Soc. London Ser. A* 342:381–402
136. Singh RN. 1997. Neurobiology of the gustatory systems of *Drosophila* and some terrestrial insects. *Microsc. Res. Tech.* 39:547–63
137. Sinoir Y. 1969. Le rôle des palpes et du labre dans le comportement de prise de nourriture chez la larve du criquet migrateur. *Ann. Nutr. Alim.* 23:167–94
138. Smedley SR, Eisner T. 1995. Sodium uptake by puddling in a moth. *Science* 270:1816–18
139. Sperling JLH, Mitchell BK. 1991. A comparative study of host recognition and the sense of taste in *Leptinotarsa*. *J. Exp. Biol.* 157:439–59
140. Städler E. 2002. Plant chemical cues important for egg deposition by herbivorous insects. In *Chemoecology of Insect Eggs and Egg Deposition*, ed. M Hilker, T Meiners, pp. 171–204. Berlin: Blackwell Sci. 410 pp.
141. Städler E, Hanson FE. 1975. Olfactory capabilities of the "gustatory" chemoreceptors of the tobacco hornworm larvae. *J. Comp. Physiol.* 104:97–102
142. Szentesi A, Bernays EA. 1984. A study of behavioural habituation to a feeding deterrent in nymphs of *Schistocerca gregaria*. *Physiol. Entomol.* 9:329–40
143. Tanaka Y, Asaoka K, Takeda S. 1994. Different feeding and gustatory responses to ecdysone and 20-hydroxyecdysone by larvae of the silkworm, *Bombyx mori*. *J. Chem. Ecol.* 20:125–33
144. Telang A, Booton V, Chapman RF, Wheeler DE. 2001. How female caterpillars accumulate their nutrient reserves. *J. Insect Physiol.* 47:1055–64
145. Thorsteinson AJ. 1960. Host selection in phytophagous insects. *A. Rev. Entomol.* 5:193–218
146. Trumper S, Simpson SJ. 1993. The regulation of salt intake by nymphs of *Locusta migratoria*. *J. Insect Physiol.* 39:857–64
147. van Drongelen W. 1979. Contact chemoreception of host plant specific chemicals in larvae of various *Yponomeuta* species (Lepidoptera). *J. Comp. Physiol.* 134:265–79
148. van Loon JJA. 1990. Chemoreception of phenolic acids and flavonoids in larvae of two species of *Pieris*. *J. Comp. Physiol. A* 166:889–99
149. van Loon JJA, Schoonhoven LM. 1999. Specialist deterrent chemoreceptors enable *Pieris* caterpillars to discriminate between chemically different deterrents. *Entomol. Exp. Appl.* 91:29–35
150. van Loon JJA, van Eeuwijk FA. 1989. Chemoreception of amino acids in larvae of two species of *Pieris*. *Physiol. Entomol.* 14:459–69
151. van Loon JJA, Wang CZ, Nielsen JK, Gols R, Qui YT. 2002. Flavonoids from cabbage are feeding stimulants for diamondback moth larvae additional to glucosinolates: chemoreception and behavior. *Entomol. Exp. Appl.* In press.
152. Varanka I. 1981. Taste reception by maxillary palps of the migratory locust. *Adv. Physiol. Sci.* 23:459–80
153. Waldbauer GP, Cohen RW, Friedman S. 1984. Self-selection of an optimal nutrient mix from defined diets by larvae of the corn-earworm, *Heliothis zea* (Boddie). *Physiol. Zool.* 57:590–97
154. Warthen JD, Morgan ED. 1985. Insect feeding deterrents. See Ref. 95a, 6:23–134
155. White PR, Chapman RF. 1990. Tarsal chemoreception in the polyphagous grasshopper *Schistocerca americana*: behavioural assays, sensilla distributions and electrophysiology. *Physiol. Entomol.* 15:105–21

156. White PR, Chapman RF, Ascoli-Christensen A. 1990. Interaction between two neurons in contact chemosensilla of the grasshopper, *Schistocerca americana*. *J. Comp. Physiol. A* 167:431–36
157. Wieczorek H. 1976. The glycoside receptor of the larvae of *Mamestra brassicae* L. (Lepidoptera, Noctuidae). *J. Comp. Physiol. A* 106:153–76
158. Wieczorek H, Koppl R. 1978. Effect of sugars on the labellar water cell receptor of the fly. *J. Comp. Physiol.* 126:131–36
159. Zacharuk RY. 1985. Antennae and sensilla. See Ref. 79a, 6:1–69

Annu. Rev. Entomol. 2003. 48:485–503
doi: 10.1146/annurev.ento.48.091801.112525

First published online as a Review in Advance on October 17, 2002

SIGNALING PATHWAYS AND PHYSIOLOGICAL FUNCTIONS OF *DROSOPHILA MELANOGASTER* FMRFAMIDE-RELATED PEPTIDES

Ruthann Nichols
Biological Chemistry Department, University of Michigan, Ann Arbor, Michigan 48109-0606; e-mail: nicholsr@umich.edu

Key Words drosulfakinin, dromyosuppressin, FaRP, gut, heart

■ **Abstract** FMRFamide-related peptides (FaRPs) contain a C-terminal RFamide but unique N-terminal extensions. They are expressed throughout the animal kingdom and affect numerous biological activities. Like other animal species, *Drosophila melanogaster* contains multiple genes that encode different FaRPs. The ease of genetic manipulations, the availability of genomic sequence data, the existence of established bioassays, and its short lifespan make *D. melanogaster* a versatile experimental organism in which to investigate peptide processing, functions, and signal transduction pathways. Here, the structures, precursor organizations, distributions, and activities of FaRPs encoded by *D. melanogaster* FMRFamide (*dFMRFamide*), myosuppressin (*Dms*), and sulfakinin (*Dsk*) genes are reviewed, and predictions are made on their signaling pathways and biological functions.

CONTENTS

0066-4170/03/0107-0485$14.00

INTRODUCTION

Peptides are a diverse and important class of messengers and hormones that transmit and regulate numerous behavioral, developmental, and physiological processes. Peptides serve as critical signal transducers and modulators in animals ranging from insects to humans. The roles that peptides play in biology suggest that aberrant peptide synthesis, expression, or signaling could result in abnormal or impaired functioning that leads to dysfunction or death. Thus, delineating how peptides act is a fundamental basic science question with application to the health sciences, from the control of insects to the development of diagnostic tests and therapeutic agents for use in humans.

Frequently, peptides can be grouped together based on a common structure. Peptides with a RFamide C terminus but with unique N-terminal extensions comprise a large family. FMRFamide was the first RFamide-containing peptide to be isolated (49); thus, this family is referred to as FMRFamide-related peptides or FaRPs. First identified in a mollusk as a cardioexcitatory peptide, FaRPs are now known to affect a wide range of processes from behavior to physiology in invertebrates and vertebrates (21, 23, 33, 53, 66).

In 1944, Berta and Ernst Scharrer (54) reported that the insect pars intercerebralis-corpora cardiaca-corpora allata system and the mammalian hypothalamo-hypophysial system are similar. Subsequently, analysis of peptides identified from insect neurosecretory cells proved that insects and humans contain some peptides that are alike (3). The similarities in peptide structures and activities provide the opportunity to conduct research in experimentally versatile lower organisms such as *Drosophila melanogaster* for application to higher animal species, including humans. On the other hand, the differences in peptide structures and signaling pathways between insects and vertebrates may serve as potential target sites for the development of safe and effective pest management tools.

D. melanogaster is a powerful scientific system in which to delineate peptide function and signaling because research in behavior and physiology can be combined with biochemistry, genetics, and molecular biology techniques in an animal species whose generation time is short and whose genome sequence is available (1). Here, the peptides, genes, spatial and temporal distributions, and activities of three FaRPs, *D. melanogaster* FMRFamide [*dFMRFamide* (34, 55)], *Drosophila* myosuppressin or dromyosuppressin [*Dms* (37)], and *Drosophila* sulfakinin or drosulfakinin [*Dsk* (45)], are reviewed. These data suggest how FaRPs transduce signals and provide insight into their functions.

FMRFamide-CONTAINING PEPTIDES

Peptides

The first FaRP reported was FMRFamide, a cardioexcitatory peptide isolated from the clam *Macrocallista nimbosa* (49). Members of a subgroup of FaRPs each contain a FMRFamide C terminus, yet their N-terminal extensions vary in structure and length. Thus, they are known as FMRFamide-containing peptides. Five FMRFamide-containing peptides are encoded in *dFMRFamide*, the *D. melanogaster* FMRFamide gene (34, 55). The peptides are DPKQDFMRFamide, TPAEDFMRFamide, SDNFMRFamide, SPKQDFMRFamide, and PDNFMRFamide. Two of the peptides, DPKQDFMRFamide and SPKQDFMRFamide, differ by only one amino acid residue, their N terminus. Likewise, SDNFMRFamide and PDNFMRFamide differ by only one amino acid residue, their N terminus. The isolation of DPKQDFMRFamide, TPAEDFMRFamide, and SDNFMRFamide from adult *D. melanogaster* has been published (34, 37). Recently, PDNFMRFamide was identified using liquid chromatography in combination with tandem mass spectrometry from the *D. melanogaster* larval central nervous system (1a).

Of the three classes of FaRPs reviewed here, the consensus structure of FMRFamide-containing peptides varies the most between animal species; only the C-terminal FMRFamide is conserved. The length and structure of the N-terminal extensions vary considerably. In addition, the number of times FMRFamide-containing peptides are encoded in a gene differs between animal species. A comparison of *D. melanogaster* FMRFamide-containing peptides to other Diptera illustrates this variation. The FMRFamide gene in *Drosophila virilis* encodes DPKQDFMRFamide, SDNFMRFamide, SPKQDFMRFamide, and PDNFMRFamide; however, it does not encode TPAEDFMRFamide (59). Fourteen different FMRFamide-containing peptides were isolated from the blow fly, *Calliphora vomitoria* (7), but only PDNFMRFamide is conserved between *D. melanogaster* and *C. vomitoria*.

Gene

Isolation of DPKQDFMRFamide led to the cloning of *dFMRFamide* (34), a gene that contains two exons of 106 bp and 1352bp, and one intron of about 2500 bp (34, 55). The *dFMRFamide* precursor is 347 amino acids in length and contains a predicted signal sequence of 20 residues. *dFMRFamide* is a single copy gene that localizes to the right arm of the second chromosome at the cytological band 46C. *dFMRFamide* encodes five copies of DPKQDFMRFamide, two of TPAEDFMRFamide, and one each of SDNFMRFamide, SPKQDFMRFamide, and PDNFMRFamide, all bounded by conventional proteolytic processing sites. The arrangement of these peptides in the polyprotein precursor encoded in *dFMRFamide* is as follows with the predicted signal peptide in italics, the proteolytic cleavage sites underlined, and the peptides in bold type. The C-terminal glycyl is processed to produce an amide in the mature peptide. *MGIALMFLLALYQMQSAIHS*EIIDTPNYAGNSLQDTDSEVSPSQDNDLVDALLGNDQTERAELEFRHPISVIGIDYSKNAVVLHFQKHGRKPRYKYDPELEAKRRSVQDMPMHFGKRQAEQLPPEGSYAGSDELEGMAKRAAMDRYGR**DPKQDFMRF**GR**DPKQDFMRF**GR**DPKQDFMRF**GR**DPKQDFMRF**GR**DPKQDFMRF**GR**TPAEDFMRF**GR**TPAEDFMRF**

G<u>R</u>**SDNFMRF**G<u>R</u>SPHEEL<u>R</u>**SPKQDFMRF**G<u>R</u>**PDNFMRF**G<u>R</u>SAPQDFVRS GKMDSNFIRFGKSLKDKAPAAPESKPVKSNQGDPGERSPVMTELFKKQE LQDQQVKNGAQATTTQDGSVEQDQFFGQ.

As in *D. melanogaster*, the *D. virilis* FMRFamide precursor contains two exons and one intron (59). The deduced precursors are similar in size and the presumed signal peptides are conserved. In addition, the *D. virilis* FMRFamide gene encodes a polyprotein precursor and, based on the positions of conventional proteolytic processing sites, may be cleaved to produce multiple peptides. The relative positions of SDNFMRFamide, SPKQDFMRFamide, and PDNFMRFamide in the precursor and the number of times they are encoded in the polyproteins are the same in both species. Thus, there is considerable conservation of gene organization between the *dFMRFamide* gene in these two distantly related drosophilids. However, there are some differences as well. For instance, compared to the five copies of DPKQDFMRFamide in *D. melanogaster dFMRFamide*, the peptide is encoded only once in *D. virilis*. In addition, *D. melanogaster* TPAEDFMRFamide is not predicted in the *D. virilis* gene.

Spatial and Temporal Distribution

Antisera generated to FMRFamide were used to show that immunoreactive material is present throughout the central nervous system including cells and processes in the brain lobes, optic lobes, and ventral ganglion (Figure 1*a*, see color insert), as well as in the gastrointestinal tract, and in the reproductive system in a bilaterally symmetric pattern (43, 65). The presence of FMRFamide staining in diverse cell types is consistent with the role of FaRPs in a variety of physiological and developmental processes. Similary, the distribution of FMRFamide immunoreactive material is widespread in other animal species (21, 48). Although these data are interesting, antisera generated to FMRFamide recognize the common C-terminal RFamide and do not discriminate between individual FaRPs. Antisera generated to antigens designed to the unique N-terminal structures of the FaRPs distinguish between these structurally similar peptides and yield data on processing and the sites of synthesis and release for individual *D. melanogaster* FMRFamide-containing peptides (27, 39, 41, 42, 44).

DPKQDFMRFamide antisera, generated to DPKQD, a sequence that represents the unique N-terminal extension of the peptide, recognize DPKQDFMRFamide but not TPAEDFMRFamide or SDNFMRFamide (38, 42). DPKQDFMRFamide antisera stain cells and processes that are a subset of FMRFamide immunoreactivity, but the DPKQDFMRFamide staining pattern does not overlap with TPAEDFMRFamide antisera or SDNFMRFamide antisera staining (27, 38, 42). In embryos, DPKQDFMRFamide is expressed in subesophageal ganglion cells and in a pair of cells in each of the three thoracic ganglia. Also stained in third instar larvae are cells in the superior protocerebrum, a pair of cells in the second thoracic ganglion, and cells in the posterior most abdominal ganglion. Additional cells in pupae that are stained include a pair of lateral protocerebral cells. DPKQDFMRFamide immunoreactivity in adults is similar to pupae; however, two additional pairs of cells in the optic lobe are stained. DPKQDFMRFamide antisera do not stain gastrointestinal tissue. Tissue incubated with DPKQDFMRFamide antisera preabsorbed with DPKQDFMRFamide resulted in a complete loss of signal. SPKQDFMRFamide, a putative dFMRFamide gene product, differs from DPKQDFMRFamide by only one amino acid; thus, DPKQDFMRFamide polyclonal antisera may recognize SPKQDFMRFamide. The level of signal intensity in tissue stained using DPKQDFMRFamide antisera preabsorbed with SPKQDFMRFamide is slightly reduced, indicating that DPKQDFMRFamide antisera have some affinity for SPKQDFMRFamide (42).

TPAEDFMRFamide antisera, generated to TPAED, a sequence that represents the unique N-terminal extension of the peptide, recognize TPAEDFMRFamide but not DPKQDFMRFamide or SDNFMRFamide (38, 41). TPAEDFMRFamide antisera stain cells and processes that are a subset of FMRFamide immunoreactivity, but the TPAEDFMRFamide staining pattern does not overlap with DPKQDFMRFamide antisera or SDNFMRFamide antisera staining (27, 38, 41). In embryos, TPAEDFMRFamide antisera stain numerous cells in each of the three thoracic ganglia and the abdominal ganglia. Staining in first instar larvae resembles that observed in embryos; however, additional cells in the subesophageal ganglion and a cluster of cells in the medial protocerebrum express TPAEDFMRFamide. Cellular expression of TPAEDFMRFamide in pupae is similar to that in larvae; however, there is an increase in the complexity in the arborization of the immunoreactive processes. Staining in adults is similar to pupae; however, a second pair of cells in the first thoracic ganglion no longer expresses TPAEDFMRFamide, and one pair of cells in the lateral protocerebrum is stained. TPAEDFMRFamide is also expressed in the gastrointestinal system at all stages of development. Compared to the larval midgut, several more cells are stained with TPAEDFMRFamide antisera than in the pupal and adult midgut. Tissue incubated with TPAEDFMRFamide antisera preabsorbed with TPAEDFMRFamide resulted in a complete loss of signal.

SDNFMRFamide antisera, generated to SDNFM, a sequence that represents the unique N-terminal extension of the peptide, recognize SDNFMRFamide but not DPKQDFMRFamide or TPAEDFMRFamide (38, 42). SDNFMRFamide antisera stain cells and processes that are a subset of FMRFamide immunoreactivity, but the SDNFMRFamide staining pattern does not overlap with DPKQDFMRFamide antisera or TPAEDFMRFamide antisera staining (27, 38, 44). In all stages of development, SDNFMRFamide antisera stain a cluster of cells in the subesophageal ganglion. In pupae, processes stained by SDNFMRFamide antisera extend into the brain lobes. In adults, the number of SDNFMRFamide-immunoreactive processes increases compared to adults. SDNFMRFamide antisera do not stain any cells in the thoracic or abdominal ganglia or in gastrointestinal tissue. Tissue incubated with SDNFMRFamide antisera preabsorbed with SDNFMRFamide resulted in a complete loss of signal. PDNFMRFamide, encoded in *dFMRFamide*, differs from SDNFMRFamide by only one amino acid; thus, SDNFMRFamide polyclonal antisera may recognize PDNFMRFamide. The level of signal intensity in tissue stained using SDNFMRFamide antisera preabsorbed with PDNFMRFamide is slightly reduced, indicating that SDNFMRFamide antisera have some affinity for PDNFMRFamide (44).

The expression patterns of the three FMRFamide-containing peptides are unique compared to one another and are a subset of FMRFamide antisera staining. The peptide-specific antisera show an increase in the amount of immunoreactive material in cells and processes located throughout the central nervous system. In some instances, the individual expression patterns of these peptides appear to be similar because of the close proximity of the stained cells. However, there is no overlap in the cells or processes that contain DPKQDFMRFamide, TPAEDFMRFamide, or SDNFMRFamide as detected by double and triple immunolabeling experiments (Figures 1*b*,*c*, see color insert) in whole mount tissue preparations (27, 38, 41–44).

Activity

Consistent with the activities of FaRPs observed in other animal species (6, 7, 21, 25, 53, 66), *D. melanogaster* FMRFamide-containing peptides affect several physiological processes including heart rate, gut motility, and synaptic activity (15, 19, 20, 40). Another similarity among species is that *D. melanogaster* FMRFamide-containing peptides require a specific N-terminal structure for biological activities (6, 21, 29, 30, 40). For example, SDNFMRFamide decreases in vivo pupal (Figure 2) and adult heart rate; however, DPKQDFMRFamide, TPAEDFMRFamide, and FMRFamide do not (40). Additional evidence for the importance

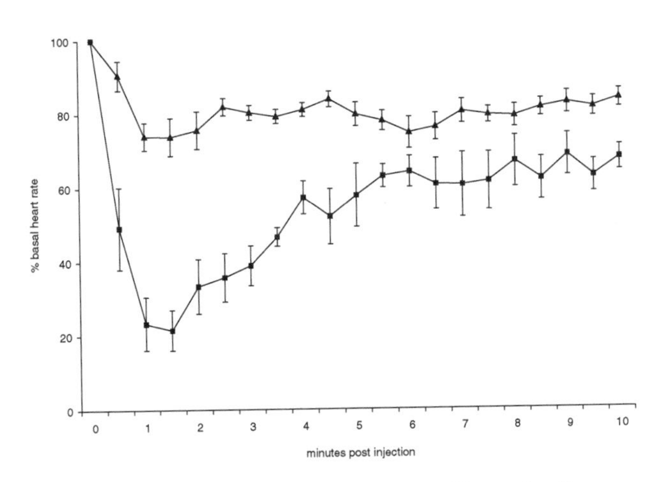

Figure 2 The effects of SDNFMRFamide and DMS on the frequency of pupal *D. melanogaster* heart contractions in vivo. Effects of SDNFMRFamide and DMS on the frequency of contractions of the *D. melanogaster* pupal heart rate measured in vivo. Heart rate is reported as a percent of basal heart rate measured in each animal prior to injection of a peptide. Saline does not have a significant effect on heart rate (data not shown). The errors are standard errors of the mean.

of the N terminus comes from the finding that TPAEDFMRFamide decreases the frequency of spontaneous contractions in the crop, an anterior portion of the gut; however, DPKQDFMRFamide and SDNFMRFamide do not (20). These data are consistent with the conclusion that the structure of the N-terminal extension is critical for activity.

MYOSUPPRESSIN PEPTIDES

Peptides

The first myosuppressin isolated was leucomyosuppressin (LMS), a myoinhibiting peptide from the cockroach, *Leucophaea maderae* (16, 17). Leucomyosuppressin was identified based on its ability to reduce the frequency of hindgut contractions. The naturally occurring myosuppressin peptide, TDVDHVFLRFamide, was isolated from adult *D. melanogaster* using a RFamide-specific radioimmunoassay (37). The *D. melanogaster* peptide was the most abundant peptide identified in the purification of RFamide-specific peptides. At the time the peptide was purified, it could not be predicted from known nucleotide sequence data, so the peptide defined a new gene. Based on the high degree of structure identity between *D. melanogaster* myosuppressin and leucomyosuppressin—only one amino acid differs, the N terminus—the fruit fly peptide was designated *Drosophila* myosuppressin or dromyosuppressin (DMS). Based on standard nomenclature used to name *D. melanogaster* genes first identified by biochemical isolation of the gene product, the dromyosuppressin gene is designated *Dms*.

PDVDHVFLRFamide, a peptide that differs from DMS by only the N-terminal amino acid residue, was isolated from the locusts, *Schistocerca gregaria* and *Locusta migratoria* (24, 25, 47, 51, 52, 58). A peptide identical in structure to DMS, TDVDHVFLRFamide, was isolated from the flesh fly, *Neobellieria bullata* (11). These three classes of myosuppressin peptides are represented by the consensus structure XDVDHVFLRFamide. Other peptides, with slightly less structure identity, can also be considered myosuppressins based on both structure and activity. One structurally similar peptide, ADVGHVFLRFamide, was isolated from the locust *L. migratoria* (25). An additional structurally related peptide, pEDVVHSFLRFamide, MasFLRFamide III, was isolated from *Manduca sexta* (21). Taken together, these myosuppressin peptides contain the consensus structure, XDVXHXFLRFamide. Myosuppressins are FaRPs owing to the presence of a RFamide C terminus. However, myosuppressin peptides contain a FLRFamide C terminus and, thus, differ from FMRFamide-containing peptides.

Gene

A partial *Dms* cDNA was amplified from adult *D. melanogaster* mRNA by RT-PCR (4) and used as a probe to isolate the *Dms* gene from a genomic library. The majority of the DMS cDNA and gene sequence agree with the data collected by the *Drosophila* Genome Project (4, 61). Analysis of the nucleotide sequence shows that a 100-amino-acid translation product contains a signal peptide of 24 amino acid residues. *Dms* is a single copy gene that localizes to the right arm of the third chromosome at the cytological band 96A1. The *Dms* transcript is approximately 1000 bp in length (4, 61). The organization of *Dms* is similar to myosuppressin genes in other animal species (5). One copy of the *D. melanogaster* myosuppressin peptide bounded by conventional dibasic proteolytic cleavage sites is encoded in the gene. The arrangement of the peptide in the polyprotein precursor encoded in *Dms* is as follows with the predicted signal peptide in italics, the proteolytic cleavage sites underlined, and the peptide in bold type. The C-terminal glycyl is processed to produce an amide in the mature peptide. *MSFAQFFVACCLAIVLLAVS NTRA*AVQGPPLCQSGIVEEMPPHIRKVCQALENSDQLTSALKSYINNEASA LVANSDDLLKNYNKR**TDVDHVFLRF**GKRR.

Spatial and Temporal Distribution

In order to determine the distribution of DMS-immunoreactive material, peptide-specific antisera were generated to the unique N-terminal structure TDVDHV (28). The common C-terminal structure, RFamide, was not included in the antigen to avoid cross-reactivity with other FaRP peptides. In embryos and larvae (Figure 3*a*), superior and medial protocerebral cells and subesophageal ganglion cells stain with DMS antisera. In pupae, cells in the optic lobes, additional cells in the superior protocerebrum, cells in the thoracic ganglia and posterior abdominal ganglion, and cells along the midline of the ventral ganglion are also stained with DMS antisera. Expression in pupae is distinguished by the loss of staining in one pair of medial protocerebral cells. In general, DMS immunoreactivity in adult brain (Figure 3*b*) is similar to that observed in pupae. DMS antisera did not stain the larval proventriculus, midgut, or hindgut; however, a pair of rectal cells is stained. In adults, staining of the gut was similar to that observed in larvae. DMS-immunoreactive processes project from rectal cells stained by DMS antisera to impinge on the oviduct and the male accessory glands (J. McCormick & R. Nichols, unpublished data). In addition, the crop was highly innervated by DMS-immunoreactive processes that extend from cells in the brain (28). The distribution of DMS in *D. melanogaster* is similar to myosuppressin-like immunoreactivity in other animal species (28a, 50).

Activity

Myosuppressins reduce the frequency of spontaneous contractions of several muscles including crop, heart, midgut, and oviduct (4, 11, 16, 17, 20, 21, 23–25, 29, 30, 51, 52). *M. sexta*–like myosuppressin also affects muscle activity, increasing the force of neurally evoked contractions in flight muscle and stimulating moth ileum contractions (21). Myosuppressin stimulates invertase release in cockroach and α-amylase release in the midgut of the cockroach, *Diploptera punctata*, and of the weevil, *Rhynchophorus ferrugineus* (14, 31).

The effect of DMS on *D. melanogaster* heart rate was measured using an in vivo microscope-based assay (67). DMS decreases in vivo heart rate in larva,

(*a*)

(*b*)

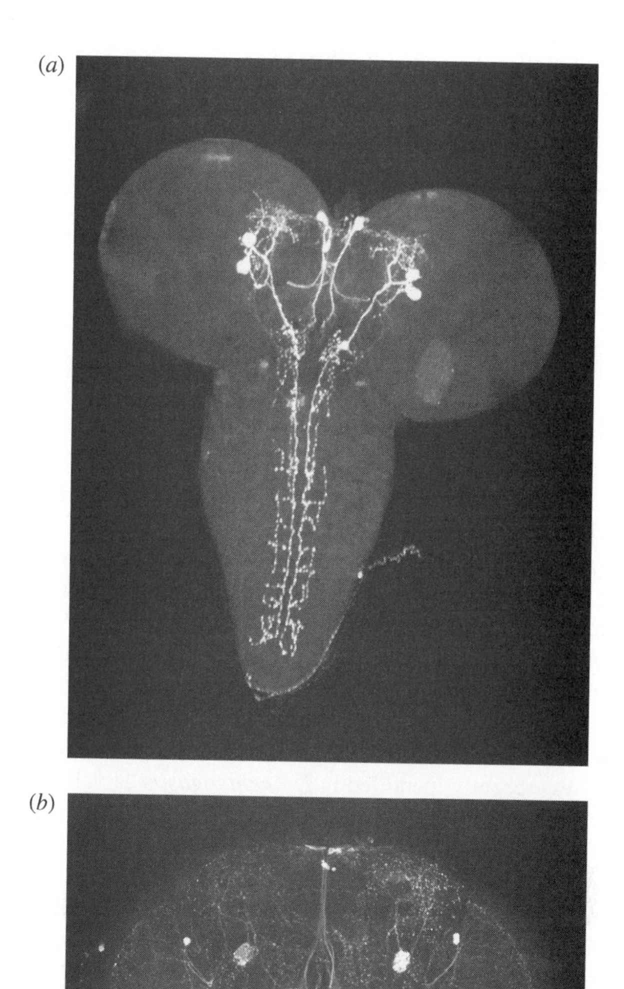

Figure 3 (*a*) Dromyosuppressin antisera staining in larval *D. melanogaster* central nervous system. A whole mount tissue preparation of a larval central nervous system was stained with DMS antisera. (*b*) Dromyosuppressin antisera staining in adult *D. melanogaster* brain. A whole mount tissue preparation of an adult brain was stained with DMS antisera.

pupa, and adult *D. melanogaster* in a dose-dependent manner (4). At 10^{-7} M in the animal, DMS decreases pupal heart rate to about 40% of basal rate (Figure 2). The threshold for the effect of DMS on in vivo pupal heart rate is approximately 10^{-11} M. DMS also decreases the frequency of spontaneous contractions of the blow fly, *Phormia regina*, crop (50). Using an in vitro bioassay, Richer et al. (50) found that 2.5×10^{-8} M DMS slowed blow fly crop motility rate to about 4% of the basal rate. At 10^{-6} M, DMS stops *D. melanogaster* crop motility (20, 29, 30) and decreases oviduct motility (R. Nichols, K. Berry, J. Fuentes & E. Walling, unpublished data).

SULFAKININ PEPTIDES

Peptides

The first sulfakinin reported was leucosulfakinin (LSK), EQFED**Y**GHMRFamide, where **Y** represents a sulfated tyrosine, isolated from the cockroach, *Leucophaea maderae* (33). LSK was identified in chromatographic fractions of cockroach brain as a peptide that increases the frequency and amplitude of hindgut contractions. A second stimulatory peptide with a C-terminal heptapeptide structure identical to LSK was designated leucosulfakinin II (LSK II), pQSDD**Y**GHMRFamide (32). The naturally occurring sulfakinin peptide, FDD**Y**GHMRFamide, was isolated from adult *D. melanogaster* (DSK I) using a RFamide-specific radioimmunoassay (36). Partial structure data were obtained for a second *D. melanogaster* sulfakinin peptide (DSK II), GGDDQFDDYGHMRFamide (R. Nichols, unpublished data). DSK II was recently identified in *D. melanogaster* larval central nervous system (1a).

Sulfakinins have been identified in other insect species including EQFDD**Y**GH MRFamide (Pea-SK I) and pQSDD**Y**GHMRFamide (Pea-SK II) from the cockroach, *Periplaneta americana* (9, 48, 63); pELASDD**Y**GHMRFamide (Lom-SK) from the locust, *L. migratoria* (57); FDDYGHMRFX (Neb-SK I) and XXEEQFD DYGHMRFX (Neb-SK II) from the flesh fly, *N. bullata* (12); and FDD**Y**GHMRF-amide (Cav-SK I) and GGEEQFDD**Y**GHMRFamide (Cav-SK II) from the blow fly, *C. vomitoria* (8). Sulfakinins have the common C-terminal structure X(E, D)DYGHMRFamide, where X represents one to six additional amino acid residues. In some sulfakinins, the tyrosine residue is sulfated. In other cases, however, the unsulfated form was isolated or the peptide was deduced from the nucleotide sequence; thus, the posttranslational modification, sulfation of a tyrosyl, cannot be predicted.

Gene

The first sulfakinin gene identified was the *D. melanogaster* sulfakinin gene (*Dsk*) (45). A genomic library was screened with oligonucleotides designed to the leucosulfakinin peptides. *Dsk* contains a 128-amino-acid translation product with a signal peptide of 20 amino acid residues. *Dsk* is a single copy gene that localizes to the right arm of the third chromosome at 81F. Northern blot analysis indicates the transcript is about 800 bp. The organization of *Dsk* is similar to the *C. vomitoria* sulfakinin gene (8). The arrangement of these peptides in the polyprotein precursor encoded in *Dsk* is as follows with the predicted signal peptide in italics, the proteolytic cleavage sites underlined, and the peptides in bold type. The C-terminal glycyl is processed to produce an amide in the mature peptide. *MPLWALAFCFLVVLPI PAQT*TSLQNAKDDRRLQELESKIGGEIDQPIANLVGPSFSLFGDRR**NQKT MSF**GRRVPLISRPIIPIELDLLMDNDDERTKAKR**FDDYGHMRF**GKR**GGD DQFDDYGHMRF**GR.

The DSK 0, NQKTMSFamide, DSK I, FDDYGHMRFamide, and DSK II, GGDDQFDDYGHMRFamide peptides are encoded once in the gene and are bounded by conventional dibasic proteolytic cleavage sites. DSK I and DSK II are FaRPs and contain the sulfakinin consensus structure. In contrast, DSK 0 is not a FaRP or a sulfakinin; thus, DSK 0 was not recognized in the RFamide-specific radioimmunoassay used to purify sulfakinins (36). Recently, however, DSK 0 was identified in adult *D. melanogaster* brains and its predicted structure was confirmed by mass spectrometry (56). In addition, putative peptides that contain MSFamide and related C-terminal motifs were recently identified in *Caenorhabditis elegans* (35).

Spatial and Temporal Distribution

Sulfakinin precursors encode multiple peptides that are expressed in neural and gut tissues (8, 9, 36, 39, 60). In order to determine the spatial and temporal distribution of *D. melanogaster* sulfakinins, antisera generated to DSK I conjugated to thyroglobulin (36, 39). However, because DSK I is a FaRP, antisera generated to DSK I recognize other RFamide-containing peptides. Thus, in order to obtain antisera specific to sulfakinins, antisera to the common C-terminal RFamide were removed by passing DSK I antisera over a FMRFamide affinity column. Antisera that did not bind FMRFamide were collected and recognized DSK I but not FMRFamide (36).

DSK II is a 5-amino-acid extension of DSK I; thus, antisera generated to DSK I recognize both DSK I and DSK II and, therefore, are referred to as DSK antisera. Antisera generated to the 5-amino-acid extension of DSK II do, however, distinguish between DSK II and DSK I and are referred to as DSK II antisera (39).

In embryos, weak but consistent DSK antisera staining is present in two pairs of superior protocerebral cells and two pairs of medial protocerebral cells (39). In larvae, DSK antisera stain cells in the superior protocerebrum, medial protocerebrum, thoracic ganglia, and the posterior most abdominal ganglion (Figure 4*a*). In pupae, additional cells in the superior protocerebrum, lateral protocerebrum, medial protocerebrum, and thoracic ganglia are stained. Immunoreactive processes that extend from the superior protocerebral cells increase in staining intensity and complexity in pupae compared to larvae. DSK antisera staining in adults is similar

(*a*)

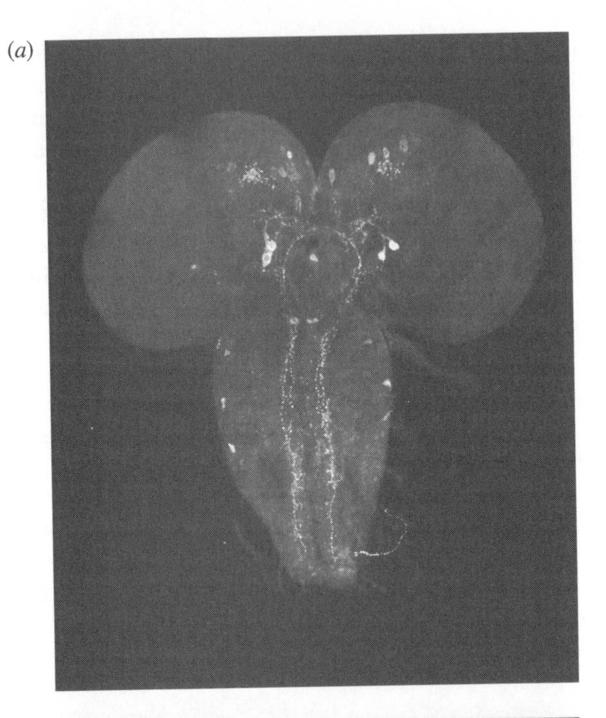

(*b*)

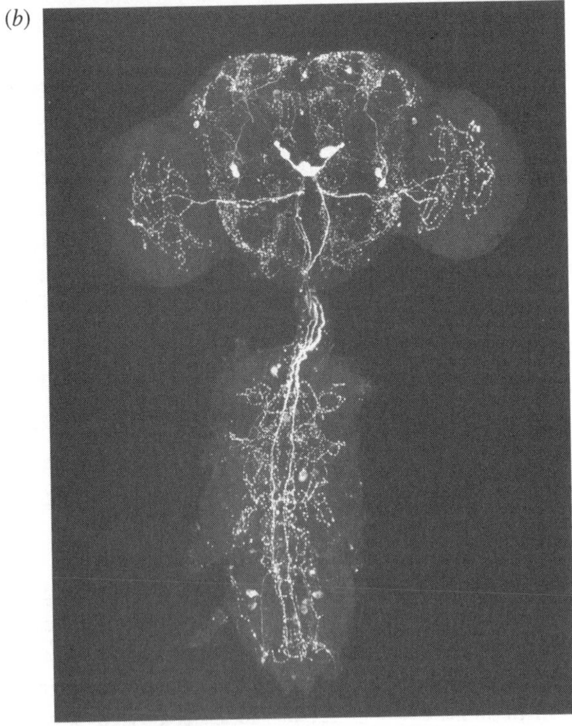

to that observed in larvae and pupae but with a vast increase in intensity of signal and complexity of immunoreactive processes in the brain and ventral nerve cord (Figure 4*b*).

In embryos, weak but consistent DSK II antisera staining is observed in cells in the superior protocerebrum and medial protocerebrum. In larvae, DSK II antisera stain cells in the superior protocerebrum, medial protocerebrum, and thoracic ganglia (39). In pupae, the same staining pattern is seen as in larvae. In adults, cells in the lateral protocerebrum and additional cells in the medial protocerebrum are also stained. The most striking difference between the staining patterns of DSK I antisera and DSK II antisera is that DSK II antisera do not stain all the cells in the superior protocerebrum and subesophageal ganglion, and that no cells in the abdominal ganglion are stained by DSK I antisera. In addition, DSK II antisera stain only the lateral protocerebral cells in the adults but not in the larvae and pupae. Additionally, DSK II antisera stain only a few processes compared to DSK antisera. Sulfakinin peptides are present in the central nervous system and gastrointestinal tissue of other animal species (8, 9).

Activity

Like other bioactive peptides, sulfakinins affect several biological processes. Sulfated drosulfakinin increases the frequency of spontaneous contractions of the *D. melanogaster* heart, measured using a semi-isolated tissue preparation (R. Nichols & B. Manoogian, unpublished data). Sulfated forms of myosuppressin peptides increase the frequency and amplitude of hindgut contractions in the cockroach *L. maderae* (31, 32) and increase the frequency of spontaneous contractions of the cockroach heart, *P. americana* (48). Sulfakinins also reduce food intake in the locust *S. gregaria* (64). Recently, a sulfakinin peptide was identified in brain extracts based on its antifeedant activity in cockroach *Blattella germanica* (26).

CONCLUSION

Three subclasses of *D. melanogaster* FaRPs—FMRFamide-containing peptides, dromyosuppressin, and drosulfakinins—are similar yet different in their structures, precursor organizations, distributions, and activities. Each contains a C-terminal RFamide, yet their N-terminal extensions differ considerably in both length and composition. Although these three subclasses of FaRPs all contain a RFamide, their consensus structures vary considerably. Only the four amino acid residues,

←

Figure 4 (*a*) Drosulfakinin antisera staining in larval *D. melanogaster* central nervous system. A whole mount tissue preparation of a larval central nervous system was stained with DSK antisera. (*b*) Drosulfakinin antisera staining in adult *D. melanogaster* central nervous system. A whole mount tissue preparation of an adult central nervous system was stained with DSK antisera.

FMRF, are present in all FMRFamide-containing peptides; however, in myosuppressin peptides, 9 of 10 amino acids are conserved. Only one residue, the N terminus, varies. The sulfakinins are intermediate of the FaRPs in conservation across animal species with the consensus structure (E,D)DYGHMRFamide. Of these three subclasses of FaRPs, sulfakinins are unique in structure because they contain a modified amino acid residue, sulfated tyrosine.

A different gene encodes each of these three subclasses of FaRPs. Peptides can be predicted in each of the three precursor peptides based on conventional proteolytic processing sites. In addition, a glycyl residue exists that can undergo processing to generate a C-terminal amide, a posttranslational modification critical for many active peptides (10) including FaRPs. The *dFMRFamide* precursor is a polyprotein that encodes five different FMRFamide-containing peptides; two are encoded multiple times, i.e., five copies of DPKQDFMRFamide and two copies of TPAEDFMRFamide. In contrast, the *Dms* precursor encodes only one copy of one peptide, DMS or TDVDHVFLRFamide. The organization of the *Dms* gene is representative of myosuppressin genes in other animal species (5). Like *dFMRFamide*, the *Dsk* precursor is a polyprotein; however, it only encodes one copy of two peptides each with the sulfakinin consensus structure. A third peptide, NQKTMSFamide, encoded in the *Dsk* precursor, is designated DSK 0 because it is 5′ relative to DSK I and DSK II and because it does not contain the sulfakinin consensus structure. The processing and predicted structure of DSK 0 was confirmed by mass spectral analysis of adult *D. melanogaster* head tissue (56). The *Dsk* gene is representative of sulfakinin genes found in other animal species; however, DSK 0 is not conserved in the sulfakinin gene of the blow fly *C. vomitoria* (8).

The three subclasses of FaRPs are unique, relative to one another, in their spatial and temporal distributions. All FMRFamide-containing peptides, myosuppressins, and sulfakinins are present in the central nervous system at all stages of development. However, they are not all expressed in gastrointestinal tissue or in reproductive tissue. There is little overlap or colocalization of these FaRPs, which is consistent with the synthesis and release of these structurally related peptides being regulated under different mechanisms. Their extensive spatial and temporal distributions suggest FaRPs transmit and modulate numerous unique biological processes.

The availability of the *D. melanogaster* genome sequence is being used in computer-based searches to "identify" numerous genes including peptides and receptors (1, 22, 61). These data have also made more scientists aware of the value of this versatile experimental organism for applied health research. Several *D. melanogaster* peptide genes were identified and described by searching the genome database (22, 61). Relatively little is known about *D. melanogaster* FaRP signal transduction pathways; however, based on studies in other animal species, FaRPs are likely to signal through conventional G protein–coupled receptors. A sulfakinin receptor was found and characterized as a G protein–coupled receptor (22). Future research directed at elucidating the peptide binding properties of orphan receptors will likely result in identifying additional *D. melanogaster* FaRP

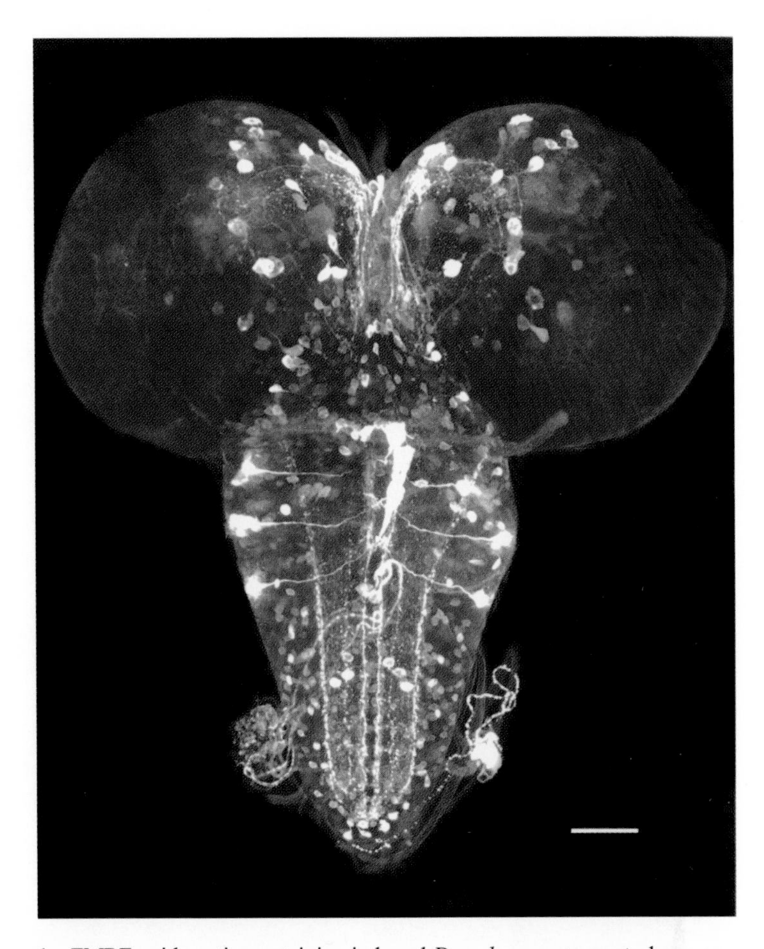

Figure 1*a* FMRFamide antisera staining in larval *D. melanogaster* central nervous system. A whole mount tissue preparation of a larval central nervous system was stained with FMRFamide antisera. The bar represents about 50 μm in length.

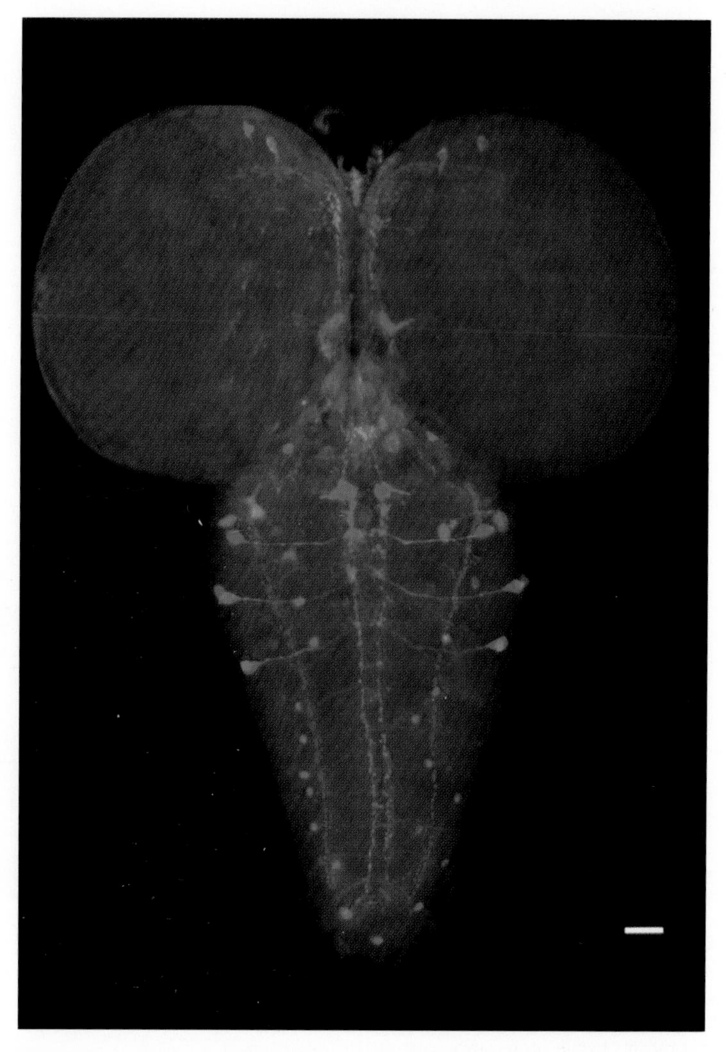

Figure 1*b* Triple immunolabeling in larval *D. melanogaster* central nervous system. A whole mount tissue preparation of a larval central nervous system was stained with three different FMRFamide-containing peptide-specific antisera: DPKQDFMRFamide-specific (*green*), TPAEDFMRFamide-specific (*blue*), and SDNFMRFamide-specific antisera (*red*).

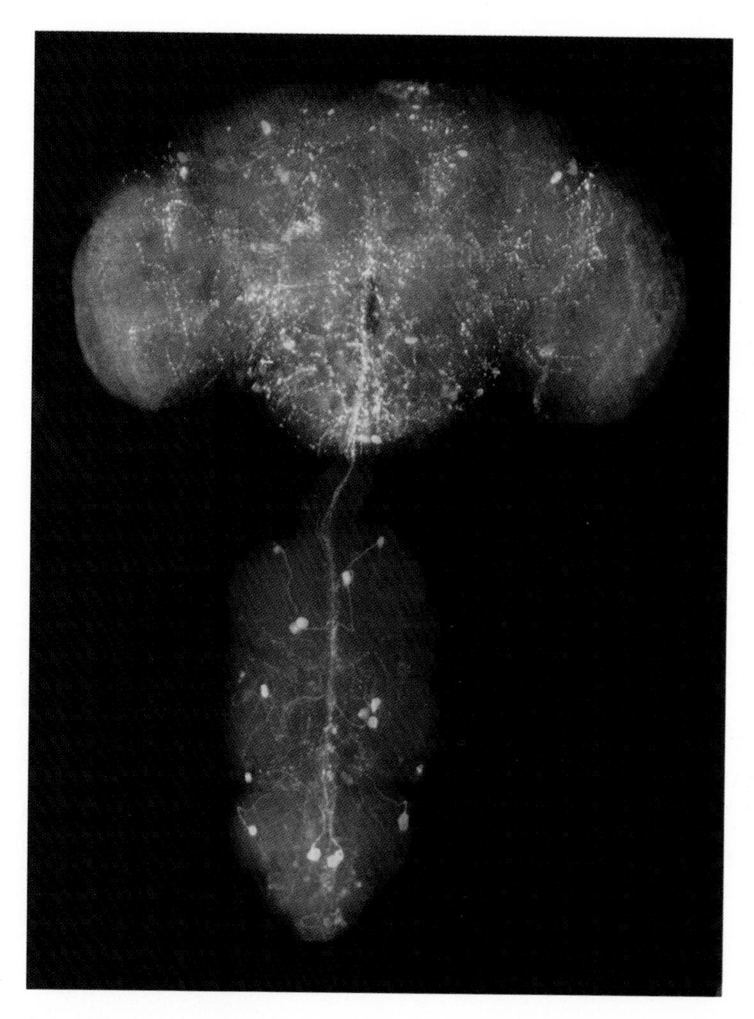

Figure 1*c* Triple immunolabeling in adult *D. melanogaster* central nervous system. A whole mount tissue preparation of an adult central nervous system was stained with three different FMRFamide-containing peptide-specific antisera: DPKQDFMRFamide-specific (*green*), TPAEDFMRFamide-specific (*blue*), and SDNFMRFamide-specific antisera (*red*).

receptors. Binding studies combined with use of mutants of signal transduction pathway components and pharmacological analysis will help elucidate additional molecular mechanisms relevant to FaRP signaling.

The synthesis and release sites of these structurally related peptides differ, which suggests FaRPs are subject to multiple and different sensory cues. In addition, structure-activity data are consistent with the existence of multiple and unique FaRP receptors or ligand-receptor binding requirements. FaRPs appear to have different effects on the frequency of heart contractions based on the site of application (19, 40, 44). In addition, FaRPs have different effects on the frequency of spontaneous contractions of the heart and the crop, an anterior portion of the gut (29, 30). Some FaRPs are similar in effect at the neuromuscular junction (15), which is not in conflict with peptide activities in heart and gut being unique because the two assays are dramatically different. Taken together, these data suggest that tissue-specific FaRP receptor subtypes exist and that FaRPs are not redundant in biological function(s).

In conclusion, investigations in several animal species suggest that FaRPs play roles as transmitters and modulators of numerous biological processes. Data indicate these structurally related peptides are individually unique in several ways including precursor processing and sites of synthesis and release, and ligand-receptor binding requirements. Like their counterparts in other animal species, *D. melanogaster* FaRPs are brain-gut peptides that affect the frequency of spontaneous contractions of different muscles and demonstrate tissue-specific activities.

ACKNOWLEDGMENTS

An NSF IBN Grant (IBN 0076615) to R. Nichols supported research from the Nichols laboratory and the writing of this review.

The *Annual Review of Entomology* is online at http://ento.annualreviews.org

LITERATURE CITED

1. Adams MD, Celniker SE, Holt RA, Evans CA, Gocayne JD, et al. 2000. The genome sequence of *Drosophila melanogaster*. *Science* 287:2185–95

1a. Baggerman G, Anja C, De Loof A, Schoofs L. 2002. Peptidomics of the larval *Drosophila melanogaster* central nervous system. *J. Biol. Chem.* In press

2. Deleted in proof

3. De Loof A, Schoofs L. 1990. Homologies between the amino acid sequences of some vertebrate peptide hormones and peptides isolated from invertebrate sources. *Comp. Biochem. Physiol. B* 95:459–68

4. Dickerson M, McCormick J, Paisley K, Nichols R. 1996. *Regulation of* Drosophila *heart rate*. Presented at Annu. Meet. Soc. Neurosci., 26th, Washington, DC

5. Donly BC, Fuse M, Orchard I, Tobe SS, Bendena WG. 1996. Characterization of the gene for leucomyosuppressin and its expression in the brain of the cockroach *Diploptera punctata*. *Insect Biochem. Mol. Biol.* 26:627–37

6. Duve H, Elia AJ, Orchard I, Johnsen AH,

Thorpe A. 1993. The effects of calliFMRFamides and other FMRFamide-related neuropeptides on the activity of the heart of the blowfly *Calliphora vomitoria*. *J. Insect Physiol.* 39:31–40
7. Duve H, Johnsen AH, Sewell JC, Scott AG, Orchard I, et al. 1992. Isolation, structure, and activity of Phe-Met-Arg-Phe-NH2 neuropeptides (designated calliFMRFamides) from the blowfly *Calliphora vomitoria*. *Proc. Natl. Acad. Sci. USA* 89:2326–30
8. Duve H, Thorpe A, Scott AG, Johnsen AH, Rehfeld JF, et al. 1995. The sulfakinins of the blowfly *Calliphora vomitoria*. Peptide isolation, gene cloning and expression studies. *Eur. J. Biochem.* 232:633–40
9. East PD, Hales DF, Cooper PD. 1997. Distribution of sulfakinin-like peptides in the central and sympathetic nervous system of the American cockroach, *Periplaneta americana* (L.) and the field cricket, *Teleogryllus commodus* (Walker). *Tissue Cell* 29:347–54
10. Eipper BA, Stoffers DA, Mains RE. 1992. The biosynthesis of neuropeptides: peptide alpha-amidation. *Annu. Rev. Neurosci.* 15:57–85
11. Fonágy A, Schoofs L, Proost P, Van Damme J, Bueds H, De Loof A. 1992. Isolation, primary structure and synthesis of neomyosuppressin, a myoinhibiting neuropeptide from the grey fleshfly, *Neobellieria bullata*. *Comp. Biochem. Physiol. C* 102:239–45
12. Fonágy A, Schoofs L, Proost P, Van Damme J, De Loof A. 1992. Isolation and primary structure of two sulfakinin-like peptides from the fleshfly, *Neobellieria bullata*. *Comp. Biochem. Physiol. C* 103:135–42
13. Friedman J, Starkman J, Nichols R. 2001. DPKQDFMRFamide expression is similar in two distantly related *Drosophila* species. *Peptides* 22:235–39
14. Fusé M, Zhang JR, Partridge E, Nachman RJ, Orchard I, et al. 1999. Effects of an allatostatin and a myosuppressin on midgut carbohydrate enzyme activity in the cockroach *Diploptera punctata*. *Peptides* 20:1285–93
15. Hewes RS, Snowdeal EC III, Saitoe M, Taghert PH. 1998. Functional redundancy of FMRFamide-related peptides at the *Drosophila* larval neuromuscular junction. *J. Neurosci.* 18:7138–51
16. Holman GM, Cook BJ, Nachman RJ. 1986. Isolation, primary structure and synthesis of leucomyosuppressin, an insect neuropeptide that inhibits spontaneous contractions of the cockroach hindgut. *Comp. Biochem. Physiol. C* 85:329–33
17. Holman GM, Cook BJ, Nachman RJ. 1986. Primary structure and synthesis of a blocked myotropic neuropeptide isolated from the cockroach, *Leucophaea maderae*. *Comp. Biochem. Physiol. C* 85:219–24
18. Johnsen AH, Duve H, Davey M, Hall M, Thorpe A. 2000. Sulfakinin neuropeptides in a crustacean. Isolation, identification and tissue localization in the tiger prawn *Penaeus monodon*. *Eur. J. Biochem.* 267:1153–60
19. Johnson E, Ringo J, Dowse H. 2000. Native and heterologous neuropeptides are cardioactive in *Drosophila melanogaster*. *J. Insect Physiol.* 46:1229–36
20. Kaminski S, Orlowski E, Berry K, Nichols R. 2002. Effects of three *Drosophila melanogaster* myotropins on the frequency of foregut contractions differ. *J. Neurogenet.* 16:125–34
21. Kingan TG, Shabanowitz J, Hunt DF, Witten JL. 1996. Characterization of two myotrophic neuropeptides in the FMRFamide family from the segmental ganglia of the moth *Manduca sexta*: candidate neurohormones and neuromodulators. *J. Exp. Biol.* 199:1095–104
22. Kubiak TM, Larsen MJ, Burton KJ, Bannow CA, Martin RA, et al. 2002. Cloning and functional expression of the first *Drosophila melanogaster* sulfakinin

receptor DSK-R1. *Biochem. Biophys. Res. Comm.* 291:313–20
23. Lange AB, Orchard I. 1998. The effects of SchistoFLRFamide on contractions of locust midgut. *Peptides* 19:459–67
24. Lange A, Orchard I, Te Brugge VA. 1991. Evidence for the involvement of a SchistoFLRFamide-like peptide in the neural control of locust oviduct. *J. Comp. Physiol. A* 168:383–92
25. Lange A, Peeff NM, Orchard I. 1994. Isolation, sequence, and bioactivity of FMRFamide-related peptides from the locust ventral nerve cord. *Peptides* 15:1089–94
26. Maestro JL, Aguilar R, Pascual N, Valero ML, Piulachs MD, et al. 2001. Screening of antifeedant activity in brain extracts led to the identification of sulfakinin as a satiety promoter in the German cockroach. Are arthropod sulfakinins homologous to vertebrate gastrins-cholecystokinins? *Eur. J. Biochem.* 268:5824–30
27. McCormick J, Lim I, Nichols R. 1999. Neuropeptide precursor processing detected by triple immunolabeling. *Cell Tissue Res.* 297:197–202
28. McCormick J, Nichols R. 1993. Spatial and temporal expression identify dromyosuppressin as a brain-gut peptide in *Drosophila melanogaster*. *J. Comp. Neurol.* 338:279–88
28a. Meola SM, Wright MS, Nichols R, Pendleton MW. 1996. Localization of myosuppressinlike peptides in the hypocerebral ganglion of two blood-feeding flies: horn fly and stable fly (Diptera:Muscidae). *J. Med. Entomol.* 33:473–81
29. Merte J, Nichols R. 2002. *Drosophila melanogaster* myotropins have unique functions and signaling pathways. *Peptides* 23:757–63
30. Merte J, Nichols R. 2002. Roles of *Drosophila melanogaster* FMRFamides in physiology: redundant or diverse? *Peptides* 23:209–20
31. Nachman RJ, Giard W, Favrel P, Suresh T, Sreekumar S, Holman GM. 1997. Insect myosuppressins and sulfakinins stimulate release of the digestive enzyme α-amylase in two invertebrates: the scallop *Pecten maximus* and insect *Rhynchophorus ferrugineus*. *Ann. N.Y. Acad. Sci.* 814:335–38
32. Nachman RJ, Holman GM, Cook BJ, Haddon WF, Ling N. 1986. Leucosulfakinin-II, a blocked sulfated insect neuropeptide with homology to cholecystokinin and gastrin. *Biochem. Biophys. Res. Commun.* 140:357–64
33. Nachman RJ, Holman GM, Haddon WF, Ling N. 1986. Leucosulfakinin, a sulfated insect neuropeptide with homology to gastrin and cholecystokinin. *Science* 234:71–73
34. Nambu JR, Murphy-Erdosh C, Andrews PC, Feistner GJ, Scheller RH. 1988. Isolation and characterization of a *Drosophila* neuropeptide gene. *Neuron* 1:55–61
35. Nathoo AN, Moeller RA, Westlund BA, Hart AC. 2001. Identification of neuropeptide-like protein gene families in *Caenorhabditis elegans* and other species. *Proc. Natl. Acad. Sci. USA* 98:14000–5
36. Nichols R. 1992. Isolation and expression of the *Drosophila* drosulfakinin neural peptide gene product, DSK-I. *Mol. Cell. Neurosci.* 3:342–47
37. Nichols R. 1992. Isolation and structural characterization of *Drosophila* TDVDHVFLRFamide and FMRFamide-containing neural peptides. *J. Mol. Neurosci.* 3:213–18
38. Nichols R, Lim I, McCormick J. 1999. Antisera to multiple antigenic peptides detect neuropeptide processing. *Neuropeptides* 33:35–40
39. Nichols R, Lim IA. 1996. Spatial and temporal immunocytochemical analysis of drosulfakinin (Dsk) gene products in the *Drosophila melanogaster* central nervous system. *Cell Tissue Res.* 283:107–16
40. Nichols R, McCormick J, Cohen M, Howe E, Jean C, et al. 1999. Differential

processing of neuropeptides influences *Drosophila* heart rate. *J. Neurogenet.* 13:89–104
41. Nichols R, McCormick J, Lim I. 1999. Regulation of *Drosophila* FMRFamide neuropeptide gene expression. *J. Neurobiol.* 39:347–58
42. Nichols R, McCormick J, Lim I, Caserta L. 1995. Cellular expression of the *Drosophila melanogaster* FMRFamide neuropeptide gene product DPKQDFMRFamide. Evidence for differential processing of the FMRFamide polypeptide precursor. *J. Mol. Neurosci.* 6:1–10
43. Nichols R, McCormick JB, Lim IA. 1999. Structure, function, and expression of *Drosophila melanogaster* FMRFamide-related peptides. *Ann. N.Y. Acad. Sci.* 897:264–72
44. Nichols R, McCormick JB, Lim IA, Starkman JS. 1995. Spatial and temporal analysis of the *Drosophila* FMRFamide neuropeptide gene product SDNFMRFamide: evidence for a restricted expression pattern. *Neuropeptides* 29:205–13
45. Nichols R, Schneuwly SA, Dixon JE. 1988. Identification and characterization of a *Drosophila* homologue to the vertebrate neuropeptide cholecystokinin. *J. Biol. Chem.* 263:12167–70
46. Deleted in proof
47. Peeff NM, Orchard I, Lange AB. 1994. Isolation, sequence, and bioactivity of PDVDHVFLRFamide and ADVGHVFLRFamide peptides from the locust central nervous system. *Peptides* 15:387–92
48. Predel R, Rapus J, Eckert M. 2001. Myoinhibitory neuropeptides in the American cockroach. *Peptides* 22:199–208
49. Price DA, Greenberg MJ. 1977. Purification and characterization of a cardioexcitatory neuropeptide from the central ganglia of a bivalve mollusc. *Prep. Biochem.* 7:261–81
50. Richer S, Stoffolano JG, Yin C-M, Nichols R. 2000. Innervation of dromyosuppressin (DMS) immunoreactive processes and effect of DMS and benzethonium chloride on the *Phormia regina* (Meigen) crop. *J. Comp. Neurol.* 421:136–42
51. Robb S, Evans P. 1994. The modulatory effect of Schisto FLRFamide on heart and skeletal muscle in the locust *Schistocerca gregaria. J. Exp. Biol.* 197:437–42
52. Robb S, Packman LC, Evans PD. 1989. Isolation, primary structure and bioactivity of Schisto-FLRF-amide, a FMRF-amide-like neuropeptide from the locust *Schistocerca gregaria. Biochem. Biophys. Res. Commun.* 160:850–56
53. Rosoff ML, Doble KE, Price DA, Li C. 1993. The flp-1 propeptide is processed into multiple, highly similar FMRFamide-like peptides in *Caenorhabditis elegans. Peptides* 14:331–38
54. Scharrer B, Scharrer E. 1944. Neurosecretion. II. A comparison between the intercerebralis-cardiacum-allatum system of invertebrates and the hypothalamo-hypophysial system of vertebrates. *Biol. Bull.* 87:242–51
55. Schneider LE, Taghert PH. 1988. Isolation and characterization of a *Drosophila* gene that encodes multiple neuropeptides related to Phe-Met-Arg-Phe-NH2 (FMRFamide). *Proc. Natl. Acad. Sci. USA* 85:1993–97
56. Schoofs L. 2002. *Peptidomics: an elegant research tool to identify differentially expressed or released neuropeptides in insects.* Presented at Invertebr. Neuropept. Conf., Hua Hin, Thail.
57. Schoofs L, Holman GM, Hayes T, DeLoof A. 1990. Isolation and identification of a sulfakinin-like peptide with sequence homolog to vertebrate gastrin and cholecystokinin, from the brain of *Locusta migratoria.* In *Chromatography and Isolation of Insect Hormones and Pheromones*, ed. AR McCaffery, ID Wilson, pp. 231–41. New York: Plenum
58. Schoofs L, Holman GM, Paemen L, Veelaert D, Amelinckx M, De Loof A. 1993. Isolation, identification, and synthesis

of PDVDHVFLRFamide (SchistoFLRFamide) in *Locusta migratoria* and its association with the male accessory glands, the salivary glands, the heart, and the oviduct. *Peptides* 14:409–21
59. Taghert PH, Schneider LE. 1990. Interspecific comparison of a *Drosophila* gene encoding FMRFamide-related neuropeptides. *J. Neurosci.* 10:1929–42
60. Tibbetts MF, Nichols R. 1993. Immunocytochemistry of sequence-related neuropeptides in *Drosophila*. *Neuropeptides* 24:321–25
61. Vanden Broeck J. 2001. Neuropeptides and their precursors in the fruitfly, *Drosophila melanogaster*. *Peptides* 22:241–54
62. Veelaert D, Devreese B, Vanden Broeck J, Yu CG, Schoofs L, et al. 1996. Isolation and characterization of schistostatin-2 (11–18) from the desert locust, *Schistocerca gregaria*: a truncated analog of schistostatin-2. *Regul. Pept.* 67:195–99
63. Veenstra JA. 1989. Isolation and structure of two gastrin/CCK-like neuropeptides from the American cockroach homologous to the leucosulfakinins. *Neuropeptides* 14:145–49
64. Wei Z, Baggerman GJ, Nachman R, Goldsworthy G, Verhaert P, et al. 2000. Sulfakinins reduce food intake in the desert locust, *Schistocerca gregaria*. *J. Insect Physiol.* 46:1259–65
65. White K, Hurteau T, Punsal P. 1986. Neuropeptide-FMRFamide-like immunoreactivity in *Drosophila*: development and distribution. *J. Comp. Neurol.* 247: 430–38
66. Yang HY, Fratta W, Majane EA, Costa E. 1985. Isolation, sequencing, synthesis, and pharmacological characterization of two brain neuropeptides that modulate the action of morphine. *Proc. Natl. Acad. Sci. USA* 82:7757–61
67. Zornik E, Paisley K, Nichols R. 1999. Neural messengers and a peptide modulate *Drosophila* heart rate. *Peptides* 20:45–51

Annu. Rev. Entomol. 2003. 48:505–19
doi: 10.1146/annurev.ento.48.091801.112621

First published online as a Review in Advance on September 3, 2002

POPULATION-LEVEL EFFECTS OF PESTICIDES AND OTHER TOXICANTS ON ARTHROPODS

John D. Stark[1] and John E. Banks[2]
[1]*Department of Entomology, Washington State University, 7612 Pioneer Way East, Puyallup, Washington 98371; e-mail: stark@puyallup.wsu.edu*
[2]*Interdisciplinary Arts and Sciences, University of Washington, Tacoma, 1900 Commerce Street, Tacoma, Washington 98402; e-mail: banksj@u.washington.edu*

Key Words ecotoxicology, demography, life tables, intrinsic rate of increase, population growth rate

■ **Abstract** New developments in ecotoxicology are changing the way pesticides and other toxicants are evaluated. An emphasis on life histories and population fitness through the use of demography, other measures of population growth rate, field studies, and modeling are being exploited to derive better estimates of pesticide impacts on both target and nontarget species than traditional lethal dose estimates. We review the state of the art in demographic toxicology, an approach to the evaluation of toxicity that uses life history parameters and other measures of population growth rate. A review of the literature revealed that 75 studies on the use of demography and similar measures of population growth rate in toxicology have been published since 1962. Of these 75 studies, the majority involved arthropods. Recent evaluations have indicated that ecotoxicological analysis based on population growth rate results in more accurate assessments of the impacts of pesticides and other toxicants because measures of population growth rate combine lethal and sublethal effects, which lethal dose/concentration estimates (LD/LC50) cannot do. We contend that to advance our knowledge of toxicant impacts on arthropods, the population growth rate approach should be widely adopted.

CONTENTS

0066-4170/03/0107-0505$14.00

INTRODUCTION

The myriad effects of pesticides and other toxicants on insects and other arthropods have been the subject of an enormous number of studies in the past several decades. Much of this large body of information is based on laboratory studies—not surprising, as lab contact tests are the preferred method of assessing pesticide impacts on natural enemies such as biological control agents (18). As a reflection of this trend, fewer than one third of the records that comprise the SELCTV database, a compilation of literature on the effects of pesticides on nontarget arthropods, are field studies (99, 100).

Another overwhelming bias in the toxicological literature, especially in invertebrate studies, is the use of acute mortality values, usually in the form of lethal dose/concentration estimates. The use of mortality and median lethal dose/concentration as toxicological endpoints comprises 95% of published studies in the SELCTV database (18), emphasizing how entrenched this methodology has been. Although the results of such estimates, coupled with the convenience and rapidity of laboratory settings, have been extremely valuable, interpretation of the data is severely limited (93). New developments in ecotoxicology are changing the way pesticide effects are evaluated. An emphasis on life histories and population fitness through the use of demography, field studies, and modeling is being employed to derive better estimates of pesticide impacts on both target and nontarget species (4, 46). We explore the differences in laboratory lethal dose/concentration studies versus population endpoints. We focus on particular aspects of these differences in perspective that have serious and often surprising consequences for how we analyze the risk that toxicants pose to insects and other arthropods and the communities in which they live.

ENDPOINTS OF TOXICOLOGICAL EFFECT

Effects on Individuals

To date, most toxicological studies involve an evaluation of the impact of toxicants on individuals or some component of individuals (e.g., cell cultures or enzyme systems). Most studies conducted by toxicologists focus on mode of action (toxicodynamics) and/or uptake and excretion rates and storage of chemicals in the body (toxicokinetics). In general, such studies are designed to reduce variation in the subject animal's age, size, life stage, weight, and overall health so effects attributable to exposure to a toxicant can be easily discerned. The time frame for evaluation of effects varies depending on the endpoints being examined and the type of study. Despite some problems described below, there are some clear advantages associated with the approach toxicologists use to evaluate toxicity in the laboratory. First, reduced variation among test animals renders the statistical power of such studies relatively high, so stronger inferences may be made (79). In contrast, toxicological studies conducted with field-collected organisms invariably involve individuals of different ages/stages, weight, and health (80), obscuring the relationship between the toxicant and its effects.

There are two major types of toxicological studies: acute- and chronic-exposure studies. Acute data are usually generated after a single, short exposure to a chemical (a few hours to a few days). Mortality is the endpoint of interest for many acute studies, although other endpoints such as enzyme kinetics or knockdown may be evaluated. In contrast, chronic exposure studies monitor the effects of repeated exposures to toxicants over longer time periods. The relative merits of each of these approaches depend, naturally, upon the type of toxicological insults being investigated. More significantly, the metrics used in each of these approaches differ in some important ways discussed next.

DOSE RESPONSE AND DETERMINATION OF THE LD_{50} FOR ACUTE EXPOSURES

The median lethal dose or concentration (LD_{50} or LC_{50}, respectively) is the most commonly used measure of acute toxic effect. For the LD_{50}, lethal doses are based on either (*a*) a known amount of toxicant per amount of body weight (milligrams toxicant/kilogram bodyweight) or (*b*) amount of toxicant per animal (milligrams toxicant/animal). In both cases, the amount of toxicant the organism receives is known precisely. In contrast, the LC_{50} is based on the amount of toxicant in an environmental medium such as water, soil, or air (milligrams toxicant/liter of water), and the amount of toxicant that enters the organism is not known. Because of the subtle differences in uncertainties associated with these two metrics, confusion persists even among toxicologists about the difference between lethal dose and lethal concentration estimates [see Robertson & Preisler (79) for their definitions].

The LD_{50} is a statistically derived measure of the relationship between dose and response and is an estimate of the dose that causes 50% mortality of a group of organisms being studied (27). Plotting the percent affected (dead) against dose or concentration yields a sigmoidal curve. No effect is observed at low concentrations, but at higher concentrations some of the organisms begin to respond. The threshold is the lowest dose that elicits a response, but eventually a dose is reached that kills all the organisms being evaluated (maximum effect). Because the dose-response relationship is sigmoidal, it is difficult to estimate the LD_{50} or the slope of the dose-response line. As a result, several methods have been developed to straighten the dose-response curve and thereby better estimate the LD_{50}. Trevan (101) proposed the first of these methods, which has been followed by a series of modifications. Most acute toxicity data are analyzed with probit or logit analysis. In the past, estimation of the LD_{50} was done by hand, but now excellent computer software is available to calculate dose-response statistics (e.g., SAS, Polo-PC, ToxCalc, and Systat).

Chronic Toxicity Studies

Chronic toxicity data are usually generated after continuous exposure to sublethal doses/concentrations over many days or even throughout a lifetime. Life span, weight gain, reproduction, cancer, and birth defects are often the endpoints of interest in chronic studies.

The highest concentration for which no effect is observed is referred to as the no observable effect concentration (NOEC) or level (NOEL). This measure can be applied to various endpoints of toxic effect, although its value has recently been called into question (56, 60). Van Straalen et al. (109) calculated NOEC values for several toxic endpoints for Collembola and oribatid mites exposed to cadmium. The challenge is determining which NOEC value should be used in establishing predictive estimates for species persistence. Clearly, the lowest NOEC value provides a natural baseline, but the relevance of each particular endpoint to the important population biology parameters of the species must be examined more thoroughly.

Lethal dose/concentration estimate data abound in the literature on pest management and environmental protection. Rather than providing an unrealistic absolute measure of the effects of toxicants on organisms of interest, the primary use of these data is simply as a comparative measure of intrinsic susceptibility. That is, in the proper context lethal dose/concentration data are extremely valuable: They provide a means of comparing the toxicity of various chemicals with a particular species or of comparing the intrinsic susceptibility among species with a particular chemical. However, there are limitations to the use of these data (93). Nonetheless, even the most recent examples of risk assessment rely heavily on the LC_{50} (89).

Sublethal Effects

Despite the singular focus of most toxicological studies on survival/mortality estimates, there is an increasing awareness of more subtle toxicant effects that warrants closer attention. Simply put, individuals that survive toxicant exposure may still sustain significant damage. Sublethal effects may be manifested as reductions in life span (94), development rates (110), fertility (36, 98), and fecundity (35, 96), changes in sex ratio (110), and changes in behavior such as feeding (24), searching (21), and oviposition (63). Thus, toxicants can exert subtle as well as overt effects that must be considered when examining their impact.

MEASURING MULTIPLE ENDPOINTS OF EFFECTS: LIFE TABLE RESPONSE EXPERIMENTS

Precisely because exposure to a toxicant can result in mortality as well as multiple sublethal effects, the use of simplistic toxicity metrics often results in underestimates of the total effects of toxicants. In particular, the traditional lethal dose/concentration estimates are not designed to measure the total effect of a toxicant but rather to analyze statistically one effect at a time.

Demography has been suggested as a more desirable means to evaluate the total effect of toxicants simply because it accounts for all effects a toxicant might have on a population (46, 49). Furthermore, demographic studies are usually conducted throughout the life span of an organism and thus provide a complete time-series portrait of toxicity. Demographic toxicological studies or life table response experiments provide a measure of effect on population growth rate. Forbes & Calow (29) have found that demographic toxicological data are superior to other types of toxicity data; such comparative arguments are the subject of a recent volume on toxicity metrics (46).

Life table response experiments are conducted by exposing individuals or groups to increasing doses or concentrations of a toxicant over their life span. Daily mortality and reproduction are recorded, and these data are then used to generate life table parameters. Because life tables are often developed in the laboratory and measurements are taken on individuals, rather than populations, a realistic measure of population growth rate is not obtained. However, the measure that is obtained, the intrinsic rate of increase, has been shown to be a more accurate measure of toxic effect than lethal concentration estimates (29).

There are two ways of expressing population growth rate: the intrinsic rate of increase (r_m) and lambda (λ) (population multiplication factor), which is the antilog of r_m. The intrinsic rate of increase is a measure of the ability of a population to increase logarithmically in an unlimited environment. The calculation of the population growth rate requires knowledge of a population's survivorship and fecundity schedule, which is usually recorded from studies of individuals or groups (10). The growth rate is then calculated by iteratively solving Equation 1:

$$1 = \sum l_x m_x e^{-r_m x}, \tag{1}$$

where X is the age of the cohort, l_x is the proportion of individuals surviving to age X, m_x is the number of females produced per female of age X, and r_m is the intrinsic rate of increase for the population. The population's sex ratio is assumed to be 1:1, and typically only females are considered in life tables. Positive values of r_m indicate exponential population increase, r_m equal to zero indicates that the population is stable, and negative values of r_m indicate that the population is declining exponentially and heading toward extinction.

Matrix algebra is another approach used to generate r_m or λ (11). In this case probabilities of survivorship and reproduction are organized into a transition matrix, which is then multiplied by a population vector containing information about the size of the population at each age X. Repeated iterations of the multiplication of the population vector and the transition matrix result in a stable population vector, so further multiplication by the transition matrix is tantamount to multiplication of the population vector by a constant. Mathematically, this may be written as

$$\lambda Z = LZ, \tag{2}$$

where L is the transition matrix, Z is the population vector at equilibrium, and λ is the population growth rate. Once the population vector has reached equilibrium, this equation can easily be solved for λ.

Real applications of demographic toxicology typically incorporate life table parameters in meaningful comparative exercises; for instance, life table parameters for unexposed populations and populations exposed to various concentrations of a toxicant/pollutant can be incorporated, and ensuing population responses can be compared. This approach is especially appealing because ecological and toxicological parameters are combined, resulting in better predictions about the effects of toxicants at the population level. Ultimately, the power of the intrinsic rate of increase (r_m) as an ecologically meaningful bioassay parameter for toxicology studies lies in the fact that this statistic is based on both survivorship (l_x) and fecundity (m_x) (2). Several excellent papers dealing with the theoretical basis for life table response experiments and/or modeling have been published (12, 29, 30, 34, 57, 84, 86). However, estimates of r_m are environment specific and must be viewed within the context of the experimental conditions.

Examples of Demographic Toxicological Studies

An examination of several scientific databases revealed that 75 studies have been published dealing with demographic toxicology or other population growth rate measures (Table 1; follow the Supplemental Material link in the online version of this chapter or at http://www.annualreviews.org/). The earliest published study in which demography was used in a toxicological context was by Marshall (70), who examined the effects of gamma radiation on *Daphnia pulex*. Notably, after Marshall's initial study in 1962 the next demographic toxicological study was not published until 1976. Five demographic toxicity studies were published in the 1970s, 27 in the 1980s, and 35 in the 1990s. Thus, even though there is a general consensus that the demographic approach provides better toxicological information than lethal dose estimates, this method has not been widely adopted. Of the 75 demographic toxicity studies published, the majority was conducted with arthropods (Table 1; follow the Supplemental Material link on the Annual Reviews homepage at http://www.annualreviews.org/). Many demographic studies are conducted with arthropods and other invertebrates because these organisms generally reproduce quickly and have relatively short life spans, especially compared with mammalian life histories. These attributes make working with arthropods much easier than working with vertebrates.

Advantages of Demography for Evaluation of Toxicity

There are important advantages inherent in using the demographic toxicological approach over traditional lethal concentration estimates. Not only is a total measure of toxic effect obtained, but other interactions not perceptible in short-term toxicity tests can be evaluated. For example, results of several studies have indicated that sublethal effects can be very subtle and affect populations at concentrations lower than the traditional concentration response curve. Studying the effects of acidic water on populations of *D. pulex*, Walton et al. (114) found that 96-h acute toxicity tests detected no effects of acidic water, but negative effects were easily discerned using an evaluation of the intrinsic rate of increase. In a later study Bechmann (6) discovered that some toxicants can affect demographic parameters well below the traditional concentration-response curve, resulting in population decline and extinction at levels previously assumed to have no effect based on endpoints for individuals.

Another phenomenon that may occur in stressed populations is increased reproduction. Population compensation is the process by which the reproductive potential of surviving individuals in a population increases because there is less competition for resources as the population is thinned (95). Removal of individuals from a population, whether through the action of toxicants, natural disasters, hunting, or fishing, results in survivors having more resources available, and thus they may reproduce at a greater rate. Offspring are also often healthier and weigh more. In fact, some populations can withstand large losses of individuals with little impact. Nicholson (76) found that losses as high as 25% had no long-term impact on a population of sheep blow flies, *Lucilia cuprina*. Similarly, losses of as high as 50% in populations of *Daphnia* resulted in only a slight decline (87).

A demographic toxicological study of the effects of the insecticide imidacloprid on the pea aphid, *Acyrthosiphon pisum* (Harris), indicated that aphids could maintain high population growth rates even when exposed to acute LC_{60} (111). Furthermore, imidacloprid caused no sublethal effects, yet lethal concentration estimates were not able to predict population growth owing to population compensation. The range of complexity evident in the results of such population-level studies is yet another indicator of the importance of broadening our perspectives on the uses and interpretations of different toxicological endpoints.

Disadvantages of Demography for Evaluation of Toxicity

The major disadvantage to the use of demographic toxicology is that development of life table data is expensive and time consuming. Furthermore, limitations exist for species that have long life spans, time to sexual maturation, and gestation periods, and low numbers of offspring. For this reason, many demographic toxicological studies are done with invertebrates; it is easy to imagine the difficulties in conducting toxicological studies over the life span of most mammals. Additionally, population growth rates derived under laboratory conditions do not reflect conditions in the wild: Density dependence, competition, immigration, emigration, predation, and/or parasitism are not usually measured in life table response experiments. However, several authors (34, 66, 85) have now considered density dependence in the context of these experiments.

Another problem with many life table response experiments is that organisms are exposed to constant toxicant concentrations when actually, many toxicants degrade in the environment (115a).

Alternative Population Growth Rate Methods

Life table response experiments can be expensive. One way to reduce cost is to use partial life tables (62). Another population growth rate method is the instantaneous rate of increase (31, 39, 48, 120). These studies are less time consuming and costly than the development of a complete life table.

The instantaneous rate of increase (r_i) is a direct measure of population growth rate, and like the intrinsic rate of increase, it integrates survivorship and fecundity. Unlike r_m, however, density-dependent regulatory mechanisms can come into play depending on how long the study is carried out and how many resources are made available to the study animals.

The instantaneous rate of increase is calculated by the following equation:

$$r_i = \ln(N_f/N_o)/\Delta T, \tag{3}$$

where N_f is the final number of animals, N_o is the initial number of animals, and ΔT is the change in time (number of days the experiment was run). Solving for r_i yields a rate of population increase or decline. Positive values of r_i indicate a growing population, $r_i = 0$ indicates a stable population, and negative r_i values indicate a population in decline and headed toward extinction. Although this is not demography in the true sense, this approach does yield a measure of population growth. This approach has been used in a toxicological context by several authors (3, 84, 91, 92, 95, 112).

POPULATION AGE/STAGE STRUCTURE AND DIFFERENTIAL SUSCEPTIBILITY AMONG LIFE STAGES

The majority of demographic studies start with eggs or neonates. A potential problem with this approach is that different life stages of an organism may exhibit different susceptibility to toxicants. Stark & Wennergren (97) developed life tables for neonate and adult aphids that had been exposed to an insecticide, Margosan-O, that was more toxic to immature stages than to adults. Results of life table response experiments indicated that populations exposed from birth were much more susceptible than populations exposed as adults. Stark & Banken (91) studied the effects of pesticides on differently structured starting populations of the two-spotted spider mite, *Tetranychus urticae* (Koch), and the pea aphid, *A. pisum* (Harris). Three structured populations were tested: The first consisted of eggs and neonates for *A. pisum* and *T. urticae*, respectively, the second consisted of the stable age distribution for each species, and the third consisted of young adult females only. Population growth rate after exposure to pesticides was significantly influenced by the starting population structure. Because different stages/ages of a species may have different susceptibilities to toxicants, it is essential that these factors be considered when estimating population susceptibility to toxicants.

CONCLUSIONS

We hope the differences highlighted here between traditional individual-based toxicity assessments and newer population-level metrics are but a starting point for further discussion among those interested in toxicological tests on arthropods and even other taxa. Ecologists have long been wary of relying on simplistic individual or single-species perspectives in their pursuit of a better understanding of critical topics such as community dynamics and multispecies interactions [e.g., Paine (77)]. Even applied entomological studies in biological control, arguably inextricably linked to issues of pesticide disturbance, have moved beyond reductionist philosophies to pursue more holistic population- and community-level properties emerging from multispecies interactions (15, 68, 69). Similarly, we hope to see continued increasing awareness of the need for the consideration of more complex (e.g., population-level, demography-based) endpoints in toxicological studies.

ACKNOWLEDGMENTS

The authors thank all the people who have contributed to the literature on demographic ecotoxicology. In particular we thank our colleagues Hal Caswell, Peter Calow, Valery Forbes, Nico van Straalen, Jan Kammenga, Peter Kareiva, Bas (SALM) Kooijman, Ryszard Laskowski, Richard Sibly, and Uno Wennergren.

The *Annual Review of Entomology* is online at http://ento.annualreviews.org

LITERATURE CITED

1. Ahmadi A. 1983. Demographic toxicology as a method for studying the dicofol-two-spotted spider mite (Acari: Tetranychidae) system. *J. Econ. Entomol.* 76: 239–42
2. Allan JD, Daniels RE. 1982. Life table evaluation of chronic exposure of *Eurytemora affinis* (Copepoda) to kepone. *Mar. Biol.* 66:179–84
3. Banken JAO, Stark JD. 1998. Multiple routes of pesticide exposure and the risk of pesticides to biological controls: a study of neem and the seven-spot lady beetle, *Coccinella septempunctata* L. *J. Econ. Entomol.* 91:1–6
4. Banks JE, Stark JD. 1998. What is ecotoxicology? An ad-hoc grab bag or an interdisciplinary science? *Integr. Biol.* 5:195–204
5. Barbour MT, Growes CG, McCulloch WL. 1989. Evaluation of the intrinsic rate of increase as an endpoint for *Ceriodaphnia* chronic tests. In *Aquatic Toxicology and Environmental Fate: Eleventh Volume*, ed. GW Suter II, MA Lewis, pp. 273–88. Philadelphia: Am. Soc. Test. Mater.
6. Bechmann RK. 1994. Use of life tables and LC_{50} tests to evaluate chronic and acute toxicity effects of copper on the marine copepod *Tisbe furcata* (Baird). *Environ. Toxicol. Chem.* 13:1509–17
7. Bengtsson G, Gunnarsson T, Rundgren S. 1985. Influence of metals on reproduction, mortality and population growth in *Onychiurus armatus* (Collembola). *J. Appl. Ecol.* 22:967–78
8. Bertram PE, Hart BA. 1979. Longevity

and reproduction of *Daphnia pulex* (de Geer) exposed to cadmium-contaminated food or water. *Environ. Pollut.* 19: 295–305

9. Boykin LS, Campbell WV. 1982. Rate of population increase of the two-spotted spider mite *Tetranychus urticae* Acari Tetranychidae on peanut *Arachis hypogaea* leaves treated with pesticides. *J. Econ. Entomol.* 75:966–71
10. Carey JR. 1993. *Applied Demography for Biologists*. Oxford, UK: Oxford Univ. Press
11. Caswell H. 1989. *Matrix Population Models*. Sunderland, MA: Sinauer
12. Caswell H. 1996. Demography meets ecotoxicology: untangling the population level effects of toxic substances. In *Ecotoxicology. A Hierarchical Treatment*, ed. MC Newman, CH Jagoe, pp. 255–82. Boca Raton, FL: Lewis
13. Chandini T. 1988. Effects of different food (*Chlorella*) concentrations on the chronic toxicity of cadmium to survivorship, growth and reproduction of *Echinisca triserialis* (Crustacea: Cladocera). *Environ. Pollut.* 54:139–54
14. Chi H. 1990. Timing of control based on the stage structure of pest populations: a simulation approach. *J. Econ. Entomol.* 83:1143–50
15. Cohen JE, Schoenly K, Heong KL, Justo H, Arida G, et al. 1994. A food web approach to evaluating the effect of insecticide spraying on insect pest population dynamics in a Philippine irrigated rice ecosystem. *J. Appl. Ecol.* 31:747–63
16. Coniglio L, Baudo R. 1989. Life-tables of *Daphnia obtuse* (Kurz) surviving exposure to toxic concentrations of chromium. *Hydrobiologia* 188/189:407–10
17. Costa SD, Barbercheck ME, Kennedy GG. 2000. Sub-lethal effects with acute and chronic exposure of Colorado potato beetle (Coleoptera: Chrysomelidae) to the δ-endotoxin of *Bacillus thuringiensis*. *J. Econ. Entomol.* 93:680–89
18. Croft BA. 1990. *Arthropod Biological Control Agents and Pesticides*. New York: Wiley
19. Crommentuijn T, Brils J, Van Straalen NM. 1993. Influence of cadmium on life-history characteristics of *Folsomia candida* (Willem). *Ecotoxicol. Environ. Saf.* 26:216–27
20. Crommentuijn T, Doodeman CJAM, Doornekamp A, van Gestel CAM. 1997. Life table study with the springtail *Folsomia candida* (Willem) exposed to cadmium, chlorpyrifos and triphenyltin hydroxide. In *Ecological Risk Assessment of Contaminants in Soil*, ed. NM van Straalen, H. Løkke, pp. 275–91. London: Chapman & Hall
21. Dabrowski ZT. 1969. Laboratory studies on the toxicity of pesticides for *Typhlodromus finlandicus* (Oud.) and *Phytoseius macropilis* (Banks) (Phytoseiidae, Acarina). *Rocz. Nauk Rol.* 95:337–69
22. Daniels RE, Allan JD. 1981. Life table evaluation of chronic exposure to a pesticide. *Can. J. Fish. Aquat. Sci.* 38:485–94
23. Day K, Kaushik NK. 1987. An assessment of the chronic toxicity of the synthetic pyrethroid, fenvalerate to *Daphnia galeata mendotae*, using life tables. *Environ. Pollut.* 44:13–26
24. Dempster JP. 1968. The sublethal effect of DDT on the rate of feeding by the ground beetle *Harpalus rufipes*. *Entomol. Exp. Appl.* 11:51–54
25. Enserink L, De la Haye M, Maas-Diepeveen H. 1993. Reproductive strategy of *Daphnia magna*: implications for chronic toxicity tests. *Aquat. Toxicol.* 25: 111–24
26. Fernandez-Casalderrey A, Ferrando MD, Andreu-Moliner E. 1993. Effects of endosulfan on survival, growth and reproduction of *Daphnia magna*. *Comp. Biochem. Physiol. C* 106:437–41
27. Finney DJ. 1971. *Probit Analysis*. Cambridge: Cambridge Univ. Press. 2nd ed.
28. Fitzmayer KM, Geiger JG, Van Den Avyle MJ. 1982. Effects of chronic exposure to simazine on the cladoceran,

Daphnia pulex. *Arch. Environ. Contam. Toxicol.* 11:603–9

29. Forbes VE, Calow P. 1999. Is the per capita rate of increase a good measure of population-level effects in ecotoxicology? *Environ. Toxicol. Chem.* 18:1544–56
30. Forbes VE, Sibly RM, Calow P. 2001. Toxicant impacts on density limited populations: a critical review of theory, practice, and results. *Ecol. Appl.* 11:1249–57
31. Frazer BD. 1972. Population dynamics and recognition of biotypes in the pea aphid (Homoptera: Aphididae). *Can. Entomol.* 104:1729–33
32. Gentile JH, Gentile SM, Hairston NG Jr, Sullican BK. 1982. The use of life-tables for evaluating the chronic toxicity of pollutants to *Mysidopsis bahia*. *Hydrobiologia* 93:179–87
33. Gentile JH, Gentile SM, Hoffman G, Heltshe JF, Hairston N Jr. 1983. The effects of a chronic mercury exposure on survival, reproduction and population dynamics of *Mysidopsis bahia*. *Environ. Toxicol. Chem.* 2:61–68
34. Grant A. 1998. Population consequences of chronic toxicity: incorporating density dependence into the analysis of life table response experiments. *Ecol. Model.* 105:325–35
35. Grapel H. 1982. Investigations on the influence of some insecticides on natural enemies of aphids. *Z. Pflanzenkr. Pflanzenschutz* 89:241–52
36. Grosch DS, Hoffman AC. 1973. The vulnerability of specific cells in the oogenetic sequence of *Bracon hebetor* to some degradation products of carbamate pesticides. *Environ. Entomol.* 2:1029–32
37. Halbach U. 1984. Population dynamics of rotifers and its consequences of ecotoxicology. *Hydrobiologia* 109:79–96
38. Halbach U, Siebert M, Westermayer M, Wissel C. 1983. Population ecology of rotifers as a bioassay tool for ecotoxicological tests in aquatic environments. *Ecotoxicol. Environ. Saf.* 7:484–513
39. Hall DJ. 1964. An experimental approach to the dynamics of a natural population of *Daphnia galeata mendotae*. *Ecology* 45:94–112
40. Deleted in proof
41. Hansen FT, Forbes VE, Forbes TL. 1999. The effects of chronic exposure to 4-*n*-nonylphenol on life history traits and population dynamics of the polychaete *Capitella* sp. I. *Ecol. Appl.* 9:482–95
42. Hansen FT, Forbes VE, Forbes TL. 1999. Using elasticity analysis of demographic models to link toxicant effects on individuals to the population level: an example. *Funct. Ecol.* 13:157–62
43. Hendriks AJ, Enserink EL. 1996. Modeling responses of single species populations to microcontaminants as a function of species size, with examples for water-fleas (*Daphnia magna*) and cormorants (*Phalacrocorax carbo*). *Ecol. Model.* 88:247–62

43a. Ibrahim YB, Yee TS. 2000. Influence of sublethal exposure to abamectin on the biological performance of *Neoseiulus longispinosus* (Acari: Phytoseiidae). *J. Econ. Entomol.* 93:1085–89

44. Janssen CR, Persoone G, Snell TW. 1994. Cyst-based toxicity tests. VIII. Short-chronic toxicity tests with the freshwater rotifer *Brachionus calciflorus*. *Aquat. Toxicol.* 28:243–58
45. Kammenga J, Laskowski R. 2000. *Demography in Ecotoxicology*. Chichester, UK: Wiley
46. Kammenga JE, Busschers M, van Straalen NM, Jepson PC, Bakker J. 1996. Stress-induced fitness reduction is not determined by the most sensitive life-cycle trait. *Funct. Ecol.* 10:106–11
47. Kammenga JE, Riksen JAG. 1996. Comparing differences in species sensitivity to toxicants: phenotypic plasticity versus concentration-response relationships. *Environ. Toxicol. Chem.* 15:1649–53

48. Kareiva P, Sahakian R. 1990. Tritrophic effects of a simple architectural mutation in pea plants. *Nature* 345:433–34
49. Kareiva P, Stark J, Wennergren U. 1996. Using demographic theory, community ecology and spatial models to illuminate ecotoxicology. *Ecotoxicology: Ecological Dimensions*, ed. DJ Baird, L Maltby, PW Greig-Smith, PET Douben, pp. 13–23. London: Chapman & Hall
50. Kjær C, Elmegaard N, Axelsen JA, Andersen PN, Seidelin N. 1998. The impact of phenology, exposure and instar susceptibility on insecticide effects on a chrysomelid beetle population. *Pestic. Sci.* 52:361–71
51. Klok C, de Roos AM. 1996. Population level consequences of toxicological influences on individual growth and reproduction in *Lumbricus rubellus* (Lumbridicae, Oligochaeta). *Ecotoxicol. Environ. Saf.* 33:118–27
52. Klok C, de Roos AM, Marinissen CY, Baveco HM, Ma WC. 1997. Assessing the effects of abiotic environmental stress on population growth in *Lumbricus rubellus* (Lumbricidae, Oligochaeta). *Soil Biol. Biochem.* 29:287–93
53. Klüttgen B, Ratte HT. 1994. Effects of different food doses on cadmium toxicity to *Daphnia magna*. *Environ. Toxicol. Chem.* 13:1619–27
54. Koivisto S, Ketola M, Walls M. 1992. Comparison of five cladoceran species in short- and long-term copper exposure. *Hydroiologia* 248:125–36
55. Kooijman SALM. 1985. Toxicity at population level. In *Multispecies Toxicity Testing*, ed. J Cairns Jr, pp. 143–64. New York: Pergamon
56. Kooijman SALM. 1996. An alternative for NOEC exists, but the standard model has to be abandoned first. *Oikos* 75:310–16
57. Kooijman SALM, Metz JAJ. 1984. On the dynamics of chemically stressed populations: the deduction on population consequences from effects on individuals. *Ecotoxicol. Environ. Saf.* 8:254–74
58. Kramarz P, Laskowski R. 1997. Effect of zinc contamination on life history parameters of a ground beetle, *Poecilus cupreus*. *Bull. Environ. Contam. Toxicol.* 59:525–30
59. Krishnan M, Chockalingam S. 1989. Toxic and sublethal effects of endosulfan and carbaryl on growth and egg production of *Moina micrura* Kurz (Cladocera: Moinidae). *Environ. Pollut.* 56:319–26
60. Laskowski R. 1995. Some good reasons to ban the use of NOEC, LOEC, and related concepts in ecotoxicology. *Oikos* 73:140–44
61. Laskowski R. 1997. Estimating fitness costs of pollution in Iteroparous invertebrates. In *Ecological Principles for Risk Assessment of Contaminants in Soil*, ed. NM van Straalen, H Løkke, pp. 305–19. London: Chapman & Hall
62. Laskowski R, Hopkin S. 1996. Effect of Zn, Cu, Pb, and Cd on fitness in snails (*Helix aspersa*). *Ecotoxicol. Environ. Saf.* 34:59–69
63. Lawrence PO. 1981. Developmental and reproductive biologies of the parasitic wasp, *Biosteres longicaudatus*, reared on hosts treated with a chitin synthesis inhibitor. *Insect Sci. Appl.* 1:403–6
64. Levin L, Caswell H, Bridges T, Dibacco C, Cabrera D, Plaia G. 1996. Demographic responses of estuarine polychaetes to pollutants: life table response experiments. *Ecol. Appl.* 6:1295–313
65. Linke-Gamenick I, Forbes VE, Mendez N. 2000. Effects of chronic fluoranthene exposure on sibling species of *Capitella* with different development modes. *Mar. Ecol. Prog. Ser.* 203:191–203
66. Linke-Gamenick I, Forbes VE, Sibly RM. 1999. Density dependent effects of a toxicant on life history traits and population dynamics of a capitellid polchaete. *Mar. Ecol. Prog. Ser.* 184:139–48
67. Longstaff BC, Desmarchelier JM. 1983.

Effects of the temperature toxicity relationships of certain pesticides upon the population growth of *Sitophilus oryzae* Coleoptera: Curculionidae. *J. Stored Prod. Res.* 19:25–30
68. Losey JE, Denno RF. 1997. The escape response of pea aphids to foliar-foraging predators: factor affecting dropping behaviour. *Ecol. Entomol.* 23:53–61
69. Losey JE, Denno RF. 1998. Positive predator-predator interactions: enhanced predation rates and synergistic suppression of aphid populations. *Ecology* 79: 2143–52
70. Marshall JS. 1962. The effects of continuous gamma radiation on the intrinsic rate of natural increase on *Daphnia pulex*. *Ecology*. 43:598–607
71. Marshall JS. 1978. Population dynamics of *Daphnia galeata mendoae* as modified by chronic cadmium stress. *J. Fish. Res. Board Can.* 35:461–69
72. Marshall JS. 1978. Field verification of cadmium toxicity to laboratory *Daphnia* populations. *Bull. Environ. Contam. Toxicol.* 20:387–93
73. Deleted in proof
74. Martínez-Jerónimo F, Villasenor R, Espinosa F, Rios G. 1993. Use of life tables and application factors for evaluating chronic toxicity of kraft mill wastes on *Daphnia magna*. *Bull. Environ. Contam. Toxicol.* 50:377–84
75. Meyer JS, Ingersoll CG, McDonald LL. 1987. Sensitivity analysis of population growth rates estimated from Cladoceran chronic toxicity tests. *Environ. Toxicol. Chem.* 6:115–26
76. Nicholson AJ. 1954. Compensatory reactions of populations to stresses, and their evolutionary significance. *Aust. J. Zool.* 2:1–8
77. Paine RT. 1966. Food web complexity and species diversity. *Am. Nat.* 100:65–75
78. Rao TR, Sarma SS. 1986. Demographic parameters of *Brachionus patulus* Muller (Rotifera) exposed to sublethal DDT concentrations at low and high food levels. *Hydrobiologia* 139:193–200
79. Robertson JL, Preisler HK. 1992. *Pesticide Bioassays with Arthropods*. Boca Raton, FL: CRC
80. Robertson JL, Worner SP. 1990. Population toxicology: suggestions for laboratory bioassays to predict pesticide efficacy. *J. Econ. Entomol.* 83:8–12
81. Rumpf S, Frampton C, Dietrich DR. 1998. Effects of conventional insecticides and insect growth regulators on fecundity and other life-table parameters of *Micromus tasmaniae* (Neuroptera: Hemerobiidae). *J. Econ. Entomol.* 91:34–40
82. Schobben JHM, van Straalen NM. 1987. A model for the calculation of the intrinsic population growth rate for a stage-structured population in ecotoxicology: applied to *Platymothrus peltifer*. *Report* (in Dutch), Dep. Ecol. Ecotoxicol. Vrije Univ., Amsterdam, The Netherlands
83. Seitz A, Ratte HT. 1991. Aquatic ecotoxicology: on the problems of extrapolation from laboratory experiments with individuals and populations to community effects in the field. *Comp. Biochem. Physiol. C* 100:301–4
84. Sibly RM. 1999. Efficient experimental designs for studying stress and population density in animal populations. *Ecol. Appl.* 9:496–503
85. Sibly RM. 2000. How environmental stress affects density dependence and carrying capacity in a marine copepod. *J. Appl. Ecol.* 37:388–97
86. Sibly RM, Calow P. 1989. A life-cycle theory of responses to stress. *Biol. J. Linn. Soc.* 37:101–16
87. Slobodkin LB, Richman S. 1956. The effect of removal of fixed percentages of the newborn on size and variability in populations of *Daphnia pulicaria* (Forbes). *Limnol. Oceanogr.* 1:209–37
88. Snell TW, Moffat D. 1992. A 2-D life cycle test with the rotifer *Brachionus*

calyciflorus. *Environ. Toxicol. Chem.* 11: 1249–57
89. Solomon KR, Baker DB, Richards RP, Dixon KR, Klaine SJ, et al. 1996. Ecological risk assessment of atrazine in North American surface waters. *Environ. Toxicol. Chem.* 15:31–76
90. Stark JD. 2001. Population-level effects of the neem insecticide, Neemix on *Daphnia pulex*. *J. Environ. Sci. Health B* 36:457–65
91. Stark JD, Banken JAO. 1999. Importance of population structure at the time of toxicant exposure. *Ecotoxicol. Environ. Saf.* 42:282–87
92. Stark JD, Banks JE. 2001. Selective pesticides: Are they less hazardous to the environment? *BioScience* 51:980–82
93. Stark JD, Jepson, PC, Mayer D. 1995. Limitations to the use of topical toxicity data for predictions of pesticide side-effects in the field. *J. Econ. Entomol.* 88:1081–88
94. Stark JD, Rangus T. 1994. Lethal and sublethal effects of the neem insecticide, Margosan-O, on the pea aphid. *Pestic. Sci.* 41:155–60
95. Stark JD, Tanigoshi L, Bounfour M, Antonelli A. 1997. Reproductive potential: its influence on the susceptibility of a species to pesticides. *Ecotoxicol. Environ. Saf.* 37:273–79
96. Stark JD, Vargas RI, Messing RH, Purcell M. 1992. Effects of cyromazine and diazinon on three economically important Hawaiian tephritid fruit flies (Diptera: Tephritidae) and their endoparasitoids (Hymenoptera: Braconidae). *J. Econ. Entomol.* 85:1687–94
97. Stark JD, Wennergren U. 1995. Can population effects of pesticides be predicted from demographic toxicological studies? *J. Econ. Entomol.* 88:1089–96
98. Stark JD, Wong TTY, Vargas RI, Thalman RK. 1992. Survival, longevity, and reproduction of tephritid fruit fly parasitoids (Hymenoptera: Braconidae) reared from fruit flies exposed to azadirachtin. *J. Econ. Entomol.* 85:1125–29
99. Theiling KM. 1987. *The SELCTV database: the susceptibility of arthropod natural enemies of agricultural pests to pesticides*. MS thesis. Oregon State Univ., Corvallis. 170 pp.
100. Theiling KM, Croft BA. 1988. Pesticide side-effects on arthropod natural enemies: a database summary. *Agric. Ecosyst. Environ.* 21:191–218
101. Trevan JW. 1927. The error of determination of toxicity. *Proc. R. Soc. London Ser. B* 101:483–14
102. Van der Hoeven N. 1990. Effect of 3,4-dichloroaniline and metavanadate on *Daphnia* populations. *Ecotoxicol. Environ. Saf.* 20:53–70
103. Van Leeuwen CJ, Luttmer WJ, Griffioen PS. 1985. The use of cohorts and populations in chronic toxicity studies with *Daphnia magna*: a cadmium example. *Ecotoxicol. Environ. Saf.* 9:26–39
104. Van Leeuwen CJ, Moberts F, Niebeek G. 1985. Aquatic toxicological aspects of dithiocarbamates and related compounds. II. Effects on survival, reproduction and growth of *Daphnia magna*. *Aquat. Toxicol.* 7:165–75
105. Van Leeuwen CJ, Neibeek G, Rijkeboer M. 1987. Effects of chemical stress on the population dynamics of *Daphnia magna* : a comparison of two test procedures. *Ecotoxicol. Environ. Saf.* 14:1–11
106. Van Leeuwen CJ, Rijkeboer M, Niebeek G. 1986. Population dynamics of *Daphnia magna* as modified by chronic bromide stress. *Hydrobiologia* 133:277–85
107. Van Straalen NM, de Goede RGM. 1987. Productivity as a population performance index in life-cycle toxicity tests. *Water Sci. Technol.* 19:13–20
108. Van Straalen NM, Kammenga JE. 1998. Assessment of ecotoxicity at the population level using demographic parameters. In *Ecotoxicology*, ed. G Schüürmann, B Markert, pp. 621–44. New York: Wiley

109. Van Straalen NM, Schobben JHM, de Goede RGM. 1989. Population consequences of cadmium toxicity in soil microarthropods. *Ecotoxicol. Environ. Saf.* 17:190–204
110. Vinson SB. 1974. Effect of an insect growth regulator on two parasitoids developing from treated tobacco budworm larvae. *J. Econ. Entomol.* 67:335–36
111. Walthall WK, Stark JD. 1997. A comparison of acute mortality and population growth rate as endpoints of toxicological effects. *Ecotoxicol. Environ. Saf.* 37:45–52
112. Walthall WK, Stark JD. 1997. Comparison of two population-level ecotoxicological endpoints: the intrinsic (rm) and instantaneous (ri) rates of increase. *Environ. Toxicol. Chem.* 16:1068–73
113. Walthall WK, Stark JD. 1999. The acute and chronic toxicity of two xanthene dyes, fluorescein sodium salt and Phloxine B, to *Daphnia pulex* (Leydig). *Environ. Pollut.* 104:207–15
114. Walton WE, Camptom SM, Allan JD, Daniels JD. 1982. The effect of acid stress on survivorship and reproduction of *Daphnia pulex* (Crustacea: Cladocera). *Can. J. Zool.* 60:573–79
115. Wennergren U, Stark JD. 2000. Modeling long-term effects of pesticides on populations: beyond just counting dead animals. *Ecol. Appl.* 10:295–302
115a. Willis GH, McDowell Ll. 1987. Pesticide persistence on foliage. *Rev. Environ. Contam. Toxicol.* 100:23–73
116. Winner RW, Farrell MP. 1976. Acute and chronic toxicity of copper to four species of *Daphnia. J. Fish Res. Board Can.* 33:1685–91
117. Winner RW, Keeling T, Yeager R, Farrell MP. 1977. Effect of food type on the acute and chronic toxicity of copper to *Daphnia magna. Freshw. Biol.* 7:343–49
118. Wong CK. 1993. Effects of chromium, copper, nickel, and zinc on longevity and reproduction of the cladoceran *Moina macro copa. Bull. Environ. Contam. Toxicol.* 50:633–39
119. Wong CK, Wong PK. 1990. Life table evaluation of the effects of cadmium exposure on the freshwater cladoceran, *Moina macrocopa. Bull. Environ. Contam. Toxicol.* 44:135–41
120. Zeng F, Pederson G, Ellsbury M, Davis F. 1993. Demographic statistics for the pea aphid (Homoptera: Aphididae) on resistant and susceptible red clovers. *J. Econ. Entomol.* 86:1852–56

Annu. Rev. Entomol. 2003. 48:521–47
doi: 10.1146/annurev.ento.48.091801.112700

First published online as a Review in Advance on October 4, 2002

BELOWGROUND HERBIVORY BY INSECTS: Influence on Plants and Aboveground Herbivores

Bernd Blossey[1] and Tamaru R. Hunt-Joshi[2]
[1]Department of Natural Resources, Fernow Hall, Cornell University, Ithaca, New York 14853; e-mail: bb22@cornell.edu
[2]Department of Biology, PO Box 871501, Arizona State University, Tempe, Arizona 85287; e-mail: trh3@cornell.edu

Key Words root herbivory, plant-herbivore interaction, biological weed control, contramensalism, plant physiology

■ **Abstract** Investigations of plant-herbivore interactions continue to be popular; however, a bias neglecting root feeders may limit our ability to understand how herbivores shape plant life histories. Root feeders can cause dramatic plant population declines, often associated with secondary stress factors such as drought or grazing. These severe impacts resulted in substantial interest in root feeders as agricultural pests and increasingly as biological weed control agents, particularly in North America. Despite logistical difficulties, establishment rates in biocontrol programs are equal or exceed those of aboveground herbivores (67.2% for aboveground herbivores, 77.5% for belowground herbivores) and root feeders are more likely to contribute to control (53.7% versus 33.6%). Models predicting root feeders would be negatively affected by competitively superior aboveground herbivores may be limited to early successional habitats or generalist root feeders attacking annual plants. In later successional habitats, root feeders become more abundant and appear to be the more potent force in driving plant performance and plant community composition. Aboveground herbivores, even at high population levels, were unable to prevent buildup of root herbivore populations and the resulting population collapse of their host plants. Significant information gaps exist about the impact of root feeders on plant physiology and secondary chemistry and their importance in natural areas, particularly in the tropics.

CONTENTS

0066-4170/03/0107-0521$14.00

INTRODUCTION

Interactions of insect herbivores with plants and herbivores as components of terrestrial food webs have fascinated scientists for generations (18, 19, 21, 22, 30, 32, 65, 74, 128). However, a recent review (136) concluded that, despite an enormous accumulation of data, our ability to evaluate the role of herbivores in shaping natural selection or communities is handicapped by lack of long-term studies and our limited ability to determine spatial variation in herbivory. A systematic propensity to study aboveground herbivores may further limit our ability to develop more predictive models, and most reviews mention belowground herbivores only in passing (30, 32).

This knowledge bias does not reflect the importance of belowground plant tissues that contribute 50%–90% of plant biomass (5, 14, 112). Root herbivores may shape plant life histories, plant fitness, or plant communities, but inherent difficulties observing and manipulating them has limited enthusiasm of researchers (5, 20). The importance of root herbivores is generally acknowledged (30), and in 1990, Brown & Gange (20) ended a review by stating, "... it is likely that in the next decade, root feeders will loose their 'forgotten status' and be recognized as an important component of plant/herbivore systems." Although recognition has been growing, particularly in weed biocontrol programs, we are still faced with a considerable lack of data about root feeders and their impact on plants. In a review of induced herbivore resistance, only 2 of 115 studies investigated the response of plants to root herbivory (74). In several well-known plant systems (*Cucurbita* producing cucurbitacin; *Nicotiana* producing nicotine), concentrations of defensive chemicals in roots exceed those in stems and leaves (74). Although roots are responsible for nicotine synthesis, only folivore-plant interactions have been investigated (74). Even for such well-recognized root herbivores as the North American periodical cicadas (149), information about induced resistance is restricted to the plant's response to oviposition into twigs (74). The majority of the cicada life cycle (13–17 years) is spent feeding belowground; however, we currently have no information about interactions between cicada nymphs and root tissues (74).

Root herbivores can be of major importance in food webs (150); an introduced European weevil (*Barypeithes pallidus*) has become widespread in northeastern North America, and adults are the most important food item for red-backed salamanders (*Plethodon cinereus*) in the spring (J. Maerz, personal communication). However, how *B. pallidus* larval feeding may affect mature trees, seedlings, or forest regeneration is unknown. Emergence of occasionally large populations of root herbivores constitutes a significant energy flux from below- to aboveground pools. In prairie grasslands in Kansas, annual bulk N precipitation ranged from 6–12 kg/ha/yr while total N input (including N_2 fixation) ranged from 11–25 kg/ha/yr. The emergence of several (annual) cicada species represented a redistribution of 16%–36% (about 4 kg/ha/yr) of the overall annual input (23).

Presence and impact of root herbivores often go unnoticed until plant populations collapse. Such impacts have been observed for conifers attacked by hepialid moths, *Korscheltellus gracilis* (83, 145, 148); vineyards attacked by grape root borer, *Vitaceae polistiformis* (3); bush lupine, *Lupinus arboreus*, attacked by ghost moth, *Hepialus californicus* (138); various examples in weed biological control such as the suppression of tansy ragwort, *Senecio jacobaea*, by the flea beetle, *Longitarsus jacobaeae* (95); and the effects of scarabaeid larvae on crops, pastures, and lawns (4, 17, 27, 101, 116, 123). Interest in root-feeding insects has increased since the last comprehensive reviews (5, 21, 103), albeit slowly. We focus our review on recent studies, i.e., those published within the past 10–15 years. Much background on root herbivores (5, 21), Japanese beetles (115), and periodical cicadas (149) will not be repeated here. Although we acknowledge the importance of mammals, nematodes, or pathogens as root feeders (5, 21, 103), we restrict our review to root-feeding insects and exclude other fossorial organisms considered in previous reviews (for example root-feeding nematodes and mammals). Our specific objectives were to review (*a*) our knowledge base (i.e., how many plant-herbivore systems have been investigated and what type of experimental designs were applied), (*b*) the representation of different habitat types and climatic regions, (*c*) how different plant life history forms (annual, biennial, perennial, grass, shrub, or tree) are represented and how their response to root herbivory may differ, (*d*) the effects of root herbivory on plants including physiology and secondary chemistry, and (*e*) the effects of root herbivory on other organisms.

THE DATABASE

We used Biosis and Agricola for initial literature surveys but discovered that entire groups of root-feeding organisms (such as crickets, cicadas, termites, and many agricultural pests) were not represented. We amended our searches by using taxonomic groups and reviewed references in published studies for additional sources. We also queried researchers in biological control to supplement our results. We limit our treatment to insects feeding on living plant tissues, thus insects feeding on dead or decaying wood or other dead plant structures are not incorporated. We defined root feeders as insects feeding on belowground plant tissues during some part of their life cycle (usually as nymphs or larvae) including species feeding on root crowns. We reviewed a total of 241 papers (follow the Supplemental Material link in the online version of this chapter or at http://www.annualreviews.org/) and selected 189 that provided information about interaction of root feeders with their host plants for further analyses. We did not include papers containing only listings or descriptions of root-feeding herbivores but retained those that considered the overall impact of root-feeding herbivores on plant communities (such as grasslands), even when they did not refer to a particular species. Many papers are collections of various experiments or use several root herbivores. Not all papers could be grouped according to our predetermined categories, thus for several analyses, the total number of studies or species may vary substantially.

The database is extremely biased toward temperate ecosystems; of 174 studies that could be grouped according to climate, only 15 (8.6%) explored tropical ecosystems (all agricultural). Future investigations are needed to assess whether this simply reflects lack of research in the tropics or whether root-feeding insects are more important or abundant in temperate ecosystems. Overall, we found information about 70 different plant species and their interaction with 178 different root herbivores. Motivations to investigate root feeders could be grouped in three major categories [basic ecological investigations (9 plants, 19 herbivores), root feeders as agricultural pests (28 plants, 71 herbivores), and root feeders as potential biological control agents (33 plants, 88 herbivores)]. Lack of basic ecological studies provides further evidence that ecologists continue to ignore belowground herbivores. The magnitude in our lack of understanding becomes clear if we consider that in North America alone, some 17,000 native plant species and at least 5000 introduced and naturalized plant species occur (102). Both agricultural and weed biocontrol scientists cannot ignore root herbivores because their work is goal oriented: controlling root pests in agriculture and finding the most potent herbivore to suppress invasive plants in weed biocontrol. Logistical difficulties in the study of root herbivores are no excuse to neglect important insect pests or to ignore potentially important biological control agents. While the most serious pests in agriculture have not changed in past decades (the literature continues to be dominated by studies on *Diabrotica* spp.), weed biocontrol programs provide new and important information about root feeders and their impact on their host plants.

The vast majority of studies (41%) target root feeders of plants found in North America (Table 1). European root feeders are comparatively well known, and this pattern is entirely driven by interest in potential biocontrol agents for invasive plants of European origin (Table 1). With the exception of few studies in Australia and New Zealand, we know virtually nothing about root feeders in natural areas of the southern hemisphere (Table 1). Among habitat types investigated,

TABLE 1 Geographic origin of root herbivore species studied in ecology, agriculture, or weed biological control (multiple entries possible)

			Biological control	
	Ecology	Agriculture	Target area	Native area
Africa	0	4	4	0
Asia	0	9	0	2
Australia (includes New Zealand, Pacific Islands, and Hawaii)	1	10	20	0
Europe	5	8	0	78
South America	0	0	0	5
Central America	4	1	0	2
North America (United States and Canada)	9	44	57	1

row crops (33.8%) and grasslands, including rangeland and turf (49.2%), dominate the database followed by forest and shrub habitat (12%), wetlands (2.7%), and orchards (2.2%). Interactions of root feeders with herbaceous perennial plants dominate (37%) followed by annual plants (31.9%), herbaceous biennials (7.5%), grasses (8.9%), trees (11.7%), and shrubs (2.8%). Studies were of variable length and ranged from days (5.5%) to months (35.6%) to years (58.9%); however, short-term studies (<5 years) were dominating. The majority of the studies (60%) involved external root feeders and mining or chewing herbivores (79.4%). Information about sucking herbivores (19.6%) is limited and we know virtually nothing about belowground gall makers (1%). Among the studies, 39.4% evaluated the impact of external attack of the water and nutrient adsorption system, i.e., fine roots of plants, 36.2% evaluated attack of taproots or rootstocks, with the remaining studies investigating herbivores in root crowns and rhizomes.

The majority of studies (76.3%) used experimental methods to establish treatments while the remaining were categorized as descriptive. Evaluating natural herbivory was the most common approach (52%) while exclusion of insect herbivores using insecticide (15.5%) and release of biological control agents (20%) were about equally common. Mowing (2%) was often part of investigations about impacts of turf-feeding grubs and was used to simulate ungulate herbivory, caging (6%) and artificial herbivory (4.5%) were occasionally attempted. Over 50% of the studies included field experiments; however, a significant portion investigated impact of root herbivores in greenhouses/growth chambers (28.6%) and the remaining (17.2%) in a common garden using potted plants. Measures to assess the impact of root herbivores on plant performance included damage to roots (60 studies) and changes in biomass (54 studies), reproduction (44 studies), and survival (45 studies), with many studies typically measuring multiple variables. Other measures included biomass of leaves and stems (27 and 23 studies, respectively), plant chemistry (20 studies), and interaction with other organisms (19 studies). We know little about the impact of root herbivores on plant physiology (10 studies), plant population dynamics (14 studies), or plant architecture (2 studies).

ROOT HERBIVORES AS PESTS

Impacts of root herbivores constitute a continuum from potentially positive to occasionally disastrous mortality. Under low attack by *Agapeta zoegana*, *Centaurea maculosa* plants showed increased flowering compared to control plants (104); in the greenhouse Kentucky blue and tall fescue provided with additional N produced more biomass if attacked by grubs than unattacked control plants (33). Low herbivore populations may not affect plant performance at all, and herbivore attack can be considered neutral (29, 33, 66); however, disastrous mass mortality may also occur (138). Communication is often made more difficult because of differences in terminology. Agricultural scientists define resistance to include antixenosis, antibiosis, and tolerance (129). Antixenosis represents plant traits conferring nonpreference of herbivores, i.e., reduced acceptance for oviposition or feeding, while antibiosis represents plant traits that adversely affect herbivore performance or survival (129). Tolerance, the ability to withstand herbivore attack, is incorporated under resistance and contrasted with susceptibility (129). This terminology has its roots in the definition of resistance put forth by Painter (114) as "the relative amount of heritable qualities possessed by the plant that influences the ultimate degree of damage done by the insect. In practical agriculture, resistance represents the ability of a certain variety to produce a larger crop of good quality than do ordinary varieties at the same level of infestation." Painter (114) also defined tolerance as "a basis of resistance in which the plant shows an ability to grow and reproduce itself or to repair injury to a marked degree in spite of supporting a population approximately equal to that damaging a susceptible host." As Berenbaum & Zangerl (11) already pointed out, the measure of plant damage may represent misleading information on resistance because tolerant plants may sustain much larger herbivore damage and yet produce equal amounts of biomass or yield. The concepts of antixenosis and antibiosis are almost entirely absent from ecology textbooks (30, 32, 74), and ecologists separate resistance (plant traits reducing herbivore populations) and tolerance (traits that allow plants to maintain performance even under sustained herbivore attack). An interesting perspective and discussion of these differences has been presented previously (2, 11, 78). In this chapter we adopt a definition of resistance as involving active or passive (antibiosis or antixenosis) mechanisms to reduce herbivore loads on plants, while tolerance is the ability to maintain plant performance under sustained herbivore attack.

Investigations on impact of corn rootworm (*Diabrotica virgifera*) on maize provide valuable data about interaction of root herbivore and plant because studies were conducted under a range of environmental conditions using various genotypes (40, 48, 126, 127, 134). The focus of these studies is the attempt to minimize insecticide input while maximizing yield. Corn rootworm damage reduced silk production and grain yields, and tolerance of damage was influenced by water and nitrogen availability and plant density (133, 134). Maize can tolerate relatively high levels of corn rootworm damage unless attacked plants were also water or nitrogen stressed (64, 127). Genotype-specific variations in root size, resistance to "lodging" (125), or the ability to regrow adventitious roots above the site of injury can affect tolerance of maize to root damage (126). Compensatory regrowth of root tissues was more pronounced under moderate than severe root damage, suggesting existence of a damage threshold for maize root herbivory. Maize varieties with higher levels of defensive chemicals (hydroxamic acid) in the roots had much greater resistance to corn rootworm (7, 8).

Root herbivores of *Brassica* crops (oil seed rape, kale, swede, canola) affect plant performance, commercial yields, and defensive chemistries. *Delia* flies damaging *Brassica* spp. roots cause significant reductions in yield, flowering, seed production, and leaf, stem, and root biomass (24, 94). Other crops such as coconut, sweet potato, groundnuts, and grapes also had significantly depressed yields when damaged by various root herbivores (1, 17, 69, 113), although amaranth was not affected (142). Root herbivory by the mealybug *Rhizoecus kondonis* caused chlorosis and stunting in alfalfa (49). Similarly, the clover root curculio, *Sitona hispidulus*, reduced alfalfa cover, survival, and height by 23.2%–38.9%, along with significant increases in weed biomass. Other fodder crops such as lucerne (*Medicago sativa*) were also negatively affected by *Sitona discoideus*, reducing yields by up to 30% (52).

Consistently, crop tolerance to root herbivory is reported to depend upon interactions with other stress factors, particularly drought. After mowing, lucerne attacked by *S. discoideus* often showed signs of severe drought stress and became abruptly dormant (51). Crop yields in sweet potato, lucerne, pigeonpea, and groundnuts were depressed when root damage was combined with water stress (6, 69), disease (17, 68, 69, 72), or both (120). Maize and turf grasses suffered greater damage by root herbivores when water and nutrients were limiting (33–35, 64, 116, 119, 127, 133). Tolerance to corn rootworm partly depended on phenological development of attacked plants (48, 64, 126), with older plants less susceptible to damage. Competition with weedy species increased stress and reduced yield of cassava plants infested with scarabaeid larvae (27). Tolerance to root damage can vary considerably among crop varieties and among grass species; moderate root damage led to overcompensation in some grass species; however, severe infestations inevitably reduced foliage biomass and aesthetic quality of turf grasses (33–35, 70, 108–110, 116, 118, 119, 122, 140).

In perennial crops, older plants often experience heavier damage probably because of increased length of exposure to herbivores (55). Plantations and orchards are often severely affected by root-feeding herbivores that reduce survival of saplings and trees and yield of orchard trees (75, 76, 82, 99, 143, 144, 148). Periodical cicadas, with their long life cycles, are often abundant in orchards and plantations and can do considerable damage (84). Large populations of scarabaeidae larvae that damaged tree roots were maintained in fir plantations by perennial grasses (primary larval food) interspersed between trees (75, 76).

EFFECT OF ROOT FEEDERS ON PLANT PHYSIOLOGY

Herbivore damage to roots can affect water and nutrient absorption, carbohydrate storage, synthesis of plant hormones, and production of secondary chemicals (54, 80). Our current knowledge was obtained examining impact of chewing herbivores on plant physiology. Chewing herbivores can be particularly damaging because they consume plant organs involved in water and nutrient adsorption or resources stored belowground. Attack can sever vascular connections between roots and shoots, causing increased water stress and disruption of resource flows. Alfalfa and lucerne plants attacked by *S. discoideus* went into dormancy and displayed symptoms typical of drought stress (106) after the first hay cut. Scarab damage to *Capsella bursa-pastoris* resulted in water stress, even when well watered (43). Similarly, damage to *Sonchus oleraceus* roots by chafer beetles resulted in a lower water content and higher nitrogen concentration in leaves, symptoms typical of water stress [(88) but see (89)]. However, mining by larvae of the weevil *Hylobius transversovittatus* inside rootstocks of a wetland perennial (purple loosestrife, *Lythrum salicaria*) did not elicit a water stress response (67). Instead, leaves of infested plants showed a consistently higher rate of carbon assimilation, perhaps triggered by changes in plant source–sink ratios. Herbivore tissue damage can greatly alter this ratio by instigating translocation of assimilates toward sink tissues. Because starch accumulation in leaves inhibits photosynthesis, increased assimilate translocation can reduce foliar starch levels below a threshold and result in increased net photosynthetic rates (111).

Although in general effects of root herbivores on plant physiology are poorly known, effects of corn rootworm (*Diabrotica* spp.) feeding on maize roots and root fly (*Delia* spp.) feeding on *Brassica* crops have been studied in unusual detail. Early larval attack by *D. virgifera* reduced CO_2 assimilation of maize and plant growth (48, 64, 126). However, there was no impact on CO_2 assimilation in maize six days after the end of the larval feeding period (when there was maximum root damage), although stomatal conductance and root system size were still reduced (48, 64, 126). This response was also dependent upon herbivore intensity, suggesting that a combination of damage severity and compensatory plant growth is important in mediating shoot growth and CO_2 assimilation response to root herbivory (124, 126). Corn rootworm herbivory did not affect stomatal conductance or net photosynthetic rates in well-watered maize; however, photosynthesis was decreased up to 23% and stomatal conductance was significantly reduced in water-stressed plants (40).

Root herbivores can have a major impact on sugar and carbohydrate storage of plants. With increasing infestation by the white grub, *Ligyrus subtropicus*, on sugarcane, foliar N, P, and K concentrations decreased, while Ca and Mg were unaffected (26). Attack may reduce sugar yield by as much as 39% through severing of the root system (26). Larval feeding by *Delia radicum* reduced total sugar content of swede, kale, and rape, but effects were host plant and genotype specific and varied from sugar to sugar (fructose, glucose, or sucrose) (63). Glucose and fructose concentrations were low or unaffected by *D. radicum* damage to rape and kale but were consistently reduced in swede, and sucrose levels were either reduced or unaffected in all crops. Larval feeding also reduced root biomass by up to 47% within three weeks of attack (63). The probability for successful larval development of *D. radicum* was not affected by sugar content of the root; however, pupal weights were positively correlated with root mass and root sugar concentrations (63). Similar results were obtained when turnip root fly, *D. floralis*, attacked swede, and host plant genotype strongly influenced the response of plants (62). In addition, *D. floralis* damage to swede, rape, and kale increased lignin content and detergent fiber content (61).

Carbohydrate contents of roots can also be greatly reduced by two chrysomelid species feeding on purple loosestrife leaves (77). Preliminary results (63) indicate that carbohydrate reserves in purple loosestrife can be depleted by the root herbivore *H. transversovittatus*. However, herbivore feeding on fine roots of annual species (such as chafer larvae) had no effect on root carbohydrate levels (90). Further investigations are needed to assess whether interspecific interactions between above- and belowground herbivores involve changes in carbohydrate storage levels. Belowground herbivores could be particularly affected if the quality and quantity of their food are reduced by aboveground herbivores.

Roots play an important part in the defensive chemistry of plants (74). Species in the Brassicaceae produce glucosinolates, defensive chemicals with anti-herbivore properties, in the epidermis and vascular cambium of roots and show higher concentrations in belowground than in aboveground tissues (54). Although rarely investigated, root herbivores can induce plants to produce defensive secondary chemicals, similar to well-known examples of induced responses to foliar herbivory (74). Vine weevil damage to spinach roots induced a 30% increase in 20-hydroxyecdysone (20-E) in root and shoot tissues (131). This compound, a phyto-ecdysteroid, can disrupt insect development and acts as a feeding deterrent to herbivores (25, 141). Citrus roots damaged by *Diaprepes abbreviatus* were induced to produce chitinases; these compounds can affect the peritrophic membrane of insect midguts and can facilitate the entry of pathogenic microbes into the hemolymph (92). In maize, strains that increased levels of hydroxamic acid in response to corn rootworm feeding developed improved resistance, produced more biomass, and had better yield than strains with lower levels of hydroxamic acid (7, 8).

An interesting aspect on how root herbivores might influence plant community dynamics and successional processes is through increased "loss" of nitrogen and carbon by attacked plants. Recent evidence suggests that herbivore feeding above- and belowground may not only affect biological activity and decomposition processes in the soil through litter and fecal inputs but also through root exudates (38). Pulse labeling experiments using $^{14}C\text{-}CO_2$ and measuring microbial activity in the soil demonstrated that above- and belowground herbivores cause carbon "leaching" of roots (41). Although these sophisticated experiments have so far only used nematodes as belowground herbivores (38), Murray & Hatch (107) demonstrated nitrogen exudation from clover root nodules attacked by *Sitona* larvae. Similar results can be expected from other root herbivores. However, these relationships are not necessarily linear, i.e., higher "sepage rates" as damage to roots increases. Highest C-flux (measured as microbial activity) occurred in low herbivory treatments, while damage above a threshold had either no or a negative effect on microbial soil activity (38). These results are strong indications for plant-initiated processes (aboveground herbivores do not cause root damage) not simply exudate from damaged plant cells. Root exudate may affect competitive hierarchies in plant communities. Attacked plants, such as clover, exude N from root nodules, which was taken up immediately by *Lolium perenne* grass roots (107), a nonhost for *Sitona* larvae. Herbivory not only transferred nutrients to a competing species, it also directly reduced the biomass of the attacked plant. Wheat seedlings planted together with clover infested with *Sitona lepidus* larvae obtained 7% of its nitrogen from clover, from either increased tissue decay or direct exudation from roots (106). Root exudates may also play a role in attracting natural enemies of root herbivores. When *Otiorhynchus sulcatus* larvae attacked roots of *Thuja occidentalis*, released chemicals attracted entomopathogenic nematodes to damaged plants (147).

Comparing Natural and Artificial Root Herbivory

In their quest to understand how herbivory affects plant performance, researchers working with aboveground herbivores have frequently used artificial damage to simulate herbivory, despite a growing body of evidence showing artificial and natural herbivory eliciting different plant responses (9). Certainly the difficulty of establishing and controlling root herbivory levels may drive investigators to choose artificial over natural root herbivory treatments. However, artificial herbivore damage to roots tends to be rather crude and rarely mimics insect herbivore damage. We found only nine studies that employed artificial root herbivore treatments, suggesting that even artificial root herbivory is an unsatisfactory method. Most importantly, there is ample evidence that artificial root herbivory elicits vastly different plant responses (biomass production, physiological and chemical changes, attraction of natural enemies) than natural root herbivory.

In experiments exploring the impact of corn rootworm (*Diabrotica* spp.) larvae on maize, mechanical root damage greatly limited the ability of roots to supply water to aerial parts of the plant (124), resulting in decreased sap flow, reduced shoot biomass production, and reduced plant height (46). Larval rootworm feeding damage was far less damaging to the water transport system of maize (124) and did not (even with 100 larvae per plant) decrease sap flow, biomass production, or plant height (46). Root volume, CO_2 assimilation, leaf stomatal conductance, and volume of adventitious roots were significantly different in treatments where maize plants were damaged using artificial herbivory compared to treatments using natural herbivory by *Diabrotica* sp. (126). Pruning of *Medicago sativa* root nodules using either scalpels or *Sitona hispidula* larvae resulted in overcompensation (shoot biomass, root biomass, and nodule biomass) 18 days post treatment with insect herbivory, eliciting a far stronger response than artificial herbivory (118). Turnip root fly (*D. floralis*) attack increased glucosinulate content of rape roots, whereas mechanical damage actually decreased it (53). Larval feeding in roots also changed the composition of glucosinulates by increasing indole glucosinulates, whereas no such changes were observed in roots that were mechanically damaged (53). However, in this study, both natural and artificial herbivory treatments increased foliar glucobrassicin contents. Similarly, levels of 20-hydroxyecdysone (an insect molting hormone synthesized in roots) in spinach (*Spinacia oleracea*) were different among plants exposed to artificial or natural herbivory by *O. sulcatus* (132).

Attack by aboveground-feeding herbivores may result in volatile emissions that attract natural enemies (79) or repel oviposition on already-attacked plants by foraging herbivores (36). We now have evidence for similar effects in response to belowground herbivore feeding. In experiments testing the attraction of

T. occidentalis roots damaged by *O. sulcatus* larvae, entomopathogenic nematodes clearly preferred roots with weevil damage even if the weevil larva was absent (147). Nematodes responded to damage over a distance of 7 cm; however, mechanical damage could not elicit the same response (147).

ROOT FEEDERS AS BIOLOGICAL WEED CONTROL AGENTS

In the past 100 years, 88 releases of 49 different root-feeding insect herbivores targeting 19 different invasive plant species have been made in 10 countries worldwide, and almost two thirds of all releases occurred in North America (United States, 32 releases; Canada, 27; Australia, 16; New Zealand, 5; six other countries, 1). Initially, root feeders were "below the radar screen" of weed biocontrol practitioners and scientists (only four releases occurred from 1902 to 1960) (Table 2). Root feeders were not even considered in scoring systems to evaluate potential biological control agents published in the mid-1980s (50). The number of releases rose sharply from 5 in the 1960s, to 8 in the 1970s, to 12 in the 1980s, and to 21 in the 1990s. However, the number of control programs using aboveground

TABLE 2 Release, establishment, and contribution to control of root feeders and aboveground herbivores in weed biocontrol programs in 20-year intervals. (Data represent summary of results/country)

	Root feeders[a] (*N* = 49 species)				**Above ground herbivores**[a] (*N* = 288 species)			
	Established		**Not established**		**Established**		**Not established**	
Year	N	%	N	%	N	%	N	%
Before 1900	—	—	—	—	2	66.7	1	33.3
1901–1920	0	0	1	100	48	73.8	17	26.2
1921–1940	2	50	2	50	58	61.7	36	38.3
1941–1960	—	—	—	—	45	57	34	43
1961–1980	25	70.2	11	29.8	185	66.3	94	33.7
1981–2002	34	89.5	4	10.5	136	73.5	49	26.5
Total[b]	61	77.5	18	22.5	474	67.2	231	32.8
Number of species contributing to control[c]	23	53.7			192	33.6		

[a]Data from (71), updated for root feeders.

[b]Excludes unknowns for aboveground herbivores ($N = 76$).

[c]Excludes root feeders released after 1996 ($N = 8$); and cases with unknown outcome for aboveground herbivores ($N = 109$ of 681). Data for aboveground herbivores from (31).

biocontrol agents continues to be almost 10 times as high (Table 2). Despite difficulties working with root feeders and generally low numbers available for release (59), establishment rates of root feeders are at least equal to those of aboveground herbivores (Table 2). Several of the obvious failures can be attributed to herbivore–host plant incompatibilities. Five of 11 species that failed to establish are sesiid moths in the genus *Chamaesphecia*, released against North American *Euphorbia* spp., where tests demonstrated the inability of these moths to utilize North American leafy spurge genotypes (146). Even more impressive is the contribution of root feeders to control of their invasive host plants. For this analysis we relied on data provided by Julien & Griffiths (71) and additional queries of biocontrol practitioners. As already discussed by McFadyen (98), for most species we have no quantitative information about their ability to suppress their host plant. We have relied on expert opinion and have included cases where a species may contribute to the control of its host plant but not provide substantial or complete control. However, this still provides useful information in comparing above- and belowground herbivores as biocontrol agents. Of 49 released species, 23 contribute to the control of their host plant. If we remove eight species, where insufficient time has elapsed since their release (released after 1996) to assess contribution to the control of their host plant from the analysis, over 50% of the released root-feeding biocontrol agents contribute to the suppression of invasive plant populations (Table 2). Only about one third of the aboveground biocontrol agents contribute to the suppression of their host plant (Table 2). Most of the successful root feeding control agents are Coleoptera, particularly Curculionidae and Chrysomelidae (Table 3), while Lepidoptera and Diptera appear difficult to work with and even after establishment many do not contribute to the suppression of their host plants (Table 3).

Overall, the biocontrol scientists we questioned agreed that logistical difficulties have prevented a more intensive focus on root feeders as biocontrol agents. However, root feeders may show greater diet restriction than folivores (15, 16, 44), making them an even better choice in biocontrol programs. Although increased logistical difficulties may be associated with host-specificity tests of root feeders, programs may gain increased specificity of control agents. Considerable financial and scientific energy is spent on researching and releasing numerous aboveground biocontrol agents; these resources may be better spent improving the ability to work with root herbivores. Significantly higher overall establishment rates and success of root-feeding herbivores in suppressing invasive plant populations demonstrate the need for increased emphasis on belowground herbivores in weed biocontrol programs.

While many weed biocontrol scientists have embraced working with root feeders, the interesting and important ecological question why root feeders (and in particular beetles) may be more successful biocontrol agents remains. Predation and parasitism rank high among reasons cited for failure in biological control (31, 81). Specialized natural enemies active in the native range will be eliminated before release of any biocontrol agent (above- or belowground feeder). The success in

TABLE 3 Number of root herbivores in Coleoptera, Lepidoptera, and Diptera released, established, and contributing to suppression of invasive plant species in weed biocontrol programs

Order/Family	Number of species released	Number of species established	Number of species contributing to control
Coleoptera	**31**	**26**	**20**
Chrysomelidae	12	9	8
Curculionidae	14	13	9
Cerambycidae	3	2	1
Buprestidae	2	2	2
Lepidoptera	**15**	**9**	**2**
Cosmopterygidae	2	2	0
Gelechiidae	1	1	0
Sesiidae	7	2	1
Tortricidae	2	2	0
Cochylidae	1	1	1
Pterophoridae	1	0	0
Noctuidae	1	1	0
Diptera	**3**	**1**	**0**
Syrphidae	2	1	1
Anthomyiidae	1	0	0

establishment and control potential of root feeders may well be a function of their safety from generalist predation and parasitism (56, 57). Root feeders, by virtue of their feeding niche, occupy a safe refuge from insect parasitism (56, 57). The relative safety of root feeders from generalist predators may make them ideally suited to build up effectively without interference by natural enemies in their introduced range. Parasitoids will be unable to suppress their herbivorous host populations if herbivores occupy sufficiently large refuges (like roots) (58). That safety is a function of feeding niche is supported by low numbers of parasitoids reared in biocontrol investigations from root feeders. An interesting situation occurs for two root feeders attacking scentless chamomile, *Tripleurospermum perforatum*. Here, two weevils have slightly different life histories as early instars with one feeding longer aboveground in the stem or leaf, whereas the other feeds entirely belowground. The species whose larvae are longer exposed aboveground are attacked by a eulophid wasp (20%–70% attack rate), while the second species escapes this attack except for an occasional encounter (2%–5%) (60). However, although root feeders may escape aerial predators and parasitoids, they are by no means safe and encounter potent entomopathogenic nematodes that may greatly reduce populations of root feeders and safeguard plants (13, 137–139). Further detailed and long-term studies are needed to assess the role of root feeders in shaping plant communities, including weed biocontrol programs, and how natural enemies may shape interactions of root feeders and their host plants.

INTERACTION OF ABOVEGROUND AND BELOWGROUND HERBIVORES

Herbivores often cause extensive damage to their host plants, and it is common that multiple herbivores attack the same plant. Interspecific competition among insects exploiting a common resource can greatly affect herbivore populations. A review of interspecific interaction of herbivorous insects (37) found that competition was frequent, often asymmetric, and particularly intense among sap feeders, wood and stem borers, and seed and fruit feeders, while mandibulate herbivores appeared less affected. Overall, of 193 pairwise interactions, only 18% did not report competitive effects, although the literature may be biased in favor of reporting competition versus reporting no effects (37). The majority of interactions involved competition of aboveground herbivores; only five studies involved investigations of interactions of spatially separated (above- and belowground) herbivores, and in all cases root feeders were adversely affected by folivores (37). Summarizing the, albeit limited, evidence about interactions of root feeders and folivores in the early 1990s, Masters et al. (90) proposed a conceptual model for interaction of root and foliar herbivores. According to this model, root feeding improves performance and fitness of folivores, whereas folivore feeding has a negative impact on root feeders. The conceptual model described root feeding as a stress that removes water and nutrient uptake tissues, creating an effect similar to drought stress, which results in increased concentrations of soluble amino acids and carbohydrates in foliage (90). These increased nutrient levels in plant tissues become available to aboveground herbivores and result in improved performance (90). Foliar herbivory, in turn, should reduce plant growth belowground, effectively limiting quality and quantity of belowground tissues, thereby decreasing growth and performance of root herbivores (90). Masters et al. (90) conclude that if indeed root quality and quantity and thus root herbivore populations are driven by aboveground herbivores, implications for plant community processes could be considerable.

A number of recent studies allows us to examine the accumulated evidence and validity of the model (90). Although spatially or temporally separated herbivores may strongly interact and compete with each other via shared host plants (37), we are particularly interested in the long-term effects of such interactions on plant individuals (survival, reproduction) and plant communities and whether root feeders are inferior competitors (37, 90). Weed biocontrol scientists are particularly interested in such interactions because of a recent surge in root feeders as biocontrol agents. If aboveground herbivores constitute superior competitors and suppress root feeder populations, emphasis on introducing a number of species attacking different plant parts, including root feeders, may be misguided and a waste of valuable resources.

Masters et al. (90) base their original model on studies using externally feeding herbivores (chafer, scarabaeid, and tipulid larvae) removing fine roots of annual plants (*Chenopodium album*, *Capsella bursa-pastoris*, *Sonchus oleraceus*), although they do not explicitly limit their model to interactions of generalist root feeders attacking annual plants. The ability of folivores to suppress growth and populations of root herbivores has been supported by a number of studies. Leaf-feeding aphids strongly suppressed populations and size of root-feeding aphids on *C. album* (100); aphids on *C. bursa-pastoris* suppressed growth of chafer larvae (43); leaf mining flies on *S. oleraceus* decreased growth of chafer larvae; increasing herbivore load aboveground further decreased performance of the belowground herbivore (87); and shoot-feeding aphids *Aphis fabae fabae* limited population buildup of root-feeding aphids *Pemphigous populitransversus* on seedlings of the perennial *Cardamine pratensis* (130). In field experiments, suppression of generalist root herbivores (scarabaeid and tipulid larvae) using insecticide was associated with lower populations of a tephritid fly *Terellia ruficauda* and its hymenopteran parasitoids on *Cirsium palustre* (91) and with lower aphid populations of *Myzus persicae* on *S. oleraceus* (88) compared to plants exposed to belowground root herbivory, although populations of belowground herbivores in these treatments were not determined. Root-feeding herbivores increased teneral weight of *M. persicae* and appeared to increase aphid fecundity (88) and pupal weight of mining flies (87) on *S. oleraceus*. Although root feeding did not affect development time of *A. fabae*, it increased fecundity, through weight gain, and longevity (43). However, no positive effect of root herbivory by *P. populitransversus* could be detected for the folivore *A. fabae* on *C. pratensis* (130). The leaf mining fly, *Chromatomyia syngenesiae*, on the annual *S. oleraceus* had higher pupal weight (a measure of fitness) when the garden chafer larvae also attacked the host plant. However, contrary to models of water stress and improved nutritional quality of aboveground plant structures as a result of root herbivory, foliar nitrogen was not increased (89).

Supporting evidence for the proposed model (90) is generally derived from early successional annual plant species (42, 87–89, 100), often using late instar generalist root feeders, and aphids or leaf miners as folivores, in experimental treatments that last only a few weeks (43, 87–89). Field experiments were conducted in early successional stages, often with plants colonizing bare ground (105, 121). While the proposed patterns of interactions (90) appear robust for annual plants, providing important information about early successional habitats, these results cannot be generalized to long-lived plants or later successional communities. Resource availability for root feeders increases as bare ground and annual species are replaced by long-lived plants that, at least in temperate climates, move substantial amounts of resources belowground. Not surprisingly, root herbivore populations increase with successional age of plant communities (21), and interactions of below- and aboveground herbivores in these "mature" communities need urgent attention.

The tephritid fly *Eurosta comma* induces galls on the rhizomes (but not roots) of several goldenrod species, including *Solidago missouriensis* (117). Galled clones produced fewer new rhizomes and allocated less mass to leaves and stems; however, attack by the folivore *Trirhabda canadensis* did not affect rhizome production or populations of *E. comma* until defoliation reached 100% (117). Ramets attacked by the root herbivore withstood or recovered better from defoliation than ungalled ramets. However, with increasing defoliation by the leaf beetle, overall gall biomass and individual gall mass were reduced, suggesting some negative effects of foliar herbivory. The Masters et al. model (90) would predict increased N in leaves of galled plants, yet foliar N levels were not significantly different between galled and ungalled clones and were unaffected by grazing intensity (117). However, ungalled rhizomes had higher N concentrations than galled ramets.

Another well-studied example of above- and belowground herbivory in a later successional community involves bush lupine, *Lupinus arboreus*, and its associated herbivores in California grasslands. The most important herbivore driving population dynamics of the system is the root-feeding ghost moth, *Hepialus californicus* (138). Although there is some evidence for the ability of an aboveground herbivore, the tussock moth, *Orgyia vetusta*, to locally and temporarily suppress populations of belowground herbivores, tussock moth attack did not prevent population increases of *H. californicus*, which caused catastrophic population crashes of bush lupine (138). Experimentally excluding belowground herbivores for three years did not affect populations and attack rates of folivores (85). Also, contrary to expectations of the Masters et al. model (90), mining by the weevil *H. transversovittatus* in rootstocks of *L. salicaria* had no effect on development or population levels of a folivore (*Galerucella calmariensis*) in large field cages (67). In addition, carbon and nitrogen concentrations in purple loosestrife leaves and stems were unaltered by *H. transversovittatus* attack (67). Intense leaf beetle damage by *G. calmariensis* reduced survival of *Hylobius* larvae mining in the rootstocks in potted plant experiments (67). However, when interactions between *Galerucella* and *Hylobius* were examined in a four-year field experiment, leaf herbivory did not affect root herbivores, not even in a year when outbreaking *Galerucella* populations defoliated *L. salicaria* (67).

Data contradicting a major prediction (foliar herbivory will reduce populations and performance of root herbivores) of Masters et al. (90) are accumulating. Simulated aboveground herbivory (by ungulates) removing 30%–70% of aboveground biomass by clipping of the perennial bunchgrass *Muhlenbergia quadridentata* did not affect chafer larval growth or survival over a six-month period (101). Instead, root feeders decreased aboveground biomass by 45%, suggesting that root herbivores can reduce resource availability for aboveground herbivores (101). Similarly, in shortgrass prairies grazed by cattle, populations of white grubs, particularly *Phyllophaga fimbripes*, were higher in grazed pastures (27). Grub feeding occasionally resulted in increased plant mortality in grazed pastures (ungrazed and slightly grazed pastures did not differ in belowground resource availability). Population explosions of Scarabaeidae, *Phyllophaga* spp., on mowed pastures were attributed to oviposition preferences of females (75). This preference was not associated with a fitness reduction or reduced survival of larvae (75). Together, these examples provide substantial evidence for lack of negative impacts of folivores on root feeders, thus contradicting predictions of Masters et al. (90). In fact, many of the examples appear to suggest that root feeders are superior competitors in later successional communities. Additional support for this finding comes from weed biocontrol programs (71), in particular the program targeting tansy ragwort,

Senecio jacobaeae. Two herbivores, the folivore cinnabar moth, *Tyria jacobaeae*, and the root herbivore, the flea beetle, *Longitarsus jacobaeae*, were introduced to North America in the 1950s and 1960s. After six years and releases of similar numbers of adults at the Cascade Head Scenic Recreation Area in Oregon (95), both species reached outbreak densities and the host plant population crashed to less than 1% of the original abundance (96). There was no inhibitory effect of the folivore on buildup of *L. jacobaeae* that occurred at peak densities and maximum population size of *T. jacobaeae* (defoliation of >90% of flowering stems creating intense resource competition). *L. jacobaeae* now limits populations of *T. jacobaeae* by maintaining the host plant at low abundance (95–97).

Root herbivores have comparatively long larval feeding periods (20), become more common in later successional habitats (18), and are slow in colonizing new host populations (3, 12, 86, 93, 138). Short-term experiments, even those that encompass an entire generation of an aboveground herbivore, may not capture interactions as they play out under field conditions. Attack by early instars of root herbivores can result in completely different plant responses than feeding by mature larvae (126). In addition, folivores likely encounter host plants already attacked by root herbivores, particularly in long-lived plants, rather than simultaneously as in experimental treatments. The sequence of herbivore colonization may have important consequences on how plants respond to tissue loss. The rather chronic nature of root herbivory versus the acute short-term attack of folivores will have important implications on how plants respond to herbivory and makes experimental studies difficult (85). Short-term experiments cannot substitute for long-term field investigations, and similar to accumulation of chronic herbivore impact over consecutive seasons (22, 39, 128), interactions of above- and belowground herbivores may have a significant temporal component. While competition for common resources of folivores and root feeders may be mediated through changes in their common host plant, data from later successional plant communities suggest that root feeders may be competitively superior and the driving force in above- and belowground plant-herbivore interactions. If future studies confirm these initial results, the importance of root feeders in shaping succession and composition of plant communities may greatly exceed the importance of aboveground herbivores.

SUMMARY

Plants are embedded in a network of interactions among other plants, above- and belowground resource availability, and above- and belowground herbivores. All these components will interact and in concert (not in isolation) determine the success and failure of plants to grow and reproduce and their role within plant communities. For further improvement in our understanding of factors shaping communities, root feeders need increased attention in ecological studies across many different habitats and climate zones. Andersen (5) suggests that due to high seasonality outside the tropics, plants use roots or other belowground organs to store assimilate. Lack of high-quality resources would certainly explain a lower abundance of root herbivores at lower latitudes, and herbivorous nematodes appear to be more common in temperate regions (5).

An important consideration for every researcher is the ability for observation, measurement, or experimental manipulation of a system. Working with root-feeding insects has a number of unique challenges. The ability to observe is often limited to the aerial life stages (adults and sometimes eggs), while direct observation of root-feeding larvae is nearly impossible without major disturbance or destruction. The often long life cycles of many root feeders require that treatments be maintained over extended periods of time (occasionally years) to accurately measure herbivore impact (12, 60, 85, 105, 112). In addition, experience has shown that establishing treatments or working with high larval densities is difficult owing to increased larval mortality, particularly in early instars (10, 28, 41, 45, 73, 112, 136). Often anticipated treatment levels were not achieved (67, 112); however, in working with older larvae, establishment rates can be exceptionally high (26, 47), often above 90%, and recovery rates are good. However, care has to be applied in the interpretation of results if only older larvae have been used because often the impact of small, medium, and large larvae over the duration of their development will be different. Herbivore impact is also likely to accumulate over time, and assessing frequency and magnitude of herbivory is important to understand the overall impact of root herbivory (39). Our review demonstrates that useful information has recently been forthcoming from weed biocontrol programs. The availability of root feeders in many weed biocontrol programs presents an excellent opportunity for in-depth and long-term investigations.

The impact of herbivore feeding on factors of interest to the investigator is best described by a damage function using a range of herbivore densities (30). Measures of plant performance such as growth, biomass allocation, reproductive effort, or yield as in many agricultural studies are likely to respond differently across a gradient of herbivory and tissue loss. The shape of the damage function reveals information about important factors such as compensation, tolerance, or damage thresholds. Unfortunately, the development of damage functions has not been forthcoming in studies of belowground insect herbivores. We have significant evidence for damage thresholds in plants attacked by root herbivores (29, 33, 34, 52, 66), with attack often going unnoticed (below the threshold) until plant populations collapse. Such widespread mortality is often associated with additional stress factors exposing impact of chronic root herbivory. Spotted knapweed (*Centaurea maculosa*) roots attacked by larvae of the moth *Agapeta zoegana* showed reduced root length (5 cm in attacked versus 30 cm in unattacked plants) (135). Attacked plants have reduced access to water below the surface area, making them more vulnerable to drought conditions that may lead to a sudden population collapse.

We found little support for the model by Masters et al. (90) that root feeders are inferior competitors if their host plant is simultaneously attacked by a folivore. With the exception of early successional communities and externally feeding larvae on annual plants, root feeders control abundance of and access to the host plant. This was particularly obvious in weed biocontrol programs where root feeders increased in abundance and their attack subsequently resulted in population crashes of their host plant despite simultaneous attack by folivores (95, 96).

This review has examined the recently accumulated research involving insect root herbivores. Although the pace has picked up, root feeders continue to be neglected in studies examining plant-herbivore interactions and the influence of root herbivory on plant community composition. Although the statement that the "forgotten status" of root feeders is gone (20) was too optimistic, root-feeding insects have become more popular, particularly in weed biocontrol programs. We anticipate that the next decade will see another increase in research targeting root feeders and their interactions with plants and aboveground herbivores. Root herbivores continue to represent significant opportunities for new discoveries in how terrestrial communities are structured.

ACKNOWLEDGMENTS

We thank our colleagues in weed biocontrol for their input in updating our references, in particular Simon Fowler (New Zealand), Rich Hansen, Eric Coombs, Mark Schwarzlaender (Unites States), Rosemarie DeClerck-Floate (Canada), John Hoffman, Costas Zachariades (South Africa), Andy Sheppard, David Briese, Tim Heard, Mic Julien, and Rachel McFadyen (Australia). Jessica McKenney helped locate papers and compile the bibliography, and two anonymous reviewers provided valuable advice on an earlier version of this manuscript.

The *Annual Review of Entomology* is online at http://ento.annualreviews.org

LITERATURE CITED

1. Abraham VA, Mohandas N. 1988. Chemical control of the white grub *Leucopholis coneophora* Burm., a pest of the coconut plant. *Trop. Agric.* 65:355–57
2. Alexander HM. 1992. Evolution of disease resistance in natural plant populations. See Ref. 41a, pp. 326–44
3. Alm SR, Williams RN, Pavuk DM, Snow WJ, Heinlein MA. 1989. Distribution and seasonal flight activity of male grape root borers (Lepidoptera: Sesiidae) in Ohio. *J. Econ. Entomol.* 82:1604–8
4. Ambethgar V, Lakshmanan V, Dinkakaran D, Selarvarajan M. 1999. Mycosis of cashew stem and root borer, *Plocaederus ferringineus* L. (Coleoptera: Cerambicidae) by *Metarhizium anisopliae* (Metsch.) Sorokin (Deuteromycotina: Moniliales) from Tamil Nadu (India). *Crop Sci. J. Entomol. Res.* 23:81–83
5. Andersen DC. 1987. Below-ground herbivory in natural communities: a review emphasizing fossorial animals. *Q. Rev. Biol.* 62:261–86
6. Anioke SC. 1996. Effect of time of planting and harvesting of sweet potato *Ipomea batatas* (L.) Lam. on yield and insect damage in South-Eastern Nigeria. *Entomon* 21:137–41
7. Assabgui RA, Annason JT, Hamilton RI. 1995. Field evaluations of hydroxamic acids as antibiosis factors in elite maize inbreds to the western corn rootworm (Coleoptera: Chrysomelidae). *J. Econ. Entomol.* 88:1482–93
8. Assabgui RA, Hamilton RI, Arnason JT.

1995. Hydroxamic acid content and plant development of maize (*Zea mays* L.) in relation to damage by the western corn rootworm, *Diabrotica virgifera virgifera* LeConte. *Can. J. Plant Sci.* 75:851–56
9. Baldwin IT. 1990. Herbivory simulations in ecological research. *Trends Ecol. Evol.* 5:91–93
10. Beavers JB. 1982. Biology of *Diaprepes abbreviatus* (Coleoptera: Curculionidae) reared on an artificial diet. *Fla. Entomol.* 65:263–69
11. Berenbaum MR, Zangerl AR. 1992. Quantification of chemical coevolution. See Ref. 41a, pp. 69–87
12. Blossey B. 1993. Herbivory below ground and biological weed control: life history of a root-boring weevil on purple loosestrife. *Oecologia* 94:380–87
13. Blossey B, Ehlers R-U. 1991. Entomopathogenic nematodes (*Heterorhabditis* spp. and *Steinernema anomali*) as potential antagonists of the biological weed control agent *Hylobius transversovittatus* (Coleoptera: Curculionidae). *J. Insect Pathol.* 58:453–54
14. Blossey B, Schat M. 1997. Performance of *Galerucella calmariensis* (Coleoptera: Chrysomelidae) on different North American populations of purple loosestrife. *Environ. Entomol.* 26:439–45
15. Blossey B, Schroeder D, Hight SD, Malecki RA. 1994. Host specificity and environmental impact of the weevil *Hylobius transversovittatus,* a biological control agent of purple loosestrife (*Lythrum salicaria*). *Weed Sci.* 42:128–33
16. Blossey B, Schroeder D, Hight SD, Malecki RA. 1994. Host specificity and environmental impact of two leaf beetles (*Galerucella calmariensis* and *G. pusilla*) for biological control of purple loosestrife (*Lythrum salicaria*). *Weed Sci.* 42:134–40
17. Braun AR, van der Fliert E. 1999. Evaluation of the impact of sweetpotato weevil (*Cylas formicarius*) and of the effectiveness of *Cylas* sex pheromone traps at the farm level in Indonesia. *Int. J. Pest. Manage.* 45:101–10
18. Brown VK, Gange AC. 1989. Differential effects of above- and below-ground insect herbivory during early plant succession. *Oikos* 54:67–76
19. Brown VK, Gange AC. 1989. Herbivory by soil-dwelling insects depresses plant species richness. *Funct. Ecol.* 3:667–71
20. Brown VK, Gange AC. 1990. Insect herbivory below ground. *Adv. Ecol. Res.* 20:1–58
21. Brown VK, Gange AC. 1992. Secondary plant succession: How is it modified by insect herbivory? *Vegetatio* 101:3–13
22. Cain ML, Carson WP, Root RB. 1991. Long-term suppression of insect herbivores increases the production and growth of *Solidago altissima* rhizomes. *Oecologia* 88:251–57
23. Callahan MA, While MR, Meyer CK, Brock BL, Charlton RE. 2000. Feeding ecology and emergence production of annual cicadas (Homoptera: Cicadidae) in tallgrass prairie. *Oecologia* 123:535–42
24. Campbell LG, Anderson AW, Dregseth R, Smith LJ. 1998. Association between sugarbeet root yield and sugarbeet root maggot (Diptera: Otitidae) damage. *J. Econ. Entomol.* 91:522–27
25. Champs F. 1991. Plant ectdysteroids and their interactions with insects. In *Ecological Chemistry and Biochemistry of Plant Terpenoids*, ed. JB Harborene, FA Tomas-Barberan, pp. 331–76. Oxford: Clarendon
26. Coale FJ, Cherry RH. 1989. Impact of white grub (*Ligyrus subtropicus* (Blatchley)) infestation on sugarcane nutrition. *J. Plant Nutr.* 12:1351–59
27. Coffin DP, Laycock WA, Lauenroth WK. 1998. Disturbance intensity and above- and belowground herbivore effects on long-term (14 y) recovery of a semiarid grassland. *Plant Ecol.* 139:221–23
28. Cordo HA, DeLoach CJ, Ferrer R, Briano J. 1995. Bionomics of *Carmenta haematica* (Ureta) (Lepidoptera: Sesiidae) which

attacks snakeweeds (*Gutierrezia* spp.) in Argentina. *Biol. Control* 5:11–24
29. Cormier D, Martel P. 1997. Effects of soil insecticide treatments on northern corn rootworm, *Diabrotica barberi* (Coleoptera: Chrysomelidae), populations and on corn yield. *Phytoprotection* 78:67–73
30. Crawley MJ. 1989. Insect herbivores and plant population dynamics. *Annu. Rev. Entomol.* 34:531–64
31. Crawley MJ. 1989. The successes and failures of weed biocontrol using insects. *Biocontrol News Inf.* 10:213–23
32. Crawley MJ. 1997. *Plant Ecology*. Oxford: Blackwell Sci. 717 pp. 2nd ed.
33. Crutchfield BA, Potter DA. 1995. Damage relationships of Japanese beetle and southern masked chafer (Coleoptera: Scarabaeidae) grubs in cool-season turfgrasses. *J. Econ. Entomol.* 88:1049–56
34. Crutchfield BA, Potter DA. 1995. Tolerance of cool-season turfgrasses to feeding by Japanese beetle and southern masked chafer (Coleoptera: Scarabaeidae). *J. Econ. Entomol.* 88:1380–87
35. Crutchfield BA, Potter DA, Powell AJ. 1995. Irrigation and nitrogen fertilization effects on white grub injury to Kentucky bluegrass and tall fescue turfgrass. *Crop Sci.* 35:1122–26
36. De Moraes CM, Mescher MC, Tumlinson JH. 2001. Caterpillar-induced nocturnal plant volatiles repel conspecific females. *Nature* 410:577–80
37. Denno RF, McClure MS, Ott JR. 1995. Interspecific interactions in phytophagous insects: competition reexamined and resurrected. *Annu. Rev. Entomol.* 40:297–331
38. Denton CS, Bardgett RD, Cook R, Hobbs PJ. 1999. Low amounts of root herbivory positively influence the rhizosphere microbial community in a temperate grassland soil. *Soil Biol. Biochem.* 31:155–65
39. Doak DF. 1992. Lifetime impacts of herbivory for a perennial plant. *Ecology* 73:2086–99
40. Dunn JP, Frommelt K. 1998. Effects of below-ground herbivory by *Diabrotica virgifera virgifera* (Coleoptera: Chrysomelidae) and soil moisture on leaf gas exchange of maize. *J. Appl. Entomol.* 122: 179–83
41. Friedli J, Bacher S. 2001. Mutualistic interaction between a weevil and a rust fungus, two parasites of the weed *Cirsium arvense*. *Oecologia* 129:571–76
41a. Fritz RS, Simms EL, eds. 1992. *Plant Resistance to Herbivores and Pathogens: Ecology, Evolution, and Genetics*. Chicago: Univ. Chicago Press
42. Ganade G, Brown VK. 1997. Effects of below-ground insects, mycorrhizal fungi and soil fertility on the establishment of *Vicia* in grassland communities. *Oecologia* 109:374–81
43. Gange AC, Brown VK. 1989. Effects of root herbivory by an insect on a foliar-feeding species, mediated through changes in the host plant. *Oecologia* 81: 38–42
44. Gassman A, Schroeder D. 1995. The search for effective biological control agents in Europe: history and lessons from leafy spurge (*Euphorbia esula* L.) and cypress spurge (*Euphorbia cyparissia* L.). *Biol. Control* 5:466–77
45. Gassman A, Tosevski I. 1994. Biology and host specificity of *Chamaesphecia hungarica* and *C. astatiformis* (Lep.: Sesiidae) two candidates for the biological control of leafy spurge (*Euphorbia esula* (Euphorbiaceae)) in North America. *Entomophaga* 39:237–45
46. Gavloski JE, Whitfield GH, Ellis CR. 1992. Effect of larvae of western corn rootworm (Coleoptera: Chrysomelidae) and of mechanical root pruning on sap flow and growth of corn. *J. Econ. Entomol.* 85:1434–41
47. Gaynor DL, Lane GA, Biggs DR, Sutherland ORW. 1986. Measurement of grass grub resistance of bean in a controlled environment. *N. Z. J. Exp. Agric.* 14:77–82
48. Godfrey LD, Meinke LJ, Wright RJ. 1993. Vegetative and reproductive biomass accumulation in field corn: response to

root injury by western corn rootworm (Coleoptera: Chrysomelidae). *J. Econ. Entomol.* 86:1557–73
49. Godfrey LD, Pickel C. 1998. Seasonal dynamics and management schemes for a subterranean mealybug, *Rhizoecus kondonis* Kuwana, pest of alfalfa. *Southw. Entomol.* 23:343–50
50. Goeden RD. 1983. Critique and revision of Harris' scoring system for selection of insect agents in biological control of weeds. *Prot. Ecol.* 5:287–301
51. Goldson SL, Bourdot GW, Proffit JR. 1987. A study of the effects of *Sitona discoideus* (Coleoptera: Curculionidae) larval feeding on the growth and development of lucerne (*Medicago sativa*). *J. Appl. Ecol.* 24:153–61
52. Goldson SL, Dyson CB, Proffitt JR, Frampton ER, Logan JA. 1985. The effect of *Sitona discoideus* Gyllenhal (Coleoptera: Curculionidae) on lucerne yields in New Zealand. *Bull. Entomol. Res.* 75:429–42
53. Griffiths DW, Birch ANE, MacFarlane-Smith WH. 1994. Induced changes in the indole glucosinolate content of oilseed and forage rape (*Brassica napus*) plants in response to either turnip root fly (*Delia floralis*) larval feeding or artificial root damage. *J. Sci. Food. Agric.* 65:171–78
54. Hara M, Fujii Y, Sasada Y, Kuboi T. 2000. cDNA cloning of radish (*Raphanus sativus*) Myrosinase and tissue-specific expression in root. *Plant Cell Physiol.* 41:1102–9
55. Harris JP, Smith BJ, Olien WC. 1994. Activity of grape root borer (Lepidoptera: Sesiidae) in southern Mississippi. *J. Econ. Entomol.* 87:1058–61
56. Hawkins BA. 1988. Species diversity in the third and fourth trophic levels: patterns and mechanisms. *J. Anim. Ecol.* 57:137–62
57. Hawkins BA. 1990. Global patterns of parasitoid assemblage size. *J. Anim. Ecol.* 59:57–72
58. Hawkins BA, Thomas MB, Hochberg ME. 1993. Refuge theory and biological control. *Science* 262:1429–31
59. Hight SD, Blossey B, Laing J, DeClerck-Floate R. 1995. Establishment of biological control agents from Europe against *Lythrum salicaria* in North America. *Environ. Entomol.* 24:967–77
60. Hinz HL, Müller-Schärer H. 2000. Suitability of two root-mining weevils for the biological control of scentless chamomile, *Tripleurospermum perforatum*, with special regard to potential non-target effects. *Bull. Entomol. Res.* 90:497–508
61. Hopkins RJ, Birch ANE, Griffiths DW, Morrison IM, McKinlay RG. 1995. Changes in the dry matter, sugar, plant fibre and lignin contents of swede, rape and kale roots in response to turnip root fly (*Delia floralis*) larval damage. *J. Sci. Food. Agric.* 69:321–28
62. Hopkins RJ, Griffiths DW, Birch ANE, McKinlay RG, Hall JE. 1993. Relationships between turnip root fly (*Delia floralis*) larval development and the sugar content of swede (*Brassica napus* ssp. *rapifera*) roots. *Ann. Appl. Biol.* 122:405–15
63. Hopkins RJ, Griffiths DW, McKinlay RG, Birch ANE. 1999. The relationship between cabbage root fly (*Delia radicum*) larval feeding and the freeze-dried matter and sugar content of *Brassica* roots. *Entomol. Exp. Appl.* 92:109–17
64. Hou X, Meinke LJ, Arkebauer TJ. 1997. Soil moisture and larval western corn rootworm injury: influence on gas exchange parameters in corn. *Agric. J.* 89:709–17
65. Huffaker CB, Kennet CE. 1959. A ten-year study of vegetational changes associated with biological control of klamath weed. *J. Range Manage.* 12:69–82
66. Hunt DWA, Lintereur G, Raffa KF. 1992. Rearing method for *Hylobius radicis* and *H. pales* (Coleoptera: Curculionidae). *J. Econ. Entomol.* 85:1873–77
67. Hunt TR. 2002. *Influence of leaf and root herbivory on purple loosestrife and the*

response of the surrounding wetland community. PhD thesis. Cornell Univ. 176 pp.
68. Jin X, Morton J, Butler L. 1992. Interactions between *Fusarium avenaceum* and *Hylastinus obscurus* (Coleoptera: Scolytidae) and their influence on root decline in red clover. *J. Econ. Entomol.* 85:1340–46
69. Johnson RA, Lamb RW, Wood TG. 1981. Termite damage and crop loss studies in Nigeria—a survey of damage to groundnuts. *Trop. Pest Manage.* 27:325–42
70. Jones A, Schalk JM, Dukes PD. 1987. Control of soil insect injury by resistance in sweet potato. *J. Am. Soc. Hortic. Sci.* 112:195–97
71. Julien MH, Griffiths MW. 1998. *Biological Control of Weeds. A World Catalogue of Agents and Their Target Weeds.* Wallingford/Oxford: CABI Int. 223 pp. 4th ed.
72. Kalb DW, Bergstrom GC, Shields EJ. 1994. Prevalence, severity and association of fungal crown and root rots with injury by the clover root curculio in New York alfalfa. *Am. Phytopathol. Soc.* 78:491–95
73. Karban R. 1985. Addition of periodical cicada nymphs to an oak forest: effects on cicada density, acorn production, and rootlet density. *J. Kans. Entomol. Soc.* 58:269–76
74. Karban R, Baldwin IT. 1997. *Induced Responses to Herbivory*. Chicago: Univ. Chicago Press. 319 pp.
75. Kard BMR, Hain FP. 1987. White grub (Coleoptera: Scarabaeidae) densities, weed control practices, and root damage to fraser fir Christmas trees in the Southern Appalachians. *J. Econ. Entomol.* 80:1072–75
76. Kard BMR, Hain FP. 1988. Influence of ground covers on white grub (Coleoptera: Scarabaeidae) populations and their feeding damage to roots of fraser fir christmas trees in the Southern Appalachians. *Environ. Entomol.* 17:63–66
77. Katovitch EJS, Becker RL, Ragsdale DW. 1999. Effect of *Galerucella* spp. on survival of purple loosestrife (*Lythrum salicaria*) roots and crowns. *Weed Sci.* 47:360–65
78. Kennedy GG, Barbour JD. 1992. Resistance variation in natural and managed systems. See Ref. 41a, pp. 13–41
79. Kessler A, Baldwin IT. 2001. Defensive function of herbivore-induced plant volatile emission in nature. *Science* 291: 2141–44
80. Larcher W. 1995. *Plant Physiological Ecology*. Berlin: Springer. 506 pp. 3rd ed.
81. Lawton JH. 1990. Biological control of plants: a review of generalisations, rules and principles using insects as agents. In *Alternatives to Chemical Controls of Weeds*, ed. C Basset, LJ Whitehouse, JA Zabkiewicz, pp. 3–17. Rotorua: New Zealand FRI Bull. 155
82. Leather SR, Day KR, Salisbury AN. 1999. The biology and ecology of the large pine weevil, *Hylobius abietis* (Coleoptera: Curculionidae): a problem of dispersal? *Bull. Entomol. Res.* 89:3–16
83. Leonard JG, Tobi DR, Parker BL. 1991. Spatial and temporal distribution of *Korscheltellus gracilis* larvae (Lepidoptera: Hepialidae) in the Green Mountains, Vermont. *Environ. Entomol.* 20:371–76
84. Lloyd M, White J. 1987. Xylem feeding by periodical cicada nymphs on pine and grass roots, with novel suggestions for pest control in conifer plantations and orchards. *Ohio J. Sci.* 87:50–54
85. Maron JL. 1998. Insect herbivory above- and belowground: individual and joint effects on plant fitness. *Ecology* 79:1281–93
86. Maron JL. 2001. Intraspecific competition and subterranean herbivory: individual and interactive effects on bush lupine. *Oikos* 92:178–86
87. Masters GJ. 1995. The effect of herbivore density on host plant mediated interactions between two insects. *Ecol. Res.* 10:125–33
88. Masters GJ. 1995. The impact of root herbivory on aphid performance: field and laboratory evidence. *Acta Oecol.* 16:135–42

89. Masters GJ, Brown VK. 1992. Plant-mediated interactions between two spatially separated insects. *Funct. Ecol.* 6:175–79
90. Masters GJ, Brown VK, Gange AC. 1993. Plant mediated interactions between above- and below-ground insect herbivores. *Oikos* 66:148–51
91. Masters GJ, Jones TH, Rogers M. 2001. Host-plant mediated effects of root herbivory on insect seed predators and their parasitoids. *Oecologia* 127:246–50
92. Mayer RT, Shapiro JP, Berdis E, Hearn CJ, McCollum TG. 1995. Citrus rootstock responses to herbivory by larvae of the sugarcane rootstock borer weevil (*Diaprepes abbreviatus*). *Physiol. Plant.* 94:164–73
93. McCabe TL, Wagner DL. 1989. The biology of *Sthenopis auratu* (Grote) (Lepidoptera: Hepialidae). *J. N. Y. Entomol. Soc.* 97:1–10
94. McDonald RS, Sears MK. 1991. Effects of root damage by cabbage maggot, *Delia radicum* (L.) (Diptera: Anthomyiidae), on yield of canola, *Brassica campestris* L., under laboratory conditions. *Can. Entomol.* 123:861–67
95. McEvoy P, Cox C, Coombs E. 1991. Successful biological control of ragwort, *Senecio jacobaea*, by introduced insects in Oregon. *Ecol. Appl.* 1:430–42
96. McEvoy PB. 1985. Depression in ragwort (*Senecio jacobaea*) abundance following introduction of *Tyria jacobaeae* and *Longitarsus jacobaeae* on the central coast of Oregon. *Proc. 6th Int. Symp. Biol. Control Weeds*, ed. ES Delfosse, pp. 57–64. Vancouver: Agric. Can.
97. McEvoy PB, Rudd NT, Cox CS, Huso M. 1993. Disturbance, competition and herbivory effects on ragwort *Senecio jacobaea* populations. *Ecol. Monogr.* 63:55–75
98. McFadyen REC. 1998. Biological control of weeds. *Annu. Rev. Entomol.* 43:369–93
99. Meshram PB, Pathak SC, Jamaluddin A. 1990. Effect of some soil insecticides in controlling the major insect pests in teak nursery. *India For.* 116:206–13
100. Moran NA, Whitham TG. 1990. Interspecific competition between root-feeding and leaf-galling aphids mediated by host-plant resistance. *Ecology* 71:1050–58
101. Morón-Rios A, Dirzo R, Jaramillo VJ. 1997. Defoliation and below-ground herbivory in the grass *Muhlenbergia quadridentata*: effects on plant performance and on the root-feeder *Phyllophaga* sp. (Coleoptera, Melolonthidae). *Oecologia* 110:237–42
102. Morse LE, Kartesz JT, Kutner LS. 1995. Native vascular plants. In *Our Living Resources: A Report to the Nation on the Distribution, Abundance, and Health of U. S. Plants, Animals, and Ecosystems*, ed. ET LaRoe, GS Farris, CE Puckett, PD Doran, MJ Mac, pp. 205–9. Washington, DC: U.S. Dep. Inter., Natl. Biol. Serv.
103. Mortimer SR, van der Putten WH, Brown VK. 1999. Insect and nematode herbivory under ground: interactions and role in vegetation succession. In *Herbivores: Between Plants and Predators*, ed. H Olff, VK Brown, RH Drent, pp. 205–38. Oxford: Blackwell Sci.
104. Müller H, Steinger T. 1990. Separate and joint effects of root herbivores, plant competition and nitrogen shortage on resource allocation and components of reproduction in *Centaurea maculosa* (Compositae). *Symp. Biol. Hung.* 39:215–24
105. Müller-Schärer H, Brown VK. 1995. Direct and indirect effects of above- and below-ground insect herbivory on plant density and performance of *Tripleurospermum perforatum* during early plant succession. *Oikos* 72:36–41
106. Murray PJ, Clements RO. 1998. Transfer of nitrogen between clover and wheat: effect of root herbivory. *Eur. J. Soil Biol.* 34:25–30
107. Murray PJ, Hatch DJ. 1994. *Sitona* weevils (Coleoptera: Curculionidae) as agents for rapid transfer of nitrogen from white clover (*Trifolium repens* L.) to perennial

ryegrass (*Lolium perenne* L.). *Ann. Appl. Biol.* 125:29–33

108. N'Guessan FK, Quisenberry SS, Croughan TP. 1994. Evaluation of rice anther culture lines for tolerance to the rice water weevil (Coleoptera: Curculionidae). *Environ. Entomol.* 23:331–36
109. N'Guessan FK, Quisenberry SS, Croughan TP. 1994. Evaluation of rice tissue culture lines for resistance to the rice water weevil (Coleoptera: Curculionidae). *J. Econ. Entomol.* 87:504–13
110. N'Guessan FK, Quisenberry SS, Thompson RA, Lindscome SD. 1994. Assessment of Louisiana rice breeding lines for tolerance to the rice water weevil (Coleoptera: Curculionidae). *J. Econ. Entomol.* 87:476–81
111. Neales TF, Incoll LD. 1968. The role of leaf photosynthesis rate by the level of assimilate concentration in the leaf: a review of the hypothesis. *Bot. Rev.* 34:107–25
112. Nötzold R, Blossey B, Newton E. 1998. The influence of belowground herbivory and plant competition on growth and biomass allocation of purple loosestrife. *Oecologia* 113:82–93
113. Olien WC, Smith BJ, Hegwood J, Patrick C. 1993. Grape root borer: a review of the life cycle and strategies for integrated control. *Hortic. Sci.* 28:1154–56
114. Painter RH. 1951. *Insect Resistance in Crop Plants.* New York: Macmillan. 520 pp.
115. Potter DA, Held DW. 2002. Biology and management of the Japanese beetle. *Annu. Rev. Entomol.* 47:175–205
116. Potter DA, Patterson CG, Redmond CT. 1992. Influence of turfgrass species and tall fescue endophyte on feeding ecology of Japanese beetle and southern masked chafer grubs (Coleoptera: Scarabaeidae). *J. Econ. Entomol.* 85:900–9
117. Preus LE, Morrow PA. 1999. Direct and indirect effects of two herbivore species on resource allocation in their changed host plant: the rhizome galler *Eurosta comma*, the folivore *Trirhabda canadensis* and *Solidago missouriensis*. *Oecologia* 119:219–26
118. Quinn MA, Hall MH. 1992. Compensatory response of a legume root-nodule system to nodule herbivory by *Sitona hispidulus*. *Entomol. Exp. Appl.* 64:167–76
119. Ramsell J, Malloch AJC, Whittaker JB. 1993. When grazed by *Tipula paludosa, Lolium perenne* is a stronger competitor of *Rumex obtusifolius*. *J. Ecol.* 81:777–86
120. Reddy MV, Yule DF, Reddy VR, Georges PJ. 1992. Attack on pigeonpea (*Cajanus cajan* (L.) Millsp.) by *Odontotermes obesus* (Rambur) and *Microtermes obesi* Holmgren (Isoptera: Microtermitinae). *Trop. Pest Manage.* 38:239–40
121. Rees M, Brown VK. 1992. Interactions between invertebrate herbivores and plant competition. *J. Ecol.* 80:353–60
122. Reilly CC, Gentry CR, McVay JR. 1987. Biochemical evidence for resistance of rootstocks to the peachtree borer and species separation of peachtree borer and lesser peachtree borer (Lepidoptera: Sesiidae) on peach trees. *J. Econ. Entomol.* 80:338–43
123. Rice ME. 1994. Damage assessment of the annual white grub, *Cyclocephala lirida* (Coleoptera: Scarabaeidae), in corn and soybean. *J. Econ. Entomol.* 87:220–22
124. Riedell WE. 1990. Rootworm and mechanical damage effects on root morphology and water relations in maize. *Crop Sci.* 30:628–31
125. Riedell WE, Evenson PD. 1993. Rootworm feeding tolerance in single-cross maize hybrids from different eras. *Crop Sci.* 33:951–55
126. Riedell WE, Reese RN. 1999. Maize morphology and shoot CO_2 assimilation after root damage by western corn rootworm larvae. *Crop Sci.* 39:1332–40
127. Riedell WE, Schumacher TE, Evenson PD. 1996. Nitrogen fertilizer management

to improve crop tolerance to corn rootworm larval feeding damage. *Agric. J.* 88:27–32

128. Root RB. 1996. Herbivore pressure on goldenrods (*Solidago altissima*): its variation and cumulative effects. *Ecology* 77:1074–87
129. Ruuth P. 1988. Resistance of cruciferous crops to turnip root fly. *J. Agric. Sci. Fin.* 60:269–79
130. Salt DT, Fenwick P, Whittaker JB. 1996. Interspecific herbivore interactions in a high CO_2 environment: root and shoot aphids feeding on *Cardamine. Oikos* 77: 326–30
131. Schmelz EA, Grebenok RJ, Galbraith DW, Bowers WS. 1998. Damage-induced accumulation of phytoecdysteroids in spinach: a rapid root response involving the octadecanoic acid pathway. *J. Chem. Ecol.* 24:339–60
132. Schmelz EA, Grebenok RJ, Galbraith DW, Bowers WS. 1999. Insect-induced synthesis of phytoecdysteroids in spinach, *Spinacia oleracea. J. Chem. Ecol.* 25:1739–57
133. Spike BP, Tollefson JJ. 1989. Relationship of plant phenology to corn yield loss resulting from western corn rootworm (Coleoptera: Chrysomelidae) larval injury, nitrogen deficiency and high plant density. *J. Econ. Entomol.* 82:226–31
134. Spike BP, Tollefson JJ. 1989. Relationship of root ratings, root size, and root regrowth to yield of corn injured by western corn rootworm (Coleoptera: Chrysomelidae). *J. Econ. Entomol.* 82:1760–63
135. Story JM, Good WR, White LJ, Smith L. 2000. Effects of the interaction of the biocontrol agent *Agapeta zoegana* L. (Lepidoptera: Cochylidae) and grass competition on spotted knapweed. *Biol. Control* 17:182–90
136. Stoyer TL, Kok LT. 1986. Field nurseries for propagating *Trichosirocalus horridus* (Coleoptera: Curculionidae), a biological control agent for *Carduus* thistles. *J. Econ. Entomol.* 79:873–76
137. Strong DR, Kaya HK, Whipple AV, Child AL, Kraig S, et al. 1996. Entomopathogenic nematodes: natural enemies of root-feeding caterpillars on bush lupine. *Oecologia* 108:167–73
138. Strong DR, Maron JL, Connors PG, Whipple A, Harrison S, Jefferies RL. 1995. High mortality, fluctuation in numbers, and heavy subterranean insect herbivory. *Oecologia* 104:85–92
139. Strong DR, Whipple AV, Child AL, Dennis B. 1999. Model selection for a subterranean trophic cascade: root-feeding caterpillars and entomopathogenic nematodes. *Ecology* 80:2750–61
140. Talekar NS, Lai R-M, Cheng K-W. 1989. Integrated control of sweetpotato weevil at Penghu Island. *Plant Prot. Bull.* 31:185–91
141. Tanaka Y, Asaoka K, Takeda S. 1994. Different feeding and gustatory responses to ecdysone and 20-hydroxyecdysone by larvae of the silkworm, *Bombix mori. J. Chem. Ecol.* 20:125–33
142. Terry IL, Lee CW. 1990. Infestation of cultivated *Amaranthus* by the weevil *Conotrachelus seniculus* in Southeastern Arizona. *Southw. Entomol.* 15:27–30
143. Thakur ML. 1988. Principal insect pests of Eucalyptus in India and their management. *India J. Entomol.* 50:39–44
144. Thakur ML, Kumar S, Negi A, Rawat DS. 1989. Chemical control of termites in Eucalyptus hybrid. *India For.* 115:733–43
145. Tobi DR, Grehan JR, Parker BL. 1993. Review of the ecological and economic significance of forest Hepialidae (Insecta: Lepidoptera). *Forest Ecol. Manage.* 56:1–12
146. Tosevski I, Gassman A, Schroeder D. 1996. Description of European *Chamaesphecia* spp. (Lepidoptera: Sesiidae) feeding on *Euphorbia* (Euphorbiaceae), and their potential for biological control of leafy spurge (*Euphorbia esula*) in North America. *Bull. Entomol. Res.* 86:703–14
147. van Tol RWHM, van der Sommen ATC, Boff MIC, van Bezooijen J, Sabelis MW,

Smits PH. 2001. Plants protect their roots by alerting the enemies of grubs. *Ecol. Lett.* 4:292–94

148. Wagner DL, Tobi DR, Wallner WE, Parker BL. 1991. *Korscheltellus gracilis* (Grote): a pest of red spruce and balsam fir roots (Lepidoptera: Hepialidae). *Can. Entomol.* 123:255–63
149. Williams KS, Simon C. 1995. The ecology, behavior, and evolution of periodical Cicadas. *Annu. Rev. Entomol.* 40:269–95
150. Williams KS, Smith KG, Stephen FM. 1993. Emergence of a 13-yr. periodical Cicadas (Cicadidae: Magicicada): phenology, mortality, and predator satiation. *Ecology* 74:1143–52

Annu. Rev. Entomol. 2003. 48:549–77
doi: 10.1146/annurev.ento.48.091801.112559

GRASSES AND GALL MIDGES: Plant Defense and Insect Adaptation

M. O. Harris,[1] J. J. Stuart,[2] M. Mohan,[3] S. Nair,[3] R. J. Lamb,[4] and O. Rohfritsch[5]

[1]*Department of Entomology, North Dakota State University, Fargo, North Dakota 58105; e-mail: marion.harris@ndsu.nodak.edu*
[2]*Department of Entomology, Purdue University, West Lafayette, Indiana 47907-1158; e-mail: jeff_stuart@entm.purdue.edu*
[3]*International Centre for Genetic Engineering and Biotechnology (ICGEB), Aruna Asaf Ali Marg, New Delhi, 110 067, India; e-mail: mohanm@del2.vsnl.net.in; suresh@icgeb.res.in*
[4]*Cereal Research Centre, Agriculture and Agri-Food Canada, Winnipeg, Manitoba R3T 2M9, Canada; e-mail: RLAMB@em.agr.ca*
[5]*Institut de Biologie Moléculaire des Plantes, C.N.R.S., Strasbourg, 67084 France; e-mail: odette.rohfritsch@libertysurf.fr*

Key Words Cecidomyiidae, rice, wheat, resistance genes, avirulence genes

■ **Abstract** The interactions of two economically important gall midge species, the rice gall midge and the Hessian fly, with their host plants, rice and wheat, respectively, are characterized by plant defense via *R* genes and insect adaptation via *avr* genes. The interaction of a third gall midge species, the orange wheat blossom midge, with wheat defense *R* genes has not yet exhibited insect adaptation. Because of the simple genetics underlying important aspects of these gall midge–grass interactions, a unique opportunity exists for integrating plant and insect molecular genetics with coevolutionary ecology. We present an overview of some genetic, physiological, behavioral, and ecological studies that will contribute to this integration and point to areas in need of study.

CONTENTS

0066-4170/03/0107-0549$14.00

INTRODUCTION

The current revolution in molecular biology will change fundamental aspects of our lives in ways we cannot yet imagine. In the production of food, we might create a "smart" plant that combines the agronomic traits desired by humans, i.e., yield and quality, with enhanced defense against a variety of enemies. Many questions arise from this possibility. Are the plant defense systems against the variety of enemies related or independent? Are there fitness costs associated with defense? And if so, are these costs associated with significant losses in yield and quality in crop plants? These questions point to a need for change in the study of plant defense. Instead of studying pairwise interactions between a host plant and a single enemy, the plant must become the focal point, with each plant enemy ultimately viewed within the larger context of all biotic and abiotic causes of plant stress.

The development of model plant systems has started to bring together the disciplines studying plant defense. Molecular biologists, plant biochemists, and plant pathologists collaborate in studies on the defense systems of the model plant *Arabidopsis* (a small cruciferous weed). The result has been significant progress toward understanding many features of plant defense including the structure and function of plant disease resistance genes (30) and the genetic architecture of disease resistance (144). Yet for some disciplines, such as entomology, this plant system has less to offer. *Arabidopsis* is without specialized insect enemies. Insects can be used to inflict damage to *Arabidopsis* in the laboratory, e.g., green peach aphid (92), but this damage is not natural. As a consequence, the lessons learned from studying *Arabidopsis*-insect interactions may not be relevant to the ecology and evolution of insect-plant interactions. Moreover, whatever is learned about the defense systems of *Arabidopsis*, a dicot, will tell us little about plant defense in monocots, such as cereal grasses. Monocots and dicots, the two major groups of flowering plants, diverged 200 million years ago (141a).

Insects provide many examples of what appears to be antagonistic insect-plant coevolution (67). Antagonistic coevolution is defined by the following events: The insect increases its fitness by attacking and successfully exploiting a plant. The reduced fitness of the attacked plant favors the selection of a novel defense that spreads through the plant population. This reduces the fitness of the insect and selects for a virulent insect genotype that spreads through the insect population. The rates at which defense and virulence reciprocally evolve depend on the genetics and ecology of the plant and insect, as well as the costs of defense and virulence. Many of the best-known examples of antagonistic coevolution come from insects with chewing mouthparts (66). Here, plant defenses include the production of chemicals that inhibit feeding and digestion as well as the production of volatile signals that attract the natural enemies of the insect pest (22). Typically, plant defense systems against chewing insects are controlled by a number of genes rather than one or two genes (13).

In spite of the many fascinating features of the interactions between plants and chewing insects, the study of plant defense against insects needs to be investigated in a greater variety of systems. A plant defense controlled by one or a small number of genes, i.e., monogenic or oligogenic resistance, would provide a much-needed entry point into the molecular basis of plant defense against insects. Likewise, insect adaptation controlled by one or a small number of genes would provide an entry point into the molecular basis of insect adaptation to plant defense. This type of genetic interaction is common in plant-pathogen interactions (1) and is described by the gene-for-gene concept (33). For dominant gene-for-gene interactions, a plant defense or resistance gene, referred to as an *R* gene, is thought to control an important step in the recognition of an elicitor produced by the pathogen avirulence (*avr*) gene. Pathogen adaptation to an *R* gene then occurs through modifications in the corresponding *avr* gene (1).

Insect-plant systems with *R* genes are thought to be rare (88) and therefore are seen as less realistic models of coevolution than systems with plant defense controlled by many genes. The fact that most insect *R* genes have been found in crop plants rather than wild plants is another argument used against these systems. However, although insect *R* genes may be rare for chewing insects, they are not rare in insect-plant interactions that involve less mobile insects that feed by other mechanisms, e.g., aphids, planthoppers, and gall midges (15, 98, 142). Although many, if not all, of the known insect *R* genes have been studied in crop plants, *R* genes have their origins in the wild progenitors of these crops (15, 105, 108).

In this review we present a case for the gall midges (Diptera: Cecidomyiidae) as useful systems in plant defense. As well as exhibiting *R* genes, gall midge–plant interactions exhibit a number of other useful features. First, the plant defense associated with *R* genes has a dramatic and easily quantifiable effect on the fitness of the insect, i.e., the first instar gall midge larva dies within days of first attack. Second, there is genetic variability in the gall midge response to *R* genes, variability associated with one or a small number of *avr* genes. Third, several plants that possess gall midge *R* genes also possess well-characterized *R* genes for other insects as well as *R* genes for pathogens (1, 15, 142). This third feature creates the potential for studies of cross-talk among pathways for insect and pathogen defense (31). Finally, a number of gall midge–plant interactions are economically important.

This means that knowledge of plant defense and gall midge virulence will have direct benefits for agriculture.

For the purposes of this review, we focus on three gall midge species that attack two economically important grasses: the rice gall midge, *Orseolia orzyae* (Wood-Mason), which attacks rice, *Oryza sativa*, the Hessian fly, *Mayetiola destructor* (Say), and the orange wheat blossom midge, *Sitodiplosis mosellana* (Géhin), which both attack wheat, *Triticum aestivum*. The three gall midges were chosen because for all three species *R* genes control important features of plant defense. Moreover, the two grasses, rice and wheat, are the two most important crop plants in the world.

A number of recent reviews of the rice gall midge, Hessian fly, and wheat midge have discussed their pest status and the use of *R* genes in agriculture (15, 108, 118). The aim of our review is to present these three gall midge–plant systems in a new light. We see in these systems an opportunity for integrating plant and insect molecular genetics, coevolutionary ecology, cross-talk among various insect and pathogen defense pathways, and applications of *R* genes in agriculture. This review comprises an overview of some of the genetic, ecological, behavioral, and physiological studies that will contribute to this integration.

THE GRASSES: RICE AND WHEAT

It is said by scientists who study grasses that the settled human societies in which science and art began to flourish were made possible by the selection of a small number of grasses for domestication and cultivation (18). More than 7000 years ago the three major cereal grasses were independently domesticated in the three centers of early agriculture: wheat in Southwest Asia, maize in the tropical highlands of Mexico, and rice in China (23). Today the cereal grasses produce three times more dry edible matter than all other major foodstuffs combined.

The economic importance of rice and wheat has made them the subjects of extensive cytogenetic and molecular studies (25, 70, 71). The late Ernie Sears developed the first set of aneuploid genetic stocks in wheat. A discovery from these genetic stocks was the ability of chromosomes of different ancestral origins to compensate for the absence of particular chromosomes. Studies on chromosome compensation resulted in the classification of the 21 chromosomes of wheat into seven homoeologous groups. Molecular genetic maps proved that homoeologous chromosomes carry similar genes. Furthermore, gene content and gene order are conserved across the grasses. The term synteny describes cases in which gene order is highly conserved across species (25).

For studies of evolution, plant molecular biology, and the application of molecular genetics to agriculture, rice is considered a better system than wheat (70, 71, 80). Rice has a relatively small genome, six times bigger than *Arabidopsis* and 40 times smaller than wheat. The entire rice genome is being sequenced (http://rgp.dna.affrc.go.jp/index.html) and a draft sequence has been published (44a, 144a). More than 2200 markers have been mapped (55). A large number of molecular and cytogenetic tools are available for the study of the rice genome (19, 70) and the genetic transformation system is well established (19). These features have made the rice genome the base genome for studying syntenic relationships among grasses (25). In terms of practical applications, the rice system has progressed faster than the wheat system. For example, pyramiding of resistance genes in rice was achieved several years ago, e.g., resistance genes for rice bacterial blight, *Xanthomonas oryzae* (60). Pyramiding of resistance genes has not yet been achieved in wheat.

GALL MIDGES AND PLANT GALLS

Within the order Diptera, the cecidomyiids together with the mosquitoes, blackflies, sandflies, and fungus gnats are classified in the suborder Nematocera (39). The Cecidomyiidae is usually divided into three subfamilies, the Lestremiinae, Porricondylinae, and Cecidomyiinae. Species in the Lestremiinae and the Porricondylinae are considered primitive and feed on fungus in decaying organic matter. The Cecidomyiinae comprise the youngest and largest subfamily. Its 3850 described species represent about 80% of all known cecidomyiid species. Some of these are fungus feeders, but most are plant feeders or predators. The greatest proportion of species in the Cecidomyiinae can be divided into two monophyletic supertribes, the Lasiopteridi and the Cecidomyiidi. Of the three gall midges we discuss in our review, the Hessian fly belongs to the Lasiopteridi, whereas the rice midge and wheat midge are Cecidomyiidi (39).

The plant-feeding cecidomyiids have several interesting biological features. The first is the production of unisexual progenies (134). Thus, most commonly, an individual female produces only daughters or only sons. The genetics underlying unisexual progenies are discussed later in this review under Gall Midge Reproductive Biology. A second interesting feature is the ability of the third instar gall midge larva to exhibit prolonged diapause (5). In the wheat midge, prolonged diapause can result in adult emergence for up to 12 years after larval feeding is completed (5).

Plant galls are a third feature associated with the plant-feeding cecidomyiids. Galls are defined as aberrant plant cells, tissues, or organs that result from stimulation from foreign organisms. An enormous variety of plant galls are stimulated by an equally enormous variety of organisms, including other insects (e.g., cynipid wasps, aphids, coccoids), mites, viruses, bacteria, fungi, and nematodes. For the gall midge larva, a critical feature of the gall is the nutritive tissue that is induced at the feeding site (112). Enhanced protein synthesis by cells in the nutritive tissue, combined with enhanced transport of nutrients from the vascular system of the plant, provide the gall midge larva with a diet rich in soluble amino acids and sugars (112). Microscopic features of the gall are discussed in greater detail later in this review under Gall Midge Larvae and Plants. In the gall midges, microscopic features of the gall are commonly accompanied by a macroscopic gall structure that bears little resemblance to the plant part from which it arose. When the structure encloses the insect, the gall is referred to as a covering gall. The rice gall midge larva, which feeds at the apical or lateral buds of the rice plant, stimulates the production of a covering gall. The Hessian fly larva, which feeds at the crown or nodes of the wheat plant, does not stimulate any significant macroscopic gall structure but is classified as a gall midge because of the nutritive tissue observed at feeding sites (O. Rohfritsch, unpublished data). The wheat midge, which feeds on the developing wheat seed, also is not associated with a significant macroscopic gall structure. Plant cells at wheat midge feeding sites have not yet been examined for aberrant growth and development.

THE THREE GALL MIDGE SPECIES

Asian Rice Gall Midge

The Asian rice gall midge was considered a minor pest of rice until the mid-1960s when the "Green Revolution" in rice occurred in Asia (102). The new high-yielding rice varieties were, and continue to be, highly susceptible to the Asian rice gall midge. Today the Asian rice gall midge is one of the four most important insect pests in Southeast Asia, South Asia, and China (118). The African rice gall midge, *Orseolia oryzivora*, is another gall midge pest of rice found in Africa. The Asian rice gall midge develops galls on several wild grasses as well as cultivated and wild *Oryza* species (A. Devi, unpublished data). Two of these wild grasses, *Cynodon dactylon* and *Isachne aristatum*, are abundant in areas adjoining paddy fields and may be important alternative hosts.

The rice gall midge begins its associations with plants when the mated female selects plants for oviposition (118). The female visits many plants during her short life, laying 100–400 eggs either on the leaf sheath or on the hairs of the ligules of the rice leaf in groups of 2–6 eggs. The newly hatched larva moves down between the leaf sheaths until it reaches the growing point of the apical or lateral buds. Feeding suppresses the development of the growth cone and stimulates the formation of a gall, which when fully formed consists of a long ivory-white tube that terminates in a solid plug of white tissue, referred to as the "silver shoot." The rice gall midge has a major impact on the fitness of a rice plant. Early attack of the plant results in stunting and profuse tillering, but few tillers produce panicles. In India, systematic screening for resistant rice genotypes was initiated in 1948 (118). Virulent rice gall midge genotypes were first reported in 1969.

Hessian Fly

The Hessian fly is a serious pest of bread wheat and durum wheat in the United States and North Africa (81, 108) but also completes its life cycle on a number of other grasses. Most of these grasses are closely related to wheat, e.g., cultivated cereals (i.e., barley, rye, triticale) and wild grasses (e.g., *Agropyron*, *Elymus*, and *Aegilops*) in 16 genera of the grass tribe Triticeae (54). Other hosts are in the grass tribe Bromeae (54). In the Mediterranean region and Europe, the Hessian fly is sympatric with a number of other *Mayetiola* species that also attack grasses, for example, the barley stem gall midge, *M. hordei* (40), the rye midge, *M. secalis* Bollow, the brome midge, *M. bromicola*, and the oat midge, *M. avena* (5).

The Hessian fly is believed to have originated in Southwest Asia, where wheat also has its origins (5). Hessian fly populations spread west to the Mediterranean region, north and west to Europe, and east to Siberia and central Asia. The Hessian fly was first reported in North America in 1777 and in New Zealand in 1888. It was during the devastating outbreak that followed the introduction of the Hessian fly into North America that its common name made its first appearance in print (61). During the Revolutionary War, the hiring by the British of mercenary soldiers from the German state of Hesse made the name Hessian the "most opprobrious Term our Language affords" (61). Today the Hessian fly is found in all major wheat-growing areas of North America.

The association of the Hessian fly with plants begins when the time-limited mated female, carrying 50–400 eggs, flies from plant to plant, laying 2 to 5 eggs on a leaf blade (52). Larvae migrate down the leaf blade, between the leaf sheaths, to protected feeding sites at the base of the plant. Successful establishment of a larva reduces the fitness of a wheat plant in several ways (15). If a seedling is attacked, the growth of the plant is stunted and the plant usually dies without producing seeds. If an older plant is attacked, fewer and smaller seeds are produced. Resistance to Hessian fly was first noted over 200 years ago (61). The first observations of Hessian fly virulence were discussed by Painter (98).

Orange Wheat Blossom Midge

The orange wheat blossom midge is a significant pest of both bread wheat and durum wheat in North America, Europe, Russia, Japan, and China (15, 76). In Europe, Russia, and Japan, the orange wheat blossom midge coexists with another gall midge that attacks the wheat head, the lemon wheat blossom midge, *Contarinia tritici* (Kirby) (5). We limit our discussion to the orange wheat blossom midge and hereafter refer to it as the wheat midge. The wheat midge is oligophagous. All seventeen species in the genus *Triticum* are hosts (138). The wheat midge also is occasionally found on rye, barley, and a number of wild grasses, e.g., *Roegneria*, *Clinelymus*, and *Aneurolepidum* in China (146).

Early literature on the distribution of the wheat midge in Eurasia refers to a more northerly distribution than the origins of wheat itself (5). The wheat midge arrived in North America in the 1800s and was established in Quebec by 1828. In 1854, major outbreaks occurred in the Northeast and Midwest. From the early 1990s up to the present time, an outbreak of wheat midge in the northern Great Plains of North America has caused important economic losses (76).

Wheat midge host selection occurs when the mated female, carrying 60–80 eggs, deposits eggs on the spikelets of wheat heads (104). Oviposition typically occurs in the days preceeding anthesis as the wheat head emerges from the flag leaf and before pollination occurs. Eggs are laid 1 to 2 eggs at a time, often on the inner surfaces of the leaf-like glumes that enclose the florets (122). Newly hatched larvae crawl 5–20 mm and feed on the surface of the primordial seed. Attack by the wheat midge is associated with a reduced proportion of well-formed wheat seeds (75). As well as yield losses, wheat midge adversely affects grain quality and important agronomic characters. Host plant resistance to wheat midge was reported in the late 1880s in North America (32). There currently are no reports of wheat midge virulence to *R* gene resistance.

GALL MIDGE REPRODUCTIVE BIOLOGY

Gall midges have a relatively long-lived immature stage specialized for feeding and a short-lived adult stage specialized for reproduction (39). The life span of adult gall midges is commonly 1–2 days and rarely longer than 3–4 days. Water is probably their only nourishment (39). In this time the fly must mate, disperse, locate hosts, and oviposit. The placement of eggs determines not only which plant the larva feeds upon but also where on the plant feeding will occur (54, 62, 63).

Genetics and Sex Determination

Like the closely related black fungus gnats (Nematocera: Sciaridae), gall midges have anomalous cytogenetics that involve both chromosome elimination and chromosome imprinting (134, 135). As a matter of gross genomic organization and chromosome behavior, these processes are particularly well conserved among the Cecidomyiinae and are typified by the Hessian fly and the rice gall midge (65, 115, 127, 128), as summarized below.

The gall midge germ line contains both E and S chromosomes. The E chromosomes are germ line limited and variable in number. Their function is unknown. The S chromosomes are found in both the germ line and the soma. These are composed of two autosomes (A1 and A2) and two X chromosomes (X1 and X2). The female somatic line is diploid for all four S chromosomes (A1A2X1X2/A1A2X1X2). The male somatic line is diploid for the autosomes but carries only the maternally derived X chromosomes (A1A2X1X2/A1A2). In the germ line, during oogenesis, the S chromosomes undergo meiosis while the E chromosomes divide mitotically. Each ovum produced contains a haploid set of S chromosomes and a full complement of E chromosomes. During spermatogenesis, there is no genetic recombination, and the E chromosomes are eliminated. All of the sperm have an identical nuclear content: a maternally derived set of S chromosomes. Thus, both sexes are homogametic and all zygotes begin development with the same chromosome complement. The somatic karyotype and the sex of the embryo are established as the E chromosomes are eliminated from the presumptive somatic nuclei. The paternally derived X1 and X2 chromosomes also may be eliminated from the soma of the embryo. If they are, the resulting somatic karyotype (A1A2X1X2/A1A2) establishes male development. If the X1 and X2 chromosomes are retained, female development ensues.

The mechanism that determines whether the paternal set of X chromosomes will be eliminated or retained clearly depends on maternal genotype, as individual females usually produce either all-female or all-male offspring (4, 97, 115, 116, 129). This mechanism has apparently enabled many gall midge species to skew the sex ratio toward females (4). A genetic model involving a single autosomal gene with maternal effects is sufficient to explain unisexual reproduction in the Hessian fly (129), and recent experiments have genetically mapped such a locus on Hessian fly chromosome A1 (J. Stuart, unpublished data). Nonetheless, other mechanisms may be involved. *Wolbachia*, for example, has been implicated in the skewed mitochondrial DNA inheritance observed in the rice gall midge (6) and may play a role in chromosome elimination.

Mating Behavior

Mating systems of gall midges are based on female-produced sex pheromones (51) produced shortly after eclosion of the female. The newly eclosed female walks several centimeters from her eclosion site, stops, and a few seconds later, exhibits a posture associated with the release of the sex pheromone. In virgin females, sex pheromone release is exhibited in a regular daily pattern that corresponds with biosynthesis of the sex pheromone (36). The Hessian fly appears to have a sex pheromone with more than one component, only one of which has been identified (38). The wheat midge has a single-component sex pheromone, recently identified as 2,7-nonanediyl dibutyrate (47). The rice gall midge is known from field-trapping studies to have a sex pheromone (51), but pheromone components have not been identified. Monogamy is common in female gall midges (39) and is promoted by several physiological and behavioral changes that occur after mating (14, 36). The newly mated female stays near her eclosion site and is inactive for a number of hours until she becomes active during the oviposition phase (54).

In contrast to females, male gall midges are constantly in motion during the period of sex pheromone release and continue to find and mate with virgin females until death occurs (14). Exposure to the volatile sex pheromone triggers upwind flight to the female, i.e., the source of the sex pheromone (51). Given the opportunity, male gall midges will mate many times and fertilize many eggs. In the Hessian fly, medium-sized males are capable of fertilizing anywhere from 2409 to 3445 eggs, i.e., the progeny of 10 to 35 females (14). There is no laboratory or field evidence that suggests that differences in rice gall midge or Hessian fly *avr* genes act as a mating barrier (7, 16). Black et al. (16) partitioned Hessian fly variance in allozyme frequencies among populations from throughout U.S. wheat-growing regions. Their results clearly indicate that gene flow is not restricted among Hessian fly genotypes carrying different sets of *avr* genes. Hessian fly genotypes carrying different sets of *avr* genes are referred to as biotypes. The large within biotype variation found by Black et al. indicates there may be restricted gene flow among local Hessian fly populations. This result indicates a high probability of inbreeding. On the other hand, the production of unisexual progenies by gall midge larvae reduces the probability of inbreeding. Inbreeding is important because it increases the rate at which new recessive *avr* mutations become homozygous, i.e., the rate at which gall midge virulence to *R* genes evolves (45).

Oviposition and Dispersal

In regard to the postmating behavior of the adult female, the Hessian fly has been studied in the greatest detail. Hessian fly females are constantly in motion during the oviposition phase (52). As well as flying, females land on plants, examine plant surfaces, oviposit eggs, and sit for only short periods before again taking flight. Given an abundance of highly stimulatory plants, the majority of flights are short flights from one oviposition site to the next, with a single longer flight occurring after a series of short flights (53, 139). Plant stimuli influence the relative frequency of short versus longer flights. If the female encounters only highly stimulatory plants, the longer flights are exhibited infrequently. If the female encounters only nonstimulatory plants, longer flights become more frequent.

Although it is commonly said that adult gall midges are weak fliers, Hessian fly females can move considerable distances while searching for oviposition sites. Females avoid flying in strong winds (140), but in light winds had an estimated diffusion rate in wheat of 660 m^2 during 2 h of ovipositional activity (141). In fields of nonhosts, estimated diffusion rates of adult females were significantly greater, i.e., 1500 m^2 during 2 h. As well as two-dimensional diffusion through the crop, Hessian fly females exhibit vertical or phototactic flights (53) that may take them out of the crop canopy and into faster moving air. McColloch (85) concluded that mated Hessian fly females, but not males or virgin females, may be carried several kilometers by the wind. After their travels, such females were fully capable of ovipositing on host plants.

During host selection, the Hessian fly female orients to both nonspecific plant stimuli and stimuli specific to grasses (37, 53). As the female approaches within several centimeters of the plant, olfactory signals stimulate landing. After landing, tactile and nonvolatile chemicals associated with leaf surface waxes stimulate one or more bouts of oviposition. Two stimulatory chemicals 1-octacosanal and 6-methoxy-2-benzoxazolinone (MBOA) have been identified (93). The latter is interesting because it is a plant defense chemical and also is the decomposition product of a well-known plant defense chemical, 2,4-dihydroxy-7-methoxy-1,4-benzoxazin-3-one (DIMBOA). In addition to plant chemicals, female Hessian flies base oviposition decisions on information about the fine structure of the leaf surface (62, 63).

Gall midge larvae benefit from many, if not most, of the decisions made by the female gall midge as she visits plants and allocates egg clutches to particular plant genotypes, physiological and developmental states, and locations within the plant (54, 64). In contrast, gall midge larvae do not benefit from the behavior of ovipositing females with respect to *R* genes. In a number of gall midge species, including the Hessian fly, rice midge, and wheat midge, adult females do not discriminate between plant genotypes that do and do not carry *R* genes (15, 54, 78, 98, 118). Various explanations have been given for this apparent flaw in parental care. For gall midges that attack perennial hosts, there may be insufficient selection pressure for the evolution of discrimination against plants carrying *R* genes (78), i.e., females only emerge in areas that contain at least some hosts that do not carry *R* genes and move only short distances before ovipositing. For gall midges that attack annual hosts, there is presumably greater selection pressure for the evolution of discrimination. Yet plants carrying *R* genes may not produce a unique signal that females ovipositing on the surface of the plant can detect (54). Moreover, interactions between virulent and avirulent larvae may reduce the selection pressure for adaptation of host selection behavior. In the Hessian fly, larvae that would normally die on a plant carrying a particular *R* gene, i.e., larvae carrying a functional *avr* gene corresponding to the plant's *R* gene, survive when larvae that do not carry the functional *avr* gene co-occupy the plant (49). Given these complicating factors, the ovipositing female would need a lot of information to exhibit optimal host selection, i.e., not only information about which *R* genes each plant carries but also information about the *avr* genes carried by her own offspring and any other larvae that co-occupy the plant.

GALL MIDGE LARVAE AND PLANTS

Plant Reactions to Attack by Virulent Larvae

In all gall midge–plant interactions studied to date, attack by a virulent gall midge larva results in the differentiation of nutritive tissue cells at the feeding site (112). This nutritive tissue provides the larva with a diet rich in soluble amino acids and sugars. Directly below the attacking larva, cells are inhibited from further growth but are kept alive and under constant stress by larval activity. These cells are induced to high proteosynthetic activity. Between these cells and the vascular bundles of the attacked organ are cells that have the features of transfer cells, i.e., numerous and large plasmodesmata, an enlarged nucleus, a dense cytoplasm, and a highly developed endoplasmic reticulum (112). Cells nearest the vascular bundles contain starch and may function as storage nutritive cells. Critical to the formation and maintenance of the nutritive tissue is the presence of an active gall midge larva. Any cessation of larval activity leads to a rapid regression of nutritive cells to cells resembling parenchyma cells. A second type of nutritive tissue is observed in other gall midge–plant interactions (112). The physiological and cytological characteristics of this nutritive tissue are similar to plant tissues associated with the developing wheat embryo (121).

Precisely how a gall midge larva stimulates the creation of the nutritive tissue and obtains its food are not known. Many features of the larva and the attacked cells point to a feeding mechanism that, as a first step, requires a change in the permeability of the plant cell wall (112, 113). In the Hessian fly, it has been proposed that the highly specialized mouthparts inject salivary substances into the plant cells (57). Once plant solutes have passed through the cell wall, the application of large quantities of saliva, produced by the highly developed salivary glands of the larva (126), presumably contributes to extraoral digestion of plant solutes. The larva then sucks up the partially digested food.

Plant Reactions to Attack by Avirulent Larvae

Among the gall midges, the rice gall midge and the Hessian fly are notable because the genetics underlying both avirulence of the gall midge larva and resistance of the plant have been studied in some detail. These genetics, as well as a biochemical model for interactions between gall midge *avr* genes and plant *R* genes, are discussed later in this review under Grass *R* Genes and Gall Midge *Avr* Genes. Unfortunately, in comparison with genetic studies, anatomical and biochemical investigations of the interactions of the rice gall midge and Hessian fly with their hosts are in their infancy. For the wheat midge–plant interaction, genetic as well as biochemical and anatomical studies have just begun.

In contrast to the virulent gall midge larvae that successfully establish and grow on hosts, avirulent gall midge larvae do not establish, do not appear to grow, and typically die within 2 to 5 days of arrival at feeding sites. Theoretically, death of avirulent gall midge larvae could result from either a lack of a plant response (i.e., evasion) or an active plant defense. The importance of the larval-induced plant nutritive tissue for virulent gall midge larvae may create an opportunity for plants to evade parasitism simply by not responding to larval stimulation and not producing a nutritive tissue. In contrast to plant resistance by evasion, plant resistance could result from an active plant defense that either directly kills gall midge larvae via toxins or indirectly kills larvae by blocking the formation of the nutritive tissue.

Anatomical and biochemical studies of gall midge–plant interactions [(96, 112); O. Rohfritsch, unpublished data], as well as genetic studies of gall midge avirulence and plant resistance (see later discussion), point to an active plant defense triggered by an interaction between the product of a gall midge *avr* gene and the product of a corresponding plant *R* gene. One such active plant defense is illustrated by the response of *Fagus silvatica* to the gall midge *Hartigiola annulipes* (O. Rohfritsch, unpublished data). On the third day of larval attack, cells at the periphery of the attacked area exhibit increased production of phenolics. During the next 24 h, cells nearer and nearer to the larva exhibit increases in phenolic production. This response appears to prevent the establishment of a nutritive tissue. Death of the larva is observed 4 days after initial attack of the plant. Subsequent to the death of the larva, tissue at and around the attack site becomes necrotic.

Induced plant defense via enhanced phenolic production also may occur in plants carrying *R* genes for the wheat midge and rice gall midge (27, 102). In the wheat midge–wheat system, resistant wheat genotypes have both a higher constitutive level of phenolics before wheat midge attack and a more rapid induction of phenolics after attack (27). There appears to be a significant cost associated with the induced component of this plant defense. Relative to an unattacked resistant seed, an attacked resistant seed has a biomass reduction of 28% (75). Phenolics also may be important for plant resistance to the rice gall midge (102). In some, but not all, rice plants resistant to the rice gall midge, browning is observed at the attack site (118). Browning is associated with the collapse of cells, especially those in the meristematic tissue of the leaf primordium. The apices of the attacked terminal shoot are severely affected by cell collapse, but the unattacked axillary shoots appear unaffected.

Another plant defense response, referred to as the hypersensitive response (HR), is exhibited by *Salix viminalis* when it is attacked by the gall midge *Dasineura marginemtorquens* (96). HR is commonly seen in pathogen-plant interactions and provides defense against the largest possible number of potential pathogens (1, 123). Classically HR has two defining features: It is an extremely rapid response, starting at the attack site within a few hours, and it involves the death of many plant cells at and around the attack site. The response of *S. viminalis* to *D. marginemtorquens* exhibits both of these features. Within 12 h of the initial attack, nearly all cell layers within the leaf are engaged in the necrotic reaction. Cell damage in the injured cells triggers biochemical reactions leading to characteristic markers of HR, e.g., callose deposition in cell walls, accumulation of polyphenols, and the production of autofluorescent compounds. Cell death blocks the attacker's access to nutrients and/or prevents penetration by the attacker.

Because necrosis and cell collapse have been observed at sites where dead larvae are found, HR responses have been claimed for resistance to the rice gall midge (11), Hessian fly (48), and wheat midge (73). Anatomical and biochemical studies will be necessary to determine if the necrosis observed in these responses is the actual cause of larval death or occurs only after some other defense mechanism has caused larval death. Preliminary studies of plant responses to avirulent Hessian fly larvae (O. Rohfritsch, unpublished data) point to a defense that is not a hypersensitive response. In plants expressing *R* genes, defense has two important features: cell wall lignification and the synthesis of phenolics. Defense occurs in a small number of cells, particularly the cells that are adjacent to cells ruptured by the larva or cells in contact with the larval cuticle. In contrast to cells involved in HR, cells defended against the Hessian fly continue to live.

The relatively large number of *R* genes available for the Hessian fly and rice gall midge will create an opportunity for testing assumptions of the elicitor-receptor model of *avr* gene–*R* gene interactions (see Grass *R* Genes and Gall Midge *Avr* Genes, below). An important assumption of this model is that the sequence of defense responses triggered by the elicitor-receptor interaction will be the same for all *avr* gene–*R* gene pairs. The best test of this assumption will require the creation of plants genetically modified to express single *R* genes, i.e., a set of plants that differ only in regard to individual *R* genes. Until such genetically modified genotypes are available, *R* genes backcrossed to a similar genetic background may provide some information. For example, in the Hessian fly–wheat system a number of *R* genes have been backcrossed to the susceptible genotype "Newton" (100). In preliminary observations, differences have been observed in the defense responses of genotypes carrying the backcrossed *R* genes, *H6*, *H9*, or *H13* (O. Rohfritsch, unpublished data). Larval death occurred earliest on plants expressing *H13* (2 to 3 days). In this genotype, there appeared to be both a constitutive defense, provided by thicker cell walls, and an induced defense, i.e., cell wall lignification and the production of phenolics. On plants expressing *H9*, larval death occurred a day later (3 to 4 days). On this genotype, perhaps because the cell walls are thinner, the larva destroyed more cells before defenses were induced. On plants expressing *H6*, larval death was further delayed (5 to 6 days). In this genotype, induced defense included hypertrophy (growth in the size of cells) as well as cell wall lignification. Even the susceptible genotype "Newton" defended itself to a small degree, i.e., a small number of cells exhibited cell wall lignification. Yet this defense was too small and too late to block the creation of the nutritive tissue. These observations suggest that the conclusion of Benhamou (9) regarding pathogen-plant interactions can be extended to gall midge–plant interactions: Both resistant and susceptible plants respond to attack by the induction of coordinated defense responses but differ in the speed at which the defense is mounted.

While observed defense responses to gall midge attack have all involved active defense, the possibility remains that some plants avoid attack by evasion, i.e., the gall midge larva dies because the plant does not respond to stimulation by the larva and therefore never creates a nutritive tissue. For example, resistance by evasion may be important in nonhost resistance (58). Nonhost resistance is expressed by all plant genotypes within a plant species (e.g., for the Hessian fly, all genotypes within oat, *Avena sativum*). Host resistance is expressed by a small number of plant genotypes within an otherwise susceptible plant species (e.g., for the Hessian fly, a number of genotypes within wheat, *T. aestivum*).

GRASS *R* GENES AND GALL MIDGE *AVR* GENES

The Gene-for-Gene Concept

The gene-for-gene concept is centered on the interplay between a major *R* gene in a host plant and a corresponding *avr* gene in a plant parasite (33), in this case, a gall midge. The *R* gene has two types of alleles: dominant, functional, resistance-conferring alleles (*R*) and recessive, nonfunctional, susceptibility alleles (*r*). Likewise, each *avr* gene has two types of alleles: dominant, functional, avirulence-conferring alleles (*A*) and recessive, nonfunctional, virulence alleles (*v*). The interaction between the *R* gene and the *avr* gene is typically manifested as either a resistant plant and dead gall midge larvae (the noncompatible interaction) or a susceptible plant and living larvae (the compatible interaction). Multiple *R* loci in the plant confer resistance to the gall midge, and for each *R* locus in the plant, a corresponding *avr* locus exists in the insect.

With few exceptions, the genetics of the Hessian fly–wheat interaction fit this model extremely well. Hessian fly resistance in wheat is usually conditioned by dominant or partially dominant alleles of major Hessian fly resistance loci, designated *H1* through *H29* [the original sources of these genes and their positions on wheat chromosomes have recently been reviewed (108)]. Hessian flies can overcome the resistance conferred by all but nine of these genes (17, 108–110). To date, the genetics of virulence to seven *R* genes, *H3*, *H5*, *H6*, *H7H8*, *H9*, *H13*, and *H25*, has been investigated in experimental matings (34, 41, 42, 44, 56, 119, 130, 145). In agreement with the gene-for-gene model, virulence to each of these genes is conditioned by mutations in simply inherited *avr* genes. Using nomenclature established by Formusoh et al. (34), these have been designated as *vH3*, *vH5*, and so on. Three X-linked *avr* genes (*vH6*, *vH9*, and *vH13*) have been mapped relative to each other and other markers on chromosome X2 (111a, 119, 130). The linkage distances between these loci are considerable (>25 cM). Two autosomal *avr* genes, *vH3* and *vH5*, have recently been mapped on chromosome A2 (S. Behura & J. Stuart, unpublished results).

The relationship between gall midge biotypes and *avr* genes is one of phenotype and genotype with respect to a limited set of *R* genes. A gall midge is assigned to a biotype after its progeny are tested for virulence on a set of plant "differentials" that contain different *R* genes. In the Hessian fly biotype nomenclature (43), individuals are assigned to a biotype based on the pattern of virulence they exhibit with respect to *H3*, *H5*, *H6*, and *H7H8* (108). For example, the Hessian fly phenotype called "biotype B" survives on wheat carrying *H3* or *H7H8*, but fails to survive on wheat carrying *H5* or *H6*. Individuals of this biotype are either homozygous or heterozygous for a dominant avirulence allele at loci *vH5* and *vH6*, and homozygous for recessive virulence alleles at loci *vH3* and *vH7H8*. In the rice gall midge, individuals are assigned to a biotype based on their virulence to rice cultivars carrying *GM1*, *GM2*, or *gm3* (118). Clearly, the use of the term biotype will become quite cumbersome as the number of possible avirulence genes and allelic combinations increases with each discovery of a new *avr* gene (26).

There are a few interesting anomalies in regard to the gene-for-gene model and the Hessian fly–wheat interaction. One *R* gene (*h4*) is recessive, and two partially dominant *R* genes (*H7* and *H8*) must be paired for either gene to express resistance fully (108). In addition, there are occasions when Hessian fly larvae survive on resistant, unstunted wheats (29, 56, 124). El Bouhssini et al. (29) discovered that the dominance and dose of the resistance allele, combined with the dominance and dose of the avirulence allele, influence the expression of this phenomenon.

Genetic evidence for a gene-for-gene relationship between rice and the Asian rice gall midge is equally compelling. Monogenetic resistance is quite common: At least seven nonallelic rice gall midge resistance genes have been identified from different indica rice varieties alone (118). Four of these, *Gm2*, *Gm4(t)*, *Gm6(t)*, and *Gm7*, have been mapped on rice chromosomes (68, 89–91, 117). Moreover, rice gall midge biotypes appear to result from mutations in *avr* genes. Thus far, 13 biotypes have been identified that differ in their ability to survive on rice plants carrying different combinations of resistance genes (118). Most importantly, experimental matings between biotypes of the rice gall midge showed that virulence to *Gm2* is inherited as an X-linked recessive allele of a single gene (7, 12). Behura et al. (7) also discovered an X-linked DNA marker linked to virulence to *Gm2*. Additional, DNA markers have been developed that distinguish between the African rice gall midge (*O. oryzivora*), the Asian rice gall midge, and the major Indian biotypes of the Asian rice gall midge (8). Interestingly, biotype-specific integration sites of *mariner*-like transposable elements have also been discovered (6). Experiments are in progress to determine if these polymorphisms are tightly linked to *avr* genes in the rice gall midge genome.

The Elicitor-Receptor Model

The biochemical basis of gene-for-gene interactions is not fully understood. However, an "elicitor-receptor" model has gained acceptance as a working hypothesis and is perhaps the simplest way to envisage how the mechanism might operate (69, 125). In this model, the *R* alleles encode "receptor" proteins capable of triggering inducible defense reactions when they bind to an appropriate ligand. The *A* alleles in the parasite encode this ligand, the so-called "elicitor" of the resistance reaction. The binding of an *A*-encoded elicitor to an *R*-encoded receptor triggers a defense response in the plant, damaging or killing the parasite. Plant genotypes lacking the dominant *R* allele fail to produce the receptor and thus do not initiate a defense response. The parasite establishes on the plant and damages it by feeding. Similarly, when the parasite lacks the functional *A* allele, no elicitor is made, and the parasite avoids detection and establishes a feeding site.

The structure of a variety of cloned cultivar-specific *R* genes that confer resistance to specific plant pathogens and nematodes has lent considerable support to this model (10, 30, 125). In fact, the *Pseudomonas syringae* resistance gene *Pto* in tomato binds directly with the *P. syringae* avirulence elicitor, *AvrPto*, and initiates disease resistance (131). However, unlike *Pto*, the majority of cloned resistance genes belongs to the large and diverse family of genes that encode proteins with C-terminal leucine-rich repeats (LRRs) and nucleotide binding site (NBS) domains. LRRs may be capable of protein-protein interaction (30), and NBS domains may play a role in signal transduction (50, 84, 125, 136, 137). In dicotyledonous plants, some NBS-LRR genes encode a domain with homology to the animal innate immunity factors, *Toll* and interleukin receptor–like genes. However, NBS-LRR genes with those domains are apparently lacking in grass genomes (99). As might be expected of genes that are utilized as recognition factors in an immune response, the NBS-LRR gene family is extensive in plants, comprising approximately 300 genes in *Arabidopsis* (87) and perhaps 1500 genes in rice (144). The genes themselves are organized in large extended clusters of orthologous genes. Examples of known NBS-LRR *R* genes include the *Xanthomonas* resistance gene, *Xa1*, in rice (143) and the nematode resistance gene, *Cre3*, in wheat (72). A case of insect resistance conferred by an NBS-LRR gene is also known: The nematode *R* gene *Mi* of tomato confers resistance to the potato aphid in plant transformation experiments (114).

Several features of gall midge *R* genes in rice and wheat indicate that they might be members of the NBS-LRR family of genes. Evidence is strongest in rice where the genome is much smaller and has been sequenced. The *R* genes, *Gm2*, *Gm6*(t), and *Gm 7*, are clustered in the rice genome (117). Each of these genes has been mapped near the same location on rice chromosome 4. *Gm*2 resides near the major *Xanthomonas* resistance NBS-LRR gene *Xa1*. A region of rice chromosome 4 (ca. 170 kb) that encompasses RFLP markers RG214 [linked to *Gm6(t)* (68)], F8 and F10 [linked to *Gm2* (90, 107)], and SA598 [linked to *Gm7* (117)] has been isolated and sequenced (S. Nair, M. Mohan & J. Stuart, unpublished data). Analysis of this sequence is currently in progress with the goal of isolating *Gm2*. Significant progress toward the identification of molecular markers linked to Hessian fly resistance genes in wheat has also been made (28, 83), and a gene (*Hfr-1*) that is upregulated by Hessian fly feeding on *H9*-resistant plants has been isolated (135a). Several Hessian fly resistance genes in wheat are clustered in the wheat genome: Of the 29 known resistance genes, 10 are located on chromosome 5A, and 2 are on chromosome 6D.

Although some common motifs are present in some bacterial *avr* genes (79, 133), it is difficult to predict what products gall midge *avr* genes may encode. To date, *avr* genes have yet to be cloned from either an insect or a nematode. Nonetheless, there are two themes in plant pathology that are likely to have considerable relevance to insect *avr* genes (21, 24, 133). First, *avr* genes in plant pathogens commonly have a dual function: an avirulence function for the plant whereby the plant recognizes the presence of the parasite, and a virulence or pathogenicity function for the parasite that enhances the parasite's ability to survive, develop, and reproduce. Second, as virulence factors, the products of the *avr* genes likely interact with defense and/or developmental pathways. Indeed, *R* genes may only indirectly interact with *avr* gene products by "guarding" the targets of the *avr* genes (82, 132).

Cloning *R* Genes

Because of synteny in cereal genomes (25), the cloning of insect *R* genes in rice may facilitate the cloning of insect *R* genes in wheat. Owing to the huge size of the wheat genome (16,000 Mbp), the map-based cloning methods that have yielded great returns in rice are inefficient when applied to wheat. However, owing to the high degree of homology between the genomes of rice and wheat (25) and the conservation of nucleotide sequences within genes, cloned *R* genes from rice can be used to isolate orthologous regions in wheat. As the rice gall midge and Hessian fly are closely related, it is likely that *R* genes in orthologous regions of rice and wheat are similar in sequence. For example, the Hessian fly resistance gene *H20* on chromosome 2B may reside in a region that is orthologous to the region of rice chromosome 4 where *Gm2* resides. Thus, when the rice gall midge *Gm2* gene is isolated and cloned, it may be useful as a probe for cloning the Hessian fly *H20* gene in wheat. This possibility should be tested by determining if *H20* maps to a region of the wheat chromosome 2B that is orthologous to the *Gm2* region of rice chromosome 4. Even if *H20* is located elsewhere, a gene similar to *Gm2* may exist in the orthologous region of wheat. Regardless, perhaps the most intriguing and practical experiments will involve the transformation of wheat cultivars with rice gall midge *R* genes to determine if they provide resistance to the Hessian fly.

GALL MIDGE RESISTANCE GENES IN AGRICULTURE

Gall midge *R* genes have played, and will continue to play, an important role in the control of gall midge pests of agriculture. Yet concerns about gall midge adaptation are real. Times to adaptation have been estimated as 8, 7, and 3 years for three Hessian fly *R* genes, *H3*, *H5*, and *H6*, respectively (35, 46). The deployment and loss of *R* genes against the rice gall midge and Hessian fly have been reviewed before (15, 20, 45, 108, 118) and will not be reviewed again here. Instead, for the rice gall midge and Hessian fly, we briefly discuss how the deployment of *R* genes may change in the future.

Rice Gall Midge and Hessian Fly

In the rice gall midge and Hessian fly systems, the status quo of sequential release of single *R* genes followed by insect adaptation is rarely viewed as optimal (20, 45). Monitoring insect virulence (i.e., biotypes) and developing cultivars with new *R* genes are costly and time consuming. Moreover, *R* genes are assumed to be limited in number and therefore should not be wasted. Insect *R* genes resemble resistance traits deployed in transgenic insecticidal cultivars in the need to both monitor insect virulence and conserve a valuable resource (46).

For the rice gall midge and Hessian fly, two important changes in the status quo are likely. First, techniques for monitoring and describing virulence in gall midge populations will change. The biotype concept (26), used for many decades to describe patterns of virulence to *R* genes in gall midge populations (15, 108), will be refined or abandoned as DNA-based techniques for identifying gall midge genotypes are developed. The rice gall midge system has made the greatest progress in this regard. DNA-based molecular markers developed using random amplified polymorphic DNA (RAPD) and amplified fragment length polymorphism (AFLP) techniques have been used to distinguish biotypes of both the Asian rice gall midge and the African rice gall midge (7, 8).

The pyramiding of two *R* genes seems the second likely change in the status quo. Simulation models predict that, under most conditions, a cultivar carrying two pyramided *R* genes will be effective for a greater number of years than the two genes deployed sequentially (142). Pyramiding of *R* genes will be aided by DNA-based marker-assisted selection (MAS) (89). Markers associated with specific *R* genes have been identified for both rice gall midge and Hessian fly (28, 94, 95, 117). The cloning of *R* genes also creates the possibility of transforming susceptible elite cultivars. Because more is known about rice midge *R* genes and because the transformation system for rice is more advanced, both pyramiding of *R* genes and transformation with *R* genes are likely to occur sooner for rice gall midge than for the Hessian fly. Once rice gall midge *R* genes are used to transform plants, their deployment in agriculture may be regulated by government agencies, as resistance traits in transgenic insecticidal cultivars have been regulated (46).

Orange Wheat Blossom Midge

The wheat midge differs from the rice gall midge and Hessian fly in that only a single *R* gene has been identified (3, 73, 86). In North America this *R* gene has not yet been deployed in agriculture but has been deployed for a number of years in large test plots in Manitoba, Canada. If cultivars carrying the *R* gene perform well in registration trials in Canada, deployment will begin in Canadian wheat fields within the next 2–5 years. Two to three years later, deployment of the *R* gene in Canadian durum wheat and North Dakota bread wheat is expected.

The loss of both rice gall midge and Hessian fly *R* genes raises concerns about the potential loss of the single wheat midge *R* gene. The wheat midge system shares many similarities with the two gall midge–grass systems. The wheat midge resistance trait is inherited as a single gene (86). Wheat midge larvae die during the early stages of attack (73). Successful defense has an induced component, a more rapid production of phenolic acids (27). In field populations, a small number of wheat midge larvae (approximately 1 in 1000) survive to the adult stage on wheat carrying the *R* gene (73). These larvae typically are of an unusually small body size and therefore are difficult to rear for tests of virulence.

A possible difference between the wheat midge and the Hessian fly concerns the number of *R* genes that may be found. The 29 Hessian fly *R* genes were relatively easy to find (15). The wheat midge attacks wheat, yet the finding of the single wheat midge *R* gene occurred after a prolonged effort in a number of countries, all of which view the wheat midge as a major pest. An evolutionary perspective on this question is that the frequency of resistance alleles in a plant species is related to the virulence of the plant's primary pest (13, 67). If this is true for gall midge *R* genes, the greater virulence of the Hessian fly on wheat may mean that there will always be more Hessian fly *R* genes than wheat midge *R* genes. On the other hand, *R* genes may be more common for gall midges that attack the early stages of plant growth, e.g., Hessian fly and rice gall midge, than for gall midges that directly attack the offspring of the plant, i.e., seeds. The sorghum midge and the sunflower midge are two other gall midge pests that attack seeds. In spite of a large effort, *R* genes and resistance traits similar to those found for rice gall midge, Hessian fly, and wheat midge have not been found for sorghum midge and sunflower midge (2, 120).

Similarities among the *R* genes against the three gall midge species, coupled with the possibility that additional *R* genes for wheat midge may not be found in the near future, lead to an inevitable conclusion. Efforts must be taken to delay or prevent the adaptation of wheat midge to the currently available and highly valuable *R* gene. Two strategies have been proposed (15). The first strategy involves pyramiding in a single cultivar, the *R* gene in combination with a resistance trait that results in reduced oviposition (74, 77). Here the assumption is that pyramiding the two unrelated resistance traits will delay gall midge adaptation to either trait (45). Combining the wheat midge *R* gene with traits for reduced oviposition will be difficult because oviposition is a highly variable process (122) and because genetic markers for the oviposition traits are not available. The second strategy applies principles of population genetics to dilute virulence in the pest by maintaining adequate frequencies of *avr* genes in the population (45, 111). Susceptible plants would be deployed as a refuge for avirulent wheat midge within a larger crop of wheat expressing the *R* gene.

Although strategies for delaying adaptation to the wheat midge *R* gene may be available, any technology that promises to help farmers in a significant way tends to create its own momentum and, in doing so, may successfully evade concerns raised by the inventors of the technology. An added problem for the wheat midge *R* gene is its deployment on both sides of the Canadian–United States border. If one country takes action to prolong the effectiveness of the wheat midge *R* gene, this effort may be undermined by inaction on the other side of the border.

CONCLUSIONS AND FUTURE DIRECTIONS

The best-known feature of gall midge–grass interactions is the genetic variability associated with both the plant's defense response to gall midge and the gall midge's response to plant defense. As in many pathogen-plant interactions, genetic variability arises from *R* genes in the plant and *avr* genes in the gall midge. In most respects, grass *R* genes and gall midge *avr* genes fit the gene-for-gene concept of coevolutionary interactions between plants and their enemies.

Although the gene-for-gene concept may ultimately prove overly simplistic (88), it currently functions as a working hypothesis for interactions between grasses and gall midges. As such, it has generated questions about antagonistic coevolution between grasses and gall midges. One such question concerns the independence of coevolving pairs of *R* and *avr* genes. If gene pairs do evolve independently, this may be reflected in the genetic architecture of *R* genes. Research to date on a small number of *R* genes in both the rice gall midge–rice system and the Hessian fly–wheat system indicates that some *R* genes are clustered in the plant genome and therefore may not evolve independently. Further genetic studies will show the extent of gall midge *R* gene clustering and the location of gall midge *R* genes relative to *R* genes for other insects, pathogens, and nematodes.

Other hypotheses generated by the gene-for-gene concept focus on the nature of the interaction between the plant and the gall midge larva. The assumption is that the plant defends itself after detecting an elicitor produced by the larva. Toward this end, molecular genetic studies on the Hessian fly have focused on mapping *avr* genes. Using both bulked segregant analysis (111a, 119, 130) and an AFLP-based linkage map (J. Stuart et al., unpublished data), *vH3*, *vH5*, *vH6*, *vH9*, and *vH13* have been positioned to specific regions of the Hessian fly polytene chromosomes. This information has permitted the initiation of a chromosome walk designed to clone *vH13*. The correspondence of candidate *avr* gene products identified by other methods can now be tested by determining whether the genes that encode those products reside on the Hessian fly polytene in positions that correspond to mapped *avr* genes.

A second assumption is that the sequence of defense responses is the same regardless of which *R* gene triggered the sequence. A first step will be the unequivocal identification of the defense mechanisms that prevent the gall midge larva from establishing a feeding site. Unequivocal identifications of plant defense mechanisms are also rare for pathogen-plant interactions (58). Microscopy studies, such as those described for the Hessian fly [(57); O. Rohfritsch, unpublished data], as well as biochemical and physiological studies will be necessary.

Although the gene-for-gene concept will remain important, studies of the ecology of gall midge–plant interactions will be necessary for studies of antagonistic coevolution. The gene-for-gene concept can create the impression that *R* genes deployed in crop plants function in a vacuum, independent of noncrop ecosystems as well as other natural mortality factors. This is plainly not the case for gall midge–plant interactions. There are many anecdotal examples of gall midge populations building up on wild hosts and then invading crop hosts (5), yet nothing is known of gall midge population dynamics on wild hosts or adult female movement between wild and crop hosts. Wild hosts also may be important reservoirs of *R* genes. This may be true for the greenbug, an aphid that attacks wheat and other cereal grasses (105). Natural mortality factors in both crop and noncrop ecosystems also may influence for antagonistic coevolution. For all three gall midge species, egg-larval parasitoids in the genus *Platygaster* cause significant mortality; yet, little is known of their biology (5, 15, 118). Studies of interactions of gall midges with other organisms (e.g., other insects, pathogens, and nematodes) that attack the plant will provide the foundations for studies on cross-talk between plant defense pathways (31, 101).

The costs of defense and virulence also are of interest. Because crop cultivars carrying gall midge *R* genes have been happily grown by farmers, the assumption is that defense via *R* genes bears little cost. Yet the costs of resistance need to be tested, preferably under a variety of ecological conditions (59, 106). The cost of virulence to gall midges is interesting because it represents the loss of function of an *avr* gene. As in pathogen-plant interactions (1, 133), gall midge *avr* genes may encode pathogenicity or virulence factors critical for the survival and growth of the larva. Does the loss of function of *avr* genes result in reduced fitness on plants that do not carry *R* genes? Assessing the costs of resistance and virulence will benefit from genetic studies of gall midge *R* genes and *avr* genes, especially when it becomes possible to precisely insert *R* genes via marker-assisted selection or transform plants with *R* genes (106).

As discussed by others (45, 46, 101, 111), the integration of genetic and ecological studies will benefit agriculture. Understanding the genetic architecture of *R* genes and *avr* genes, as well as the function of these genes, will improve our chances of finding, selecting, and deploying insect *R* genes in a more intelligent manner. Mapping and cloning of *R* genes will make it easier to create new resistant cultivars through marker-assisted breeding (89) or genetic transformation. Finally, a detailed understanding of a small number of gall midge–plant interactions will help us make smarter decisions when novel *R* genes are deployed against gall midges or other insects. Gall midge *R* genes ought to be recognized as valuable resources that, if used in an informed and creative manner, will provide good control of a number of devastating insect pests for many years to come.

ACKNOWLEDGMENTS

Many thanks to John Weiland, Ray Gagné, Bill Black, Kirk Hartel, Stig Larsson, Jennifer Brockmann, Carla Jordon, Karin Gross, and Kirk Anderson for helpful comments on earlier versions of the manuscript.

The *Annual Review of Entomology* is online at http://ento.annualreviews.org

LITERATURE CITED

1. Agrios G. 1997. *Plant Pathology*. San Diego: Academic. 635 pp.
2. Anderson M, Brewer G. 1991. Mechanisms of hybrid sunflower resistance to the sunflower midge (Diptera: Cecidomyiidae). *J. Econ. Entomol.* 84:1060–67
3. Barker P, McKenzie R. 1996. Possible sources of resistance to the wheat midge in wheat. *Can. J. Plant Sci.* 76:689–95
4. Barnes HF. 1931. The sex ratio at the time of emergence and the occurrence of unisexual families in the gall midges (Diptera: Cecidomyiidae). *J. Genet.* 14:225–34
5. Barnes HF. 1956. *Gall Midges of Cereals*. London: Lockwood
6. Behura SK, Sahu SC, Mohan M, Nair S. 2001. *Wolbachia* in the Asian rice gall midge, *Orseolia oryzae* (Wood-Mason): correlation between host mitotypes and infection status. *Insect Mol. Biol.* 10:163–71

6a. Behura SK, Nair, S, Mohan M. 2001. Polymorphisms flanking the *mariner* integration sites in the rice gall midge (*Orseolia oryzae* Wood Mason) genome are biotype specific. *Genome* 44:947–54

7. Behura SK, Sahu SC, Nair S, Mohan M. 2000. An AFLP marker that differentiates biotypes of the Indian gall midge (*Orseolia oryzae*, Wood-Mason) is linked to sex and avirulence. *Mol. Gen. Genet.* 263:328–34
8. Behura SK, Sahu SC, Rajamani S, Devi A, Mago R. 1999. Differentiation of Asian rice gall midge, *Orseolia oryzae* (Wood-Mason), biotypes by sequence characterized amplified regions (SCARs). *Insect Mol. Biol.* 8:391–97
9. Benhamou N. 1996. Elicitor-induced plant defence pathways. *Trends Plant Sci.* 1:233–40
10. Bent AF. 1996. Plant disease resistance genes: function meets structure. *Plant Cell* 8:1757–71
11. Bentur JS, Kalode MB. 1996. Hypersensitive reaction and induced resistance in rice against the Asian rice gall midge *Orseolia oryzae*. *Entomol. Exp. Appl.* 78:77–81
12. Bentur JS, Pasalu IC, Kalode MB. 1992. Inheritance of virulence in rice gall midge (*Orseolia oryzae*). *Indian J. Agric. Sci.* 62:492–93
13. Berenbaum M, Zangerl A. 1998. Chemical phenotype matching between a plant and its insect herbivore. *Proc. Natl. Acad. Sci. USA* 95:13743–48
14. Bergh J, Harris MO, Rose S. 1992. Factors inducing mated behavior in female Hessian flies (Diptera: Cecidomyiidae). *Ann. Entomol. Soc. Am.* 85:224–33
15. Berzonsky W, Shanower T, Lamb R, McKenzie R, Ding H, et al. 2002. Breeding wheat for resistance to insects. *Plant Breed. Rev.* 22:221–97

16. Black WC IV, Hatchett JH, Krchma LJ. 1990. Allozyme variation among populations of the Hessian fly (*Mayetiola destructor*) in the United States. *J. Hered.* 81:331–37
17. Buntin GD, Bruckner PL, Johnson JW, Foster JE. 1990. Effectiveness of selected genes for Hessian fly resistance in wheat. *J. Agric. Entomol.* 7:284–91
17a. Chapman G, ed. 1992. *Grass Evolution and Domestication*. Cambridge, UK: Cambridge Univ. Press. 390 pp.
18. Chapman G. 1992. Preface. See Ref. 17a, pp. xv–xvii
19. Christou P. 1997. Rice transformation: bombardment. In *Oryza: From Molecule to Plant*, T Sasaki, G Moore, pp. 197–203. Netherlands: Kluwer. 254 pp.
20. Cox TS, Hatchett JH. 1986. Genetic model for wheat/Hessian fly (Diptera: Cecidomyiidae) interaction: strategies for deployment of resistance genes in wheat cultivars. *Environ. Entomol.* 15: 24–31
21. Cruz CMV, Bai J, Ona I, Leung H, Nelson RJ. 2000. Predicting durability of a disease resistance gene based on an assessment of the fitness loss and epidemiological consequences of avirulence gene mutation. *Proc. Natl. Acad. Sci. USA* 97:13500–5
22. De Moares C, Lewis W, Pare P, Alborn H, Tumlinson J. 1998. Herbivore-infested plants selectively attract parasitoids. *Nature* 393:570–72
23. de Wet J. 1992. The three phases of cereal domestication. See Ref. 17a, pp. 176–98
24. De Wit PJGM. 1992. Molecular characterization of gene-for-gene systems in plant-fungus interactions and the application of avirulence genes in the control of plant pathogens. *Annu. Rev. Phytopathol.* 30:391–418
25. Devos KM, Gale MD. 1997. Comparative genetics in the grasses. *Plant Mol. Biol.* 35:3–15
26. Diehl SR, Bush GL. 1984. An evolutionary and applied perspective of insect biotypes. *Annu. Rev. Entomol.* 29:471–504
27. Ding H, Lamb R, Ames N. 2000. Inducible production of phenolic acids in wheat and antibiotic resistance to *Sitodiplosis mosellana*. *J. Chem. Ecol.* 26:969–85
28. Dweikat I, Ohm H, Patterson F, Cambron S. 1997. Identification of RAPD markers for 11 Hessian fly resistance genes in wheat. *Theor. Appl. Genet.* 94:419
29. El Bouhssini ME, Hatchett JH, Cox TS, Wilde GE. 2001. Genotypic interaction between resistance genes in wheat and virulence genes in the Hessian fly *Mayetiola destructor* (Diptera: Cecidomyiidae). *Bull. Entomol. Res.* 91: 327–31
30. Ellis J, Dodds P, Pryor T. 2000. Structure, function and evolution of plant resistance genes. *Curr. Opin. Plant Biol.* 3:278–84
31. Felton G, Korth K. 2000. Trade-offs between pathogen and herbivore resistance. *Curr. Opin. Plant Biol.* 3:309–14
32. Fletcher J. 1888. *Report of the entomologist and botanist*. Gov. Can., Ottawa. 152 pp.
33. Flor HH. 1946. Genetics of pathogenicity in *Melampsora lini*. *J. Agric. Res.* 73:335–57
34. Formusoh ES, Hatchett JH, Black WC IV, Stuart JJ. 1996. Sex-linked inheritance of virulence against wheat resistance gene *H9* in the Hessian fly (Diptera: Cecidomyiidae). *Ann. Entomol. Soc. Am.* 89:428–44
35. Foster JE, Ohm H, Patterson F, Taylor P. 1991. Effectiveness of deploying single gene resistances in wheat for controlling damage by the Hessian fly (Diptera: Cecidomyiidae). *Environ. Entomol.* 20:964–69
36. Foster SP, Bergh J, Rose S, Harris MO. 1991. Aspects of pheromone biosynthesis in the Hessian fly, *Mayetiola destructor* (Say). *J. Insect Physiol.* 12:899–906
37. Foster SP, Harris MO. 1992. Foliar chemicals of wheat and related grasses influencing oviposition by Hessian fly,

Mayetiola destructor. *J. Chem. Ecol.* 18:1965–80

38. Foster SP, Harris MO, Millar J. 1991. Identification of the sex pheromone of the Hessian fly, *Mayetiola destructor* (Say). *Naturwissenschaften* 78:130–31
39. Gagné R. 1989. *The Plant-Feeding Gall Midges of North America*. Ithaca: Cornell Univ. Press
40. Gagné R, Hatchett JH, Lhaloui S, El Bouhssini ME. 1991. Hessian fly and barley stem gall midge, two different species of *Mayetiola* (Diptera: Cecidomyiidae) in Morocco. *Ann. Entomol. Soc. Am.* 84:436–43
41. Gallun RL. 1977. Genetic basis of Hessian fly epidemics. *Ann. NY Acad. Sci.* 287:222–29
42. Gallun RL. 1978. Genetics of biotypes B and C of the Hessian fly. *Ann. Entomol. Soc. Am.* 71:481–86
43. Gallun RL, Deay HO, Cartwright WB. 1961. Four races of Hessian fly selected and developed from an Indiana population. *Purdue Agric. Res. Stn. Bull.* 732:1–8
44. Gallun RL, Hatchett JH. 1969. Genetic evidence of elimination of chromosomes in the Hessian fly. *Ann. Entomol. Soc. Am.* 62:1095–101

44a. Goff SA, Ricke D, Lan TH, Presting G, Wang R, et al. 2002. A draft sequence of the rice genome (*Oryza sativa* L. ssp. japonica). *Science* 296:92–100

45. Gould F. 1986. Simulation models for predicting durability of insect-resistant germ plasm: Hessian fly (Diptera: Cecidomyiidae)-resistant winter wheat. *Environ. Entomol.* 15:11–23
46. Gould F. 1998. Sustainability of transgenic insecticidal cultivars: integrating pest genetics and ecology. *Annu. Rev. Entomol.* 43:701–26
47. Gries R, Gries G, Khaskin G, King S, Olfert O, et al. 2000. Sex pheromone of the orange wheat blossom midge, *Sitodiplosis mosellana*. *Naturwissenschaften* 87:450–54
48. Grover P. 1995. Hypersensitive response of wheat to the Hessian fly. *Entomol. Exp. Appl.* 74:283–94
49. Grover P, Shukle RH, Foster JE. 1989. Interactions of Hessian fly (Diptera: Cecidomyiidae) biotypes on resistant wheat. *Environ. Entomol.* 18:687–90
50. Hammond-Kosack KE, Jones JDG. 1997. Plant disease resistance genes. *Annu. Rev. Plant Physiol. Plant Mol. Biol.* 48:575–607
51. Harris MO, Foster SP. 1999. Gall midges. In *Pheromones of Non-Lepidopteran Insects in Agriculture*, ed. J Hardie, A Minks, pp. 27–49. Oxford: CAB Int. 466 pp.
52. Harris MO, Rose S. 1989. Temporal changes in the egglaying behavior of the Hessian fly, *Mayetiola destructor*. *Entomol. Exp. Appl.* 53:17–29
53. Harris MO, Rose S, Malsch P. 1993. The role of vision in the host-plant finding behaviour of the Hessian fly. *Physiol. Entomol.* 18:31–42
54. Harris MO, Sandanayaka M, Griffin W. 2001. Oviposition preferences of the Hessian fly and their consequences for the survival and reproductive potential of offspring. *Ecol. Entomol.* 26:473–86
55. Harushima Y, Yano M, Shomura A, Sato M, Shimano T. 1998. A high-density rice genetic linkage map with 2275 markers using a single F2 population. *Genetics* 148:479–94
56. Hatchett JH, Gallun RL. 1970. Genetics of the ability of the Hessian fly, *Mayetiola destructor*, to survive on wheats having different genes for resistance. *Ann. Entomol. Soc. Am.* 63:1400–7
57. Hatchett JH, Kreitner GL, Elzinga RJ. 1990. Larval mouthparts and feeding mechanism of the Hessian fly (Diptera: Cecidomyiidae). *Ann. Entomol. Soc. Am.* 83:1137–47
58. Heath M. 2000. Nonhost resistance and nonspecific plant defenses. *Curr. Opin. Plant Biol.* 3:315–19
59. Heil M, Baldwin I. 2002. Fitness costs

of induced resistance: emerging experimental support for a slippery concept. *Trends Plant Sci.* 7:61–67
60. Huang N, Angeles E, Domingo J, Magpantay G, Singh S. 1997. Pyramiding of bacterial blight resistance genes in rice: marker-aided selection using RFLP and PCR. *Theor. Appl. Genet.* 95:313–20
61. Hunter B. 2001. *Rage for Grain: Flour Milling in the Mid-Atlantic, 1750–1815.* PhD thesis. Univ. Delaware. 322 pp.
62. Kanno H, Harris MO. 2000. Leaf physical and chemical features influence the selection of plant genotypes by the Hessian fly. *J. Chem. Ecol.* 26:2335–54
63. Kanno H, Harris MO. 2000. Physical features of grass leaves influence the placement of eggs within the plant by the Hessian fly. *Entomol. Exp. Appl.* 96:69–80
64. Kanno H, Harris MO. 2002. Avoidance of occupied hosts by the Hessian fly: oviposition behaviour and consequences for larval survival. *Ecol. Entomol.* 27:177–88
65. Kar B. 1999. *Cytogenetics of the rice gall midge,* Orseolia oryzae *Wood-Mason.* PhD thesis. Utkal Univ., Bhubneswar, Orissa, India. 105 pp.
66. Karban R, Baldwin I. 1997. *Induced Responses to Herbivory.* Chicago: Univ. Chicago Press
67. Kareiva P. 1999. Coevolutionary arms races: is victory possible? *Proc. Natl. Acad. Sci. USA* 96:8–10
68. Katiyar SK, Tan Y, Huang B, Chandel G, Xu Y, et al. 2001. Molecular mapping of gene *Gm-6(t)* which confers resistance against four biotypes of Asian rice gall midge in China. *Theor. Appl. Genet.* 103:953–61
69. Keen NT. 1990. Gene-for-gene complementarity in plant-pathogen interactions. *Annu. Rev. Genet.* 24:447–63
70. Kochert G. 1992. Rice as a model system. See Ref. 17a, pp. 290–315
71. Lagudah E, Appels R. 1992. Wheat as a model system. See Ref. 17a, pp. 225–65
72. Lagudah ES, Moullet O, Appels R. 1997. Map-based cloning of a gene sequence encoding a nucleotide-binding domain and a leucine-rich region at the *Cre3* nematode resistance locus of wheat. *Genome* 40:659–65
73. Lamb R, McKenzie R, Wise I, Barker P, Smith M, Olfert O. 2000. Resistance to wheat midge, *Sitodiplosis mosellana* (Diptera: Cecidomyiidae), in spring wheat (Gramineae). *Can. Entomol.* 132: 591–95
74. Lamb R, Smith M, Wise I, Clarke P, Clarke J. 2001. Oviposition deterrence to *Sitodiplosis mosellana* (Diptera: Cecidomyiidae): a source of resistance for durum wheat (Gramineae). *Can. Entomol.* 133:579–91
75. Lamb R, Tucker J, Wise I, Smith M. 2000. Trophic interaction between *Sitodiplosis mosellana* (Diptera: Cecidomyiidae) and spring wheat: implications for seed production. *Can. Entomol.* 132:607–25
76. Lamb R, Wise I, Olfert O, Gavloski J, Barker P. 1999. Distribution and seasonal abundance of the wheat midge, *Sitodiplosis mosellana* (Diptera: Cecidomyiidae), in Manitoba. *Can. Entomol.* 131:387–98
77. Lamb R, Wise I, Smith M, McKenzie R, Thomas J, Olfert O, et al. 2002. Oviposition deterrence against *Sitodiplosis mosellana* (Diptera: Cecidomyiidae) in spring wheat (Gramineae). *Can. Entomol.* 134:85–96
78. Larsson S, Ekbom B. 1995. Oviposition mistakes in herbivorous insects: confusion or a step towards a new host plant? *Oikos* 72:155–60
79. Leach JE, White FF. 1996. Bacterial avirulence genes. *Annu. Rev. Phytopathol.* 34:153–79
80. Lemaux P, Qualset C. 2001. Advances in technology for wheat breeding. *Israel J. Plant Sci.* 49:105–15
81. Lhaloui S, Hatchett JH, Wilde GE. 1996. Evaluation of New Zealand barleys for resistance to *Mayetiola destructor* and

M. hordei (Diptera: Cecidomyiidae) and the effect of temperature on resistance expression to Hessian fly. *J. Econ. Entomol.* 89:562–67
82. Luderer R, Joosten MHAJ. 2001. Avirulence proteins of plant pathogens: determinants of victory and defeat. *Mol. Plant Pathol.* 2:355–64
83. Ma Z-Q, Gill BS, Sorrells ME, Tanksley SD. 1993. RFLP markers linked to two Hessian fly resistance genes in wheat (*Triticum aestivum*, L.) from *Triticum tauschii* (Coss.) Schmal. *Theor. Appl. Genet.* 85:750–54
84. Martin GM. 1999. Functional analysis of plant disease resistance genes and their downstream effectors. *Curr. Opin. Plant Biol.* 2:273–79
85. McColloch J. 1917. Wind as a factor in the dispersion of the Hessian fly. *J. Econ. Entomol.* 10:162–70
86. McKenzie R, Lamb R, Aung T, Wise I, Barker P, Olfert O. 2002. Inheritance of resistance to wheat midge, *Sitodiplosis mosellana*, in spring wheat. *Plant Breed.* In press
87. Meyers BC, Dickerman AW, Michelmore RW, Sivaramakrishan S, Sobral BW, Young ND. 1999. Plant disease resistance genes encode members of an ancient and diverse protein family within the nucleotide-binding superfamily. *Plant J.* 20:317–32
88. Mitchell-Olds T, Bergelson J. 2000. Biotic interactions: genomics and coevolution. *Curr. Opin. Plant Biol.* 3:273–77
89. Mohan M, Nair S, Bhagwat A, Krishna TG, Yano M. 1997. Genome mapping, molecular markers and marker-assisted selection in crop plants. *Mol. Breed.* 3:87–103
90. Mohan M, Nair SN, Bentur JS, Rao UP, Bennett J. 1994. RFLP and RAPD mapping of the rice *Gm2* gene that confers resistance to biotype 1 of gall midge (*Orselia oryzae*). *Theor. Appl. Genet.* 87:782–88
91. Mohan M, Sathyanarayanan PV, Kumar A, Srivastava MN, Nair S. 1997. Molecular mapping of a resistance-specific PCR-based marker linked to a gall midge resistance gene (*Gm4t*) in rice. *Theor. Appl. Genet.* 95:777–82
92. Moran P, Thompson G. 2001. Molecular responses to aphid feeding in *Arabidopsis* in relation to plant defense pathways. *Plant Physiol.* 125:1074–85
93. Morris B, Foster S, Harris M. 2000. Identification of 1-octacosanal and 6-methoxy-2-benzoxazolinone (MBOA) from wheat as ovipositional stimulants for the Hessian fly. *J. Chem. Ecol.* 26:859–73
94. Nair S, Bentur JS, Rao PU, Mohan M. 1995. DNA markers tightly linked to a gall midge resistance gene (*Gm2*) are potentially useful for marker-aided selection in rice breeding. *Theor. Appl. Genet.* 91:68–73
95. Nair S, Kumar A, Srivastava MN, Mohan M. 1996. PCR-based DNA markers linked to a gall midge resistance gene, *Gm4t* has potential for marker-aided selection in rice. *Theor. Appl. Genet.* 92:660–65
96. Ollerstam O, Rohfritsch O, Hoglund S, Larsson S. 2002. A rapid hypersensitive response associated with resistance in the willow *Salix viminalis* against the gall midge *Dasineura marginemtorquens*. *Entomol. Exp. Appl.* 102:153–62
97. Painter RH. 1930. Observations on the biology of the Hessian fly. *J. Econ. Entomol.* 23:326–28
98. Painter RH. 1951. *Insect Resistance in Crop Plants*. New York: Macmillan
99. Pan Q, Wendel J, Fluhr R. 2000. Divergent evolution of plant NBS-LRR resistance gene homologues in dicot and cereal genomes. *J. Mol. Evol.* 50:203–13
100. Patterson F, Maas F, Foster JE, Ratcliffe RH, Cambron S, et al. 1994. Registration of eight Hessian fly resistant common winter wheat germplasm lines (Carol,

Erin, Flynn, Iris, Joy, Karen, Lola, and Molly). *Crop Sci.* 34:314–15
101. Paul N, Hatcher P, Taylor J. 2000. Coping with multiple enemies: an integration of molecular and ecological perspectives. *Trends Plant Sci.* 5:220–25
102. Peries I. 1994. Studies on host-plant resistance in rice to gall midge *Orseolia oryzae*. In *Plant Galls*, ed. M Williams, 49:231–43. Oxford: Clarendon
103. Deleted in proof
104. Pivnick K, Labbé E. 1993. Daily patterns of activity of females of the orange wheat blossom midge, *Sitodiplosis mosellana* (Gehin) (Diptera: Cecidomyiidae). *Can. Entomol.* 125:725–36
105. Porter D, Burd J, Shufran K, Webster J, Teetes G. 1997. Greenbug (Homoptera: Aphididae) biotypes: selected by resistant cultivars or preadapted opportunists? *J. Econ. Entomol.* 90:1055–65
106. Purrington C. 2000. Costs of resistance. *Curr. Opin. Plant Biol.* 3:305–8
107. Rajyashri KR, Nair S, Ohmido N, Fukui K, Kurata N, et al. 1998. Isolation and FISH mapping of yeast artificial chromosomes (YACs) encompassing an allele of the Gm2 gene for gall midge resistance in rice. *Theor. Appl. Genet.* 97:507–14
108. Ratcliffe RH, Hatchett JH. 1997. Biology and genetics of the Hessian fly and resistance in wheat. In *New Developments in Entomology*, ed. K Bondari, pp. 47–56. Trivandurm, India: Res. Signpost, Sci. Inf. Guild
109. Ratcliffe RH, Ohm HW, Patterson FL, Cambron SE, Safranski GG. 1996. Response of resistance genes *H9–H19* in wheat to Hessian fly (Diptera: Cecidomyiidae) laboratory biotypes and field populations from the eastern United States. *J. Econ. Entomol.* 89:1309–17
110. Ratcliffe RH, Safranski GG, Patterson FL, Ohm HW, Taylor PL. 1994. Biotype status of Hessian fly (Diptera: Cecidomyiidae) populations from the eastern United States and their response to 14 Hessian fly resistance genes. *J. Econ. Entomol.* 87:1113–21
111. Rausher M. 2001. Co-evolution and plant resistance to natural enemies. *Nature* 411:857–64
111a. Rider SD, Sun W, Ratcliffe RH, Stuart JJ. 2002. Chromosome landing near avirulence gene *vH13* in the Hessian fly. *Genome*. In press
112. Rohfritsch O. 1992. Patterns in gall development. In *Biology of Insect-Induced Galls*, ed. J Shorthouse, O Rohfritsch, pp. 60–86. New York: Oxford Univ. Press
113. Rohfritsch O. 1999. A so-called 'rudimentary gall' induced by the gall midge *Physemocecis hartigi* on leaves of *Tilia intermedia*. *Can. J. Bot.* 77:460–47
114. Rossi M, Googin RL, Milligan SB, Kaloshian I, Ullman DE, Williamson VM. 1998. The nematode resistance gene *Mi* of tomato confers resistance against the potato aphid. *Proc. Natl. Acad. Sci. USA* 95:9750–54
115. Sahu SC, Bose L, Pani J, Rajamani S, Mathur KC. 1996. Karyotype of rice gall midge, *Orseolia oryzae* Wood-Mason. *Curr. Sci.* 70:874
116. Sain M, Kalode MB. 1988. Production of unisexual progeny in rice gall midge, *Orseolia oryzae* (Wood-Mason). *Curr. Sci.* 57:860–61
117. Sardesai N, Kumar A, Rajyashri KR, Nair S, Mohan M. 2002. Identification and mapping of an AFLP marker linked to *Gm7*, a gall midge resistance gene and its conversion to a SCAR marker for its utility in marker-aided selection in rice. *Theor. Appl. Genet.* In press
118. Sardesai N, Rajyashri KR, Behura SK, Nair S, Mohan M. 2001. Genetic, physiological and molecular interactions of rice and its major dipteran pest, gall midge. *Plant Cell Tissue Org. Cult.* 64:115–31
119. Schulte SJ, Rider SD, Hatchett JH, Stuart JJ. 1999. Molecular genetic mapping of three X-linked avirulence genes, *vH6*, *vH9* and *vH13*, in the Hessian fly. *Genome* 42:821–28

120. Sharma H, Mukuru S, Prasad K, Manyasa E, Pande S. 1999. Identification of stable sources of resistance in sorghum to midge and their reaction to leaf diseases. *Crop Prot.* 18:29–37
121. Smart M, O'Brien T. 1983. The development of the wheat embryo in relation to the neighbouring tissues. *Protoplasma* 114:1–13
122. Smith M, Lamb R. 2001. Factors influencing oviposition by *Sitodiplosis mosellana* (Diptera: Cecidomyiidae) on wheat spikes (Gramineae). *Can. Entomol.* 133:533–48
123. Somssich I, Hahlbrock K. 1998. Pathogen defence in plants—a paradigm of biological complexity. *Trends Plant Sci.* 3:86–90
124. Sosa O. 1981. Biotypes J and L of the Hessian fly discovered in an Indiana wheat field. *J. Econ. Entomol.* 74:180–82
125. Staskawicz BJ, Ausubel FM, Baker BJ, Ellis JG, Jones JDG. 1995. Molecular genetics of plant disease resistance. *Science* 268:661–67
126. Stuart JJ, Hatchett JH. 1987. Morphogenesis and cytology of the salivary gland of the Hessian fly, *Mayetiola destructor* (Diptera: Cecidomyiidae). *Ann. Entomol. Soc. Am.* 80:475–82
127. Stuart JJ, Hatchett JH. 1988. Cytogenetics of the Hessian fly, *Mayetiola destructor* (Say). I. Mitotic karyotype analysis and polytene chromosome correlations. *J. Hered.* 79:184–89
128. Stuart JJ, Hatchett JH. 1988. Cytogenetics of the Hessian fly, *Mayetiola destructor* (Say). II. Inheritance and behavior of somatic and germ-line-limited chromosomes. *J. Hered.* 79:190–99
129. Stuart JJ, Hatchett JH. 1991. Genetics of sex determination in the Hessian fly, *Mayetiola destructor*. *J. Hered.* 82:43–52
130. Stuart JJ, Schulte SJ, Hall PS, Mayer KM. 1998. Genetic mapping of Hessian fly avirulence gene *vH6* using bulked segregant analysis. *Genome* 41:702–8
131. Tang X, Frederick RD, Zhou J, Halterman DA, Jia Y, et al. 1996. Initiation of plant disease resistance by physical interaction of *AvrPto* and the *Pto* kinase. *Science* 274:2060–63
132. Van der Biezen E, Jones JDG. 1998. Plant disease-resistance proteins and the gene-for-gene concept. *Trends Biochem. Sci.* 23:454–56
133. White FF, Yang B, Johnson LB. 2000. Prospects for understanding avirulence gene function. *Curr. Opin. Plant Biol.* 3:291–98
134. White MJD. 1950. *Cytological studies on gall midges (Cecidomyiidae)*. Univ. Texas Publ. 5007:5–80
135. White MJD. 1973. *Animal Cytology and Evolution*. Cambridge: Cambridge Univ. Press

135a. Williams CE, Collier CC, Nemacheck JA, Liang C, Cambron SE. 2002. A lectin-like gene responds systemically to attempted feeding by avirulent first-instar Hessian fly larvae. *J. Chem. Ecol.* 28:1411–28

136. Williamson VM. 1998. Root-knot nematode resistance genes in tomato and their potential for future use. *Annu. Rev. Phytopathol.* 36:227–93
137. Williamson VM, Hussey RS. 1996. Nematode pathogenesis and resistance in plants. *Plant Cell* 8:1101–15
138. Wise I, Lamb R, Smith M. 2001. Domestication of wheats (Gramineae) and their susceptibility to herbivory by *Sitodiplosis mosellana* (Diptera: Cecidomyiidae). *Can. Entomol.* 133:255–67
139. Withers T, Harris MO. 1996. Foraging for oviposition sites in the Hessian fly (*Mayetiola destructor*): random and nonrandom aspects of movement. *Ecol. Entomol.* 21:382–95
140. Withers T, Harris MO. 1997. The influence of wind on Hessian fly flight and egglaying behavior. *Environ. Entomol.* 26:327–33

141. Withers T, Harris MO, Madie C. 1997. Dispersal of mated female Hessian flies in field arrays of host and non-host plants. *Environ. Entomol.* 26:1247–57
141a. Wolfe KH, Gouy M, Yang YW, Sharp PM, Li WH. 1989. Date of the monocot-dicot divergence estimated from chloroplast DNA sequence data. *Proc. Natl. Acad. Sci. USA* 86:6201
142. Yencho G, Cohen M, Byrne P. 2000. Applications of tagging and mapping insect resistance loci in plants. *Annu. Rev. Entomol.* 45:393–422
143. Yoshimura S, Yamanouchi U, Katayose Y, Toki S, Wang Z-X. 1998. Expression of *Xa1*, a bacterial blight-resistance gene in rice, is induced by bacterial inoculation. *Proc. Natl. Acad. Sci. USA.* 95:1663–68
144. Young ND. 2000. The genetic architecture of resistance. *Curr. Opin. Plant Biol.* 3:285–90
144a. Yu J, Wang J, Wang GKS, Li S, et al. 2002. A draft sequence of the rice genome (*Oryza sativa* L. ssp. indica). *Science* 296:79–92
145. Zantoko L, Shukle RH. 1997. Genetics of virulence in the Hessian fly to resistance gene *H13* in wheat. *J. Hered.* 88:120–23
146. Zheng S. 1965. *Wheat Midge*. Beijing: Agricultural Press

Annu. Rev. Entomol. 2003. 48:579–602
doi: 10.1146/annurev.ento.48.091801.112749

First published online as a Review in Advance on September 25, 2002

ANALYSIS AND FUNCTION OF TRANSCRIPTIONAL REGULATORY ELEMENTS: Insights from *Drosophila*

David N. Arnosti
Department of Biochemistry and Molecular Biology and Program in Genetics, Michigan State University, East Lansing, Michigan 48824-1319; e-mail: arnosti@msu.edu

Key Words transcription, promoter, enhancer, transcriptional regulation

■ **Abstract** Analysis of gene expression is assuming an increasingly important role in elucidating the molecular basis of insect biology. Transcriptional regulation of gene expression is directed by a variety of *cis*-acting DNA elements that control spatial and temporal patterns of expression. This review summarizes current knowledge about properties of transcriptional regulatory elements, based largely on research in *Drosophila melanogaster*, and outlines ways that new technologies are providing tools to facilitate the study of transcriptional regulatory elements in other insects.

CONTENTS

INTRODUCTION

Transcriptional regulation is woven into the entire fabric of biology; thus, knowledge of gene expression is a critical component of understanding many issues relevant to entomology, including pesticide and disease resistance, behavior, evolution, and development. These and other topics have been studied with the tools of molecular biology, allowing the identification of relevant structural genes; however, the control of these genes is often poorly understood. Transcriptional regulation is effected by *cis*-acting DNA sequences that direct the assembly of the protein machinery responsible for transcription. Alterations in these *cis*-acting regulatory sequences are thought to play driving roles in morphological diversification and evolution of developmental mechanisms (106). In addition, quantitative trait loci that contribute to adaptations often correspond to genetic alterations in putative *cis*-acting regulatory elements rather than protein-coding regions of genes (68).

Often the entomologist will be presented with sequences of genomic DNA that may contain transcriptional regulatory sequences, but functional understanding of these DNA elements can be hampered by the lack of genetic and molecular genetic tools. Functional analysis of insect transcriptional regulatory regions is based largely on work in *Drosophila melanogaster*. Fortunately, general properties of eukaryotic transcriptional regulation are highly conserved, especially among metazoans; thus, lessons learned from *Drosophila* can often be directly applied to other insects. The extensive analysis of *Drosophila* transcriptional "wiring diagrams" should provide the basis for accelerated analysis of *cis*-acting elements in other organisms.

This review focuses on the *cis*-acting DNA sequences, rather than the *trans*-acting transcriptional machinery, to assist in answering questions about subjects such as the size of a regulatory region, the possibility of coordinate regulation of closely spaced genes, the nature and importance of basal promoter sequences, and how to predict the transcriptional output of a given promoter. We are still some way from answering all these questions without the use of empirical tests, but consideration of general properties of transcriptional control regions identified in *Drosophila* will allow investigators to identify features that might be important for their system. In addition, new bioinformatic approaches promise to directly "read" transcriptional regulatory information from the genome, at least partly circumventing the need to experimentally determine the function of *cis*-acting sequences. This possibility is especially important to those working in genetically less tractable systems.

For more information on the transcriptional machinery, the reader is directed to recent reviews on RNA polymerase II transcription (58, 80, 84, 116), chromatin and chromatin remodeling machinery (90, 115), transcriptional activators and repressors (26, 58), boundary elements (4, 36), more-specialized elements such as molecular memory modules (4, 69), and heterochromatin (42).

FEATURES OF TRANSCRIPTIONAL REGULATORY REGIONS

Three general types of *cis*-acting elements control the activity of RNA polymerase II–transcribed (i.e., protein coding) genes (Figure 1): (*a*) basal promoter sequences near the transcriptional initiation site. These elements provide a binding site for RNA polymerase II and the basal transcriptional machinery that acts on most promoters. (*b*) Enhancer elements that contain binding sites for sequence-specific transcription activators and repressors, which regulate levels of gene activity. (*c*) Boundary elements or insulators, which can functionally separate regulatory elements. Rather than acting in separate roles, the three types of elements are functionally interrelated, and these complexities contribute to the specificity of gene regulation. The term "promoter" is often used to indicate all regulatory sequences associated with a gene, including enhancer sequences and the basal promoter. More specialized elements, such as those that interact with Polycomb and Trithorax regulatory proteins, are not thought to be associated with most genes and are discussed in recent reviews (31, 69).

Transcriptional regulatory information can be located entirely within 100 bp of the transcriptional initiation site, as with testis-specific promoters, or in some

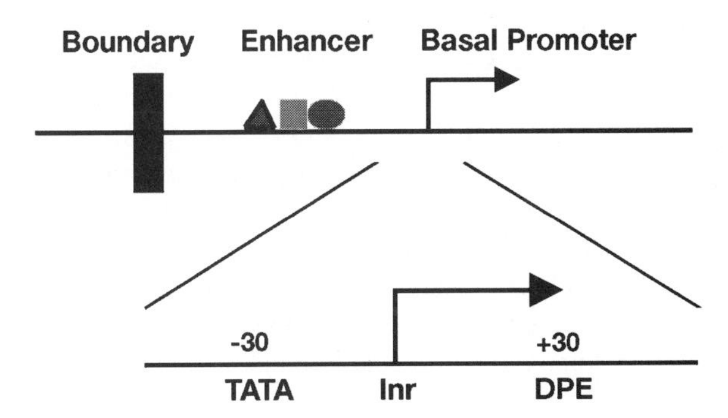

Figure 1 *Cis*-acting transcriptional control elements. Basal promoter includes ~100 bp of sequence flanking the initiation site, onto which general transcriptional machinery loads. Enhancer sequences (ranging in size from 50 bp to several hundred basepairs) are located at variable distances from +1; these bind to sequence-specific transcriptional regulators that control output levels of basal promoter. Boundary elements, found near some genes, insulate the gene from local chromatin effects and can screen the gene from the influence of distal enhancers. The inset shows the disposition of three sequence elements of basal promoters (see text).

cases distributed over a large region, as in the *bithorax* complex, which contains three HOX genes in a 300-kbp region replete with regulatory elements (34, 60). In general, larger regulatory regions are required for diverse tissue-specific and temporal patterns of expression because in most cases individual enhancers are required for separate portions of an expression pattern. Genes dedicated to a single function, such as tissue-specific structural genes, often contain all information for tissue and temporal specificity in a small regulatory element of a few hundred basepairs.

When no other information is available, the region within a few kilobasepairs 5′ of the initiation site is often the first place examined for regulatory information. For many genes, such as those encoding heat shock proteins, glue proteins, some immune-response proteins, cytochrome P450 enzymes, and chorion proteins, it appears that all necessary regulatory information is present within 1 kbp of the basal promoter (30, 57, 59, 78, 107). For a number of important regulatory genes, however, essential transcriptional information is found tens of kilobasepairs from the initiation site (33, 60, 75).

How representative are *cis*-regulatory elements in *Drosophila* for other insects and for metazoans in general? Given the high degree of conservation of the transcriptional machinery in all eukaryotes (2, 58), the ability of regulatory elements to function in heterologous systems (53, 77, 86), and the common occurrence of long-distance regulatory elements in metazoans, it is clear that in most aspects *Drosophila* is an appropriate model system. One area in which this system was not believed to be representative was in DNA methylation. Until recently, *D. melanogaster* was generally thought to lack 5-methylcytosine, an important modified nucleotide in mammalian genomes that functions in transcriptional regulatory phenomena such as imprinting and certain types of repression. DNA methylation in other insects such as aphid, cricket, and mosquito have been previously reported, leading to the question of whether *Drosophila* is atypical in this respect (37, 109). Recent genomic analyses showing the presence of genes homologous to DNA methyltransferase and methyl CpG binding proteins in *D. melanogaster* have prompted a reexamination of this issue (47, 109). Using new, more sensitive techniques, low levels of 5-methylcytosine have been identified, primarily during early stages of embryogenesis (37, 67). The possible functional role of cytosine methylation in *Drosophila* remains unknown, as in other insects; thus, we do not know whether some types of transcriptional regulation via DNA methylation will be accurately modeled in the fly.

BASAL PROMOTER ELEMENTS

A basal promoter can be defined as the ~100 bp of sequence surrounding the transcriptional initiation site that comes into close contact with the general transcriptional machinery. This region contains TATA sequences (consensus TATAAA) centered at −30, Initiator (Inr) sequences at +1 (consensus TCAGT), and downstream promoter elements (DPE) at +30 (consensus A/GGA/TC/TGT) (Figure 1).

Not all basal promoters contain all three elements. In a survey of 205 promoters with accurately mapped start sites, approximately half had a recognizable TATA sequence, almost half had a DPE, and about one third had neither (56). Initiator-like sequences were found in approximately one quarter of arthropod promoters surveyed (24). Identification of the transcriptional initiation site requires mapping the 5′ end of a transcribed RNA by primer extension, S1 nuclease protection assays, or cloning of cDNAs, especially those isolated based on the mRNA 5′ cap structure (104). Computer identification of basal promoters in genomic sequences has met with limited success, but a current effort to map ~2000 *Drosophila* transcription start sites from oligo-capped cDNA libraries should provide a much larger basis for development of bioinformatic methods [(81); U. Ohler, personal communication].

The elements of the basal promoter provide nucleation sites for binding by basal transcriptional machinery. The TATA sequence interacts with the TATA binding protein (TBP), a crucial part of the basal transcription machinery that helps anchor the RNA polymerase and basal transcription factors at the promoter. TBP is a subunit of the multicomponent TFIID general transcription factor, which contains about 10 TBP-associated proteins (TAFs) (2). The initiator region contacts TAFs and the RNA polymerase itself (23, 87). The DPE was identified by its contacts with TAF proteins and has been suggested to function as an alternative docking site for the TFIID factor (19).

The functional role of these basal promoter elements appears to be strongly context dependent. In vitro studies in mammalian systems indicated that TATA and Inr elements can be functionally redundant (121), but in vivo analysis of the *hsp70* promoter indicated that TATA function was indispensable and was not redundant with the Inr (117). On this promoter, Inr and DPE have important but largely redundant activities, so mutation of either element individually has little effect (117). In contrast to the *hsp70* promoter, loss of the conserved Initiator element of the TATA-less testis-specific β tubulin promoter has a stronger effect, decreasing transcription by one half (92). On the TATA-less *vermillion* promoter, a TATA element can substitute for a DPE (32), suggesting equivalence of function, but a number of embryonic enhancers tested can discriminate between otherwise identical basal promoters containing either a TATA element or a DPE (20, 83). An additional indication of functional distinctions between basal promoters is that TATA-containing promoters are repressed by the NC2 (Dr1/Drap1) basal transcription factor, whereas DPE-containing promoters are activated by the same factor in in vitro assays (113).

Basal promoter regions are also sometimes bound by sequence-specific transcription factors that play important roles in regulation of individual genes. Binding of the Zn finger Ovo protein to the initiator region of the *otu* gene is important for ovary-specific expression (64). The TATA-less *dpp* core promoter extending from −22 to +6 is bound by a sequence-specific factor in the embryo, and this core promoter is by itself sufficient to direct a qualitatively correct pattern of expression (96). GAGA factor binding sites located within the *even-skipped* basal promoter confer enhancer-blocking activities upon this element (82). The temporal regulation of two promoters of Adh has been linked to direct binding of the inactive promoter by the AEF-1 repressor protein (88). For the most part, it is not known whether sequence-specific factors have the same function when bound to the basal promoter as when they bind to more distal sequences. Often sequence-specific transcription factor binding sites are found close to the basal promoter, but in contrast to the examples cited above, the exact placement of sequence-specific regulators with respect to the transcriptional initiation site is not critical; thus, these binding sites can be considered as enhancer elements.

The general conclusion to be drawn from these studies is that basal promoter sequences and elements cannot be assumed to be generic and interchangeable, a consideration to be taken into account in the design of transgenic vectors. Although in cell culture diverse basal promoters can be functionally interchangeable (51), in some cases the structure of basal promoters can be important for the function of endogenous genes (99) (see Shared or Exclusive Enhancer-Promoter Interactions, below). Alternative forms of TBP protein have been found in *Drosophila*; these proteins are expressed in tissue-specific manners and have alternative sequence specificity, suggesting that alternative basal transcriptional complexes may nucleate on distinct types of basal promoters (110).

BOUNDARY ELEMENTS

Boundary elements or insulators are regulatory regions typically several hundred basepairs in length that bind to sequence-specific factors, establishing a barrier between regulatory elements on one side and basal promoters on the other and insulating genes from repressive effects of heterochromatin-mediated silencing. These elements have been found in both *Drosophila* and vertebrates and play important roles in the regulation of HOX-complex genes in *Drosophila* and imprinted genes in vertebrates. Recent reviews discuss the functional properties of these elements and some of the proteins that interact with them (4, 36). It is clear that the *Fab* and *Mcp* boundary elements of the *bithorax* complex are critical for the proper regulation of this complex locus, separating the activation or repression signals established on distinct portions of the locus (76). However, most genes do not have the complex *cis*-regulatory design of these genes, and it appears that boundary elements do not play a decisive role in the regulation of most genes. Functional studies of these elements have therefore focused on loci into which an element has been inserted by a transposon or by the investigator (4, 36, 97).

The heterochromatic insulating properties of elements bound by the cellular Su(Hw) protein have been exploited to make transformation vectors that are less susceptible to position effects on gene expression (8). Although their effectiveness outside of *Drosophila* has not been demonstrated, boundary-element function is in some cases conserved even between *Drosophila* and vertebrates. It is therefore likely that Su(Hw) insulators will work in other insect systems.

ENHANCERS

Definition of the Term Enhancer and Functional Analysis

Enhancers were originally defined as viral, and later as cellular, DNA sequences that increased expression of a linked gene in an orientation- and distance-independent manner (7, 13). In contemporary usage, an enhancer can refer to any discrete (usually less than 1 kbp) element that binds sequence-specific transcription factors acting in a positive or negative manner. Usually more than one single type of transcription factor binds to an enhancer, but artificial multimerized binding sites for a single factor can act like an enhancer (16). For the purposes of this discussion I consider a functional cluster of transcription factor binding sites as an enhancer, regardless of the distance to the initiation site.

Enhancers have been suggested to function through two distinct pathways: through remodeling of chromatin, thereby facilitating or interfering with binding of the transcriptional machinery, and through direct interactions with the general transcriptional machinery (13). Both mechanisms appear to be important in vivo. The result of these activities can be to trip a gene from an inactive to an active state (on/off effect) or to modulate transcription levels (rheostat effect) (4).

Identification of enhancers was achieved originally through genetic approaches such as classic mutations that altered expression of HOX genes in the *bithorax* complex (60). With the advent of transgenic technology in *Drosophila*, numerous "promoter bashing" experiments allowed the direct identification of regulatory regions by functional analysis (45, 95, 102). Recent progress in bioinformatics and the completion of the *Drosophila* genome has facilitated this work by identifying potential regulatory regions based on conserved clusters of binding sites that in some cases are evolutionarily conserved. Putative factor binding sites have been traditionally identified through biochemical approaches, such as DNAseI protection or gel-mobility shift assays (122). Functional in vivo assays, in which individual elements are tested and the effects of mutating binding sites are assayed, are still required to verify these predictions, however.

Alternative approaches have been used, especially where transgenic methods are impractical, including transfection of reporter constructs in cell culture. These experiments have been useful in functional analysis for some types of compact regulatory elements, such as those associated with heat shock, cytochrome P450, and vitellogenin genes (30, 57, 72). This approach is less satisfactory with promoters that require cell-type-specific factors or input from signal transduction pathways. In vitro transcriptional assays have also been extensively employed, but as with cell culture assays, in vitro transcription assays are more successful in modeling the potential activities of a single factor rather than the activity of a complex enhancer. Less commonly, regulatory regions have also been studied using biolistic transformation of insect tissues, electroporation, and introduction of genes on viral vectors [(54) and references therein]. With the development of efficient broad-spectrum transformation systems, analysis of *cis*-regulatory regions in transgenic organisms should be the benchmark for understanding enhancer function (5).

Models of Enhancer Activity

Two distinct models of enhancer action have been proposed to ascribe different computational functional roles to the enhancer (Figure 2). In the first, the "enhanceosome" model, the arrangement of binding sites within the enhancer is critical to dictating the correct output of the element, so the enhancer acts as a molecular computer, leading to a single output directed to the general machinery (25, 103). In the best-studied example, the regulatory element controlling the human β interferon promoter, cooperative interactions between architectural proteins and sequence-specific activators assemble a monolithic complex that subsequently recruits coactivators required to fire the promoter (74) (Figure 2*A*). In the second model the enhancer acts as an information display, or "billboard," which is then read and interpreted by consecutive interactions with the basal machinery. In the case of a billboard enhancer, exact binding site locations are less critical, and both activating and repressing states can be represented at the same time within an enhancer (55). It is not cooperative assembly of a higher-order structure, but successive interactions with the basal transcriptional machinery, and the biochemical consequences of these multiple interactions, that dictates the output of the enhancer

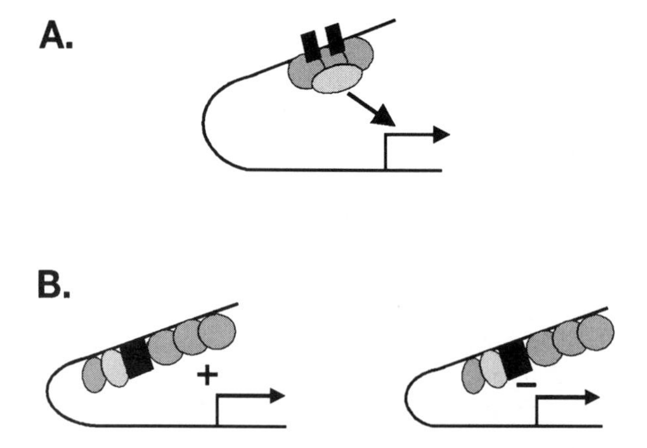

Figure 2 Alternative models of enhancer action. (*A*) "Enhanceosome" complex of carefully positioned, cooperatively bound transcriptional regulators provides a single output to the basal promoter. (*B*) "Billboard" or information display enhancer contains loosely constrained transcription factor sites binding proteins that interact with the basal machinery in a variety of conformations. The subelements of the enhancer can represent conflicting signals, which are deciphered by consecutive interactions with the basal machinery. (*Left*) Activators send positive output to promoter, resulting in transcriptional initiation; (*right*) negative signal output from a repressor.

(Figure 2*B*). The billboard enhancer model appears to more accurately describe many developmentally active enhancers, whose exact composition of binding sites is subject to rapid change in evolutionary time, even as the overall output remains constant (65, 66). Considering these functional differences, a distinction should be made between the terms *enhanceosome*, which implies important cooperative assembly processes within the enhancer-bound proteins, and *enhancer*, which may or may not function in this manner.

Many enhancers activated by signal transduction pathways require more than one type of activator to achieve correct tissue specificity. The requirement for multiple signal inputs is suggested to reflect a common enhancer design scheme, whereby a field of cells is made competent for expression by the presence of a widely expressed activator (e.g., field-selector gene) that by itself is a weak activator (9). Only in those nuclei receiving an input from a signal transduction cascade are additional activating proteins bound to the enhancer, allowing the gene to fire. To achieve greater specificity, in the absence of this signal corepressors bind in place of the coactivators, suppressing the weak activity of the widely expressed general activator. In the case of the Notch signaling pathway, the DNA binding protein Su(H) can interact either with a fragment of Notch as a coactivator or with Hairless as a corepressor (9). Synergistic interactions with the widely expressed Scalloped/Vestigial activator proteins provide a strong transcriptional output. The exact positioning of these transcription factors within a native enhancer is not critical, for an arbitrary arrangement of two binding sites each for Su(H) and Sd is sufficient to mimic the expression pattern of a native element (39).

Pattern Generation by Enhancers

Enhancers can act as information integrators or simply passively replicate a particular pattern of gene expression established previously (4). It is difficult to predict a priori whether tissue and temporal specificity of transcription is associated with a large, complex regulatory element or a simple element. In the case of the *eve* pair ruled gene, the iterated seven-stripe blastoderm expression pattern reflects the integration of positional and temporal information by five separable elements comprising 4 kbp of DNA bound by over seven distinct types of transcriptional regulator (33, 101). The large amount of *cis*-regulatory sequence is required to program cells as either *eve* expressing or *eve* nonexpressing, an essential step in the establishment of segmentation. In contrast, after cells have differentiated, expression of tissue-specific factors can recapitulate this tissue-specific pattern by binding to simple *cis* elements. For example, small autoregulatory elements in *eve* and *ftz* are sufficient to drive the multistripe pattern in later embryos, in response to earlier stripe patterning, and a short element containing multimerized Pax6/eyeless factor sites is sufficient to drive gene expression in an eye-specific pattern (49, 94, 95). In experimental settings, precise tissue-specific gene expression can be achieved using transgenes containing five binding sites for the yeast Gal4 activator, which is expressed under the control of tissue-specific enhancers (16).

The biochemistry of gene activation does not require more than one single type of transcription factor; Pax6/eyeless and Gal4 alone can drive robust expression. More typically, though, several proteins are involved in regulation, even on small tissue-specific transcriptional switches. A minimal regulatory element from the *yp1/2* ovarian-specific enhancer contains binding sites for four distinct proteins within ~40 bp: a tissue-specific activator that synergizes with a ubiquitously expressed activator, a constitutive repressor that represses the gene in nonovarian tissues, and a female-specific activator that displaces the repressor (1). A 125-bp larval fat body–specific element for the *Fbp1* gene contains a timing element in the form of an ecdysone receptor binding site that permits expression in response to proper levels of hormone, an AEF-1 repressor binding site that keeps the gene off in nonfat body tissues, and redundant GATAb protein sites that provide both activation and antirepression activities (17).

These minimal elements often represent only the pattern-specific part of the entire control region, with other sequences playing a role in overall signal amplification. In one case, the proximal element of the mosquito vitellogenin regulatory region binds to four distinct activators and is sufficient to drive low-level expression of the element in the correct stage and tissue-specific manner, but upstream sequences with additional factor binding sites are required to achieve wild-type levels of expression (53). Similar regulatory designs with separate specificity/signal amplification modules have been described in the sea urchin (3).

Shared or Exclusive Enhancer-Promoter Interactions

Enhancers can activate basal promoters over long distances, thus raising the question of how regulatory information is targeted to the correct gene. In some cases, regulatory sequences lie between divergently transcribed genes, which may share *cis*-regulatory elements. In other cases, regulation may be selective. An example of shared regulation is seen in the case of the divergent yolk protein 1 and 2 genes (*yp1* and *yp2*), which are separated by 1225-bp-containing fat body and ovary-specific enhancers (15). Both genes are simultaneously fired in each tissue by the same regulatory elements. In some cases in which an enhancer engages multiple promoters, promoter competition occurs, but in the case of the *yp1* and *yp2* genes, access to enhancers does not appear to be rate limiting; rather, a poorly understood promoter dependence was noted, in which damage to one basal promoter reduced the activity of the other (97).

Divergently transcribed genes may also share regulatory sequences, but in a temporally separated fashion. The divergently transcribed *Pig1* and *Sgs4* genes are expressed in the larval salivary gland in a sequential fashion, with shut-off of *Pig1* transcription accompanying upregulation of the *Sgs4* gene (59, 79). These genes are separated by a 838-bp region that contains multiple factor binding sites for activators and at least one inhibitory factor. One cluster of activators is capable of activating both genes simultaneously, but an additional element allows transcriptional activity to be directed to the Pig1 promoter until mid third instar (79).

Then transcriptional activity is focused on *Sgs4* until an ecdysone-regulated loss of one activator protein inhibits *Sgs4* expression (89).

Some divergently transcribed genes do not share regulatory information at all. Such enhancer-promoter specificity can be dictated by the elements of the basal promoter, including TATA, DPE, and Inr elements. Distinct types of basal promoters allow enhancers near the *dpp* and *gsb* genes to selectively activate the correct promoter and not neighboring genes (62, 75). A molecular basis of such specificity has been suggested from studies of the AE-1 enhancer, which normally activates only the *ftz* promoter and not the divergent *scr* promoter. This enhancer's specificity appears to be dependent on the presence of a TATA sequence within the activated gene (83). A recent transgenic comparison of DPE- or TATA-containing basal promoters identified endogenous enhancers that activated only DPE-containing promoters, as well as enhancers that activated only a TATA-containing promoter. A majority of the enhancers tested activated both genes, however, suggesting that most enhancers activate disparate promoters (20).

In addition to basal promoter specificity, restriction of enhancer activity to one promoter over another might also be effected by placement of a boundary element between the two genes. Experimentally, this has been observed for the *yp1* and *yp2* genes using Su(Hw) elements (97). In another case, boundary element–like activity mediated by the *eve* basal promoter was also observed (82). Aside from these experimental situations, it is not known whether boundary elements are frequently utilized to screen promoters from unwanted enhancer attention. In the best understood cases of biological activity of boundary elements in the *bithorax* complex, the activity of these elements seems critical to separate differentially regulated enhancers (76).

Enhancer Redundancy

The billboard model of enhancers suggests that there are multiple configurations in which enhancer-bound factors can interact with the basal transcriptional machinery. Consistent with this model, functional analysis of many *cis*-regulatory elements suggests that redundant regulatory information is represented in discrete, separable sequences. In the case of yolk protein genes, for example, the 125-bp enhancer that gives fat body activity can be deleted from the upstream region without measurably affecting expression, and a deletion that removes most of the intergenic region only had subtle effects on expression (86, 97). *Krüppel* expression in the central domain of the embryo is directed by two adjacent enhancers that each independently drive the correct pattern of gene expression (45). In the case of the *eve* stripe 2 enhancer, removal of activator binding sites in the context of a 5.2-kbp element had weak and variable effects in contrast to the more drastic effects noted when a minimal stripe 2 element was assayed (101, 102). A similar redundancy in enhancer design was noted with enhancers that drive the blastoderm stripe pattern of expression of *ftz*, where mutations in individual activator sites had little measurable effect on activity, although mutation of multiple sites abolished activity (40). Repressor binding sites also function in a redundant manner, as in the case of multiple Brinker binding sites within the *zen* promoter (91). Regulatory mutations that involve a single protein binding site are in general rarely detected, suggesting that most enhancers are built to withstand a fair degree of change without suffering catastrophic loss of activity. In addition to the likely functional redundancy built into many systems, our inability to measure subtle changes in transcriptional output of altered elements may also explain the apparent robustness of *cis*-regulatory elements to alteration (105).

In contrast to this redundancy, some complex regulatory regions do show strong dependence on a few sites. The regulatory region of the *diptericin* antibacterial gene, for instance, contains conserved binding sites for Rel-type transcriptional activators (73). Although the sites are not sufficient to confer activation, loss of two of these sites in a 2.2-kbp enhancer abolished induction of a reporter in response to bacterial challenge, consistent with an enhanceosome type of model (73). Both insect and vertebrate genes utilize Rel domain proteins in innate immune responses, and enhancers such as this that are designed for sudden inducibility might tend to function as enhanceosomes (103).

SOME SPECIAL PROMOTERS

Modular Enhancers and Short-Range Repression

Many developmentally regulated genes have independently acting modular control elements. An initial indication of the modularity of transcriptional regulatory regions came from the identification of mutant alleles of the *hairy* gene, which is normally expressed in a seven-stripe pattern in the embryo (46). Deletions upstream of the structural gene caused a loss of individual stripes rather than a general effect on all regions of expression. Subsequent dissection of *hairy* regulatory regions revealed the presence of individual modular enhancer elements, each of which is subject to the action of activators and repressor proteins, a finding that has been replicated with many other genes, including *even-skipped* (33, 101). One surprising finding from studies of *eve* enhancers is that in an individual nucleus at a given moment, one enhancer can be repressed while an adjacent enhancer actively signals the promoter. The reason for this functional independence is that the silenced enhancers are repressed by proteins that work over a short range (<100 bp), so distances between enhancers of a few hundred bases are sufficient to prevent improper cross-regulation (38). Artificially juxtaposing two normally distant enhancers causes improper repression of one element by repressors bound to the other (100). Repressors that act over long distances have also been identified, but it is not yet clear how their activities are integrated into modular control regions (26, 38).

Compact Testis-Specific Promoters

Unlike the uniformly compact promoters associated with RNA Pol I and III transcription units, RNA Pol II–regulated promoters active in most tissues are variable in size. An exception to this observation is formed by a number of testis-specific genes with unusually small transcriptional control elements, often extending no more than 100 nt 5′ of the initiation site. The expression of the testis-specific $\beta 2$ tubulin gene depends on a 14-bp element at −40 that contributes tissue specificity and on two small elements just 5′ and 3′ of +1 that are important for overall activity of the promoter (92). As is the case with other testis-specific genes, the nature of the proteins that might interact with these functional motifs is not known, although binding activities have been identified in gel-shift assays (92). Like other testis-specific promoters, the $\beta 2$ tubulin gene contains a conserved Inr but lacks a TATA box. The 14-bp element does not activate a heterologous TATA-containing basal promoter, suggesting that a distinct type of basal machinery architecture might be used on this class of gene. This notion is supported by the finding that the testis-specific TAFII80 and TAFII110 homologs encoded by *cannonball* and *no-hitter*, respectively, are required for wild-type expression of a number of testis-specific genes (2, 44). Cannonball and No-hitter may function as part of a tissue-specific form of TFIID or, like other TAFII's, they may function as part of a histone acetylation complex (90, 115).

In some cases, the close spacing of testis-specific genes would indicate small *cis*-acting regions. For example, a cluster of four adenylyl cyclase genes had intergenic spacings of only 166, 84, and 39 bp (21). However, it is important to consider that some regulatory elements map within transcribed regions or within the sequence of upstream genes. In $\beta 2$ tubulin and the H1-like *don juan* genes, there are transcriptional control elements immediately downstream of the initiation site, overlapping sequences that play a role in translational control of the mRNA (14, 92). In some cases, control elements are located within transcribed regions of upstream genes, such as the 53-bp element responsible for male-specific transcription of *gonadal* found at −330 bp within the coding sequence of an upstream gene not expressed in male germline and a control element for male-specific transcription of *Janus B* located within an exon of the upstream *Janus A* gene (93, 119).

Promoters in Heterochromatin

Most protein-coding genes are located in euchromatic regions, but a small number of genes such as *light*, *rolled*, and *concertina* are embedded in heterochromatin, chromosomal regions with a more compact structure and rich in repetitive sequences (42, 111). Little is known about the transcriptional regulation of such genes, except that in contrast to most genes, which are repressed by proximity to heterochromatin, these are fully functional, and in the case of *light* are positively regulated by heterochromatin (41). The *D. melanogaster light* gene has an unusual pattern of transcriptional initiation, with start sites spread over a region of ~200 nucleotides. In *D. virilis*, by contrast, in which the *light* gene is located in euchromatin, transcription initiates at a single site. When introduced into *D. melanogaster*, the *D. virilis* gene can rescue a *light* mutant, which indicates that the overall regulation of the genes has been conserved despite the differences in chromatin context and basal promoter function (120). Future work should identify the specific regulatory features that allow this gene to function in a heterochromatic context.

runt: A Disperse Regulatory Element

Similar to *eve* and *ftz*, initial embryonic patterning of the *runt* gene is under control of individual, modular stripe enhancers. Unlike *eve* and *ftz*, however, later expression of *runt* is not regulated by a compact autoregulatory element, but by a large region, over 5 kbp in size, that cannot be subdivided into independent elements (52). This large regulatory element appears to be unusual, but traditional approaches to promoter analysis have concentrated on those elements that are compact (<1 kbp); thus, the true frequency of disperse elements may be higher than generally appreciated. Evolutionary conservation of this disperse *runt* element is low, suggesting that interacting functional subelements might be widely distributed within this regulatory region (114).

EVOLUTION OF *CIS*-REGULATORY ELEMENTS

The rates of change in transcriptional regulatory regions during evolution can differ markedly from those in protein-coding sequences. In some cases the sequences of transcriptional regulatory regions are considerably altered, yet the transcriptional output remains the same, indicating that different arrangements of transcription factors can have the same function (65, 66). The constraints on regulatory regions are considerably more relaxed than those on protein-coding regions, probably owing in part to the redundant, flexible nature of transcriptional regions. In other cases, transcriptional regulatory regions themselves appear to be the driving force behind evolutionary changes, as for instance in the altered expression patterns of HOX genes correlated with diverse limb development schemes in arthropods (6). Duplicated genes can diverge in function solely on the basis of transcriptional regulation, as demonstrated in the case of the homologous transcription factors Paired, Gooseberry, and Gooseberry neuro. These proteins have essentially identical activities but differ in developmental roles because of their genes' distinct transcriptional control regions (63).

Functional Analysis

Conservation or divergence in transcriptional control regions has been tested both functionally and by sequence analysis. One of the first demonstrations of conservation of transcriptional regulation between different insect orders was the finding that a chorion gene cluster from *Bombyx mori* is correctly regulated in *Drosophila* in a sex- and tissue-specific manner, despite the great evolutionary distance between lepidopterans and dipterans, the lack of homology between the chorion genes of the two species, and the distinct gene arrangement (77). Subsequent tests have shown that the vitellogenin gene from the mosquito *Aedes aegyti* is regulated in the correct stage and tissue-specific manner in *Drosophila* (although not in a female-specific manner), despite the great differences in yolk proteins and vitellogenesis between these organisms (53). Other heterologous genes correctly regulated in *Drosophila* include gut-specific protease genes from the mosquito *Anopheles gambiae* and the black fly *Simulium vittatum* [(98, 118) and references therein].

Sequence Analysis

Given these often remarkable similarities in the function of regulatory regions from divergent organisms, is it possible to identify similar sequences by computer alignment? The picture is mixed. A comparison of yolk-protein regulatory regions among *D. melanogaster* and Hawaiian drosophilids demonstrated that there have been extensive changes in the short region between the divergently transcribed genes, so much so that there was no clear overall alignment between previously mapped enhancer elements in *D. melanogaster* and similar areas in *D. grimshawi* (86). Only short, interspersed sequences were found, but transgenic analysis confirmed that these sequences were functionally conserved (86). The extent of divergence between yolk protein genes may be an extreme, however, because of selective evolutionary pressures on vitellogenesis, which leads to subtle changes in regulation that are not readily evident in an experimental setting.

Much closer similarities were noted in an analysis of enhancers of the conserved pair rule gene *ftz* in *D. hydei*. In sequence alignments with previously characterized enhancers from *D. melanogaster*, conserved blocks hundreds of basepairs in size with 50%–65% overall identity were noted. These enhancers gave a *ftz*-type expression pattern in *D. melanogaster* (50, 70). A recent comparative analysis of 100 kbp of known or suspected regulatory DNA from 40 loci between *D. melanogaster* and *D. virilis* (~40 million years of separation, similar to *D. hydei*) found that overall about 25% of the sequence was conserved, with most of the conserved sequences present as small blocks of median length of 19 bp, and a substantial fraction up to 40–60 bp in length (11). Thus, even between divergent drosophilids, there appears to be sufficient conservation to identify regulatory blocks in many cases and to identify divergent features of enhancer design. For instance, comparison of the *even-skipped* stripe 2 enhancer from an array of *Drosophila* species revealed that overall organization and general size of this enhancer was maintained and many binding sites were conserved (66). A specific binding site for the Bicoid activator that played an important role in *D. melanogaster* was not conserved, however. A recent deletion appears to have brought a nearby Giant repressor site closer to this recently acquired Bicoid activator site, an example of the "tuning" of enhancer output (43). Thus, whereas both enhancers have the same overall output when tested in *D. melanogaster*, a chimeric enhancer containing half *D. pseudoobscura* and half *D. melanogaster* sequences gave improper regulatory output, evidence of changes in enhancer design (65).

The greater degree of divergence between regulatory regions in more distantly related species complicates efforts to locate enhancers by sequence alignment alone, even when functional aspects are similar. In some cases, lack of conservation reflects a fundamental change in regulatory strategy. An 8.7-kbp fragment 5′ of *hairy* from the beetle *Tribolium casteneum* drives a portion of the conserved striped pattern in central regions of *D. melanogaster*, but there are no obvious conserved regulatory regions. A more detailed functional analysis of individual conserved sequence clusters might reveal conserved functional elements, but the failure to recapitulate some portions of the stripe pattern in the fly is probably due to fundamental differences in anterior and terminal patterning in *Tribolium* and *Drosophila* (29). Preliminary analysis of putative regulatory elements in the *Tribolium* HOX complex reveals a similar lack of conservation between these two species, suggesting that with current approaches direct functional testing will remain critical for identification and analysis of possibly conserved regulatory regions (18).

NEW TECHNOLOGIES

Genomics

Our current understanding of transcriptional switch elements derives from empirical experiments involving the dissection of basal promoters and enhancers and from in vivo tests. The most powerful studies, such as those conducted on complex regulatory regions such as the *bithorax* complex and *even-skipped*, are initiated with genetic characterization of *trans*-activating factors, identification of *cis*-regulatory elements, mapping of factor binding sites within those regions, mutagenesis of those putative binding sites, and testing of the elements in transgenic organisms. This approach requires good genetic and transgenic tools; thus, most studies of this type have been restricted to *Drosophila*.

Applications of genomics and bioinformatics associated with the *Drosophila* genomic project, as well as more extensive phylogenetic comparisons and advances in transgenesis, will facilitate the analysis of *cis*-regulatory elements, enabling work to proceed in other, less genetically tractable systems. Comprehensive analysis of gene expression in *Drosophila* is being conducted by gene array analysis and in situ hybridization. Gene arrays have been used to track expression on an organismal level and more recently in a tissue-specific basis in both wild-type and mutant backgrounds (48, 61). As examples, recent gene array analysis of the *Drosophila* immune response, mesoderm-specific genes, and metamorphosis will provide a baseline set of data for comparison in other species (27, 35, 112). A systematic analysis of embryonic expression patterns by in situ hybridization now underway at the Berkeley Drosophila Genome Project will add a spatial component to the temporal information gained from gene array studies (108). From these studies we can expect to develop a comprehensive knowledge of the *Drosophila* transcriptome, including genetic pathways. This large body of empirical data will give researchers models for gene regulation by which to measure homologs in other insect species, possibly revealing conserved or divergent features of regulatory elements or expression patterns.

Bioinformatics

Bioinformatic analysis can be used to identify binding sites of known or novel factors. This identification is carried out in two ways. For factors with well-characterized binding sites, a consensus binding site, or a position weight matrix representing a population of known binding sites can be used to search genomic sequences. Relatively simple search paradigms, such as searching for clusters of a binding motif within 400 bp, have been successful in producing useful information. A genomic search for clustered Dorsal protein binding sites identified 15 putative Dorsal-regulated enhancers, some of which were assayed by transgenic analysis (71). A search for clusters of activator and repressor binding sites in the 1-Mbp region surrounding the *even-skipped* gene successfully identified previously functionally mapped transcriptional control elements flanking *eve* and identified a novel enhancer that proved to be a control element for the *giant* gene (12).

When the *trans*-acting factors for genes are not known, one can search for overrepresented motifs near genes that show similar expression patterns. Such an approach was successfully used to identify repeated motifs common to eight odorant receptor proteins expressed primarily in maxilla. Mutagenesis of the regulatory regions confirmed that the elements are important for the tissue-specific regulation of these genes (22). Alternatively, one can search a known regulatory region using statistical methods for motifs that are overrepresented; a recent study showed that this technique located sites of known regulatory importance in *Drosophila* enhancers (85).

Phylogenetic Analysis

Phylogenetic comparisons are powerful tools to identify regions of genomic DNA that likely contain regulatory information. The imminent or completed sequencing of additional dipteran genomes, such as that of *D. pseudoobscura* and *An. gambiae* will facilitate analysis of *cis*-acting control regions on a global scale, so the putative regulatory regions of any conserved gene will be available for bioinformatic and functional analysis. It remains to be seen how close the conservation of regulatory regions will be in more evolutionarily distant insects and whether the genetic control networks are sufficiently similar to allow direct modeling of entire enhancers from dipterans. At least at the level of individual factor binding sites, genes with similar regulatory profiles, such as those encoding rhodopsins, antimicrobial peptides, and cytochrome P450s, appear to use the same transcription factors; thus, at least short motifs should be conserved (10, 27, 30).

Transgenesis in Other Systems

Genomic and bioinformatic tools can accelerate identification and characterization of *cis* elements, but models of transcriptional regulatory regions still require validation with functional tests. Recent advances in insect transgenesis (5) offer the possibility of testing the function of putative regulatory regions in most species. The lepidopteran PiggyBac vector has been reported to successfully transform many different insect orders, including Coleoptera, Lepidoptera, and Diptera (10, 29). The use of eye-specific green fluorescent protein marker genes to identify transgenic progeny allows transformation of wild-type animals, obviating the need to identify mutant strains with phenotypes that can be rescued. In addition to studying regulatory elements, transgenesis has the potential to generate mutant phenotypes using RNA interference (28), allowing genetic analysis in nonmodel system insects. A future paradigm for characterizing the regulation of a novel gene in a nonmodel insect system would be identification of possible homologs in a completed genome, examination of possible evolutionarily conserved regulatory regions and binding sites for *trans*-acting factors, experimental testing of *cis* elements on transformation vectors, and disruption of putative regulatory factors through RNA interference.

CONCLUDING REMARKS

Genome-wide analysis of transcriptional programs in *Drosophila* will contribute greatly to our understanding of basic transcriptional mechanisms used in all metazoans by providing for the first time a comprehensive overview of all transcription factors, a genome-wide comparison of transcriptional elements, and their conservation in disparate species. Comparisons with *Drosophila* do not eliminate the requirement for direct functional analysis of transcriptional elements in species of interest, but new advances will facilitate the design of these experiments, and work in previously nonmodel organisms will undoubtedly make important contributions to basic science. A better understanding of insect *cis*-acting regulatory regions will also contribute to many areas relevant to entomology, including identification of useful promoters for engineering resistance in disease vectors, identifying the molecular basis of pesticide resistance, and understanding evolution mediated by changes in transcriptional control elements.

ACKNOWLEDGMENTS

Special thanks to S. Brown, T. Ip, B. Lemaitre, U. Ohler, A. Raikhel, D. Tautz, and B. Wakimoto for helpful suggestions and sharing information prior to publication.

The *Annual Review of Entomology* is online at http://ento.annualreviews.org

LITERATURE CITED

1. An W, Wensink PC. 1995. Integrating sex- and tissue-specific regulation within a single *Drosophila* enhancer. *Genes Dev.* 9:256–66
2. Aoyagi N, Wassarman DA. 2000. Genes encoding *Drosophila melanogaster* RNA polymerase II general transcription factors: diversity in TFIIA and TFIID components contributes to gene-specific transcriptional regulation. *J. Cell Biol.* 150: F45–50
3. Arnone MI, Davidson EH. 1997. The hardwiring of development: organization and function of genomic regulatory systems. *Development* 124:1851–64
4. Arnosti DN. 2002. Design and function of transcriptional switches in *Drosophila*. *Insect Biochem. Mol. Biol.* 32(10):1257–73
5. Atkinson PW, Pinkerton AC, O'Brochta DA. 2001. Genetic transformation systems in insects. *Annu. Rev. Entomol.* 46: 317–46
6. Averof M, Patel NH. 1997. Crustacean appendage evolution associated with changes in *Hox* gene expression. *Nature* 388:682–86
7. Banerji J, Rusconi S, Schaffner W. 1981. Expression of a beta-globin gene is enhanced by remote SV40 DNA sequences. *Cell* 27:299–308
8. Barolo S, Carver LA, Posakony JW. 2000. GFP and beta-galactosidase transformation vectors for promoter/enhancer analysis in *Drosophila*. *Biotechniques* 29:726–32
9. Barolo S, Posakony JW. 2002. Three habits of highly effective signaling pathways; principles of transcriptional control by developmental cell signaling. *Genes Dev.* 16:1167–81
10. Berghammer AJ, Klingler M, Wimmer EA. 1999. A universal marker for transgenic insects. *Nature* 402:370–71
11. Bergman CM, Kreitman M. 2001. Analysis of conserved noncoding DNA in *Drosophila* reveals similar constraints in intergenic and intronic sequences. *Genome Res.* 11:1335–45
12. Berman BP, Nibu Y, Pfeiffer BD, Tomancak P, Celniker SE, et al. 2002. Exploiting transcription factor binding site clustering to identify *cis*-regulatory modules involved in pattern formation in the *Drosophila* genome. *Proc. Natl. Acad. Sci. USA* 99:757–62
13. Blackwood EM, Kadonaga JT. 1998. Going the distance: a current view of enhancer action. *Science* 281:60–63
14. Blümer N, Schreiter K, Hempel L, Santel A, Hollmann M, et al. 2002. A new translational repression element and unusual transcriptional control regulate expression of *don juan* during *Drosophila* spermatogenesis. *Mech. Dev.* 110:97–112
15. Bownes M. 1994. The regulation of the yolk protein genes, a family of sex differentiation genes in *Drosophila melanogaster*. *BioEssays* 16:745–52
16. Brand AH, Perrimon N. 1993. Targeted gene expression as a means of altering cell fates and generating dominant phenotypes. *Development* 118:401–15
17. Brodu V, Mugat B, Fichelson P, Lepesant JA, Antoniewski C. 2001. A UAS site substitution approach to the in vivo dissection of promoters: interplay between the GATAb activator and the AEF-1 repressor at a *Drosophila* ecdysone response unit. *Development* 128:2593–602
18. Brown SJ, Fellers JP, Shippy TD, Richardson EA, Maxwell M, et al. 2002. Sequence of the *Tribolium castaneum* homeotic complex. The region corresponding to the *Drosophila melanogaster* Antennapedia complex. *Genetics* 160:1067–74
19. Burke TW, Willy PJ, Kutach AK, Butler JE, Kadonaga JT. 1998. The DPE,

a conserved downstream core promoter element that is functionally analogous to the TATA box. *Cold Spring Harbor Symp. Quant. Biol.* 63:75–82

20. Butler JE, Kadonaga JT. 2001. Enhancer-promoter specificity mediated by DPE or TATA core promoter motifs. *Genes Dev.* 15:2515–19
21. Cann MJ, Chung E, Levin LR. 2000. A new family of adenylyl cyclase genes in the male germline of *Drosophila melanogaster*. *Dev. Genes Evol.* 210:200–6
22. Carlson J. 2002. *Odor and taste receptors: genetics and e-genetics*. Presented at Annu. Drosophila Res. Conf., 43rd, San Diego, CA
23. Chalkley GE, Verrijzer CP. 1999. DNA binding site selection by RNA polymerase II TAFs: a TAF(II)250-TAF(II)150 complex recognizes the initiator. *EMBO J.* 18:4835–45
24. Cherbas L, Cherbas P. 1993. The arthropod initiator: the capsite consensus plays an important role in transcription. *Insect Biochem. Mol. Biol.* 23:81–90
25. Courey AJ. 2001. Cooperativity in transcriptional control. *Curr. Biol.* 11:R250–52
26. Courey AJ, Jia S. 2001. Transcriptional repression: the long and the short of it. *Genes Dev.* 15:2786–96
27. De Gregorio E, Spellman PT, Rubin GM, Lemaitre B. 2001. Genome-wide analysis of the *Drosophila* immune response by using oligonucleotide microarrays. *Proc. Natl. Acad. Sci. USA* 98:12590–95
28. Denell R, Shippy T. 2001. Comparative insect developmental genetics: phenotypes without mutants. *BioEssays* 23: 379–82
29. Eckert C, Wolff C, Wimmer E, Tautz D. 2002. Functional analysis of the regulatory region of the pair-rule gene *hairy* in *Tribolium* suggests regulatory divergence in spite of conserved expression. Submitted
30. Feyereisen R. 1999. Insect P450 enzymes. *Annu. Rev. Entomol.* 44:507–33
31. Francis NJ, Kingston RE. 2001. Mechanisms of transcriptional memory. *Nat. Rev. Mol. Cell. Biol.* 2:409–21
32. Fridell YW, Searles LL. 1992. In vivo transcriptional analysis of the TATA-less promoter of the *Drosophila melanogaster vermilion* gene. *Mol. Cell. Biol.* 12:4571–77
33. Fujioka M, Emi-Sarker Y, Yusibova GL, Goto T, Jaynes JB. 1999. Analysis of an *even-skipped* rescue transgene reveals both composite and discrete neuronal and early blastoderm enhancers, and multi-stripe positioning by gap gene repressor gradients. *Development* 126:2527–38
34. Fuller MT. 1993. Spermatogenesis. In *The Development of* Drosophila melanogaster, ed. M Bate, A Martinez Arias, 1:71–147. New York: Cold Spring Harbor Press
35. Furlong EE, Andersen EC, Null B, White KP, Scott MP. 2001. Patterns of gene expression during *Drosophila* mesoderm development. *Science* 293:1629–33
36. Gerasimova TI, Corces VG. 2001. Chromatin insulators and boundaries: effects on transcription and nuclear organization. *Annu. Rev. Genet.* 35:193–208
37. Gowher H, Leismann O, Jeltsch A. 2000. DNA of *Drosophila melanogaster* contains 5-methylcytosine. *EMBO J.* 19: 6918–23
38. Gray S, Levine M. 1996. Transcriptional repression in development. *Curr. Opin. Cell. Biol.* 8:358–64
39. Guss KA, Nelson CE, Hudson A, Kraus ME, Carroll SB. 2001. Control of a genetic regulatory network by a selector gene. *Science* 292:1164–67
40. Han W, Yu Y, Su K, Kohanski RA, Pick L. 1998. A binding site for multiple transcriptional activators in the *fushi tarazu* proximal enhancer is essential for gene expression in vivo. *Mol. Cell. Biol.* 18: 3384–94
41. Hearn MG, Hedrick A, Grigliatti TA, Wakimoto BT. 1991. The effect of modifiers of position-effect variegation on

the variegation of heterochromatic genes of *Drosophila melanogaster*. *Genetics* 128:785–97

42. Henikoff S. 2000. Heterochromatin function in complex genomes. *Biochim. Biophys. Acta* 1470:1–8
43. Hewitt GF, Strunk BS, Margulies C, Priputin T, Wang X-D, et al. 1999. Transcriptional repression by the *Drosophila* giant protein: *cis* element positioning provides an alternative means of interpreting an effector gradient. *Development* 126:1201–10
44. Hiller MA, Lin TY, Wood C, Fuller MT. 2001. Developmental regulation of transcription by a tissue-specific TAF homolog. *Genes Dev.* 15:1021–30
45. Hoch M, Schröder C, Seifert E, Jäckle H. 1990. *cis*-acting control elements for *Krüppel* expression in the *Drosophila* embryo. *EMBO J.* 9:2587–95
46. Hooper KL, Parkhurst SM, Ish-Horowicz D. 1989. Spatial control of hairy protein expression during embryogenesis. *Development* 107:489–504
47. Hung MS, Karthikeyan N, Huang B, Koo HC, Kiger J, Shen CJ. 1999. *Drosophila* proteins related to vertebrate DNA (5-cytosine) methyltransferases. *Proc. Natl. Acad. Sci. USA* 96:11940–45
48. Imam FB, Johnson EJ, Arbeitman M, Furlong E, Null B, et al. 2002. *A transcriptional view of* Drosophila *development*. Presented at Annu. Drosophila Res. Conf., 43rd, San Diego, CA
49. Jiang J, Hoey T, Levine M. 1991. Autoregulation of a segmentation gene in *Drosophila*: combinatorial interaction of the *even-skipped* homeo box protein with a distal enhancer element. *Genes Dev.* 5: 265–77
50. Jost W, Yu Y, Pick L, Preiss A, Maier D. 1995. Structure and regulation of the *fushi tarazu* gene from *Drosophila hydei*. *Roux's Arch. Dev. Biol.* 205:160–70
51. Kermekchiev M, Pettersson M, Matthias P, Schaffner W. 1991. Every enhancer works with every promoter for all the combinations tested: could new regulatory pathways evolve by enhancer shuffling? *Gene Exp.* 1:71–81
52. Klingler M, Soong J, Butler B, Gergen JP. 1996. Disperse versus compact elements for the regulation of *runt* stripes in *Drosophila*. *Dev. Biol.* 177:73–84
53. Kokoza VA, Martin D, Mienaltowski MJ, Ahmed A, Morton CM, Raikhel AS. 2001. Transcriptional regulation of the mosquito vitellogenin gene via a blood meal-triggered cascade. *Gene* 274:47–65
54. Kravariti L, Thomas J, Sourmeli S, Rodakis GC, Mauchamp B, et al. 2001. The biolistic method as a tool for testing the differential activity of putative silkmoth chorion gene promoters. *Insect Biochem. Mol. Biol.* 31:473–79
55. Kulkarni M, Arnosti DN. 2002. Parallel signaling by transcriptional enhancers. Submitted
56. Kutach AK, Kadonaga JT. 2000. The downstream promoter element DPE appears to be as widely used as the TATA box in *Drosophila* core promoters. *Mol. Cell. Biol.* 20:4754–64
57. Lawson R, Mestril R, Schiller P, Voellmy R. 1984. Expression of heat shock-beta-galactosidase hybrid genes in cultured *Drosophila* cells. *Mol. Gen. Genet.* 198: 116–24
58. Lee TI, Young RA. 2000. Transcription of eukaryotic protein-coding genes. *Annu. Rev. Genet.* 34:77–137
59. Lehmann M. 1996. Drosophila *Sgs* genes: stage and tissue specificity of hormone responsiveness. *BioEssays* 18:47–54
60. Lewis EB. 1998. The bithorax complex: the first fifty years. *Int. J. Dev. Biol.* 42: 403–15
61. Li TR, White KP. 2002. *Developmental gene expression profiles of individual tissues and organs in* Drosophila. Presented at Annu. Drosophila Res. Conf., 43rd, San Diego, CA
62. Li X, Noll M. 1994. Compatibility between enhancers and promoters determines the transcriptional specificity of

gooseberry and *gooseberry neuro* in the *Drosophila* embryo. *EMBO J.* 13:400–6

63. Li X, Noll M. 1994. Evolution of distinct developmental functions of three *Drosophila* genes by acquisition of different *cis*-regulatory regions. *Nature* 367: 83–87
64. Lu J, Oliver B. 2001. Drosophila OVO regulates ovarian tumor transcription by binding unusually near the transcription start site. *Development* 128:1671–86
65. Ludwig MZ, Bergman C, Patel NH, Kreitman M. 2000. Evidence for stabilizing selection in a eukaryotic enhancer element. *Nature* 403:564–67
66. Ludwig MZ, Kreitman M. 1998. Functional analysis of *eve* stripe 2 enhancer evolution in *Drosophila*: rules governing conservation and change. *Development* 125:949–58
67. Lyko F, Ramsahoye BH, Jaenisch R. 2000. DNA methylation in *Drosophila melanogaster*. *Nature* 408:538–40
68. Mackay TF. 2001. Quantitative trait loci in Drosophila. *Nat. Rev. Genet.* 2:11–20
69. Mahmoudi T, Verrijzer CP. 2001. Chromatin silencing and activation by Polycomb and trithorax group proteins. *Oncogene* 20:3055–66
70. Maier D, Preiss A, Powell JR. 1990. Regulation of the segmentation gene *fushi tarazu* has been functionally conserved in *Drosophila*. *EMBO J.* 9:3957–66
71. Markstein M, Markstein P, Markstein V, Levine MS. 2002. Genome-wide analysis of clustered Dorsal binding sites identifies putative target genes in the *Drosophila* embryo. *Proc. Natl. Acad. Sci. USA* 99:763–68
72. Martin D, Wang SF, Raikhel AS. 2001. The vitellogenin gene of the mosquito *Aedes aegypti* is a direct target of ecdysteroid receptor. *Mol. Cell. Endocrinol.* 173:75–86
73. Meister M, Braun A, Kappler C, Reichhart JM, Hoffmann JA. 1994. Insect immunity. A transgenic analysis in *Drosophila* defines several functional domains in the *diptericin* promoter. *EMBO J.* 13:5958–66
74. Merika M, Thanos D. 2001. Enhanceosomes. *Curr. Opin. Genet. Dev.* 11:205–8
75. Merli C, Bergstrom DE, Cygan JA, Blackman RK. 1996. Promoter specificity mediates the independent regulation of neighboring genes. *Genes Dev.* 10:1260–70
76. Mihaly J, Hogga I, Barges S, Galloni M, Mishra RK, et al. 1998. Chromatin domain boundaries in the Bithorax complex. *Cell. Mol. Life Sci.* 54:60–70
77. Mitsialis SA, Kafatos FC. 1985. Regulatory elements controlling chorion gene expression are conserved between flies and moths. *Nature* 317:453–56
78. Mitsialis SA, Spoerel N, Leviten M, Kafatos FC. 1987. A short 5'-flanking DNA region is sufficient for developmentally correct expression of moth chorion genes in *Drosophila*. *Proc. Natl. Acad. Sci. USA* 84:7987–91
79. Mougneau E, von Seggern D, Fowler T, Rosenblatt J, Jongens T, et al. 1993. A transcriptional switch between the *Pig-1* and *Sgs-4* genes of *Drosophila melanogaster*. *Mol. Cell. Biol.* 13:184–95
80. Näär AM, Lemon BD, Tjian R. 2001. Transcriptional coactivator complexes. *Annu. Rev. Biochem.* 70:475–501
81. Ohler U. 2000. Promoter prediction on a genomic scale: the *Adh* experience. *Genome Res.* 10:539–42
82. Ohtsuki S, Levine M. 1998. GAGA mediates the enhancer blocking activity of the *eve* promoter in the *Drosophila* embryo. *Genes Dev.* 12:3325–30
83. Ohtsuki S, Levine M, Cai HN. 1998. Different core promoters possess distinct regulatory activities in the *Drosophila* embryo. *Genes Dev.* 12:547–56
84. Orphanides G, Reinberg D. 2002. A unified theory of gene expression. *Cell* 108:439–51
85. Papatsenko DA, Makeev VJ, Lifanov AP, Regnier M, Nazina AG, Desplan C. 2002. Extraction of functional binding sites from unique regulatory regions:

the *Drosophila* early developmental enhancers. *Genome Res.* 12:470–81
86. Piano F, Parisi MJ, Karess R, Kambysellis MP. 1999. Evidence for redundancy but not *trans* factor-*cis* element coevolution in the regulation of *Drosophila* Yp genes. *Genetics* 152:605–16
87. Purnell BA, Emanuel PA, Gilmour DS. 1994. TFIID sequence recognition of the initiator and sequences farther downstream in *Drosophila* class II genes. *Genes Dev.* 8:830–42
88. Ren B, Maniatis T. 1998. Regulation of *Drosophila Adh* promoter switching by an initiator-targeted repression mechanism. *EMBO J.* 17:1076–86
89. Renault N, King-Jones K, Lehmann M. 2001. Downregulation of the tissue-specific transcription factor Fork head by *Broad-Complex* mediates a stage-specific hormone response. *Development* 128: 3729–37
90. Roth SY, Denu JM, Allis CD. 2001. Histone acetyltransferases. *Annu. Rev. Biochem.* 70:81–120
91. Rushlow C, Colosimo PF, Lin MC, Xu M, Kirov N. 2001. Transcriptional regulation of the *Drosophila* gene *zen* by competing Smad and Brinker inputs. *Genes Dev.* 15:340–51
92. Santel A, Kaufmann J, Hyland R, Renkawitz-Pohl R. 2000. The initiator element of the *Drosophila* $\beta 2$ tubulin gene core promoter contributes to gene expression in vivo but is not required for male germ-cell specific expression. *Nucleic Acids Res.* 28:1439–46
93. Schulz RA, Xie XL, Miksch JL. 1990. *cis*-acting sequences required for the germ line expression of the *Drosophila gonadal* gene. *Dev. Biol.* 140:455–58
94. Sheng G, Thouvenot E, Schmucker D, Wilson DS, Desplan C. 1997. Direct regulation of rhodopsin 1 by *Pax-6/eyeless* in *Drosophila*: evidence for a conserved function in photoreceptors. *Genes Dev.* 11:1122–31
95. Schier AF, Gehring WJ. 1993. Analysis of a *fushi tarazu* autoregulatory element: multiple sequence elements contribute to enhancer activity. *EMBO J.* 12:1111–19
96. Schwyter DH, Huang JD, Dubnicoff T, Courey AJ. 1995. The *decapentaplegic* core promoter region plays an integral role, in the spatial control of transcription. *Mol. Cell. Biol.* 15:3960–68
97. Scott KS, Geyer PK. 1995. Effects of the su(Hw) insulator protein on the expression of the divergently transcribed *Drosophila* yolk protein genes. *EMBO J.* 14:6258–67
98. Skavdis G, Siden-Kiamos I, Muller HM, Crisanti A, Louis C. 1996. Conserved function of *Anopheles gambiae* midgut-specific promoters in the fruitfly. *EMBO J.* 15:344–50
99. Smale ST. 2001. Core promoters: active contributors to combinatorial gene regulation. *Genes Dev.* 15:2503–8
100. Small S, Arnosti DN, Levine M. 1993. Spacing ensures autonomous expression of different stripe enhancers in the *even-skipped* promoter. *Development* 119:767–72
101. Small S, Blair A, Levine M. 1992. Regulation of *even-skipped* stripe 2 in the *Drosophila* embryo. *EMBO J.* 11:4047–57
102. Stanojevic D, Small S, Levine M. 1991. Regulation of a segmentation stripe by overlapping activators and repressors in the *Drosophila* embryo. *Science* 254: 1385–87
103. Struhl K. 2001. Gene regulation. A paradigm for precision. *Science* 293:1054–55
104. Suzuki Y, Tsunoda T, Sese J, Taira H, Mizushima-Sugano J, et al. 2001. Identification and characterization of the potential promoter regions of 1031 kinds of human genes. *Genome Res.* 11:677–84
105. Tautz D. 2000. A genetic uncertainty problem. *Trends Genet.* 16:475–77
106. Tautz D. 2000. Evolution of transcriptional regulation. *Curr. Opin. Genet. Dev.* 10:575–79

107. Tingvall TO, Roos E, Engstrom Y. 2001. The GATA factor Serpent is required for the onset of the humoral immune response in *Drosophila* embryos. *Proc. Natl. Acad. Sci. USA* 98:3884–88
108. Tomancak P, Beaton A, Weiszmann R, Kwan E, Celniker S, Rubin GM. 2002. *Patterns of gene expression in development-embryology meets bioinformatics*. Presented at the Annu. Drosophila Res. Conf., 43rd, San Diego, CA
109. Tweedie S, Ng HH, Barlow AL, Turner BM, Hendrich B, Bird A. 1999. Vestiges of a DNA methylation system in *Drosophila melanogaster*? *Nat. Genet.* 23:389–90
110. Verrijzer CP. 2001. Transcription factor IID: not so basal after all. *Science* 293: 2010–11
111. Weiler KS, Wakimoto BT. 1995. Heterochromatin and gene expression in *Drosophila. Annu. Rev. Genet.* 29:577–605
112. White KP, Rifkin SA, Hurban P, Hogness DS. 1999. Microarray analysis of *Drosophila* development during metamorphosis. *Science* 286:2179–84
113. Willy PJ, Kobayashi R, Kadonaga JT. 2000. A basal transcription factor that activates or represses transcription. *Science* 290:982–85
114. Wolff C, Pepling M, Gergen P, Klingler M. 1999. Structure and evolution of a pair-rule interaction element: *runt* regulatory sequences in *D. melanogaster* and *D. virilis*. *Mech. Dev.* 80:87–99
115. Workman JL, Kingston RE. 1998. Alteration of nucleosome structure as a mechanism of transcriptional regulation. *Annu. Rev. Biochem.* 67:545–79
116. Woychik NA, Hampsey M. 2002. The RNA polymerase II machinery: structure illuminates function. *Cell* 108:453–63
117. Wu CH, Madabusi L, Nishioka H, Emanuel P, Sypes M, et al. 2001. Analysis of core promoter sequences located downstream from the TATA element in the *hsp70* promoter from *Drosophila melanogaster*. *Mol. Cell. Biol.* 21:1593–602
118. Xiong B, Jacobs-Lorena M. 1995. Gut-specific transcriptional regulatory elements of the carboxypeptidase gene are conserved between black flies and *Drosophila*. *Proc. Natl. Acad. Sci. USA* 92:9313–17
119. Yanicostas C, Lepesant JA. 1990. Transcriptional and translational *cis*-regulatory sequences of the spermatocyte-specific *Drosophila janusB* gene are located in the 3′ exonic region of the overlapping *janusA* gene. *Mol. Gen. Genet.* 224:450–58
120. Yasuhara JC, DeCrease C, Slade D, Wakimoto BT. 2002. *A comparative study of heterochromatic and euchromatic* light *genes in* Drosophila *species*. Presented at the 43rd Annu. Drosophila Res. Conf., San Diego, CA
121. Zenzie-Gregory B, O'Shea-Greenfield A, Smale ST. 1992. Similar mechanisms for transcription initiation mediated through a TATA box or an initiator element. *J. Biol. Chem.* 267:2823–30
122. Zhu J, Kokoza V, Raikhel A. 2002. Analysis of stage- and tissue-specific gene expression in insect vectors. In *The Biology of Disease Vectors*, pp. xx–xx, ed. WC Marquardt. Niwot: Univ. Colorado Press. 2nd ed.

Subject Index

Cumulative Indexes

CONTRIBUTING AUTHORS, VOLUMES 39–48

CHAPTER TITLES, VOLUMES 39–48

Acarines, Arachnids, and Other Noninsect Arthropods

Agricultural Entomology

Behavior

Biological Control

Bionomics (See also Ecology)

Ecology (See also Bionomics; Behavior)

Forest Entomology

Genetics

Historical and Other

Insecticides and Toxicology

Medical and Veterinary Entomology

Miscellaneous

Morphology

Systematics, Evolution, and Biogeography

Vectors of Plant Pathogens